Loukas Grafakos

Classical Fourier Analysis

Second Edition

Springer

Loukas Grafakos
Department of Mathematics
University of Missouri
loukas@math.missouri.edu
Columbia, MO 65211
USA

ISSN: 0072-5285
ISBN 978-1-4419-1855-0
DOI: 10.1007/978-0-387-09432-8
e-ISBN 978-0-387-09432-8

Mathematics Subject Classification (2000): 42-xx 42-02

springer.com

Grafakos: Classical Fourier Analysis, 2nd ed

Softcover reprint of the hardcover 2nd edition 2008

First Indian Reprint 2012
Second Indian Reprint 2013

ISBN 978-81-322-0406-0

This edition is published by Springer (India) Private Limited, (a part of Springer Science+Business Media), Registered Office: 7th Floor, Vijaya Building, 17 Barakhamba Road, New Delhi – 110 001, India.

Printed in India by Rashtriya Printers, Delhi.

To Suzanne

Preface

The great response to the publication of the book *Classical and Modern Fourier Analysis* has been very gratifying. I am delighted that Springer has offered to publish the second edition of this book in two volumes: *Classical Fourier Analysis, 2nd Edition,* and *Modern Fourier Analysis, 2nd Edition.*

These volumes are mainly addressed to graduate students who wish to study Fourier analysis. This first volume is intended to serve as a text for a one-semester course in the subject. The prerequisite for understanding the material herein is satisfactory completion of courses in measure theory, Lebesgue integration, and complex variables.

The details included in the proofs make the exposition longer. Although it will behoove many readers to skim through the more technical aspects of the presentation and concentrate on the flow of ideas, the fact that details are present will be comforting to some. The exercises at the end of each section enrich the material of the corresponding section and provide an opportunity to develop additional intuition and deeper comprehension. The historical notes of each chapter are intended to provide an account of past research but also to suggest directions for further investigation. The appendix includes miscellaneous auxiliary material needed throughout the text.

A web site for the book is maintained at

http://math.missouri.edu/∼loukas/FourierAnalysis.html

I am solely responsible for any misprints, mistakes, and historical omissions in this book. Please contact me directly (loukas@math.missouri.edu) if you have corrections, comments, suggestions for improvements, or questions.

Columbia, Missouri,
April 2008

Loukas Grafakos

Acknowledgments

I am very fortunate that several people have pointed out errors, misprints, and omissions in the first edition of this book. Others have clarified issues I raised concerning the material it contains. All these individuals have provided me with invaluable help that resulted in the improved exposition of the present second edition. For these reasons, I would like to express my deep appreciation and sincere gratitude to the following people:

Marco Annoni, Pascal Auscher, Andrew Bailey, Dmitriy Bilyk, Marcin Bownik, Leonardo Colzani, Simon Cowell, Mita Das, Geoffrey Diestel, Yong Ding, Jacek Dziubanski, Wei He, Petr Honzík, Heidi Hulsizer, Philippe Jaming, Svante Janson, Ana Jiménez del Toro, John Kahl, Cornelia Kaiser, Nigel Kalton, Kim Jin Myong, Doowon Koh, Elena Koutcherik, Enrico Laeng, Sungyun Lee, Qifan Li, Chin-Cheng Lin, Liguang Liu, Stig-Olof Londen, Diego Maldonado, José María Martell, Mieczyslaw Mastylo, Parasar Mohanty, Carlo Morpurgo, Andrew Morris, Mihail Mourgoglou, Virginia Naibo, Hiro Oh, Marco Peloso, Maria Cristina Pereyra, Carlos Pérez, Humberto Rafeiro, Maria Carmen Reguera Rodríguez, Alexander Samborskiy, Andreas Seeger, Steven Senger, Sumi Seo, Christopher Shane, Shu Shen, Yoshihiro Sawano, Vladimir Stepanov, Erin Terwilleger, Rodolfo Torres, Suzanne Tourville, Ignacio Uriarte-Tuero, Kunyang Wang, Huoxiong Wu, Takashi Yamamoto, and Dachun Yang.

For their valuable suggestions, corrections, and other important assistance at different stages in the preparation of the first edition of this book, I would like to offer my deepest gratitude to the following individuals:

Georges Alexopoulos, Nakhlé Asmar, Bruno Calado, Carmen Chicone, David Cramer, Geoffrey Diestel, Jakub Duda, Brenda Frazier, Derrick Hart, Mark Hoffmann, Steven Hofmann, Helge Holden, Brian Hollenbeck, Petr Honzík, Alexander Iosevich, Tunde Jakab, Svante Janson, Ana Jiménez del Toro, Gregory Jones, Nigel Kalton, Emmanouil Katsoprinakis, Dennis Kletzing, Steven Krantz, Douglas Kurtz, George Lobell, Xiaochun Li, José María Martell, Antonios Melas, Keith Mersman, Stephen Montgomety-Smith, Andrea Nahmod, Nguyen Cong Phuc, Krzysztof Oleszkiewicz, Cristina Pereyra, Carlos Pérez, Daniel Redmond, Jorge Rivera-Noriega, Dmitriy Ryabogin, Christopher Sansing, Lynn Savino Wendel, Shih-Chi Shen,

Roman Shvidkoy, Elias Stein, Atanas Stefanov, Terence Tao, Erin Terwilleger, Christoph Thiele, Rodolfo Torres, Deanie Tourville, Nikolaos Tzirakis Don Vaught, Igor Verbitsky, Brett Wick, James Wright, and Linqiao Zhao.

I would also like to thank all reviewers who provided me with an abundance of meaningful remarks, corrections, and suggestions for improvements. Finally, I would like to thank Springer editor Mark Spencer, Springer's digital product support personnel Frank Ganz and Frank McGuckin, and copyeditor David Kramer for their invaluable assistance during the preparation of this edition.

Contents

Chapter 1
L^p Spaces and Interpolation

Many quantitative properties of functions are expressed in terms of their integrability to a power. For this reason it is desirable to acquire a good understanding of spaces of functions whose modulus to a power p is integrable. These are called Lebesgue spaces and are denoted by L^p. Although an in-depth study of Lebesgue spaces falls outside the scope of this book, it seems appropriate to devote a chapter to reviewing some of their fundamental properties.

The emphasis of this review is basic interpolation between Lebesgue spaces. Many problems in Fourier analysis concern boundedness of operators on Lebesgue spaces, and interpolation provides a framework that often simplifies this study. For instance, in order to show that a linear operator maps L^p to itself for all $1 < p < \infty$, it is sufficient to show that it maps the (smaller) Lorentz space $L^{p,1}$ into the (larger) Lorentz space $L^{p,\infty}$ for the same range of p's. Moreover, some further reductions can be made in terms of the Lorentz space $L^{p,1}$. This and other considerations indicate that interpolation is a powerful tool in the study of boundedness of operators.

Although we are mainly concerned with L^p subspaces of Euclidean spaces, we discuss in this chapter L^p spaces of arbitrary measure spaces, since they represent a useful general setting. Many results in the text require working with general measures instead of Lebesgue measure.

1.1 L^p and Weak L^p

Let X be a measure space and let μ be a positive, not necessarily finite, measure on X. For $0 < p < \infty$, $L^p(X,\mu)$ denotes the set of all complex-valued μ-measurable functions on X whose modulus to the pth power is integrable. $L^\infty(X,\mu)$ is the set of all complex-valued μ-measurable functions f on X such that for some $B > 0$, the set $\{x : |f(x)| > B\}$ has μ-measure zero. Two functions in $L^p(X,\mu)$ are considered equal if they are equal μ-almost everywhere. The notation $L^p(\mathbf{R}^n)$ is reserved for the space $L^p(\mathbf{R}^n, |\cdot|)$, where $|\cdot|$ denotes n-dimensional Lebesgue measure. Lebesgue measure on $\mathbf{R}^n$ is also denoted by dx. Within context and in the absence of ambi-

L. Grafakos, *Classical Fourier Analysis, Second Edition*,
DOI: 10.1007/978-0-387-09432-8_1,

guity, $L^p(X,\mu)$ is simply written as L^p. The space $L^p(\mathbf{Z})$ equipped with counting measure is denoted by $\ell^p(\mathbf{Z})$ or simply ℓ^p.

For $0<p<\infty$, we define the L^p quasinorm of a function f by

$$\big\|f\big\|_{L^p(X,\mu)} = \left(\int_X |f(x)|^p\,d\mu(x)\right)^{\frac{1}{p}} \tag{1.1.1}$$

and for $p=\infty$ by

$$\big\|f\big\|_{L^\infty(X,\mu)} = \text{ess.sup}\,|f| = \inf\big\{B>0\,:\,\mu(\{x: |f(x)|>B\})=0\big\}. \tag{1.1.2}$$

It is well known that Minkowski's (or the triangle) inequality

$$\big\|f+g\big\|_{L^p(X,\mu)} \le \big\|f\big\|_{L^p(X,\mu)} + \big\|g\big\|_{L^p(X,\mu)} \tag{1.1.3}$$

holds for all f, g in $L^p = L^p(X,\mu)$, whenever $1\le p\le\infty$. Since in addition $\big\|f\big\|_{L^p(X,\mu)}=0$ implies that $f=0$ (μ-a.e.), the L^p spaces are normed linear spaces for $1\le p\le\infty$. For $0<p<1$, inequality (1.1.3) is reversed when $f,g\ge 0$. However, the following substitute of (1.1.3) holds:

$$\big\|f+g\big\|_{L^p(X,\mu)} \le 2^{(1-p)/p}\big(\big\|f\big\|_{L^p(X,\mu)} + \big\|g\big\|_{L^p(X,\mu)}\big), \tag{1.1.4}$$

and thus $L^p(X,\mu)$ is a quasinormed linear space. See also Exercise 1.1.5. For all $0<p\le\infty$, it can be shown that every Cauchy sequence in $L^p(X,\mu)$ is convergent, and hence the spaces $L^p(X,\mu)$ are complete. For the case $0<p<1$ we refer to Exercise 1.1.8. Therefore, the L^p spaces are Banach spaces for $1\le p\le\infty$ and quasi-Banach spaces for $0<p<1$. For any $p\in(0,\infty)\setminus\{1\}$ we use the notation $p'=\frac{p}{p-1}$. Moreover, we set $1'=\infty$ and $\infty'=1$, so that $p''=p$ for all $p\in(0,\infty]$. Hölder's inequality says that for all $p\in[1,\infty]$ and all measurable functions f,g on (X,μ) we have

$$\big\|fg\big\|_{L^1} \le \big\|f\big\|_{L^p}\big\|g\big\|_{L^{p'}}.$$

It is a well-known fact that the dual $(L^p)^*$ of L^p is isometric to $L^{p'}$ for all $1\le p<\infty$. Furthermore, the L^p norm of a function can be obtained via duality when $1\le p\le\infty$ as follows:

$$\big\|f\big\|_{L^p} = \sup_{\|g\|_{L^{p'}}=1}\left|\int_X fg\,d\mu\right|.$$

For the endpoint cases $p=1$, $p=\infty$, see Exercise 1.4.12(a), (b).

1.1.1 The Distribution Function

Definition 1.1.1. For f a measurable function on X, the *distribution function* of f is the function d_f defined on $[0,\infty)$ as follows:

$$d_f(\alpha) = \mu(\{x \in X : |f(x)| > \alpha\}). \tag{1.1.5}$$

The distribution function d_f provides information about the size of f but not about the behavior of f itself near any given point. For instance, a function on $\mathbf{R}^n$ and each of its translates have the same distribution function. It follows from Definition 1.1.1 that d_f is a decreasing function of α (not necessarily strictly).

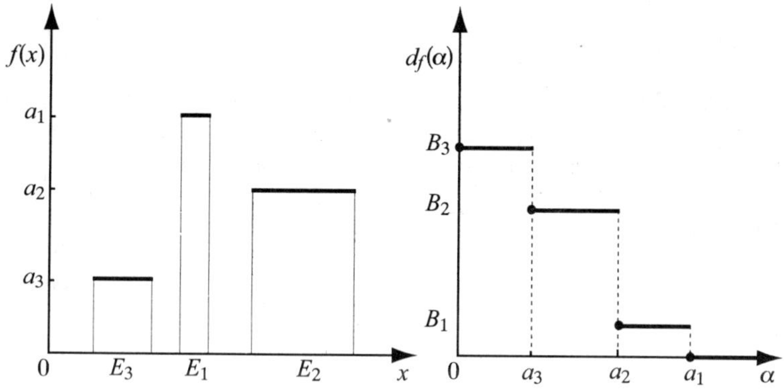

Fig. 1.1 The graph of a simple function $f = \sum_{k=1}^{3} a_k \chi_{E_k}$ and its distribution function $d_f(\alpha)$. Here $B_j = \sum_{k=1}^{j} \mu(E_k)$.

Example 1.1.2. Recall that simple functions are finite linear combinations of characteristic functions of sets of finite measure. For pedagogical reasons we compute the distribution function d_f of a nonnegative simple function

$$f(x) = \sum_{j=1}^{N} a_j \chi_{E_j}(x),$$

where the sets E_j are pairwise disjoint and $a_1 > \cdots > a_N > 0$. If $\alpha \geq a_1$, then clearly $d_f(\alpha) = 0$. However, if $a_2 \leq \alpha < a_1$ then $|f(x)| > \alpha$ precisely when $x \in E_1$, and in general, if $a_{j+1} \leq \alpha < a_j$, then $|f(x)| > \alpha$ precisely when $x \in E_1 \cup \cdots \cup E_j$. Setting

$$B_j = \sum_{k=1}^{j} \mu(E_k),$$

we have

$$d_f(\alpha) = \sum_{j=0}^{N} B_j \chi_{[a_{j+1}, a_j)}(\alpha),$$

where $a_0 = \infty$ and $B_0 = a_{N+1} = 0$. Figure 1.1 illustrates this example when $N = 3$.

We now state a few simple facts about the distribution function d_f.

Proposition 1.1.3. *Let f and g be measurable functions on (X,μ). Then for all $\alpha,\beta>0$ we have*

(1) $|g|\le|f|$ μ-a.e. implies that $d_g\le d_f$;

(2) $d_{cf}(\alpha)=d_f(\alpha/|c|)$, for all $c\in\mathbf{C}\setminus\{0\}$;

(3) $d_{f+g}(\alpha+\beta)\le d_f(\alpha)+d_g(\beta)$;

(4) $d_{fg}(\alpha\beta)\le d_f(\alpha)+d_g(\beta)$.

Proof. The simple proofs are left to the reader. □

Knowledge of the distribution function d_f provides sufficient information to evaluate the L^p norm of a function f precisely. We state and prove the following important description of the L^p norm in terms of the distribution function.

Proposition 1.1.4. *For f in $L^p(X,\mu)$, $0<p<\infty$, we have*

$$\|f\|_{L^p}^p = p\int_0^\infty \alpha^{p-1} d_f(\alpha)\,d\alpha\,. \tag{1.1.6}$$

Proof. Indeed, we have

$$\begin{aligned}
p\int_0^\infty \alpha^{p-1} d_f(\alpha)\,d\alpha &= p\int_0^\infty \alpha^{p-1}\int_X \chi_{\{x:\,|f(x)|>\alpha\}}\,d\mu(x)\,d\alpha\\
&= \int_X\int_0^{|f(x)|} p\alpha^{p-1}\,d\alpha\,d\mu(x)\\
&= \int_X |f(x)|^p\,d\mu(x)\\
&= \|f\|_{L^p}^p,
\end{aligned}$$

where we used Fubini's theorem in the second equality. This proves (1.1.6). □

Notice that the same argument yields the more general fact that for any increasing continuously differentiable function φ on $[0,\infty)$ with $\varphi(0)=0$ we have

$$\int_X \varphi(|f|)\,d\mu = \int_0^\infty \varphi'(\alpha)d_f(\alpha)\,d\alpha\,. \tag{1.1.7}$$

Definition 1.1.5. For $0<p<\infty$, the space *weak* $L^p(X,\mu)$ is defined as the set of all μ-measurable functions f such that

$$\begin{aligned}
\|f\|_{L^{p,\infty}} &= \inf\Big\{C>0:\, d_f(\alpha)\le \frac{C^p}{\alpha^p} \quad \text{for all} \quad \alpha>0\Big\} &(1.1.8)\\
&= \sup\big\{\gamma d_f(\gamma)^{1/p}:\,\gamma>0\big\} &(1.1.9)
\end{aligned}$$

is finite. The space *weak-*$L^\infty(X,\mu)$ is by definition $L^\infty(X,\mu)$.

The reader should check that (1.1.9) and (1.1.8) are in fact equal. The weak L^p spaces are denoted by $L^{p,\infty}(X,\mu)$. Two functions in $L^{p,\infty}(X,\mu)$ are considered equal if they are equal μ-a.e. The notation $L^{p,\infty}(\mathbf{R}^n)$ is reserved for $L^{p,\infty}(\mathbf{R}^n, |\cdot|)$. Using Proposition 1.1.3 (2), we can easily show that

$$\big\|kf\big\|_{L^{p,\infty}} = |k|\big\|f\big\|_{L^{p,\infty}}, \tag{1.1.10}$$

for any complex nonzero constant k. The analogue of (1.1.3) is

$$\big\|f+g\big\|_{L^{p,\infty}} \le c_p\big(\big\|f\big\|_{L^{p,\infty}} + \big\|g\big\|_{L^{p,\infty}}\big), \tag{1.1.11}$$

where $c_p = \max(2, 2^{1/p})$, a fact that follows from Proposition 1.1.3 (3), taking both α and β equal to $\alpha/2$. We also have that

$$\big\|f\big\|_{L^{p,\infty}(X,\mu)} = 0 \Rightarrow f = 0 \qquad \mu\text{-a.e.} \tag{1.1.12}$$

In view of (1.1.10), (1.1.11), and (1.1.12), $L^{p,\infty}$ is a quasinormed linear space for $0 < p < \infty$.

The weak L^p spaces are larger than the usual L^p spaces. We have the following:

Proposition 1.1.6. *For any $0 < p < \infty$ and any f in $L^p(X,\mu)$ we have $\big\|f\big\|_{L^{p,\infty}} \le \big\|f\big\|_{L^p}$; hence $L^p(X,\mu) \subseteq L^{p,\infty}(X,\mu)$.*

Proof. This is just a trivial consequence of Chebyshev's inequality:

$$\alpha^p d_f(\alpha) \le \int_{\{x:|f(x)|>\alpha\}} |f(x)|^p \, d\mu(x). \tag{1.1.13}$$

The integral in (1.1.13) is at most $\big\|f\big\|_{L^p}^p$ and using (1.1.9) we obtain that $\big\|f\big\|_{L^{p,\infty}} \le \big\|f\big\|_{L^p}$. □

The inclusion $L^p \subseteq L^{p,\infty}$ is strict. For example, on $\mathbf{R}^n$ with the usual Lebesgue measure, let $h(x) = |x|^{-\frac{n}{p}}$. Obviously, h is not in $L^p(\mathbf{R}^n)$ but h is in $L^{p,\infty}(\mathbf{R}^n)$ with $\big\|h\big\|_{L^{p,\infty}(\mathbf{R}^n)} = v_n$, where v_n is the measure of the unit ball of $\mathbf{R}^n$.

It is not immediate from their definition that the weak L^p spaces are complete with respect to the quasinorm $\big\|\cdot\big\|_{L^{p,\infty}}$. The completeness of these spaces is proved in Theorem 1.4.11, but it is also a consequence of Theorem 1.1.13, proved in this section.

1.1.2 Convergence in Measure

Next we discuss some convergence notions. The following notion is important in probability theory.

Definition 1.1.7. Let f, f_n, $n = 1, 2, \dots$, be measurable functions on the measure space (X,μ). The sequence f_n is said to *converge in measure* to f if for all $\varepsilon > 0$ there exists an $n_0 \in \mathbf{Z}^+$ such that

$$n > n_0 \implies \mu(\{x \in X : |f_n(x) - f(x)| > \varepsilon\}) < \varepsilon. \tag{1.1.14}$$

Remark 1.1.8. The preceding definition is equivalent to the following statement:

$$\text{For all } \varepsilon > 0 \quad \lim_{n\to\infty} \mu(\{x \in X : |f_n(x) - f(x)| > \varepsilon\}) = 0. \tag{1.1.15}$$

Clearly (1.1.15) implies (1.1.14). To see the converse given $\varepsilon > 0$, pick $0 < \delta < \varepsilon$ and apply (1.1.14) for this δ. There exists an $n_0 \in \mathbf{Z}^+$ such that

$$\mu(\{x \in X : |f_n(x) - f(x)| > \delta\}) < \delta$$

holds for $n > n_0$. Since

$$\mu(\{x \in X : |f_n(x) - f(x)| > \varepsilon\}) \le \mu(\{x \in X : |f_n(x) - f(x)| > \delta\}),$$

we conclude that

$$\mu(\{x \in X : |f_n(x) - f(x)| > \varepsilon\}) < \delta$$

for all $n > n_0$. Let $n \to \infty$ to deduce that

$$\limsup_{n\to\infty} \mu(\{x \in X : |f_n(x) - f(x)| > \varepsilon\}) \le \delta. \tag{1.1.16}$$

Since (1.1.16) holds for all $0 < \delta < \varepsilon$, (1.1.15) follows by letting $\delta \to 0$.

Convergence in measure is a weaker notion than convergence in either L^p or $L^{p,\infty}$, $0 < p \le \infty$, as the following proposition indicates:

Proposition 1.1.9. *Let $0 < p \le \infty$ and f_n, f be in $L^{p,\infty}(X,\mu)$.*

(1) If f_n, f are in L^p and $f_n \to f$ in L^p, then $f_n \to f$ in $L^{p,\infty}$.
(2) If $f_n \to f$ in $L^{p,\infty}$, then f_n converges to f in measure.

Proof. Fix $0 < p < \infty$. Proposition 1.1.6 gives that for all $\varepsilon > 0$ we have

$$\mu(\{x \in X : |f_n(x) - f(x)| > \varepsilon\}) \le \frac{1}{\varepsilon^p} \int_X |f_n - f|^p \, d\mu.$$

This shows that convergence in L^p implies convergence in weak L^p. The case $p = \infty$ is tautological.

Given $\varepsilon > 0$ find an n_0 such that for $n > n_0$, we have

$$\big\|f_n - f\big\|_{L^{p,\infty}} = \sup_{\alpha > 0} \alpha \mu(\{x \in X : |f_n(x) - f(x)| > \alpha\})^{\frac{1}{p}} < \varepsilon^{\frac{1}{p}+1}.$$

Taking $\alpha = \varepsilon$, we conclude that convergence in $L^{p,\infty}$ implies convergence in measure. □

Example 1.1.10. Fix $0 < p < \infty$. On $[0,1]$ define the functions

$$f_{k,j} = k^{1/p} \chi_{(\frac{j-1}{k}, \frac{j}{k})}, \qquad k \ge 1,\ 1 \le j \le k.$$

Consider the sequence $\{f_{1,1}, f_{2,1}, f_{2,2}, f_{3,1}, f_{3,2}, f_{3,3}, \dots\}$. Observe that

$$|\{x : f_{k,j}(x) > 0\}| = 1/k.$$

Therefore, $f_{k,j}$ converges to 0 in measure. Likewise, observe that

$$\|f_{k,j}\|_{L^{p,\infty}} = \sup_{\alpha>0} \alpha |\{x : f_{k,j}(x) > \alpha\}|^{1/p} \geq \sup_{k\geq 1} \frac{(k-1/k)^{1/p}}{k^{1/p}} = 1,$$

which implies that $f_{k,j}$ does not converge to 0 in $L^{p,\infty}$.

It turns out that every sequence convergent in $L^p(X,\mu)$ or in $L^{p,\infty}(X,\mu)$ has a subsequence that converges a.e. to the same limit.

Theorem 1.1.11. *Let f_n and f be complex-valued measurable functions on a measure space (X,μ) and suppose that f_n converges to f in measure. Then some subsequence of f_n converges to f μ-a.e.*

Proof. For all $k = 1,2,\dots$ choose inductively n_k such that

$$\mu(\{x \in X : |f_{n_k}(x) - f(x)| > 2^{-k}\}) < 2^{-k} \tag{1.1.17}$$

and such that $n_1 < n_2 < \cdots < n_k < \cdots$. Define the sets

$$A_k = \{x \in X : |f_{n_k}(x) - f(x)| > 2^{-k}\}.$$

Equation (1.1.17) implies that

$$\mu\Big(\bigcup_{k=m}^{\infty} A_k\Big) \leq \sum_{k=m}^{\infty} \mu(A_k) \leq \sum_{k=m}^{\infty} 2^{-k} = 2^{1-m} \tag{1.1.18}$$

for all $m = 1,2,3,\dots$. It follows from (1.1.18) that

$$\mu\Big(\bigcup_{k=1}^{\infty} A_k\Big) \leq 1 < \infty. \tag{1.1.19}$$

Using (1.1.18) and (1.1.19), we conclude that the sequence of the measures of the sets $\{\bigcup_{k=m}^{\infty} A_k\}_{m=1}^{\infty}$ converges as $m \to \infty$ to

$$\mu\Big(\bigcap_{m=1}^{\infty}\bigcup_{k=m}^{\infty} A_k\Big) = 0. \tag{1.1.20}$$

To finish the proof, observe that the null set in (1.1.20) contains the set of all $x \in X$ for which $f_{n_k}(x)$ does not converge to $f(x)$. □

In many situations we are given a sequence of functions and we would like to extract a convergent subsequence. One way to achieve this is via the next theorem, which is a useful variant of Theorem 1.1.11. We first give a relevant definition.

Definition 1.1.12. We say that a sequence of measurable functions $\{f_n\}$ on the measure space (X,μ) is *Cauchy in measure* if for every $\varepsilon > 0$, there exists an $n_0 \in \mathbf{Z}^+$ such that for $n,m > n_0$ we have

$$\mu(\{x \in X : |f_m(x) - f_n(x)| > \varepsilon\}) < \varepsilon.$$

Theorem 1.1.13. *Let (X,μ) be a measure space and let $\{f_n\}$ be a complex-valued sequence on X that is Cauchy in measure. Then some subsequence of f_n converges μ-a.e.*

Proof. The proof is very similar to that of Theorem 1.1.11. For all $k = 1,2,\dots$ choose n_k inductively such that

$$\mu(\{x \in X : |f_{n_k}(x) - f_{n_{k+1}}(x)| > 2^{-k}\}) < 2^{-k} \tag{1.1.21}$$

and such that $n_1 < n_2 < \cdots < n_k < n_{k+1} < \cdots$. Define

$$A_k = \{x \in X : |f_{n_k}(x) - f_{n_{k+1}}(x)| > 2^{-k}\}.$$

As shown in the proof of Theorem 1.1.11, (1.1.21) implies that

$$\mu\bigg(\bigcap_{m=1}^{\infty}\bigcup_{k=m}^{\infty} A_k\bigg) = 0. \tag{1.1.22}$$

For $x \notin \bigcup_{k=m}^{\infty} A_k$ and $i \ge j \ge j_0 \ge m$ (and j_0 large enough) we have

$$|f_{n_i}(x) - f_{n_j}(x)| \le \sum_{l=j}^{i-1} |f_{n_l}(x) - f_{n_{l+1}}(x)| \le \sum_{l=j}^{i-1} 2^{-l} \le 2^{1-j} \le 2^{1-j_0}.$$

This implies that the sequence $\{f_{n_i}(x)\}_i$ is Cauchy for every x in the set $(\bigcup_{k=m}^{\infty} A_k)^c$ and therefore converges for all such x. We define a function

$$f(x) = \begin{cases} \lim_{j\to\infty} f_{n_j}(x) & \text{when } x \notin \bigcap_{m=1}^{\infty}\bigcup_{k=m}^{\infty} A_k\,, \\ 0 & \text{when } x \in \bigcap_{m=1}^{\infty}\bigcup_{k=m}^{\infty} A_k\,. \end{cases}$$

Then $f_{n_j} \to f$ almost everywhere. □

1.1.3 A First Glimpse at Interpolation

It is a useful fact that if a function f is in $L^p(X,\mu)$ and in $L^q(X,\mu)$, then it also lies in $L^r(X,\mu)$ for all $p < r < q$. The usefulness of the spaces $L^{p,\infty}$ can be seen from the following sharpening of this statement:

Proposition 1.1.14. *Let $0 < p < q \le \infty$ and let f in $L^{p,\infty}(X,\mu) \cap L^{q,\infty}(X,\mu)$. Then f is in $L^r(X,\mu)$ for all $p < r < q$ and*

$$\|f\|_{L^r} \le \left(\frac{r}{r-p}+\frac{r}{q-r}\right)^{\frac{1}{r}} \|f\|_{L^{p,\infty}}^{\frac{\frac{1}{r}-\frac{1}{q}}{\frac{1}{p}-\frac{1}{q}}} \|f\|_{L^{q,\infty}}^{\frac{\frac{1}{p}-\frac{1}{r}}{\frac{1}{p}-\frac{1}{q}}}, \tag{1.1.23}$$

with the suitable interpretation when $q=\infty$.

Proof. Let us take first $q<\infty$. We know that

$$d_f(\alpha) \le \min\left(\frac{\|f\|_{L^{p,\infty}}^p}{\alpha^p}, \frac{\|f\|_{L^{q,\infty}}^q}{\alpha^q}\right). \tag{1.1.24}$$

Set

$$B=\left(\frac{\|f\|_{L^{q,\infty}}^q}{\|f\|_{L^{p,\infty}}^p}\right)^{\frac{1}{q-p}}. \tag{1.1.25}$$

We now estimate the L^r norm of f. By (1.1.24), (1.1.25), and Proposition 1.1.4 we have

$$\begin{aligned}
\|f\|_{L^r(X,\mu)}^r &= r\int_0^\infty \alpha^{r-1} d_f(\alpha)\,d\alpha \\
&\le r\int_0^\infty \alpha^{r-1}\min\left(\frac{\|f\|_{L^{p,\infty}}^p}{\alpha^p}, \frac{\|f\|_{L^{q,\infty}}^q}{\alpha^q}\right)d\alpha \\
&= r\int_0^B \alpha^{r-1-p}\|f\|_{L^{p,\infty}}^p\,d\alpha + r\int_B^\infty \alpha^{r-1-q}\|f\|_{L^{q,\infty}}^q\,d\alpha \\
&= \frac{r}{r-p}\|f\|_{L^{p,\infty}}^p B^{r-p} + \frac{r}{q-r}\|f\|_{L^{q,\infty}}^q B^{r-q} \\
&= \left(\frac{r}{r-p}+\frac{r}{q-r}\right)\left(\|f\|_{L^{p,\infty}}^p\right)^{\frac{q-r}{q-p}}\left(\|f\|_{L^{q,\infty}}^q\right)^{\frac{r-p}{q-p}}.
\end{aligned} \tag{1.1.26}$$

Observe that the integrals converge, since $r-p>0$ and $r-q<0$.

The case $q=\infty$ is easier. Since $d_f(\alpha)=0$ for $\alpha>\|f\|_{L^\infty}$ we need to use only the inequality $d_f(\alpha)\le\alpha^{-p}\|f\|_{L^{p,\infty}}^p$ for $\alpha\le\|f\|_{L^\infty}$ in estimating the first integral in (1.1.26). We obtain

$$\|f\|_{L^r}^r \le \frac{r}{r-p}\|f\|_{L^{p,\infty}}^p\|f\|_{L^\infty}^{r-p},$$

which is nothing other than (1.1.23) when $q=\infty$. This completes the proof. □

Note that (1.1.23) holds with constant 1 if $L^{p,\infty}$ and $L^{q,\infty}$ are replaced by L^p and L^q, respectively. It is often convenient to work with functions that are only locally in some L^p space. This leads to the following definition.

Definition 1.1.15. For $0<p<\infty$, the space $L^p_{\mathrm{loc}}(\mathbf{R}^n,|\cdot|)$ or simply $L^p_{\mathrm{loc}}(\mathbf{R}^n)$ is the set of all Lebesgue-measurable functions f on $\mathbf{R}^n$ that satisfy

$$\int_K |f(x)|^p\,dx<\infty \tag{1.1.27}$$

for any compact subset K of $\mathbf{R}^n$. Functions that satisfy (1.1.27) with $p = 1$ are called *locally integrable* functions on $\mathbf{R}^n$.

The union of all $L^p(\mathbf{R}^n)$ spaces for $1 \le p \le \infty$ is contained in $L^1_{\mathrm{loc}}(\mathbf{R}^n)$. More generally, for $0 < p < q < \infty$ we have the following:

$$L^q(\mathbf{R}^n) \subseteq L^q_{\mathrm{loc}}(\mathbf{R}^n) \subseteq L^p_{\mathrm{loc}}(\mathbf{R}^n).$$

Functions in $L^p(\mathbf{R}^n)$ for $0 < p < 1$ may not be locally integrable. For example, take $f(x) = |x|^{-n-\alpha}\chi_{|x|\le 1}$, which is in $L^p(\mathbf{R}^n)$ when $p < n/(n+\alpha)$, and observe that f is not integrable over any open set in $\mathbf{R}^n$ containing the origin.

Exercises

1.1.1. Suppose f and f_n are measurable functions on (X,μ). Prove that
(a) d_f is right continuous on $[0,\infty)$.
(b) If $|f| \le \liminf_{n\to\infty} |f_n|$ μ-a.e., then $d_f \le \liminf_{n\to\infty} d_{f_n}$.
(c) If $|f_n| \uparrow |f|$, then $d_{f_n} \uparrow d_f$.
[*Hint:* Part (a): Let t_n be a decreasing sequence of positive numbers that tends to zero. Show that $d_f(\alpha_0 + t_n) \uparrow d_f(\alpha_0)$ using a convergence theorem. Part (b): Let $E = \{x \in X : |f(x)| > \alpha\}$ and $E_n = \{x \in X : |f_n(x)| > \alpha\}$. Use that $\mu\big(\bigcap_{n=m}^\infty E_n\big) \le \liminf_{n\to\infty} \mu(E_n)$ and $E \subseteq \bigcup_{m=1}^\infty \bigcap_{n=m}^\infty E_n$ μ-a.e.]

1.1.2. (*Hölder's inequality*) Let $0 < p, p_1, \dots, p_k \le \infty$, where $k \ge 2$, and let f_j be in $L^{p_j} = L^{p_j}(X,\mu)$. Assume that

$$\frac{1}{p} = \frac{1}{p_1} + \cdots + \frac{1}{p_k}.$$

(a) Show that the product $f_1 \cdots f_k$ is in L^p and that

$$\big\|f_1 \cdots f_k\big\|_{L^p} \le \big\|f_1\big\|_{L^{p_1}} \cdots \big\|f_k\big\|_{L^{p_k}}.$$

(b) When no p_j is infinite, show that if equality holds in part (a), then it must be the case that $c_1|f_1|^{p_1} = \cdots = c_k|f_k|^{p_k}$ a.e. for some $c_j \ge 0$.
(c) Let $0 < q < 1$. For $r < 0$ and $g > 0$ almost everywhere, let $\big\|g\big\|_{L^r} = \big\|g^{-1}\big\|_{L^{|r|}}^{-1}$. Show that for $f \ge 0$, $g > 0$ a.e. we have

$$\big\|fg\big\|_{L^1} \ge \big\|f\big\|_{L^q}\big\|g\big\|_{L^{q'}}.$$

1.1.3. Let (X,μ) be a measure space.
(a) If f is in $L^{p_0}(X,\mu)$ for some $p_0 < \infty$, prove that

$$\lim_{p\to\infty} \big\|f\big\|_{L^p} = \big\|f\big\|_{L^\infty}.$$

(b) (*Jensen's inequality*) Suppose that $\mu(X) = 1$. Show that

$$\|f\|_{L^p} \geq \exp\left(\int_X \log|f(x)|\,d\mu(x)\right)$$

for all $0 < p < \infty$.
(c) If $\mu(X) = 1$ and f is in some $L^{p_0}(X,\mu)$ for some $p_0 > 0$, then

$$\lim_{p\to 0}\|f\|_{L^p} = \exp\left(\int_X \log|f(x)|\,d\mu(x)\right)$$

with the interpretation $e^{-\infty} = 0$.
[*Hint:* Part (a): Given $0 < \varepsilon < \|f\|_{L^\infty}$, find a measurable set $E \subseteq X$ of positive measure such that $|f(x)| \geq \|f\|_{L^\infty} - \varepsilon$ for all $x \in E$. Then $\|f\|_{L^p} \geq (\|f\|_{L^\infty} - \varepsilon)\mu(E)^{1/p}$ and thus $\liminf_{p\to\infty}\|f\|_{L^p} \geq \|f\|_{L^\infty} - \varepsilon$. Part (b) is a direct consequence of Jensen's inequality $\int_X \log|h|\,d\mu \leq \log\left(\int_X |h|\,d\mu\right)$. Part (c): Fix a sequence $0 < p_n < p_0$ such that $p_n \downarrow 0$ and define

$$h_n(x) = \frac{1}{p_0}(|f(x)|^{p_0} - 1) - \frac{1}{p_n}(|f(x)|^{p_n} - 1).$$

Use that $\frac{1}{p}(t^p - 1) \downarrow \log t$ as $p \downarrow 0$ for all $t > 0$. The Lebesgue monotone convergence theorem yields $\int_X h_n\,d\mu \uparrow \int_X h\,d\mu$, hence $\int_X \frac{1}{p_n}(|f|^{p_n} - 1)\,d\mu \downarrow \int_X \log|f|\,d\mu$, where the latter could be $-\infty$. Use

$$\exp\left(\int_X \log|f|\,d\mu\right) \leq \left(\int_X |f|^{p_n}\,d\mu\right)^{\frac{1}{p_n}} \leq \exp\left(\int_X \frac{1}{p_n}(|f|^{p_n} - 1)\,d\mu\right)$$

to complete the proof.]

1.1.4. Let a_j be a sequence of positive reals. Show that
(a) $\left(\sum_{j=1}^\infty a_j\right)^\theta \leq \sum_{j=1}^\infty a_j^\theta$, for any $0 \leq \theta \leq 1$.
(b) $\sum_{j=1}^\infty a_j^\theta \leq \left(\sum_{j=1}^\infty a_j\right)^\theta$, for any $1 \leq \theta < \infty$.
(c) $\left(\sum_{j=1}^N a_j\right)^\theta \leq N^{\theta-1}\sum_{j=1}^N a_j^\theta$, when $1 \leq \theta < \infty$.
(d) $\sum_{j=1}^N a_j^\theta \leq N^{1-\theta}\left(\sum_{j=1}^N a_j\right)^\theta$, when $0 \leq \theta \leq 1$.

1.1.5. Let $\{f_j\}_{j=1}^N$ be a sequence of $L^p(X,\mu)$ functions.
(a) (*Minkowski's inequality*) For $1 \leq p \leq \infty$ show that

$$\Big\|\sum_{j=1}^N f_j\Big\|_{L^p} \leq \sum_{j=1}^N \|f_j\|_{L^p}.$$

(b) (*Minkowski's inequality*) For $0 < p < 1$ and $f_j \geq 0$ prove that

$$\sum_{j=1}^{N} \|f_j\|_{L^p} \le \|\sum_{j=1}^{N} f_j\|_{L^p} .$$

(c) For $0 < p < 1$ show that

$$\|\sum_{j=1}^{N} f_j\|_{L^p} \le N^{\frac{1-p}{p}} \sum_{j=1}^{N} \|f_j\|_{L^p} .$$

(d) The constant $N^{\frac{1-p}{p}}$ in part (c) is best possible.
[*Hint:* Part (c): Use Exercise 1.1.4(c). Part (d): Take $\{f_j\}_{j=1}^N$ to be characteristic functions of disjoint sets with the same measure.]

1.1.6. (*Minkowski's integral inequality*) Let $1 \le p < \infty$. Let F be a measurable function on the product space $(X,\mu) \times (T,\nu)$, where μ, ν are σ-finite. Show that

$$\left[\int_T \left(\int_X |F(x,t)|\,d\mu(x)\right)^p d\nu(t)\right]^{\frac{1}{p}} \le \int_X \left[\int_T |F(x,t)|^p\,d\nu(t)\right]^{\frac{1}{p}} d\mu(x) .$$

Moreover, prove that when $0 < p < 1$, then the preceding inequality is reversed.

1.1.7. Let $f_1, \dots, f_N$ be in $L^{p,\infty}(X,\mu)$.
(a) Prove that for $1 \le p < \infty$ we have

$$\|\sum_{j=1}^{N} f_j\|_{L^{p,\infty}} \le N \sum_{j=1}^{N} \|f_j\|_{L^{p,\infty}} .$$

(b) Show that for $0 < p < 1$ we have

$$\|\sum_{j=1}^{N} f_j\|_{L^{p,\infty}} \le N^{\frac{1}{p}} \sum_{j=1}^{N} \|f_j\|_{L^{p,\infty}} .$$

[*Hint:* Use that $\mu(\{|f_1 + \cdots + f_N| > \alpha\}) \le \sum_{j=1}^N \mu(\{|f_j| > \alpha/N\})$ and Exercise 1.1.4(a) and (c).]

1.1.8. Let $0 < p < \infty$. Prove that $L^p(X,\mu)$ is a complete quasinormed space. This means that every quasinorm Cauchy sequence is quasinorm convergent.
[*Hint:* Let f_n be a Cauchy sequence in L^p. Pass to a subsequence $\{n_i\}_i$ such that $\|f_{n_{i+1}} - f_{n_i}\|_{L^p} \le 2^{-i}$. Then the series $f = f_{n_1} + \sum_{i=1}^\infty (f_{n_{i+1}} - f_{n_i})$ converges in L^p.]

1.1.9. Let (X,μ) be a measure space with $\mu(X) < \infty$. Suppose that a sequence of measurable functions f_n on X converges to f μ-a.e. Prove that f_n converges to f in measure.
[*Hint:* For $\varepsilon > 0$, $\{x \in X : f_n(x) \to f(x)\} \subseteq \bigcup_{m=1}^\infty \bigcap_{n=m}^\infty \{x \in X : |f_n(x) - f(x)| < \varepsilon\}$.]

1.1.10. Given a measurable function f on (X,μ) and $\gamma > 0$, define $f_\gamma = f\chi_{|f|>\gamma}$ and $f^\gamma = f - f_\gamma = f\chi_{|f|\le\gamma}$.
(a) Prove that

$$d_{f_\gamma}(\alpha) = \begin{cases} d_f(\alpha) & \text{when} \quad \alpha > \gamma, \\ d_f(\gamma) & \text{when} \quad \alpha \le \gamma, \end{cases}$$

$$d_{f^\gamma}(\alpha) = \begin{cases} 0 & \text{when} \quad \alpha \ge \gamma, \\ d_f(\alpha) - d_f(\gamma) & \text{when} \quad \alpha < \gamma. \end{cases}$$

(b) If $f \in L^p(X,\mu)$ then

$$\begin{aligned} \|f_\gamma\|_{L^p}^p &= p\int_\gamma^\infty \alpha^{p-1} d_f(\alpha)\,d\alpha + \gamma^p d_f(\gamma), \\ \|f^\gamma\|_{L^p}^p &= p\int_0^\gamma \alpha^{p-1} d_f(\alpha)\,d\alpha - \gamma^p d_f(\gamma), \\ \int_{\gamma<|f|\le\delta} |f|^p\,d\mu &= p\int_\gamma^\delta d_f(\alpha)\alpha^{p-1}\,d\alpha - \delta^p d_f(\delta) + \gamma^p d_f(\gamma). \end{aligned}$$

(c) If f is in $L^{p,\infty}(X,\mu)$ prove that f^γ is in $L^q(X,\mu)$ for any $q > p$ and f_γ is in $L^q(X,\mu)$ for any $q < p$. Thus $L^{p,\infty} \subseteq L^{p_0} + L^{p_1}$ when $0 < p_0 < p < p_1 \le \infty$.

1.1.11. Let (X,μ) be a measure space and let E be a subset of X with $\mu(E) < \infty$. Assume that f is in $L^{p,\infty}(X,\mu)$ for some $0 < p < \infty$.
(a) Show that for $0 < q < p$ we have

$$\int_E |f(x)|^q\,d\mu(x) \le \frac{p}{p-q}\mu(E)^{1-\frac{q}{p}}\|f\|_{L^{p,\infty}}^q.$$

(b) Conclude that if $\mu(X) < \infty$ and $0 < q < p$, then

$$L^p(X,\mu) \subseteq L^{p,\infty}(X,\mu) \subseteq L^q(X,\mu).$$

[*Hint:* Part (a): Use $\mu\big(E\cap\{|f|>\alpha\}\big) \le \min\big(\mu(E), \alpha^{-p}\|f\|_{L^{p,\infty}}^p\big)$.]

1.1.12. (*Normability of weak L^p for $p > 1$*) Let (X,μ) be a measure space and let $0 < p < \infty$. Pick $0 < r < p$ and define

$$|||\, f\,|||_{L^{p,\infty}} = \sup_{0<\mu(E)<\infty} \mu(E)^{-\frac{1}{r}+\frac{1}{p}}\left(\int_E |f|^r\,d\mu\right)^{\frac{1}{r}},$$

where the supremum is taken over all measurable subsets E of X of finite measure.
(a) Use Exercise 1.1.11 with $q = r$ to conclude that

$$|||\, f\,|||_{L^{p,\infty}} \le \left(\frac{p}{p-r}\right)^{\frac{1}{r}}\|f\|_{L^{p,\infty}}$$

for all f in $L^{p,\infty}(X,\mu)$.
(b) Take $E = \{|f| > \alpha\}$ to deduce that $\|f\|_{L^{p,\infty}} \le \||f\||_{L^{p,\infty}}$ for all f in $L^{p,\infty}(X,\mu)$.
(c) Show that $L^{p,\infty}(X,\mu)$ is metrizable for all $0 < p < \infty$ and normable when $p > 1$ (by picking $r = 1$).
(d) Use the characterization of the weak L^p quasinorm obtained in parts (a) and (b) to prove Fatou's theorem for this space: For all measurable functions g_n on X we have

$$\big\| \liminf_{n\to\infty} |g_n| \big\|_{L^{p,\infty}} \le C_p \liminf_{n\to\infty} \big\| g_n \big\|_{L^{p,\infty}}$$

for some constant C_p that depends only on $p \in (0,\infty)$.

1.1.13. Consider the $N!$ functions on the line

$$f_\sigma = \sum_{j=1}^{N} \frac{N}{\sigma(j)} \chi_{[\frac{j-1}{N},\frac{j}{N})},$$

where σ is a permutation of the set $\{1,2,\dots,N\}$.
(a) Show that each f_σ satisfies $\|f_\sigma\|_{L^{1,\infty}} = 1$.
(b) Show that $\big\|\sum_{\sigma\in S_N} f_\sigma\big\|_{L^{1,\infty}} = N!\big(1+\frac{1}{2}+\cdots+\frac{1}{N}\big)$.
(c) Conclude that the space $L^{1,\infty}(\mathbf{R})$ is not normable.
(d) Use a similar argument to prove that $L^{1,\infty}(\mathbf{R}^n)$ is not normable by considering the functions

$$f_\sigma(x_1,\dots,x_n) = \sum_{j_1=1}^{N}\cdots\sum_{j_n=1}^{N} \frac{N^n}{\sigma(\tau(j_1,\dots,j_n))} \chi_{[\frac{j_1-1}{N},\frac{j_1}{N})}(x_1)\cdots\chi_{[\frac{j_n-1}{N},\frac{j_n}{N})}(x_n),$$

where σ is a permutation of the set $\{1,2,\dots,N^n\}$ and τ is a fixed injective map from the set of all n-tuples of integers with coordinates $1 \le j \le N$ onto the set $\{1,2,\dots,N^n\}$. One may take

$$\tau(j_1,\dots,j_n) = j_1 + N(j_2-1) + N^2(j_3-1) + \cdots + N^{n-1}(j_n-1),$$

for instance.

1.1.14. Let $0 < p < 1$, $0 < s < \infty$ and let (X,μ) be a measure space.
(a) Let f be a measurable function on X. Show that

$$\int_{|f|\le s} |f|\,d\mu \le \frac{s^{1-p}}{1-p}\|f\|_{L^{p,\infty}}^p.$$

(b) Let f_j, $1 \le j \le m$, be measurable functions on X. Show that

$$\Big\| \max_{1\le j\le m} |f_j| \Big\|_{L^{p,\infty}}^p \le \sum_{j=1}^{m} \|f_j\|_{L^{p,\infty}}^p.$$

(c) Conclude that

$$\|f_1+\cdots+f_m\|_{L^{p,\infty}}^p \le \frac{2-p}{1-p}\sum_{j=1}^m \|f_j\|_{L^{p,\infty}}^p .$$

The latter estimate is referred to as the p-normability of weak L^p for $p<1$.
[*Hint:* Part (c): First obtain the estimate

$$d_{f_1+\cdots+f_m}(\alpha) \le \mu(\{|f_1+\cdots+f_m|>\alpha, \max|f_j|\le\alpha\}) + d_{\max|f_j|}(\alpha)$$

for all $\alpha>0$.]

1.1.15. (*Hölder's inequality for weak spaces*) Let f_j be in $L^{p_j,\infty}$ of a measure space X where $0<p_j<\infty$ and $1\le j\le k$. Let

$$\frac{1}{p} = \frac{1}{p_1}+\cdots+\frac{1}{p_k}.$$

Prove that

$$\|f_1\cdots f_k\|_{L^{p,\infty}} \le p^{-\frac{1}{p}}\prod_{j=1}^k p_j^{\frac{1}{p_j}}\prod_{j=1}^k \|f_j\|_{L^{p_j,\infty}}.$$

[*Hint:* Take $\|f_j\|_{L^{p_j,\infty}}=1$ for all j. Control $d_{f_1\cdots f_k}(\alpha)$ by

$$\begin{aligned}&\mu(\{|f_1|>\alpha/s_1\})+\cdots+\mu(\{|f_{k-1}|>s_{k-2}/s_{k-1}\})+\mu(\{|f_k|>s_{k-1}\})\\ &\le (s_1/\alpha)^{p_1}+(s_2/s_1)^{p_2}+\cdots+(s_{k-1}/s_{k-2})^{p_{k-1}}+(1/s_{k-1})^{p_k}.\end{aligned}$$

Set $x_1=s_1/\alpha$, $x_2=s_2/s_1,\dots,x_k=1/s_{k-1}$. Minimize $x_1^{p_1}+\cdots+x_k^{p_k}$ subject to the constraint $x_1\cdots x_k=1/\alpha$.]

1.1.16. Let $0<p_0<p<p_1\le\infty$ and let $\frac{1}{p}=\frac{1-\theta}{p_0}+\frac{\theta}{p_1}$ for some $\theta\in[0,1]$. Prove the following:

$$\begin{aligned}\|f\|_{L^p} &\le \|f\|_{L^{p_0}}^{1-\theta}\|f\|_{L^{p_1}}^{\theta},\\ \|f\|_{L^{p,\infty}} &\le \|f\|_{L^{p_0,\infty}}^{1-\theta}\|f\|_{L^{p_1,\infty}}^{\theta}.\end{aligned}$$

1.1.17. (*Loomis and Whitney* [178]) Follow the steps below to prove the *isoperimetric inequality*. For $n\ge2$ and $1\le j\le n$ define the projection maps $\pi_j:\mathbf{R}^n\to\mathbf{R}^{n-1}$ by setting for $x=(x_1,\dots,x_n)$,

$$\pi_j(x) = (x_1,\dots,x_{j-1},x_{j+1},\dots,x_n),$$

with the obvious interpretations when $j=1$ or $j=n$.
(a) For maps $f_j:\mathbf{R}^{n-1}\to\mathbf{C}$ prove that

$$\Lambda(f_1,\dots,f_n) = \int_{\mathbf{R}^n}\prod_{j=1}^n |f_j\circ\pi_j|\,dx \le \prod_{j=1}^n \|f_j\|_{L^{n-1}(\mathbf{R}^{n-1})}.$$

(b) Let Ω be a compact set with a rectifiable boundary in $\mathbf{R}^n$ where $n \geq 2$. Show that there is a constant c_n independent of Ω such that

$$|\Omega| \leq c_n |\partial\Omega|^{\frac{n}{n-1}},$$

where the expression $|\partial\Omega|$ denotes the $(n-1)$-dimensional surface measure of the boundary of Ω.
[*Hint:* Part (a): Use induction starting with $n = 2$. Then write

$$\begin{aligned} \Lambda(f_1,\dots,f_n) &\leq \int_{\mathbf{R}^{n-1}} P(x_1,\dots,x_{n-1}) |f_n(\pi_n(x))| \, dx_1 \cdots dx_{n-1} \\ &\leq \|P\|_{L^{\frac{n-1}{n-2}}(\mathbf{R}^{n-1})} \|f_n \circ \pi_n\|_{L^{n-1}(\mathbf{R}^{n-1})}, \end{aligned}$$

where $P(x_1,\dots,x_{n-1}) = \int_{\mathbf{R}} |f_1(\pi_1(x)) \cdots f_{n-1}(\pi_{n-1}(x))| \, dx_n$, and apply the induction hypothesis to the $n-1$ functions

$$\left[\int_{\mathbf{R}} f_j(\pi_j(x))^{n-1} \, dx_n\right]^{\frac{1}{n-2}},$$

for $j = 1,\dots,n-1$, to obtain the required conclusion. Part (b): Specialize part (a) to the case $f_j = \chi_{\pi_j[\Omega]}$ to obtain

$$|\Omega| \leq |\pi_1[\Omega]|^{\frac{1}{n-1}} \cdots |\pi_n[\Omega]|^{\frac{1}{n-1}}$$

and then use that $|\pi_j[\Omega]| \leq \frac{1}{2}|\partial\Omega|$.]

1.2 Convolution and Approximate Identities

The notion of convolution can be defined on measure spaces endowed with a group structure. It turns out that the most natural environment to define convolution is the context of topological groups. Although the focus of this book is harmonic analysis on Euclidean spaces, we develop the notion of convolution on general groups. This allows us to study this concept on $\mathbf{R}^n$, $\mathbf{Z}^n$, and $\mathbf{T}^n$, in a unified way. Moreover, since the basic properties of convolutions and approximate identities do not require commutativity of the group operation, we may assume that the underlying groups are not necessarily abelian. Thus, the results in this section can be also applied to nonabelian structures such as the Heisenberg group.

1.2.1 Examples of Topological Groups

A topological group G is a Hausdorff topological space that is also a group with law

$$(x,y) \mapsto xy \tag{1.2.1}$$

such that the maps $(x,y) \mapsto xy$ and $x \mapsto x^{-1}$ are continuous.

Example 1.2.1. The standard examples are provided by the spaces $\mathbf{R}^n$ and $\mathbf{Z}^n$ with the usual topology and the usual addition of n-tuples. Another example is the space $\mathbf{T}^n$ defined as follows:

$$\mathbf{T}^n = \underbrace{[0,1] \times \cdots \times [0,1]}_{n \text{ times}}$$

with the usual topology and group law addition of n-tuples mod 1, that is,

$$(x_1,\dots,x_n) + (y_1,\dots,y_n) = ((x_1+y_1) \bmod 1, \dots, (x_n+y_n) \bmod 1).$$

Let G be a locally compact group. It is known that G possesses a positive measure λ on the Borel sets that is nonzero on all nonempty open sets and is left invariant, meaning that

$$\lambda(tA) = \lambda(A), \tag{1.2.2}$$

for all measurable sets A and all $t \in G$. Such a measure λ is called a (left) *Haar measure* on G. For a constructive proof of the existence of Haar measure we refer to Lang [168, §16.3]. Furthermore, Haar measure is unique up to positive multiplicative constants. If G is abelian then any left Haar measure on G is a constant multiple of any given *right Haar measure* on G, the latter meaning right invariant [i.e., $\lambda(At) = \lambda(A)$, for all measurable $A \subseteq G$ and $t \in G$].

Example 1.2.2. Let $G = \mathbf{R}^* = \mathbf{R} \setminus \{0\}$ with group law the usual multiplication. It is easy to verify that the measure $\lambda = dx/|x|$ is invariant under multiplicative translations, that is,

$$\int_{-\infty}^{\infty} f(tx)\,\frac{dx}{|x|} = \int_{-\infty}^{\infty} f(x)\,\frac{dx}{|x|},$$

for all f in $L^1(G,\mu)$ and all $t \in \mathbf{R}^*$. Therefore, $dx/|x|$ is a Haar measure. [Taking $f = \chi_A$ gives $\lambda(tA) = \lambda(A)$.]

Example 1.2.3. Similarly, on the multiplicative group $G = \mathbf{R}^+$, a Haar measure is dx/x.

Example 1.2.4. Counting measure is a Haar measure on the group $\mathbf{Z}^n$ with group operation the usual addition.

Example 1.2.5. The *Heisenberg group* $\mathbf{H}^n$ is the set $\mathbf{C}^n \times \mathbf{R}$ with the group operation

$$(z_1,\dots,z_n,t)(w_1,\dots,w_n,s) = \Big(z_1+w_1,\dots,z_n+w_n,t+s+2\,\mathrm{Im}\sum_{j=1}^n z_j\overline{w_j}\Big).$$

It can easily be seen that the identity element e of this group is $0 \in \mathbf{C}^n \times \mathbf{R}$ and $(z_1,\dots,z_n,t)^{-1} = (-z_1,\dots,-z_n,-t)$. Topologically the Heisenberg group is identified with $\mathbf{C}^n \times \mathbf{R}$, and both left and right Haar measure on $\mathbf{H}^n$ is Lebesgue measure. The norm

$$|(z_1,\dots,z_n,t)| = \left[\left(\sum_{j=1}^{n}|z_j|^2\right)^2 + t^2\right]^{\frac{1}{4}}$$

introduces balls $B_r(x) = \{y \in \mathbf{H}^n : |y^{-1}x| < r\}$ on the Heisenberg group that are quite different from Euclidean balls. For x close to the origin, the balls $B_r(x)$ are not far from being Euclidean, but for x far away from $e = 0$ they look like slanted truncated cylinders. The Heisenberg group can be naturally identified as the boundary of the unit ball in $\mathbf{C}^n$ and plays an important role in quantum mechanics.

1.2.2 Convolution

Throughout the rest of this section, fix a locally compact group G and a left invariant Haar measure λ on G. The spaces $L^p(G,\lambda)$ and $L^{p,\infty}(G,\lambda)$ are simply denoted by $L^p(G)$ and $L^{p,\infty}(G)$.

Left invariance of λ is equivalent to the fact that for all $t \in G$ and all $f \in L^1(G)$,

$$\int_G f(tx)\,d\lambda(x) = \int_G f(x)\,d\lambda(x). \tag{1.2.3}$$

Equation (1.2.3) is a restatement of (1.2.2) if f is a characteristic function. For a general $f \in L^1(G)$ it follows by linearity and approximation.

We are now ready to define the operation of convolution.

Definition 1.2.6. Let f, g be in $L^1(G)$. Define the *convolution* $f * g$ by

$$(f*g)(x) = \int_G f(y)g(y^{-1}x)\,d\lambda(y). \tag{1.2.4}$$

For instance, if $G = \mathbf{R}^n$ with the usual additive structure, then $y^{-1} = -y$ and the integral in (1.2.4) is written as

$$(f*g)(x) = \int_{\mathbf{R}^n} f(y)g(x-y)\,dy.$$

Remark 1.2.7. The right-hand side of (1.2.4) is defined a.e., since the following double integral converges absolutely:

$$\begin{aligned}
&\int_G\int_G |f(y)||g(y^{-1}x)|\,d\lambda(y)\,d\lambda(x)\\
&= \int_G\int_G |f(y)||g(y^{-1}x)|\,d\lambda(x)\,d\lambda(y)\\
&= \int_G |f(y)|\int_G |g(y^{-1}x)|\,d\lambda(x)\,d\lambda(y)\\
&= \int_G |f(y)|\int_G |g(x)|\,d\lambda(x)\,d\lambda(y) \qquad \text{by (1.2.2)}\\
&= \|f\|_{L^1(G)}\|g\|_{L^1(G)} < +\infty.
\end{aligned}$$

The change of variables $z = x^{-1}y$ yields that (1.2.4) is in fact equal to

$$(f * g)(x) = \int_G f(xz)g(z^{-1})\,d\lambda(z), \tag{1.2.5}$$

where the substitution of $d\lambda(y)$ by $d\lambda(z)$ is justified by left invariance.

Example 1.2.8. On $\mathbf{R}$ let $f(x) = 1$ when $-1 \le x \le 1$ and zero otherwise. We see that $(f*f)(x)$ is equal to the length of the intersection of the intervals $[-1,1]$ and $[x-1,x+1]$. It follows that $(f*f)(x) = 2-|x|$ for $|x| \le 2$ and zero otherwise. Observe that $f*f$ is a smoother function than f. Similarly, we obtain that $f*f*f$ is a smoother function than $f*f$.

There is an analogous calculation when g is the characteristic function of the unit disk $B(0,1)$ in $\mathbf{R}^2$. A simple computation gives

$$\begin{aligned}(g*g)(x) &= \big|B(0,1)\cap B(x,1)\big| = \int_{-\sqrt{1-\frac14|x|^2}}^{+\sqrt{1-\frac14|x|^2}} \Big(2\sqrt{1-t^2}-|x|\Big)\,dt \\ &= 2\arcsin\left(\sqrt{1-\tfrac14|x|^2}\right) - |x|\sqrt{1-\tfrac14|x|^2}\end{aligned}$$

when $x = (x_1,x_2)$ in $\mathbf{R}^2$ satisfies $|x| \le 2$, while $(g*g)(x) = 0$ if $|x| \ge 2$.

A calculation similar to that in Remark 1.2.7 yields that

$$\big\|f*g\big\|_{L^1(G)} \le \big\|f\big\|_{L^1(G)}\big\|g\big\|_{L^1(G)}, \tag{1.2.6}$$

that is, the convolution of two integrable functions is also an integrable function with L^1 norm less than or equal to the product of the L^1 norms.

Proposition 1.2.9. *For all f, g, h in $L^1(G)$, the following properties are valid:*

(1) $f*(g*h) = (f*g)*h$ (associativity)
(2) $f*(g+h) = f*g+f*h$ and $(f+g)*h = f*h+g*h$ (distributivity)

Proof. The easy proofs are omitted. □

Proposition 1.2.9 implies that $L^1(G)$ is a (not necessarily commutative) Banach algebra under the convolution product.

1.2.3 Basic Convolution Inequalities

The most fundamental inequality involving convolutions is the following.

Theorem 1.2.10. ***(Minkowski's inequality)*** *Let $1 \le p \le \infty$. For f in $L^p(G)$ and g in $L^1(G)$ we have*

$$\big\|g*f\big\|_{L^p(G)} \le \big\|g\big\|_{L^1(G)}\big\|f\big\|_{L^p(G)}. \tag{1.2.7}$$

Proof. Estimate (1.2.7) follows directly from Exercise 1.1.6. Here we give a direct proof. We may assume that $1 < p < \infty$, since the cases $p = 1$ and $p = \infty$ are simple. Clearly, we have

$$|(g * f)(x)| \le \int_G |f(y^{-1}x)|\,|g(y)|\,d\lambda(y). \tag{1.2.8}$$

Apply Hölder's inequality in (1.2.8) with respect to the measure $|g(y)|\,d\lambda(y)$ to the functions $y \mapsto f(y^{-1}x)$ and 1 with exponents p and $p' = p/(p-1)$, respectively. We obtain

$$|(g * f)(x)| \le \left(\int_G |f(y^{-1}x)|^p |g(y)|\,d\lambda(y)\right)^{\frac{1}{p}} \left(\int_G |g(y)|\,d\lambda(y)\right)^{\frac{1}{p'}}. \tag{1.2.9}$$

Taking L^p norms of both sides of (1.2.9) we deduce

$$\begin{aligned}
\|g * f\|_{L^p} &\le \left(\|g\|_{L^1}^{p-1} \int_G \int_G |f(y^{-1}x)|^p |g(y)|\,d\lambda(y)\,d\lambda(x)\right)^{\frac{1}{p}} \\
&= \left(\|g\|_{L^1}^{p-1} \int_G \int_G |f(y^{-1}x)|^p\,d\lambda(x)|g(y)|\,d\lambda(y)\right)^{\frac{1}{p}} \\
&= \left(\|g\|_{L^1}^{p-1} \int_G \int_G |f(x)|^p\,d\lambda(x)|g(y)|\,d\lambda(y)\right)^{\frac{1}{p}} \qquad \text{by (1.2.3)} \\
&= \left(\|f\|_{L^p}^p \|g\|_{L^1} \|g\|_{L^1}^{p-1}\right)^{\frac{1}{p}} = \|f\|_{L^p} \|g\|_{L^1},
\end{aligned}$$

where the second equality follows by Fubini's theorem. The proof is complete. □

Remark 1.2.11. Theorem 1.2.10 may fail for nonabelian groups if $g * f$ is replaced by $f * g$ in (1.2.7). Note, however, that if

$$\|g\|_{L^1} = \|\widetilde{g}\|_{L^1}, \tag{1.2.10}$$

where $\widetilde{g}(x) = g(x^{-1})$, then (1.2.7) holds when the quantity $\|g * f\|_{L^p(G)}$ is replaced by $\|f * g\|_{L^p(G)}$. To see this, observe that if (1.2.10) holds, then we can use (1.2.5) to conclude that if f in $L^p(G)$ and g in $L^1(G)$, then

$$\|f * g\|_{L^p(G)} \le \|g\|_{L^1(G)} \|f\|_{L^p(G)}. \tag{1.2.11}$$

If the left Haar measure satisfies

$$\lambda(A) = \lambda(A^{-1}) \tag{1.2.12}$$

for all measurable $A \subseteq G$, then (1.2.10) holds and thus (1.2.11) is satisfied for all g in $L^1(G)$. This is, for instance, the case for the Heisenberg group $\mathbf{H}^n$.

Minkowski's inequality (1.2.11) is only a special case of Young's inequality in which the function g can be in any space $L^r(G)$ for $1 \le r \le \infty$.

Theorem 1.2.12. *(Young's inequality) Let $1 \le p,q,r \le \infty$ satisfy*

$$\frac{1}{q}+1=\frac{1}{p}+\frac{1}{r}. \tag{1.2.13}$$

Then for all f in $L^p(G)$ and all g in $L^r(G)$ satisfying $\|g\|_{L^r(G)}=\|\widetilde{g}\|_{L^r(G)}$ we have

$$\|f*g\|_{L^q(G)} \le \|g\|_{L^r(G)}\|f\|_{L^p(G)}. \tag{1.2.14}$$

Proof. Young's inequality is proved in a way similar to Minkowski's inequality. We do a suitable splitting of the product $|f(y)||g(y^{-1}x)|$ and apply Hölder's inequality. Observe that when $r<\infty$, the hypotheses on the indices imply that

$$\frac{1}{r'}+\frac{1}{q}+\frac{1}{p'}=1\,, \qquad \frac{p}{q}+\frac{p}{r'}=1\,, \qquad \frac{r}{q}+\frac{r}{p'}=1\,.$$

Using Hölder's inequality with exponents r', q, and p', we obtain

$$\begin{aligned}
|(f*g)(x)| &\le \int_G |f(y)|\,|g(y^{-1}x)|\,d\lambda(y)\\
&\le \int_G |f(y)|^{\frac{p}{r'}}\left(|f(y)|^{\frac{p}{q}}|g(y^{-1}x)|^{\frac{r}{q}}\right)|g(y^{-1}x)|^{\frac{r}{p'}}\,d\lambda(y)\\
&\le \|f\|_{L^p}^{\frac{p}{r'}}\left(\int_G |f(y)|^p|g(y^{-1}x)|^r\,d\lambda(y)\right)^{\frac{1}{q}}\left(\int_G |g(y^{-1}x)|^r\,d\lambda(y)\right)^{\frac{1}{p'}}\\
&= \|f\|_{L^p}^{\frac{p}{r'}}\left(\int_G |f(y)|^p|g(y^{-1}x)|^r\,d\lambda(y)\right)^{\frac{1}{q}}\left(\int_G |\widetilde{g}(x^{-1}y)|^r\,d\lambda(y)\right)^{\frac{1}{p'}}\\
&= \left(\int_G |f(y)|^p|g(y^{-1}x)|^r\,d\lambda(y)\right)^{\frac{1}{q}}\|f\|_{L^p}^{\frac{p}{r'}}\|\widetilde{g}\|_{L^r}^{\frac{r}{p'}},
\end{aligned}$$

where we used left invariance. Now take L^q norms (in x) and apply Fubini's theorem to deduce that

$$\begin{aligned}
\|f*g\|_{L^q} &\le \|f\|_{L^p}^{\frac{p}{r'}}\|\widetilde{g}\|_{L^r}^{\frac{r}{p'}}\left(\int_G\int_G |f(y)|^p|g(y^{-1}x)|^r\,d\lambda(x)\,d\lambda(y)\right)^{\frac{1}{q}}\\
&= \|f\|_{L^p}^{\frac{p}{r'}}\|\widetilde{g}\|_{L^r}^{\frac{r}{p'}}\|f\|_{L^p}^{\frac{p}{q}}\|g\|_{L^r}^{\frac{r}{q}}\\
&= \|g\|_{L^r}\|f\|_{L^p},
\end{aligned}$$

using the hypothesis on g. Finally, note that if $r=\infty$, the assumptions on p and q imply that $p=1$ and $q=\infty$, in which case the required inequality trivially holds. □

We now give a version of Theorem 1.2.12 for weak L^p spaces. Theorem 1.2.13 is improved in Section 1.4.

Theorem 1.2.13. *(Young's inequality for weak type spaces) Let G be a locally compact group with left Haar measure λ that satisfies (1.2.12). Let $1 \le p < \infty$ and*

$1 < q, r < \infty$ satisfy

$$\frac{1}{q} + 1 = \frac{1}{p} + \frac{1}{r}. \tag{1.2.15}$$

Then there exists a constant $C_{p,q,r} > 0$ such that for all f in $L^p(G)$ and g in $L^{r,\infty}(G)$ we have

$$\|f * g\|_{L^{q,\infty}(G)} \le C_{p,q,r} \|g\|_{L^{r,\infty}(G)} \|f\|_{L^p(G)}. \tag{1.2.16}$$

Proof. The proof is based on a suitable splitting of the function g. Let M be a positive real number to be chosen later. Define $g_1 = g\chi_{|g| \le M}$ and $g_2 = g\chi_{|g|>M}$. In view of Exercise 1.1.10(a) we have

$$d_{g_1}(\alpha) = \begin{cases} 0 & \text{if } \alpha \ge M, \\ d_g(\alpha) - d_g(M) & \text{if } \alpha < M, \end{cases} \tag{1.2.17}$$

$$d_{g_2}(\alpha) = \begin{cases} d_g(\alpha) & \text{if } \alpha > M, \\ d_g(M) & \text{if } \alpha \le M. \end{cases} \tag{1.2.18}$$

Proposition 1.1.3 gives

$$d_{f*g}(\alpha) \le d_{f*g_1}(\alpha/2) + d_{f*g_2}(\alpha/2), \tag{1.2.19}$$

and thus it suffices to estimate the distribution functions of $f * g_1$ and $f * g_2$. Since g_1 is the "small" part of g, it is in L^s for any $s > r$. In fact, we have

$$\begin{aligned} \int_G |g_1(x)|^s \, d\lambda(x) &= s \int_0^\infty \alpha^{s-1} d_{g_1}(\alpha)\, d\alpha \\ &= s \int_0^M \alpha^{s-1} (d_g(\alpha) - d_g(M))\, d\alpha \\ &\le s \int_0^M \alpha^{s-1-r} \|g\|_{L^{r,\infty}}^r \, d\alpha - s \int_0^M \alpha^{s-1} d_g(M)\, d\alpha \\ &= \frac{s}{s-r} M^{s-r} \|g\|_{L^{r,\infty}}^r - M^s d_g(M), \end{aligned} \tag{1.2.20}$$

when $s < \infty$.

Similarly, since g_2 is the "large" part of g, it is in L^t for any $t < r$, and

$$\begin{aligned} \int_G |g_2(x)|^t \, d\lambda(x) &= t \int_0^\infty \alpha^{t-1} d_{g_2}(\alpha)\, d\alpha \\ &= t \int_0^M \alpha^{t-1} d_g(M)\, d\alpha + t \int_M^\infty \alpha^{t-1} d_g(\alpha)\, d\alpha \\ &\le M^t d_g(M) + t \int_M^\infty \alpha^{t-1-r} \|g\|_{L^{r,\infty}}^r \, d\alpha \\ &\le M^{t-r} \|g\|_{L^{r,\infty}}^r + \frac{t}{r-t} M^{t-r} \|g\|_{L^{r,\infty}}^r \\ &= \frac{r}{r-t} M^{t-r} \|g\|_{L^{r,\infty}}^r. \end{aligned} \tag{1.2.21}$$

Since $1/r = 1/p' + 1/q$, it follows that $1 < r < p'$. Select $t = 1$ and $s = p'$. Hölder's inequality and (1.2.20) give

$$|(f * g_1)(x)| \leq \|f\|_{L^p} \|g_1\|_{L^{p'}} \leq \|f\|_{L^p} \left(\frac{p'}{p'-r} M^{p'-r} \|g\|_{L^{r,\infty}}^r \right)^{\frac{1}{p'}}, \tag{1.2.22}$$

when $p' < \infty$, while

$$|(f * g_1)(x)| \leq \|f\|_{L^p} M \tag{1.2.23}$$

if $p' = \infty$. Choose an M such that the right-hand side of (1.2.22) if $p' < \infty$, or (1.2.23) if $p' = \infty$, is equal to $\alpha/2$. For instance, choose

$$M = (\alpha^{p'} 2^{-p'} r q^{-1} \|f\|_{L^p}^{-p'} \|g\|_{L^{r,\infty}}^{-r})^{1/(p'-r)}$$

if $p' < \infty$ and $M = \alpha/(2\|f\|_{L^1})$ if $p' = \infty$. For this choice of M we have that

$$d_{f * g_1}(\alpha/2) = 0.$$

Next by Theorem 1.2.10 and (1.2.21) with $t = 1$ we obtain

$$\|f * g_2\|_{L^p} \leq \|f\|_{L^p} \|g_2\|_{L^1} \leq \|f\|_{L^p} \frac{r}{r-1} M^{1-r} \|g\|_{L^{r,\infty}}^r. \tag{1.2.24}$$

For the value of M chosen, using (1.2.24) and Chebyshev's inequality, we obtain

$$\begin{aligned} d_{f*g}(\alpha) &\leq d_{f*g_2}(\alpha/2) \\ &\leq (2\|f * g_2\|_{L^p} \alpha^{-1})^p \\ &\leq (2r\|f\|_{L^p} M^{1-r} \|g\|_{L^{r,\infty}}^r (r-1)^{-1} \alpha^{-1})^p \\ &= C_{p,q,r}^q \alpha^{-q} \|f\|_{L^p}^q \|g\|_{L^{r,\infty}}^q, \end{aligned} \tag{1.2.25}$$

which is the required inequality. This proof gives that the constant $C_{p,q,r}$ blows up like $(r-1)^{-p/q}$ as $r \to 1$. □

Example 1.2.14. Theorem 1.2.13 may fail at some endpoints:

(1) $r = 1$ and $1 \leq p = q \leq \infty$. On **R** take $g(x) = 1/|x|$ and $f = \chi_{[0,1]}$. Clearly, g is in $L^{1,\infty}$ and f in L^p for all $1 \leq p \leq \infty$, but the convolution of f and g is identically equal to infinity on the interval $[0,1]$. Therefore, (1.2.16) fails in this case.

(2) $q = \infty$ and $1 < r = p' < \infty$. On **R** let $f(x) = (|x|^{1/p} \log|x|)^{-1}$ for $|x| \geq 2$ and zero otherwise, and also let $g(x) = |x|^{-1/r}$. We see that $(f * g)(x) = \infty$ for $|x| \leq 1$. Thus (1.2.16) fails in this case also.

(3) $r = q = \infty$ and $p = 1$. Then inequality (1.2.16) trivially holds.

1.2.4 Approximate Identities

We now introduce the notion of approximate identities. The Banach algebra $L^1(G)$ may not have a unit element, that is, an element f_0 such that

$$f_0 * f = f = f * f_0 \tag{1.2.26}$$

for all $f \in L^1(G)$. In particular, this is the case when $G = \mathbf{R}$; in fact, the only f_0 that satisfies (1.2.26) for all $f \in L^1(\mathbf{R})$ is not a function but the Dirac delta distribution, introduced in Chapter 2. It is reasonable therefore to introduce the notion of approximate unit or identity, a family of functions k_ε with the property $k_\varepsilon * f \to f$ in L^1 as $\varepsilon \to 0$.

Definition 1.2.15. An *approximate identity* (as $\varepsilon \to 0$) is a family of $L^1(G)$ functions k_ε with the following three properties:

(i) There exists a constant $c > 0$ such that $\|k_\varepsilon\|_{L^1(G)} \le c$ for all $\varepsilon > 0$.
(ii) $\int_G k_\varepsilon(x)\,d\lambda(x) = 1$ for all $\varepsilon > 0$.
(iii) For any neighborhood V of the identity element e of the group G we have $\int_{V^c} |k_\varepsilon(x)|\,d\lambda(x) \to 0$ as $\varepsilon \to 0$.

The construction of approximate identities on general locally compact groups G is beyond the scope of this book and is omitted. We refer to Hewitt and Ross [125] for details. In this book we are interested only in groups with Euclidean structure, where approximate identities exist in abundance. See the following examples.

Sometimes we think of approximate identities as sequences $\{k_n\}_n$. In this case property (iii) holds as $n \to \infty$. It is best to visualize approximate identities as sequences of positive functions k_n that spike near 0 in such a way that the signed area under the graph of each function remains constant (equal to one) but the support shrinks to zero. See Figure 1.2.

Example 1.2.16. On $\mathbf{R}^n$ let $k(x)$ be an integrable function with integral one. Let $k_\varepsilon(x) = \varepsilon^{-n} k(\varepsilon^{-1}x)$. It is straightforward to see that $k_\varepsilon(x)$ is an approximate identity. Property (iii) follows from the fact that

$$\int_{|x| \ge \delta/\varepsilon} |k(x)|\,dx \to 0$$

as $\varepsilon \to 0$ for δ fixed.

Example 1.2.17. On $\mathbf{R}$ let $P(x) = (\pi(x^2+1))^{-1}$ and $P_\varepsilon(x) = \varepsilon^{-1}P(\varepsilon^{-1}x)$ for $\varepsilon > 0$. Since P_ε and P have the same L^1 norm and

$$\int_{-\infty}^{+\infty} \frac{1}{x^2+1}\,dx = \lim_{x \to +\infty} \big[\arctan(x) - \arctan(-x)\big] = (\pi/2) - (-\pi/2) = \pi\,,$$

property (ii) is satisfied. Property (iii) follows from the fact that

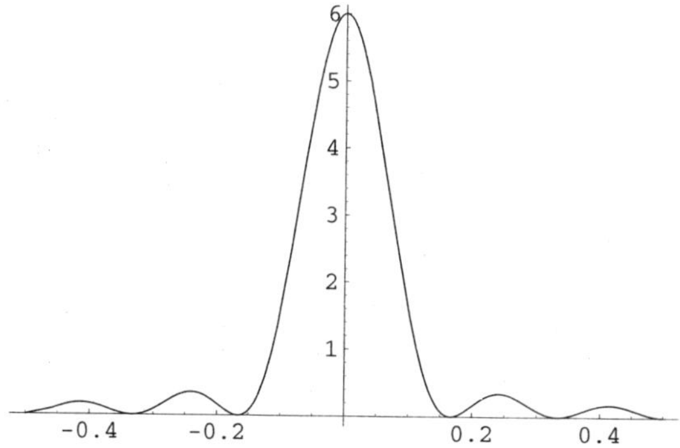

Fig. 1.2 The Fejér kernel F_5 plotted on the interval $[-\frac{1}{2}, \frac{1}{2}]$.

$$\frac{1}{\pi}\int_{|x|\geq\delta}\frac{1}{\varepsilon}\frac{1}{(x/\varepsilon)^2+1}\,dx = 1-\frac{2}{\pi}\arctan(\delta/\varepsilon)\to 0 \qquad \text{as } \varepsilon\to 0,$$

for all $\delta > 0$. The function P_ε is called the *Poisson kernel.*

Example 1.2.18. On the circle group $\mathbf{T}^1$ let

$$F_N(t) = \sum_{j=-N}^{N}\left(1-\frac{|j|}{N+1}\right)e^{2\pi i j t} = \frac{1}{N+1}\left(\frac{\sin(\pi(N+1)t)}{\sin(\pi t)}\right)^2. \tag{1.2.27}$$

To check the previous equality we use that

$$\sin^2(x) = (2-e^{2ix}-e^{-2ix})/4,$$

and we carry out the calculation. F_N is called the *Fejér kernel.* To see that the sequence $\{F_N\}_N$ is an approximate identity, we check conditions (i), (ii), and (iii) in Definition 1.2.15. Property (iii) follows from the expression giving F_N in terms of sines, while property (i) follows from the expression giving F_N in terms of exponentials. Property (ii) is identical to property (i), since F_N is nonnegative.

Next comes the basic theorem concerning approximate identities.

Theorem 1.2.19. *Let k_ε be an approximate identity on a locally compact group G with left Haar measure λ.*

(1) If $f\in L^p(G)$ for $1\leq p<\infty$, then $\big\|k_\varepsilon * f - f\big\|_{L^p(G)}\to 0$ as $\varepsilon\to 0$.
(2) When $p=\infty$, the following is valid: If f is continuous in a neighborhood of a a compact subset K of G, then $\big\|k_\varepsilon * f - f\big\|_{L^\infty(K)}\to 0$ as $\varepsilon\to 0$.

Proof. We start with the case $1\leq p<\infty$. We recall that continuous functions with compact support are dense in L^p of locally compact Hausdorff spaces equipped with measures arising from nonnegative linear functionals (see Hewitt and Ross [125], Theorem 12.10). For a continuous function with compact support g we have $|g(h^{-1}x)-g(x)|^p\leq(2\big\|g\big\|_{L^\infty})^p$ for h in a relatively compact neighborhood of the origin e, and by the Lebesgue dominated convergence theorem we obtain

$$\int_G |g(h^{-1}x) - g(x)|^p \, d\lambda(x) \to 0 \tag{1.2.28}$$

as $h \to e$. Now approximate a given f in $L^p(G)$ by a continuous function with compact support g to deduce that

$$\int_G |f(h^{-1}x) - f(x)|^p \, d\lambda(x) \to 0 \qquad \text{as} \qquad h \to e. \tag{1.2.29}$$

Because of (1.2.29), given a $\delta > 0$ there exists a neighborhood V of e such that

$$h \in V \implies \int_G |f(h^{-1}x) - f(x)|^p \, d\lambda(x) < \left(\frac{\delta}{2c}\right)^p, \tag{1.2.30}$$

where c is the constant that appears in Definition 1.2.15 (i). Since k_ε has integral one for all $\varepsilon > 0$, we have

$$\begin{aligned}(k_\varepsilon * f)(x) - f(x) &= (k_\varepsilon * f)(x) - f(x) \int_G k_\varepsilon(y) \, d\lambda(y) \\ &= \int_G (f(y^{-1}x) - f(x)) k_\varepsilon(y) \, d\lambda(y) \\ &= \int_V (f(y^{-1}x) - f(x)) k_\varepsilon(y) \, d\lambda(y) \\ &\qquad + \int_{V^c} (f(y^{-1}x) - f(x)) k_\varepsilon(y) \, d\lambda(y).\end{aligned} \tag{1.2.31}$$

Now take L^p norms in x in (1.2.31). In view of (1.2.30),

$$\begin{aligned}&\left\| \int_V (f(y^{-1}x) - f(x)) k_\varepsilon(y) \, d\lambda(y) \right\|_{L^p(G, d\lambda(x))} \\ &\qquad \le \int_V \|f(y^{-1}x) - f(x)\|_{L^p(G, d\lambda(x))} |k_\varepsilon(y)| \, d\lambda(y) \\ &\qquad \le \int_V \frac{\delta}{2c} |k_\varepsilon(y)| \, d\lambda(y) < \frac{\delta}{2},\end{aligned} \tag{1.2.32}$$

while

$$\begin{aligned}&\left\| \int_{V^c} (f(y^{-1}x) - f(x)) k_\varepsilon(y) \, d\lambda(y) \right\|_{L^p(G, d\lambda(x))} \\ &\qquad \le \int_{V^c} 2\|f\|_{L^p(G)} |k_\varepsilon(y)| \, d\lambda(y) < \frac{\delta}{2},\end{aligned} \tag{1.2.33}$$

provided we have that

$$\int_{V^c} |k_\varepsilon(x)| \, d\lambda(x) < \frac{\delta}{4\|f\|_{L^p}}. \tag{1.2.34}$$

Choose $\varepsilon_0 > 0$ such that (1.2.34) is valid for $\varepsilon < \varepsilon_0$ by property (iii). Now (1.2.32) and (1.2.33) imply the required conclusion.

The case $p = \infty$ follows similarly. Since f is uniformly continuous on K, given $\delta > 0$ find a neighborhood V of $e \in G$ such that

$$h \in V \implies |f(h^{-1}x) - f(x)| < \frac{\delta}{2c} \qquad \text{for all } x \in K, \tag{1.2.35}$$

where c is as in Definition 1.2.15(i), and then find an $\varepsilon_0 > 0$ such that for $0 < \varepsilon < \varepsilon_0$ we have

$$\int_{V^c} |k_\varepsilon(y)|\, d\lambda(y) < \frac{\delta}{4\|f\|_{L^\infty}}. \tag{1.2.36}$$

Using (1.2.35) and (1.2.36), we easily conclude that

$$\begin{aligned}
&\sup_{x\in K} |(k_\varepsilon * f)(x) - f(x)| \\
&\le \int_V |k_\varepsilon(y)| \sup_{x\in K} |f(y^{-1}x) - f(x)|\, d\lambda(y) + \int_{V^c} |k_\varepsilon(y)| \sup_{x\in K} |f(y^{-1}x) - f(x)|\, d\lambda(y) \\
&\le \frac{\delta}{2} + \frac{\delta}{2} = \delta,
\end{aligned}$$

which shows that $k_\varepsilon * f$ converges uniformly to f on K as $\varepsilon \to 0$. □

Remark 1.2.20. Observe that if Haar measure satisfies (1.2.12), then the conclusion of Theorem 1.2.19 also holds for $f * k_\varepsilon$.

A simple modification in the proof of Theorem 1.2.19 yields the following variant.

Theorem 1.2.21. *Let k_ε be a family of functions on a locally compact group G that satisfies properties (i) and (iii) of Definition 1.2.15 and also*

$$\int_G k_\varepsilon(x)\, d\lambda(x) = a \in \mathbf{C}, \qquad \textit{for all } \varepsilon > 0.$$

Let $f \in L^p(G)$ for some $1 \le p \le \infty$.

(a) If $1 \le p < \infty$, then $\|k_\varepsilon * f - af\|_{L^p(G)} \to 0$ as $\varepsilon \to 0$.

(b) If $p = \infty$ and f is continuous on a compact $K \subseteq G$, then

$$\|k_\varepsilon * f - af\|_{L^\infty(K)} \to 0$$

as $\varepsilon \to 0$.

Remark 1.2.22. With the notation of Theorem 1.2.21, if f is continuous and tends to zero at infinity, then $\|k_\varepsilon * f - af\|_{L^\infty(G)} \to 0$. To see this, simply observe that outside a compact subset of G, both $k_\varepsilon * f$, af have small L^∞ norm, while inside a compact subset of G, uniform convergence holds.

Exercises

1.2.1. Let G be a locally compact group and let f,g in $L^1(G)$ be supported in the subsets A and B of G, respectively. Prove that $f*g$ is supported in the algebraic product set AB.

1.2.2. For a function f on a locally compact group G and $t \in G$, let ${}^t f(x) = f(tx)$ and $f^t(x) = f(xt)$. Show that

$$ {}^t f * g = {}^t(f*g) \qquad \text{and} \qquad f*g^t = (f*g)^t $$

whenever $f,g \in L^1(G)$, equipped with left Haar measure.

1.2.3. Let G be a locally compact group with left Haar measure. Let $f \in L^p(G)$ and $\widetilde{g} \in L^{p'}(G)$, where $1 < p < \infty$; recall that $\widetilde{g}(x) = g(x^{-1})$. For $t,x \in G$, let ${}^t g(x) = g(tx)$. Show that for any $\varepsilon > 0$ there exists a relatively compact symmetric neighborhood of the origin U such that $u \in U$ implies $\big\| {}^u\widetilde{g} - \widetilde{g} \big\|_{L^{p'}(G)} < \varepsilon$ and therefore

$$ |(f*g)(v) - (f*g)(w)| < \big\|f\big\|_{L^p}\, \varepsilon $$

whenever $vw^{-1} \in U$.

1.2.4. Let G be a locally compact group and let $1 \le p \le \infty$. Let $f \in L^p(G)$ and μ be a finite Borel measure on G with total variation $\big\|\mu\big\|$. Define

$$ (\mu * f)(x) = \int_G f(y^{-1}x)\, d\mu(y)\,. $$

Show that if μ is an absolutely continuous measure, then the preceding definition extends (1.2.4). Prove that $\big\|\mu * f\big\|_{L^p(G)} \le \big\|\mu\big\| \big\|f\big\|_{L^p(G)}$.

1.2.5. Show that a Haar measure λ for the multiplicative group of all positive real numbers is

$$ \lambda(A) = \int_0^\infty \chi_A(t)\, \frac{dt}{t}\,. $$

1.2.6. Let $G = \mathbf{R}^2 \setminus \{(0,y) : y \in \mathbf{R}\}$ with group operation $(x,y)(z,w) = (xz, xw+y)$. [Think of G as the group of all 2×2 matrices with bottom row $(0,1)$ and nonzero top left entry.] Show that a left Haar measure on G is

$$ \lambda(A) = \int_{-\infty}^{+\infty} \int_{-\infty}^{+\infty} \chi_A(x,y)\, \frac{dx\,dy}{x^2}\,, $$

while a right Haar measure on G is

$$ \rho(A) = \int_{-\infty}^{+\infty} \int_{-\infty}^{+\infty} \chi_A(x,y)\, \frac{dx\,dy}{|x|}\,. $$

1.2.7. (*Hardy [118], [119]*) Use Theorem 1.2.10 to prove that

$$\left(\int_0^\infty \left(\frac{1}{x}\int_0^x |f(t)|\,dt\right)^p dx\right)^{\frac{1}{p}} \le \frac{p}{p-1}\|f\|_{L^p(0,\infty)},$$

$$\left(\int_0^\infty \left(\int_x^\infty |f(t)|\,dt\right)^p dx\right)^{\frac{1}{p}} \le p\left(\int_0^\infty |f(t)|^p t^p\,dt\right)^{\frac{1}{p}},$$

when $1 < p < \infty$.
[*Hint:* On the multiplicative group $(\mathbf{R}^+, \frac{dt}{t})$ consider the convolution of the function $|f(x)|x^{\frac{1}{p}}$ with the function $x^{-\frac{1}{p'}}\chi_{[1,\infty)}$ and the convolution of the function $|f(x)|x^{1+\frac{1}{p}}$ with $x^{\frac{1}{p}}\chi_{(0,1]}$.]

1.2.8. (*G. H. Hardy*) Let $0 < b < \infty$ and $1 \le p < \infty$. Prove that

$$\left(\int_0^\infty \left(\int_0^x |f(t)|\,dt\right)^p x^{-b-1}\,dx\right)^{\frac{1}{p}} \le \frac{p}{b}\left(\int_0^\infty |f(t)|^p t^{p-b-1}\,dt\right)^{\frac{1}{p}},$$

$$\left(\int_0^\infty \left(\int_x^\infty |f(t)|\,dt\right)^p x^{b-1}\,dx\right)^{\frac{1}{p}} \le \frac{p}{b}\left(\int_0^\infty |f(t)|^p t^{p+b-1}\,dt\right)^{\frac{1}{p}}.$$

[*Hint:* On the multiplicative group $(\mathbf{R}^+, \frac{dt}{t})$ consider the convolution of the function $|f(x)|x^{1-\frac{b}{p}}$ with $x^{-\frac{b}{p}}\chi_{[1,\infty)}$ and of the function $|f(x)|x^{1+\frac{b}{p}}$ with $x^{\frac{b}{p}}\chi_{(0,1]}$.]

1.2.9. On $\mathbf{R}^n$ let $T(f) = f * K$, where K is a positive L^1 function and f is in L^p, $1 \le p \le \infty$. Prove that the operator norm of $T : L^p \to L^p$ is equal to $\|K\|_{L^1}$.
[*Hint:* Clearly, $\|T\|_{L^p\to L^p} \le \|K\|_{L^1}$. Conversely, fix $0 < \varepsilon < 1$ and let N be a positive integer. Let $\chi_N = \chi_{B(0,N)}$ and for any $R > 0$ let $K_R = K\chi_{B(0,R)}$, where $B(x,R)$ is the ball of radius R centered at x. Observe that for $|x| \le (1-\varepsilon)N$, we have $B(0,N\varepsilon) \subseteq B(x,N)$; thus $\int_{\mathbf{R}^n} \chi_N(x-y)K_{N\varepsilon}(y)\,dy = \int_{\mathbf{R}^n} K_{N\varepsilon}(y)\,dy = \|K_{N\varepsilon}\|_{L^1}$. Then

$$\frac{\|K * \chi_N\|_{L^p}^p}{\|\chi_N\|_{L^p}^p} \ge \frac{\|K_{N\varepsilon} * \chi_N\|_{L^p(B(0,(1-\varepsilon)N)}^p}{\|\chi_N\|_{L^p}^p} \ge \|K_{N\varepsilon}\|_{L^1}^p (1-\varepsilon)^n.$$

Let $N \to \infty$ first and then $\varepsilon \to 0$.]

1.2.10. On the multiplicative group $(\mathbf{R}^+, \frac{dt}{t})$ let $T(f) = f * K$, where K is a positive L^1 function and f is in L^p, $1 \le p \le \infty$. Prove that the operator norm of $T : L^p \to L^p$ is equal to the L^1 norm of K. Deduce that the constants $p/(p-1)$ and p/b are sharp in Exercises 1.2.7 and 1.2.8.
[*Hint:* Adapt the idea of Exercise 1.2.9 to this setting.]

1.2.11. Let $Q_j(t) = c_j(1-t^2)^j$ for $t \in [-1,1]$ and zero elsewhere, where c_j is chosen such that $\int_{-1}^1 Q_j(t)\,dt = 1$ for all $j = 1,2,\dots$.
(a) Show that $c_j < \sqrt{j}$.

(b) Use part (a) to show that $\{Q_j\}_j$ is an approximate identity on $\mathbf{R}$ as $j \to \infty$.
(c) Given a continuous function f on $\mathbf{R}$ that vanishes outside the interval $[-1,1]$, show that $f * Q_j$ converges to f uniformly on $[-1,1]$.
(d) (*Weierstrass*) Prove that every continuous function on $[-1,1]$ can be approximated uniformly by polynomials.
[*Hint:* Part (a): Consider the integral $\int_{|t|\le n^{-1/2}} Q_j(t)\,dt$. Part (d): Consider the function $g(t) = f(t) - f(-1) - \frac{t+1}{2}(f(1) - f(-1))$.]

1.2.12. (Christ and Grafakos [51]) Let $F \ge 0$, $G \ge 0$ be measurable functions on the sphere $\mathbf{S}^{n-1}$ and let $K \ge 0$ be a measurable function on $[-1,1]$. Prove that

$$\int_{\mathbf{S}^{n-1}}\int_{\mathbf{S}^{n-1}} F(\theta)G(\varphi)K(\theta\cdot\varphi)\,d\varphi\,d\theta \le C\|F\|_{L^p(\mathbf{S}^{n-1})}\|G\|_{L^{p'}(\mathbf{S}^{n-1})},$$

where $1 \le p \le \infty$, $\theta\cdot\varphi = \sum_{j=1}^n \theta_j\varphi_j$ and $C = \int_{\mathbf{S}^{n-1}} K(\theta\cdot\varphi)\,d\varphi$, which is independent of θ. Moreover, show that C is the best possible constant in the preceding inequality. Using duality, compute the norm of the linear operator

$$F(\theta) \mapsto \int_{\mathbf{S}^{n-1}} F(\theta)K(\theta\cdot\varphi)\,d\varphi$$

from $L^p(\mathbf{S}^{n-1})$ to itself.
[*Hint:* Observe that $\int_{\mathbf{S}^{n-1}}\int_{\mathbf{S}^{n-1}} F(\theta)G(\varphi)K(\theta\cdot\varphi)\,d\varphi\,d\theta$ is bounded by the quantity $\left\{\int_{\mathbf{S}^{n-1}}\left[\int_{\mathbf{S}^{n-1}} F(\theta)K(\theta\cdot\varphi)\,d\theta\right]^p d\varphi\right\}^{\frac{1}{p}}\|G\|_{L^{p'}(\mathbf{S}^{n-1})}$. Apply Hölder's inequality to the functions F and 1 with respect to the measure $K(\theta\cdot\varphi)\,d\theta$ to deduce that $\int_{\mathbf{S}^{n-1}} F(\theta)K(\theta\cdot\varphi)\,d\theta$ is controlled by

$$\left(\int_{\mathbf{S}^{n-1}} F(\theta)^p K(\theta\cdot\varphi)\,d\theta\right)^{1/p}\left(\int_{\mathbf{S}^{n-1}} K(\theta\cdot\varphi)\,d\theta\right)^{1/p'}.$$

Use Fubini's theorem to bound the latter by

$$\|F\|_{L^p(\mathbf{S}^{n-1})}\|G\|_{L^{p'}(\mathbf{S}^{n-1})}\int_{\mathbf{S}^{n-1}} K(\theta\cdot\varphi)\,d\varphi.$$

Note that equality is attained if and only if both F and G are constants.]

1.3 Interpolation

The theory of interpolation of operators is vast and extensive. In this section we are mainly concerned with a couple of basic interpolation results that appear in a variety of applications and constitute the foundation of the field. These results are the *Marcinkiewicz interpolation theorem* and the *Riesz–Thorin interpolation theorem*. These theorems are traditionally proved using real and complex variables techniques, respectively. A byproduct of the Riesz–Thorin interpolation theorem, *Stein's theorem on interpolation of analytic families of operators*, has also proved to be an important and useful tool in many applications and is presented at the end of the section.

We begin by setting up the background required to formulate the results of this section. Let (X,μ) and (Y,ν) be two measure spaces. Suppose we are given a linear operator T, initially defined on the set of simple functions on X, such that for all f simple on X, $T(f)$ is a ν-measurable function on Y. Let $0<p<\infty$ and $0<q<\infty$. If there exists a constant $C_{p,q}>0$ such that for all simple functions f on X we have

$$\big\|T(f)\big\|_{L^q(Y,\nu)} \le C_{p,q}\big\|f\big\|_{L^p(X,\mu)}, \tag{1.3.1}$$

then by density, T admits a unique bounded extension from $L^p(X,\mu)$ to $L^q(Y,\nu)$. This extension is also denoted by T. Operators that map L^p to L^q are called of *strong type* (p,q) and operators that map L^p to $L^{q,\infty}$ are called *weak type* (p,q).

1.3.1 Real Method: The Marcinkiewicz Interpolation Theorem

Definition 1.3.1. Let T be an operator defined on a linear space of complex-valued measurable functions on a measure space (X,μ) and taking values in the set of all complex-valued finite almost everywhere measurable functions on a measure space (Y,ν). Then T is called *linear* if for all f, g and all $\lambda\in\mathbf{C}$, we have

$$T(f+g)=T(f)+T(g) \qquad \text{and} \qquad T(\lambda f)=\lambda T(f). \tag{1.3.2}$$

T is called *sublinear* if for all f, g and all $\lambda\in\mathbf{C}$, we have

$$|T(f+g)|\le|T(f)|+|T(g)| \qquad \text{and} \qquad |T(\lambda f)|=|\lambda||T(f)|. \tag{1.3.3}$$

T is called *quasilinear* if for all f, g and all $\lambda\in\mathbf{C}$, we have

$$|T(f+g)|\le K(|T(f)|+|T(g)|) \qquad \text{and} \qquad |T(\lambda f)|=|\lambda||T(f)| \tag{1.3.4}$$

for some constant $K>0$. Sublinearity is a special case of quasilinearity.

We begin with the first interpolation theorem.

Theorem 1.3.2. *Let (X,μ) and (Y,ν) be measure spaces and let $0<p_0<p_1\le\infty$. Let T be a sublinear operator defined on the space $L^{p_0}(X)+L^{p_1}(X)$ and taking values in the space of measurable functions on Y. Assume that there exist two positive constants A_0 and A_1 such that*

$$\big\|T(f)\big\|_{L^{p_0,\infty}(Y)} \le A_0\big\|f\big\|_{L^{p_0}(X)} \qquad \textit{for all } f\in L^{p_0}(X), \tag{1.3.5}$$

$$\big\|T(f)\big\|_{L^{p_1,\infty}(Y)} \le A_1\big\|f\big\|_{L^{p_1}(X)} \qquad \textit{for all } f\in L^{p_1}(X). \tag{1.3.6}$$

Then for all $p_0<p<p_1$ and for all f in $L^p(X)$ we have the estimate

$$\|T(f)\|_{L^p(Y)} \le A\|f\|_{L^p(X)}, \tag{1.3.7}$$

where

$$A = 2\left(\frac{p}{p-p_0}+\frac{p}{p_1-p}\right)^{\frac{1}{p}} A_0^{\frac{\frac{1}{p}-\frac{1}{p_1}}{\frac{1}{p_0}-\frac{1}{p_1}}} A_1^{\frac{\frac{1}{p_0}-\frac{1}{p}}{\frac{1}{p_0}-\frac{1}{p_1}}}. \tag{1.3.8}$$

Proof. Assume first that $p_1 < \infty$. Fix f a function in $L^p(X)$ and $\alpha > 0$. We split $f = f_0^\alpha + f_1^\alpha$, where f_0^α is in L^{p_0} and f_1^α is in L^{p_1}. The splitting is obtained by cutting $|f|$ at height $\delta\alpha$ for some $\delta > 0$ to be determined later. Set

$$f_0^\alpha(x) = \begin{cases} f(x) & \text{for } |f(x)| > \delta\alpha, \\ 0 & \text{for } |f(x)| \le \delta\alpha, \end{cases}$$

$$f_1^\alpha(x) = \begin{cases} f(x) & \text{for } |f(x)| \le \delta\alpha, \\ 0 & \text{for } |f(x)| > \delta\alpha. \end{cases}$$

It can be checked easily that f_0^α (the unbounded part of f) is an L^{p_0} function and that f_1^α (the bounded part of f) is an L^{p_1} function. Indeed, since $p_0 < p$, we have

$$\|f_0^\alpha\|_{L^{p_0}}^{p_0} = \int_{|f|>\delta\alpha} |f|^p |f|^{p_0-p}\, d\mu(x) \le (\delta\alpha)^{p_0-p}\|f\|_{L^p}^p$$

and similarly, since $p < p_1$,

$$\|f_1^\alpha\|_{L^{p_1}}^{p_1} \le (\delta\alpha)^{p_1-p}\|f\|_{L^p}^p.$$

By the sublinearity property (1.3.3) we obtain that

$$|T(f)| \le |T(f_0^\alpha)| + |T(f_1^\alpha)|,$$

which implies

$$\{x : |T(f)(x)| > \alpha\} \subseteq \{x : |T(f_0^\alpha)(x)| > \alpha/2\} \cup \{x : |T(f_1^\alpha)(x)| > \alpha/2\},$$

and therefore

$$d_{T(f)}(\alpha) \le d_{T(f_0^\alpha)}(\alpha/2) + d_{T(f_1^\alpha)}(\alpha/2). \tag{1.3.9}$$

Hypotheses (1.3.5) and (1.3.6) together with (1.3.9) now give

$$d_{T(f)}(\alpha) \le \frac{A_0^{p_0}}{(\alpha/2)^{p_0}} \int_{|f|>\delta\alpha} |f(x)|^{p_0}\, d\mu(x) + \frac{A_1^{p_1}}{(\alpha/2)^{p_1}} \int_{|f|\le\delta\alpha} |f(x)|^{p_1}\, d\mu(x).$$

In view of the last estimate and Proposition 1.1.4, we obtain that

$$
\begin{aligned}
\|T(f)\|_{L^p}^p &\le p(2A_0)^{p_0}\int_0^\infty \alpha^{p-1}\alpha^{-p_0}\int_{|f|>\delta\alpha}|f(x)|^{p_0}\,d\mu(x)\,d\alpha\\
&\quad + p(2A_1)^{p_1}\int_0^\infty \alpha^{p-1}\alpha^{-p_1}\int_{|f|\le\delta\alpha}|f(x)|^{p_1}\,d\mu(x)\,d\alpha\\
&= p(2A_0)^{p_0}\int_X |f(x)|^{p_0}\int_0^{\frac{1}{\delta}|f(x)|}\alpha^{p-1-p_0}\,d\alpha\,d\mu(x)\\
&\quad + p(2A_1)^{p_1}\int_X |f(x)|^{p_1}\int_{\frac{1}{\delta}|f(x)|}^\infty \alpha^{p-1-p_1}\,d\alpha\,d\mu(x)\\
&= \frac{p(2A_0)^{p_0}}{p-p_0}\frac{1}{\delta^{p-p_0}}\int_X |f(x)|^{p_0}|f(x)|^{p-p_0}\,d\mu(x)\\
&\quad + \frac{p(2A_1)^{p_1}}{p_1-p}\frac{1}{\delta^{p-p_1}}\int_X |f(x)|^{p_1}|f(x)|^{p-p_1}\,d\mu(x)\\
&= p\left(\frac{(2A_0)^{p_0}}{p-p_0}\frac{1}{\delta^{p-p_0}} + \frac{(2A_1)^{p_1}}{p_1-p}\delta^{p_1-p}\right)\|f\|_{L^p}^p,
\end{aligned}
$$

and the convergence of the integrals in α is justified from $p_0 < p < p_1$. We pick $\delta > 0$ such that

$$
(2A_0)^{p_0}\frac{1}{\delta^{p-p_0}} = (2A_1)^{p_1}\delta^{p_1-p},
$$

and observe that the last displayed constant is equal to the pth power of the constant in (1.3.8). We have therefore proved the theorem when $p_1 < \infty$.

We now consider the case $p_1 = \infty$. Write $f = f_0^\alpha + f_1^\alpha$, where

$$
f_0^\alpha(x) = \begin{cases} f(x) & \text{for } |f(x)| > \gamma\alpha,\\ 0 & \text{for } |f(x)| \le \gamma\alpha,\end{cases}
$$

$$
f_1^\alpha(x) = \begin{cases} f(x) & \text{for } |f(x)| \le \gamma\alpha,\\ 0 & \text{for } |f(x)| > \gamma\alpha.\end{cases}
$$

We have

$$
\|T(f_1^\alpha)\|_{L^\infty} \le A_1\|f_1^\alpha\|_{L^\infty} \le A_1\gamma\alpha = \alpha/2,
$$

provided we choose $\gamma = (2A_1)^{-1}$. It follows that the set $\{x : |T(f_1^\alpha)(x)| > \alpha/2\}$ has measure zero. Therefore,

$$
d_{T(f)}(\alpha) \le d_{T(f_0^\alpha)}(\alpha/2).
$$

Since T maps L^{p_0} to $L^{p_0,\infty}$, it follows that

$$
d_{T(f_0^\alpha)}(\alpha/2) \le \frac{(2A_0)^{p_0}\|f_0^\alpha\|_{L^{p_0}}^{p_0}}{\alpha^{p_0}} = \frac{(2A_0)^{p_0}}{\alpha^{p_0}}\int_{|f|>\gamma\alpha}|f(x)|^{p_0}\,d\mu(x). \qquad (1.3.10)
$$

Using (1.3.10) and Proposition 1.1.4, we obtain

$$\begin{aligned}\|T(f)\|_{L^p}^p &= p\int_0^\infty \alpha^{p-1} d_{T(f)}(\alpha)\,d\alpha\\ &\le p\int_0^\infty \alpha^{p-1} d_{T(f_0^\alpha)}(\alpha/2)\,d\alpha\\ &\le p\int_0^\infty \alpha^{p-1}\frac{(2A_0)^{p_0}}{\alpha^{p_0}}\int_{|f|>\alpha/(2A_1)} |f(x)|^{p_0}\,d\mu(x)\,d\alpha\\ &= p(2A_0)^{p_0}\int_X |f(x)|^{p_0}\int_0^{2A_1|f(x)|}\alpha^{p-p_0-1}\,d\alpha\,d\mu(x)\\ &= \frac{p(2A_1)^{p-p_0}(2A_0)^{p_0}}{p-p_0}\int_X |f(x)|^p\,d\mu(x).\end{aligned}$$

This proves the theorem with constant

$$A = 2\left(\frac{p}{p-p_0}\right)^{\frac{1}{p}} A_1^{1-\frac{p_0}{p}} A_0^{\frac{p_0}{p}}. \tag{1.3.11}$$

Observe that when $p_1 = \infty$, the constant in (1.3.11) coincides with that in (1.3.8). □

Remark 1.3.3. If T is a linear operator (instead of sublinear), then we can relax the hypotheses of Theorem 1.3.2 by assuming that (1.3.5) and (1.3.6) hold for all simple functions f on X. Then the functions f_0^α and f_1^α constructed in the proof are also simple, and we conclude that (1.3.7) holds for all simple functions f on X. By density, T has a unique extension on $L^p(X)$ that also satisfies (1.3.7).

1.3.2 Complex Method: The Riesz–Thorin Interpolation Theorem

The next interpolation theorem assumes stronger endpoint estimates, but yields a more natural bound on the norm of the operator on the intermediate spaces. Unfortunately, it is mostly applicable for linear operators and in some cases for sublinear operators (often via a linearization process). It does not apply to quasilinear operators without some loss in the constant. A short history of this theorem is discussed at the end of this chapter.

Theorem 1.3.4. *Let (X,μ) and (Y,ν) be two measure spaces. Let T be a linear operator defined on the set of all simple functions on X and taking values in the set of measurable functions on Y. Let $1 \le p_0, p_1, q_0, q_1 \le \infty$ and assume that*

$$\begin{aligned}\big\|T(f)\big\|_{L^{q_0}} &\le M_0\big\|f\big\|_{L^{p_0}},\\ \big\|T(f)\big\|_{L^{q_1}} &\le M_1\big\|f\big\|_{L^{p_1}},\end{aligned} \tag{1.3.12}$$

for all simple functions f on X. Then for all $0<\theta<1$ we have

$$\big\|T(f)\big\|_{L^q} \le M_0^{1-\theta}M_1^{\theta}\big\|f\big\|_{L^p} \tag{1.3.13}$$

for all simple functions f on X, where

$$\frac{1}{p} = \frac{1-\theta}{p_0} + \frac{\theta}{p_1} \qquad \textit{and} \qquad \frac{1}{q} = \frac{1-\theta}{q_0} + \frac{\theta}{q_1}. \tag{1.3.14}$$

By density, T has a unique extension as a bounded operator from $L^p(X,\mu)$ to $L^q(Y,\nu)$ for all p and q as in (1.3.14).

We note that in many applications, T may be defined on $L^{p_0}+L^{p_1}$, in which case hypothesis (1.3.12) and conclusion (1.3.13) can be stated in terms of functions in the corresponding Lebesgue spaces.

Proof. Let

$$f = \sum_{k=1}^{m} a_k e^{i\alpha_k} \chi_{A_k}$$

be a simple function on X, where $a_k > 0$, α_k are real, and A_k are pairwise disjoint subsets of X with finite measure.

We need to control

$$\big\|T(f)\big\|_{L^q(Y,\nu)} = \sup\left|\int_Y T(f)(x)g(x)\,d\nu(x)\right|,$$

where the supremum is taken over all simple functions g on Y with $L^{q'}$ norm less than or equal to 1. Write

$$g = \sum_{j=1}^{n} b_j e^{i\beta_j} \chi_{B_j}, \tag{1.3.15}$$

where $b_j > 0$, β_j are real, and B_j are pairwise disjoint subsets of Y with finite measure. Let

$$P(z) = \frac{p}{p_0}(1-z) + \frac{p}{p_1}z \qquad \text{and} \qquad Q(z) = \frac{q'}{q_0'}(1-z) + \frac{q'}{q_1'}z. \tag{1.3.16}$$

For z in the closed strip $\overline{S} = \{z \in \mathbf{C} : 0 \le \operatorname{Re} z \le 1\}$, define

$$F(z) = \int_Y T(f_z)(x)\,g_z(x)\,d\nu(x),$$

where

$$f_z = \sum_{k=1}^{m} a_k^{P(z)} e^{i\alpha_k} \chi_{A_k}, \quad g_z = \sum_{j=1}^{n} b_j^{Q(z)} e^{i\beta_j} \chi_{B_j}. \tag{1.3.17}$$

By linearity,

$$F(z) = \sum_{k=1}^{m}\sum_{j=1}^{n} a_k^{P(z)} b_j^{Q(z)} e^{i\alpha_k} e^{i\beta_j} \int_Y T(\chi_{A_k})(x)\,\chi_{B_j}(x)\,d\nu(x),$$

and hence F is analytic in z, since $a_k, b_j > 0$.

Let us now consider a $z \in \overline{S}$ with $\mathrm{Re}\, z = 0$. By the disjointness of the sets A_k we have $\|f_z\|_{L^{p_0}}^{p_0} = \|f\|_{L^p}^{p}$, since $|a_k^{P(z)}| = a_k^{\frac{p}{p_0}}$. Similarly, by the disjointness of the sets B_j we have that $\|g_z\|_{L^{q_0'}}^{q_0'} = \|g\|_{L^{q'}}^{q'}$, since $|b_j^{Q(z)}| = b_j^{\frac{q'}{q_0'}}$.

By the same token, when $\mathrm{Re}\, z = 1$, we have $\|f_z\|_{L^{p_1}}^{p_1} = \|f\|_{L^p}^{p}$ and $\|g_z\|_{L^{q_1'}}^{q_1'} = \|g\|_{L^{q'}}^{q'}$. Hölder's inequality and the hypothesis now give

$$\begin{aligned} |F(z)| &\leq \|T(f_z)\|_{L^{q_0}} \|g_z\|_{L^{q_0'}} \\ &\leq M_0 \|f_z\|_{L^{p_0}} \|g_z\|_{L^{q_0'}} = M_0 \|f\|_{L^p}^{\frac{p}{p_0}} \|g\|_{L^{q'}}^{\frac{q'}{q_0'}}, \end{aligned} \tag{1.3.18}$$

when $\mathrm{Re}\, z = 0$. Similarly, we obtain

$$|F(z)| \leq M_1 \|f\|_{L^p}^{\frac{p}{p_1}} \|g\|_{L^{q'}}^{\frac{q'}{q_1'}}, \tag{1.3.19}$$

when $\mathrm{Re}\, z = 1$.

We state the following lemma, known as *Hadamard's three lines lemma*, whose proof we postpone until the end of this section.

Lemma 1.3.5. *Let F be analytic in the open strip $S = \{z \in \mathbf{C} : 0 < \mathrm{Re}\ z < 1\}$, continuous and bounded on its closure, such that $|F(z)| \leq B_0$ when $\mathrm{Re}\ z = 0$ and $|F(z)| \leq B_1$ when $\mathrm{Re}\ z = 1$, where $0 < B_0, B_1 < \infty$. Then $|F(z)| \leq B_0^{1-\theta} B_1^{\theta}$ when $\mathrm{Re}\ z = \theta$, for any $0 \leq \theta \leq 1$.*

Returning to the proof of Theorem 1.3.4, we observe that F is analytic in the open strip S and continuous on its closure. Also, F is bounded on the closed unit strip (by some constant that depends on f and g). Therefore, (1.3.18), (1.3.19), and Lemma 1.3.5 give

$$\begin{aligned} |F(z)| &\leq \Big(M_0 \|f\|_{L^p}^{\frac{p}{p_0}} \|g\|_{L^{q'}}^{\frac{q'}{q_0'}} \Big)^{1-\theta} \Big(M_1 \|f\|_{L^p}^{\frac{p}{p_1}} \|g\|_{L^{q'}}^{\frac{q'}{q_1'}} \Big)^{\theta} \\ &= M_0^{1-\theta} M_1^{\theta} \|f\|_{L^p} \|g\|_{L^{q'}}, \end{aligned}$$

when $\mathrm{Re}\, z = \theta$. Observe that $P(\theta) = Q(\theta) = 1$ and hence

$$F(\theta) = \int_Y T(f)\, g\, d\nu .$$

Taking the supremum over all simple functions g on Y with $L^{q'}$ norm less than or equal to one, we conclude the proof of the theorem. □

We now give an application of the theorem just proved.

Example 1.3.6. One may prove Young's inequality (Theorem 1.2.12) using the Riesz–Thorin interpolation theorem (Theorem 1.3.4). Fix a function g in L^r and let $T(f) = f * g$. Since $T : L^1 \to L^r$ with norm at most $\|g\|_{L^r}$ and $T : L^{r'} \to L^\infty$ with norm at most $\|g\|_{L^r}$, Theorem 1.3.4 gives that T maps L^p to L^q with norm at most the quantity $\|g\|_{L^r}^{\theta}\|g\|_{L^r}^{1-\theta} = \|g\|_{L^r}$, where

$$\frac{1}{p} = \frac{1-\theta}{1} + \frac{\theta}{r'} \quad \text{and} \quad \frac{1}{q} = \frac{1-\theta}{r} + \frac{\theta}{\infty}. \tag{1.3.20}$$

Finally, observe that equations (1.3.20) give (1.2.13).

1.3.3 Interpolation of Analytic Families of Operators

Theorem 1.3.4 can now be extended to the case in which the interpolated operators are allowed to vary. In particular, if a family of operators depends analytically on a parameter z, then the proof of this theorem can be adapted to work in this setting.

We now describe the setup for this theorem. Let (X,μ) and (Y,ν) be measure spaces. Suppose that for every z in the closed strip $\overline{S} = \{z \in \mathbf{C} : 0 \le \operatorname{Re} z \le 1\}$ there is an associated linear operator T_z defined on the space of simple functions on X and taking values in the space of measurable functions on Y such that

$$\int_Y |T_z(f)\, g|\, d\nu < \infty \tag{1.3.21}$$

whenever f and g are simple functions on X and Y, respectively. The family $\{T_z\}_z$ is said to be *analytic* if the function

$$z \mapsto \int_Y T_z(f)\, g\, d\nu \tag{1.3.22}$$

is analytic in the open strip $S = \{z \in \mathbf{C} : 0 < \operatorname{Re} z < 1\}$ and continuous on its closure. Finally, the analytic family is of *admissible growth* if there is a constant $a < \pi$ and a constant $C_{f,g}$ such that

$$e^{-a|\operatorname{Im} z|} \log \left| \int_Y T_z(f)\, g\, d\nu \right| \le C_{f,g} < \infty \tag{1.3.23}$$

for all z satisfying $0 \le \operatorname{Re} z \le 1$. The extension of the Riesz–Thorin interpolation theorem is now stated.

Theorem 1.3.7. *Let T_z be an analytic family of linear operators of admissible growth. Let $1 \le p_0, p_1, q_0, q_1 \le \infty$ and suppose that M_0 and M_1 are positive functions on the real line such that*

$$\sup_{-\infty < y < +\infty} e^{-b|y|} \log M_j(y) < \infty \tag{1.3.24}$$

for $j=0,1$ and some $b<\pi$. Let $0<\theta<1$ satisfy

$$\frac{1}{p}=\frac{1-\theta}{p_0}+\frac{\theta}{p_1} \qquad \text{and} \qquad \frac{1}{q}=\frac{1-\theta}{q_0}+\frac{\theta}{q_1}. \tag{1.3.25}$$

Suppose that

$$\big\|T_{iy}(f)\big\|_{L^{q_0}} \le M_0(y)\big\|f\big\|_{L^{p_0}}, \tag{1.3.26}$$

$$\big\|T_{1+iy}(f)\big\|_{L^{q_1}} \le M_1(y)\big\|f\big\|_{L^{p_1}}, \tag{1.3.27}$$

for all simple functions f on X. Then

$$\big\|T_\theta(f)\big\|_{L^q} \le M(\theta)\big\|f\big\|_{L^p} \qquad \text{when } 0<\theta<1 \tag{1.3.28}$$

for all simple functions f on X, where for $0<t<1$,

$$M(t)=\exp\left\{\frac{\sin(\pi t)}{2}\int_{-\infty}^{\infty}\left[\frac{\log M_0(y)}{\cosh(\pi y)-\cos(\pi t)}+\frac{\log M_1(y)}{\cosh(\pi y)+\cos(\pi t)}\right]dy\right\}.$$

By density, T_θ has a unique extension as a bounded operator from $L^p(X,\mu)$ to $L^q(Y,\nu)$ for all p and q as in (1.3.25).

As expected, the proof of the previous theorem is based on an extension of Lemma 1.3.5.

Lemma 1.3.8. *Let F be analytic on the open strip $S=\{z\in \mathbf{C}: 0<\operatorname{Re} z<1\}$ and continuous on its closure such that*

$$\sup_{z\in\overline{S}} e^{-a|\operatorname{Im} z|}\log|F(z)| \le A<\infty \tag{1.3.29}$$

for some fixed A and $a<\pi$. Then

$$|F(x)|\le \exp\left\{\frac{\sin(\pi x)}{2}\int_{-\infty}^{\infty}\left[\frac{\log|F(iy)|}{\cosh(\pi y)-\cos(\pi x)}+\frac{\log|F(1+iy)|}{\cosh(\pi y)+\cos(\pi x)}\right]dy\right\}$$

whenever $0<x<1$.

Assuming Lemma 1.3.8, we prove Theorem 1.3.7.

Proof. As in the proof of Theorem 1.3.4, we work with simple functions f on X and g on Y. Fix $0<\theta<1$ and also fix simple functions f,g such that $\big\|f\big\|_{L^p}=1=\big\|g\big\|_{L^{q'}}$. Let

$$f=\sum_{k=1}^{m} a_k e^{i\alpha_k}\chi_{A_k} \quad \text{and} \quad g=\sum_{j=1}^{n} b_j e^{i\beta_j}\chi_{B_j},$$

where $a_k>0$, $b_j>0$, α_k, β_k are real, A_k are pairwise disjoint subsets of X with finite measure, and B_j are pairwise disjoint subsets of Y with finite measure. Let f_z and g_z be as in the proof of Theorem 1.3.4. Define

$$F(z) = \int_Y T_z(f_z)\, g_z\, d\nu\,. \tag{1.3.30}$$

It follows from the assumptions about $\{T_z\}_z$ that $F(z)$ is an analytic function that satisfies the hypotheses of Lemma 1.3.8. Moreover,

$$\big\|f_{iy}\big\|_{L^{p_0}}^{p_0} = \big\|f\big\|_{L^p}^{p} = 1 = \big\|g\big\|_{L^{q'}}^{q'} = \big\|g_{iy}\big\|_{L^{q_0'}}^{q_0'} \quad \text{when } y \in \mathbf{R}\,, \tag{1.3.31}$$

$$\big\|f_{1+iy}\big\|_{L^{p_1}}^{p_1} = \big\|f\big\|_{L^p}^{p} = 1 = \big\|g\big\|_{L^{q'}}^{q'} = \big\|g_{1+iy}\big\|_{L^{q_1'}}^{q_1'} \quad \text{when } y \in \mathbf{R}\,. \tag{1.3.32}$$

Hölder's inequality, (1.3.31), and the hypothesis (1.3.26) now give

$$\begin{aligned} |F(iy)| &\le \big\|T_{iy}(f_{iy})\big\|_{L^{q_0}} \big\|g_{iy}\big\|_{L^{q_0'}} \\ &\le M_0(y) \big\|f_{iy}\big\|_{L^{p_0}} \big\|g_{iy}\big\|_{L^{q_0'}} = M_0(y) \end{aligned}$$

for all y real. Similarly, (1.3.32), and (1.3.27) imply

$$\begin{aligned} |F(1+iy)| &\le \big\|T_{1+iy}(f_{1+iy})\big\|_{L^{q_1}} \big\|g_{1+iy}\big\|_{L^{q_1'}} \\ &\le M_1(y) \big\|f_{1+iy}\big\|_{L^{p_1}} \big\|g_{1+iy}\big\|_{L^{q_1'}} = M_1(y)\,. \end{aligned}$$

for all $y \in \mathbf{R}$. Therefore, the hypotheses of Lemma 1.3.8 are satisfied. We conclude that

$$\left| \int_Y T_\theta(f)\, g\, d\nu \right| = |F(\theta)| \le M(\theta)\,, \tag{1.3.33}$$

where $M(x)$ is the function given in the hypothesis of the theorem.

Taking the supremum over all simple functions g on Y with $L^{q'}$ norm equal to one, we conclude the proof of the theorem. □

1.3.4 Proofs of Lemmas 1.3.5 and 1.3.8

Proof of Lemma 1.3.5. Define analytic functions

$$G(z) = F(z)(B_0^{1-z} B_1^z)^{-1} \qquad \text{and} \qquad G_n(z) = G(z) e^{(z^2-1)/n}\,.$$

Since F is bounded on the closed unit strip and $B_0^{1-z} B_1^z$ is bounded from below, we conclude that G is bounded by some constant M on the closed strip. Also, G is bounded by one on its boundary. Since

$$|G_n(x+iy)| \le M e^{-y^2/n} e^{(x^2-1)/n} \le M e^{-y^2/n}\,,$$

we deduce that $G_n(x+iy)$ converges to zero uniformly in $0 \le x \le 1$ as $|y| \to \infty$. Select $y(n) > 0$ such that for $|y| \ge y(n)$, $|G_n(x+iy)| \le 1$ uniformly in $x \in [0,1]$. By the maximum principle we obtain that $|G_n(z)| \le 1$ in the rectangle $[0,1] \times [-y(n), y(n)]$; hence $|G_n(z)| \le 1$ everywhere in the closed strip. Letting $n \to \infty$, we conclude that $|G(z)| \le 1$ in the closed strip.

□

Having disposed of the proof of Lemma 1.3.5, we end this section with a proof of Lemma 1.3.8.

Proof of Lemma 1.3.8. Recall the Poisson integral formula

$$U(z) = \frac{1}{2\pi} \int_{-\pi}^{+\pi} U(Re^{i\varphi}) \frac{R^2 - \rho^2}{|Re^{i\varphi} - \rho e^{i\theta}|^2} d\varphi, \qquad z = \rho e^{i\theta}, \tag{1.3.34}$$

which is valid for a harmonic function U defined on the unit disk $D = \{z : |z| < 1\}$ when $|z| < R < 1$. See Rudin [229, p. 258].

Consider now a subharmonic function u on D that is continuous on the circle $|\zeta| = R < 1$. When $U = u$, the right side of (1.3.34) defines a harmonic function on the set $\{z \in \mathbf{C} : |z| < R\}$ that coincides with u on the circle $|\zeta| = R$. The maximum principle for subharmonic functions (Rudin [229, p. 362]) implies that for $|z| < R < 1$ we have

$$u(z) \le \frac{1}{2\pi} \int_{-\pi}^{+\pi} u(Re^{i\varphi}) \frac{R^2 - \rho^2}{|Re^{i\varphi} - \rho e^{i\theta}|^2} d\varphi, \qquad z = \rho e^{i\theta}. \tag{1.3.35}$$

This is valid for all subharmonic functions u on D that are continuous on the circle $|\zeta| = R$ when $\rho < R < 1$.

It is not difficult to verify that

$$h(\zeta) = \frac{1}{\pi i} \log\left(i \frac{1+\zeta}{1-\zeta} \right)$$

is a conformal map from D onto the strip $S = (0,1) \times \mathbf{R}$. Indeed, $i(1+\zeta)/(1-\zeta)$ lies in the upper half-plane and the preceding complex logarithm is a well defined holomorphic function that takes the upper half-plane onto the strip $\mathbf{R} \times (0, \pi)$. Since $F \circ h$ is a holomorphic function on D, $\log|F \circ h|$ is a subharmonic function on D. Applying (1.3.35) to the function $z \mapsto \log|F(h(z))|$, we obtain

$$\log|F(h(z))| \le \frac{1}{2\pi} \int_{-\pi}^{+\pi} \log|F(h(Re^{i\varphi}))| \frac{R^2 - \rho^2}{R^2 - 2\rho R \cos(\theta - \varphi) + \rho^2} d\varphi \tag{1.3.36}$$

when $z = \rho e^{i\varphi}$ and $|z| = \rho < R$. Observe that when $|\zeta| = 1$ and $\zeta \ne \pm 1$, $h(\zeta)$ has real part zero or one. It follows from the hypothesis that

$$\log|F(h(\zeta))| \le Ae^{a|\operatorname{Im} h(\zeta)|} = Ae^{a\left|\operatorname{Im} \frac{1}{\pi i} \log\left(i\frac{1+\zeta}{1-\zeta}\right)\right|} \le Ae^{\frac{a}{\pi}\left|\log\left|\frac{1+\zeta}{1-\zeta}\right|\right|}.$$

Therefore, $\log|F(h(\zeta))|$ is bounded by a multiple of $|1+\zeta|^{-a/\pi}+|1-\zeta|^{-a/\pi}$, which is integrable over the set $|\zeta|=1$, since $a<\pi$. Fix now $z=\rho e^{i\theta}$ with $\rho<R$ and let $R\to 1$ in (1.3.36). The Lebesgue dominated convergence theorem gives that

$$\log|F(h(\rho e^{i\theta}))| \le \frac{1}{2\pi}\int_{-\pi}^{+\pi} \log|F(h(e^{i\varphi}))| \frac{1-\rho^2}{1-2\rho\cos(\theta-\varphi)+\rho^2}\, d\varphi. \quad (1.3.37)$$

Setting $x=h(\rho e^{i\theta})$, we obtain that

$$\rho e^{i\theta} = h^{-1}(x) = \frac{e^{\pi i x}-i}{e^{\pi i x}+i} = -i\frac{\cos(\pi x)}{1+\sin(\pi x)} = \left\{\frac{\cos(\pi x)}{1+\sin(\pi x)}\right\} e^{-i(\pi/2)},$$

from which it follows that $\rho=(\cos(\pi x))/(1+\sin(\pi x))$ and $\theta=-(\pi/2)$, when $0<x\le\frac{1}{2}$, while $\rho=-(\cos(\pi x))/(1+\sin(\pi x))$ and $\theta=\pi/2$, when $\frac{1}{2}\le x<1$. In either case we easily deduce that

$$\frac{1-\rho^2}{1-2\rho\cos(\theta-\varphi)+\rho^2} = \frac{\sin(\pi x)}{1+\cos(\pi x)\sin(\varphi)}.$$

Using this we write (1.3.37) as

$$\log|F(x)| \le \frac{1}{2\pi}\int_{-\pi}^{\pi} \frac{\sin(\pi x)}{1+\cos(\pi x)\sin(\varphi)} \log|F(h(e^{i\varphi}))|\, d\varphi. \quad (1.3.38)$$

We now change variables. On the interval $[-\pi,0)$ we use the change of variables $iy=h(e^{i\varphi})$ or, equivalently, $e^{i\varphi}=-\tanh(\pi y)-i\,\mathrm{sech}(\pi y)$. Observe that as φ ranges from $-\pi$ to 0, y ranges from $+\infty$ to $-\infty$. Furthermore, $d\varphi=-\pi\,\mathrm{sech}(\pi y)\,dy$. We have

$$\begin{aligned} &\frac{1}{2\pi}\int_{-\pi}^{0} \frac{\sin(\pi x)}{1+\cos(\pi x)\sin(\varphi)} \log|F(h(e^{i\varphi}))|\, d\varphi \\ &\qquad = \frac{1}{2}\int_{-\infty}^{\infty} \frac{\sin(\pi x)}{\cosh(\pi y)-\cos(\pi x)} \log|F(iy)|\, dy. \end{aligned} \quad (1.3.39)$$

On the interval $(0,\pi]$ we use the change of variables $1+iy=h(e^{i\varphi})$ or, equivalently, $e^{i\varphi}=-\tanh(\pi y)+i\,\mathrm{sech}(\pi y)$. Observe that as φ ranges from 0 to π, y ranges from $-\infty$ to $+\infty$. Furthermore, $d\varphi=\pi\,\mathrm{sech}(\pi y)\,dy$. Similarly, we obtain

$$\begin{aligned} &\frac{1}{2\pi}\int_{0}^{\pi} \frac{\sin(\pi x)}{1+\cos(\pi x)\sin(\varphi)} \log|F(h(e^{i\varphi}))|\, d\varphi \\ &\qquad = \frac{1}{2}\int_{-\infty}^{+\infty} \frac{\sin(\pi x)}{\cosh(\pi y)+\cos(\pi x)} \log|F(1+iy)|\, dy. \end{aligned} \quad (1.3.40)$$

Now add (1.3.39) and (1.3.40) and use (1.3.38) to conclude the proof. □

Exercises

1.3.1. Generalize Theorem 1.3.2 to the situation in which T is quasilinear, that is, it satisfies for some $K > 0$,

$$|T(\lambda f)| = |\lambda|\,|T(f)| \quad \text{and} \quad |T(f+g)| \le K(|T(f)| + |T(g)|),$$

for all $\lambda \in \mathbf{C}$, and all f, g in the domain of T. Prove that in this case, the constant A in (1.3.7) can be taken to be K times the constant in (1.3.8).

1.3.2. Let $1 < p < r \le \infty$ and suppose that T is a linear operator that maps L^1 to $L^{1,\infty}$ with norm A_0 and L^r to L^r with norm A_1. Prove that T maps L^p to L^p with norm at most

$$8\,(p-1)^{-\frac{1}{p}} A_0^{\frac{\frac{1}{p}-\frac{1}{r}}{1-\frac{1}{r}}} A_1^{\frac{1-\frac{1}{p}}{1-\frac{1}{r}}}.$$

[*Hint:* First interpolate between L^1 and L^r using Theorem 1.3.2 and then interpolate between $L^{\frac{p+1}{2}}$ and L^r using Theorem 1.3.4.]

1.3.3. Let $0 < p_0 < p < p_1 \le \infty$ and let T be an operator as in Theorem 1.3.2 that also satisfies

$$|T(f)| \le T(|f|),$$

for all $f \in L^{p_0} + L^{p_1}$.
(a) If $p_0 = 1$ and $p_1 = \infty$, prove that T maps L^p to L^p with norm at most

$$\frac{p}{p-1} A_0^{\frac{1}{p}} A_1^{1-\frac{1}{p}}.$$

(b) More generally, if $p_0 < p < p_1 = \infty$, prove that the norm of T from L^p to L^p is at most

$$p^{1+\frac{1}{p}} \left[\frac{B(p_0+1, p-p_0)}{p_0^{p_0}(p-p_0)^{p-p_0}}\right]^{\frac{1}{p}} A_0^{\frac{p_0}{p}} A_1^{1-\frac{p_0}{p}},$$

where $B(s,t) = \int_0^1 x^{s-1}(1-x)^{t-1}\,dx$ is the usual Beta function.
(c) When $0 < p_0 < p_1 < \infty$, then the norm of T from L^p to L^p is at most

$$\min_{0<\lambda<1} p^{\frac{1}{p}} \left(\frac{B(p-p_0, p_0+1)}{(1-\lambda)^{p_0}} + \frac{\frac{p_1-p+1}{p_1-p}}{\lambda^{p_1}}\right)^{\frac{1}{p}} A_0^{\frac{\frac{1}{p}-\frac{1}{p_1}}{\frac{1}{p_0}-\frac{1}{p_1}}} A_1^{\frac{\frac{1}{p_0}-\frac{1}{p}}{\frac{1}{p_0}-\frac{1}{p_1}}}.$$

[*Hint:* Parts (a), (b): The hypothesis $|T(f)| \le T(|f|)$ reduces matters to nonnegative functions. For $f \ge 0$ and for fixed $\alpha > 0$ write $f = f_0 + f_1$, where $f_0 = f - \lambda\alpha/A_1$ when $f \ge \lambda\alpha/A_1$ and zero otherwise for some $0 < \lambda < 1$. Then we have that $|\{|T(f)| > \alpha\}| \le |\{|T(f_0)| > (1-\lambda)\alpha\}|$. When $p_1 < \infty$ write $f = f_0 + f_1$, where $f_0 = f - \delta\alpha$ when $f \ge \delta\alpha$ and zero otherwise. Use that $|\{|T(f)| > \alpha\}| \le |\{|T(f_0)| > (1-\lambda)\alpha\}| + |\{|T(f_1)| > \lambda\alpha\}|$ and optimize over $\delta > 0$.]

1.3.4. Let $0 < \alpha, \beta < \pi$. Let T_z be a family of linear operators defined on the strip $S_{a,b} = \{z \in \mathbf{C} : a \le \operatorname{Re} z \le b\}$ that is analytic on the interior of $S_{a,b}$, in the sense of (1.3.22), continuous on its closure, and satisfies for all $z \in S_{a,b}$,

$$e^{-\alpha|\operatorname{Im} z|/(b-a)} \log \left| \int_Y T_z(f) g \, d\nu \right| \le C_{f,g} < \infty .$$

Let $1 \le p_0, q_0, p_1, q_1 \le \infty$. Suppose that T_{a+iy} maps $L^{p_0}(X)$ to $L^{q_0}(Y)$ with bound $M_0(y)$ and T_{b+iy} maps $L^{p_1}(X)$ to $L^{q_1}(Y)$ with bound $M_1(y)$, where

$$\sup_{-\infty<y<\infty} e^{-\beta|y|/(b-a)} \log M_j(y) < \infty, \quad j = 0, 1.$$

Then for $a < t < b$, T_t maps $L^p(X)$ to $L^q(Y)$, where

$$\frac{1}{p} = \frac{\frac{b-t}{b-a}}{p_0} + \frac{\frac{t-a}{b-a}}{p_1} \quad \text{and} \quad \frac{1}{q} = \frac{\frac{b-t}{b-a}}{q_0} + \frac{\frac{t-a}{b-a}}{q_1} .$$

1.3.5. (*Stein [251]*) On $\mathbf{R}^n$ let $K_\lambda(x_1, \dots, x_n)$ be the function

$$\frac{\pi^{\frac{n-1}{2}} \Gamma(\lambda+1)}{\Gamma(\lambda + \frac{n+1}{2})} \int_{-1}^{+1} e^{2\pi i s (x_1^2 + \cdots + x_n^2)^{1/2}} (1-s^2)^{\lambda + \frac{n-1}{2}} \, ds ,$$

where λ is a complex number. Let T_λ be the operator given by convolution with K_λ. Show that T_λ maps $L^p(\mathbf{R}^n)$ to itself for $\operatorname{Re}\lambda > (n-1)|\frac{1}{2} - \frac{1}{p}|$.
[*Hint:* Using the result in Appendix B.5, show that when $\operatorname{Re}\lambda = 0$, T_λ maps $L^2(\mathbf{R}^n)$ to itself with norm 1. Using the estimates in Appendices B.6 and B.7, conclude that T_λ maps $L^1(\mathbf{R}^n)$ to itself with an appropriate constant when $\operatorname{Re}\lambda = (n-1)/2 + \delta$ (for $\delta > 0$) and then appeal to Theorem 1.3.7.]

1.3.6. Under the same hypotheses as in Theorem 1.3.7, prove the stronger conclusion

$$\|T_z(f)\|_{L^q} \le B(z) \|f\|_{L^p}$$

for z in the open strip $S = (0,1) \times \mathbf{R}$, where

$$B(t+is) = \exp\left\{ \frac{\sin(\pi t)}{2} \int_{-\infty}^{\infty} \left[\frac{\log M_0(y)}{\cosh(\pi(y-s)) - \cos(\pi t)} + \frac{\log M_1(y)}{\cosh(\pi(y-s)) + \cos(\pi t)} \right] dy \right\}.$$

[*Hint:* Apply Theorem 1.3.7 to the analytic family $\widetilde{T_z} = T_{z+is}$.]

1.3.7. (*Yano [294]*) Let (X,μ) and (Y,ν) be two measure spaces with $\mu(X) < \infty$ and $\nu(Y) < \infty$. Let T be a sublinear operator that maps $L^p(X)$ to $L^p(Y)$ for every $1 < p \le 2$ with norm $\|T\|_{L^p \to L^p} \le A(p-1)^{-\alpha}$ for some fixed $A, \alpha > 0$. Prove that for all f measurable on X we have

$$\int_Y |T(f)|\,d\nu \le 6A(1+\nu(Y))^{\frac{1}{2}}\left[\int_X |f|(\log_2^+|f|)^\alpha\,d\mu + C_\alpha + \mu(X)^{\frac{1}{2}}\right],$$

where $C_\alpha = \sum_{k=1}^\infty k^\alpha (2/3)^k$. This result provides an example of *extrapolation*.
[*Hint:* Write

$$f = \sum_{k=0}^\infty f\chi_{S_k},$$

where $S_k = \{2^k \le |f| < 2^{k+1}\}$ when $k \ge 1$ and $S_0 = \{|f| < 2\}$. Using Hölder's inequality and the hypotheses on T, obtain that

$$\int_Y |T(f\chi_{S_k})|\,d\nu \le 2A\nu(Y)^{\frac{1}{k+1}} 2^k k^\alpha \mu(S_k)^{\frac{k}{k+1}}$$

for $k \ge 1$. Note that for $k \ge 1$ we have $\nu(Y)^{\frac{1}{k+1}} \le \max(1,\nu(Y))^{\frac{1}{2}}$ and consider the cases $\mu(S_k) \ge 3^{-k-1}$ and $\mu(S_k) \le 3^{-k-1}$ when summing in $k \ge 1$. The term with $k = 0$ is easier.]

1.3.8. Prove that for $0 < x < 1$ we have

$$\frac{1}{2}\int_{-\infty}^{+\infty} \frac{\sin(\pi x)}{\cosh(\pi y) + \cos(\pi x)}\,dy = x,$$
$$\frac{1}{2}\int_{-\infty}^{+\infty} \frac{\sin(\pi x)}{\cosh(\pi y) - \cos(\pi x)}\,dy = 1 - x,$$

and conclude that Lemma 1.3.8 is indeed an extension of Lemma 1.3.5.
[*Hint:* In the first integral write $\cosh(\pi y) = \frac{1}{2}(e^{\pi y} + e^{-\pi y})$. Then use the change of variables $z = e^{\pi y}$.]

1.4 Lorentz Spaces

Suppose that f is a measurable function on a measure space (X,μ). It would be desirable to have another function f^* defined on $[0,\infty)$ that is decreasing and *equidistributed* with f. By this we mean

$$d_f(\alpha) = d_{f^*}(\alpha) \tag{1.4.1}$$

for all $\alpha \ge 0$. This is achieved via a simple construction discussed in this section.

1.4.1 Decreasing Rearrangements

Definition 1.4.1. Let f be a complex-valued function defined on X. The *decreasing rearrangement* of f is the function f^* defined on $[0,\infty)$ by

$$f^*(t) = \inf\{s > 0 : \; d_f(s) \le t\}. \tag{1.4.2}$$

We adopt the convention $\inf \emptyset = \infty$, thus having $f^*(t) = \infty$ whenever $d_f(\alpha) > t$ for all $\alpha \ge 0$. Observe that f^* is decreasing and supported in $[0, \mu(X)]$.

Before we proceed with properties of the function f^*, we work out three examples.

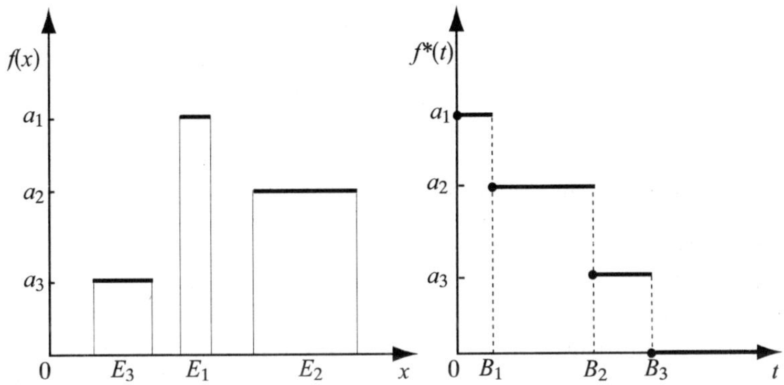

Fig. 1.3 The graph of a simple function $f(x)$ and its decreasing rearrangement $f^*(t)$.

Example 1.4.2. Consider the simple function of Example 1.1.2,

$$f(x) = \sum_{j=1}^{N} a_j \chi_{E_j}(x),$$

where the sets E_j have finite measure and are pairwise disjoint and $a_1 > \cdots > a_N$. We saw in Example 1.1.2 that

$$d_f(\alpha) = \sum_{j=0}^{N} B_j \chi_{[a_{j+1}, a_j)}(\alpha),$$

where

$$B_j = \sum_{i=1}^{j} \mu(E_i)$$

and $a_{N+1} = B_0 = 0$ and $a_0 = \infty$. Observe that for $B_0 \le t < B_1$, the smallest $s > 0$ with $d_f(s) \le t$ is a_1. Similarly, for $B_1 \le t < B_2$, the smallest $s > 0$ with $d_f(s) \le t$ is a_2. Arguing this way, it is not difficult to see that

$$f^*(t) = \sum_{j=1}^{N} a_j \chi_{[B_{j-1}, B_j)}(t).$$

Example 1.4.3. On $(\mathbf{R}^n, dx)$ let

$$f(x) = \frac{1}{1+|x|^p}, \qquad 0 < p < \infty.$$

A computation shows that

$$d_f(\alpha) = \begin{cases} v_n(\frac{1}{\alpha} - 1)^{\frac{n}{p}} & \text{if } \alpha < 1, \\ 0 & \text{if } \alpha \geq 1, \end{cases}$$

and therefore

$$f^*(t) = \frac{1}{(t/v_n)^{p/n} + 1},$$

where v_n is the volume of the unit ball in $\mathbf{R}^n$.

Example 1.4.4. Again on $(\mathbf{R}^n, dx)$ let $g(x) = 1 - e^{-|x|^2}$. We can easily see that $d_g(\alpha) = 0$ if $\alpha \geq 1$ and $d_g(\alpha) = \infty$ if $\alpha < 1$. We conclude that $g^*(t) = 1$ for all $t \geq 0$. This example indicates that although quantitative information is preserved, significant qualitative information is lost in passing from a function to its decreasing rearrangement.

It is clear from the previous examples that f^* is continuous from the right and decreasing. The following are some properties of the function f^*.

Proposition 1.4.5. *For f, g, f_n μ-measurable, $k \in \mathbf{C}$, and $0 \leq t, s, t_1, t_2 < \infty$ we have*

(1) $f^*(d_f(\alpha)) \leq \alpha$ *whenever* $\alpha > 0$.

(2) $d_f(f^*(t)) \leq t$.

(3) $f^*(t) > s$ *if and only if* $t < d_f(s)$; *that is,* $\{t \geq 0 : f^*(t) > s\} = [0, d_f(s))$.

(4) $|g| \leq |f|$ μ*-a.e. implies that* $g^* \leq f^*$ *and* $|f|^* = f^*$.

(5) $(kf)^* = |k| f^*$.

(6) $(f+g)^*(t_1 + t_2) \leq f^*(t_1) + g^*(t_2)$.

(7) $(fg)^*(t_1 + t_2) \leq f^*(t_1) g^*(t_2)$.

(8) $|f_n| \uparrow |f|$ μ*-a.e. implies* $f_n^* \uparrow f^*$.

(9) $|f| \leq \liminf\limits_{n\to\infty} |f_n|$ μ*-a.e. implies* $f^* \leq \liminf\limits_{n\to\infty} f_n^*$.

(10) f^* *is right continuous on* $[0, \infty)$.

(11) $t \leq \mu(\{|f| \geq f^*(t)\})$ *if* $\mu(\{|f| \geq f^*(t) - c\}) < \infty$ *for some* $c > 0$.

(12) $d_f = d_{f^*}$.

(13) $(|f|^p)^* = (f^*)^p$ *when* $0 < p < \infty$.

(14) $\displaystyle\int_X |f|^p \, d\mu = \int_0^\infty f^*(t)^p \, dt$ *when* $0 < p < \infty$.

(15) $\|f\|_{L^\infty} = f^*(0)$.

(16) $\sup\limits_{t>0} t^q f^*(t) = \sup\limits_{\alpha>0} \alpha \left(d_f(\alpha)\right)^q$ *for* $0 < q < \infty$.

Proof. Property (1): The set $A = \{s > 0 : d_f(s) \le d_f(\alpha)\}$ contains α and thus $f^*(d_f(\alpha)) = \inf A \le \alpha$.

Property (2): Let $s_n \in \{s > 0 : d_f(s) \le t\}$ be such that $s_n \downarrow f^*(t)$. Then $d_f(s_n) \le t$, and the right continuity of d_f (Exercise 1.1.1(a)) implies that $d_f(f^*(t)) \le t$.

Property (3): If $s < f^*(t) = \inf\{u > 0 : d_f(u) \le t\}$, then $s \notin \{u > 0 : d_f(u) \le t\}$ which gives $d_f(s) > t$. Conversely, if for some $t < d_f(s)$ we had $f^*(t) \le s$, applying d_f and using property (2) would yield the contradiction $d_f(s) \le d_f(f^*(t)) \le t$.

Properties (4) and (5) are left to the reader.

Properties (6) and (7): Let $A = \{s_1 > 0 : d_f(s_1) \le t_1\}$, $B = \{s_2 > 0 : d_g(s_2) \le t_2\}$, $P = \{s > 0 : d_{fg}(s) \le t_1 + t_2\}$, and $S = \{s > 0 : d_{f+g}(s) \le t_1 + t_2\}$. Then $A + B \subseteqq S$ and $A \cdot B \subseteqq P$; thus $(f+g)^*(t_1+t_2) = \inf S \le s_1 + s_2$ and $(fg)^*(t_1+t_2) = \inf P \le s_1 s_2$ are valid for all $s_1 \in A$ and $s_2 \in B$. Taking the infimum over all $s_1 \in A$ and $s_2 \in B$ yields the conclusions.

Property (8): It follows from the definition of decreasing rearrangements that $f_n^* \le f_{n+1}^* \le f^*$ for all n. Let $h = \lim_{n\to\infty} f_n^*$; then obviously $h \le f^*$. Since $f_n^* \le h$, we have $d_{f_n}(h(t)) \le d_{f_n}(f_n^*(t)) \le t$, which implies, in view of Exercise 1.1.1(c), that $d_f(h(t)) \le t$ by letting $n \to \infty$. It follows that $f^* \le h$, hence $h = f^*$.

Property (9): Set $F_n = \inf_{m\ge n} |f_m|$ and $h = \liminf_{n\to\infty} |f_n| = \sup_{n\ge 1} F_n$. Since $F_n \uparrow h$, property (8) yields that $F_n^* \uparrow h^*$ as $n \to \infty$. By hypothesis we have $|f| \le h$, hence $f^* \le h^* = \sup_n F_n^*$. Since $F_n \le |f_m|$ for $m \ge n$, it follows that $F_n^* \le f_m^*$ for $m \ge n$; thus $F_n^* \le \inf_{m\ge n} f_m^*$. Putting these facts together, we obtain $f^* \le h^* \le \sup_n \inf_{m\ge n} f_m^* = \liminf_{n\to\infty} f_n^*$.

Property (10): If $f^*(t_0) = 0$, then $f^*(t) = 0$ for all $t > t_0$ and thus f^* is right continuous at t_0. Suppose $f^*(t_0) > 0$. Pick α such that $0 < \alpha < f^*(t_0)$ and let $\{t_n\}_{n=1}^\infty$ be a sequence of real numbers decreasing to zero. The definition of f^* yields that $d_f(f^*(t_0) - \alpha) > t_0$. Since $t_n \downarrow 0$, there is an $n_0 \in \mathbf{Z}^+$ such that $d_f(f^*(t_0) - \alpha) > t_0 + t_n$ for all $n \ge n_0$. Property (3) yields that for all $n \ge n_0$ we have $f^*(t_0) - \alpha < f^*(t_0 + t_n)$, and since the latter is at most $f^*(t_0)$, the right continuity of f^* follows.

Property (11): The definition of f^* yields that the set $A_n = \{|f| > f^*(t) - c/n\}$ has measure $\mu(A_n) > t$. The sets A_n form a decreasing sequence as n increases and $\mu(A_1) < \infty$ by assumption. Consequently, $\{|f| \ge f^*(t)\} = \bigcap_{n=1}^\infty A_n$ has measure greater than or equal to t.

Property (12): This is immediate for nonnegative simple functions in view of Examples 1.1.2 and 1.4.2. For an arbitrary measurable function f, find a sequence of nonnegative simple functions f_n such that $f_n \uparrow |f|$ and apply (9).

Property (13): It follows from $d_{|f|^p}(\alpha) = d_f(\alpha^{1/p}) = d_{f^*}(\alpha^{1/p}) = d_{(f^*)^p}(\alpha)$ for all $\alpha > 0$.

Property (14): This is a consequence of property (12) and of Proposition 1.1.4.

Property (15): This is a restatement of (1.1.2).

Property (16): Given $\alpha > 0$, pick ε satisfying $0 < \varepsilon < \alpha$. Property (3) yields $f^*(d_f(\alpha) - \varepsilon) > \alpha$, which implies that

$$\sup_{t>0} t^q f^*(t) \geq (d_f(\alpha)-\varepsilon)^q f^*(d_f(\alpha)-\varepsilon) > (d_f(\alpha)-\varepsilon)^q \alpha .$$

We first let $\varepsilon \to 0$ and then take the supremum over all $\alpha > 0$ to obtain one direction. Conversely, given $t > 0$, pick $0 < \varepsilon < f^*(t)$. Property (3) yields $d_f(f^*(t)-\varepsilon) > t$. This implies that $\sup_{\alpha>0} \alpha (d_f(\alpha))^q \geq (f^*(t)-\varepsilon)(d_f(f^*(t)-\varepsilon))^q > (f^*(t)-\varepsilon)t^q$. We first let $\varepsilon \to 0$ and then take the supremum over all $t > 0$ to obtain the opposite direction of the claimed equality. □

1.4.2 Lorentz Spaces

Having disposed of the basic properties of decreasing rearrangements of functions, we proceed with the definition of the Lorentz spaces.

Definition 1.4.6. Given f a measurable function on a measure space (X,μ) and $0 < p,q \leq \infty$, define

$$\left\| f \right\|_{L^{p,q}} = \begin{cases} \left(\displaystyle\int_0^\infty \left(t^{\frac{1}{p}} f^*(t) \right)^q \frac{dt}{t} \right)^{\frac{1}{q}} & \text{if } q < \infty, \\ \displaystyle\sup_{t>0} t^{\frac{1}{p}} f^*(t) & \text{if } q = \infty. \end{cases}$$

The set of all f with $\left\| f \right\|_{L^{p,q}} < \infty$ is denoted by $L^{p,q}(X,\mu)$ and is called the *Lorentz space* with indices p and q.

As in L^p and in weak L^p, two functions in $L^{p,q}(X,\mu)$ are considered equal if they are equal μ-almost everywhere. Observe that the previous definition implies that $L^{\infty,\infty} = L^\infty$, $L^{p,\infty} =$ weak L^p in view of Proposition 1.4.5 (16) and that $L^{p,p} = L^p$.

Remark 1.4.7. Observe that for all $0 < p,r < \infty$ and $0 < q \leq \infty$ we have

$$\left\| |g|^r \right\|_{L^{p,q}} = \left\| g \right\|_{L^{pr,qr}}^r . \tag{1.4.3}$$

On $\mathbf{R}^n$ let $\delta^\varepsilon(f)(x) = f(\varepsilon x)$, $\varepsilon > 0$, be the dilation operator. It is straightforward that $d_{\delta^\varepsilon(f)}(\alpha) = \varepsilon^{-n} d_f(\alpha)$ and $(\delta^\varepsilon(f))^*(t) = f^*(\varepsilon^n t)$. It follows that Lorentz norms satisfy the following dilation identity:

$$\left\| \delta^\varepsilon(f) \right\|_{L^{p,q}} = \varepsilon^{-n/p} \left\| f \right\|_{L^{p,q}} . \tag{1.4.4}$$

Next, we calculate the Lorentz norms of a simple function.

Example 1.4.8. Using the notation of Example 1.4.2, when $0 < p,q < \infty$ we have

$$\left\| f \right\|_{L^{p,q}} = \left(\frac{p}{q} \right)^{\frac{1}{q}} \left[a_1^q B_1^{\frac{q}{p}} + a_2^q \left(B_2^{\frac{q}{p}} - B_1^{\frac{q}{p}} \right) + \cdots + a_N^q \left(B_N^{\frac{q}{p}} - B_{N-1}^{\frac{q}{p}} \right) \right]^{\frac{1}{q}} ,$$

and also

$$\|f\|_{L^{p,\infty}} = \sup_{1\le j\le N} a_j B_j^{\frac{1}{p}}\,.$$

The preceding expression for $\|f\|_{L^{p,q}}$ is also valid when $p=\infty$, but in this case it is equal to infinity if at least one a_j is strictly positive. We conclude that the only simple function with finite $L^{\infty,q}$ norm is the zero function. For this reason we have that $L^{\infty,q}=\{0\}$ for every $0<q<\infty$.

Proposition 1.4.9. *For $0<p<\infty$ and $0<q\le\infty$, we have the identity*

$$\|f\|_{L^{p,q}} = p^{\frac{1}{q}}\left(\int_0^\infty \left[d_f(s)^{\frac{1}{p}}s\right]^q \frac{ds}{s}\right)^{\frac{1}{q}}. \tag{1.4.5}$$

Proof. The case $q=\infty$ is statement (16) in Proposition 1.4.5, and we may therefore concentrate on the case $q<\infty$. If f is the simple function of Example 1.1.2, then

$$d_f(s) = \sum_{j=1}^{N} B_j \chi_{[a_{j+1},a_j)}(s)$$

with the understanding that $a_{N+1}=0$. Using the this formula and identity in Example 1.4.8, we obtain the validity of (1.4.5) for simple functions. In general, given a measurable function f, find a sequence of nonnegative simple functions such that $f_n\uparrow|f|$ a.e. Then $d_{f_n}\uparrow d_f$ (Exercise 1.1.1(c)) and $f_n^*\uparrow f^*$ (Proposition 1.4.5 (8)). Using the Lebesgue monotone convergence theorem we deduce (1.4.5). □

Since $L^{p,p}\subseteq L^{p,\infty}$, one may wonder whether these spaces are nested. The next result shows that for any fixed p, the Lorentz spaces $L^{p,q}$ increase as the exponent q increases.

Proposition 1.4.10. *Suppose $0<p\le\infty$ and $0<q<r\le\infty$. Then there exists a constant $c_{p,q,r}$ (which depends on p, q, and r) such that*

$$\|f\|_{L^{p,r}} \le c_{p,q,r}\|f\|_{L^{p,q}}\,. \tag{1.4.6}$$

In other words, $L^{p,q}$ is a subspace of $L^{p,r}$.

Proof. We may assume $p<\infty$, since the case $p=\infty$ is trivial. We have

$$\begin{aligned} t^{1/p}f^*(t) &= \left\{\frac{q}{p}\int_0^t [s^{1/p}f^*(t)]^q\frac{ds}{s}\right\}^{1/q} \\ &\le \left\{\frac{q}{p}\int_0^t [s^{1/p}f^*(s)]^q\frac{ds}{s}\right\}^{1/q} \qquad \text{since } f^* \text{ is decreasing,} \\ &\le \left(\frac{q}{p}\right)^{1/q}\|f\|_{L^{p,q}}\,. \end{aligned}$$

Hence, taking the supremum over all $t>0$, we obtain

$$\|f\|_{L^{p,\infty}} \le \left(\frac{q}{p}\right)^{1/q} \|f\|_{L^{p,q}}. \tag{1.4.7}$$

This establishes (1.4.6) in the case $r = \infty$. Finally, when $r < \infty$, we have

$$\|f\|_{L^{p,r}} = \left\{ \int_0^\infty [t^{1/p} f^*(t)]^{r-q+q} \frac{dt}{t} \right\}^{1/r} \le \|f\|_{L^{p,\infty}}^{(r-q)/r} \|f\|_{L^{p,q}}^{q/r}. \tag{1.4.8}$$

Inequality (1.4.7) combined with (1.4.8) gives (1.4.6) with $c_{p,q,r} = (q/p)^{(r-q)/rq}$. □

Unfortunately, the functionals $\|\cdot\|_{L^{p,q}}$ do not satisfy the triangle inequality. For instance, consider the functions $f(t) = t$ and $g(t) = 1-t$ defined on $[0,1]$. Then $f^*(\alpha) = g^*(\alpha) = (1-\alpha)\chi_{[0,1]}(\alpha)$. A simple calculation shows that the triangle inequality for these functions with respect to the norm $\|\cdot\|_{L^{p,q}}$ would be equivalent to

$$\frac{p}{q} \le 2^q \frac{\Gamma(q+1)\Gamma(q/p)}{\Gamma(q+1+q/p)},$$

which fails in general. However, since for all $t > 0$ we have

$$(f+g)^*(t) \le f^*(t/2) + g^*(t/2),$$

the estimate

$$\|f+g\|_{L^{p,q}} \le c_{p,q}\big(\|f\|_{L^{p,q}} + \|g\|_{L^{p,q}}\big), \tag{1.4.9}$$

where $c_{p,q} = 2^{1/p}\max(1, 2^{(1-q)/q})$, is a consequence of (1.1.4). Also, if $\|f\|_{L^{p,q}} = 0$ then we must have $f = 0$ μ-a.e. Therefore, $L^{p,q}$ is a quasinormed space for all $0 < p,q \le \infty$. Is this space complete with respect to its quasinorm? The next theorem answers this question.

Theorem 1.4.11. *Let (X,μ) be a measure space. Then for all $0 < p,q \le \infty$, the spaces $L^{p,q}(X,\mu)$ are complete with respect to their quasinorm and they are therefore quasi-Banach spaces.*

Proof. We consider only the case $p < \infty$. First we note that convergence in $L^{p,q}$ implies convergence in measure. When $q = \infty$, this is proved in Proposition 1.1.9. When $q < \infty$, in view of Proposition 1.4.5 (16) and (1.4.7), it follows that

$$\sup_{t>0} t^{1/p} f^*(t) = \sup_{\alpha>0} \alpha d_f(\alpha)^{1/p} \le \left(\frac{q}{p}\right)^{1/q} \|f\|_{L^{p,q}}$$

for all $f \in L^{p,q}$, from which the same conclusion follows. Let $\{f_n\}$ be a Cauchy sequence in $L^{p,q}$. Then $\{f_n\}$ is Cauchy in measure, and hence it has a subsequence $\{f_{n_k}\}$ that converges almost everywhere to some f by Theorem 1.1.13. Fix k_0 and apply property (9) in Proposition 1.4.5. Since $|f - f_{n_{k_0}}| = \lim_{k\to\infty} |f_{n_k} - f_{n_{k_0}}|$, it follows that

$$(f - f_{n_{k_0}})^*(t) \le \liminf_{k\to\infty} (f_{n_k} - f_{n_{k_0}})^*(t). \tag{1.4.10}$$

Raise (1.4.10) to the power q, multiply by $t^{q/p}$, integrate with respect to dt/t over $(0,\infty)$, and apply Fatou's lemma to obtain

$$\left\|f - f_{n_{k_0}}\right\|_{L^{p,q}}^q \le \liminf_{k\to\infty} \left\|f_{n_k} - f_{n_{k_0}}\right\|_{L^{p,q}}^q. \tag{1.4.11}$$

Now let $k_0 \to \infty$ in (1.4.11) and use the fact that $\{f_n\}$ is Cauchy to conclude that f_{n_k} converges to f in $L^{p,q}$. It is a general fact that if a Cauchy sequence has a convergent subsequence in a quasinormed space, then the sequence is convergent to the same limit. It follows that f_n converges to f in $L^{p,q}$. □

Remark 1.4.12. It can be shown that the spaces $L^{p,q}$ are normable when p, q are bigger than 1; see Exercise 1.4.3. Therefore, these spaces can be normed to become Banach spaces.

It is natural to ask whether simple functions are dense in $L^{p,q}$. This is in fact the case when $q \neq \infty$.

Theorem 1.4.13. *Simple functions are dense in $L^{p,q}(X,\mu)$ when $0 < q < \infty$.*

Proof. Let $f \in L^{p,q}(X,\mu)$. Assume without loss of generality that $f \ge 0$. Given $n = 1,2,3,\dots$, we find a simple function $f_n \ge 0$ such that

$$f_n(x) = 0$$

when $f(x) \le 1/n$, and

$$f(x) - \frac{1}{n} \le f_n(x) \le f(x)$$

when $f(x) > 1/n$, except on a set of measure less than $1/n$. It follows that

$$\mu(\{x \in X : |f(x) - f_n(x)| > 1/n\}) < 1/n;$$

hence $(f - f_n)^*(t) \le 1/n$ for $t \ge 1/n$. Thus

$$(f - f_n)^*(t) \to 0 \quad \text{as } n \to \infty \text{ and} \quad f_n^*(t) \le f^*(t) \quad \text{for all } t > 0.$$

Since $(f - f_n)^*(t) \le 2f^*(t/2)$, an application of the Lebesgue dominated convergence theorem gives that $\left\|f_n - f\right\|_{L^{p,q}} \to 0$ as $n \to \infty$. □

Remark 1.4.14. One may wonder whether simple functions are dense in $L^{p,\infty}$. This turns out to be false for all $0 < p \le \infty$. However, if X is σ-finite, countable linear combinations of characteristic functions of sets with finite measure are dense in $L^{p,\infty}(X,\mu)$. We call such functions *countably simple*. See Exercise 1.4.4 for details.

1.4.3 Duals of Lorentz Spaces

Given a quasi-Banach space Z with norm $\|\cdot\|_Z$, its dual Z^* is defined as the space of all continuous linear functionals T on Z equipped with the norm

$$\|T\|_{Z^*} = \sup_{\|x\|_Z=1} |T(x)|.$$

Observe that the dual of a quasi-Banach space is always a Banach space.

We are now considering the following question: What are the dual spaces $(L^{p,q})^*$ of $L^{p,q}$? The answer to this question presents some technical difficulties for general measure spaces. In this exposition we restrict our attention to σ-finite nonatomic measure spaces, where the situation is simpler.

Definition 1.4.15. A subset A of a measure space (X,μ) is called an *atom* if $\mu(A) > 0$ and every subset B of A has measure either equal to zero or equal to $\mu(A)$. A measure space (X,μ) is called *nonatomic* if it contains no atoms. In other words, X is nonatomic if and only if for any $A \subseteq X$ with $\mu(A) > 0$, there exists a proper subset $B \subsetneq A$ with $\mu(B) > 0$ and $\mu(A \setminus B) > 0$.

For instance, $\mathbf{R}$ with Lebesgue measure is nonatomic, but any measure space with counting measure is atomic. Nonatomic spaces have the property that every measurable subset of them with strictly positive measure contains subsets of any given measure smaller than the measure of the original subset. See Exercise 1.4.5.

Definition 1.4.16. A measure space is called *σ-finite* if there is a sequence of measurable sets K_N with $\mu(K_N) < \infty$ such that

$$\bigcup_{N=1}^{\infty} K_N = X.$$

For instance, $\mathbf{R}^n$ equipped with Lebesgue measure is a σ-finite measure space. So is $\mathbf{Z}^n$ with the usual counting measure.

Theorem 1.4.17. *Suppose that (X,μ) is a nonatomic σ-finite measure space. Then*

(i)	$(L^{p,q})^* = \{0\}$,	*when* $0 < p < 1$, $0 < q \le \infty$,
(ii)	$(L^{p,q})^* = L^\infty$,	*when* $p = 1$, $0 < q \le 1$,
(iii)	$(L^{p,q})^* = \{0\}$,	*when* $p = 1$, $1 < q < \infty$,
(iv)	$(L^{p,q})^* \neq \{0\}$,	*when* $p = 1$, $q = \infty$,
(v)	$(L^{p,q})^* = L^{p',\infty}$,	*when* $1 < p < \infty$, $0 < q \le 1$,
(vi)	$(L^{p,q})^* = L^{p',q'}$,	*when* $1 < p < \infty$, $1 < q < \infty$,
(vii)	$(L^{p,q})^* \neq \{0\}$,	*when* $1 < p < \infty$, $q = \infty$,
(viii)	$(L^{p,q})^* \neq \{0\}$,	*when* $p = q = \infty$.

Proof. Since X is σ-finite, we have $X = \bigcup_{N=1}^{\infty} K_N$, where K_N is an increasing sequence of sets with $\mu(K_N) < \infty$. Given $T \in (L^{p,q})^*$, where $0 < p < \infty$ and $0 < q \le \infty$, consider the measure $\sigma(E) = T(\chi_E)$. Since σ satisfies $|\sigma(E)| \le (p/q)^{1/q}\|T\|\mu(E)^{1/p}$ when $q < \infty$ and $|\sigma(E)| \le \|T\|\mu(E)^{1/p}$, it follows that σ is absolutely continuous with respect to the measure μ. By the Radon–Nikodym theorem, there exists a complex-valued measurable function g (which satisfies $\int_{K_N} |g|\,d\mu < \infty$ for all N) such that

$$\sigma(E) = T(\chi_E) = \int_X g\chi_E\,d\mu\,. \tag{1.4.12}$$

Linearity implies that (1.4.12) holds for any simple function on X. The continuity of T and the density of the simple functions on $L^{p,q}$ (when $q < \infty$) gives

$$T(f) = \int_X g f\,d\mu \tag{1.4.13}$$

for every $f \in L^{p,q}$. We now examine each case (i)–(viii) separately.

(i) We first consider the case $0 < p < 1$. Let $f = \sum_n a_n\chi_{E_n}$ be a simple function on X (take f to be countably simple when $q = \infty$). If X is nonatomic, we can split each E_n as a union of N disjoint sets E_{jn} each having measure $N^{-1}\mu(E_n)$. Let $f_j = \sum_n a_n\chi_{E_{jn}}$. We see that $\|f_j\|_{L^{p,q}} = N^{-1/p}\|f\|_{L^{p,q}}$. Now if $T \in (L^{p,q})^*$, it follows that

$$|T(f)| \le \sum_{j=1}^N |T(f_j)| \le \|T\| \sum_{j=1}^N \|f_j\|_{L^{p,q}} \le \|T\|N^{1-1/p}\|f\|_{L^{p,q}}\,.$$

Let $N \to \infty$ and use that $p < 1$ to obtain that $T = 0$.

(ii) We now consider the case $p = 1$ and $0 < q \le 1$. Clearly, every $g \in L^\infty$ gives a bounded linear functional on $L^{1,q}$, since

$$\left|\int_X f g\,d\mu\right| \le \|g\|_{L^\infty}\|f\|_{L^1} \le C_q\|g\|_{L^\infty}\|f\|_{L^{1,q}}\,.$$

Conversely, suppose that $T \in (L^{1,q})^*$ where $q \le 1$. The function g given in (1.4.12) satisfies

$$\left|\int_E g\,d\mu\right| \le \|T\|\,\mu(E)$$

for all $E \subseteq K_N$, and hence $|g| \le \|T\|$ μ-a.e. on every K_N. See Rudin [229, p. 31] (Theorem 1.40) for a proof of this fact. It follows that $\|g\|_{L^\infty} \le \|T\|$ and hence $(L^{1,q})^* = L^\infty$.

(iii) Let us now take $p = 1$, $1 < q < \infty$, and suppose that $T \in (L^{1,q})^*$. Then

$$\left|\int_X f g\,d\mu\right| \le \|T\|\;\|f\|_{L^{1,q}}, \tag{1.4.14}$$

where g is the function in (1.4.13). We show that $g = 0$ a.e. Suppose that $|g| \ge \delta$ on some set E_0 with $\mu(E_0) > 0$. Let $f = \overline{g}|g|^{-1}\chi_{E_0}h$, where $h \ge 0$. Then (1.4.14) implies that

$$\|h\|_{L^1(E_0)} \le \|T\|\delta^{-1}\|h\|_{L^{1,q}(E_0)}$$

for all $h \geq 0$. Since X is nonatomic, this can't happen unless $T = 0$. See Exercise 1.4.8.

(iv) In the case $p = 1$, $q = \infty$ something interesting happens. Since every continuous linear functional on $L^{1,\infty}$ extends to a continuous linear functional on $L^{1,q}$ for $1 < q < \infty$, it must necessarily vanish on all simple functions by part (iii). However, $(L^{1,\infty})^*$ contains nontrivial linear functionals. For details we refer to the articles of Cwikel and Fefferman [63], [64].

(v) We now take up the case $p > 1$ and $0 < q \leq 1$. Using Exercise 1.4.1(b) and Proposition 1.4.10, we see that if $g \in L^{p',\infty}$, then

$$\begin{aligned}\left|\int_X fg\,d\mu\right| &\leq \int_0^\infty t^{\frac{1}{p}} f^*(t) t^{\frac{1}{p'}} g^*(t)\,\frac{dt}{t}\\ &\leq \|f\|_{L^{p,1}}\|g\|_{L^{p',\infty}}\\ &\leq C_{p,q}\|f\|_{L^{p,q}}\|g\|_{L^{p',\infty}}.\end{aligned}$$

Conversely, suppose that $T \in (L^{p,q})^*$ when $1 < p < \infty$ and $0 < q \leq 1$. Let g satisfy (1.4.13). Taking $f = \overline{g}|g|^{-1}\chi_{|g|>\alpha}$ and using that

$$\left|\int_X fg\,d\mu\right| \leq \|T\|\|f\|_{L^{p,q}},$$

we obtain that

$$\alpha\mu(\{|g| > \alpha\}) \leq (p/q)^{1/q}\|T\|\mu(\{|g| > \alpha\})^{\frac{1}{p}}.$$

It follows that $\|g\|_{L^{p',\infty}} \approx \|T\|$.

(vi) Using Exercise 1.4.1(b) and Hölder's inequality, we obtain

$$\left|\int_X fg\,d\mu\right| \leq \int_0^\infty t^{\frac{1}{p}} f^*(t) t^{\frac{1}{p'}} g^*(t)\,\frac{dt}{t} \leq \|f\|_{L^{p,q}}\|g\|_{L^{p',q'}};$$

thus every $g \in L^{p',q'}$ gives a bounded linear functional on $L^{p,q}$. Conversely, let T be in $(L^{p,q})^*$. By (1.4.13), T is given by integration against a locally integrable function g. It remains to prove that $g \in L^{p',q'}$. For all f in $L^{p,q}(X)$ we have

$$\int_0^\infty f^*(t)g^*(t)\,dt = \sup_{h:\,d_h=d_f}\left|\int_X hg\,d\mu\right| \leq \|T\|\,\|f\|_{L^{p,q}}, \tag{1.4.15}$$

where the equality is a consequence of the fact that X is nonatomic (see Exercise 1.4.5). Pick a function f on X such that

$$f^*(t) = \int_{t/2}^\infty s^{\frac{q'}{p'}-1} g^*(s)^{q'-1}\,\frac{ds}{s}. \tag{1.4.16}$$

This can be achieved again by Exercise 1.4.5. The fact that the integral in (1.4.16) converges is a consequence of the observation that the function f^* defined in

(1.4.16) lies in the space $L^q(0,\infty)$ with respect to the measure $t^{q/p-1}dt$. This follows from the inequality

$$\begin{aligned}\|f\|_{L^{p,q}} &= \left(\int_0^\infty t^{\frac{q}{p}}\left[\int_{t/2}^\infty s^{\frac{q'}{p'}-1} g^*(s)^{q'-1}\frac{ds}{s}\right]^q \frac{dt}{t}\right)^{\frac{1}{q}}\\ &\le C_1(p,q)\left(\int_0^\infty (t^{\frac{1}{p'}} g^*(t))^{q'}\frac{dt}{t}\right)^{\frac{1}{q}}\\ &= C_1(p,q)\|g\|_{L^{p',q'}}^{q'/q} < \infty,\end{aligned}$$

which is a consequence of Hardy's second inequality in Exercise 1.2.8 with $b = q/p$. Using (1.4.15), we conclude that

$$\int_0^\infty f^*(t)g^*(t)\,dt \le \|T\|\ \|f\|_{L^{p,q}} \le C_1(p,q)\|T\|\ \|g\|_{L^{p',q'}}^{q'/q}. \tag{1.4.17}$$

On the other hand, we have

$$\begin{aligned}\int_0^\infty f^*(t)g^*(t)\,dt &\ge \int_0^\infty \int_{t/2}^t s^{\frac{q'}{p'}-1} g^*(s)^{q'-1}\frac{ds}{s}\,g^*(t)\,dt\\ &\ge \int_0^\infty g^*(t)^{q'}\int_{t/2}^t s^{\frac{q'}{p'}-1}\frac{ds}{s}\,dt \\ &= C_2(p,q)\|g\|_{L^{p',q'}}^{q'}.\end{aligned} \tag{1.4.18}$$

Combining (1.4.17) and (1.4.18), we obtain $\|g\|_{L^{p',q'}} \le C(p,q)\|T\|$. This estimate is valid only when we have a priori knowledge that $\|g\|_{L^{p',q'}} < \infty$. Suitably modifying the preceding proof and using that $\|g\|_{L^{p',q'}(K_N)} < \infty$, we obtain that $\|g\|_{L^{p',q'}(K_N)} \le C(p,q)\|T\|$ for all $N = 1,2,\ldots$. Letting $N\to\infty$, we obtain the required conclusion.

(vii) For a complete characterization of this space, we refer to the article of Cwikel [62].

(viii) The dual of L^∞ can be identified with the set of all bounded finitely additive set functions. See Dunford and Schwartz [77]. □

Remark 1.4.18. Some parts of Theorem 1.4.17 are false if X is atomic. For instance, the dual of $\ell^p(\mathbf{Z})$ contains l^∞ when $0<p<1$ and thus it is not $\{0\}$.

1.4.4 The Off-Diagonal Marcinkiewicz Interpolation Theorem

We now present the main result of this section, the off-diagonal extension of Marcinkiewicz's interpolation theorem (Theorem 1.3.2). Recall that an operator T is called quasilinear if it satisfies

$$|T(\lambda f)| = |\lambda|\,|T(f)| \qquad \text{and} \qquad |T(f+g)| \le K(|T(f)|+|T(g)|),$$

for some $K > 0$, $\lambda \in \mathbf{C}$, and all functions f, g in the domain of T. To avoid trivialities, we assume that $K \geq 1$.

Theorem 1.4.19. *Let $0 < r \leq \infty$, $0 < p_0 \neq p_1 \leq \infty$, and $0 < q_0 \neq q_1 \leq \infty$ and let (X,μ) and (Y,ν) be two measure spaces. Let T be either a quasilinear operator defined on $L^{p_0}(X)+L^{p_1}(X)$ and taking values in the set of measurable functions on Y or a linear operator defined on the set of simple functions on X and taking values as before. Assume that for some $M_0, M_1 < \infty$ the following (restricted) weak type estimates hold:*

$$\big\|T(\chi_A)\big\|_{L^{q_0,\infty}} \leq M_0\,\mu(A)^{1/p_0}\,, \tag{1.4.19}$$

$$\big\|T(\chi_A)\big\|_{L^{q_1,\infty}} \leq M_1\,\mu(A)^{1/p_1}\,, \tag{1.4.20}$$

for all measurable subsets A of X with $\mu(A) < \infty$. Fix $0 < \theta < 1$ and let

$$\frac{1}{p} = \frac{1-\theta}{p_0} + \frac{\theta}{p_1} \qquad \text{and} \qquad \frac{1}{q} = \frac{1-\theta}{q_0} + \frac{\theta}{q_1}\,. \tag{1.4.21}$$

Then there exists a constant M, which depends on K, p_0, p_1, q_0, q_1, M_0, M_1, r, and θ, such that for all functions f in the domain of T and in $L^{p,r}(X)$ we have

$$\big\|T(f)\big\|_{L^{q,r}} \leq M\big\|f\big\|_{L^{p,r}}\,. \tag{1.4.22}$$

We note that $L^{p,\infty} \subseteq L^{p_0}+L^{p_1}$ (Exercise 1.1.10(c)), and thus T is well defined on $L^{p,r}$ for all $r \leq \infty$. If $r < \infty$ and T is linear and defined on the set of simple functions on X, then T has a unique extension that satisfies (1.4.22) for all f in $L^{p,r}(X)$, since simple functions are dense in this space.

Before we give the proof of Theorem 1.4.19, we state and prove a lemma that is interesting on its own.

Lemma 1.4.20. *Let $0 < p < \infty$ and $0 < q \leq \infty$. Let T be either a quasilinear operator defined on $L^p(X,\mu)$ and taking values in the set of measurable functions of a measure space (Y,ν), or a linear operator initially defined on the space of simple functions on X and taking values as before. Suppose that there exists a constant $L > 0$ such that for all $A \subseteq X$ of finite measure we have*

$$\big\|T(\chi_A)\big\|_{L^{q,\infty}} \leq L\mu(A)^{1/p}. \tag{1.4.23}$$

Fix $\alpha_0 < q$ with $0 < \alpha_0 \leq \frac{\log 2}{\log 2K}$. Then for all $0 < \alpha \leq \alpha_0$ there exists a constant $C(p,q,K,\alpha) > 0$ (depending only on the parameters indicated) such that for all functions f in $L^{p,\alpha}(X)$ that lie in the domain of T, we have the estimate

$$\big\|T(f)\big\|_{L^{q,\infty}} \leq C(p,q,K,\alpha)L\,\big\|f\big\|_{L^{p,\alpha}}\,. \tag{1.4.24}$$

Lemma 1.4.20 is saying that if a quasilinear operator satisfies a $L^{p,1} \to L^{q,\infty}$ estimate uniformly on all characteristic functions, then it must map a Lorentz space $L^{p,\alpha}$ to $L^{q,\infty}$ for some $\alpha < 1$.

Proof. It suffices to prove Lemma 1.4.20 for $f \geq 0$, since we can express a general function f as

$$f = (f_1 - f_2) + i(f_3 - f_4),$$

where $f_j \geq 0$, and use quasilinearity.

It follows from the Aoki–Rolewicz theorem (Exercise 1.4.6) that for all $f_1, \dots, f_m$ we have the pointwise inequality

$$\begin{aligned} |T(f_1 + \cdots + f_m)| &\leq 4 \left(\sum_{j=1}^{m} |T(f_j)|^{\alpha_1} \right)^{\frac{1}{\alpha_1}} \\ &\leq 4 \left(\sum_{j=1}^{m} |T(f_j)|^{\alpha} \right)^{\frac{1}{\alpha}}, \end{aligned} \tag{1.4.25}$$

where $0 < \alpha \leq \alpha_1$ and α_1 satisfies the equation

$$(2K)^{\alpha_1} = 2.$$

The second inequality in (1.4.25) is a simple consequence of the fact that $\alpha \leq \alpha_1$. Fix $\alpha_0 > 0$ with

$$\alpha_0 \leq \alpha_1 = \frac{\log 2}{\log 2K} \qquad \text{and} \qquad \alpha_0 < q.$$

This ensures that the quasinormed space $L^{q/\alpha,\infty}$ is normable when $\alpha \leq \alpha_0$. In fact, Exercise 1.1.12 gives that the space $L^{s,\infty}$ is normable as long as $s > 1$ and for some equivalent norm $|||\, f \,|||_{L^{s,\infty}}$ we have

$$\|f\|_{L^{s,\infty}} \leq |||\, f \,|||_{L^{s,\infty}} \leq \frac{s}{s-1} \|f\|_{L^{s,\infty}}.$$

Next we claim that for any $f \geq 0$ we have

$$\|T(f\chi_A)\|_{L^{q,\infty}} \leq C(q,\alpha) L \mu(A)^{1/p} \|f\|_{L^\infty}. \tag{1.4.26}$$

To prove (1.4.26) first observe that multiplying by a suitable constant, we may assume that $f \leq 1$. Write

$$f(x) = \sum_{j=1}^{\infty} d_j(x) 2^{-j}$$

in binary expansion, where $d_j(x) = 0$ or 1. Let

$$B_j = \{x \in A : \, d_j(x) = 1\}.$$

Then $B_j \subseteq A$ and the function $f\chi_A$ can be written as the sum

$$\sum_{j=1}^{\infty} 2^{-j} \chi_{B_j}.$$

We use (1.4.25) once and (1.4.3) twice in the following argument. We have

$$\begin{aligned}
\big\|T(f\chi_A)\big\|_{L^{q,\infty}} &\le 4\Big\|\Big(\sum_{j=1}^{\infty}2^{-j\alpha}|T(\chi_{B_j})|^{\alpha}\Big)^{\frac{1}{\alpha}}\Big\|_{L^{q,\infty}}\\
&= 4\Big\|\sum_{j=1}^{\infty}2^{-j\alpha}|T(\chi_{B_j})|^{\alpha}\Big\|_{L^{q/\alpha,\infty}}^{\frac{1}{\alpha}}\\
&\le 4\,\Big|\Big|\Big|\sum_{j=1}^{\infty}2^{-j\alpha}|T(\chi_{B_j})|^{\alpha}\Big|\Big|\Big|_{L^{q/\alpha,\infty}}^{\frac{1}{\alpha}}\\
&\le 4\Big(\sum_{j=1}^{\infty}2^{-j\alpha}\,\Big|\Big|\Big|\,|T(\chi_{B_j})|^{\alpha}\Big|\Big|\Big|_{L^{q/\alpha,\infty}}\Big)^{\frac{1}{\alpha}}\\
&\le 4\Big(\frac{q}{q-\alpha}\Big)^{\frac{1}{\alpha}}\Big(\sum_{j=1}^{\infty}2^{-j\alpha}\Big\||T(\chi_{B_j})|^{\alpha}\Big\|_{L^{q/\alpha,\infty}}\Big)^{\frac{1}{\alpha}}\\
&= 4\Big(\frac{q}{q-\alpha}\Big)^{\frac{1}{\alpha}}\Big(\sum_{j=1}^{\infty}2^{-j\alpha}\|T(\chi_{B_j})\|_{L^{q,\infty}}^{\alpha}\Big)^{\frac{1}{\alpha}}\\
&\le 4\Big(\frac{q}{q-\alpha}\Big)^{\frac{1}{\alpha}}L\Big(\sum_{j=1}^{\infty}2^{-j\alpha}\mu(B_j)^{\alpha/p}\Big)^{\frac{1}{\alpha}}\\
&\le 2\Big(\frac{q}{q-\alpha}\Big)^{\frac{1}{\alpha}}(1-2^{-\alpha})^{-\frac{1}{\alpha}}L\mu(A)^{1/p},
\end{aligned}$$

since $B_j \subseteq A$. This establishes (1.4.26) with

$$C(q,\alpha)=2\Big(\frac{q}{q-\alpha}\Big)^{\frac{1}{\alpha}}(1-2^{-\alpha})^{-\frac{1}{\alpha}}.$$

Now write the function f as

$$f=\sum_{n=-\infty}^{\infty}f\chi_{A_n},$$

where A_n are measurable sets defined by

$$A_n=\{x\in X:\ f^*(2^{n+1})<|f(x)|\le f^*(2^n)\}. \tag{1.4.27}$$

Observe that

$$\begin{aligned}
\mu(A_n) &= \big|\{t\in\mathbf{R}:\ f^*(2^{n+1})<f^*(t)\le f^*(2^n)\}\big|\\
&= \big|[2^n,2^{n+1}]\big|\\
&= 2^n,
\end{aligned}$$

since f and f^* are equidistributed. Next we have

$$
\begin{aligned}
\big\|T(f)\big\|_{L^{q,\infty}} &\le 4\bigg\|\Big(\sum_{n=-\infty}^{\infty}|T(f\chi_{A_n})|^{\alpha}\Big)^{\frac{1}{\alpha}}\bigg\|_{L^{q,\infty}}\\
&= 4\bigg\|\sum_{n=-\infty}^{\infty}|T(f\chi_{A_n})|^{\alpha}\bigg\|_{L^{q/\alpha,\infty}}^{\frac{1}{\alpha}}\\
&\le 4\,\bigg|\bigg|\bigg|\sum_{n=-\infty}^{\infty}|T(f\chi_{A_n})|^{\alpha}\bigg|\bigg|\bigg|_{L^{q/\alpha,\infty}}^{\frac{1}{\alpha}}\\
&\le 4\Big(\sum_{n=-\infty}^{\infty}\big|\big|\big||T(f\chi_{A_n})|^{\alpha}\big|\big|\big|_{L^{q/\alpha,\infty}}\Big)^{\frac{1}{\alpha}}\\
&\le 4\Big(\frac{q}{q-\alpha}\Big)^{\frac{1}{\alpha}}\Big(\sum_{n=-\infty}^{\infty}\big\||T(f\chi_{A_n})|^{\alpha}\big\|_{L^{q/\alpha,\infty}}\Big)^{\frac{1}{\alpha}}\\
&\le 4\Big(\frac{q}{q-\alpha}\Big)^{\frac{1}{\alpha}}\Big(\sum_{n=-\infty}^{\infty}\big\|T(f\chi_{A_n})\big\|_{L^{q,\infty}}^{\alpha}\Big)^{\frac{1}{\alpha}}\\
&\le 8\Big(\frac{q}{q-\alpha}\Big)^{\frac{2}{\alpha}}(1-2^{-\alpha})^{-\frac{1}{\alpha}}L\Big(\sum_{n=-\infty}^{\infty}f^*(2^n)^{\alpha}2^{n\alpha/p}\Big)^{\frac{1}{\alpha}}\\
&\le 8\Big(\frac{q}{q-\alpha}\Big)^{\frac{2}{\alpha}}(1-2^{-\alpha})^{-\frac{1}{\alpha}}(\log 2)^{\frac{1}{\alpha}}L\big\|f\big\|_{L^{p,\alpha}}.
\end{aligned}
$$

Taking into account the splitting $f=f_1-f_2+if_3-if_4$, where $f_j\ge 0$, we conclude the proof of the lemma with constant

$$
C(p,q,K,\alpha)=C_pK^2\Big(\frac{q}{q-\alpha}\Big)^{2/\alpha}(\log 2)^{\frac{1}{\alpha}}(1-2^{-\alpha})^{-\frac{1}{\alpha}}. \tag{1.4.28}
$$

Recall that we have been assuming that $\alpha<\min\big(\frac{\log 2}{\log 2K},q\big)$ throughout. □

We now continue with the proof of Theorem 1.4.19.

Proof. We assume that $p_0<p_1$, since if $p_0>p_1$ we may simply reverse the roles of p_0 and p_1. We first consider the case $p_1<\infty$. Lemma 1.4.20 implies that

$$
\begin{aligned}
\big\|T(f)\big\|_{L^{q_0,\infty}} &\le A_0\big\|f\big\|_{L^{p_0,m}},\\
\big\|T(f)\big\|_{L^{q_1,\infty}} &\le A_1\big\|f\big\|_{L^{p_1,m}},
\end{aligned} \tag{1.4.29}
$$

where $m=\frac{1}{2}\min\big(q_0,q_1,\frac{\log 2}{\log 2K}\big)$, $A_0=C(p_0,q_0,K,m)M_0$, $A_1=C(p_1,q_1,K,m)M_1$, and $C(p,q,K,\alpha)$ is as in (1.4.28).

Fix a function f in $L^{p,r}$. Split $f=f^t+f_t$ as follows:

$$
f^t(x)=\begin{cases} f(x) & \text{if } |f(x)|>f^*(t^{\gamma}),\\ 0 & \text{if } |f(x)|\le f^*(t^{\gamma}),\end{cases}
$$

$$f_t(x) = \begin{cases} 0 & \text{if } |f(x)| > f^*(t^\gamma), \\ f(x) & \text{if } |f(x)| \le f^*(t^\gamma), \end{cases}$$

where γ is the following nonzero real number:

$$\gamma = \frac{\frac{1}{q_0} - \frac{1}{q}}{\frac{1}{p_0} - \frac{1}{p}} = \frac{\frac{1}{q} - \frac{1}{q_1}}{\frac{1}{p} - \frac{1}{p_1}}.$$

Next, observe that the following inequalities are valid:

$$(f^t)^*(s) \le \begin{cases} f^*(s) & \text{if } 0 < s < t^\gamma, \\ 0 & \text{if } s \ge t^\gamma, \end{cases}$$

$$(f_t)^*(s) \le \begin{cases} f^*(t^\gamma) & \text{if } 0 < s < t^\gamma, \\ f^*(s) & \text{if } s \ge t^\gamma. \end{cases}$$

It follows from these inequalities that f^t lies in $L^{p_0,m}$ and f_t lies in $L^{p_1,m}$ for all $t > 0$.
The sublinearity of the operator T and (1.4.9) imply

$$\begin{aligned} \big\|T(f)\big\|_{L^{q,r}} &= \big\|t^{\frac{1}{q}} T(f)^*(t)\big\|_{L^r(\frac{dt}{t})} \\ &\le K\big\|t^{\frac{1}{q}}(|T(f_t)| + |T(f^t)|)^*(t)\big\|_{L^r(\frac{dt}{t})} \\ &\le K\big\|t^{\frac{1}{q}} T(f_t)^*(\tfrac{t}{2}) + t^{\frac{1}{q}} T(f^t)^*(\tfrac{t}{2})\big\|_{L^r(\frac{dt}{t})} \\ &\le K a_r\Big(\big\|t^{\frac{1}{q}} T(f_t)^*(\tfrac{t}{2})\big\|_{L^r(\frac{dt}{t})} + \big\|t^{\frac{1}{q}} T(f^t)^*(\tfrac{t}{2})\big\|_{L^r(\frac{dt}{t})}\Big) \\ &\le K a_r\Big(\big\|t^{\frac{1}{q}-\frac{1}{q_0}} t^{\frac{1}{q_0}} T(f_t)^*(\tfrac{t}{2})\big\|_{L^r(\frac{dt}{t})} \\ &\qquad\qquad + \big\|t^{\frac{1}{q}-\frac{1}{q_1}} t^{\frac{1}{q_1}} T(f^t)^*(\tfrac{t}{2})\big\|_{L^r(\frac{dt}{t})}\Big), \end{aligned} \tag{1.4.30}$$

where

$$a_r = \begin{cases} 1 & \text{when } r \ge 1, \\ 2^{(1-r)/r} & \text{when } r \le 1. \end{cases}$$

It follows from (1.4.29) that

$$t^{\frac{1}{q_0}} T(f^t)^*(\tfrac{t}{2}) \le 2^{\frac{1}{q_0}} \sup_{s>0} s^{\frac{1}{q_0}} T(f^t)^*(s) \le 2^{\frac{1}{q_0}} A_0 \big\|f^t\big\|_{L^{p_0,m}}, \tag{1.4.31}$$

$$t^{\frac{1}{q_1}} T(f_t)^*(\tfrac{t}{2}) \le 2^{\frac{1}{q_1}} \sup_{s>0} s^{\frac{1}{q_1}} T(f_t)^*(s) \le 2^{\frac{1}{q_1}} A_1 \big\|f_t\big\|_{L^{p_1,m}}, \tag{1.4.32}$$

for all $t > 0$. Now use (1.4.31) and (1.4.32) to estimate (1.4.30) by

$$K a_r 2^{\frac{1}{q_0}} A_0 \left\| t^{\frac{1}{q}-\frac{1}{q_0}} \big\|f^t\big\|_{L^{p_0,m}} \right\|_{L^r(\frac{dt}{t})} + K a_r 2^{\frac{1}{q_1}} A_1 \left\| t^{\frac{1}{q}-\frac{1}{q_1}} \big\|f_t\big\|_{L^{p_1,m}} \right\|_{L^r(\frac{dt}{t})},$$

which is the same as

$$
\begin{aligned}
&Ka_r 2^{\frac{1}{q_0}} A_0 \left\| t^{-\gamma(\frac{1}{p_0}-\frac{1}{p})} \| f^t \|_{L^{p_0,m}} \right\|_{L^r(\frac{dt}{t})} \\
&\qquad + Ka_r 2^{\frac{1}{q_1}} A_1 \left\| t^{\gamma(\frac{1}{p}-\frac{1}{p_1})} \| f_t \|_{L^{p_1,m}} \right\|_{L^r(\frac{dt}{t})} .
\end{aligned}
\tag{1.4.33}
$$

Next, we change variables $u = t^\gamma$ in the first term of (1.4.33) to obtain

$$
\begin{aligned}
&Ka_r 2^{\frac{1}{q_0}} A_0 \left\| t^{-\gamma(\frac{1}{p_0}-\frac{1}{p})} \| f^t \|_{L^{p_0,m}} \right\|_{L^r(\frac{dt}{t})} \\
&\quad \le Ka_r \frac{2^{\frac{1}{q_0}} A_0}{|\gamma|^{1/r}} \left\| u^{-(\frac{1}{p_0}-\frac{1}{p})} \left(\int_0^u f^*(s)^m s^{\frac{m}{p_0}} \frac{ds}{s} \right)^{\frac{1}{m}} \right\|_{L^r(\frac{du}{u})} \\
&\quad \le Ka_r \frac{2^{\frac{1}{q_0}} A_0}{m|\gamma|^{1/r}} \frac{r}{r(\frac{1}{p_0}-\frac{1}{p})} \left(\int_0^\infty (s^{\frac{1}{p_0}} f^*(s))^r s^{-r(\frac{1}{p_0}-\frac{1}{p})} \frac{ds}{s} \right)^{\frac{1}{r}} \\
&\quad = Ka_r \frac{2^{\frac{1}{q_0}} A_0}{m|\gamma|^{1/r}} \frac{1}{\frac{1}{p_0}-\frac{1}{p}} \| f \|_{L^{p,r}} ,
\end{aligned}
$$

where the last inequality is a consequence of Hardy's first inequality in Exercise 1.2.8 with $p = r/m \ge 1$ and $b = (1/p_0 - 1/p)r$.

Similarly, change variables $u = t^\gamma$ in the second term of (1.4.33) to obtain

$$
\begin{aligned}
&Ka_r 2^{\frac{1}{q_1}} A_1 \left\| t^{\gamma(\frac{1}{p}-\frac{1}{p_1})} \| f_t \|_{L^{p_1,m}} \right\|_{L^r(\frac{dt}{t})} \\
&\le \frac{Ka_r 2^{\frac{1}{q_1}} A_1}{|\gamma|^{1/r}} \left\| u^{\frac{1}{p}-\frac{1}{p_1}} \left[\int_0^u f^*(u)^m s^{\frac{m}{p_1}} \frac{ds}{s} + \int_u^\infty f^*(s)^m s^{\frac{m}{p_1}} \frac{ds}{s} \right]^{\frac{1}{m}} \right\|_{L^r(\frac{du}{u})} \\
&\le \frac{Ka_r^2 2^{\frac{1-m}{m}} 2^{\frac{1}{q_1}} A_1}{|\gamma|^{1/r}} \left\{ \frac{p_1}{m} \left\| u^{\frac{1}{p}-\frac{1}{p_1}} f^*(u) u^{\frac{1}{p_1}} \right\|_{L^r(\frac{du}{u})} \right. \\
&\qquad\qquad \left. + \left\| u^{\frac{1}{p}-\frac{1}{p_1}} \left(\int_u^\infty f^*(s)^m s^{\frac{m}{p_1}} \frac{ds}{s} \right)^{\frac{1}{m}} \right\|_{L^r(\frac{du}{u})} \right\} \\
&\le \frac{Ka_r^2 2^{\frac{1-m}{m}} 2^{\frac{1}{q_1}} A_1}{|\gamma|^{1/r}} \left\{ \frac{p_1}{m} \| f \|_{L^{p,r}} + \frac{r}{mr(\frac{1}{p}-\frac{1}{p_1})} \left\| u^{r(\frac{1}{p}-\frac{1}{p_1})} f^*(u)^r u^{\frac{r}{p_1}} \right\|_{L^r(\frac{du}{u})} \right\} \\
&= \frac{Ka_r^2 2^{\frac{1-m}{m}} 2^{\frac{1}{q_1}} A_1}{|\gamma|^{1/r}} \left\{ \frac{p_1}{m} + \frac{1}{m(\frac{1}{p}-\frac{1}{p_1})} \right\} \| f \|_{L^{p,r}} ,
\end{aligned}
$$

where the last inequality above is Hardy's second inequality in Exercise 1.2.8 with $p = r/m \ge 1$ and $b = (1/p - 1/p_1)r$.

We have now shown that

$$\big\|T(f)\big\|_{L^{q,r}} \le M\big\|f\big\|_{L^{p,r}}$$

with constant

$$M = Ka_r \frac{2^{\frac{1}{q_0}} + 2^{\frac{1}{q_1}}}{m|\gamma|^{1/r}} \left(\frac{A_0}{\frac{1}{p_0} - \frac{1}{p}} + a_r 2^{\frac{1-m}{m}} A_1 \Big(p_1 + \frac{1}{\frac{1}{p} - \frac{1}{p_1}} \Big) \right). \tag{1.4.34}$$

We have been tacitly assuming that $r < \infty$. The remaining case is a simple consequence of the result just proved by letting $r \to \infty$, in which case $a_r \to 1$ and $|\gamma|^{1/r} \to 1$.

We now turn to the case $p_1 = \infty$. Hypotheses (1.4.19) and (1.4.20) together with Exercise 1.1.16 imply that

$$\big\|T(\chi_A)\big\|_{L^{q,\infty}} \le M_0^{1-\theta} M_1^{\theta} \mu(A)^{1/p}$$

for all $0 \le \theta \le 1$. We select $\lambda \in (0,1)$ such that the indices $p = p_\lambda$ and $q = q_\lambda$ defined by (1.4.21) when $\theta = \lambda$ satisfy $p_0 < p < p_\lambda < \infty$ and q_λ is strictly between q_0 and q_1. Then apply the case $p_1 < \infty$ just proved with p_0, q_0 as before and $p_1 = p_\lambda$ and $q_1 = q_\lambda$. The result follows with M as in (1.4.34) except that p_1 is replaced by p_λ and q_1 by q_λ. □

Corollary 1.4.21. *Let T be as in the statement of Theorem 1.4.19 and let $0 < p_0 \neq p_1 \le \infty$ and $0 < q_0 \neq q_1 \le \infty$. If T maps L^{p_0} to $L^{q_0,\infty}$ and L^{p_1} to $L^{q_1,\infty}$, and for some $0 < \theta < 1$ we have*

$$\frac{1}{p} = \frac{1-\theta}{p_0} + \frac{\theta}{p_1}, \qquad \frac{1}{q} = \frac{1-\theta}{q_0} + \frac{\theta}{q_1}, \qquad \textit{and} \qquad p \le q,$$

then T satisfies the strong type estimate $\big\|T(f)\big\|_{L^q} \le C\big\|f\big\|_{L^p}$ for all functions f in the domain of T. Moreover, if T is linear, then it has a bounded extension from $L^p(X,\mu)$ to $L^q(Y,\nu)$.

Proof. Take $r = q$ in the previous theorem. □

Definition 1.4.22. Let $0 < p,q \le \infty$. We call an operator T of *restricted weak type* (p,q) if it satisfies

$$\big\|T(\chi_A)\big\|_{L^{q,\infty}} \le C\mu(A)^{1/p}$$

for all subsets A of a measure space (X,μ) with finite measure. Using this terminology, Corollary 1.4.21 says that if a quasilinear operator T is of restricted weak types (p_0,q_0) and (p_1,q_1) for some $p_0 \neq p_1$ and $q_0 \neq q_1$, then it is bounded from L^p to L^q when $p \le q$.

We now give examples to indicate why the assumptions $p_0 \neq p_1$ and $q_0 \neq q_1$ cannot be dropped in Theorem 1.4.19.

Example 1.4.23. Let $X = Y = \mathbf{R}$ and

$$T(f)(x) = |x|^{-1/2} \int_0^1 f(t)\,dt\,.$$

Then $\alpha|\{x:\ |T(\chi_A)(x)| > \alpha\}|^{1/2} = 2^{1/2}|A \cap [0,1]|$ and thus T is of restricted weak types $(1,2)$ and $(3,2)$. But observe that T does not map $L^2 = L^{2,2}$ to $L^{q,2}$. Thus Theorem 1.4.19 fails if the assumption $q_0 \neq q_1$ is dropped. The dual operator

$$S(f)(x) = \chi_{[0,1]}(x) \int_{-\infty}^{+\infty} f(t)|t|^{-1/2}\,dt$$

satisfies $\alpha|\{x:\ |S(\chi_A)(x)| > \alpha\}|^{1/q} \leq c|A|^{1/2}$ when $q = 1$ or 3, and thus it furnishes an example of an operator of restricted weak types $(2,1)$ and $(2,3)$ that is not L^2 bounded. Thus Theorem 1.4.19 fails if the assumption $p_0 \neq p_1$ is dropped.

As an application of Theorem 1.4.19, we give the following strengthening of Theorem 1.2.13.

Theorem 1.4.24. ***(Young's inequality for weak type spaces)*** *Let G be a locally compact group with left Haar measure λ that satisfies (1.2.12) for all measurable subsets A of G. Let $1 < p,q,r < \infty$ satisfy*

$$\frac{1}{q} + 1 = \frac{1}{p} + \frac{1}{r}\,. \tag{1.4.35}$$

Then there exists a constant $B_{pqr} > 0$ such that for all f in $L^p(G)$ and g in $L^{r,\infty}(G)$ we have

$$\big\|f * g\big\|_{L^q(G)} \leq B_{pqr}\big\|g\big\|_{L^{r,\infty}(G)}\big\|f\big\|_{L^p(G)}\,. \tag{1.4.36}$$

Proof. We fix $1 < p,q < \infty$. Since p and q range in an open interval, we can find $p_0 < p < p_1$, $q_0 < q < q_1$, and $0 < \theta < 1$ such that (1.4.21) and (1.4.35) hold. Let $T(f) = f * g$, defined for all functions f on G. By Theorem 1.2.13, T extends to a bounded operator from L^{p_0} to $L^{p_1,\infty}$ and from L^{q_0} to $L^{q_1,\infty}$. It follows from the Marcinkiewicz interpolation theorem that T extends to a bounded operator from $L^p(G)$ to $L^q(G)$. □

Exercises

1.4.1. (a) Let g be a nonnegative simple function on (X,μ) and let A be a measurable subset of X. Prove that

$$\int_A g\,d\mu \leq \int_0^{\mu(A)} g^*(t)\,dt.$$

(b) (*G. H. Hardy and J. E. Littlewood*) For f and g measurable on (X,μ), prove that

$$\int_X |f(x)g(x)|\,d\mu(x) \le \int_0^\infty f^*(t)g^*(t)\,dt.$$

Compare this result to the classical Hardy–Littlewood result asserting that if $a_j, b_j > 0$, the sum $\sum_j a_j b_j$ is greatest when both a_j and b_j are rearranged in decreasing order (for this see Hardy, Littlewood, and Pólya [122, p. 261]).

1.4.2. Prove that if $f \in L^{q_0,\infty} \cap L^{q_1,\infty}$ for some $0 < q_0 < q_1 \le \infty$, then $f \in L^{q,s}$ for all $0 < s \le \infty$ and $q_0 < q < q_1$.

1.4.3. (*Hunt [134]*) Given $0 < p, q < \infty$, fix an $r = r(p,q) > 0$ such that $r \le 1$, $r \le q$ and $r < p$. For $t \le \mu(X)$ define

$$f^{**}(t) = \sup_{\mu(E)\ge t} \left(\frac{1}{\mu(E)} \int_E |f|^r\,d\mu \right)^{1/r},$$

while for $t > \mu(X)$ (if $\mu(X) < \infty$) let

$$f^{**}(t) = \left(\frac{1}{t} \int_X |f|^r\,d\mu \right)^{1/r}.$$

Also define

$$|||\, f \,|||_{L^{p,q}} = \left(\int_0^\infty \left(t^{\frac{1}{p}} f^{**}(t)\right)^q \frac{dt}{t} \right)^{\frac{1}{q}}.$$

(The function f^{**} and the functional $f \to |||\, f \,|||_{L^{p,q}}$ depend on r.)
(a) Prove that the inequality

$$(((f+g)^{**})(t))^r \le (f^{**}(t))^r + (g^{**}(t))^r$$

is valid for all $t \ge 0$. Since $r \le q$, conclude that the functional

$$f \to |||\, f \,|||_{L^{p,q}}^r$$

is subadditive and hence it is a norm when $r = 1$ (this is possible only if $p > 1$).
(b) Show that for all f we have

$$\|f\|_{L^{p,q}} \le |||\, f \,|||_{L^{p,q}} \le \left(\frac{p}{p-r} \right)^{1/r} \|f\|_{L^{p,q}}.$$

(c) In conjunction with Exercise 1.1.12, conclude that $L^{p,q}$ is metrizable whenever $0 < p, q \le \infty$ and also normable when $1 < p < \infty$ and $1 \le q \le \infty$.

1.4.4. (a) Show that on a σ-finite measure space (X,μ) the set of countable linear combinations of simple functions is dense in $L^{p,\infty}(X)$.
(b) Prove that simple functions are not dense in $L^{p,\infty}(\mathbf{R})$ for any $0 < p \le \infty$.
[*Hint:* Part (b): Show that the function $h(x) = x^{-1/p}\chi_{x>0}$ cannot be approximated by a sequence of simple functions $L^{p,\infty}$. To see this, partition the inter-

val $(0,\infty)$ into small subintervals of length $\varepsilon > 0$ and let f_ε be the step function $\sum_{-[1/\varepsilon]}^{[1/\varepsilon]} f(k\varepsilon)\chi_{[k\varepsilon,(k+1)\varepsilon]}(x)$. Show that for some $c > 0$ we have $\|f_\varepsilon - f\|_{L^{p,\infty}} \geq c$.]

1.4.5. Let (X,μ) be a nonatomic measure space. Prove the following facts:
(a) If $A_0 \subseteq A_1 \subseteq X$, $0 < \mu(A_1) < \infty$, and $\mu(A_0) \leq t \leq \mu(A_1)$, then there exists an $E_t \subseteq A_1$ with $\mu(E_t) = t$.
(b) Given $\varphi(t)$ continuous and decreasing on $[0,\infty)$, there exists a measurable function f on X with $f^*(t) = \varphi(t)$ for all $t > 0$.
(c) Given $A \subseteq X$ with $0 < \mu(A) < \infty$ and g an integrable function on X, there exists a subset $\widetilde{A}$ of X with $\mu(\widetilde{A}) = \mu(A)$ such that

$$\int_{\widetilde{A}} g\,d\mu = \int_0^{\mu(A)} g^*(s)\,ds.$$

(d) Given f and g measurable functions on X, we have

$$\sup_{h:\, d_h = d_f} \left| \int_X h g\,d\mu \right| = \int_0^\infty f^*(s) g^*(s)\,ds,$$

where the supremum is taken over all h equidistributed with f.
[*Hint:* Part (a): Reduce matters to the situation in which $A_0 = \emptyset$. Consider first the case that for all $A \subseteq X$ there exists a subset B of X satisfying $\frac{1}{10}\mu(A) \leq \mu(B) \leq \frac{9}{10}\mu(A)$. Then we can find subsets of A_1 of measure in any arbitrarily small interval, and by continuity the required conclusion follows. Next consider the case in which there is a subset A_1 of X such that every $B \subseteq A_1$ satisfies $\mu(B) < \frac{1}{10}\mu(A_1)$ or $\mu(B) > \frac{9}{10}\mu(A_1)$. Without loss of generality, normalize μ so that $\mu(A_1) = 1$. Let $\mu_1 = \sup\{\mu(C) : C \subseteq A_1, \mu(C) < \frac{1}{10}\}$ and pick $B_1 \subseteq A_1$ such that $\frac{1}{2}\mu_1 \leq \mu(B_1) \leq \mu_1$. Set $A_2 = A_1 \setminus B_1$ and define $\mu_2 = \sup\{\mu(C) : C \subseteq A_2, \mu(C) < \frac{1}{10}\}$. Continue in this way and define sets $A_1 \supseteq A_2 \supseteq A_3 \supseteq \cdots$ and numbers $\frac{1}{10} \geq \mu_1 \geq \mu_2 \geq \mu_3 \geq \cdots$. If $C \subseteq A_{n+1}$ with $\mu(C \cup A_{n+1}) < \frac{1}{10}$, then $C \cup B_n \subseteq A_n$ with $\mu(C \cup B_n) < \frac{1}{5} < \frac{9}{10}$, and hence by assumption we must have $\mu(C \cup B_n) < \frac{1}{10}$. Conclude that $\mu_{n+1} \leq \frac{1}{2}\mu_n$ and that $\mu(A_n) \geq \frac{4}{5}$ for all $n = 1,2,\dots$. Then the set $\bigcap_{n=1}^\infty A_n$ must be an atom. Part (b): First show that when d is a simple right continuous decreasing function on $[0,\infty)$ there exists a measurable f on X such that $f^* = d$. For general continuous functions, use approximation. Part (c): Let $t = \mu(A)$ and define $A_1 = \{x : |g(x)| > g^*(t)\}$ and $A_2 = \{x : |g(x)| \geq g^*(t)\}$. Then $A_1 \subseteq A_2$ and $\mu(A_1) \leq t \leq \mu(A_2)$. Pick $\widetilde{A}$ such that $A_1 \subseteq \widetilde{A} \subseteq A_2$ and $\mu(\widetilde{A}) = t = \mu(A)$ by part (a). Then $\int_{\widetilde{A}} g\,d\mu = \int_X g\chi_{\widetilde{A}}\,d\mu = \int_0^\infty (g\chi_{\widetilde{A}})^*\,ds = \int_0^{\mu(\widetilde{A})} g^*(s)\,ds$. Part (d): Let $f = \sum_{j=1}^N a_j \chi_{A_j}$ where $a_1 > a_2 > \cdots > a_N > 0$ and the A_j are pairwise disjoint. Write f as $\sum_{j=1}^N b_j \chi_{B_j}$, where $b_j = (a_j - a_{j+1})$ and $B_j = A_1 \cup \cdots \cup A_j$. Pick $\widetilde{B_j}$ as in part (c). Then $\widetilde{B_1} \subseteq \cdots \subseteq \widetilde{B_N}$ and the function $f_1 = \sum_{j=1}^N b_j \chi_{\widetilde{B_j}}$ has the same distribution function as f. It follows from part (c) that $\int_X f_1 g\,d\mu = \int_0^\infty f^*(s)g^*(s)\,ds$. The case of a general function f follows from that in which f is simple using Exercise 1.4.1 and approximation.]

1.4.6. (*Aoki [5]/ Rolewicz [224]*) Let $\|\cdot\|$ be a nonnegative functional on a vector space X that satisfies

$$\|x+y\| \le K(\|x\|+\|y\|)$$

for all x and y in X. (To avoid trivialities, assume that $K \ge 1$.) Then for α defined by the equation

$$(2K)^\alpha = 2 \qquad (\alpha \le 1),$$

we have

$$\|x_1+\cdots+x_n\|^\alpha \le 4\left(\|x_1\|^\alpha+\cdots+\|x_n\|^\alpha\right)$$

for all $n=1,2,\dots$ and all $x_1, x_2, \dots, x_n$ in X.
[*Hint:* Quasilinearity implies that $\|x_1+\cdots+x_n\| \le \max_{1\le j\le n}[(2K)^j\|x_j\|]$ for all $x_1,\dots,x_n$ in X (use that $K\ge 1$). Define $H:\ X\to \mathbf{R}$ by setting $H(0)=0$ and $H(x)=2^{j/\alpha}$ if $2^{j-1}<\|x\|^\alpha\le 2^j$. Then $\|x\|\le H(x)\le 2^{1/\alpha}\|x\|$ for all $x\in X$. Prove by induction that $\|x_1+\cdots+x_n\|^\alpha \le 2(H(x_1)^\alpha+\cdots+H(x_n)^\alpha)$. Suppose that this statement is true when $n=m$. To show its validity for $n=m+1$, without loss of generality assume that $\|x_1\|\ge\|x_2\|\ge\cdots\ge\|x_{m+1}\|$. Then $H(x_1)\ge H(x_2)\ge\cdots\ge H(x_{m+1})$. Assume that all the $H(x_j)$'s are distinct. Then since $H(x_j)^\alpha$ are distinct powers of 2, they must satisfy $H(x_j)^\alpha \le 2^{-j+1}H(x_1)^\alpha$. Then

$$\begin{aligned}
\|x_1+\cdots+x_{m+1}\|^\alpha &\le \Big[\max_{1\le j\le m+1}(2K)^j\|x_j\|\Big]^\alpha\\
&\le \Big[\max_{1\le j\le m+1}(2K)^j H(x_j)\Big]^\alpha\\
&\le \Big[\max_{1\le j\le m+1}(2K)^j 2^{1/\alpha}2^{-j/\alpha}H(x_1)\Big]^\alpha\\
&= 2H(x_1)^\alpha\\
&\le 2\big(H(x_1)^\alpha+\cdots+H(x_{m+1})^\alpha\big).
\end{aligned}$$

We now consider the case that $H(x_j)=H(x_{j+1})$ for some $1\le j\le m$. Then for some integer r we must have $2^{r-1}<\|x_{j+1}\|^\alpha\le\|x_j\|^\alpha\le 2^r$ and $H(x_j)=2^{r/\alpha}$. Next note that

$$\|x_j+x_{j+1}\|^\alpha \le K^\alpha(\|x_j\|+\|x_{j+1}\|)^\alpha \le K^\alpha(22^r)^\alpha \le 2^{r+1}.$$

This implies

$$H(x_j+x_{j+1})^\alpha \le 2^{r+1} = 2^r+2^r = H(x_j)^\alpha + H(x_{j+1})^\alpha.$$

Now apply the inductive hypothesis to $x_1,\dots,x_{j-1},x_j+x_{j+1},x_{j+1},\dots,x_m$ and use the previous inequality to obtain the required conclusion.]

1.4.7. (*Stein and Weiss [264]*) Let (X,μ) and (Y,ν) be measure spaces. Let Z be a Banach space of complex-valued measurable functions on Y. Assume that Z is closed under absolute values and satisfies $\|f\|_Z = \||f|\|_Z$. Suppose that T is a linear operator defined on the space of measurable functions on (X,μ) and taking values in Z. Suppose that for some constant $A>0$ we have the restricted weak type estimate

$$\big\|T(\chi_E)\big\|_Z \le A\mu(E)^{1/p}$$

for all E measurable subsets of X and some $0 < p < \infty$. Then there is a constant $C(p) > 0$ such that

$$\big\|T(f)\big\|_Z \le C(p)A\big\|f\big\|_{L^{p,1}}$$

for all f in the domain of T.
[*Hint:* Let $f = \sum_{j=1}^N a_j\chi_{E_j} \ge 0$, where $a_1 > a_2 > \cdots > a_N > 0$, $\mu(E_j) < \infty$ pairwise disjoint. Let $F_j = E_1 \cup \cdots \cup E_j$, $B_0 = 0$, and $B_j = \mu(F_j)$ for $j \ge 1$. Write $f = \sum_{j=1}^N (a_j - a_{j+1})\chi_{F_j}$, where $a_{N+1} = 0$. Then

$$\begin{aligned}
\big\|T(f)\big\|_Z &= \big\||T(f)|\big\|_Z \\
&\le \sum_{j=1}^N (a_j - a_{j+1})\big\|T(\chi_{F_j})\big\|_Z \\
&\le A\sum_{j=1}^N (a_j - a_{j+1})(\mu(F_j))^{1/p} \\
&= A\sum_{j=0}^{N-1} a_{j+1}(B_{j+1}^{1/p} - B_j^{1/p}) \\
&= p^{-1}A\big\|f\big\|_{L^{p,1}},
\end{aligned}$$

where the penultimate equality follows summing by parts; see Appendix F.]

1.4.8. Let $0 < p < \infty$ and $0 < q_1 < q_2 \le \infty$. Let $\alpha, \beta, q > 0$.
(a) Show that the function $f(t) = t^{-\alpha}(\log t^{-1})^{-\beta}\chi_{(0,1)}(t)$ lies in $L^{p,q}(\mathbf{R})$ if and only if either $p > 1/\alpha$ or both $p = 1/\alpha$ and $q > 1/\beta$.
(b) Show that the function $t^{-\frac{1}{p}}(\log t^{-1})^{-\frac{1}{q_1}}\chi_{(0,1)}(t)$ lies in $L^{p,q_2}(\mathbf{R})$ but not in $L^{p,q_1}(\mathbf{R})$.
(c) On $\mathbf{R}^n$ construct examples to show that $L^{p,q_1} \subsetneqq L^{p,q_2}$.
(d) On a general nonatomic measure space (X,μ) prove that there *does not* exist a constant $C(p,q_1,q_2) > 0$ such that for all f in $L^{p,q_2}(X)$ the following is valid:

$$\big\|f\big\|_{L^{p,q_1}} \le C(p,q_1,q_2)\big\|f\big\|_{L^{p,q_2}}.$$

1.4.9. (*Stein and Weiss [263]*) Let $L^p(\omega)$ denote the weighted L^p space with measure $\omega(x)dx$. Let T be a sublinear operator that maps

$$\begin{aligned}
T:\ L^{p_0}(\omega_0) &\to L^{q_0,\infty}(w),\\
T:\ L^{p_1}(\omega_1) &\to L^{q_1,\infty}(w),
\end{aligned}$$

for some $p_0 \ne p_1$, where $0 < p_0, p_1, q_0, q_1 \le \infty$ and $\omega_0, \omega_1, \omega$ are positive functions. Suppose that

$$\frac{1}{p_\theta} = \frac{1-\theta}{p_0} + \frac{\theta}{p_1}, \qquad \frac{1}{q_\theta} = \frac{1-\theta}{q_0} + \frac{\theta}{q_1}.$$

Then T maps

$$L^{p_\theta}\left(\omega_0^{\frac{1-\theta}{p_0}p_\theta}\omega_1^{\frac{\theta}{p_1}p_\theta}\right) \to L^{q_\theta,p_\theta}(\omega).$$

[*Hint:* Define

$$L(f) = (\omega_1/\omega_0)^{\frac{1}{p_1-p_0}} f$$

and observe that for each $\theta \in [0,1]$, L maps

$$L^{p_\theta}\left(\omega_0^{\frac{1-\theta}{p_0}p_\theta}\omega_1^{\frac{\theta}{p_1}p_\theta}\right) \to L^{p_\theta}\left((\omega_0^{p_1}\omega_1^{-p_0})^{\frac{1}{p_1-p_0}}\right)$$

isometrically. Then apply the classical Marcinkiewicz interpolation theorem to the sublinear operator $T\circ L^{-1}$, and the required conclusion easily follows.]

1.4.10. (*Kalton [147]/Stein and Weiss [266]*) Let λ_n be a sequence of positive numbers with $\sum_n \lambda_n \le 1$ and $\sum_n \lambda_n \log(\frac{1}{\lambda_n}) = K < \infty$.
(a) Let f_n be a sequence of complex-valued functions in $L^{1,\infty}(X)$ such that $\|f_n\|_{L^{1,\infty}} \le 1$ uniformly in n. Prove that $\sum_n \lambda_n f_n$ lies in $L^{1,\infty}(X)$ with norm at most $2(K+2)$. (This property is referred to as the *logconvexity* of $L^{1,\infty}$.)
(b) Let T_n be a sequence of sublinear operators that map $L^1(X)$ to $L^{1,\infty}(Y)$ with norms $\|T_n\|_{L^1\to L^{1,\infty}} \le B$ uniformly in n. Use part (a) to prove that $\sum_n \lambda_n T_n$ maps $L^1(X)$ to $L^{1,\infty}(Y)$ with norm at most $2B(K+2)$.
(c) Given $\delta > 0$ pick $0 < \varepsilon < \delta$ and use the simple estimate

$$\mu(\{\sum_{n=1}^\infty 2^{-\delta n} f_n > \alpha\}) \le \sum_{n=1}^\infty \mu(\{2^{-\delta n} f_n > (2^\varepsilon - 1)2^{-\varepsilon n}\alpha\})$$

to obtain a simple proof of the statements in part (a) and (b) when $\lambda_n = 2^{-\delta n}$, $n = 1,2,\dots$, and zero otherwise.
[*Hint:* Part (a): For fixed $\alpha > 0$, write $f_n = u_n + v_n + w_n$, where $u_n = f_n \chi_{|f_n|\le \frac{\alpha}{2}}$, $v_n = f_n\chi_{|f_n|>\frac{\alpha}{2\lambda_n}}$, and $w_n = f_n\chi_{\frac{\alpha}{2}<|f_n|\le\frac{\alpha}{2\lambda_n}}$. Let $u = \sum_n \lambda_n u_n$, $v = \sum_n \lambda_n v_n$, and $w = \sum_n \lambda_n w_n$. Clearly $|u| \le \alpha/2$. Also $\{v \ne 0\} \subseteq \bigcup_n \{|f_n| > \frac{\alpha}{2\lambda_n}\}$; hence $\mu(\{v\ne 0\}) \le \frac{2}{\alpha}$. Finally,

$$\begin{aligned}\int_X |w|\,d\mu &\le \sum_n \lambda_n \int_X |f_n|\chi_{\frac{\alpha}{2}<|f_n|\le\frac{\alpha}{2\lambda_n}}\,d\mu\\ &\le \sum_n \lambda_n\left[\int_{\alpha/2}^{\alpha/(2\lambda_n)} d_{f_n}(\beta)\,d\beta + \int_0^{\alpha/2} d_{f_n}(\alpha/2)\,d\beta\right]\\ &\le K+1.\end{aligned}$$

Using $\mu(\{|u+v+w| > \alpha\}) \le \mu(\{|u| > \alpha/2\}) + \mu(\{|v| \ne 0\}) + \mu(\{|w| > \alpha/2\})$, deduce the conclusion.]

1.4.11. Construct a sequence of functions f_k in $L^{1,\infty}(\mathbf{R}^n)$ and a function $f \in L^{1,\infty}$ such that $\|f_k - f\|_{L^\infty} \to 0$ but $\|f_k\|_{L^{1,\infty}} \to \infty$ as $k \to \infty$.

1.4.12. (a) Suppose that X is a quasi-Banach space and let X^* be its dual (which is always a Banach space). Prove that for all $T \in X^*$ we have

$$\|T\|_{X^*} = \sup_{\substack{x\in X\\ \|x\|_X\le 1}} |T(x)|.$$

(b) Now suppose that X is a Banach space. Use the Hahn–Banach theorem to prove that for every $x \in X$ we have

$$\|x\|_X = \sup_{\substack{T\in X^*\\ \|T\|_{X^*}\le 1}} |T(x)|.$$

Observe that this result may fail for quasi-Banach spaces. For example, if $X = L^{1,\infty}$, every linear functional on X^* vanishes on the set of simple functions.
(c) Take $X = L^{p,1}$ and $X^* = L^{p',\infty}$. Then for $1 < p < \infty$ both of these spaces are normable. Conclude that

$$\|f\|_{L^{p,1}} \approx \sup_{\|g\|_{L^{p',\infty}}\le 1} \left|\int_X fg\,d\mu\right|,$$

$$\|f\|_{L^{p,\infty}} \approx \sup_{\|g\|_{L^{p',1}}\le 1} \left|\int_X fg\,d\mu\right|.$$

1.4.13. Let $0 < p,q < \infty$. Prove that any function in $L^{p,q}(X,\mu)$ can be written as

$$f = \sum_{n=-\infty}^{+\infty} c_n f_n,$$

where f_n is a function bounded by $2^{-n/p}$, supported on a set of measure 2^n, and the sequence $\{c_k\}_k$ lies in ℓ^q and satisfies

$$2^{-\frac{1}{p}}(\log 2)^{\frac{1}{q}} \big\|\{c_k\}_k\big\|_{\ell^q} \le \|f\|_{L^{p,q}} \le \big\|\{c_k\}_k\big\|_{\ell^q} 2^{\frac{1}{p}}(\log 2)^{\frac{1}{q}}.$$

[*Hint:* Let $c_n = 2^{n/p} f^*(2^n)$ and $f_n = c_n^{-1} f \chi_{A_n}$ where A_n is as in (1.4.27).]

1.4.14. (*T. Tao*) Let $0 < p < \infty$, $0 < \gamma < 1$, $A > 0$, and let f be a measurable function on a measure space (X,μ).
(a) Suppose that $\|f\|_{L^{p,\infty}} \le A$. Then for every measurable set E of finite measure there exists a measurable subset E' of E with $\mu(E') \ge \gamma\mu(E)$ such that

$$\left|\int_{E'} f\,d\mu\right| \le C_\gamma A\,\mu(E)^{1-\frac{1}{p}},$$

where $C_\gamma = (1-\gamma)^{-1/p}$.
(b) Conversely, if the last condition holds for some $C_\gamma, A < \infty$ and all measurable

subsets E of finite measure, then $\|f\|_{L^{p,\infty}} \le c_\gamma A$, where $c_\gamma = C_\gamma 4^{1/p}\gamma^{-1}\sqrt{2}$.
(c) Conclude that

$$\|f\|_{L^{p,\infty}} \approx \sup_{\substack{E\subseteq X\\ 0<\mu(E)<\infty}} \inf_{\substack{E'\subseteq E\\ \mu(E')\ge\frac{1}{2}\mu(E)}} \mu(E)^{-1+\frac{1}{p}}\left|\int_{E'} f\,d\mu\right|.$$

[*Hint:* Part (a): Take $E' = E\setminus\{|f| > A(1-\gamma)^{-\frac{1}{p}}\mu(E)^{-\frac{1}{p}}\}$. Part (b): Given $\alpha > 0$, note that the set $\{|f| > \alpha\}$ is contained in

$$\{\mathrm{Re}\, f > \tfrac{\alpha}{\sqrt{2}}\}\cup\{\mathrm{Im}\, f > \tfrac{\alpha}{\sqrt{2}}\}\cup\{\mathrm{Re}\, f < -\tfrac{\alpha}{\sqrt{2}}\}\cup\{\mathrm{Im}\, f < -\tfrac{\alpha}{\sqrt{2}}\}.$$

For E any of the preceding four sets, let E' be a subset of it with measure at least $\gamma\mu(E)$ such as in the hypothesis. Then $\left|\int_{E'} f\,d\mu\right| \ge \frac{\alpha}{\sqrt{2}}\gamma\mu(E)$, which gives $\|f\|_{L^{p,\infty}} \le 4^{1/p}\gamma^{-1}C_\gamma\sqrt{2}A$.]

1.4.15. Given a linear operator T defined on the set of measurable functions on a measure space (X,μ) and taking values in the set of measurable functions on a measure space (Y,ν), define its "transpose" T^t via the identity

$$\int_Y T(f)\,g\,d\nu = \int_X T^t(g)\,f\,d\mu$$

for all measurable functions f on X and g on Y, whenever the integrals converge. Let T be such a linear operator given in the form

$$T(f)(y) = \int_X K(y,x)f(x)\,d\mu(x),$$

where K is measurable and bounded by some constant $M > 0$. Suppose that T maps $L^1(X)$ to $L^{1,\infty}(Y)$ with norm $\|T\|$, and T^t maps $L^1(Y)$ to $L^{1,\infty}(X)$ with norm $\|T^t\|$. Show that for all $1 < p < \infty$ there exists a constant C_p that depends only on p and is independent of M such that T maps $L^p(X)$ to $L^p(Y)$ with norm

$$\|T\|_{L^p(X)\to L^p(Y)} \le C_p\|T\|^{\frac{1}{p}}\|T^t\|^{1-\frac{1}{p}}.$$

[*Hint:* For $R > 0$, let $\mathscr{B}_R$ be the set of all (A,B), where A is a measurable subset of X with $\mu(A) \le R$ and B is a measurable subset of Y with $\nu(B) \le R$. Let $\mathscr{B}_{R,M}$ be the set of all (A,B) in $\mathscr{B}_R$ such that $|K(x,y)| \le M$ for all $x \in A$ and $y \in B$. Also let $M_p = M_p(R,M) < \infty$ be the smallest constant such that for all $(A,B) \in \mathscr{B}_{R,M}$ we have $\left|\int_B T(\chi_A)\,d\nu\right| \le M_p\mu(A)^{\frac{1}{p}}\nu(B)^{\frac{1}{p'}}$. Let $\delta > 0$ and $(A,B) \in \mathscr{B}_{R,M}$. If $\mu(A) \le \delta\nu(B)$, use Exercise 1.4.14 to find a B' with $\nu(B') \ge \frac{1}{2}\nu(B)$ such that $\left|\int_{B'} T(\chi_A)\,d\nu\right| \le c\|T\|\mu(A) \le c\,\delta^{\frac{1}{p'}}\|T\|\mu(A)^{\frac{1}{p}}\nu(B)^{\frac{1}{p'}}$. Then $\nu(B\setminus B') \le \frac{1}{2}\nu(B)$ and we have

$$\left|\int_{B\setminus B'} T(\chi_A)\,d\nu\right| \le M_p 2^{-\frac{1}{p'}}\mu(A)^{\frac{1}{p}}\nu(B)^{\frac{1}{p'}}.$$

Summing, we obtain $M_p \le M_p 2^{-\frac{1}{p'}} + c\delta^{\frac{1}{p'}}\|T\|$. Whenever $\nu(B) \le \delta^{-1}\mu(A)$, write $\left|\int_B T(\chi_A)\,d\nu\right| = \left|\int_A T^t(\chi_B)\,d\mu\right|$ and use Exercise 1.4.14 to find a set A' with $\mu(A') \ge \frac{1}{2}\mu(A)$ and $\left|\int_{A'} T^t(\chi_B)\,d\mu\right| \le c\|T^t\|\nu(B)$. Argue similarly to obtain $M_p \le M_p 2^{-\frac{1}{p}} + c\delta^{-\frac{1}{p}}\|T^t\|$. Pick a suitable δ to optimize both expressions. Obtain that M_p is independent of R and M. Considering $B_+ = B \cap \{T(\chi_A) > 0\}$ and $B_- = B \cap \{T(\chi_A) < 0\}$, obtain that $\int_B |T(\chi_A)|\,d\nu \le 2M_p\mu(A)^{\frac{1}{p}}\nu(B)^{\frac{1}{p'}}$ for $(A,B) \in \mathscr{B}_{R,M}$. Use Fatou's lemma to remove the restriction that $(A,B) \in \mathscr{B}_{R,M}$. Finally, use the characterization of $\|\cdot\|_{L^{p,\infty}}$ obtained in Exercise 1.1.12 with $r = 1$ to conclude that $\|T(\chi_A)\|_{L^{p,\infty}} \le C_p\|T\|^{\frac{1}{p}}\|T^t\|^{1-\frac{1}{p}}\mu(A)^{\frac{1}{p}}$.]

1.4.16. (*Bourgain [29]*) Let $0 < p_0 < p_1 < \infty$ and $0 < \alpha, \beta, A, B < \infty$. Suppose that a family of sublinear operators T_k is of restricted weak type (p_0, p_0) with constant $A2^{-k\alpha}$ and of restricted weak type (p_1, p_1) with constant $B2^{-k\beta}$ for all $k \in \mathbf{Z}$. Show that there is a constant $C = C(\alpha, \beta, p_0, p_1)$ such that $\sum_{k\in\mathbf{Z}} T_k$ is of restricted weak type (p,p) with constant $CA^{1-\theta}B^{\theta}$, where $\theta = \alpha/(\alpha+\beta)$ and

$$\frac{1}{p} = \frac{1-\theta}{p_0} + \frac{\theta}{p_1}.$$

[*Hint:* Estimate $\mu(\{|T(\chi_E)| > \lambda\})$ by the sum $\sum_{k\ge k_0}\mu(\{|T_k(\chi_E)| > c\lambda 2^{\alpha'(k_0-k)}\}) + \sum_{k\le k_0}\mu(\{|T_k(\chi_E)| > c\lambda 2^{\beta'(k-k_0)}\})$, where c is a suitable constant and $0 < \alpha' < \alpha$, $\beta < \beta' < \infty$. Apply the restricted weak type (p_0,p_0) hypothesis on each term of the first sum, the restricted weak type (p_1,p_1) hypothesis on each term of the second sum, and choose k_0 to optimize the resulting expression.]

APPENDIX: SOME MULTILINEAR INTERPOLATION

Multilinear maps are defined on products on linear spaces and take values in another linear space. We are interested in the situation that these linear spaces are function spaces. Let $(X_1,\mu_1),\dots,(X_m,\mu_m)$ be measure spaces, let $\mathscr{D}_j$ be spaces of measurable functions on X_j, and let T be a map defined on $\mathscr{D}_1 \times \cdots \times \mathscr{D}_m$ and taking values in the set of measurable functions on another measure space (Z,σ). Then T is called multilinear if for all f_j, g_j in $\mathscr{D}_j$ and all scalars λ we have

$$\begin{aligned} |T(f_1,\dots,\lambda f_j,\dots,f_m)| &= |\lambda|\,|T(f_1,\dots,f_j,\dots,f_m)|, \\ T(f_1,\dots,f_j+g_j,\dots,f_m) &= T(f_1,\dots,f_j,\dots,f_m) + T(f_1,\dots,g_j,\dots,f_m). \end{aligned}$$

If $\mathscr{D}_j$ are dense subspaces of $L^{p_j}(X_j,\mu_j)$ and T is a multlinear map defined on $\prod_{j=1}^m \mathscr{D}_j$ and satisfies

$$\|T(f_1,\dots,f_m)\|_{L^p(Z)} \le C\|f_1\|_{L^{p_1}(X_1)}\cdots\|f_m\|_{L^{p_m}(X_m)},$$

for all $f_j \in \mathscr{D}_j$, then T has a bounded extension from $L^{p_1} \times \cdots \times L^{p_m} \to Z$. The norm of a multilinear map $T : L^{p_1} \times \cdots \times L^{p_m} \to Z$ is the smallest constant C such that the preceding inequality holds and is denoted by

$$\big\|T\big\|_{L^{p_1}\times\cdots\times L^{p_m}\to Z}.$$

Suppose that T is defined on $\prod_{j=1}^m \mathscr{D}_j$, where each $\mathscr{D}_j$ contains the simple functions. We say that T is quasimultilinear if there is a $K>0$ such that for all $1\le j\le m$, all f_j, g_j in $\mathscr{D}_j$, and all $\lambda \in \mathbf{C}$ we have

$$\begin{aligned}|T(f_1,\dots,\lambda f_j,\dots,f_m)| &= |\lambda|\,|T(f_1,\dots,f_j,\dots,f_m)|,\\ |T(f_1,\dots,f_j+g_j,\dots,f_m)| &\le K\big(|T(f_1,\dots,f_j,\dots,f_m)|+|T(f_1,\dots,g_j,\dots,f_m)|\big).\end{aligned}$$

In the special case in which $K=1$, T is called multisublinear.

1.4.17. Let T be a multilinear map defined on the set of simple functions of the product of m measure spaces $(X_1,\mu_1)\times\cdots\times(X_m,\mu_m)$ and taking values in the set of measurable functions on another measure space (Z,σ). Let $1\le p_{jk}\le\infty$ for $1\le k\le m$ and $j\in\{0,1\}$ and also let $1\le p_j\le\infty$ for $j\in\{0,1\}$. Suppose that T satisfies

$$\big\|T(f_1,\dots,f_m)\big\|_{L^{p_j}}\le M_j\big\|f_1\big\|_{L^{p_{j1}}}\cdots\big\|f_m\big\|_{L^{p_{jm}}},\qquad j=0,1,$$

for all simple functions f_k on X_k. Let $(1/q,1/q_1,\dots,1/q_m)$ lie on the open line segment joining $(1/p_0,1/p_{01},\dots,1/p_{0m})$ and $(1/p_1,1/p_{11},\dots,1/p_{1m})$ in $\mathbf{R}^{m+1}$. Then for some $0<\theta<1$ we have

$$\frac{1}{q}=\frac{1-\theta}{p_0}+\frac{\theta}{p_1},\qquad \frac{1}{q_k}=\frac{1-\theta}{p_{0k}}+\frac{\theta}{p_{1k}},\qquad 1\le k\le m.$$

Prove that T has a bounded extension from $L^{q_1}\times\cdots\times L^{q_m}$ to L^q that satisfies

$$\big\|T(f_1,\dots,f_m)\big\|_{L^q}\le M_0^{1-\theta}M_1^{\theta}\big\|f_1\big\|_{L^{q_1}}\cdots\big\|f_m\big\|_{L^{q_m}}$$

for all $f_k\in L^{q_k}(X_k)$.
[*Hint:* Adapt the proof of Theorem 1.3.4.]

1.4.18. Let $(X_1,\mu_1),\dots,(X_m,\mu_m)$ be measure spaces, let $\mathscr{D}_j$ be spaces of measurable functions on X_j that contain the simple functions, and let T be a quasi-multilinear map defined on $\mathscr{D}_1\times\cdots\times\mathscr{D}_m$ that takes values in the set of measurable functions on another measure space (Z,σ). Let $0<p_{jk}\le\infty$ for $1\le j\le m+1$ and $1\le k\le m$, and also let $0<p_j\le\infty$ for $1\le j\le m+1$. Suppose that for all $1\le j\le m+1$, T satisfies

$$\big\|T(\chi_{E_1},\dots,\chi_{E_m})\big\|_{L^{p_j,\infty}}\le M\mu_1(E_1)^{\frac{1}{p_{j1}}}\cdots\mu_m(E_m)^{\frac{1}{p_{jm}}}$$

for all sets E_k of finite μ_k measure. Assume that the system

$$\begin{pmatrix} 1/p_{11} & 1/p_{12} & \dots & 1/p_{1m} & 1 \\ 1/p_{21} & 1/p_{22} & \dots & 1/p_{2m} & 1 \\ \vdots & \vdots & \vdots & \vdots & \vdots \\ 1/p_{m1} & 1/p_{m2} & \dots & 1/p_{mm} & 1 \\ 1/p_{(m+1)1} & 1/p_{(m+1)2} & \dots & 1/p_{(m+1)m} & 1 \end{pmatrix} \begin{pmatrix} \sigma_1 \\ \sigma_2 \\ \vdots \\ \sigma_m \\ \tau \end{pmatrix} = \begin{pmatrix} 1/p_1 \\ 1/p_2 \\ \vdots \\ 1/p_m \\ 1/p_{m+1} \end{pmatrix}$$

has a *unique* solution $(\sigma_1,\dots,\sigma_m,\tau) \in \mathbf{R}^{m+1}$ with *not all* $\sigma_j = 0$. (This assumption implies that the determinant of the displayed square matrix is nonzero.) Suppose that the point $(1/q,1/q_1,\dots,1/q_m)$ lies in the open convex hull of the $m+1$ points $(1/p_j,1/p_{j1},\dots,1/p_{jm})$ in $\mathbf{R}^{m+1}$, $1 \le j \le m+1$. Let $0 < t_k, t \le \infty$ satisfy

$$\sum_{\sigma_k \neq 0} \frac{1}{t_k} \ge \frac{1}{t}.$$

Prove that there exists a constant C that depends only on the p_{jk}'s, q_k's, p_j's, and on K (but not on M) such that for all f_j in $\mathscr{D}_j$ we have

$$\big\|T(f_1,\dots,f_m)\big\|_{L^{q,t}} \le CM\big\|f_1\big\|_{L^{q_1,t_1}} \cdots \big\|f_m\big\|_{L^{q_m,t_m}}.$$

[*Hint:* Split the functions f_j as in the proof of Theorem 1.4.19. For simplicity, you may want to prove this result only when $m = 2$.]

1.4.19. (*O' Neil [207]*) Show that

$$\big\|f * g\big\|_{L^{r,s}} \le C_{p,q,s_1,s_2}\big\|f\big\|_{L^{p,s_1}}\big\|g\big\|_{L^{q,s_2}},$$

whenever $1 < p,q,r < \infty$, $0 < s_1,s_2 \le \infty$, $\frac{1}{p} + \frac{1}{q} = \frac{1}{r} + 1$, and $\frac{1}{s_1} + \frac{1}{s_2} = \frac{1}{s}$. Also deduce Hölder's inequality for Lorentz spaces,

$$\big\|fg\big\|_{L^{r,s}} \le C_{p,q,s_1,s_2}\big\|f\big\|_{L^{p,s_1}}\big\|g\big\|_{L^{q,s_2}},$$

where now $0 < p,q,r \le \infty$, $0 < s_1,s_2 \le \infty$, $\frac{1}{p} + \frac{1}{q} = \frac{1}{r}$, and $\frac{1}{s_1} + \frac{1}{s_2} = \frac{1}{s}$.
[*Hint:* Use Exercise 1.4.17.]

1.4.20. (*Grafakos and Tao [112]*) Suppose that T is a multilinear operator of the form

$$T(f_1,\dots,f_m)(y) = \int_{X_1} \cdots \int_{X_m} K(x_1,\dots,x_m,y) f_1(x_1)\cdots f_m(x_m)\, d\mu_1(x_1)\cdots d\mu_m(x_m),$$

where the kernel K is bounded by some constant M. The jth transpose of T is the m-linear operator whose kernel is obtained from K by interchanging the variables x_j and y. Suppose that T and all of its transposes map $L^1(X_1) \times \cdots \times L^1(X_m)$ to $L^{1/m,\infty}(Y)$. Conclude that T maps $L^{p_1}(X_1) \times \cdots \times L^{p_m}(X_m)$ to $L^p(Y)$ when $1 < p_j \le \infty$, $p < \infty$, and $1/p = 1/p_1 + \cdots + 1/p_m$ with a bound independent of the kernel K.
[*Hint:* Take $p > 1$ and use the same idea as in Exercise 1.4.15. The full range of p's follows by interpolation.]

HISTORICAL NOTES

The modern theory of measure and integration was founded with the publication of Lebesgue's dissertation [169]; see also [170]. The theory of the Lebesgue integral reshaped the course of integration. The spaces $L^p([a,b])$, $1 < p < \infty$, were first investigated by Riesz [217], who obtained many important properties of them. A rigorous treatise of harmonic analysis on general groups can be found in the book of Hewitt and Ross [125]. The best possible constant C_{pqr} in Young's inequality $\|f*g\|_{L^r(\mathbf{R}^n)} \le C_{pqr}\|f\|_{L^p(\mathbf{R}^n)}\|g\|_{L^q(\mathbf{R}^n)}$, $\frac{1}{p}+\frac{1}{q}=\frac{1}{r}+1$, $1<p,q,r<\infty$, was shown by Beckner [16] to be $C_{pqr} = (B_pB_qB_{r'})^n$, where $B_p^2 = p^{1/p}(p')^{-1/p'}$.

Theorem 1.3.2 first appeared without proof in Marcinkiewicz's brief note [187]. After his death in World War II, this theorem seemed to have escaped attention until Zygmund reintroduced it in [302]. This reference presents the more difficult off-diagonal version of the theorem, derived by Zygmund. Stein and Weiss [264] strengthened Zygmund's theorem by assuming that the initial estimates are of restricted weak type whenever $1 \le p_0, p_1, q_0, q_1 \le \infty$. The extension of this result to the case $0 < p_0, p_1, q_0, q_1 < 1$ as presented in Theorem 1.4.19 is due to the author; the critical Lemma 1.4.20 was suggested by Kalton. Equivalence of restricted weak type $(1,1)$ and weak type $(1,1)$ properties for certain maximal multipliers was obtained by Moon [201]. The following partial converse of Theorem 1.2.13 is due to Stepanov [268]: If a convolution operator maps $L^1(\mathbf{R}^n)$ to $L^{q,\infty}(\mathbf{R}^n)$ for some $1 < q < \infty$ then its kernel must be in $L^{q,\infty}$.

The extrapolation result of Exercise 1.3.7 is due to Yano [294]; see also Zygmund [304, pp. 119–120]. We refer to Carro [47] for a generalization. See also the related work of Soria [250] and Tao [274].

The original version of Theorem 1.3.4 was proved by Riesz [220] in the context of bilinear forms. This version is called the Riesz convexity theorem, since it says that the logarithm of the function $M(\alpha,\beta) = \inf_{x,y}\left|\sum_{j=1}^n\sum_{k=1}^m a_{jk}x_jy_k\right| \|x\|_{\ell^{1/\alpha}}^{-1}\|y\|_{\ell^{1/\beta}}^{-1}$ (where the infimum is taken over all sequences $\{x_j\}_{j=1}^n$ in $\ell^{1/\alpha}$ and $\{y_k\}_{k=1}^m$ in $\ell^{1/\beta}$) is a convex function of (α,β) in the triangle $0 \le \alpha,\beta \le 1$, $\alpha+\beta \ge 1$. Riesz's student Thorin [278] extended this triangle to the unit square $0 \le \alpha,\beta \le 1$ and generalized this theorem by replacing the maximum of a bilinear form with the maximum of the modulus of an entire function in many variables. After the end of World War II, Thorin published his thesis [279], building the subject and giving a variety of applications. The original proof of Thorin was rather long, but a few years later, Tamarkin and Zygmund [272] gave a very elegant short proof using the maximum modulus principle in a more efficient way. Today, this theorem is referred to as the Riesz–Thorin interpolation theorem.

Calderón [34] elaborated the complex-variables proof of the Riesz–Thorin theorem into a general method of interpolation between Banach spaces. The complex interpolation method can also be defined for pairs of quasi-Banach spaces, although certain complications arise in this setting; however, the Riesz–Thorin theorem is true for pairs of L^p spaces (with the "correct" geometric mean constant) for all $0 < p \le \infty$ and also for Lorentz spaces. In this setting, duality cannot be used, but a well-developed theory of analytic functions with values in quasi-Banach spaces is crucial. We refer to the articles of Kalton [148] and [149] for details. Complex interpolation for sublinear maps is also possible; see the article of Calderón and Zygmund [38]. Interpolation for analytic families of operators (Theorem 1.3.7) is due to Stein [251]. The critical Lemma 1.3.8 used in the proof was previously obtained by Hirschman [126].

The fact that nonatomic measure spaces contain subsets of all possible measures is classical. An extension of this result to countably additive vector measures with values in finite-dimensional Banach spaces was obtained by Lyapunov [183]; for a proof of this fact, see Diestel and Uhl [75, p. 264]. The Aoki–Rolewicz theorem (Exercise 1.4.6) was proved independently by Aoki [5] and Rolewicz [224]. For a proof of this fact and a variety of its uses in the context of quasi-Banach spaces we refer to the book of Kalton, Peck, and Roberts [150].

Decreasing rearrangements of functions were introduced by Hardy and Littlewood [123]; the authors attribute their motivation to understanding cricket averages. The $L^{p,q}$ spaces were introduced by Lorentz in [179] and in [180]. A general treatment of Lorentz spaces is given in the article of Hunt [134]. The normability of the spaces $L^{p,q}$ (which holds exactly when $1 < p \le \infty$

and $1 \le q \le \infty$) can be traced back to general principles obtained by Kolmogorov [160]. The introduction of the function f^{**}, which was used in Exercise 1.4.3, to explicitly define a norm on the normable spaces $L^{p,q}$ is due to Calderón [34]. These spaces appear as intermediate spaces in the general interpolation theory of Calderón [34] and in that of Lions and Peetre [172]. The latter was pointed out by Peetre [211]. For a systematic study of the duals of Lorentz spaces we refer to Cwikel [62] and Cwikel and Fefferman [63], [64]. An extension of the Marcinkiewicz interpolation theorem to Lorentz spaces was obtained by Hunt [133]. Standard references on interpolation include the books of Bennett and Sharpley [20], Bergh and Löfström [22], Sadosky [232], and Chapter 5 in Stein and Weiss [265].

Multilinear complex interpolation (cf. Exercise 1.4.16) is a straightforward adaptation of the linear one (cf. Theorem 1.3.4); see Zygmund [304, p. 106] and Berg and Löfström [22]. The multilinear real interpolation method is more involved. References on the subject include (in chronological order) the articles of Strichartz [269], Sharpley [241] and [242], Zafran [297], Christ [48], Janson [139], and Grafakos and Kalton [105]. The latter contains, in particular, the proof of Exercise 1.4.17.

Chapter 2
Maximal Functions, Fourier Transform, and Distributions

We have already seen that the convolution of a function with a fixed density is a smoothing operation that produces a certain average of the function. Averaging is an important operation in analysis and naturally arises in many situations. The study of averages of functions is better understood and simplified by the introduction of the maximal function. This is defined as the largest average of a function over all balls containing a fixed point. Maximal functions play a key role in differentiation theory, where they are used in obtaining almost everywhere convergence for certain integral averages. Although maximal functions do do not preserve qualitative information about the given functions, they maintain crucial quantitative information, a fact of great importance in the subject of Fourier analysis.

Another important operation we study in this chapter is the Fourier transform. This is as fundamental to Fourier analysis as marrow is to the human bone. It is the father of all oscillatory integrals and a powerful transformation that carries a function from its spatial domain to its frequency domain. By doing this, it inverts the function's localization properties. Then magically, if applied one more time, it gives back the function composed with a reflection. More important, it transforms our point of view in harmonic analysis. It changes convolution to multiplication, translation to modulation, and expanding dilation to shrinking dilation, while its decay at infinity encodes information about the local smoothness of the function. The study of the Fourier transform also motivates the launch of a thorough study of general oscillatory integrals. We take a quick look at this topic with emphasis on one-dimensional results.

Distributions changed our view of analysis as they furnished a mathematical framework for many operations that did not exactly qualify to be called functions. These operations found their mathematical place in the world of functionals applied to smooth functions (called test functions). These functionals also introduced the correct interpretation for many physical objects, such as the Dirac delta function. Distributions quickly became an indispensable tool in analysis and brought a broader perspective.

L. Grafakos, *Classical Fourier Analysis, Second Edition*,
DOI: 10.1007/978-0-387-09432-8_2,

2.1 Maximal Functions

Given a Lebesgue measurable subset A of $\mathbf{R}^n$, we denote by $|A|$ its Lebesgue measure. For $x \in \mathbf{R}^n$ and $r > 0$, we denote by $B(x,r)$ the open ball of radius r centered at x. We also use the notation $aB(x,\delta) = B(x,a\delta)$, for $a > 0$, for the ball with the same center and radius $a\delta$. Given $\delta > 0$ and f a locally integrable function on $\mathbf{R}^n$, let

$$\operatorname*{Avg}_{B(x,\delta)} |f| = \frac{1}{|B(x,\delta)|} \int_{B(x,\delta)} |f(y)|\,dy$$

denote the average of $|f|$ over the ball of radius δ centered at x.

2.1.1 The Hardy–Littlewood Maximal Operator

Definition 2.1.1. The function

$$\mathcal{M}(f)(x) = \sup_{\delta>0} \operatorname*{Avg}_{B(x,\delta)} |f| = \sup_{\delta>0} \frac{1}{v_n \delta^n} \int_{|y|<\delta} |f(x-y)|\,dy$$

is called the *centered Hardy–Littlewood maximal function* of f.

Obviously we have $\mathcal{M}(f) = \mathcal{M}(|f|) \geq 0$; thus the maximal function is a positive operator. Information concerning cancellation of the function f is lost by passing to $\mathcal{M}(f)$. We show later that $\mathcal{M}(f)$ pointwise controls f (i.e., $\mathcal{M}(f) \geq |f|$ almost everywhere). Note that $\mathcal{M}$ maps L^∞ to itself, that is, we have

$$\big\|\mathcal{M}(f)\big\|_{L^\infty} \leq \big\|f\big\|_{L^\infty}.$$

Let us compute the Hardy–Littlewood maximal function of a specific function.

Example 2.1.2. On $\mathbf{R}$, let f be the characteristic function of the interval $[a,b]$. For $x \in (a,b)$, clearly $\mathcal{M}(f) = 1$. For $x \geq b$, a simple calculation shows that the largest average of f over all intervals $(x-\delta, x+\delta)$ is obtained when $\delta = x-a$. Similarly, when $x \leq a$, the largest average is obtained when $\delta = b-x$. Therefore,

$$\mathcal{M}(f)(x) = \begin{cases} (b-a)/2|x-b| & \text{when } x \leq a, \\ 1 & \text{when } x \in (a,b), \\ (b-a)/2|x-a| & \text{when } x \geq b. \end{cases}$$

Observe that $\mathcal{M}(f)$ has a jump at $x = a$ and $x = b$ equal to one-half that of f.

$\mathcal{M}$ is a sublinear operator and never vanishes. In fact, we have that if $\mathcal{M}(f)(x_0) = 0$ for some $x_0 \in \mathbf{R}^n$, then $f = 0$ a.e. Moreover, if f is compactly supported, say in $|x| \leq R$, then

$$\mathcal{M}(f)(x) \geq \frac{\|f\|_{L^1}}{v_n} \frac{1}{(|x|+R)^n}, \tag{2.1.1}$$

for $|x| \geq R$, where v_n is the volume of the unit ball in $\mathbf{R}^n$. Equation (2.1.1) implies that $\mathcal{M}(f)$ *is never* in $L^1(\mathbf{R}^n)$ if $f \neq 0$ a.e., a strong property that reflects a certain behavior of the maximal function. In fact, if g is in L^1_{loc} and $\mathcal{M}(g)$ is in $L^1(\mathbf{R}^n)$, then $g = 0$ a.e. To see this, use (2.1.1) with $g_R(x) = g(x)\chi_{|x|\leq R}$ to conclude that $g_R(x) = 0$ for almost all x in the ball of radius $R > 0$. Thus $g = 0$ a.e. in $\mathbf{R}^n$. However, it is true that $\mathcal{M}(f)$ is in $L^{1,\infty}$ when f is in L^1.

A related analogue of $\mathcal{M}(f)$ is its uncentered version $M(f)$, defined as the supremum of all averages of f over all open balls containing a given point.

Definition 2.1.3. The *uncentered Hardy–Littlewood maximal function* of f,

$$M(f)(x) = \sup_{\substack{\delta>0 \\ |y-x|<\delta}} \operatorname*{Avg}_{B(y,\delta)} |f|,$$

is defined as the supremum of the averages of $|f|$ over all open balls $B(y,\delta)$ that contain the point x.

Clearly $\mathcal{M}(f) \leq M(f)$; in other words, M is a larger operator than $\mathcal{M}$. However, $M(f) \leq 2^n \mathcal{M}(f)$ and the boundedness properties of M are identical to those of $\mathcal{M}$.

Example 2.1.4. On $\mathbf{R}$, let f be the characteristic function of the interval $I = [a,b]$. For $x \in (a,b)$, clearly $M(f)(x) = 1$. For $x > b$, a calculation shows that the largest average of f over all intervals $(y-\delta, y+\delta)$ that contain x is obtained when $\delta = \frac{1}{2}(x-a)$ and $y = \frac{1}{2}(x+a)$. Similarly, when $x < a$, the largest average is obtained when $\delta = \frac{1}{2}(b-x)$ and $y = \frac{1}{2}(b+x)$. We conclude that

$$M(f)(x) = \begin{cases} (b-a)/|x-b| & \text{when } x \leq a, \\ 1 & \text{when } x \in (a,b), \\ (b-a)/|x-a| & \text{when } x \geq b. \end{cases}$$

Observe that M does not have a jump at $x = a$ and $x = b$ and that it is comparable to the function $\big(1 + \frac{\operatorname{dist}(x,I)}{|I|}\big)^{-1}$.

We are now ready to obtain some basic properties of maximal functions. We need the following simple covering lemma.

Lemma 2.1.5. *Let $\{B_1, B_2, \dots, B_k\}$ be a finite collection of open balls in $\mathbf{R}^n$. Then there exists a finite subcollection $\{B_{j_1}, \dots, B_{j_l}\}$ of pairwise disjoint balls such that*

$$\sum_{r=1}^{l} |B_{j_r}| \geq 3^{-n} \Big| \bigcup_{i=1}^{k} B_i \Big|. \tag{2.1.2}$$

Proof. Let us reindex the balls so that

$$|B_1| \geq |B_2| \geq \cdots \geq |B_k|.$$

Let $j_1 = 1$. Having chosen $j_1, j_2, \dots, j_i$, let j_{i+1} be the least index $s > j_i$ such that $\bigcup_{m=1}^{i} B_{j_m}$ is disjoint from B_s. Since we have a finite number of balls, this process will terminate, say after l steps. We have now selected pairwise disjoint balls $B_{j_1}, \dots, B_{j_l}$. If some B_m was not selected, that is, $m \notin \{j_1, \dots, j_l\}$, then B_m must intersect a selected ball B_{j_r} for some $j_r < m$. Then B_m has smaller size than B_{j_r} and we must have $B_m \subseteq 3B_{j_r}$. This shows that the union of the unselected balls is contained in the union of the triples of the selected balls. Therefore, the union of all balls is contained in the union of the triples of the selected balls. Thus

$$\left|\bigcup_{i=1}^{k} B_i\right| \leq \left|\bigcup_{r=1}^{l} 3B_{j_r}\right| \leq \sum_{r=1}^{l} |3B_{j_r}| = 3^n \sum_{r=1}^{l} |B_{j_r}|,$$

and the required conclusion follows. □

We are now ready to prove the main theorem concerning the boundedness of the centered and uncentered maximal functions $\mathcal{M}$ and M, respectively.

Theorem 2.1.6. *The uncentered Hardy–Littlewood maximal function maps $L^1(\mathbf{R}^n)$ to $L^{1,\infty}(\mathbf{R}^n)$ with constant at most 3^n and also $L^p(\mathbf{R}^n)$ to $L^p(\mathbf{R}^n)$ for $1 < p < \infty$ with constant at most $3^{n/p} p(p-1)^{-1}$. The same is true for the centered maximal operator $\mathcal{M}$.*

We note that operators that map L^1 to $L^{1,\infty}$ are said to be *weak type* $(1,1)$.

Proof. Since $M(f) \geq \mathcal{M}(f)$, we have

$$\{x \in \mathbf{R}^n : |\mathcal{M}(f)(x)| > \alpha\} \subseteq \{x \in \mathbf{R}^n : |M(f)(x)| > \alpha\},$$

and therefore it suffices to show that

$$|\{x \in \mathbf{R}^n : |M(f)(x)| > \alpha\}| \leq 3^n \frac{\|f\|_{L^1}}{\alpha}. \tag{2.1.3}$$

We claim that the set

$$E_\alpha = \{x \in \mathbf{R}^n : |M(f)(x)| > \alpha\}$$

is open. Indeed, for $x \in E_\alpha$, there is an open ball B_x that contains x such that the average of $|f|$ over B_x is strictly bigger than α. Then the uncentered maximal function of any other point in B_x is also bigger than α, and thus B_x is contained in E_α. This proves that E_α is open.

Let K be a compact subset of E_α. For each $x \in K$ there exists an open ball B_x containing the point x such that

$$\int_{B_x} |f(y)|\,dy > \alpha |B_x|. \tag{2.1.4}$$

Observe that $B_x \subset E_\alpha$ for all x. By compactness there exists a finite subcover $\{B_{x_1},\dots,B_{x_k}\}$ of K. Using Lemma 2.1.5 we find a subcollection of pairwise disjoint balls $B_{x_{j_1}},\dots,B_{x_{j_l}}$ such that (2.1.2) holds. Using (2.1.4) and (2.1.2) we obtain

$$|K| \le \Big|\bigcup_{i=1}^{k} B_{x_i}\Big| \le 3^n \sum_{i=1}^{l} |B_{x_{j_i}}| \le \frac{3^n}{\alpha}\sum_{i=1}^{l}\int_{B_{x_{j_i}}} |f(y)|\,dy \le \frac{3^n}{\alpha}\int_{E_\alpha}|f(y)|\,dy,$$

since all the balls $B_{x_{j_i}}$ are disjoint and contained in E_α. Taking the supremum over all compact $K \subseteq E_\alpha$ and using the inner regularity of Lebesgue measure, we deduce (2.1.3). We have now proved that M maps $L^1 \to L^{1,\infty}$ with constant 3^n. It is a trivial fact that M maps $L^\infty \to L^\infty$ with constant 1. Since M is well defined and finite a.e. on $L^1 + L^\infty$, it is also on $L^p(\mathbf{R}^n)$ for $1 < p < \infty$. The Marcinkiewicz interpolation theorem (Theorem 1.3.2) implies that M maps $L^p(\mathbf{R}^n)$ to $L^p(\mathbf{R}^n)$ for all $1 < p < \infty$. Using Exercise 1.3.3, we obtain the following estimate for the operator norm of M on $L^p(\mathbf{R}^n)$:

$$\|M\|_{L^p\to L^p} \le \frac{p\,3^{\frac{n}{p}}}{p-1}. \tag{2.1.5}$$

Observe that a direct application of Theorem 1.3.2 would give the slightly worse bound of $2\big(\frac{p}{p-1}\big)^{\frac{1}{p}}3^{\frac{n}{p}}$. □

Remark 2.1.7. The previous proof gives a bound on the operator norm of M on $L^p(\mathbf{R}^n)$ that grows exponentially with the dimension. One may wonder whether this bound could be improved to a better one that does not grow exponentially in the dimension n, as $n \to \infty$. This is not possible; see Exercise 2.1.8.

Example 2.1.8. Let $R > 0$. Then there are dimensional constants c_n and c_n' such that

$$\frac{c_n' R^n}{(|x|+R)^n} \le M(\chi_{B(0,R)})(x) \le \frac{c_n R^n}{(|x|+R)^n}. \tag{2.1.6}$$

Since these functions are not integrable over $\mathbf{R}^n$, it follows that M does not map $L^1(\mathbf{R}^n)$ to $L^1(\mathbf{R}^n)$.

Next we estimate $M(M(\chi_{B(0,R)}))(x)$. First we write

$$\frac{R^n}{(|x|+R)^n} \le \chi_{B(0,R)} + \sum_{k=0}^{\infty}\frac{R^n}{(R+2^kR)^n}\chi_{B(0,2^{k+1}R)\setminus B(0,2^kR)}.$$

Using the upper estimate in (2.1.6) and the sublinearity of M, we obtain

$$\begin{aligned} M\Big(\frac{R^n}{(|\cdot|+R)^n}\Big)(x) &\le M(\chi_{B(0,R)})(x) + \sum_{k=0}^{\infty}\frac{1}{(1+2^k)^n}M(\chi_{B(0,2^{k+1}R)})(x) \\ &\le \frac{c_n R^n}{(|x|+R)^n} + \sum_{k=0}^{\infty}\frac{1}{(1+2^k)^n}\frac{c_n\,(2^{k+1}R)^n}{(|x|+2^{k+1}R)^n} \\ &\le \frac{C_n\log(e+|x|/R)}{(1+|x|/R)^n}, \end{aligned}$$

where the last estimate follows by summing separately over k satisfying $2^{k+1} \le |x|/R$ and $2^{k+1} \ge |x|/R$. Note that the presence of the logarithm does not affect the L^p boundedness of this function when $p > 1$.

2.1.2 Control of Other Maximal Operators

We now study some properties of the Hardy–Littlewood maximal function. We begin with a notational definition that we plan to use throughout this book.

Definition 2.1.9. Given a function g on $\mathbf{R}^n$ and $\varepsilon > 0$, we denote by g_ε the following function:

$$g_\varepsilon(x) = \varepsilon^{-n} g(\varepsilon^{-1}x). \tag{2.1.7}$$

As observed in Example 1.2.16, if g is an integrable function with integral equal to 1, then the family defined by (2.1.7) is an approximate identity. Therefore, convolution with g_ε is an averaging operation. The Hardy–Littlewood maximal function $\mathcal{M}(f)$ is obtained as the supremum of the averages of a function f with respect to the dilates of the kernel $k = v_n^{-1}\chi_{B(0,1)}$ in $\mathbf{R}^n$; here v_n is the volume of the unit ball $B(0,1)$. Indeed, we have

$$\begin{aligned}\mathcal{M}(f)(x) &= \sup_{\varepsilon>0} \frac{1}{v_n\varepsilon^n}\int_{\mathbf{R}^n} |f(x-y)|\,\chi_{B(0,1)}\Big(\frac{y}{\varepsilon}\Big)\,dy \\ &= \sup_{\varepsilon>0}(|f| * k_\varepsilon)(x).\end{aligned}$$

Note that the function $k = v_n^{-1}\chi_{B(0,1)}$ has integral equal to 1, and the operation given by convolution with k_ε is indeed an averaging operation.

It turns out that the Hardy–Littlewood maximal function controls the averages of a function with respect to any radially decreasing L^1 function. Recall that a function f on $\mathbf{R}^n$ is called *radial* if $f(x) = f(y)$ whenever $|x| = |y|$. Note that a radial function f on $\mathbf{R}^n$ has the form $f(x) = \varphi(|x|)$ for some function φ on $\mathbf{R}^+$. We have the following result.

Theorem 2.1.10. *Let $k \ge 0$ be a function on $[0,\infty)$ that is continuous except at a finite number of points. Suppose that $K(x) = k(|x|)$ is an integrable function on $\mathbf{R}^n$ that satisfies*

$$K(x) \ge K(y), \quad \textit{whenever } |x| \le |y| \tag{2.1.8}$$

(i.e., k is decreasing). Then the following estimate is true:

$$\sup_{\varepsilon>0}(|f| * K_\varepsilon)(x) \le \big\|K\big\|_{L^1}\mathcal{M}(f)(x) \tag{2.1.9}$$

for all locally integrable functions f on $\mathbf{R}^n$.

Proof. We prove (2.1.9) when K is radial, satisfies (2.1.8), and is compactly supported and continuous. When this case is established, select a sequence K_j of radial, compactly supported, continuous functions that increase to K as $j \to \infty$. This is possible, since the function k is continuous except at a finite number of points. If (2.1.9) holds for each K_j, passing to the limit implies that (2.1.9) also holds for K. Next, we observe that it suffices to prove (2.1.9) for $x = 0$. When this case is established, replacing $f(t)$ by $f(t+x)$ implies that (2.1.9) holds for all x.

Let us now fix a radial, continuous, and compactly supported function K with support in the ball $B(0,R)$, satisfying (2.1.8). Also fix an $f \in L^1_{\mathrm{loc}}$ and take $x = 0$. Let e_1 be the vector $(1,0,0,\dots,0)$ on the unit sphere $\mathbf{S}^{n-1}$. Polar coordinates give

$$\int_{\mathbf{R}^n} |f(y)|K_\varepsilon(-y)\,dy = \int_0^\infty \int_{\mathbf{S}^{n-1}} |f(r\theta)|K_\varepsilon(re_1)r^{n-1}\,d\theta\,dr. \tag{2.1.10}$$

Define functions

$$\begin{aligned} F(r) &= \int_{\mathbf{S}^{n-1}} |f(r\theta)|\,d\theta\,, \\ G(r) &= \int_0^r F(s)s^{n-1}\,ds\,, \end{aligned}$$

where $d\theta$ denotes surface measure on $\mathbf{S}^{n-1}$. Using these functions, (2.1.10), and integration by parts, we obtain

$$\begin{aligned} \int_{\mathbf{R}^n} |f(y)|K_\varepsilon(y)\,dy &= \int_0^{\varepsilon R} F(r)r^{n-1}K_\varepsilon(re_1)\,dr \\ &= G(\varepsilon R)K_\varepsilon(\varepsilon R e_1) - G(0)K_\varepsilon(0) - \int_0^{\varepsilon R} G(r)\,dK_\varepsilon(re_1) \\ &= \int_0^\infty G(r)\,d(-K_\varepsilon(re_1))\,, \end{aligned} \tag{2.1.11}$$

where two of the integrals are of Lebesgue–Stieltjes type and we used our assumptions that $G(0) = 0$, $K_\varepsilon(0) < \infty$, $G(\varepsilon R) < \infty$, and $K_\varepsilon(\varepsilon R e_1) = 0$. Let v_n be the volume of the unit ball in $\mathbf{R}^n$. Since

$$G(r) = \int_0^r F(s)s^{n-1}\,ds = \int_{|y|\le r} |f(y)|\,dy \le \mathcal{M}(f)(0)v_n r^n\,,$$

it follows that the expression in (2.1.11) is dominated by

$$\begin{aligned} \mathcal{M}(f)(0)v_n \int_0^\infty r^n d(-K_\varepsilon(re_1)) &= \mathcal{M}(f)(0)\int_0^\infty n v_n r^{n-1} K_\varepsilon(re_1)\,dr \\ &= \mathcal{M}(f)(0)\big\|K\big\|_{L^1}\,. \end{aligned}$$

Here we used integration by parts and the fact that the surface measure of the unit sphere $\mathbf{S}^{n-1}$ is equal to nv_n. See Appendix A.3. The theorem is now proved. □

Remark 2.1.11. Theorem 2.1.10 can be generalized as follows. If K is an L^1 function on $\mathbf{R}^n$ whose absolute value is bounded above by some continuous integrable radial function K_0 that satisfies (2.1.8), then (2.1.9) holds with $\big\|K\big\|_{L^1}$ replaced by

$\|K_0\|_{L^1}$. Such a K_0 is called a *radial decreasing majorant* of K. This observation is formulated as the following corollary.

Corollary 2.1.12. *If a function φ has an integrable radially decreasing majorant Φ, then the estimate*

$$\sup_{t>0} |(f * \varphi_t)(x)| \le \|\Phi\|_{L^1} \mathcal{M}(f)(x)$$

is valid for all locally integrable functions f on $\mathbf{R}^n$.

Example 2.1.13. Let

$$P(x) = \frac{c_n}{(1+|x|^2)^{\frac{n+1}{2}}},$$

where c_n is a constant such that

$$\int_{\mathbf{R}^n} P(x)\,dx = 1.$$

The function P is called the *Poisson kernel*. We define L^1 dilates P_t of the Poisson kernel P by setting

$$P_t(x) = t^{-n} P(t^{-1}x)$$

for $t > 0$. It is straightforward to verify that when $n \ge 2$,

$$\frac{d^2}{dt^2} P_t + \sum_{j=1}^{n} \partial_j^2 P_t = 0,$$

that is, $P_t(x_1,\dots,x_n)$ is a *harmonic function* of the variables $(x_1,\dots,x_n,t)$. Therefore, for $f \in L^p(\mathbf{R}^n)$, $1 \le p < \infty$, the function

$$u(x,t) = (f * P_t)(x)$$

is harmonic in $\mathbf{R}^{n+1}_+$ and converges to $f(x)$ in $L^p(dx)$ as $t \to 0$, since $\{P_t\}_{t>0}$ is an approximate identity. If we knew that $f * P_t$ converged to f a.e. as $t \to 0$, then we could say that $u(x,t)$ solves the *Dirichlet problem*

$$\begin{aligned} \sum_{j=1}^{n+1} \partial_j^2 u &= 0 \qquad \text{on } \mathbf{R}^{n+1}_+, \\ u(x,0) &= f(x) \qquad \text{a.e. on } \mathbf{R}^n. \end{aligned} \tag{2.1.12}$$

Solving the Dirichlet problem (2.1.12) motivates the study of the almost everywhere convergence of the expressions $f * P_t$. This is discussed in the next subsection.

Let us now compute the value of the constant c_n. Denote by ω_{n-1} the surface area of $\mathbf{S}^{n-1}$. Using polar coordinates, we obtain

$$\begin{aligned}
\frac{1}{c_n} &= \int_{\mathbf{R}^n} \frac{dx}{(1+|x|^2)^{\frac{n+1}{2}}} \\
&= \omega_{n-1} \int_0^\infty \frac{r^{n-1}}{(1+r^2)^{\frac{n+1}{2}}} dr \\
&= \omega_{n-1} \int_0^{\pi/2} (\sin\varphi)^{n-1}\, d\varphi \qquad\qquad (r = \tan\varphi) \\
&= \frac{2\pi^{\frac{n}{2}}}{\Gamma(\frac{n}{2})} \frac{1}{2} \frac{\Gamma(\frac{n}{2})\Gamma(\frac{1}{2})}{\Gamma(\frac{n+1}{2})} \\
&= \frac{\pi^{\frac{n+1}{2}}}{\Gamma(\frac{n+1}{2})},
\end{aligned}$$

where we used the formula for ω_{n-1} in Appendix A.3 and an identity in Appendix A.4. We conclude that

$$c_n = \frac{\Gamma(\frac{n+1}{2})}{\pi^{\frac{n+1}{2}}}$$

and that the Poisson kernel on $\mathbf{R}^n$ is given by

$$P(x) = \frac{\Gamma(\frac{n+1}{2})}{\pi^{\frac{n+1}{2}}} \frac{1}{(1+|x|^2)^{\frac{n+1}{2}}}. \tag{2.1.13}$$

Theorem 2.1.10 implies that the solution of the Dirichlet problem (2.1.12) is pointwise bounded by the Hardy–Littlewood maximal function of f.

2.1.3 Applications to Differentiation Theory

We continue this section by obtaining some applications of the boundedness of the Hardy–Littlewood maximal function in differentiation theory.

We now show that the weak type $(1,1)$ property of the Hardy–Littlewood maximal function implies almost everywhere convergence for a variety of families of functions. We deduce this from the more general fact that a certain weak type property for the supremum of a family of linear operators implies almost everywhere convergence.

Here is our setup. Let (X,μ), (Y,ν) be measure spaces and let $0 < p \le \infty$, $0 < q < \infty$. Suppose that D is a dense subspace of $L^p(X,\mu)$. This means that for all $f \in L^p$ and all $\delta > 0$ there exists a $g \in D$ such that $\|f-g\|_{L^p} < \delta$. Suppose that for every $\varepsilon > 0$, T_ε is a linear operator defined on $L^p(X,\mu)$ with values in the set of measurable functions on Y. Define a sublinear operator

$$T_*(f)(x) = \sup_{\varepsilon>0} |T_\varepsilon(f)(x)|. \tag{2.1.14}$$

We have the following.

Theorem 2.1.14. *Let $0 < p < \infty$, $0 < q < \infty$, and T_ε and T_* as previously. Suppose that for some $B > 0$ and all $f \in L^p(X)$ we have*

$$\big\|T_*(f)\big\|_{L^{q,\infty}} \le B\big\|f\big\|_{L^p} \tag{2.1.15}$$

and that for all $f \in D$,

$$\lim_{\varepsilon\to 0} T_\varepsilon(f) = T(f) \tag{2.1.16}$$

exists and is finite ν-a.e. (and defines a linear operator on D). Then for all functions f in $L^p(X,\mu)$ the limit (2.1.16) exists and is finite ν-a.e., and defines a linear operator T on $L^p(X)$ (uniquely extending T defined on D) that satisfies

$$\big\|T(f)\big\|_{L^{q,\infty}} \le B\big\|f\big\|_{L^p}\,. \tag{2.1.17}$$

Proof. Given f in L^p, we define the *oscillation* of f:

$$O_f(y) = \limsup_{\varepsilon\to 0}\limsup_{\theta\to 0}|T_\varepsilon(f)(y) - T_\theta(f)(y)|\,.$$

We would like to show that for all $f \in L^p$ and $\delta > 0$,

$$\nu(\{y \in Y:\, O_f(y) > \delta\}) = 0\,. \tag{2.1.18}$$

Once (2.1.18) is established, given $f \in L^p(X)$, we obtain that $O_f(y) = 0$ for ν-almost all y, which implies that $T_\varepsilon(f)(y)$ is Cauchy for ν-almost all y, and it therefore converges ν-a.e. to some $T(f)(y)$ as $\varepsilon \to 0$. The operator T defined this way on $L^p(X)$ is linear and extends T defined on D.

To approximate O_f we use density. Given $\eta > 0$, find a function $g \in D$ such that $\big\|f-g\big\|_{L^p} < \eta$. Since $T_\varepsilon(g) \to T(g)$ ν-a.e, it follows that $O_g = 0$ ν-a.e. Using this fact and the linearity of the T_ε's, we conclude that

$$O_f(y) \le O_g(y) + O_{f-g}(y) = O_{f-g}(y) \qquad \nu\text{-a.e.}$$

Now for any $\delta > 0$ we have

$$\begin{aligned}
\nu(\{y \in Y:\, O_f(y) > \delta\}) &\le \nu(\{y \in Y:\, O_{f-g}(y) > \delta\}) \\
&\le \nu(\{y \in Y:\, 2T_*(f-g)(y) > \delta\}) \\
&\le \big(2B\big\|f-g\big\|_{L^p}/\delta\big)^q \\
&\le (2B\eta/\delta)^q\,.
\end{aligned}$$

Letting $\eta \to 0$, we deduce (2.1.18). We conclude that $T_\varepsilon(f)$ is a Cauchy sequence, and hence it converges ν-a.e. to some $T(f)$. Since $|T(f)| \le |T_*(f)|$, the conclusion (2.1.17) of the theorem follows easily. □

We now derive some applications. First we return to the issue of almost everywhere convergence of the expressions $f * P_y$, where P is the Poisson kernel.

Example 2.1.15. Fix $1 \le p < \infty$ and $f \in L^p(\mathbf{R}^n)$. Let

$$P(x) = \frac{\Gamma(\frac{n+1}{2})}{\pi^{\frac{n+1}{2}}} \frac{1}{(1+|x|^2)^{\frac{n+1}{2}}}$$

be the Poisson kernel on $\mathbf{R}^n$ and let $P_\varepsilon(x) = \varepsilon^{-n}P(\varepsilon^{-1}x)$. We deduce from the previous theorem that the family $f * P_\varepsilon$ converges to f a.e. Let D be the set of all continuous functions with compact support on $\mathbf{R}^n$. Since the family $(P_\varepsilon)_{\varepsilon>0}$ is an approximate identity, Theorem 1.2.19 (2) implies that for f in D we have that $f * P_\varepsilon \to f$ uniformly on compact subsets of $\mathbf{R}^n$ and hence a.e. In view of Theorem 2.1.10, the supremum of the family of linear operators $T_\varepsilon(f) = f * P_\varepsilon$ is controlled by the Hardy–Littlewood maximal function, and thus it maps L^p to $L^{p,\infty}$ for $1 \le p < \infty$. Theorem 2.1.14 now gives that $f * P_\varepsilon$ converges to f a.e. for all $f \in L^p$.

Here is another application of Theorem 2.1.14. We refer to Exercise 2.1.10 for others.

Corollary 2.1.16. ***(Lebesgue's differentiation theorem)*** *For any locally integrable function f on $\mathbf{R}^n$ we have*

$$\lim_{r\to 0} \frac{1}{|B(x,r)|} \int_{B(x,r)} f(y)\,dy = f(x) \tag{2.1.19}$$

for almost all x in $\mathbf{R}^n$. Consequently we have $|f| \le \mathcal{M}(f)$ a.e.

Proof. Since $\mathbf{R}^n$ is the union of the balls $B(0,N)$ for $N = 1,2,3\dots$, it suffices to prove the required conclusion for almost all x inside the ball $B(0,N)$. Then we may take f supported in a larger ball, thus working with f integrable over the whole space. Let T_ε be the operator given with convolution with k_ε, where $k = v_n^{-1}\chi_{B(0,1)}$. We know that the corresponding maximal operator T_* is controlled by the the centered Hardy–Littlewood maximal function $\mathcal{M}$, which maps L^1 to $L^{1,\infty}$. It is straightforward to verify that (2.1.19) holds for all continuous functions f with compact support. Since the set of these functions is dense in L^1, and T_* maps L^1 to $L^{1,\infty}$, Theorem 2.1.14 implies that (2.1.19) holds for a general f in L^1. □

The following corollaries were inspired by Example 2.1.15.

Corollary 2.1.17. ***(Differentiation theorem for approximate identities)*** *Let K be an L^1 function on $\mathbf{R}^n$ with integral 1 that has a continuous integrable radially decreasing majorant. Then $f * K_\varepsilon \to f$ a.e. as $\varepsilon \to 0$ for all $f \in L^p(\mathbf{R}^n)$, $1 \le p < \infty$.*

Proof. It follows from Example 1.2.16 that K_ε is an approximate identity. Theorem 1.2.19 now implies that $f * K_\varepsilon \to f$ uniformly on compact sets when f is continuous. Let D be the space of all continuous functions with compact support. Then $f * K_\varepsilon \to f$ a.e. for $f \in D$. It follows from Corollary 2.1.12 that $T_*(f) = \sup_{\varepsilon>0} |f * K_\varepsilon|$ maps L^p to $L^{p,\infty}$ for $1 \le p < \infty$. Using Theorem 2.1.14, we conclude the proof of the corollary. □

Remark 2.1.18. Fix $f \in L^p(\mathbf{R}^n)$ for some $1 \le p < \infty$. Theorem 1.2.19 implies that $f * K_\varepsilon$ converges to f in L^p and hence some subsequence $f * K_{\varepsilon_n}$ of $f * K_\varepsilon$ converges to f a.e. as $n \to \infty$, ($\varepsilon_n \to 0$). Compare this result with Corollary 2.1.17, which gives a.e. convergence for the whole family $f * K_\varepsilon$ as $\varepsilon \to 0$.

Corollary 2.1.19. *(Differentiation theorem for multiples of approximate identities) Let K be a function on $\mathbf{R}^n$ that has an integrable radially decreasing majorant. Let $a = \int_{\mathbf{R}^n} K(x)\,dx$. Then for all $f \in L^p(\mathbf{R}^n)$ and $1 \le p < \infty$, $(f * K_\varepsilon)(x) \to af(x)$ for almost all $x \in \mathbf{R}^n$ as $\varepsilon \to 0$.*

Proof. Use Theorem 1.2.21 instead of Theorem 1.2.19 in the proof of Corollary 2.1.17. □

The following application of the Lebesgue differentiation theorem uses a simple *stopping-time argument.* This is the sort of argument in which a selection procedure stops when it is exhausted at a certain scale and is then repeated at the next scale. A certain refinement of the following proposition is of fundamental importance in the study of singular integrals given in Chapter 4.

Proposition 2.1.20. *Given a nonnegative integrable function f on $\mathbf{R}^n$ and $\alpha > 0$, there exist disjoint open cubes Q_j such that for almost all $x \in \big(\bigcup_j Q_j\big)^c$ we have $f(x) \le \alpha$ and*

$$\alpha < \frac{1}{|Q_j|}\int_{Q_j} f(t)\,dt \le 2^n\alpha. \tag{2.1.20}$$

Proof. The proof provides an excellent paradigm of a stopping-time argument. Start by decomposing $\mathbf{R}^n$ as a union of cubes of equal size, whose interiors are disjoint, and whose diameter is so large that $|Q|^{-1}\int_Q f(x)\,dx \le \alpha$ for every Q in this mesh. This is possible since f is integrable and $|Q|^{-1}\int_Q f(x)\,dx \to 0$ as $|Q| \to \infty$. Call the union of these cubes $\mathscr{E}_0$.

Divide each cube in the mesh into 2^n congruent cubes by bisecting each of the sides. Call the new collection of cubes $\mathscr{E}_1$. Select a cube Q in $\mathscr{E}_1$ if

$$\frac{1}{|Q|}\int_Q f(x)\,dx > \alpha \tag{2.1.21}$$

and call the set of all selected cubes $\mathscr{S}_1$. Now subdivide each cube in $\mathscr{E}_1 \setminus \mathscr{S}_1$ into 2^n congruent cubes by bisecting each of the sides as before. Call this new collection of cubes $\mathscr{E}_2$. Repeat the same procedure and select a family of cubes $\mathscr{S}_2$ that satisfy (2.1.21). Continue this way ad infinitum and call the cubes in $\bigcup_{m=1}^\infty \mathscr{S}_m$ "selected." If Q was selected, then there exists Q_1 in $\mathscr{E}_{m-1}$ containing Q that was not selected at the $(m-1)th$ step for some $m \ge 1$. Therefore,

$$\alpha < \frac{1}{|Q|}\int_Q f(x)\,dx \le 2^n \frac{1}{|Q_1|}\int_{Q_1} f(x)\,dx \le 2^n\alpha .$$

Now call F the closure of the complement of the union of all selected cubes. If $x \in F$, then there exists a sequence of cubes containing x whose diameter shrinks down to zero such that the average of f over these cubes is less than or equal to α. By Corollary 2.1.16, it follows that $f(x) \le \alpha$ almost everywhere in F. This proves the proposition. □

In the proof of Proposition 2.1.20 it was not crucial to assume that f was defined on all $\mathbf{R}^n$, but only on a cube. We now give a local version of this result.

Corollary 2.1.21. *Let $f \ge 0$ be an integrable function over a cube Q in $\mathbf{R}^n$ and let $\alpha \ge \frac{1}{|Q|}\int_Q f\,dx$. Then there exist disjoint open subcubes Q_j of Q such that for almost all $x \in Q \setminus \bigcup_j Q_j$ we have $f \le \alpha$ and (2.1.20) holds for all j.*

Proof. This easily follows by a simple modification of Proposition 2.1.20 in which $\mathbf{R}^n$ is replaced by the fixed cube Q. □

See Exercise 2.1.4 for an application of Proposition 2.1.20 involving maximal functions.

Exercises

2.1.1. A positive Borel measure μ on $\mathbf{R}^n$ is called *inner regular* if for any open subset U of $\mathbf{R}^n$ we have $\mu(U) = \sup\{\mu(K) : K \subseteqq U,\ K \text{ compact}\}$ and μ is called *locally finite* if $\mu(B) < \infty$ for all balls B.
(a) Let μ be a positive inner regular locally finite measure on $\mathbf{R}^n$ that satisfies the following *doubling condition*: There exists a constant $D(\mu) > 0$ such that for all $x \in \mathbf{R}^n$ and $r > 0$ we have

$$\mu(3B(x,r)) \le D(\mu)\,\mu(B(x,r)).$$

For $f \in L^1_{\mathrm{loc}}(\mathbf{R}^n,\mu)$ define the uncentered maximal function $M_\mu(f)$ with respect to μ by

$$M_\mu(f)(x) = \sup_{r>0} \sup_{\substack{z:\,|z-x|<r \\ \mu(B(z,r))\neq 0}} \frac{1}{\mu(B(z,r))}\int_{B(z,r)} f(y)\,d\mu(y).$$

Show that M_μ maps $L^1(\mathbf{R}^n,\mu)$ to $L^{1,\infty}(\mathbf{R}^n,\mu)$ with constant at most $D(\mu)$ and $L^p(\mathbf{R}^n,\mu)$ to itself with constant at most $2\big(\frac{p}{p-1}\big)^{\frac{1}{p}} D(\mu)^{\frac{1}{p}}$.
(b) Obtain as a consequence a differentiation theorem analogous to Corollary 2.1.16.
[*Hint:* Part (a): For $f \in L^1(\mathbf{R}^n,\mu)$ show that the set $E_\alpha = \{M_\mu(f) > \alpha\}$ is open. Then use the argument of the proof of Theorem 2.1.6 and the inner regularity of μ.]

2.1.2. On $\mathbf{R}$ consider the maximal function M_μ of Exercise 2.1.1.
(a) (*W. H. Young*) Prove the following covering lemma. Given a finite set $\mathscr{F}$ of open intervals in $\mathbf{R}$, prove that there exist two subfamilies each consisting of pairwise disjoint intervals such that the union of the intervals in the original family is equal to the union of the intervals of both subfamilies. Use this result to show that the

maximal function M_μ of Exercise 2.1.1 maps $L^1(\mu) \to L^{1,\infty}(\mu)$ with constant at most 2.
(b) (*Grafakos and Kinnunen [107]*) Prove that for any σ-finite positive measure μ on $\mathbf{R}$, $\alpha > 0$, and $f \in L^1_{\text{loc}}(\mathbf{R}, \mu)$ we have

$$\frac{1}{\alpha}\int_A |f|\,d\mu - \mu(A) \le \frac{1}{\alpha}\int_{\{|f|>\alpha\}} |f|\,d\mu - \mu(\{|f|>\alpha\}).$$

Use this result and part (a) to prove that for all $\alpha > 0$ and all locally integrable f we have

$$\mu(\{|f|>\alpha\}) + \mu(\{M_\mu(f)>\alpha\}) \le \frac{1}{\alpha}\int_{\{|f|>\alpha\}} |f|\,d\mu + \frac{1}{\alpha}\int_{\{M_\mu(f)>\alpha\}} |f|\,d\mu$$

and note that equality is obtained when $\alpha = 1$ and $f(x) = |x|^{-1/p}$.
(c) Conclude that M_μ maps $L^p(\mu)$ to $L^p(\mu)$, $1 < p < \infty$, with bound at most the unique positive solution A_p of the equation

$$(p-1)x^p - px^{p-1} - 1 = 0.$$

(d) (*Grafakos and Montgomery-Smith [109]*) If μ is the Lebesgue measure show that for $1 < p < \infty$ we have

$$\big\|M\big\|_{L^p \to L^p} = A_p,$$

where A_p is the unique positive solution of the equation in part (c).
[*Hint:* Part (a): Select a subset $\mathscr{G}$ of $\mathscr{F}$ with minimal cardinality such that $\bigcup_{J\in\mathscr{G}} J = \bigcup_{I\in\mathscr{F}} I$. Part (d): One direction follows from part (c). Conversely, $M(|x|^{-1/p})(1) = \frac{p}{p-1}\frac{\gamma^{1/p'}+1}{\gamma+1}$, where γ is the unique positive solution of the equation $\frac{p}{p-1}\frac{\gamma^{1/p'}+1}{\gamma+1} = \gamma^{-1/p}$. Conclude that $M(|x|^{-1/p})(1) = A_p$ and that $M(|x|^{-1/p}) = A_p|x|^{-1/p}$. Since this function is not in L^p, consider the family $f_\varepsilon(x) = |x|^{-1/p}\min(|x|^{-\varepsilon}, |x|^{\varepsilon})$, $\varepsilon > 0$, and show that $M(f_\varepsilon)(x) \ge (1+\gamma^{\frac{1}{p'}+\varepsilon})(1+\gamma)^{-1}(\frac{1}{p'}+\varepsilon)^{-1} f_\varepsilon(x)$ for $0 < \varepsilon < p'$.]

2.1.3. Define the centered Hardy–Littlewood maximal function $\mathcal{M}_c$ and the uncentered Hardy–Littlewood maximal function M_c using cubes with sides parallel to the axes instead of balls in $\mathbf{R}^n$. Prove that

$$v_n(n/2)^{n/2} \le \frac{M(f)}{M_c(f)} \le 2^n/v_n, \quad v_n(n/2)^{n/2} \le \frac{\mathcal{M}(f)}{\mathcal{M}_c(f)} \le 2^n/v_n,$$

where v_n is the volume of the unit ball in $\mathbf{R}^n$. Conclude that $\mathcal{M}_c$ and M_c are weak type $(1,1)$ and they map $L^p(\mathbf{R}^n)$ to itself for $1 < p \le \infty$.

2.1.4. (a) Prove the estimate:

$$|\{x \in \mathbf{R}^n : M(f)(x) > 2\alpha\}| \le \frac{3^n}{\alpha}\int_{\{|f|>\alpha\}} |f(y)|\,dy$$

and conclude that M maps L^p to $L^{p,\infty}$ with norm at most $2\cdot 3^{n/p}$ for $1\le p<\infty$. Deduce that if $f\log^+(2|f|)$ is integrable over a ball B, then $M(f)$ is integrable over the same ball B.
(b) (*Wiener [291], Stein [255]*) Apply Proposition 2.1.20 to $|f|$ and $\alpha>0$ and Exercise 2.1.3 to show that with $c_n=(n/2)^{n/2}v_n$ we have

$$|\{x\in\mathbf{R}^n:\, M(f)(x)>c_n\,\alpha\}|\ge\frac{2^{-n}}{\alpha}\int_{\{|f|>\alpha\}}|f(y)|\,dy.$$

(c) Suppose that f is integrable and supported in a ball $B(0,\rho)$. Show that for x in $B(0,2\rho)\setminus B(0,\rho)$ we have $\mathcal{M}(f)(x)\le\mathcal{M}(\rho^2|x|^{-2}x)$. Conclude that

$$\int_{B(0,2\rho)}\mathcal{M}(f)\,dx\le(4^n+1)\int_{B(0,\rho)}\mathcal{M}(f)\,dx$$

and from this deduce a similar inequality for $M(f)$.
(d) Suppose that f is integrable and supported in a ball B and that $M(f)$ is integrable over B. Let $\lambda_0=2^n|B|^{-1}\|f\|_{L^1}$. Use part (b) to prove that $f\log^+(\lambda_0^{-1}c_n|f|)$ is integrable over B.
[*Hint:* Part (a): Write $f=f\chi_{|f|>\alpha}+f\chi_{|f|\le\alpha}$. Part (c): Let $x'=\rho^2|x|^{-2}x$ for some $\rho<|x|<2\rho$. Show that for $R>|x|-\rho$, we have that

$$\int_{B(x,R)}|f(z)|\,dz\le\int_{B(x',R)}|f(z)|\,dz$$

by showing that $B(x,R)\cap B(0,\rho)\subset B(x',R)$. Part (d): For $x\notin 2B$ we have $M(f)(x)\le\lambda_0$, hence $\int_{2B}M(f)(x)\,dx\ge\int_{\lambda_0}^\infty|\{x\in 2B:\, M(f)(x)>\alpha\}|\,d\alpha.$]

2.1.5. (*A. Kolmogorov*) Let S be a sublinear operator that maps $L^1(\mathbf{R}^n)$ to $L^{1,\infty}(\mathbf{R}^n)$ with norm B. Suppose that $f\in L^1(\mathbf{R}^n)$. Prove that for any set A of finite Lebesgue measure and for all $0<q<1$ we have

$$\int_A|S(f)(x)|^q\,dx\le(1-q)^{-1}B^q|A|^{1-q}\|f\|_{L^1}^q,$$

and in particular, for the Hardy–Littlewood maximal operator,

$$\int_A M(f)(x)^q\,dx\le(1-q)^{-1}3^{nq}|A|^{1-q}\|f\|_{L^1}^q.$$

[*Hint:* Use the identity

$$\int_A|S(f)(x)|^q\,dx=\int_0^\infty q\,\alpha^{q-1}|\{x\in A:\,S(f)(x)>\alpha\}|\,d\alpha$$

and estimate the last measure by $\min(|A|,\frac{B}{\alpha}\|f\|_{L^1})$.]

2.1.6. Let $M_s(f)(x)$ be the supremum of the averages of $|f|$ over all rectangles with sides parallel to the axes containing x. The operator M_s is called the *strong maximal function*.
(a) Prove that M_s maps $L^p(\mathbf{R}^n)$ to itself.
(b) Show that the operator norm of M_s is A_p^n, where A_p is as in Exercise 2.1.2(c).
(c) Prove that M_s is not weak type (1,1).

2.1.7. Prove that if

$$|\varphi(x_1,\dots,x_n)| \le A(1+|x_1|)^{-1-\varepsilon}\cdots(1+|x_n|)^{-1-\varepsilon}$$

for some $A,\varepsilon>0$, and $\varphi_{t_1,\dots,t_n}(x)=t_1^{-1}\cdots t_n^{-1}\varphi(t_1^{-1}x_1,\dots,t_n^{-1}x_n)$, then the maximal operator

$$f\mapsto \sup_{t_1,\dots,t_n>0}|f*\varphi_{t_1,\dots,t_n}|$$

is pointwise controlled by the strong maximal function.

2.1.8. Prove that for any fixed $1<p<\infty$, the operator norm of M on $L^p(\mathbf{R}^n)$ tends to infinity as $n\to\infty$.
[*Hint:* Let f_0 be the characteristic function of the unit ball in $\mathbf{R}^n$. Consider the averages $|B_x|^{-1}\int_{B_x} f_0\,dy$, where $B_x=B\big(\frac{1}{2}(|x|-|x|^{-1})\frac{x}{|x|},\frac{1}{2}(|x|+|x|^{-1})\big)$ for $|x|>1$.]

2.1.9. (a) In $\mathbf{R}^2$ let $M_0(f)(x)$ be the maximal function obtained by taking the supremum of the averages of $|f|$ over all rectangles (of arbitrary orientation) containing x. Prove that M_0 is not bounded on $L^p(\mathbf{R}^n)$ for $p<2$ and conclude that M_0 is not weak type $(1,1)$.
(b) Let $M_{00}(f)(x)$ be the maximal function obtained by taking the supremum of the averages of $|f|$ over all rectangles in $\mathbf{R}^2$ of arbitrary orientation but fixed eccentricity containing x. (The eccentricity of a rectangle is the ratio of its longer side to its shorter side.) Using a covering lemma, show that M_{00} is weak type $(1,1)$ with a bound proportional to the square of the eccentricity.
(c) On $\mathbf{R}^n$ define a maximal function by taking the supremum of the averages of $|f|$ over all products of intervals $I_1\times\cdots\times I_n$ containing a point x with $|I_2|=a_2|I_1|,\dots,|I_n|=a_n|I_1|$ and $a_2,\dots,a_n>0$ fixed. Show that this maximal function is weak type $(1,1)$ with bound independent of the numbers $a_2,\dots,a_n$.
[*Hint:* Part (b): Let b be the eccentricity. If two rectangles with the same eccentricity intersect, then the smaller one is contained in the bigger one scaled $4b$ times. Then use an argument similar to that in Lemma 2.1.5.]

2.1.10. (a) Let p,q,X,Y be as in Theorem 2.1.14. Assume that T_ε is a family of quasilinear operators defined on $L^p(X)$ [i.e., $|T_\varepsilon(f+g)|\le K(|T_\varepsilon(f)|+|T_\varepsilon(g)|)$ for all $f,g\in L^p(X)$] such that $\lim_{\varepsilon\to 0}T_\varepsilon(f)=0$ for all f in some dense subspace D of $L^p(X)$. Use the argument of Theorem 2.1.14 to prove that $\lim_{\varepsilon\to 0}T_\varepsilon(f)=0$ for all f in $L^p(X)$.
(b) Use the result in part (a) to prove the following improvement of the Lebesgue differentiation theorem: Let $f\in L^p_{\mathrm{loc}}(\mathbf{R}^n)$ for some $1\le p<\infty$. Then for almost all $x\in\mathbf{R}^n$ we have

$$\lim_{\substack{|B|\to 0\\ B\ni x}} \frac{1}{|B|}\int_B |f(y)-f(x)|^p\,dy = 0,$$

where the limit is taken over all open balls B containing x.
[*Hint:* Define

$$T_\varepsilon(f)(x) = \sup_{B(z,\varepsilon)\ni x}\left(\frac{1}{|B(z,\varepsilon)|}\int_{B(z,\varepsilon)} |f(y)-f(x)|^p\,dy\right)^{1/p}$$

and observe that $T_*(f) = \sup_{\varepsilon>0} T_\varepsilon(f) \le |f| + M(|f|^p)^{\frac{1}{p}}$. Use Theorem 2.1.14.]

2.1.11. On $\mathbf{R}$ define the right and left maximal functions M_R and M_L as follows:

$$\begin{aligned} M_L(f)(x) &= \sup_{r>0}\frac{1}{r}\int_{x-r}^{x} |f(t)|\,dt,\\ M_R(f)(x) &= \sup_{r>0}\frac{1}{r}\int_{x}^{x+r} |f(t)|\,dt. \end{aligned}$$

(a) (*Riesz's sunrise lemma [218]*) Show that

$$\begin{aligned} |\{x\in\mathbf{R}: M_L(f)(x)>\alpha\}| &= \frac{1}{\alpha}\int_{\{M_L(f)>\alpha\}} |f(t)|\,dt,\\ |\{x\in\mathbf{R}: M_R(f)(x)>\alpha\}| &= \frac{1}{\alpha}\int_{\{M_R(f)>\alpha\}} |f(t)|\,dt. \end{aligned}$$

(b) Conclude that M_L and M_R map L^p to L^p with norm at most $p/(p-1)$ for $1<p<\infty$.
(c) Construct examples to show that the operator norms of M_L and M_R on L^p are exactly $p/(p-1)$ for $1<p<\infty$.
(d) (*K. L. Phillips*) Prove that $M = \max(M_R, M_L)$.
(e) (*J. Duoandikoetxea*) Let $N = \min(M_R, M_L)$. Since

$$M(f)^p + N(f)^p = M_L(f)^p + M_R(f)^p,$$

integrate over the line and use the following consequence of part (a),

$$\int_{\mathbf{R}} M_L(f)^p + M_R(f)^p\,dx = \frac{p}{p-1}\int_{\mathbf{R}} |f|\left(M(f)^{p-1} + N(f)^{p-1}\right)dx,$$

to prove that

$$(p-1)\|M(f)\|_{L^p}^p - p\|f\|_{L^p}\|M(f)\|_{L^p}^{p-1} - \|f\|_{L^p}^p \le 0.$$

This provides an alternative proof of the result in Exercise 2.1.2(c).

2.1.12. A cube $Q = [a_1 2^k, (a_1+1)2^k)\times\cdots\times[a_n 2^k, (a_n+1)2^k)$ on $\mathbf{R}^n$ is called *dyadic* if $k, a_1,\dots,a_n\in\mathbf{Z}$. Observe that either two dyadic cubes are disjoint or one contains the other. Define the *dyadic maximal function*

$$M_d(f)(x) = \sup_{Q \ni x} \frac{1}{|Q|} \int_Q f(y)\, dy,$$

where the supremum is taken over all dyadic cubes Q containing x.
(a) Prove that M_d maps L^1 to $L^{1,\infty}$ with constant at most one, that is, show that for all $\alpha > 0$ and $f \in L^1(\mathbf{R}^n)$ we have

$$|\{x \in \mathbf{R}^n : M_d(f)(x) > \alpha\}| \le \alpha^{-1} \int_{\{M_d(f) > \alpha\}} f(t)\, dt.$$

(b) Conclude that M_d maps $L^p(\mathbf{R}^n)$ to itself with constant at most $p/(p-1)$.

2.1.13. Observe that the proof of Theorem 2.1.6 yields the estimate

$$\lambda |\{M(f) > \lambda\}|^{\frac{1}{p}} \le 3^n |\{M(f) > \lambda\}|^{-1+\frac{1}{p}} \int_{\{M(f) > \lambda\}} |f(y)|\, dy$$

for $\lambda > 0$ and f locally integrable. Use the result of Exercise 1.1.12(a) to prove that the Hardy–Littlewood maximal operator M maps the space $L^{p,\infty}(\mathbf{R}^n)$ to itself for $1 < p < \infty$.

2.1.14. Let $K(x) = (1 + |x|)^{-n-\delta}$ be defined on $\mathbf{R}^n$. Prove that there exists a constant $C_{n,\delta}$ such that for all $\varepsilon_0 > 0$ we have the estimate

$$\sup_{\varepsilon > \varepsilon_0} (|f| * K_\varepsilon)(x) \le C_{n,\delta} \sup_{\varepsilon > \varepsilon_0} \frac{1}{\varepsilon^n} \int_{|y-x| \le \varepsilon} |f(y)|\, dy,$$

for all f locally integrable on $\mathbf{R}^n$.
[*Hint:* Apply only a minor modification to the proof of Theorem 2.1.10.]

2.2 The Schwartz Class and the Fourier Transform

In this section we introduce the single most important tool in harmonic analysis, the Fourier transform. It is often the case that the Fourier transform is introduced as an operation on L^1 functions. In this exposition we first define the Fourier transform on a smaller class, the space of Schwartz functions, which turns out to be a very natural environment. Once the basic properties of the Fourier transform are derived, we extend its definition to other spaces of functions.

We begin with some preliminaries. Given $x = (x_1, \dots, x_n) \in \mathbf{R}^n$, we set $|x| = (x_1^2 + \cdots + x_n^2)^{1/2}$. The first partial derivative of a function f on $\mathbf{R}^n$ with respect to the jth variable x_j is denoted by $\partial_j f$ while the mth partial derivative with respect to the jth variable is denoted by $\partial_j^m f$. A *multi-index* α is an ordered n-tuple of nonnegative integers. For a multi-index $\alpha = (\alpha_1, \dots, \alpha_n)$, $\partial^\alpha f$ denotes the derivative $\partial_1^{\alpha_1} \cdots \partial_n^{\alpha_n} f$. If $\alpha = (\alpha_1, \dots, \alpha_n)$ is a multi-index, $|\alpha| = \alpha_1 + \cdots + \alpha_n$ denotes its size and $\alpha! = \alpha_1! \cdots \alpha_n!$ denotes the product of the factorials of its entries. The number $|\alpha|$ indicates the *total order of differentiation* of $\partial^\alpha f$. The space of functions in $\mathbf{R}^n$ all of whose derivatives of order at most $N \in \mathbf{Z}^+$ are continuous is denoted by $\mathscr{C}^N(\mathbf{R}^n)$ and the space of all *infinitely differentiable functions* on $\mathbf{R}^n$ by $\mathscr{C}^\infty(\mathbf{R}^n)$. The space of $\mathscr{C}^\infty$ functions with compact support on $\mathbf{R}^n$ is denoted by $\mathscr{C}_0^\infty(\mathbf{R}^n)$. This space is nonempty; see Exercise 2.2.1(a).

For $x \in \mathbf{R}^n$ and $\alpha = (\alpha_1, \dots, \alpha_n)$ a multi-index, we set $x^\alpha = x_1^{\alpha_1} \cdots x_n^{\alpha_n}$. It is a simple fact to verify that

$$|x^\alpha| \le c_{n,\alpha} |x|^{|\alpha|}, \tag{2.2.1}$$

for some constant that depends on the dimension n and on α. In fact, $c_{n,\alpha}$ is the maximum of the continuous function $(x_1, \dots, x_n) \mapsto |x_1^{\alpha_1} \cdots x_n^{\alpha_n}|$ on the sphere $\mathbf{S}^{n-1} = \{x \in \mathbf{R}^n : |x| = 1\}$. The converse inequality in (2.2.1) fails. However, the following substitute of the converse of (2.2.1) is of great use: for $k \in \mathbf{Z}^+$ we have

$$|x|^k \le C_{n,k} \sum_{|\beta|=k} |x^\beta|. \tag{2.2.2}$$

To prove (2.2.2), take $1/C_{n,k}$ to be the minimum of the function

$$x \mapsto \sum_{|\beta|=k} |x^\beta|$$

on $\mathbf{S}^{n-1}$; this minimum is positive since this function has no zeros on $\mathbf{S}^{n-1}$.

We end the preliminaries by noting the validity of the one-dimensional Leibniz rule

$$\frac{d^m}{dt^m}(fg) = \sum_{k=0}^{m} \binom{m}{k} \frac{d^k f}{dt^k} \frac{d^{m-k} g}{dt^{m-k}}, \tag{2.2.3}$$

for all $\mathscr{C}^m$ functions f, g on $\mathbf{R}$, and its multidimensional analogue

$$\partial^\alpha (fg) = \sum_{\beta \le \alpha} \binom{\alpha_1}{\beta_1} \cdots \binom{\alpha_n}{\beta_n} (\partial^\beta f)(\partial^{\alpha-\beta} g), \tag{2.2.4}$$

for f, g in $\mathscr{C}^{|\alpha|}(\mathbf{R}^n)$ for some multi-index α, where the notation $\beta \le \alpha$ in (2.2.4) means that β ranges over all multi-indices satisfying $0 \le \beta_j \le \alpha_j$ for all $1 \le j \le n$. We observe that identity (2.2.4) is easily deduced by repeated application of (2.2.3), which in turn is obtained by induction.

2.2.1 The Class of Schwartz Functions

We now introduce the class of *Schwartz functions* on $\mathbf{R}^n$. Roughly speaking, a function is Schwartz if it is smooth and all of its derivatives decay faster than the reciprocal of any polynomial at infinity. More precisely, we give the following definition.

Definition 2.2.1. A $\mathscr{C}^\infty$ complex-valued function f on $\mathbf{R}^n$ is called a Schwartz function if for every pair of multi-indices α and β there exists a positive constant $C_{\alpha,\beta}$ such that

$$\rho_{\alpha,\beta}(f) = \sup_{x\in\mathbf{R}^n} |x^\alpha \partial^\beta f(x)| = C_{\alpha,\beta} < \infty. \tag{2.2.5}$$

The quantities $\rho_{\alpha,\beta}(f)$ are called the *Schwartz seminorms* of f. The set of all Schwartz functions on $\mathbf{R}^n$ is denoted by $\mathscr{S}(\mathbf{R}^n)$.

Example 2.2.2. The function $e^{-|x|^2}$ is in $\mathscr{S}(\mathbf{R}^n)$ but $e^{-|x|}$ is not, since it fails to be differentiable at the origin. The $\mathscr{C}^\infty$ function $g(x) = (1+|x|^4)^{-a}$, $a > 0$, is not in $\mathscr{S}$ since it decays only like the reciprocal of a fixed polynomial at infinity. The set of all smooth functions with compact support, $\mathscr{C}_0^\infty(\mathbf{R}^n)$, is contained in $\mathscr{S}(\mathbf{R}^n)$.

Remark 2.2.3. If f_1 is in $\mathscr{S}(\mathbf{R}^n)$ and f_2 is in $\mathscr{S}(\mathbf{R}^m)$, then the function of $m+n$ variables $f_1(x_1,\dots,x_n)f_2(x_{n+1},\dots,x_{n+m})$ is in $\mathscr{S}(\mathbf{R}^{n+m})$. If f is in $\mathscr{S}(\mathbf{R}^n)$ and $P(x)$ is a polynomial of n variables, then $P(x)f(x)$ is also in $\mathscr{S}(\mathbf{R}^n)$. If α is a multi-index and f is in $\mathscr{S}(\mathbf{R}^n)$, then $\partial^\alpha f$ is in $\mathscr{S}(\mathbf{R}^n)$. Also note that

$$f \in \mathscr{S}(\mathbf{R}^n) \iff \sup_{x\in\mathbf{R}^n} |\partial^\alpha(x^\beta f(x))| < \infty \qquad \text{for all multi-indices } \alpha,\ \beta.$$

Remark 2.2.4. The following alternative characterization of Schwartz functions is very useful. A $\mathscr{C}^\infty$ function f is in $\mathscr{S}(\mathbf{R}^n)$ if and only if for all positive integers N and all multi-indices α there exists a positive constant $C_{\alpha,N}$ such that

$$|(\partial^\alpha f)(x)| \le C_{\alpha,N}(1+|x|)^{-N}. \tag{2.2.6}$$

The simple proofs are omitted. We now discuss convergence in $\mathscr{S}(\mathbf{R}^n)$.

Definition 2.2.5. Let f_k, f be in $\mathscr{S}(\mathbf{R}^n)$ for $k = 1,2,\dots$. We say that the sequence f_k converges to f in $\mathscr{S}(\mathbf{R}^n)$ if for all multi-indices α and β we have

$$\rho_{\alpha,\beta}(f_k - f) = \sup_{x\in\mathbf{R}^n} |x^\alpha(\partial^\beta(f_k - f))(x)| \to 0 \qquad \text{as} \quad k\to\infty.$$

For instance, for any fixed $x_0 \in \mathbf{R}^n$, $f(x+x_0/k) \to f(x)$ in $\mathscr{S}(\mathbf{R}^n)$ for any f in $\mathscr{S}(\mathbf{R}^n)$ as $k\to\infty$.

This notion of convergence is compatible with a topology on $\mathscr{S}(\mathbf{R}^n)$ under which the operations $(f,g) \mapsto f+g$, $(a,f) \to af$, and $f \mapsto \partial^\alpha f$ are continuous for all complex scalars a and multi-indices α ($f,g \in \mathscr{S}(\mathbf{R}^n)$). A subbasis for open sets containing 0 in this topology is

$$\{f \in \mathscr{S} : \rho_{\alpha,\beta}(f) < r\},$$

for all α, β multi-indices and all $r \in \mathbf{Q}^+$. Observe the following: If $\rho_{\alpha,\beta}(f) = 0$, then $f = 0$. This means that $\mathscr{S}(\mathbf{R}^n)$ is a locally convex topological vector space equipped with the family of seminorms $\rho_{\alpha,\beta}$ that separate points. We refer to Reed and Simon [215] for the pertinent definitions. Since the origin in $\mathscr{S}(\mathbf{R}^n)$ has a countable base, this space is metrizable. In fact, the following is a metric on $\mathscr{S}(\mathbf{R}^n)$:

$$d(f,g)=\sum_{j=1}^{\infty}2^{-j}\frac{\rho_j(f-g)}{1+\rho_j(f-g)},$$

where ρ_j is an enumeration of all the seminorms $\rho_{\alpha,\beta}$, α and β multi-indices. One may easily verify that $\mathscr{S}$ is complete with respect to the metric d. Indeed, a Cauchy sequence $\{h_j\}_j$ in $\mathscr{S}$ would have to be Cauchy in L^∞ and therefore it would converge uniformly to some function h. The same is true for the sequences $\{\partial^\beta h_j\}_j$ and $\{x^\alpha h_j(x)\}_j$, and the limits of these sequences can be shown to be the functions $\partial^\beta h$ and $x^\alpha h(x)$, respectively. It follows that the sequence $\{h_j\}$ converges to h in $\mathscr{S}$. Therefore, $\mathscr{S}(\mathbf{R}^n)$ is a *Fréchet space (complete metrizable locally convex space).*

We note that convergence in $\mathscr{S}$ is stronger than convergence in all L^p. We have the following.

Proposition 2.2.6. *Let f, f_k, $k=1,2,3,\dots$, be in $\mathscr{S}(\mathbf{R}^n)$. If $f_k\to f$ in $\mathscr{S}$ then $f_k\to f$ in L^p for all $0<p\le\infty$. Moreover, there exists a $C_{p,n}>0$ such that*

$$\big\|\partial^\beta f\big\|_{L^p}\le C_{p,n}\sum_{|\alpha|\le[(n+1)/p]+1}\rho_{\alpha,\beta}(f) \tag{2.2.7}$$

for all f for which the right-hand side is finite.

Proof. Observe that

$$\begin{aligned}\big\|\partial^\beta f\big\|_{L^p} &\le\left(\int_{|x|\le 1}|\partial^\beta f(x)|^p\,dx+\int_{|x|\ge 1}|x|^{n+1}|\partial^\beta f(x)|^p|x|^{-(n+1)}\,dx\right)^{1/p}\\ &\le\left(v_n\big\|\partial^\beta f\big\|_{L^\infty}^p+\sup_{|x|\ge 1}|x|^{n+1}|\partial^\beta f(x)|^p\int_{|x|\ge 1}|x|^{-(n+1)}\,dx\right)^{1/p}\\ &\le C_{p,n}\Big(\big\|\partial^\beta f\big\|_{L^\infty}+\sup_{x\in\mathbf{R}^n}\big(|x|^{[(n+1)/p]+1}|\partial^\beta f(x)|\big)\Big).\end{aligned}$$

Now set $m=[(n+1)/p]+1$ and use (2.2.2) to obtain

$$|x|^m|\partial^\beta f(x)|\le C_{n,m}\sum_{|\alpha|=m}|x^\alpha\partial^\beta f(x)|.$$

Thus the L^p norm of the Schwartz function $\partial^\beta f$ is controlled by a constant multiple of a sum of some $\rho_{\alpha,0}$ seminorms of it. Conclusion (2.2.7) now follows immediately. This shows that convergence in $\mathscr{S}$ implies convergence in L^p. □

We now show that the Schwartz class is closed under certain operations.

Proposition 2.2.7. *Let f, g be in $\mathscr{S}(\mathbf{R}^n)$. Then fg and $f*g$ are in $\mathscr{S}(\mathbf{R}^n)$. Moreover,*

$$\partial^\alpha(f*g)=(\partial^\alpha f)*g=f*(\partial^\alpha g) \tag{2.2.8}$$

for all multi-indices α.

Proof. Fix f and g in $\mathscr{S}(\mathbf{R}^n)$. Let e_j be the unit vector $(0,\dots,1,\dots,0)$ with 1 in the jth entry and zeros in all the other entries. Since

$$\frac{f(y+he_j)-f(y)}{h}-(\partial_j f)(y)\to 0 \tag{2.2.9}$$

as $h\to 0$, and since the expression in (2.2.9) is pointwise bounded by some constant depending on f, the integral of the expression in (2.2.9) with respect to the measure $g(x-y)\,dy$ converges to zero as $h\to 0$ by the Lebesgue dominated convergence theorem. This proves (2.2.8) when $\alpha=(0,\dots,1,\dots,0)$. The general case follows by repeating the previous argument and using induction.

We now show that the convolution of two functions in $\mathscr{S}$ is also in $\mathscr{S}$. For each $N>n$ we have

$$|(f*g)(x)|\le C_N\int_{\mathbf{R}^n}(1+|x-y|)^{-N}(1+|y|)^{-N}dy\,. \tag{2.2.10}$$

The part of the integral in (2.2.10) over the set $\{y:\ \frac{1}{2}|x|\le|y-x|\}$ is bounded by

$$\int_{|y-x|\ge\frac{1}{2}|x|}(1+\tfrac{1}{2}|x|)^{-N}(1+|y|)^{-N}dy\le B_N(1+|x|)^{-N},$$

where B_N is a constant depending on N and on the dimension. When $\frac{1}{2}|x|\ge|y-x|$ we have that $|y|\ge\frac{1}{2}|x|$, and it follows that the part of the integral in (2.2.10) over the set $\{y:\ |y-x|\le\frac{1}{2}|x|\}$ is bounded by

$$\int_{|y-x|\le\frac{1}{2}|x|}(1+|x-y|)^{-N}(1+\tfrac{1}{2}|x|)^{-N}dy\le B_N(1+|x|)^{-N}.$$

This shows that $f*g$ decays like $(1+|x|)^{-N}$ at infinity, but since $N>n$ is arbitrary it follows that $f*g$ decays faster than the reciprocal of any polynomial.

Since $\partial^\alpha(f*g)=(\partial^\alpha f)*g$, replacing f by $\partial^\alpha f$ in the previous argument, we also conclude that all the derivatives of $f*g$ decay faster than the reciprocal of any polynomial at infinity. Using (2.2.6), we conclude that $f*g$ is in $\mathscr{S}$. Finally, the fact that fg is in $\mathscr{S}$ follows directly from Leibniz's rule (2.2.4) and (2.2.6). □

2.2.2 *The Fourier Transform of a Schwartz Function*

The Fourier transform is often introduced as an operation on L^1. In that setting, problems of convergence arise when certain manipulations of functions are performed. Also, Fourier inversion requires the additional assumption that the Fourier transform is in L^1. Here we initially introduce the Fourier transform on the space of Schwartz functions. The rapid decay of Schwartz functions at infinity allows us to develop its fundamental properties without encountering any convergence problems. The Fourier transform is a homeomorphism of the Schwartz class and Fourier inversion holds in it. For these reasons, this class is a natural environment for it.

For $x = (x_1, \dots, x_n)$, $y = (y_1, \dots, y_n)$ in $\mathbf{R}^n$ we use the notation

$$x \cdot y = \sum_{j=1}^{n} x_j y_j .$$

Definition 2.2.8. Given f in $\mathscr{S}(\mathbf{R}^n)$ we define

$$\widehat{f}(\xi) = \int_{\mathbf{R}^n} f(x) e^{-2\pi i x \cdot \xi} \, dx .$$

We call $\widehat{f}$ the Fourier transform of f.

Example 2.2.9. Let $f(x) = e^{-\pi |x|^2}$ defined on $\mathbf{R}$. Then $\widehat{f}(\xi) = f(\xi)$. First observe that the function

$$s \mapsto \int_{-\infty}^{+\infty} e^{-\pi (x+is)^2} dx, \qquad s \in \mathbf{R},$$

is constant. Indeed, its derivative is

$$\int_{-\infty}^{+\infty} -2\pi i (x+is) e^{-\pi (x+is)^2} dx = \int_{-\infty}^{+\infty} i \frac{d}{dx} (e^{-\pi (x+is)^2}) \, dx = 0 .$$

The computation of the Fourier transform of $f(x) = e^{-\pi |x|^2}$ relies on simple completion of squares. We have

$$\begin{aligned}
\int_{\mathbf{R}^n} e^{-\pi |x|^2} e^{-2\pi i \sum_{j=1}^n x_j \xi_j} dx &= \int_{\mathbf{R}^n} e^{-\pi \sum_{j=1}^n (x_j + i\xi_j)^2} e^{\pi \sum_{j=1}^n (i\xi_j)^2} dx \\
&= \left(\int_{-\infty}^{+\infty} e^{-\pi x^2} dx \right)^n e^{-\pi |\xi|^2} \\
&= e^{-\pi |\xi|^2},
\end{aligned}$$

where we used that

$$\int_{-\infty}^{+\infty} e^{-x^2} dx = \sqrt{\pi}, \tag{2.2.11}$$

a fact that can be found in Appendix A.1.

Remark 2.2.10. It follows from the definition of the Fourier transform that if f is in $\mathscr{S}(\mathbf{R}^n)$ and g is in $\mathscr{S}(\mathbf{R}^m)$, then

$$[f(x_1, \dots, x_n) g(x_{n+1}, \dots, x_{n+m})]\widehat{\;} = \widehat{f}(\xi_1, \dots, \xi_n) \widehat{g}(\xi_{n+1}, \dots, \xi_{n+m}),$$

where the first $\widehat{\;}$ denotes the Fourier transform on $\mathbf{R}^{n+m}$. In other words, the Fourier transform preserves separation of variables.

Combining this observation with the result in Example 2.2.9, we conclude that the function $f(x) = e^{-\pi |x|^2}$ defined on $\mathbf{R}^n$ is equal to its Fourier transform.

We now continue with some properties of the Fourier transform. Before we do this we introduce some notation. For a measurable function f on $\mathbf{R}^n$, $x \in \mathbf{R}^n$, and $a > 0$ we define the *translation*, *dilation*, and *reflection* of f by

$$\begin{aligned} \tau^y(f)(x) &= f(x-y) \\ \delta^a(f)(x) &= f(ax) \\ \widetilde{f}(x) &= f(-x). \end{aligned} \tag{2.2.12}$$

Also recall the notation $f_a = a^{-n}\delta^{1/a}(f)$ introduced in Definition 2.1.9.

Proposition 2.2.11. *Given f, g in $\mathscr{S}(\mathbf{R}^n)$, $y \in \mathbf{R}^n$, $b \in \mathbf{C}$, α a multi-index, and $t > 0$, we have*

(1) $\|\widehat{f}\|_{L^\infty} \le \|f\|_{L^1}$,

(2) $\widehat{f+g} = \widehat{f} + \widehat{g}$,

(3) $\widehat{bf} = b\widehat{f}$,

(4) $\widehat{\widetilde{f}} = \widetilde{\widehat{f}}$,

(5) $\widehat{\overline{f}} = \overline{\widetilde{\widehat{f}}}$,

(6) $\widehat{\tau^y(f)}(\xi) = e^{-2\pi i y\cdot\xi}\widehat{f}(\xi)$,

(7) $(e^{2\pi i x\cdot y}f(x))\widehat{\ }(\xi) = \tau^y(\widehat{f})(\xi)$,

(8) $(\delta^t(f))\widehat{\ } = t^{-n}\delta^{t^{-1}}(\widehat{f}) = (\widehat{f})_t$,

(9) $(\partial^\alpha f)\widehat{\ }(\xi) = (2\pi i\xi)^\alpha \widehat{f}(\xi)$,

(10) $(\partial^\alpha \widehat{f})(\xi) = ((-2\pi i x)^\alpha f(x))\widehat{\ }(\xi)$,

(11) $\widehat{f} \in \mathscr{S}$,

(12) $\widehat{f*g} = \widehat{f}\,\widehat{g}$,

(13) $\widehat{f\circ A}(\xi) = \widehat{f}(A\xi)$, *where A is an orthogonal matrix and ξ is a column vector.*

Proof. Property (1) follows from Definition 2.2.8 and implies that the Fourier transform is always bounded. Properties (2)–(5) are trivial. Properties (6)–(8) require a suitable change of variables but they are omitted. Property (9) is proved by integration by parts (which is justified by the rapid decay of the integrands):

$$\begin{aligned} (\partial^\alpha f)\widehat{\ }(\xi) &= \int_{\mathbf{R}^n} (\partial^\alpha f)(x) e^{-2\pi i x\cdot\xi}\,dx \\ &= (-1)^{|\alpha|}\int_{\mathbf{R}^n} f(x)(-2\pi i\xi)^\alpha e^{-2\pi i x\cdot\xi}\,dx \\ &= (2\pi i\xi)^\alpha \widehat{f}(\xi). \end{aligned}$$

To prove (10), let $\alpha = (0,\dots,1,\dots,0)$, where all entries are zero except for the jth entry, which is 1. Since

$$\frac{e^{-2\pi i x\cdot(\xi+he_j)} - e^{-2\pi i x\cdot\xi}}{h} - (-2\pi i x_j)e^{-2\pi i x\cdot\xi} \to 0 \tag{2.2.13}$$

as $h \to 0$ and the preceding function is bounded by $C|x|$ for all h and ξ, the Lebesgue dominated convergence theorem implies that the integral of the function in (2.2.13) with respect to the measure $f(x)dx$ converges to zero. Thus we have proved (10) for $\alpha = (0,\dots,1,\dots,0)$. For other α's use induction. To prove (11) we use (9), (10), and (1) in the following way:

$$\big\|x^\alpha(\partial^\beta \widehat{f})(x)\big\|_{L^\infty} = \frac{(2\pi)^{|\beta|}}{(2\pi)^{|\alpha|}}\big\|(\partial^\alpha(x^\beta f(x)))^{\widehat{}}\big\|_{L^\infty} \le \frac{(2\pi)^{|\beta|}}{(2\pi)^{|\alpha|}}\big\|\partial^\alpha(x^\beta f(x))\big\|_{L^1} < \infty.$$

Identity (12) follows from the following calculation:

$$\begin{aligned}
\widehat{f*g}(\xi) &= \int_{\mathbf{R}^n}\int_{\mathbf{R}^n} f(x-y)g(y)e^{-2\pi i x\cdot\xi}\,dy\,dx \\
&= \int_{\mathbf{R}^n}\int_{\mathbf{R}^n} f(x-y)g(y)e^{-2\pi i(x-y)\cdot\xi}e^{-2\pi i y\cdot\xi}\,dy\,dx \\
&= \int_{\mathbf{R}^n} g(y)\int_{\mathbf{R}^n} f(x-y)e^{-2\pi i(x-y)\cdot\xi}dx\,e^{-2\pi i y\cdot\xi}\,dy \\
&= \widehat{f}(\xi)\widehat{g}(\xi),
\end{aligned}$$

where the application of Fubini's theorem is justified by the absolute convergence of the integrals. Finally, we prove (13). We have

$$\begin{aligned}
\widehat{f\circ A}(\xi) &= \int_{\mathbf{R}^n} f(Ax)e^{-2\pi i x\cdot\xi}\,dx \\
&= \int_{\mathbf{R}^n} f(y)e^{-2\pi i A^{-1}y\cdot\xi}\,dy \\
&= \int_{\mathbf{R}^n} f(y)e^{-2\pi i A^t y\cdot\xi}\,dy \\
&= \int_{\mathbf{R}^n} f(y)e^{-2\pi i y\cdot A\xi}\,dy \\
&= \widehat{f}(A\xi),
\end{aligned}$$

where we used the change of variables $y = Ax$ and the fact that $|\det A| = 1$. □

Corollary 2.2.12. *The Fourier transform of a radial function is radial. Products and convolutions of radial functions are radial.*

Proof. Let ξ_1, ξ_2 in $\mathbf{R}^n$ with $|\xi_1| = |\xi_2|$. Then for some orthogonal matrix A we have $A\xi_1 = \xi_2$. Since f is radial, we have $f = f\circ A$. Then

$$\widehat{f}(\xi_2) = \widehat{f}(A\xi_1) = \widehat{f\circ A}(\xi_1) = \widehat{f}(\xi_1),$$

where we used (13) in Proposition 2.2.11 to justify the second equality. Products and convolutions of radial functions are easily seen to be radial. □

2.2.3 The Inverse Fourier Transform and Fourier Inversion

We now define the inverse Fourier transform.

Definition 2.2.13. Given a Schwartz function f, we define

$$f^\vee(x) = \widehat{f}(-x),$$

for all $x \in \mathbf{R}^n$. The operation

$$f \mapsto f^\vee$$

is called the *inverse Fourier transform.*

It is straightforward that the inverse Fourier transform shares the same properties as the Fourier transform. One may want to list (and prove) properties for the inverse Fourier transform analogous to those in Proposition 2.2.11.

We now investigate the relation between the Fourier transform and the inverse Fourier transform. In the next theorem, we prove that one is the inverse operation of the other. This property is referred to as *Fourier inversion.*

Theorem 2.2.14. *Given f, g, and h in $\mathscr{S}(\mathbf{R}^n)$, we have*

(1) $\displaystyle\int_{\mathbf{R}^n} f(x)\widehat{g}(x)\,dx = \int_{\mathbf{R}^n} \widehat{f}(x)g(x)\,dx,$

*(2) **(Fourier Inversion)*** $\quad (\widehat{f})^\vee = f = (f^\vee)\widehat{\ },$

*(3) **(Parseval's relation)*** $\displaystyle\int_{\mathbf{R}^n} f(x)\overline{h}(x)\,dx = \int_{\mathbf{R}^n} \widehat{f}(\xi)\overline{\widehat{h}(\xi)}\,d\xi,$

*(4) **(Plancherel's identity)*** $\quad \|f\|_{L^2} = \|\widehat{f}\|_{L^2} = \|f^\vee\|_{L^2},$

(5) $\displaystyle\int_{\mathbf{R}^n} f(x)g(x)\,dx = \int_{\mathbf{R}^n} \widehat{f}(x)g^\vee(x)\,dx.$

Proof. (1) follows immediately from the definition of the Fourier transform and Fubini's theorem. To prove (2) we use (1) with

$$g(\xi) = e^{2\pi i\xi\cdot t}e^{-\pi|\varepsilon\xi|^2}.$$

By Proposition 2.2.11 (7) and (8) and Example 2.2.9, we have that

$$\widehat{g}(x) = \frac{1}{\varepsilon^n}e^{-\pi|(x-t)/\varepsilon|^2},$$

which is an approximate identity. Now (1) gives

$$\int_{\mathbf{R}^n} f(x)\varepsilon^{-n}e^{-\pi\varepsilon^{-2}|x-t|^2}\,dx = \int_{\mathbf{R}^n} \widehat{f}(\xi)e^{2\pi i\xi\cdot t}e^{-\pi|\varepsilon\xi|^2}\,d\xi\,. \tag{2.2.14}$$

Now let $\varepsilon \to 0$ in (2.2.14). The left-hand side of (2.2.14) converges to $f(t)$ uniformly on compact sets by Theorem 1.2.19. The right-hand side of (2.2.14) converges to $(\widehat{f})^\vee(t)$ as $\varepsilon \to 0$ by the Lebesgue dominated convergence theorem. We conclude that $(\widehat{f})^\vee = f$ on $\mathbf{R}^n$. Replacing f by $\widetilde{f}$ and using the result just proved, we conclude that $(f^\vee)\widehat{\ } = f$.

To prove (3), use (1) with $g = \overline{\widehat{h}}$ and the fact that $\widehat{g} = \overline{h}$, which is a consequence of Proposition 2.2.11 (5) and Fourier inversion. Plancherel's identity is a trivial consequence of (3). (Sometimes the polarized identity (3) is also referred to as Plancherel's identity.) Finally, (5) easily follows from (1) and (2). □

Next we have the following simple corollary of Theorem 2.2.14.

Corollary 2.2.15. *The Fourier transform is a homeomorphism from $\mathscr{S}(\mathbf{R}^n)$ onto itself.*

Proof. The continuity of the Fourier transform (and its inverse) follows from Exercise 2.2.2, while Fourier inversion yields that this map is bijective. □

2.2.4 The Fourier Transform on $L^1 + L^2$

We have defined the Fourier transform on $\mathscr{S}(\mathbf{R}^n)$. We now extend this definition to the space $L^1(\mathbf{R}^n) + L^2(\mathbf{R}^n)$.

We begin by observing that the Fourier transform given in Definition 2.2.8,

$$\widehat{f}(\xi) = \int_{\mathbf{R}^n} f(x)e^{-2\pi i x\cdot\xi}\,dx\,,$$

makes sense as a convergent integral for functions $f \in L^1(\mathbf{R}^n)$. This allows us to extend the definition of the Fourier transform on L^1. Moreover, this operator satisfies properties (1)–(8) as well as (12) and (13) in Proposition 2.2.11, with f, g integrable. We also define the inverse Fourier transform on L^1 by setting $f^\vee(x) = \widehat{f}(-x)$ for $f \in L^1(\mathbf{R}^n)$ and we note that analogous properties hold for it. One problem in this generality is that when f is integrable, one may not necessarily have $(\widehat{f})^\vee = f$ a.e. This inversion is possible when $\widehat{f}$ is also integrable; see Exercise 2.2.6.

The integral defining the Fourier transform does not converge absolutely for functions in $L^2(\mathbf{R}^n)$; however, the Fourier transform has a natural definition in this space accompanied by an elegant theory. In view of the result in Exercise 2.2.8, the Fourier transform is an L^2 isometry on $L^1 \cap L^2$, which is a dense subspace of L^2. By density, there is a unique bounded extension of the Fourier transform on L^2. Let us denote this extension by $\mathscr{F}$. Then $\mathscr{F}$ is also an isometry on L^2, i.e.,

$$\|\mathscr{F}(f)\|_{L^2} = \|f\|_{L^2}$$

for all $f \in L^2(\mathbf{R}^n)$, and any sequence of functions $f_N \in L^1(\mathbf{R}^n) \cap L^2(\mathbf{R}^n)$ converging to a given f in $L^2(\mathbf{R}^n)$ satisfies

$$\|\widehat{f_N} - \mathscr{F}(f)\|_{L^2} \to 0, \tag{2.2.15}$$

as $N \to \infty$. In particular, the sequence of functions $f_N(x) = f(x)\chi_{|x|\le N}$ yields that

$$\widehat{f_N}(\xi) = \int_{|x|\le N} f(x)e^{-2\pi i x\cdot\xi}\,dx \tag{2.2.16}$$

converges to $\mathscr{F}(f)(\xi)$ in L^2 as $N \to \infty$. If f is both integrable and square integrable, the expressions in (2.2.16) also converge to $\widehat{f}(\xi)$ pointwise. Also, in view of Theorem 1.1.11 and (2.2.15), there is a subsequence of $\widehat{f_N}$ that converges to $\mathscr{F}(f)$ pointwise a.e. Consequently, for f in $L^1(\mathbf{R}^n) \cap L^2(\mathbf{R}^n)$ the expressions $\widehat{f}$ and $\mathscr{F}(f)$ coincide pointwise a.e. For this reason we often adopt the notation $\widehat{f}$ to denote the Fourier transform of functions f in L^2 as well.

In a similar fashion, we let $\mathscr{F}'$ be the isometry on $L^2(\mathbf{R}^n)$ that extends the operator $f \mapsto f^\vee$, which is an L^2 isometry on $L^1 \cap L^2$; the last statement follows by adapting the result of Exercise 2.2.8 to the inverse Fourier transform. Since $\varphi^\vee(x) = \widehat{\varphi}(-x)$ for φ in the Schwartz class, which is dense in L^2 (Exercise 2.2.5), it follows that $\mathscr{F}'(f)(x) = \mathscr{F}(f)(-x)$ for all $f \in L^2$ and $x \in \mathbf{R}^n$. The operators $\mathscr{F}$ and $\mathscr{F}'$ are L^2-isometries that satisfy $\mathscr{F}' \circ \mathscr{F} = \mathscr{F} \circ \mathscr{F}' = \mathrm{Id}$ on the Schwartz space. By density this identity also holds for L^2 functions and implies that $\mathscr{F}$ and $\mathscr{F}'$ are injective and surjective mappings from L^2 to itself; consequently, $\mathscr{F}'$ coincides with the inverse operator $\mathscr{F}^{-1}$ of $\mathscr{F} : L^2 \to L^2$, and Fourier inversion

$$f = \mathscr{F}^{-1} \circ \mathscr{F}(f) = \mathscr{F} \circ \mathscr{F}^{-1}(f)$$

holds on L^2.

Having set down the basic facts concerning the action of the Fourier transform on L^1 and L^2, it is now a simple matter to extend its definition on L^p for $1 < p < 2$. For functions $f \in L^p(\mathbf{R}^n)$, $1 < p < 2$, we define $\widehat{f} = \widehat{f_1} + \widehat{f_2}$, where $f_1 \in L^1$, $f_2 \in L^2$, and $f = f_1 + f_2$; we may take, for instance, $f_1 = f\chi_{|f|>1}$ and $f_2 = f\chi_{|f|\le 1}$. The definition of $\widehat{f}$ is independent of the choice of f_1 and f_2, for if $f_1 + f_2 = h_1 + h_2$ for $f_1, h_1 \in L^1$ and $f_2, h_2 \in L^2$, we have $f_1 - h_1 = h_2 - f_2 \in L^1 \cap L^2$. Since these functions are equal on $L^1 \cap L^2$, their Fourier transforms are also equal, and we obtain $\widehat{f_1} - \widehat{h_1} = \widehat{h_2} - \widehat{f_2}$, which yields $\widehat{f_1 + f_2} = \widehat{h_1 + h_2}$. We have the following result concerning the action of the Fourier transform on L^p.

Proposition 2.2.16. ***(Hausdorff–Young inequality)*** *For every function f in $L^p(\mathbf{R}^n)$ we have the estimate*

$$\|\widehat{f}\|_{L^{p'}} \le \|f\|_{L^p}$$

whenever $1 \le p \le 2$.

Proof. This follows easily from Theorem 1.3.4. Interpolate between the estimates $\|\widehat{f}\|_{L^\infty} \le \|f\|_{L^1}$ (Proposition 2.2.11 (1)) and $\|\widehat{f}\|_{L^2} \le \|f\|_{L^2}$ to obtain $\|\widehat{f}\|_{L^{p'}} \le \|f\|_{L^p}$. We conclude that the Fourier transform is a bounded operator from $L^p(\mathbf{R}^n)$ to $L^{p'}(\mathbf{R}^n)$ with norm at most 1 when $1 \le p \le 2$. □

Next, we are concerned with the behavior of the Fourier transform at infinity.

Proposition 2.2.17. ***(Riemann–Lebesgue lemma)*** *For a function f in $L^1(\mathbf{R}^n)$ we have that*

$$|\widehat{f}(\xi)| \to 0 \qquad \text{as} \qquad |\xi| \to \infty.$$

Proof. Consider the function $\chi_{[a,b]}$ on $\mathbf{R}$. A simple computation gives

$$\widehat{\chi_{[a,b]}}(\xi) = \frac{e^{-2\pi i \xi a} - e^{-2\pi i \xi b}}{2\pi i \xi},$$

which tends to zero as $|\xi| \to \infty$. Likewise, if $g = \prod_{j=1}^n \chi_{[a_j,b_j]}$ on $\mathbf{R}^n$, then

$$\widehat{g}(\xi) = \prod_{j=1}^{n} \frac{e^{-2\pi i \xi_j a_j} - e^{-2\pi i \xi_j b_j}}{2\pi i \xi_j},$$

which also tends to zero as $|\xi| \to \infty$ in $\mathbf{R}^n$.

To prove the assertion, approximate in the L^1 norm a general integrable function f on $\mathbf{R}^n$ by a finite sum h of "step functions" like g and use

$$|\widehat{f}(\xi)| \le |\widehat{f}(\xi) - \widehat{h}(\xi)| + |\widehat{h}(\xi)| \le \|f-h\|_{L^1} + |\widehat{h}(\xi)|.$$

□

We end this section with an example that illustrates some of the practical uses of the Fourier transform.

Example 2.2.18. We are asked to find a function $f(x_1,x_2,x_3)$ on $\mathbf{R}^3$ that satisfies the partial differential equation

$$f(x) + \partial_1^2\partial_2^2\partial_3^4 f(x) + 4i\partial_1^2 f(x) + \partial_2^7 f(x) = e^{-\pi|x|^2}.$$

Taking the Fourier transform on both sides of this identity and using Proposition 2.2.11 (2), (9) and the result of Example 2.2.9, we obtain

$$\widehat{f}(\xi)\Big[1 + (2\pi i\xi_1)^2(2\pi i\xi_2)^2(2\pi i\xi_3)^4 + 4i(2\pi i\xi_1)^2 + (2\pi i\xi_2)^7\Big] = e^{-\pi|\xi|^2}.$$

Let $p(\xi) = p(\xi_1,\xi_2,\xi_3)$ be the polynomial inside the square brackets. We observe that $p(\xi)$ has no real zeros and we may therefore write

$$\widehat{f}(\xi) = e^{-\pi|\xi|^2}p(\xi)^{-1} \implies f(x) = \big(e^{-\pi|\xi|^2}p(\xi)^{-1}\big)^\vee(x).$$

In general, let $P(\xi) = \sum_{|\alpha|\le N} C_\alpha \xi^\alpha$ be a polynomial in $\mathbf{R}^n$ with constant complex coefficients C_α indexed by multi-indices α. If $P(2\pi i\xi)$ has no real zeros, and u is in $\mathscr{S}(\mathbf{R}^n)$, then the partial differential equation

$$P(\partial)f = \sum_{|\alpha|\le N} C_\alpha \partial^\alpha f = u$$

is solved as before to give

$$f = \big(\widehat{u}(\xi)P(2\pi i\xi)^{-1}\big)^{\vee}.$$

Since $P(2\pi i\xi)$ has no real zeros and $u \in \mathscr{S}(\mathbf{R}^n)$, the function $\widehat{u}(\xi)P(2\pi i\xi)^{-1}$ is smooth and therefore a Schwartz function. Then f is also in $\mathscr{S}(\mathbf{R}^n)$ by Proposition 2.2.11 (11).

Exercises

2.2.1. (a) Construct a Schwartz function with compact support.
(b) Construct a $\mathscr{C}_0^\infty(\mathbf{R}^n)$ function equal to 1 on the annulus $1 \le |x| \le 2$ and vanishing off the annulus $1/2 \le |x| \le 4$.
(c) Construct a nonnegative nonzero Schwartz function f whose Fourier transform is nonnegative and compactly supported.
[*Hint:* Part (a): Try the construction in dimension one first using the $\mathscr{C}^\infty$ function $\eta(x) = e^{-1/x}$ for $x > 0$ and $\eta(x) = 0$ for $x < 0$. Part (c): Take $f = |\varphi * \widetilde{\varphi}|^2$, where $\widehat{\varphi}$ is odd, real-valued, and compactly supported.]

2.2.2. If $f_k, f \in \mathscr{S}(\mathbf{R}^n)$ and $f_k \to f$ in $\mathscr{S}(\mathbf{R}^n)$, then $\widehat{f_k} \to \widehat{f}$ and $f_k^\vee \to f^\vee$ in $\mathscr{S}(\mathbf{R}^n)$.

2.2.3. Find the *spectrum* (i.e., the set of all *eigenvalues* of the Fourier transform), that is, all complex numbers λ for which there exist nonzero functions f such that

$$\widehat{f} = \lambda f.$$

[*Hint:* Apply the Fourier transform three times to the preceding identity. Consider the functions $xe^{-\pi x^2}$, $(a+bx^2)e^{-\pi x^2}$, and $(cx+dx^3)e^{-\pi x^2}$ for suitable a,b,c,d to show that all fourth roots of unity are indeed eigenvalues of the Fourier transform.]

2.2.4. Use the idea of the proof of Proposition 2.2.7 to show that if the functions f, g defined on $\mathbf{R}^n$ satisfy $|f(x)| \le A(1+|x|)^{-M}$ and $|g(x)| \le B(1+|x|)^{-N}$ for some $M,N > n$, then

$$|(f*g)(x)| \le ABC(1+|x|)^{-L},$$

where $L = \min(N,M)$ and $C = C(N,M) > 0$.

2.2.5. (a) Show that $\mathscr{C}_0^\infty(\mathbf{R}^n)$ is dense on $L^p(\mathbf{R}^n)$ for $1 \le p < \infty$. Conclude that $\mathscr{S}(\mathbf{R}^n)$ is also dense on L^p spaces.

(b) Prove that these spaces are also dense in L^p for $0 < p < 1$.
[*Hint:* When $1 \le p < \infty$ you may convolve with an approximate identity. For $0 < p < 1$ you may approximate a compactly supported step function with a smooth function.]

2.2.6. (a) Prove that if $f \in L^1$, then $\widehat{f}$ is uniformly continuous on $\mathbf{R}^n$.
(b) Prove that for $f \in L^1$ and $g \in \mathscr{S}$ we have

$$\int_{\mathbf{R}^n} f(x)\widehat{g}(x)\,dx = \int_{\mathbf{R}^n} \widehat{f}(x)g(x)\,dx.$$

(c) Take $\widehat{g}(x) = \varepsilon^{-n}e^{-\pi\varepsilon^{-2}|x-t|^2}$ in (b) and let $\varepsilon \to 0$ to prove that if f and $\widehat{f}$ are both in L^1, then $(\widehat{f})^\vee = f$ a.e. This fact is called *Fourier inversion on* L^1.

2.2.7. (a) Prove that if f is continuous at 0, then

$$\lim_{\varepsilon\to 0}\int_{\mathbf{R}^n} \widehat{f}(x)e^{-\pi|\varepsilon x|^2}\,dx = f(0).$$

(b) Prove that if $f \in L^1(\mathbf{R}^n)$, $\widehat{f} \ge 0$, and f is continuous at zero, then $\widehat{f}$ is in L^1 and therefore Fourier inversion $f(0) = \|\widehat{f}\|_{L^1}$ holds at zero and $f = (\widehat{f})^\vee$ a.e. in general.
[*Hint:* Part (a): Take $g(x) = e^{-\pi|\varepsilon x|^2}$ in Exercise 2.2.6(b).]

2.2.8. (a) Given f in $L^1(\mathbf{R}^n) \cap L^2(\mathbf{R}^n)$, prove that

$$\|\widehat{f}\|_{L^2} = \|f\|_{L^2}.$$

[*Hint:* Let $h = f * \overline{\widetilde{f}}$, where $\widetilde{f}(x) = f(-x)$ and the bar indicates complex conjugation. Then $h \in L^1(\mathbf{R}^n)$, $\widehat{h} = |\widehat{f}|^2 \ge 0$, and h is continuous at zero. Exercise 2.2.7(b) yields $\|\widehat{f}\|_{L^2}^2 = \|\widehat{h}\|_{L^1} = h(0) = \int_{\mathbf{R}^n} f(x)\overline{\widetilde{f}(-x)}\,dx = \|f\|_{L^2}^2$.]

2.2.9. (a) Prove that for all $0 < \varepsilon < t < \infty$ we have

$$\left|\int_\varepsilon^t \frac{\sin(\xi)}{\xi}\,d\xi\right| \le 4.$$

(b) If f is an odd L^1 function on the line, conclude that for all $t > \varepsilon > 0$ we have

$$\left|\int_\varepsilon^t \frac{\widehat{f}(\xi)}{\xi}\,d\xi\right| \le 4\|f\|_{L^1}.$$

(c) Let $g(\xi)$ be a continuous odd function that is equal to $1/\log(\xi)$ for $\xi \ge 2$. Show that there does not exist an L^1 function whose Fourier transform is g.

2.2.10. Let f be in $L^1(\mathbf{R})$. Prove that

$$\int_{-\infty}^{+\infty} f(x)\,dx = \int_{-\infty}^{+\infty} f(x-1/x)\,dx.$$

2.2.11. (a) Use Exercise 2.2.10 with $f(x) = e^{-tx^2}$ to obtain the *subordination* identity

$$e^{-2t} = \frac{1}{\sqrt{\pi}} \int_0^\infty e^{-y-t^2/y} \frac{dy}{\sqrt{y}}, \qquad \text{where } t > 0.$$

(b) Set $t = \pi|x|$ and integrate with respect to $e^{-2\pi i \xi \cdot x} dx$ to prove that

$$(e^{-2\pi|x|})^\frown(\xi) = \frac{\Gamma(\frac{n+1}{2})}{\pi^{\frac{n+1}{2}}} \frac{1}{(1+|\xi|^2)^{\frac{n+1}{2}}}.$$

This calculation gives the Fourier transform of the Poisson kernel.

2.2.12. Let $1 \le p \le \infty$ and let p' be its dual index.
(a) Prove that Schwartz functions f on the line satisfy the estimate

$$\|f\|_{L^\infty}^2 \le 2\|f\|_{L^p}\|f'\|_{L^{p'}}.$$

(b) Prove that all Schwartz functions f on $\mathbf{R}^n$ satisfy the estimate

$$\|f\|_{L^\infty}^2 \le 2 \sum_{\alpha+\beta=(1,\dots,1)} \|\partial^\alpha f\|_{L^p} \|\partial^\beta f\|_{L^{p'}},$$

where the sum is taken over all multi-indices α and β whose sum is $(1,1,\dots,1)$.
[*Hint:* Part (a): Write $f(x)^2 = \int_{-\infty}^x \frac{d}{dt} f(t)^2 \, dt$.]

2.2.13. The *uncertainty principle* says that the position and the momentum of a particle cannot be simultaneously localized. Prove the following inequality, which presents a quantitative version of this principle:

$$\|f\|_{L^2(\mathbf{R}^n)}^2 \le \frac{4\pi}{n} \inf_{y \in \mathbf{R}^n} \left[\int_{\mathbf{R}^n} |x-y|^2 |f(x)|^2 \, dx \right]^{\frac{1}{2}} \inf_{z \in \mathbf{R}^n} \left[\int_{\mathbf{R}^n} |\xi - z|^2 |\widehat{f}(\xi)|^2 \, d\xi \right]^{\frac{1}{2}},$$

where f is a Schwartz function on $\mathbf{R}^n$ (or an L^2 function with sufficient decay at infinity).
[*Hint:* Let y be in $\mathbf{R}^n$. Start with

$$\|f\|_{L^2}^2 = \frac{1}{n} \int_{\mathbf{R}^n} f(x) \overline{f(x)} \sum_{j=1}^n \partial_j (x_j - y_j) \, dx,$$

integrate by parts, apply the Cauchy–Schwarz inequality, Plancherel's identity, and the identity $\sum_{j=1}^n |\widehat{\partial_j f}(\xi)|^2 = 4\pi^2 |\xi|^2 |\widehat{f}(\xi + z)|^2$ for all $\xi, z \in \mathbf{R}^n$.]

2.2.14. Let $-\infty < \alpha < \frac{n}{2} < \beta < +\infty$. Prove the validity of the following inequality:

$$\|g\|_{L^1(\mathbf{R}^n)} \le C \big\| |x|^\alpha g(x) \big\|_{L^2(\mathbf{R}^n)}^{\frac{\beta-n/2}{\beta-\alpha}} \big\| |x|^\beta g(x) \big\|_{L^2(\mathbf{R}^n)}^{\frac{n/2-\alpha}{\beta-\alpha}}$$

for some constant $C = C(n,\alpha,\beta)$ independent of g.
$\big[$*Hint:* First prove $\|g\|_{L^1} \le C\big\||x|^\alpha g(x)\big\|_{L^2} + \big\||x|^\beta g(x)\big\|_{L^2}$ and then replace $g(x)$ by $g(\lambda x)$ for some suitable $\lambda > 0$.$\big]$

2.3 The Class of Tempered Distributions

The fundamental idea of the theory of distributions is that it is generally easier to work with linear functionals acting on spaces of "nice" functions than to work with "bad" functions directly. The set of "nice" functions we consider is closed under the basic operations in analysis, and these operations are extended to distributions by duality. This wonderful interpretation has proved to be an indispensable tool that has clarified many situations in analysis.

2.3.1 Spaces of Test Functions

We recall the space $\mathscr{C}_0^\infty(\mathbf{R}^n)$ of all smooth functions with compact support, and $\mathscr{C}^\infty(\mathbf{R}^n)$ of all smooth functions on $\mathbf{R}^n$. We are mainly interested in the three spaces of "nice" functions on $\mathbf{R}^n$ that are nested as follows:

$$\mathscr{C}_0^\infty(\mathbf{R}^n) \subseteq \mathscr{S}(\mathbf{R}^n) \subseteq \mathscr{C}^\infty(\mathbf{R}^n).$$

Here $\mathscr{S}(\mathbf{R}^n)$ is the space of Schwartz functions introduced in Section 2.2.

Definition 2.3.1. We define convergence of sequences in these spaces. We say that

$$f_k \to f \text{ in } \mathscr{C}^\infty \iff f_k, f \in \mathscr{C}^\infty \text{ and } \lim_{k\to\infty} \sup_{|x|\le N} |\partial^\alpha (f_k - f)(x)| = 0$$
$$\forall\, \alpha \text{ multi-indices and all } N = 1,2,\dots.$$

$$f_k \to f \text{ in } \mathscr{S} \iff f_k, f \in \mathscr{S} \text{ and } \lim_{k\to\infty} \sup_{x\in\mathbf{R}^n} |x^\alpha \partial^\beta (f_k - f)(x)| = 0$$
$$\forall\, \alpha,\beta \text{ multi-indices.}$$

$$f_k \to f \text{ in } \mathscr{C}_0^\infty \iff f_k, f \in \mathscr{C}_0^\infty, \text{ support}(f_k) \subseteq B \text{ for all } k,\ B \text{ compact},$$
$$\text{and } \lim_{k\to\infty} \big\|\partial^\alpha (f_k - f)\big\|_{L^\infty} = 0\ \forall\, \alpha \text{ multi-indices.}$$

It follows that convergence in $\mathscr{C}_0^\infty(\mathbf{R}^n)$ implies convergence in $\mathscr{S}(\mathbf{R}^n)$, which in turn implies convergence in $\mathscr{C}^\infty(\mathbf{R}^n)$.

Example 2.3.2. Let φ be a nonzero $\mathscr{C}_0^\infty$ function on $\mathbf{R}$. We call such functions *smooth bumps*. Define the sequence of smooth bumps $\varphi_k(x) = \varphi(x-k)/k$. Then $\varphi_k(x)$ does not converge to zero in $\mathscr{C}_0^\infty(\mathbf{R})$, even though φ_k (and all of its derivatives) converge to zero uniformly. Furthermore, we see that φ_k does not converge to any function in $\mathscr{S}(\mathbf{R})$. Clearly $\varphi_k \to 0$ in $\mathscr{C}^\infty(\mathbf{R})$.

The space $\mathscr{C}^\infty(\mathbf{R}^n)$ is equipped with the family of seminorms

$$\widetilde{\rho}_{\alpha,N}(f) = \sup_{|x|\le N} |(\partial^\alpha f)(x)|, \qquad \alpha \text{ a multi-index}, N = 1,2,\dots. \tag{2.3.1}$$

It can be shown that $\mathscr{C}^\infty(\mathbf{R}^n)$ is complete with respect to this countable family of seminorms, i.e., it is a Fréchet space. However, it is true that $\mathscr{C}_0^\infty(\mathbf{R}^n)$ is not complete with respect to the topology generated by this family of seminorms.

The topology of $\mathscr{C}_0^\infty$ given in Definition 2.3.1 is the *inductive limit topology,* and under this topology it can be seen that $\mathscr{C}_0^\infty$ is complete. Indeed, $\mathscr{C}_0^\infty(\mathbf{R}^n)$ is a countable union of spaces $\bigcup_{k=1}^\infty \mathscr{C}_0^\infty(B(0,k))$ and each of these spaces is complete with respect to the topology generated by the family of seminorms $\widetilde{\rho}_{\alpha,N}$; hence so is $\mathscr{C}_0^\infty(\mathbf{R}^n)$. Nevertheless, $\mathscr{C}_0^\infty(\mathbf{R}^n)$ is not metrizable. We refer to Reed and Simon [215] for details on the topologies of these spaces.

2.3.2 *Spaces of Functionals on Test Functions*

The dual spaces (i.e., the spaces of continuous linear functionals on the sets of test functions) we introduced is denoted by

$$\begin{aligned}
(\mathscr{C}_0^\infty(\mathbf{R}^n))' &= \mathscr{D}'(\mathbf{R}^n),\\
(\mathscr{S}(\mathbf{R}^n))' &= \mathscr{S}'(\mathbf{R}^n),\\
(\mathscr{C}^\infty(\mathbf{R}^n))' &= \mathscr{E}'(\mathbf{R}^n).
\end{aligned}$$

By definition of the topologies on the dual spaces, we have

$$\begin{aligned}
T_k \to T \quad \text{in } \mathscr{D}' &\iff T_k, T \in \mathscr{D}' \text{ and } T_k(f) \to T(f) \text{ for all } f \in \mathscr{C}_0^\infty.\\
T_k \to T \quad \text{in } \mathscr{S}' &\iff T_k, T \in \mathscr{S}' \text{ and } T_k(f) \to T(f) \text{ for all } f \in \mathscr{S}.\\
T_k \to T \quad \text{in } \mathscr{E}' &\iff T_k, T \in \mathscr{E}' \text{ and } T_k(f) \to T(f) \text{ for all } f \in \mathscr{C}^\infty.
\end{aligned}$$

The dual spaces are nested as follows:

$$\mathscr{E}'(\mathbf{R}^n) \subseteq \mathscr{S}'(\mathbf{R}^n) \subseteq \mathscr{D}'(\mathbf{R}^n).$$

Definition 2.3.3. Elements of the space $\mathscr{D}'(\mathbf{R}^n)$ are called *distributions*. Elements of $\mathscr{S}'(\mathbf{R}^n)$ are called *tempered distributions*. Elements of the space $\mathscr{E}'(\mathbf{R}^n)$ are called *distributions with compact support.*

Before we discuss some examples, we give alternative characterizations of distributions, which are very useful from the practical point of view. The action of a distribution u on a test function f is represented in either one of the following two ways:

$$\langle u, f\rangle = u(f).$$

Proposition 2.3.4. *(a) A linear functional u on* $\mathscr{C}_0^\infty(\mathbf{R}^n)$ *is a distribution if and only if for every compact* $K \subseteq \mathbf{R}^n$*, there exist* $C > 0$ *and an integer m such that*

$$\left|\langle u, f\rangle\right| \le C \sum_{|\alpha|\le m} \|\partial^\alpha f\|_{L^\infty}, \quad \textit{for all } f \in \mathscr{C}^\infty \textit{ with support in } K. \tag{2.3.2}$$

(b) A linear functional u on $\mathscr{S}(\mathbf{R}^n)$ *is a tempered distribution if and only if there exist* $C > 0$ *and k, m integers such that*

$$\left|\langle u, f\rangle\right| \le C \sum_{\substack{|\alpha|\le m\\ |\beta|\le k}} \rho_{\alpha,\beta}(f), \qquad \textit{for all } f \in \mathscr{S}(\mathbf{R}^n). \tag{2.3.3}$$

(c) A linear functional u on $\mathscr{C}^\infty(\mathbf{R}^n)$ *is a distribution with compact support if and only if there exist* $C > 0$ *and N, m integers such that*

$$\left|\langle u, f\rangle\right| \le C \sum_{|\alpha|\le m} \widetilde{\rho}_{\alpha,N}(f), \qquad \textit{for all } f \in \mathscr{C}^\infty(\mathbf{R}^n), \tag{2.3.4}$$

where $\rho_{\alpha,\beta}$ *and* $\widetilde{\rho}_{\alpha,N}$ *are defined in (2.2.5) and (2.3.1).*

Proof. We prove only (2.3.3), since the proofs of (2.3.2) and (2.3.4) are similar. It is clear that (2.3.3) implies continuity of u. Conversely, it was pointed out in Section 2.2 that the family of sets $\{f \in \mathscr{S}(\mathbf{R}^n) : \rho_{\alpha,\beta}(f) < \delta\}$, where α, β are multi-indices and $\delta > 0$, forms a subbasis for the topology of $\mathscr{S}$. Thus if u is a continuous functional on $\mathscr{S}$, there exist integers k, m and a $\delta > 0$ such that

$$|\alpha| \le m,\ |\beta| \le k, \quad \text{and } \rho_{\alpha,\beta}(f) < \delta \implies \left|\langle u, f\rangle\right| \le 1. \tag{2.3.5}$$

We see that (2.3.3) follows from (2.3.5) with $C = 1/\delta$. □

Examples 2.3.5. We now discuss some important examples.

1. The *Dirac mass* at the origin δ_0. This is defined by

$$\langle \delta_0, f\rangle = f(0).$$

   We claim that δ_0 is in $\mathscr{E}'$. To see this we observe that if $f_k \to f$ in $\mathscr{C}^\infty$ then $\langle \delta_0, f_k\rangle \to \langle \delta_0, f\rangle$. The Dirac mass at a point $a \in \mathbf{R}^n$ is defined similarly by the equation

$$\langle \delta_a, f\rangle = f(a).$$

2. Some functions g can be thought of as distributions via the identification $g \mapsto L_g$, where L_g is the functional

$$L_g(f) = \int_{\mathbf{R}^n} f(x)g(x)\,dx.$$

Here are some examples: The function 1 is in $\mathscr{S}'$ but not in $\mathscr{E}'$. Compactly supported integrable functions are in $\mathscr{E}'$. The function $e^{|x|^2}$ is in $\mathscr{D}'$ but not in $\mathscr{S}'$.

3. Functions in L^1_{loc} are distributions. To see this, first observe that if $g \in L^1_{\mathrm{loc}}$, then the integral
$$L_g(f) = \int_{\mathbf{R}^n} f(x)g(x)\,dx$$
is well defined for all $f \in \mathscr{D}$, and then note that $f_k \to f$ in $\mathscr{D}$ implies that $L_g(f_k) \to L_g(f)$.
4. Functions in L^p, $1 \le p \le \infty$, are tempered distributions, but they are not in $\mathscr{E}'$ unless they have compact support.
5. Any finite Borel measure μ is a tempered distribution via the identification
$$L_\mu(f) = \int_{\mathbf{R}^n} f(x)\,d\mu(x).$$
To see this, observe that $f_k \to f$ in $\mathscr{S}$ implies that $L_\mu(f_k) \to L_\mu(f)$. Finite Borel measures may not be distributions with compact support. Lebesgue measure is also a tempered distribution.
6. Every function g that satisfies $|g(x)| \le C(1+|x|)^k$, for some real number k, is a tempered distribution. To see this, observe that
$$L_g(f) \le \sup_{x\in\mathbf{R}^n} (1+|x|)^m |f(x)| \int_{\mathbf{R}^n} (1+|x|)^{k-m} dx,$$
where $m > n+k$ and the expression $\sup_{x\in\mathbf{R}^n} |(1+|x|)^m f(x)|$ is bounded by a sum of $\rho_{\alpha,\beta}$ seminorms in the Schwartz space.
7. The function $\log|x|$ is a tempered distribution. The integral of this function against Schwartz functions is well defined. More generally, any function that is integrable in a neighborhood of the origin and satisfies $|g(x)| \le C(1+|x|)^k$ for $|x| \ge M$ is a tempered distribution.
8. Here is an example of a compactly supported distribution on $\mathbf{R}$ that is neither a function nor a measure:
$$\langle u, f\rangle = \lim_{\varepsilon\to 0} \int\limits_{\varepsilon\le|x|\le 1} f(x)\frac{dx}{x} = \lim_{\varepsilon\to 0} \int\limits_{\varepsilon\le|x|\le 1} (f(x)-f(0))\frac{dx}{x}.$$
We have that $|\langle u, f\rangle| \le 2\|f'\|_{L^\infty}$ and that if $f_n \to f$ in $\mathscr{C}^\infty$, then $\langle u, f_n\rangle \to \langle u, f\rangle$.

2.3.3 The Space of Tempered Distributions

Having set down the basic definitions of distributions, we now focus our study on the space of tempered distributions. These distributions are the most useful in harmonic analysis. The main reason for this is that the subject is concerned with boundedness of translation-invariant operators, and every such bounded operator from $L^p(\mathbf{R}^n)$ to $L^q(\mathbf{R}^n)$ is given by convolution with a tempered distribution. This fact is shown in Section 2.5.

Suppose that f and g are Schwartz functions and α a multi-index. Integrating by parts $|\alpha|$ times, we obtain

$$\int_{\mathbf{R}^n} (\partial^\alpha f)(x) g(x)\, dx = (-1)^{|\alpha|} \int_{\mathbf{R}^n} f(x) (\partial^\alpha g)(x)\, dx. \tag{2.3.6}$$

If we wanted to define the derivative of a tempered distribution u, we would have to give a definition that extends the definition of the derivative of the function and that satisfies (2.3.6) for g in $\mathscr{S}'$ and $f \in \mathscr{S}$ if the integrals in (2.3.6) are interpreted as actions of distributions on functions. We simply use equation (2.3.6) to define the derivative of a distribution.

Definition 2.3.6. Let $u \in \mathscr{S}'$ and α a multi-index. Define

$$\langle \partial^\alpha u, f \rangle = (-1)^{|\alpha|} \langle u, \partial^\alpha f \rangle . \tag{2.3.7}$$

If u is a function, the derivatives of u in the sense of distributions are called *distributional derivatives*.

In view of Theorem 2.2.14, it is natural to give the following:

Definition 2.3.7. Let $u \in \mathscr{S}'$. We define the Fourier transform $\widehat{u}$ and the inverse Fourier transform $u^\vee$ of a tempered distribution u by

$$\langle \widehat{u}, f \rangle = \langle u, \widehat{f} \rangle \qquad \text{and} \qquad \langle u^\vee, f \rangle = \langle u, f^\vee \rangle, \tag{2.3.8}$$

for all f in $\mathscr{S}$.

Example 2.3.8. We observe that $\widehat{\delta_0} = 1$. More generally, for any multi-index α we have

$$(\partial^\alpha \delta_0)\widehat{\ } = (2\pi i x)^\alpha .$$

To see this, observe that for all $f \in \mathscr{S}$ we have

$$\begin{aligned}
\langle (\partial^\alpha \delta_0)\widehat{\ }, f \rangle &= \langle \partial^\alpha \delta_0 , \widehat{f} \rangle \\
&= (-1)^{|\alpha|} \langle \delta_0 , \partial^\alpha \widehat{f} \rangle \\
&= (-1)^{|\alpha|} \langle \delta_0 , ((-2\pi i x)^\alpha f(x))\widehat{\ } \rangle \\
&= (-1)^{|\alpha|} ((-2\pi i x)^\alpha f(x))\widehat{\ }(0) \\
&= (-1)^{|\alpha|} \int_{\mathbf{R}^n} (-2\pi i x)^\alpha f(x)\, dx \\
&= \int_{\mathbf{R}^n} (2\pi i x)^\alpha f(x)\, dx .
\end{aligned}$$

This calculation indicates that $(\partial^\alpha \delta_0)\widehat{\ }$ can be identified with the function $(2\pi i x)^\alpha$.

Example 2.3.9. Recall that for $x_0 \in \mathbf{R}^n$, $\delta_{x_0}(f) = \langle \delta_{x_0}, f \rangle = f(x_0)$. Then

$$\langle \widehat{\delta_{x_0}}, h \rangle = \langle \delta_{x_0}, \widehat{h} \rangle = \widehat{h}(x_0) = \int_{\mathbf{R}^n} h(x) e^{-2\pi i x \cdot x_0}\, dx, \qquad h \in \mathscr{S}(\mathbf{R}^n),$$

that is, $\widehat{\delta_{x_0}}$ can be identified with the function $x \mapsto e^{-2\pi i x \cdot x_0}$. In particular, $\widehat{\delta_0} = 1$.

Example 2.3.10. The function $e^{|x|^2}$ is not in $\mathscr{S}'(\mathbf{R}^n)$ and therefore its Fourier transform is not defined as a distribution. However, the Fourier transform of any locally integrable function with polynomial growth at infinity is defined as a tempered distribution.

Now observe that the following are true whenever f, g are in $\mathscr{S}$.

$$\begin{aligned} \int_{\mathbf{R}^n} g(x) f(x-t)\, dx &= \int_{\mathbf{R}^n} g(x+t) f(x)\, dx, \\ \int_{\mathbf{R}^n} g(ax) f(x)\, dx &= \int_{\mathbf{R}^n} g(x) a^{-n} f(a^{-1}x)\, dx, \\ \int_{\mathbf{R}^n} \widetilde{g}(x) f(x)\, dx &= \int_{\mathbf{R}^n} g(x) \widetilde{f}(x)\, dx, \end{aligned} \tag{2.3.9}$$

for all $t \in \mathbf{R}^n$ and $a > 0$. Recall now the definitions of τ^t, δ^a, and $\widetilde{\ }$ given in (2.2.12). Motivated by (2.3.9), we give the following:

Definition 2.3.11. The *translation* $\tau^t(u)$, the *dilation* $\delta^a(u)$, and the *reflection* $\widetilde{u}$ of a tempered distribution u are defined as follows:

$$\langle \tau^t(u), f \rangle = \langle u, \tau^{-t}(f) \rangle, \tag{2.3.10}$$

$$\langle \delta^a(u), f \rangle = \langle u, a^{-n} \delta^{1/a}(f) \rangle, \tag{2.3.11}$$

$$\langle \widetilde{u}, f \rangle = \langle u, \widetilde{f} \rangle, \tag{2.3.12}$$

for all $t \in \mathbf{R}^n$ and $a > 0$. Let A be an invertible matrix. The composition of a distribution u with an invertible matrix A is the distribution

$$\langle u^A, \varphi \rangle = |\det A|^{-1} \langle u, \varphi^{A^{-1}} \rangle, \tag{2.3.13}$$

where $\varphi^{A^{-1}}(x) = \varphi(A^{-1}x)$.

It is easy to see that the operations of translation, dilation, reflection, and differentiation are continuous on tempered distributions.

Example 2.3.12. The Dirac mass at the origin δ_0 is equal to its reflection, while $\delta^a(\delta_0) = a^{-n}\delta_0$. Also, $\tau^x(\delta_0) = \delta_x$ for any $x \in \mathbf{R}^n$.

Now observe that for f, g, and h in $\mathscr{S}$ we have

$$\int_{\mathbf{R}^n} (h * g)(x) f(x)\, dx = \int_{\mathbf{R}^n} g(x) (\widetilde{h} * f)(x)\, dx. \tag{2.3.14}$$

Motivated by (2.3.14), we define the convolution of a function with a tempered distribution as follows:

Definition 2.3.13. Let $u \in \mathscr{S}'$ and $h \in \mathscr{S}$. Define the convolution $h * u$ by

$$\langle h * u, f \rangle = \langle u, \widetilde{h} * f \rangle, \qquad f \in \mathscr{S}. \tag{2.3.15}$$

Example 2.3.14. Let $u = \delta_{x_0}$ and $f \in \mathscr{S}$. Then $f * \delta_{x_0}$ is the function $x \mapsto f(x - x_0)$, for when $h \in \mathscr{S}$, we have

$$\langle f * \delta_{x_0}, h \rangle = \langle \delta_{x_0}, \widetilde{f} * h \rangle = (\widetilde{f} * h)(x_0) = \int_{\mathbf{R}^n} f(x - x_0) h(x)\, dx.$$

It follows that convolution with δ_0 is the identity operator.

We now define the product of a function and a distribution.

Definition 2.3.15. Let $u \in \mathscr{S}'$ and let h be a $\mathscr{C}^\infty$ function that has at most polynomial growth at infinity and the same is true for all of its derivatives. This means that it satisfies $|(\partial^\alpha h)(x)| \le C(1 + |x|)^{k_\alpha}$ for all α and some $k_\alpha > 0$. Then define the product hu of h and u by

$$\langle hu, f \rangle = \langle u, hf \rangle, \qquad f \in \mathscr{S}. \tag{2.3.16}$$

Note that hf is in $\mathscr{S}$ and thus (2.3.16) is well defined. The product of an arbitrary $\mathscr{C}^\infty$ function with a tempered distribution is not defined.

We observe that if a function g is supported in a set K, then for all $f \in \mathscr{C}_0^\infty(K^c)$ we have

$$\int_{\mathbf{R}^n} f(x) g(x)\, dx = 0. \tag{2.3.17}$$

Moreover, the support of g is the intersection of all closed sets K with the property (2.3.17) for all f in $\mathscr{C}_0^\infty(K^c)$. Motivated by the preceding observation we give the following:

Definition 2.3.16. Let u be in $\mathscr{D}'(\mathbf{R}^n)$. The *support* of u ($\operatorname{supp} u$) is the intersection of all closed sets K with the property

$$\varphi \in \mathscr{C}^\infty(\mathbf{R}^n), \qquad \operatorname{supp} \varphi \subseteq K^c \implies \langle u, \varphi \rangle = 0. \tag{2.3.18}$$

Distributions with compact support are exactly those whose support (as defined in the previous definition) is a compact set. To prove this assertion, we start with a distribution u with compact support as defined in Definition 2.3.3. Then there exist $C, N, m > 0$ such that (2.3.4) holds. For a smooth function f whose support is contained in $B(0,N)^c$, the expression on the right in (2.3.4) vanishes and we must therefore have $\langle u, f \rangle = 0$. This shows that the support of u is bounded, and since it is already closed (as an intersection of closed sets), it must be compact. Conversely, if the support of u as defined in Definition 2.3.16 is a compact set, then there exists an $N > 0$ such that $\operatorname{supp} u$ is contained in $B(0,N)$. We take a smooth function η that is equal to 1 on $B(0,N)$ and vanishes off $B(0,N+1)$. Then the support of $f(1-\eta)$ does not meet the support of u, and we must have

$$\langle u,f\rangle = \langle u,f\eta\rangle + \langle u,f(1-\eta)\rangle = \langle u,f\eta\rangle .$$

Taking m to be the integer that corresponds to the compact set $K=\overline{B(0,N+1)}$ in (2.3.2), and using that the L^∞ norm of $\partial^\alpha(f\eta)$ is controlled by a finite sum of seminorms $\widetilde{\rho}_{\alpha,N+1}(f)$ with $|\alpha|\le m$, we obtain the validity of (2.3.4).

Example 2.3.17. The support of Dirac mass at x_0 is the set $\{x_0\}$.

Along the same lines, we give the following definition:

Definition 2.3.18. We say that a distribution u in $\mathscr{D}'(\mathbf{R}^n)$ coincides with the function h on an open set Ω if

$$\langle u,f\rangle = \int_{\mathbf{R}^n} f(x)h(x)\,dx \qquad \text{for all } f \text{ in } \mathscr{C}_0^\infty(\Omega). \tag{2.3.19}$$

When (2.3.19) occurs we often say that u agrees with h away from Ω^c.

This definition implies that the support of the distribution $u-h$ is contained in the set Ω^c.

Example 2.3.19. The distribution $|x|^2+\delta_{a_1}+\delta_{a_2}$, where a_1, a_2 are in $\mathbf{R}^n$, coincides with the function $|x|^2$ on any open set not containing the points a_1 and a_2. Also, the distribution in Example 2.3.5 (8) coincides with the function $x^{-1}\chi_{|x|\le 1}$ away from the origin in the real line.

Having ended the streak of definitions regarding operations with distributions, we now discuss properties of convolutions and Fourier transforms.

Theorem 2.3.20. *If $u\in\mathscr{S}'$ and $\varphi\in\mathscr{S}$, then $\varphi * u$ is a $\mathscr{C}^\infty$ function. Moreover, for all multi-indices α there exist constants $C_\alpha, k_\alpha>0$ such that*

$$|\partial^\alpha(\varphi * u)(x)| \le C_\alpha(1+|x|)^{k_\alpha} .$$

Furthermore, if u has compact support, then $f*u$ is a Schwartz function.

Proof. Let ψ be in $\mathscr{S}(\mathbf{R}^n)$. We have

$$\begin{aligned}(\varphi * u)(\psi) = u(\widetilde{\varphi}*\psi) &= u\left(\int_{\mathbf{R}^n}\widetilde{\varphi}(\cdot - y)\psi(y)\,dy\right)\\ &= u\left(\int_{\mathbf{R}^n}\tau^y(\widetilde{\varphi})(\cdot)\psi(y)\,dy\right)\\ &= \int_{\mathbf{R}^n} u(\tau^y(\widetilde{\varphi}))\,\psi(y)\,dy,\end{aligned} \tag{2.3.20}$$

where the last step is justified by the continuity of u and by the fact that the Riemann sums of the integral in (2.3.20) converge to that integral in the topology of $\mathscr{S}$, a fact proved later. This calculation identifies the function $\varphi * u$ as

$$(\varphi * u)(x) = u\big(\tau^x(\widetilde{\varphi})\big). \tag{2.3.21}$$

We now show that $(\varphi * u)(x)$ is a $\mathscr{C}^\infty$ function. Let $e_j = (0,\dots,1,\dots,0)$ with 1 in the jth entry and zero elsewhere. Then

$$\frac{\tau^{-he_j}(\varphi * u)(x) - (\varphi * u)(x)}{h} = u\left(\frac{\tau^{-he_j}(\tau^x(\widetilde{\varphi})) - \tau^x(\widetilde{\varphi})}{h}\right) \to u(\tau^x(\partial_j\widetilde{\varphi}))$$

by the continuity of u and the fact that $\big(\tau^{-he_j}(\tau^x(\widetilde{\varphi})) - \tau^x(\widetilde{\varphi})\big)/h$ tends to $\tau^x(\partial_j\widetilde{\varphi})$ in $\mathscr{S}$ as $h \to 0$. See Exercise 2.3.5(a). The same calculation for higher-order derivatives shows that $\varphi * u \in \mathscr{C}^\infty$ and that $\partial^\gamma(\varphi * u) = (\partial^\gamma\varphi) * u$ for all multi-indices γ. It follows from (2.3.3) that for some C, m, and k we have

$$\begin{aligned}
|\partial^\alpha(\varphi * u)(x)| &\le C \sum_{\substack{|\gamma|\le m\\ |\beta|\le k}} \sup_{y\in\mathbf{R}^n} |y^\gamma \tau^x(\partial^{\alpha+\beta}\widetilde{\varphi})(y)| \\
&= C \sum_{\substack{|\gamma|\le m\\ |\beta|\le k}} \sup_{y\in\mathbf{R}^n} |(x+y)^\gamma (\partial^{\alpha+\beta}\widetilde{\varphi})(y)| \\
&\le C_m \sum_{|\beta|\le k} \sup_{y\in\mathbf{R}^n} (|x|^m + |y|^m)|(\partial^{\alpha+\beta}\widetilde{\varphi})(y)|,
\end{aligned} \tag{2.3.22}$$

and this clearly implies that $\partial^\alpha(\varphi * u)$ grows at most polynomially at infinity.

We now indicate why $\varphi * u$ is Schwartz whenever u has compact support. Applying estimate (2.3.4) to the function $y \mapsto \varphi(x-y)$ yields that

$$\big|\langle u, \varphi(x-\cdot)\rangle\big| = |(\varphi * u)(x)| \le C \sum_{|\alpha|\le m} \sup_{|y|\le N} |\partial_y^\alpha \varphi(x-y)|$$

for some constants C, m, N. Since

$$|\partial_y^\alpha \varphi(x-y)| \le C_{\alpha,M}(1+|x-y|)^{-M} \le C_{\alpha,M,N}(1+|x|)^{-M}$$

for $|x| \ge 2N$, it follows that $\varphi * u$ decays rapidly at infinity. Since $\partial^\gamma(\varphi * u) = (\partial^\gamma\varphi) * u$, the same argument yields that all the derivatives of $\varphi * u$ decay rapidly at infinity; hence $\varphi * u$ is a Schwartz function. Incidentally, this argument actually shows that any Schwartz seminorm of $\varphi * u$ is controlled by a finite sum of Schwartz seminorms of φ.

We now return to the point left open concerning the convergence of the Riemann sums in (2.3.20) in the topology of $\mathscr{S}(\mathbf{R}^n)$. For each $N = 1, 2, \dots$, consider a partition of $[-N,N]^n$ into $(2N^2)^n$ cubes Q_m of side length $1/N$ and let y_m be the center of each Q_m. For multi-indices α, β, we must show that

$$D_N(x) = \sum_{m=1}^{(2N^2)^n} x^\alpha \partial_x^\beta \widetilde{\varphi}(x-y_m)\psi(y_m)|Q_m| - \int_{\mathbf{R}^n} x^\alpha \partial_x^\beta \widetilde{\varphi}(x-y)\psi(y)\,dy$$

converges to zero in $L^\infty(\mathbf{R}^n)$ as $N \to \infty$. We have

$$\begin{aligned} x^\alpha \partial_x^\beta \widetilde{\varphi}(x-y_m)\psi(y_m)|Q_m| &- \int_{Q_m} x^\alpha \partial_x^\beta \widetilde{\varphi}(x-y)\psi(y)\,dy \\ &= \int_{Q_m} x^\alpha (y-y_m)\cdot \nabla_y\{\partial_x^\beta \widetilde{\varphi}(x-y)\psi(y)\}(\xi)\,dy \end{aligned}$$

for some $\xi = y+\theta(y_m-y)$, where $\theta \in [0,1]$. It follows that $|y| \le |\xi| + \sqrt{n}/N \le |\xi|+1$ for $N \ge \sqrt{n}$. It is easy to see that the last integrand is at most

$$C\,|x|^{|\alpha|}\,\frac{\sqrt{n}}{N}\,\frac{1}{(1+|x-\xi|)^M}\,\frac{1}{(2+|\xi|)^M}$$

for M large (pick $M > 2|\alpha|$), which in turn is at most

$$C'\,|x|^{|\alpha|}\,\frac{\sqrt{n}}{N}\,\frac{1}{(1+|x|)^{M/2}}\,\frac{1}{(2+|\xi|)^{M/2}} \le C'\,|x|^{|\alpha|}\,\frac{\sqrt{n}}{N}\,\frac{1}{(1+|x|)^{M/2}}\,\frac{1}{(1+|y|)^{M/2}}\,.$$

Inserting this estimate for the integrand in the last displayed integral, we obtain

$$|D_N(x)| \le \frac{C''}{N}\,\frac{|x|^{|\alpha|}}{(1+|x|)^{M/2}} \int_{[-N,N]^n} \frac{dy}{(1+|y|)^{M/2}} + \int_{([-N,N]^n)^c} |x^\alpha \partial_x^\beta \widetilde{\varphi}(x-y)\psi(y)|\,dy\,.$$

But the integrand in the last integral is controlled by

$$\frac{C'''|x|^{|\alpha|}}{(1+|x-y|)^M}\,\frac{dy}{(1+|y|)^M} \le \frac{C'''|x|^{|\alpha|}}{(1+|x|)^{M/2}}\,\frac{dy}{(1+|y|)^{M/2}}\,.$$

Using these estimates it is now easy to see that $\lim_{N\to\infty}\sup_{x\in\mathbf{R}^n}|D_N(x)| = 0$. □

Next we have the following important result regarding distributions with compact support:

Theorem 2.3.21. *If u is in $\mathscr{E}'(\mathbf{R}^n)$, then $\widehat{u}$ is a real analytic function on $\mathbf{R}^n$. Moreover, $\widehat{u}$ has a holomorphic extension on $\mathbf{C}^n$. In particular, $\widehat{u}$ is a $\mathscr{C}^\infty$ function. Furthermore, $\widehat{u}$ and all of its derivatives have polynomial growth at infinity.*

Proof. Given a distribution u with compact support and a polynomial $p(\xi)$, the action of u on the $\mathscr{C}^\infty$ function $\xi \mapsto p(\xi)e^{-2\pi i x\cdot\xi}$ is a well defined function of x, which we denote by $u(p(\cdot)e^{-2\pi i x\cdot(\cdot)})$. Here x is an element of $\mathbf{R}^n$ or $\mathbf{C}^n$.

It is straightforward to verify that the function of $z = (z_1,\dots,z_n)$

$$F(z) = u\big(e^{-2\pi i(\cdot)\cdot z}\big)$$

defined on $\mathbf{C}^n$ is holomorphic, in fact entire. Indeed, the continuity and linearity of u and the fact that $(e^{-2\pi i\xi_j h}-1)/h \to -2\pi i\xi_j$ in $\mathscr{C}^\infty(\mathbf{R}^n)$ as $h\to 0$ in the complex plane imply that F is differentiable and its derivative with respect to z_j is the action of the distribution u to the $\mathscr{C}^\infty$ function

$$\xi \mapsto (-2\pi i\xi_j)e^{-2\pi i\sum_{j=1}^n \xi_j z_j}.$$

By induction it follows that for all multi-indices α we have

$$\partial_{z_1}^{\alpha_1}\cdots\partial_{z_n}^{\alpha_n}F = u\big((-2\pi i(\cdot))^\alpha e^{-2\pi i\sum_{j=1}^n(\cdot)z_j}\big).$$

Since F is entire, its restriction on $\mathbf{R}^n$, i.e., $F(x_1,\dots,x_n)$, where $x_j=\operatorname{Re} z_j$, is real analytic. Also, as a consequence of (2.3.4), this restriction and all of its derivatives have polynomial growth at infinity.

Now for f in $\mathscr{S}(\mathbf{R}^n)$ we have

$$\langle\widehat{u},f\rangle = \langle u,\widehat{f}\rangle = u\left(\int_{\mathbf{R}^n} f(x)e^{-2\pi i x\cdot\xi}\,dx\right) = \int_{\mathbf{R}^n} f(x)u(e^{-2\pi i x\cdot(\cdot)})\,dx,$$

provided we can justify the passage of u inside the integral. The reason for this is that the Riemann sums of the integral of $f(x)e^{-2\pi i x\cdot\xi}$ over $\mathbf{R}^n$ converge to it in the topology of $\mathscr{C}^\infty$, and thus the linear functional u can be interchanged with the integral. We conclude that the tempered distribution $\widehat{u}$ can be identified with the real analytic function $x\mapsto F(x)$ whose derivatives have polynomial growth at infinity.

To justify the fact concerning the convergence of the Riemann sums, we argue as in the proof of the previous theorem. For each $N=1,2,\dots$, consider a partition of $[-N,N]^n$ into $(2N^2)^n$ cubes Q_m of side length $1/N$ and let y_m be the center of each Q_m. For a multi-index α let

$$D_N(\xi) = \sum_{m=1}^{(2N^2)^n} f(y_m)(-2\pi i y_m)^\alpha e^{-2\pi i y_m\cdot\xi}|Q_m| - \int_{\mathbf{R}^n} f(x)(-2\pi i x)^\alpha e^{-2\pi x\cdot\xi}\,dx.$$

We must show that for every $M>0$, $\sup_{|\xi|\le M}|D_N(\xi)|$ converges to zero as $N\to\infty$. Setting $g(x)=f(x)(-2\pi i x)^\alpha$, we write

$$D_N(\xi) = \sum_{m=1}^{(2N^2)^n}\int_{Q_m}\big[g(y_m)e^{-2\pi i y_m\cdot\xi}-g(x)e^{-2\pi i x\cdot\xi}\big]\,dx + \int_{([-N,N]^n)^c} g(x)e^{-2\pi x\cdot\xi}\,dx.$$

Using the mean value theorem, we bound the absolute value of the expression inside the square brackets by

$$\big(|\nabla g(z_m)|+2\pi|\xi|\,|g(z_m)|\big)\frac{\sqrt{n}}{N} \le \frac{C_K(1+|\xi|)}{(1+|z_m|)^K}\frac{\sqrt{n}}{N},$$

for some point z_m in the cube Q_m. Since

$$\sum_{m=1}^{(2N^2)^n} \int_{Q_m} \frac{C_K(1+|\xi|)}{(1+|z_m|)^K} \le C'_K(1+M) < \infty$$

for $|\xi| \le M$, it follows that $\sup_{|\xi|\le M} |D_N(\xi)| \to 0$ as $N \to \infty$. □

Next we give a proposition that extends the properties of the Fourier transform to tempered distributions.

Proposition 2.3.22. *Given u, v in $\mathscr{S}'(\mathbf{R}^n)$, $f \in \mathscr{S}$, $y \in \mathbf{R}^n$, b a complex scalar, α a multi-index, and $a > 0$, we have*

(1) $\widehat{u+v} = \widehat{u} + \widehat{v}$,

(2) $\widehat{bu} = b\widehat{u}$,

(3) *If* $u_j \to u$ *in* $\mathscr{S}'$, *then* $\widehat{u}_j \to \widehat{u}$ *in* $\mathscr{S}'$,

(4) $(\widetilde{u})\widehat{\ } = (\widehat{u})\widetilde{\ }$,

(5) $(\tau^y(u))\widehat{\ } = e^{-2\pi i y\cdot\xi}\widehat{u}$,

(6) $(e^{2\pi i x\cdot y}u)\widehat{\ } = \tau^y(\widehat{u})$,

(7) $(\delta^a(u))\widehat{\ } = (\widehat{u})_a = a^{-n}(\delta^{a^{-1}}(\widehat{u}))$,

(8) $(\partial^\alpha u)\widehat{\ } = (2\pi i\xi)^\alpha \widehat{u}$,

(9) $\partial^\alpha \widehat{u} = ((-2\pi i x)^\alpha u)\widehat{\ }$,

(10) $(\widehat{u})^\vee = u$,

(11) $\widehat{f*u} = \widehat{f}\,\widehat{u}$,

(12) $\widehat{fu} = \widehat{f} * \widehat{u}$,

(13) ***(Leibniz's rule)*** $\partial_j^m(fu) = \sum_{k=0}^m \binom{m}{k}(\partial_j^k f)(\partial_j^{m-k}u)$, $m \in \mathbf{Z}^+$,

(14) ***(Leibniz's rule)*** $\partial^\alpha(fu) = \sum_{\gamma_1=0}^{\alpha_1}\cdots\sum_{\gamma_n=0}^{\alpha_n}\binom{\alpha_1}{\gamma_1}\cdots\binom{\alpha_n}{\gamma_n}(\partial^\gamma f)(\partial^{\alpha-\gamma}u)$,

(15) *If* u_k, $u \in L^p(\mathbf{R}^n)$ *and* $u_k \to u$ *in* L^p $(1 \le p \le \infty)$, *then* $u_k \to u$ *in* $\mathscr{S}'(\mathbf{R}^n)$. *Therefore, convergence in* $\mathscr{S}$ *implies convergence in* L^p, *which in turn implies convergence in* $\mathscr{S}'(\mathbf{R}^n)$.

Proof. All the statements can be proved easily using duality and the corresponding statements for Schwartz functions. □

We continue with an application of Theorem 2.3.21.

Proposition 2.3.23. *Given* $u \in \mathscr{S}'(\mathbf{R}^n)$, *there exists a sequence of* $\mathscr{C}_0^\infty$ *functions* f_k *such that* $f_k \to u$ *in the sense of tempered distributions; in particular,* $\mathscr{C}_0^\infty(\mathbf{R}^n)$ *is dense in* $\mathscr{S}'(\mathbf{R}^n)$.

Proof. Fix a function in $\mathscr{C}_0^\infty(\mathbf{R}^n)$ with $\varphi(x) = 1$ in a neighborhood of the origin. Let $\varphi_k(x) = \delta^{1/k}(\varphi)(x) = \varphi(x/k)$. It follows from Exercise 2.3.5(b) that for $u \in \mathscr{S}'(\mathbf{R}^n)$, $\varphi_k u \to u$ in $\mathscr{S}'$. By Proposition 2.3.22 (3), we have that the map $u \mapsto (\varphi_k \widehat{u})^\vee$ is continuous on $\mathscr{S}'(\mathbf{R}^n)$. Now Theorem 2.3.21 gives that $(\varphi_k \widehat{u})^\vee$ is a $\mathscr{C}^\infty$ function and therefore $\varphi_j(\varphi_k \widehat{u})^\vee$ is in $\mathscr{C}_0^\infty(\mathbf{R}^n)$. As observed, $\varphi_j(\varphi_k \widehat{u})^\vee \to (\varphi_k \widehat{u})^\vee$ in $\mathscr{S}'$ when k is fixed and $j \to \infty$. Exercise 2.3.5(c) gives that the diagonal sequence $\varphi_k(\varphi_k f)\widehat{\ }$ converges to $f\widehat{\ }$ in $\mathscr{S}$ as $k \to \infty$ for all $f \in \mathscr{S}$. Using duality and Exercise 2.2.2, we conclude that the sequence of $\mathscr{C}_0^\infty$ functions $\varphi_k(\varphi_k \widehat{u})^\vee$ converges to u in $\mathscr{S}'$ as $k \to \infty$. □

2.3.4 The Space of Tempered Distributions Modulo Polynomials

Definition 2.3.24. We define $\mathscr{P}$ to be set of all polynomials of n real variables,

$$\sum_{|\beta| \le m} c_\beta x^\beta = \sum_{\substack{\beta_j \in \mathbf{Z}^+ \cup \{0\} \\ \beta_1 + \cdots + \beta_n \le m}} c_{(\beta_1, \dots, \beta_n)} x_1^{\beta_1} \cdots x_n^{\beta_n},$$

with complex coefficients c_β and m an arbitrary integer. We then define an equivalence relation $\sim$ on $\mathscr{S}'(\mathbf{R}^n)$ by setting

$$u \sim v \iff u - v \in \mathscr{P}.$$

The space of all resulting equivalence classes is denoted by $\mathscr{S}'/\mathscr{P}$.

To avoid cumbersome notation, two elements u, v of the same equivalence class in $\mathscr{S}'/\mathscr{P}$ are identified, and in this case we write $u = v$ in $\mathscr{S}'/\mathscr{P}$. Note that for $u, v \in \mathscr{S}'/\mathscr{P}$ we have

$$u = v \quad \text{in } \mathscr{S}'/\mathscr{P} \iff \begin{cases} \langle \widehat{u}, \phi \rangle = \langle \widehat{v}, \phi \rangle & \text{for all } \phi \in \mathscr{S}(\mathbf{R}^n) \\ \text{with support contained in } \mathbf{R}^n \setminus \{0\}. \end{cases} \tag{2.3.23}$$

Proposition 2.3.25. *Let $\mathscr{S}_\infty(\mathbf{R}^n)$ be the space of all Schwartz functions φ that satisfy*

$$\int_{\mathbf{R}^n} x^\gamma \varphi(x)\, dx = 0$$

for all multi-indices γ. Then $\mathscr{S}_\infty(\mathbf{R}^n)$ is a subspace of $\mathscr{S}(\mathbf{R}^n)$ that inherits the same topology as $\mathscr{S}(\mathbf{R}^n)$ and whose dual is $\mathscr{S}'(\mathbf{R}^n)/\mathscr{P}$, that is,

$$(\mathscr{S}_\infty(\mathbf{R}^n))' = \mathscr{S}'(\mathbf{R}^n)/\mathscr{P}.$$

Proof. Consider the map J that takes an element u of $\mathscr{S}'(\mathbf{R}^n)$ to the equivalence class in $\mathscr{S}'(\mathbf{R}^n)/\mathscr{P}$ that contains it. The kernel of this map is $\mathscr{P}$ and the claimed identification follows. □

We write $u_j \to u$ in $\mathscr{S}'(\mathbf{R}^n)/\mathscr{P}$ if and only if u_j, u are elements of $\mathscr{S}'(\mathbf{R}^n)/\mathscr{P}$ and we have

$$\langle u_j, \varphi \rangle \to \langle u, \varphi \rangle$$

as $j \to \infty$ for all φ in $\mathscr{S}_\infty(\mathbf{R}^n)$.

Exercises

2.3.1. Show that a positive measure μ that satisfies

$$\int_{\mathbf{R}^n} \frac{d\mu(x)}{(1+|x|)^k} < +\infty,$$

for some $k > 0$, can be identified with a tempered distribution. Show that if we think of Lebesgue measure as a tempered distribution, then it coincides with the constant function 1 also interpreted as a tempered distribution.

2.3.2. Let $\varphi, f \in \mathscr{S}(\mathbf{R}^n)$, and for $\varepsilon > 0$ let $\varphi_\varepsilon(x) = \varepsilon^{-n}\varphi(\varepsilon^{-1}x)$. Prove that $\varphi_\varepsilon * f \to bf$ in $\mathscr{S}$, where b is the integral of φ.

2.3.3. Prove that for all $a > 0$, $u \in \mathscr{S}'(\mathbf{R}^n)$, and $f \in \mathscr{S}(\mathbf{R}^n)$ we have

$$\delta^a(f) * \delta^a(u) = a^{-n}\delta^a(f * u).$$

2.3.4. (a) Prove that the derivative of $\chi_{[a,b]}$ is $\delta_a - \delta_b$.

(b) Compute $\partial_j \chi_{B(0,1)}$ on $\mathbf{R}^2$.

(c) Compute the Fourier transforms of the locally integrable functions $\sin x$ and $\cos x$.

(d) Prove that the derivative of the distribution $\log|x| \in \mathscr{S}'(\mathbf{R})$ is the distribution

$$u(\varphi) = \lim_{\varepsilon\to 0} \int_{\varepsilon \le |x|} \varphi(x) \frac{dx}{x}.$$

2.3.5. Let $f \in \mathscr{S}(\mathbf{R}^n)$ and let $\varphi \in \mathscr{C}_0^\infty$ be identically equal to 1 in a neighborhood of origin. Define $\varphi_k(x) = \varphi(x/k)$ as in the proof of Proposition 2.3.23.
(a) Prove that $(\tau^{-he_j}(f) - f)/h \to \partial_j f$ in $\mathscr{S}$ as $h \to 0$.
(b) Prove that $\varphi_k f \to f$ in $\mathscr{S}$ as $k \to \infty$.
(c) Prove that the sequence $\varphi_k(\varphi_k f)\widehat{\ }$ converges to $\widehat{f}$ in $\mathscr{S}$ as $k \to \infty$.

2.3.6. Use Theorem 2.3.21 to show that there does not exist a nonzero $\mathscr{C}_0^\infty$ function whose Fourier transform is also a $\mathscr{C}_0^\infty$ function.

2.3.7. Let $f \in L^p(\mathbf{R}^n)$ for some $1 \le p \le \infty$. Show that the sequence of functions

$$g_N(\xi) = \int_{B(0,N)} f(x) e^{-2\pi i x\cdot\xi}\, dx$$

converges to $\widehat{f}$ in $\mathscr{S}'$.

2.3.8. Let $(c_k)_{k\in\mathbf{Z}^n}$ be a sequence that satisfies $|c_k| \le A(1+|k|)^M$ for all k and some fixed M and $A>0$. Let δ_k denote Dirac mass at the integer k. Show that the sequence of distributions

$$\sum_{|k|\le N} c_k\delta_k$$

converges to some tempered distribution u in $\mathscr{S}'(\mathbf{R}^n)$ as $N\to\infty$. Also show that $\widehat{u}$ is the $\mathscr{S}'$ limit of the sequence of functions

$$h_N(\xi) = \sum_{|k|\le N} c_k e^{-2\pi i\xi\cdot k}.$$

2.3.9. A distribution in $\mathscr{S}'(\mathbf{R}^n)$ is called *homogeneous of degree* $\gamma\in\mathbf{C}$ if

$$\langle u, \delta^\lambda(f)\rangle = \lambda^{-n-\gamma}\langle u, f\rangle, \qquad \text{for all } \lambda>0.$$

(a) Prove that this definition agrees with the usual definition for functions.
(b) Show that δ_0 is homogeneous of degree $-n$.
(c) Prove that if u is homogeneous of degree γ, then $\partial^\alpha u$ is homogeneous of degree $\gamma-|\alpha|$.
(d) Show that u is homogeneous of degree γ if and only if $\widehat{u}$ is homogeneous of degree $-n-\gamma$.

2.3.10. Show that the functions e^{inx} and e^{-inx} converge to zero in $\mathscr{S}'$ and $\mathscr{D}'$ as $n\to\infty$. Conclude that multiplication of distributions is not a continuous operation even when it is defined. What is the limit of $\sqrt{n}(1+n|x|^2)^{-1}$ in $\mathscr{D}'(\mathbf{R})$ as $n\to\infty$?

2.3.11. (*S. Bernstein*) Let f be a bounded function on $\mathbf{R}^n$ with $\widehat{f}$ supported in the ball $B(0,R)$. Prove that for all multi-indices α there exist constants $C_{\alpha,n}$ (depending only on α and on the dimension n) such that

$$\big\|\partial^\alpha f\big\|_{L^\infty} \le C_{\alpha,n}R^{|\alpha|}\big\|f\big\|_{L^\infty}.$$

$\big[$*Hint:* Write $f = f * h_{1/R}$, where h is a Schwartz function h in $\mathbf{R}^n$ whose Fourier transform is equal to one on the ball $B(0,1)$ and vanishes outside the ball $B(0,2)$.$\big]$

2.3.12. Let $\widehat{\Phi}$ be a $\mathscr{C}_0^\infty$ function that is equal to 1 in $B(0,1)$ and let $\widehat{\Theta}$ be a $\mathscr{C}^\infty$ function that is equal to 1 in a neighborhood of infinity and vanishes in a neighborhood of zero. Prove the following.
(a) For all u in $\mathscr{S}'(\mathbf{R}^n)$ we have

$$\left(\widehat{\Phi}(\xi/2^N)\widehat{u}\right)^\vee \to u \quad \text{in } \mathscr{S}'(\mathbf{R}^n) \text{ as } N\to\infty.$$

(b) For all u in $\mathscr{S}'(\mathbf{R}^n)$ we have

$$\left(\widehat{\Theta}(\xi/2^N)\widehat{u}\right)^\vee \to 0 \quad \text{in } \mathscr{S}'(\mathbf{R}^n) \text{ as } N\to\infty.$$

(c) The convergence in part (b) also holds in the topology of $\mathscr{S}'(\mathbf{R}^n)/\mathscr{P}$.
[*Hint:* Prove first the corresponding assertions for functions φ in $\mathscr{S}$ or $\mathscr{S}_\infty$ with convergence in the topology of these spaces.]

2.3.13. Prove that there exists a function in L^p for $2 < p < \infty$ whose distributional Fourier transform is not a locally integrable function.
[*Hint:* Assume the converse. Then for all $f \in L^p(\mathbf{R}^n)$, $\widehat{f}$ is locally integrable and hence the map $f \mapsto \widehat{f}$ is a well defined linear operator from $L^p(\mathbf{R}^n)$ to $L^1(B(0,1))$ (i.e. $\|\widehat{f}\|_{L^1(B(0,1))} < \infty$ for all $f \in L^p(\mathbf{R}^n)$). Use the closed graph theorem to deduce that $\|\widehat{f}\|_{L^1(B(0,1))} \le C\|f\|_{L^p(\mathbf{R}^n)}$ for some $C < \infty$. To violate this inequality whenever $p > 2$, take $f_N(x) = (1+iN)^{-n/2}e^{-\pi(1+iN)^{-1}|x|^2}$ and let $N \to \infty$, noting that $\widehat{f_N}(\xi) = e^{-\pi|\xi|^2(1+iN)}$.]

2.4 More About Distributions and the Fourier Transform

In this section we discuss further properties of distributions and Fourier transforms and bring up certain connections that arise between harmonic analysis and partial differential equations.

2.4.1 Distributions Supported at a Point

We begin with the following characterization of distributions supported at a single point.

Proposition 2.4.1. *If $u \in \mathscr{S}'(\mathbf{R}^n)$ is supported in the singleton $\{x_0\}$, then there exists an integer k and complex numbers a_α such that*

$$u = \sum_{|\alpha| \le k} a_\alpha \partial^\alpha \delta_{x_0}.$$

Proof. Without loss of generality we may assume that $x_0 = 0$. By (2.3.3) we have that for some C, m, and k,

$$\left|\langle u, f\rangle\right| \le C \sum_{\substack{|\alpha|\le m \\ |\beta|\le k}} \sup_{x\in\mathbf{R}^n} |x^\alpha(\partial^\beta f)(x)| \qquad \text{for all } f \in \mathscr{S}(\mathbf{R}^n).$$

We now prove that if $\varphi \in \mathscr{S}$ satisfies

$$(\partial^\alpha \varphi)(0) = 0 \qquad \text{for all } |\alpha| \le k, \tag{2.4.1}$$

then $\langle u, \varphi\rangle = 0$. To see this, fix a φ satisfying (2.4.1) and let $\zeta(x)$ be a smooth function on $\mathbf{R}^n$ that is equal to 1 when $|x| \ge 2$ and equal to zero for $|x| \le 1$. Let

$\zeta^\varepsilon(x) = \zeta(x/\varepsilon)$. Then, using (2.4.1) and the continuity of the derivatives of φ at the origin, it is not hard to show that $\rho_{\alpha,\beta}(\zeta^\varepsilon\varphi - \varphi) \to 0$ as $\varepsilon \to 0$ for all $|\alpha| \le m$ and $|\beta| \le k$. Then

$$|\langle u, \varphi\rangle| \le |\langle u, \zeta^\varepsilon\varphi\rangle| + |\langle u, \zeta^\varepsilon\varphi - \varphi\rangle| \le 0 + C \sum_{\substack{|\alpha|\le m \\ |\beta|\le k}} \rho_{\alpha,\beta}(\zeta^\varepsilon\varphi - \varphi) \to 0$$

as $\varepsilon \to 0$. This proves our assertion.

Now let $f \in \mathscr{S}(\mathbf{R}^n)$. Let η be a $\mathscr{C}_0^\infty$ function on $\mathbf{R}^n$ that is equal to 1 in a neighborhood of the origin. Write

$$f(x) = \eta(x)\Big(\sum_{|\alpha|\le k} \frac{(\partial^\alpha f)(0)}{\alpha!} x^\alpha + h(x)\Big) + (1-\eta(x))f(x), \tag{2.4.2}$$

where $h(x) = O(x^{k+1})$ as $|x| \to 0$. Then ηh satisfies (2.4.1) and hence $\langle u, \eta h\rangle = 0$ by the claim. Also,

$$\langle u, ((1-\eta)f)\rangle = 0$$

by our hypothesis. Applying u to both sides of (2.4.2), we obtain

$$\langle u, f\rangle = \sum_{|\alpha|\le k} \frac{(\partial^\alpha f)(0)}{\alpha!} u(x^\alpha \eta(x)) = \sum_{|\alpha|\le k} a_\alpha (\partial^\alpha \delta_0)(f),$$

with $a_\alpha = (-1)^{|\alpha|} u(x^\alpha \eta(x))/\alpha!$. This proves the proposition. □

An immediate consequence is thc following result.

Corollary 2.4.2. *Let $u \in \mathscr{S}'(\mathbf{R}^n)$. If $\widehat{u}$ is supported in the singleton $\{\xi_0\}$, then u is a finite linear combination of functions $(-2\pi i\xi)^\alpha e^{2\pi i\xi\cdot\xi_0}$, where α is a multi-index. In particular, if $\widehat{u}$ is supported at the origin, then u is a polynomial.*

Proof. Proposition 2.4.1 gives that $\widehat{u}$ is a linear combination of derivatives of Dirac masses at ξ_0. Then Proposition 2.3.22 (8) yields the required conclusion. □

2.4.2 The Laplacian

The *Laplacian* Δ is a partial differential operator acting on tempered distributions on $\mathbf{R}^n$ as follows:

$$\Delta(u) = \sum_{j=1}^n \partial_j^2 u.$$

Solutions of Laplace's equation $\Delta(u) = 0$ are called *harmonic* distributions. We have the following:

Corollary 2.4.3. *Let $u \in \mathscr{S}'(\mathbf{R}^n)$ satisfy $\Delta(u) = 0$. Then u is a polynomial.*

Proof. Taking Fourier transforms, we obtain that $\widehat{\Delta(u)} = 0$. Therefore,

$$-4\pi^2|\xi|^2\widehat{u} = 0 \qquad \text{in } \mathscr{S}'.$$

This implies that $\widehat{u}$ is supported at the origin, and by Corollary 2.4.2 it follows that u must be polynomial. □

Liouville's classical theorem that every bounded harmonic function must be constant is a consequence of Corollary 2.4.3. See Exercise 2.4.2.

Next we would like to compute the fundamental solutions of Laplace's equation in $\mathbf{R}^n$. A distribution is called a *fundamental solution* of a partial differential operator L if we have $L(u) = \delta_0$. The following result gives the fundamental solution of the Laplacian.

Proposition 2.4.4. *For $n \geq 3$ we have*

$$\Delta(|x|^{2-n}) = -(n-2)\frac{2\pi^{n/2}}{\Gamma(n/2)}\delta_0\,, \tag{2.4.3}$$

while for $n = 2$,

$$\Delta(\log|x|) = 2\pi\delta_0\,. \tag{2.4.4}$$

Proof. We use Green's identity

$$\int_\Omega v\Delta(u) - u\Delta(v)\,dx = \int_{\partial\Omega}\left(v\frac{\partial u}{\partial\nu} - u\frac{\partial v}{\partial\nu}\right)ds\,,$$

where Ω is an open set in $\mathbf{R}^n$ with smooth boundary and $\partial v/\partial\nu$ denotes the derivative of v with respect to the outer unit normal vector. Take $\Omega = \mathbf{R}^n \setminus \overline{B(0,\varepsilon)}$, $v = |x|^{2-n}$, and $u = f$ a $\mathscr{C}_0^\infty(\mathbf{R}^n)$ function in the previous identity. The normal derivative of $f(r\theta)$ is the derivative with respect to the radial variable r. Observe that $\Delta(|x|^{2-n}) = 0$ for $x \neq 0$. We obtain

$$\int_{|x|>\varepsilon}\Delta(f)(x)|x|^{2-n}\,dx = -\int\limits_{|r\theta|=\varepsilon}\left(\varepsilon^{2-n}\frac{\partial f}{\partial r} - f(r\theta)\frac{\partial r^{2-n}}{\partial r}\right)d\theta. \tag{2.4.5}$$

Now observe two things: first, that for some $C = C(f)$ we have

$$\left|\int_{|r\theta|=\varepsilon}\frac{\partial f}{\partial r}\,d\theta\right| \leq C\varepsilon^{n-1};$$

second, that

$$\int_{|r\theta|=\varepsilon} f(r\theta)\varepsilon^{1-n}\,d\theta \to \omega_{n-1}f(0)$$

as $\varepsilon \to 0$. Letting $\varepsilon \to 0$ in (2.4.5), we obtain that

$$\lim_{\varepsilon\to 0}\int_{|x|>\varepsilon}\Delta(f)(x)|x|^{2-n}\,dx = -(n-2)\omega_{n-1}f(0)\,,$$

which implies (2.4.3) in view of the formula for ω_{n-1} given in Appendix A.3.

The proof of (2.4.4) is identical. The only difference is that the quantity $\partial r^{2-n}/\partial r$ in (2.4.5) is replaced by $\partial \log r/\partial r$. □

2.4.3 Homogeneous Distributions

The fundamental solutions of the Laplacian are locally integrable functions on $\mathbf{R}^n$ and also homogeneous of degree $2-n$ when $n \geq 3$. Since homogeneous distributions often arise in applications, it is desirable to pursue their study. Here we do not undertake such a study in depth, but we discuss a few important examples.

Definition 2.4.5. For $z \in \mathbf{C}$ we define a distribution u_z as follows:

$$\langle u_z, f\rangle = \int_{\mathbf{R}^n} \frac{\pi^{\frac{z+n}{2}}}{\Gamma\big(\frac{z+n}{2}\big)} |x|^z f(x)\, dx. \tag{2.4.6}$$

Clearly the u_z's coincide with the locally integrable functions

$$\pi^{\frac{z+n}{2}} \Gamma\big(\tfrac{z+n}{2}\big)^{-1} |x|^z$$

when $\operatorname{Re} z > -n$ and the definition makes sense only for that range of z's. It follows from its definition that u_z is a homogeneous distribution of degree z.

We would like to extend the definition of u_z for $z \in \mathbf{C}$. Let $\operatorname{Re} z > -n$ first. Fix N to be a positive integer. Given $f \in \mathscr{S}(\mathbf{R}^n)$, write the integral in (2.4.6) as follows:

$$\begin{aligned} &\int_{|x|<1} \frac{\pi^{\frac{z+n}{2}}}{\Gamma(\frac{z+n}{2})} \left\{ f(x) - \sum_{|\alpha|\leq N} \frac{(\partial^\alpha f)(0)}{\alpha!} x^\alpha \right\} |x|^z \, dx \\ &\quad + \int_{|x|>1} \frac{\pi^{\frac{z+n}{2}}}{\Gamma(\frac{z+n}{2})} f(x)|x|^z \, dx + \int_{|x|<1} \frac{\pi^{\frac{z+n}{2}}}{\Gamma(\frac{z+n}{2})} \sum_{|\alpha|\leq N} \frac{(\partial^\alpha f)(0)}{\alpha!} x^\alpha |x|^z \, dx. \end{aligned}$$

The preceding expression is equal to

$$\begin{aligned} &\int_{|x|<1} \frac{\pi^{\frac{z+n}{2}}}{\Gamma(\frac{z+n}{2})} \left\{ f(x) - \sum_{|\alpha|\leq N} \frac{(\partial^\alpha f)(0)}{\alpha!} x^\alpha \right\} |x|^z \, dx \\ &\quad + \int_{|x|>1} \frac{\pi^{\frac{z+n}{2}}}{\Gamma(\frac{z+n}{2})} f(x)|x|^z \, dx \\ &\quad + \sum_{|\alpha|\leq N} \frac{(\partial^\alpha f)(0)}{\alpha!} \frac{\pi^{\frac{z+n}{2}}}{\Gamma(\frac{z+n}{2})} \int_{r=0}^{1} \int_{\mathbf{S}^{n-1}} (r\theta)^\alpha \, r^{z+n-1} \, dr\, d\theta, \end{aligned}$$

where we switched to polar coordinates in the penultimate integral. Now set

$$\begin{aligned} b(n,\alpha,z) &= \frac{\pi^{\frac{z+n}{2}}}{\Gamma(\frac{z+n}{2})}\frac{1}{\alpha!}\bigg(\int_{\mathbf{S}^{n-1}}\theta^\alpha\,d\theta\bigg)\int_{r=0}^{1} r^{|\alpha|+n+z-1}\,dr \\ &= \frac{\pi^{\frac{z+n}{2}}}{\Gamma(\frac{z+n}{2})}\frac{\frac{1}{\alpha!}\int_{\mathbf{S}^{n-1}}\theta^\alpha\,d\theta}{|\alpha|+z+n}, \end{aligned}$$

where $\alpha = (\alpha_1,\dots,\alpha_n)$ is a multi-index. These coefficients are zero when at least one α_j is odd. Consider now the case that all the α_j's are even; then $|\alpha|$ is also even. The function $\Gamma(\frac{z+n}{2})$ has simple poles at the points

$$z=-n,\ z=-(n+2),\ z=-(n+4), \qquad \text{and so on;}$$

see Appendix A.5. These poles cancel exactly the poles of the function

$$z \mapsto (|\alpha|+z+n)^{-1}$$

at $z=-n-|\alpha|$ when $|\alpha|$ is an even integer in $[0,N]$. We therefore have

$$\begin{aligned} \langle u_z, f\rangle &= \int_{|x|\geq 1}\frac{\pi^{\frac{z+n}{2}}}{\Gamma(\frac{z+n}{2})}f(x)|x|^z\,dx + \sum_{|\alpha|\leq N} b(n,\alpha,z)\big\langle \partial^\alpha\delta_0, f\big\rangle \\ &\quad + \int_{|x|<1}\frac{\pi^{\frac{z+n}{2}}}{\Gamma(\frac{z+n}{2})}\bigg\{f(x)-\sum_{|\alpha|\leq N}\frac{(\partial^\alpha f)(0)}{\alpha!}x^\alpha\bigg\}|x|^z\,dx. \end{aligned} \tag{2.4.7}$$

Both integrals converge absolutely when $\operatorname{Re} z > -N-n-1$, since the expression inside the curly brackets above is bounded by a constant multiple of $|x|^{N+1}$, and the resulting function of z in (2.4.7) is a well defined analytic function in the range $\operatorname{Re} z > -N-n-1$.

Since N was arbitrary, $\langle u_z, f\rangle$ has an analytic extension to all of $\mathbf{C}$. Therefore, u_z is a distribution-valued entire function of z.

Next we would like to calculate the Fourier transform of u_z. We know by Exercise 2.3.9 that $\widehat{u_z}$ is a homogeneous distribution of degree $-n-z$. The choice of constant in the definition of u_z was made to justify the following result:

Theorem 2.4.6. *For all $z \in \mathbf{C}$ we have $\widehat{u_z} = u_{-n-z}$.*

Proof. The idea of the proof is straightforward. First we show that for a certain range of z's we have

$$\int_{\mathbf{R}^n}|\xi|^z\widehat{\varphi}(\xi)\,d\xi = C(n,z)\int_{\mathbf{R}^n}|x|^{-n-z}\varphi(x)\,dx, \tag{2.4.8}$$

for some fixed constant $C(n,z)$ and all $\varphi \in \mathscr{S}(\mathbf{R}^n)$. Next we pick a specific φ to evaluate the constant $C(n,z)$. Then we use analytic continuation to extend the validity of (2.4.8) for all z's. Use polar coordinates by setting $\xi = \rho\varphi$ and $x = r\theta$ in (2.4.8). We have

$$
\begin{aligned}
&\int_{\mathbf{R}^n} |\xi|^z \widehat{\varphi}(\xi)\,d\xi \\
&= \int_0^\infty \rho^{z+n-1} \int_0^\infty \int_{\mathbf{S}^{n-1}} \varphi(r\theta) \left(\int_{\mathbf{S}^{n-1}} e^{-2\pi i r\rho(\theta\cdot\varphi)} d\varphi \right) d\theta\, r^{n-1}\, dr\, d\rho \\
&= \int_0^\infty \left(\int_0^\infty \sigma(r\rho)\rho^{z+n-1}\, d\rho \right) \left(\int_{\mathbf{S}^{n-1}} \varphi(r\theta)\, d\theta \right) r^{n-1}\, dr \\
&= C(n,z) \int_0^\infty r^{-z-n} \left(\int_{\mathbf{S}^{n-1}} \varphi(r\theta)\, d\theta \right) r^{n-1}\, dr \\
&= C(n,z) \int_{\mathbf{R}^n} |x|^{-n-z} \varphi(x)\, dx,
\end{aligned}
$$

where we set

$$
\sigma(t) = \int_{\mathbf{S}^{n-1}} e^{-2\pi i t(\theta\cdot\varphi)}\, d\varphi = \int_{\mathbf{S}^{n-1}} e^{-2\pi i t(\varphi_1)}\, d\varphi, \tag{2.4.9}
$$

$$
C(n,z) = \int_0^\infty \sigma(t) t^{z+n-1}\, dt, \tag{2.4.10}
$$

and the second equality in (2.4.9) is a consequence of rotational invariance. It remains to prove that the integral in (2.4.10) converges for some range of z's.

If $n = 1$, then

$$
\sigma(t) = \int_{\mathbf{S}^0} e^{-2\pi i t\varphi}\, d\varphi = e^{-2\pi i t} + e^{2\pi i t} = 2\cos(2\pi t)
$$

and the integral in (2.4.10) converges conditionally for $-1 < \operatorname{Re} z < 0$.

Let us therefore assume that $n \ge 2$. Since $|\sigma(t)| \le \omega_{n-1}$, the integral converges near zero when $-n < \operatorname{Re} z$. Let us study the behavior of $\sigma(t)$ for t large. Using the formula in Appendix D.2 and the definition of Bessel functions in Appendix B.1, we write

$$
\sigma(t) = \int_{-1}^{1} e^{2\pi i t s} \omega_{n-2} \left(\sqrt{1-s^2}\right)^{n-2} \frac{ds}{\sqrt{1-s^2}} = c_n J_{\frac{n-2}{2}}(2\pi t),
$$

for some constant c_n. Using the asymptotics for Bessel functions (Appendix B.7), we obtain that $|\sigma(t)| \le c t^{-1/2}$ when $n - 2 > -1/2$ and $t \ge 1$. In either case the integral in (2.4.10) converges absolutely near infinity when $\operatorname{Re} z + n - 1 - 1/2 < -1$, i.e., when $\operatorname{Re} z < -n + 1/2$.

We have now proved that when $-n < \operatorname{Re} z < -n + 1/2$ we have

$$
\widehat{u_z} = C(n,z) u_{-n-z}
$$

for some constant $C(n,z)$ that we wish to compute. Insert the function $\varphi(x) = e^{-\pi|x|^2}$ in (2.4.8). Example 2.2.9 gives that this function is equal to its Fourier transform. Use polar coordinates to write

$$
\omega_{n-1} \int_0^\infty r^{z+n-1} e^{-\pi r^2}\, dr = C(n,z)\omega_{n-1} \int_0^\infty r^{-z-n+n-1} e^{-\pi r^2}\, dr.
$$

Change variables $s = \pi r^2$ and use the definition of the gamma function to obtain that

$$C(n,z) = \frac{\Gamma(\frac{z+n}{2})}{\Gamma(-\frac{z}{2})}\frac{\pi^{-\frac{z+n}{2}}}{\pi^{\frac{z}{2}}}.$$

It follows that $\widehat{u_z} = u_{-n-z}$ for the range of z's considered.

At this point observe that for every $f \in \mathscr{S}(\mathbf{R}^n)$, the function $z \mapsto \langle \widehat{u_z} - u_{-z-n}, f\rangle$ is entire and vanishes for $-n < \operatorname{Re} z < -n+1/2$. Therefore, it must vanish everywhere and the theorem is proved. $\square$

Homogeneous distributions were introduced in Exercise 2.3.9. We already saw that the Dirac mass on $\mathbf{R}^n$ is a homogeneous distribution of degree $-n$. There is another important example of a homogeneous distributions of degree $-n$, which we now discuss.

Let Ω be an integrable function on the sphere $\mathbf{S}^{n-1}$ with integral zero. Define a tempered distribution W_Ω on $\mathbf{R}^n$ by setting

$$\langle W_\Omega, f\rangle = \lim_{\varepsilon\to 0}\int_{|x|\geq\varepsilon}\frac{\Omega(x/|x|)}{|x|^n}f(x)\,dx. \tag{2.4.11}$$

We check that W_Ω is a well defined tempered distribution on $\mathbf{R}^n$. Indeed, since $\Omega(x/|x|)/|x|^n$ has integral zero over all annuli centered at the origin, we obtain

$$\begin{aligned}
|\langle W_\Omega, \varphi\rangle| &= \left|\lim_{\varepsilon\to 0}\int_{\varepsilon\leq|x|\leq 1}\frac{\Omega(x/|x|)}{|x|^n}(\varphi(x)-\varphi(0))\,dx + \int_{|x|\geq 1}\frac{\Omega(x/|x|)}{|x|^n}\varphi(x)\,dx\right| \\
&\leq \|\nabla\varphi\|_{L^\infty}\int_{|x|\leq 1}\frac{|\Omega(x/|x|)|}{|x|^{n-1}}\,dx + \Big(\sup_{x\in\mathbf{R}^n}|x|\,|\varphi(x)|\Big)\int_{|x|\geq 1}\frac{|\Omega(x/|x|)|}{|x|^{n+1}}\,dx \\
&\leq C_1\|\nabla\varphi\|_{L^\infty}\|\Omega\|_{L^1(\mathbf{S}^{n-1})} + C_2\sum_{|\alpha|\leq 1}\|\varphi(x)x^\alpha\|_{L^\infty}\|\Omega\|_{L^1(\mathbf{S}^{n-1})},
\end{aligned}$$

for suitable constants C_1 and C_2 in view of (2.2.2).

One can verify that $W_\Omega \in \mathscr{S}'(\mathbf{R}^n)$ is a homogeneous distribution of degree $-n$ just like the Dirac mass at the origin. It is an interesting fact that all homogeneous distributions on $\mathbf{R}^n$ of degree $-n$ that coincide with a smooth function away from the origin arise in this way. We have the following result.

Proposition 2.4.7. *Suppose that m is a $\mathscr{C}^\infty$ function on $\mathbf{R}^n\setminus\{0\}$ that is homogeneous of degree zero. Then there exist a scalar b and a $\mathscr{C}^\infty$ function Ω on $\mathbf{S}^{n-1}$ with integral zero such that*

$$m^\vee = b\,\delta_0 + W_\Omega, \tag{2.4.12}$$

where W_Ω denotes the distribution defined in (2.4.11).

To prove this result we need the following proposition, whose proof we postpone until the end of this section.

Proposition 2.4.8. *Suppose that u is a $\mathscr{C}^\infty$ function on $\mathbf{R}^n\setminus\{0\}$ that is homogeneous of degree $z\in\mathbf{C}$. Then $\widehat{u}$ is a $\mathscr{C}^\infty$ function on $\mathbf{R}^n\setminus\{0\}$.*

We now prove Proposition 2.4.7 using Proposition 2.4.8.

Proof. Let a be the integral of the smooth function m over $\mathbf{S}^{n-1}$. The function $m-a$ is homogeneous of degree zero and thus locally integrable on $\mathbf{R}^n$; hence it can be thought of as a distribution that we call $\widehat{u}$ (the Fourier transform of a tempered distribution u). Since $\widehat{u}$ is a $\mathscr{C}^\infty$ function on $\mathbf{R}^n \setminus \{0\}$, Proposition 2.4.8 implies that u is also a $\mathscr{C}^\infty$ function on $\mathbf{R}^n \setminus \{0\}$. Let Ω be the restriction of u on $\mathbf{S}^{n-1}$. Then Ω is a well defined $\mathscr{C}^\infty$ function on $\mathbf{S}^{n-1}$. Since u is a homogeneous function of degree $-n$ that coincides with the smooth function Ω on $\mathbf{S}^{n-1}$, it follows that $u(x) = \Omega(x/|x|)/|x|^n$ for x in $\mathbf{R}^n \setminus \{0\}$.

We show that Ω has mean value zero over $\mathbf{S}^{n-1}$. Pick a nonnegative, radial, smooth, and nonzero function ψ on $\mathbf{R}^n$ supported in the annulus $1 < |x| < 2$. Switching to polar coordinates, we write

$$\begin{aligned}
\langle u, \psi \rangle &= \int_{\mathbf{R}^n} \frac{\Omega(x/|x|)}{|x|^n} \psi(x)\, dx = c_\psi \int_{\mathbf{S}^{n-1}} \Omega(\theta)\, d\theta, \\
\langle u, \psi \rangle &= \langle \widehat{u}, \widehat{\psi} \rangle = \int_{\mathbf{R}^n} (m(\xi) - a) \widehat{\psi}(\xi)\, d\xi = c'_\psi \int_{\mathbf{S}^{n-1}} (m(\theta) - a)\, d\theta = 0,
\end{aligned}$$

and thus Ω has mean value zero over $\mathbf{S}^{n-1}$ (since $c_\psi \neq 0$).

We can now legitimately define the distribution W_Ω, which coincides with the function $\Omega(x/|x|)/|x|^n$ on $\mathbf{R}^n \setminus \{0\}$. But the distribution u also coincides with this function on $\mathbf{R}^n \setminus \{0\}$. It follows that $u - W_\Omega$ is supported at the origin. Proposition 2.4.1 now gives that $u - W_\Omega$ is a sum of derivatives of Dirac masses. Since both distributions are homogeneous of degree $-n$, it follows that

$$u - W_\Omega = c\delta_0 .$$

But $u = (m-a)^\vee = m^\vee - a\delta_0$, and thus $m^\vee = (c+a)\delta_0 + W_\Omega$. This proves the proposition. $\square$

We now turn to the proof of Proposition 2.4.8.

Proof. Let $u \in \mathscr{S}'$ be homogeneous of degree z and $\mathscr{C}^\infty$ on $\mathbf{R}^n \setminus \{0\}$. We need to show that $\widehat{u}$ is $\mathscr{C}^\infty$ away from the origin. We prove that $\widehat{u}$ is $\mathscr{C}^M$ for all M. Fix $M \in \mathbf{Z}^+$ and let α be any multi-index such that

$$|\alpha| > n + M + \operatorname{Re} z . \tag{2.4.13}$$

Pick a $\mathscr{C}^\infty$ function φ on $\mathbf{R}^n$ that is equal to 1 when $|x| \geq 2$ and equal to zero for $|x| \leq 1$. Write $u_0 = (1-\varphi)u$ and $u_\infty = \varphi u$. Then

$$\partial^\alpha u = \partial^\alpha u_0 + \partial^\alpha u_\infty \qquad \text{and thus} \qquad \widehat{\partial^\alpha u} = \widehat{\partial^\alpha u_0} + \widehat{\partial^\alpha u_\infty},$$

where the operations are performed in the sense of distributions. Since u_0 is compactly supported, Theorem 2.3.21 implies that $\widehat{\partial^\alpha u_0}$ is $\mathscr{C}^\infty$. Now Leibniz's rule gives that

$$\partial^\alpha u_\infty = v + \varphi \partial^\alpha u,$$

where v is a smooth function supported in the annulus $1 \le |x| \le 2$. Then $\widehat{v}$ is $\mathscr{C}^\infty$ and we need to show only that $\widehat{\varphi \partial^\alpha u}$ is $\mathscr{C}^M$. The function $\varphi \partial^\alpha u$ is actually $\mathscr{C}^\infty$, and by the homogeneity of $\partial^\alpha u$ (Exercise 2.3.9(c)) we obtain that $(\partial^\alpha u)(x) = |x|^{-|\alpha|+z}(\partial^\alpha u)(x/|x|)$. Since φ is supported away from zero, it follows that

$$|\varphi(x)(\partial^\alpha u)(x)| \le \frac{C_\alpha}{(1+|x|)^{|\alpha|-\operatorname{Re} z}} \tag{2.4.14}$$

for some $C_\alpha > 0$. It is now straightforward to see that if a function satisfies (2.4.14), then its Fourier transform is $\mathscr{C}^M$ whenever (2.4.13) is satisfied. See Exercise 2.4.1.

We conclude that $\widehat{\partial^\alpha u_\infty}$ is a $\mathscr{C}^M$ function whenever (2.4.13) is satisfied; thus so is $\widehat{\partial^\alpha u}$. Since $\widehat{\partial^\alpha u}(\xi) = (2\pi i \xi)^\alpha \widehat{u}(\xi)$, we deduce smoothness for $\widehat{u}$ away from the origin. Let $\xi \neq 0$. Pick a neighborhood V of ξ that does not meet the jth coordinate axis for some $1 \le j \le n$. Then $\eta_j \neq 0$ when $\eta \in V$. Let α be the multi-index $(0,\dots,M,\dots,0)$ with M in the jth coordinate and zeros elsewhere. Then $(2\pi i \eta_j)^M \widehat{u}(\eta)$ is a $\mathscr{C}^M$ function on V, and thus so is $\widehat{u}(\eta)$, since we can divide by η_j^M. We conclude that $\widehat{u}(\xi)$ is $\mathscr{C}^M$ on $\mathbf{R}^n \setminus \{0\}$. Since M is arbitrary, the conclusion follows. □

We end this section with an example that illustrates the usefulness of some of the ideas discussed in this section.

Example 2.4.9. Let η be a smooth function on $\mathbf{R}^n$ that is equal to 1 on the set $|x| \ge 1/2$ and vanishes on the set $|x| \le 1/4$. Let $0 < \operatorname{Re}(\alpha) < n$. Let

$$g(\xi) = \left(\eta(x)|x|^{-\alpha}\right)\widehat{\ }(\xi).$$

The function g decays faster than the reciprocal of any polynomial at infinity and

$$g(\xi) - \frac{\pi^{\alpha-\frac{n}{2}}\Gamma(\frac{n-\alpha}{2})}{\Gamma(\frac{\alpha}{2})}|\xi|^{\alpha-n}$$

is a $\mathscr{C}^\infty$ function on $\mathbf{R}^n$. Therefore, g is integrable on $\mathbf{R}^n$. This example indicates the interplay between the smoothness of a function and the decay of its Fourier transform. The smoothness of the function $\eta(x)|x|^{-\alpha}$ near zero is reflected by the decay of g near infinity. Moreover, the function $\eta(x)|x|^{-\alpha}$ is not affected by the bump η near infinity, and this results in a behavior of $g(\xi)$ near zero similar to that of $(|x|^{-\alpha})\widehat{\ }(\xi)$.

To see these assertions, first observe that $\partial^\gamma(\eta(x)|x|^{-\alpha})$ is integrable and thus $(-2\pi i \xi)^\gamma g(\xi)$ is bounded if γ is large enough. This gives the decay of g near infinity. We now use Theorem 2.4.6 to obtain

$$g(\xi) = \frac{\pi^{\alpha-\frac{n}{2}}\Gamma(\frac{n-\alpha}{2})}{\Gamma(\frac{\alpha}{2})}|\xi|^{\alpha-n} + \widehat{\varphi}(\xi),$$

where $\widehat{\varphi}(\xi) = \big((\eta(x)-1)|x|^{-\alpha}\big)^{\widehat{}}(\xi)$, which is $\mathscr{C}^\infty$ as the Fourier transform of a compactly supported distribution.

Exercises

2.4.1. Suppose that a function f satisfies the estimate

$$|f(x)| \le \frac{C_\alpha}{(1+|x|)^N},$$

for some $N > n$. Then $\widehat{f}$ is $\mathscr{C}^M$ when $1 \le M \le [N-n]$.

2.4.2. Use Corollary 2.4.3 to prove Liouville's theorem that every bounded harmonic function on $\mathbf{R}^n$ must be a constant. Derive as a consequence the *fundamental theorem of algebra*, stating that every polynomial on $\mathbf{C}$ must have a complex root.

2.4.3. Prove that e^x is not in $\mathscr{S}'(\mathbf{R})$ but that $e^x e^{ie^x}$ is in $\mathscr{S}'(\mathbf{R})$.

2.4.4. Show that the Schwartz function $x \mapsto \operatorname{sech}(\pi x)$, $x \in \mathbf{R}$, coincides with its Fourier transform.
[*Hint:* Integrate the function e^{iaz} over the rectangular contour with corners $(-R,0)$, $(R,0)$, $(R,i\pi)$, and $(-R,i\pi)$.]

2.4.5. (*Ismagilov [137]*) Construct an uncountable family of linearly independent Schwartz functions f_a such that $|f_a| = |f_b|$ and $|\widehat{f_a}| = |\widehat{f_b}|$ for all f_a and f_b in the family.
[*Hint:* Let w be a smooth nonzero function whose Fourier transform is supported in the interval $[-1/2,1/2]$ and let φ be a real-valued smooth nonconstant periodic function with period 1. Then take $f_a(x) = w(x)e^{i\varphi(x-a)}$ for $a \in \mathbf{R}$.]

2.4.6. Let P_y be the Poisson kernel defined in (2.1.13). Prove that for $f \in L^p(\mathbf{R}^n)$, $1 \le p < \infty$, the function

$$(x,y) \mapsto (P_y * f)(x)$$

is a harmonic function on $\mathbf{R}^{n+1}_+$. Use the Fourier transform and Exercise 2.2.11 to prove that $(P_{y_1} * P_{y_2})(x) = P_{y_1+y_2}(x)$ for all $x \in \mathbf{R}^n$.

2.4.7. (a) For a fixed $x_0 \in \mathbf{R}^n$, show that the function

$$v(x;x_0) = \frac{1-|x|^2}{|x-x_0|^n}$$

is harmonic on $\mathbf{R}^n \setminus \{x_0\}$.
(b) For fixed $x_0 \in \mathbf{S}^{n-1}$, prove that the family of functions $\theta \mapsto v(\theta; rx_0)$, $0 < r < 1$, defined on the sphere satisfies

$$\lim_{r\uparrow 1}\int_{\substack{\theta\in\mathbf{S}^{n-1}\\ |\theta-x_0|>\delta}} v(\theta;rx_0)\,d\theta=0$$

uniformly in x_0. The function $v(\theta;rx_0)$ is called the *Poisson kernel for the sphere.*
(c) Let f be a continuous function on $\mathbf{S}^{n-1}$. Prove that the function

$$u(x)=\frac{1}{\omega_{n-1}}(1-|x|^2)\int_{\mathbf{S}^{n-1}}\frac{f(\theta)}{|x-\theta|^n}\,d\theta$$

solves the Dirichlet problem $\Delta(u)=0$ on $|x|<1$ and $u=f$ on $|x|=1$.

2.4.8. Fix a real number λ, $0<\lambda<n$.
(a) Prove that

$$\int_{\mathbf{S}^n}|\xi-\eta|^{-\lambda}\,d\xi=2^{n-\lambda}\frac{\pi^{\frac{n}{2}}\Gamma(\frac{n-\lambda}{2})}{\Gamma(n-\frac{\lambda}{2})}.$$

(b) Prove that

$$\int_{\mathbf{R}^n}|x-y|^{-\lambda}(1+|x|^2)^{\frac{\lambda}{2}-n}\,dx=2^{n-\lambda}\frac{\pi^{\frac{n}{2}}\Gamma(\frac{n-\lambda}{2})}{\Gamma(n-\frac{\lambda}{2})}(1+|y|^2)^{-\frac{\lambda}{2}}.$$

[*Hint:* Use the stereographic projection in Appendix D.6.]

2.4.9. Prove the following *beta integral identity:*

$$\int_{\mathbf{R}^n}\frac{dt}{|x-t|^{\alpha_1}|y-t|^{\alpha_2}}=\pi^{\frac{n}{2}}\frac{\Gamma(\frac{n-\alpha_1}{2})\Gamma(\frac{n-\alpha_2}{2})\Gamma(\frac{\alpha_1+\alpha_2-n}{2})}{\Gamma(\frac{\alpha_1}{2})\Gamma(\frac{\alpha_2}{2})\Gamma(n-\frac{\alpha_1+\alpha_2}{2})}|x-y|^{n-\alpha_1-\alpha_2},$$

where $0<\alpha_1,\alpha_2<n$, $\alpha_1+\alpha_2>n$.

2.4.10. (a) Prove that if a function f on $\mathbf{R}^n$ ($n\ge 3$) is constant on the spheres $r\mathbf{S}^{n-1}$ for all $r>0$, then so is its Fourier transform.
(b) If a function on $\mathbf{R}^n$ ($n\ge 2$) is constant on all $(n-2)$–dimensional spheres orthogonal to $e_1=(1,0,\dots,0)$, then its Fourier transform possesses the same property.

2.4.11. (*Grafakos and Morpurgo [108]*) Suppose that $0<d_1,d_2,d_3<n$ satisfy $d_1+d_2+d_3=2n$. Prove that for any distinct $x,y,z\in\mathbf{R}^n$ we have the identity

$$\int_{\mathbf{R}^n}|x-t|^{-d_2}|y-t|^{-d_3}|z-t|^{-d_1}\,dt$$
$$=\pi^{\frac{n}{2}}\prod_{j=1}^{3}\frac{\Gamma(n-\frac{d_j}{2})}{\Gamma(\frac{d_j}{2})}|x-y|^{d_1-n}|y-z|^{d_2-n}|z-x|^{d_3-n}.$$

[*Hint:* Reduce matters to the case that $z=0$ and $y=e_1$. Then take the Fourier transform in x and use Exercise 2.4.10.]

2.4.12. (a) Integrate the function e^{iz^2} over the contour consisting of the three pieces $P_1 = \{(x,0) : 0 \le x \le R\}$, $P_2 = \{(R\cos\theta, R\sin\theta) : 0 \le \theta \le \frac{\pi}{4}\}$, and $P_3 = \{(t,t) :$ t between $\frac{R\sqrt{2}}{2}$ and $0\}$ to obtain the *Fresnel integral identity*:

$$\lim_{R\to\infty} \int_{-R}^{+R} e^{ix^2}\,dx = \frac{\sqrt{2\pi}}{2}(1+i)\,.$$

(b) Use the result in part (a) to show that the Fourier transform of the function $e^{i\pi|x|^2}$ in $\mathbf{R}^n$ is equal to $e^{i\frac{\pi n}{4}} e^{-i\pi|\xi|^2}$.

[*Hint:* Part (a): On P_2 we have $e^{-R^2\sin(2\theta)} \le e^{-\frac{4}{\pi}R^2\theta}$, and the integral over P_2 tends to 0. Part (b): Try first $n = 1$.]

2.5 Convolution Operators on L^p Spaces and Multipliers

In this section we study the class of operators that commute with translations. We prove in this section that bounded operators that commute with translations must be of convolution type. Convolution operators arise in many situations, and we would like to know under what circumstances they are bounded between L^p spaces.

2.5.1 Operators That Commute with Translations

Definition 2.5.1. A vector space X of measurable functions on $\mathbf{R}^n$ is called *closed under translations* if for $f \in X$ we have $\tau^z(f) \in X$ for all $z \in \mathbf{R}^n$. Let X and Y be vector spaces of measurable functions on $\mathbf{R}^n$ that are closed under translations. Let also T be an operator from X to Y. We say that T *commutes with translations* or is *translation-invariant* if

$$T(\tau^y(f)) = \tau^y(T(f))$$

for all $f \in X$ and all $y \in \mathbf{R}^n$.

It is automatic to see that convolution operators commute with translations. One of the main goals of this section is to prove the converse: every bounded linear operator that commutes with translations is of convolution type. We have the following:

Theorem 2.5.2. *Suppose $1 \le p,q \le \infty$. Suppose T is a bounded linear operator from $L^p(\mathbf{R}^n)$ to $L^q(\mathbf{R}^n)$ that commutes with translations. Then there exists a unique tempered distribution v such that*

$$T(f) = f * v \qquad \text{for all } f \in \mathscr{S}.$$

The theorem is a consequence of the following two results:

Lemma 2.5.3. *Under the hypotheses of Theorem 2.5.2 and for $f \in \mathscr{S}(\mathbf{R}^n)$, the distributional derivatives of $T(f)$ are L^q functions that satisfy*

$$\partial^\alpha T(f) = T(\partial^\alpha f), \qquad \textit{for all multi-indices } \alpha. \tag{2.5.1}$$

Lemma 2.5.4. *Let $1 \le q \le \infty$ and let $h \in L^q(\mathbf{R}^n)$. If all distributional derivatives $\partial^\alpha h$ are also in L^q, then h is almost everywhere equal to a continuous function H satisfying*

$$|H(0)| \le C_{n,q} \sum_{|\alpha| \le n+1} \big\|\partial^\alpha h\big\|_{L^q}. \tag{2.5.2}$$

Proof. Assuming Lemmas 2.5.3 and 2.5.4, we prove Theorem 2.5.2.

Define a linear functional u on $\mathscr{S}$ by setting

$$\langle u, f\rangle = T(f)(0).$$

By (2.5.1), (2.5.2), (2.2.7), and the boundedness of T, we have

$$\begin{aligned} |\langle u, f\rangle| &\le C_{n,q} \sum_{|\alpha|\le n+1} \big\|\partial^\alpha T(f)\big\|_{L^q} \\ &\le C_{n,q} \sum_{|\alpha|\le n+1} \big\|T(\partial^\alpha f)\big\|_{L^q} \\ &\le C_{n,q}\big\|T\big\|_{L^p\to L^q} \sum_{|\alpha|\le n+1} \big\|\partial^\alpha f\big\|_{L^p} \\ &\le C_{n,q}\big\|T\big\|_{L^p\to L^q} \sum_{|\alpha|,|\beta|\le N} \rho_{\alpha,\beta}(f), \end{aligned}$$

which implies that u is in $\mathscr{S}'$. We now set $v = \widetilde{u}$ and we claim that $T(f) = f * v$ for $f \in \mathscr{S}$. To see this, by Theorem 2.3.20 and by the translation invariance of T, we have

$$\begin{aligned} (f*\widetilde{u})(x) &= \langle \widetilde{u}, \tau^x(\widetilde{f})\rangle = \langle u, \tau^{-x}(f)\rangle \\ &= T(\tau^{-x}(f))(0) = \tau^{-x}(T(f))(0) \\ &= T(f)(x) \end{aligned}$$

whenever $f \in \mathscr{S}(\mathbf{R}^n)$. This proves the theorem. □

We now return to Lemmas 2.5.3 and 2.5.4. We begin with Lemma 2.5.3.

Proof. Let $\alpha = (0,\dots,1,\dots,0)$, where 1 is in the jth entry. Let $f, g \in \mathscr{S}$. Since

$$\frac{\tau^{-he_j}(g) - g}{h} - \partial_j g \to 0 \quad \text{in } \mathscr{S} \text{ as } h \to 0, \tag{2.5.3}$$

it follows that (2.5.3) converges to zero in L^p and thus

$$T\left(\frac{\tau^{-he_j}(g) - g}{h} - \partial_j g\right) \to 0 \quad \text{in } L^q \text{ as } h \to 0. \tag{2.5.4}$$

Therefore, (2.5.4) converges to zero when integrated against the function $\partial_j g$.

We now have

$$\begin{aligned}
\langle \partial_j T(f), g\rangle &= -\int_{\mathbf{R}^n} T(f)\,\partial_j g\,dx \\
&= -\lim_{h\to 0}\int_{\mathbf{R}^n} T(f)\Big(\frac{\tau^{-he_j}-I}{h}(g)\Big)\,dx \\
&= \lim_{h\to 0}\int_{\mathbf{R}^n} \Big(\Big(\frac{I-\tau^{he_j}}{h}\Big)\circ T(f)\Big)\,g\,dx \\
&= \lim_{h\to 0}\int_{\mathbf{R}^n} T\Big(\frac{I-\tau^{he_j}}{h}(f)\Big)\,g\,dx \\
&= \int_{\mathbf{R}^n} T(\partial_j f)\,g\,dx,
\end{aligned}$$

where we used the fact that T commutes with translations and (2.5.4). This shows that $\partial_j T(f) = T(\partial_j f)$. The general case follows by induction on $|\alpha|$. □

We now prove Lemma 2.5.4.

Proof. Let $R \ge 1$. Fix a $\mathscr{C}_0^\infty$ function φ_R that is equal to 1 in the ball $|x| \le R$ and equal to zero when $|x| \ge 2R$. Since h is in $L^q(\mathbf{R}^n)$, it follows that $\varphi_R h$ is in $L^1(\mathbf{R}^n)$. We show that $\widehat{\varphi_R h}$ is also in L^1. We begin with the inequality

$$1 \le C_n(1+|x|)^{-(n+1)} \sum_{|\alpha|\le n+1} |(-2\pi i x)^\alpha|, \tag{2.5.5}$$

which is trivial for $|x| \le 2$ and follows from (2.2.2) when $|x| \ge 2$. Now multiply (2.5.5) by $|\widehat{\varphi_R h}(x)|$ to obtain

$$\begin{aligned}
|\widehat{\varphi_R h}(x)| &\le C_n(1+|x|)^{-(n+1)} \sum_{|\alpha|\le n+1} |(-2\pi i x)^\alpha \widehat{\varphi_R h}(x)| \\
&\le C_n(1+|x|)^{-(n+1)} \sum_{|\alpha|\le n+1} \big\|(\partial^\alpha(\varphi_R h))^\wedge\big\|_{L^\infty} \\
&\le C_n(1+|x|)^{-(n+1)} \sum_{|\alpha|\le n+1} \big\|\partial^\alpha(\varphi_R h)\big\|_{L^1} \\
&\le C_n(2^n R^n v_n)^{1/q'}(1+|x|)^{-(n+1)} \sum_{|\alpha|\le n+1} \big\|\partial^\alpha(\varphi_R h)\big\|_{L^q} \\
&\le C_{n,R}(1+|x|)^{-(n+1)} \sum_{|\alpha|\le n+1} \big\|\partial^\alpha h\big\|_{L^q},
\end{aligned}$$

where we used Leibniz's rule and the fact that all derivatives of φ_R are bounded by constants (depending on R).

Integrate the previously displayed inequality with respect to x to obtain

$$\big\|\widehat{\varphi_R h}\big\|_{L^1} \le C_{R,n} \sum_{|\alpha|\le n+1} \big\|\partial^\alpha h\big\|_{L^q} < \infty. \tag{2.5.6}$$

Therefore, Fourier inversion holds for $\varphi_R h$ (see Exercise 2.2.6). This implies that $\varphi_R h$ is equal a.e. to a continuous function, namely the inverse Fourier transform of its Fourier transform. Since $\varphi_R = 1$ on the ball $B(0,R)$, we conclude that h is a.e. equal to a continuous function in this ball. Since $R > 0$ was arbitrary, it follows that h is a.e. equal to a continuous function on $\mathbf{R}^n$, which we denote by H. Finally, (2.5.2) is a direct consequence of (2.5.6) with $R = 1$, since $|H(0)| \le \|\widehat{\varphi_1 h}\|_{L^1}$. □

2.5.2 The Transpose and the Adjoint of a Linear Operator

We briefly discuss the notions of the transpose and the adjoint of a linear operator. We first recall real and complex inner products. For f, g measurable functions on $\mathbf{R}^n$, we define the *complex inner product*

$$\langle f \,|\, g \rangle = \int_{\mathbf{R}^n} f(x)\overline{g(x)}\,dx$$

whenever the integral converges absolutely. We reserve the notation

$$\langle f, g \rangle = \int_{\mathbf{R}^n} f(x)g(x)\,dx$$

for the *real inner product* on $L^2(\mathbf{R}^n)$ and also for the action of a distribution f on a test function g. (This notation also makes sense when a distribution f coincides with a function.)

Let $1 \le p, q \le \infty$. For a bounded linear operator T from $L^p(X,\mu)$ to $L^q(Y,\nu)$ we denote by T^* its *adjoint operator* defined by

$$\langle T(f) \,|\, g \rangle = \int_Y T(f)\,\overline{g}\,d\nu = \int_X f\,\overline{T^*(g)}\,d\mu = \langle f \,|\, T^*(g) \rangle \tag{2.5.7}$$

for f in $L^p(X,\mu)$ and g in $L^{q'}(Y,\nu)$ (or in a dense subspace of it). We also define the *transpose* of T as the unique operator T^t that satisfies

$$\langle T(f), g \rangle = \int_{\mathbf{R}^n} T(f)\,g\,dx = \int_{\mathbf{R}^n} f\,T^t(g)\,dx = \langle f, T^t(g) \rangle$$

for all $f \in L^p(X,\mu)$ and all $g \in L^{q'}(Y,\nu)$.

If T is an integral operator of the form

$$T(f)(x) = \int_X K(x,y)f(y)\,d\mu(y),$$

then T^* and T^t are also integral operators with kernels $K^*(x,y) = \overline{K(y,x)}$ and $K^t(x,y) = K(y,x)$, respectively. If T has the form $T(f) = (\widehat{f}m)^\vee$, that is, it is given by multiplication on the Fourier transform by a (complex-valued) function $m(\xi)$, then T^* is given by multiplication on the Fourier transform by the function $\overline{m(\xi)}$. Indeed for f,g in $\mathscr{S}(\mathbf{R}^n)$ we have

$$\begin{aligned}\int_{\mathbf{R}^n} f\,\overline{T^*(g)}\,dx &= \int_{\mathbf{R}^n} T(f)\,\overline{g}\,dx \\ &= \int_{\mathbf{R}^n} \widehat{T(f)}\,\overline{\widehat{g}}\,d\xi \\ &= \int_{\mathbf{R}^n} \widehat{f}\,m\,\overline{\widehat{g}}\,d\xi \\ &= \int_{\mathbf{R}^n} f\,\overline{(\overline{m}\widehat{g})^\vee}\,dx.\end{aligned}$$

A similar argument (using Theorem 2.2.14 (5)) gives that if T is given by multiplication on the Fourier transform by the function $m(\xi)$, then T^t is given by multiplication on the Fourier transform by the function $m(-\xi)$. Since the complex-valued functions $\overline{m(\xi)}$ and $m(-\xi)$ may be different, the operators T^* and T^t may be different in general. Also, if $m(\xi)$ is real-valued, then T is *self-adjoint* (i.e., $T = T^*$) while if $m(\xi)$ is even, then T is *self-transpose* (i.e., $T = T^t$).

2.5.3 The Spaces $\mathscr{M}^{p,q}(\mathbf{R}^n)$

Definition 2.5.5. Given $1 \le p,q \le \infty$, we denote by $\mathscr{M}^{p,q}(\mathbf{R}^n)$ the set of all bounded linear operators from $L^p(\mathbf{R}^n)$ to $L^q(\mathbf{R}^n)$ that commute with translations.

By Theorem 2.5.2 we have that every T in $\mathscr{M}^{p,q}$ is given by convolution with a tempered distribution. We introduce a norm $\|\cdot\|$ on $\mathscr{M}^{p,q}$ by setting

$$\|T\|_{\mathscr{M}^{p,q}} = \|T\|_{L^p \to L^q},$$

that is, the norm of T in $\mathscr{M}^{p,q}$ is the operator norm of T as an operator from L^p to L^q. It is a known fact that under this norm, $\mathscr{M}^{p,q}$ is a complete normed space (i.e., a Banach space).

Next we show that when $p > q$ the set $\mathscr{M}^{p,q}$ consists of only one element, namely the zero operator $T = 0$. This means that the only interesting classes of operators arise when $p \le q$.

Theorem 2.5.6. *$\mathscr{M}^{p,q} = \{0\}$ whenever $1 \le q < p < \infty$.*

Proof. Let f be a nonzero $\mathscr{C}_0^\infty$ function and let $h \in \mathbf{R}^n$. We have

$$\|\tau^h(T(f)) + T(f)\|_{L^q} = \|T(\tau^h(f) + f)\|_{L^q} \le \|T\|_{L^p \to L^q}\|\tau^h(f) + f\|_{L^p}.$$

Now let $|h| \to \infty$ and use Exercise 2.5.1. We conclude that

$$2^{\frac{1}{q}}\|T(f)\|_{L^q} \le \|T\|_{L^p \to L^q} 2^{\frac{1}{p}}\|f\|_{L^p},$$

which is impossible if $q < p$ unless T is the zero operator. □

Next we have a theorem concerning the duals of the spaces $\mathscr{M}^{p,q}(\mathbf{R}^n)$.

Theorem 2.5.7. *Let $1 < p \le q < \infty$ and $T \in \mathscr{M}^{p,q}(\mathbf{R}^n)$. Then T can be defined on $L^{q'}(\mathbf{R}^n)$, coinciding with its previous definition on the subspace $L^p(\mathbf{R}^n) \cap L^{q'}(\mathbf{R}^n)$ of $L^p(\mathbf{R}^n)$, so that it maps $L^{q'}(\mathbf{R}^n)$ to $L^{p'}(\mathbf{R}^n)$ with norm*

$$\big\|T\big\|_{L^{q'}\to L^{p'}} = \big\|T\big\|_{L^p\to L^q}. \tag{2.5.8}$$

(Recall $\infty' = 1$.) In other words, we have the following isometric identification of spaces:

$$\mathscr{M}^{q',p'}(\mathbf{R}^n) = \mathscr{M}^{p,q}(\mathbf{R}^n).$$

Proof. We first observe that if $T : L^p \to L^q$ is given by convolution with $u \in \mathscr{S}'$, then $T^* : L^{q'} \to L^{p'}$ is given by convolution with $\overline{\widetilde{u}} \in \mathscr{S}'$. Indeed, for f in $L^p(\mathbf{R}^n)$ and g in $L^{q'}(\mathbf{R}^n)$ we have

$$\begin{aligned} \int_{\mathbf{R}^n} f\,\overline{T^*(g)}\,dx &= \int_{\mathbf{R}^n} T(f)\,\overline{g}\,dx \\ &= \int_{\mathbf{R}^n} (f*u)\,\overline{g}\,dx \\ &= \int_{\mathbf{R}^n} f\,(\overline{g}*\widetilde{u})\,dx \\ &= \int_{\mathbf{R}^n} f\,\overline{g*\overline{\widetilde{u}}}\,dx. \end{aligned}$$

Therefore T^* is given by convolution with $\overline{\widetilde{u}}$. Moreover, T^* is well defined on $L^{q'}$. Using the simple identity

$$\overline{f*\overline{\widetilde{u}}} = (\widetilde{\overline{f}}*u)^{\sim}, \qquad f \in L^{q'}, \tag{2.5.9}$$

it follows that T is also well defined on $L^{q'}$. It remains to show that T (convolution with u) and T^* (convolution with $\overline{\widetilde{u}}$) map $L^{q'}$ to $L^{p'}$ with the same norm. But this easily follows from (2.5.9), which implies that

$$\frac{\big\|f*\overline{\widetilde{u}}\big\|_{L^{p'}}}{\big\|f\big\|_{L^{q'}}} = \frac{\big\|\widetilde{\overline{f}}*u\big\|_{L^{p'}}}{\big\|\widetilde{\overline{f}}\big\|_{L^{q'}}},$$

for all $f \in L^{q'}$, $f \neq 0$. We conclude that $\big\|T^*\big\|_{L^{q'}\to L^{p'}} \big\|T\big\|_{L^{q'}\to L^{p'}}$ and therefore $\big\|T\big\|_{L^p\to L^q} = \big\|T\big\|_{L^{q'}\to L^{p'}}$. □

We next focus attention on the spaces $\mathscr{M}^{p,q}(\mathbf{R}^n)$ whenever $p = q$. These spaces are of particular interest, since they include the singular integral operators, which we study in Chapter 4.

2.5.4 Characterizations of $\mathscr{M}^{1,1}(\mathbf{R}^n)$ and $\mathscr{M}^{2,2}(\mathbf{R}^n)$

It would be desirable to have a characterization of the spaces $\mathscr{M}^{p,p}$ in terms of properties of the convolving distribution. Unfortunately, this is unknown at present (it is not clear whether it is possible) except for certain cases.

Theorem 2.5.8. *An operator T is in $\mathscr{M}^{1,1}(\mathbf{R}^n)$ if and only if it is given by convolution with a finite Borel (complex-valued) measure. In this case, the norm of the operator is equal to the total variation of the measure.*

Proof. If T is given with convolution with a finite Borel measure μ, then clearly T maps L^1 to itself and $\|T\|_{L^1\to L^1} \le \|\mu\|_{\mathscr{M}}$, where $\|\mu\|_{\mathscr{M}}$ is the total variation of μ.

Conversely, let T be an operator bounded from L^1 to L^1. By Theorem 2.5.2, T is given by convolution with a tempered distribution u. Let

$$f_\varepsilon(x) = \varepsilon^{-n} e^{-\pi|x/\varepsilon|^2}.$$

Since the functions f_ε are uniformly bounded in L^1, it follows from the boundedness of T that $f_\varepsilon * u$ are also uniformly bounded in L^1. Since L^1 is naturally embedded in the space of finite Borel measures, which is the dual of the space C_{00} of continuous functions that tend to zero at infinity, we obtain that the family $f_\varepsilon * u$ lies in a fixed multiple of the unit ball of C_{00}^*. By the Banach–Alaoglu theorem, this is a weak* compact set. Therefore, some subsequence of $f_\varepsilon * u$ converges in the weak* topology to a measure μ. That is, for some $\varepsilon_k \to 0$ and all $g \in C_{00}(\mathbf{R}^n)$ we have

$$\lim_{k\to\infty} \int_{\mathbf{R}^n} g(x)(f_{\varepsilon_k} * u)(x)\,dx = \int_{\mathbf{R}^n} g(x)\,d\mu(x). \tag{2.5.10}$$

We claim that $u = \mu$. To see this, fix $g \in \mathscr{S}$. Equation (2.5.10) implies that

$$\langle u, \widetilde{f_{\varepsilon_k}} * g\rangle = \langle u, f_{\varepsilon_k} * g\rangle \to \langle \mu, g\rangle$$

as $k \to \infty$. Exercise 2.3.2 gives that $g * f_{\varepsilon_k}$ converges to g in $\mathscr{S}$. Therefore,

$$\langle u, f_{\varepsilon_k} * g\rangle \to \langle u, g\rangle.$$

It follows from (2.5.10) that $\langle u, g\rangle = \langle \mu, g\rangle$, and since g was arbitrary, $u = \mu$.

Next, (2.5.10) implies that for all $g \in C_{00}$ we have

$$\left|\int_{\mathbf{R}^n} g(x)\,d\mu(x)\right| \le \|g\|_{L^\infty} \sup_k \|f_{\varepsilon_k} * u\|_{L^1} \le \|g\|_{L^\infty} \|T\|_{L^1\to L^1}. \tag{2.5.11}$$

The Riesz representation theorem gives that the norm of the functional

$$g \mapsto \int_{\mathbf{R}^n} g(x)\,d\mu(x)$$

on C_{00} is exactly $\|\mu\|_{\mathscr{M}}$. It follows from (2.5.11) that $\|T\|_{L^1\to L^1} \geq \|\mu\|_{\mathscr{M}}$. Since the reverse inequality is obvious, we conclude that $\|T\|_{L^1\to L^1} = \|\mu\|_{\mathscr{M}}$. □

Operators given by convolution with finite complex-valued Borel measures obviously map $L^\infty(\mathbf{R}^n)$ to $L^\infty(\mathbf{R}^n)$; hence $\mathscr{M}^{1,1}(\mathbf{R}^n)$ is a subspace of $\mathscr{M}^{\infty,\infty}(\mathbf{R}^n)$. But there may exist bounded linear operators on L^∞ that commute with translations that are not given by a convolution. The following example captures a strange behavior in the case $p = q = \infty$.

Example 2.5.9. Let X be the space of all bounded complex-valued functions on the real line such that

$$\Phi(f) = \lim_{R\to+\infty} \frac{1}{R}\int_0^R f(t)\,dt$$

exists. Then Φ is a bounded linear functional on X that has a bounded extension $\widetilde{\Phi}$ on L^∞ by the Hahn–Banach theorem. We may view $\widetilde{\Phi}$ as a bounded linear operator from $L^\infty(\mathbf{R})$ to the space of constant functions, which is contained in $L^\infty(\mathbf{R})$. We note that $\widetilde{\Phi}$ commutes with translations, since for all $f \in L^\infty(\mathbf{R})$ and $x \in \mathbf{R}$ we have

$$\widetilde{\Phi}(\tau^x(f)) - \tau^x(\widetilde{\Phi}(f)) = \widetilde{\Phi}(\tau^x(f)) - \widetilde{\Phi}(f) = \widetilde{\Phi}(\tau^x(f) - f) = \Phi(\tau^x(f) - f) = 0,$$

where the last two equalities follow from the fact that for bounded functions f the expression $\frac{1}{R}\int_0^R f(t-x) - f(t)\,dt$ has limit zero as $R \to \infty$. Since the operator $\widetilde{\Phi}$ vanishes on all test functions, it is not given by convolution.

We now study the case $p = 2$. We have the following theorem:

Theorem 2.5.10. *An operator T is in $\mathscr{M}^{2,2}(\mathbf{R}^n)$ if and only if it is given by convolution with some $u \in \mathscr{S}'$ whose Fourier transform $\widehat{u}$ is an L^∞ function. In this case the norm of $T : L^2 \to L^2$ is equal to $\|\widehat{u}\|_{L^\infty}$.*

Proof. If $\widehat{u} \in L^\infty$, Plancherel's theorem gives

$$\int_{\mathbf{R}^n} |f*u|^2\,dx = \int_{\mathbf{R}^n} |\widehat{f}(\xi)\widehat{u}(\xi)|^2\,d\xi \leq \|\widehat{u}\|_{L^\infty}^2 \|\widehat{f}\|_{L^2}^2 ;$$

therefore, $\|T\|_{L^2\to L^2} \leq \|\widehat{u}\|_{L^\infty}$, and hence T is in $\mathscr{M}^{2,2}(\mathbf{R}^n)$.

Now suppose that $T \in \mathscr{M}^{2,2}(\mathbf{R}^n)$ is given by convolution with a tempered distribution u. We show that $\widehat{u}$ is a bounded function. For $R > 0$ let φ_R be a $\mathscr{C}_0^\infty$ function supported inside the ball $B(0,2R)$ and equal to one on the ball $B(0,R)$. The product of the function φ_R with the distribution $\widehat{u}$ is $\varphi_R\widehat{u} = ((\varphi_R)^\vee * u)\widehat{\ } = T(\varphi_R^\vee)\widehat{\ }$, which is an L^2 function. Since the L^2 function $\varphi_R\widehat{u}$ coincides with the distribution $\widehat{u}$ on the set $B(0,R)$, it follows that $\widehat{u}$ is in $L^2(B(0,R))$ for all $R > 0$ and therefore it is in L^2_{loc}. If $f \in L^\infty(\mathbf{R}^n)$ has compact support, the function $f\widehat{u}$ is in L^2, and therefore Plancherel's theorem and the boundedness of T give

$$\int_{\mathbf{R}^n} |f(x)\widehat{u}(x)|^2\,dx = \int_{\mathbf{R}^n} |T(f^\vee)(x)|^2\,dx \leq \|T\|_{L^2\to L^2}^2 \int_{\mathbf{R}^n} |f(x)|^2\,dx .$$

We conclude that for all bounded functions with compact support f we have

$$\int_{\mathbf{R}^n} \big(\|T\|_{L^2\to L^2}^2 - |\widehat{u}(x)|^2\big)|f(x)|^2\,dx \ge 0.$$

Taking $f = |B(x,r)|^{-1}\chi_{B(x,r)}$ for $r > 0$ and using Corollary 2.1.16, we obtain that $\|T\|_{L^2\to L^2}^2 - |\widehat{u}(x)|^2 \ge 0$ for almost all x. Hence $\widehat{u}$ is in L^∞ and $\|\widehat{u}\|_{L^\infty} \le \|T\|_{L^2\to L^2}$. Combining this with the estimate $\|T\|_{L^2\to L^2} \le \|\widehat{u}\|_{L^\infty}$, which holds if $\widehat{u} \in L^\infty$, we deduce that $\|T\|_{L^2\to L^2} = \|\widehat{u}\|_{L^\infty}$. □

2.5.5 *The Space of Fourier Multipliers* $\mathscr{M}_p(\mathbf{R}^n)$

We have now characterized all convolution operators that map L^2 to L^2. Suppose now that T is in $\mathscr{M}^{p,p}$, where $1 < p < 2$. As discussed in Theorem 2.5.7, T also maps $L^{p'}$ to $L^{p'}$. Since $p < 2 < p'$, by Theorem 1.3.4, it follows that T also maps L^2 to L^2. Thus T is given by convolution with a tempered distribution whose Fourier transform is a bounded function.

Definition 2.5.11. Given $1 \le p < \infty$, we denote by $\mathscr{M}_p(\mathbf{R}^n)$ the space of all bounded functions m on $\mathbf{R}^n$ such that the operator

$$T_m(f) = (\widehat{f}m)^\vee, \qquad f \in \mathscr{S},$$

is bounded on $L^p(\mathbf{R}^n)$ (or is initially defined in a dense subspace of $L^p(\mathbf{R}^n)$ and has a bounded extension on the whole space). The norm of m in $\mathscr{M}_p(\mathbf{R}^n)$ is defined by

$$\|m\|_{\mathscr{M}_p} = \|T_m\|_{L^p\to L^p}. \tag{2.5.12}$$

Definition 2.5.11 implies that $m \in \mathscr{M}_p$ if and only if $T_m \in \mathscr{M}^{p,p}$. Elements of the space $\mathscr{M}_p$ are called L^p *multipliers* or L^p *Fourier multipliers*. It follows from Theorem 2.5.10 that $\mathscr{M}_2$, the set of all L^2 multipliers, is L^∞. Theorem 2.5.8 implies that $\mathscr{M}_1(\mathbf{R}^n)$ is the set of the Fourier transforms of finite Borel measures that is usually denoted by $\mathscr{M}(\mathbf{R}^n)$. Theorem 2.5.7 states that a bounded function m is an L^p multiplier if and only if it is an $L^{p'}$ multiplier, and in this case

$$\|m\|_{\mathscr{M}_p} = \|m\|_{\mathscr{M}_{p'}}, \qquad 1 < p < \infty.$$

It is a consequence of Theorem 1.3.4 that the normed spaces $\mathscr{M}_p$ are nested, that is, for $1 \le p \le q \le 2$ we have

$$\mathscr{M}_1 \subseteq \mathscr{M}_p \subseteq \mathscr{M}_q \subseteq \mathscr{M}_2 = L^\infty.$$

Moreover, if $m \in \mathscr{M}_p$ and $1 \le p \le 2 \le p'$, Theorem 1.3.4 gives

$$\left\|T_m\right\|_{L^2\to L^2} \le \left\|T_m\right\|^{\frac{1}{2}}_{L^p\to L^p}\left\|T_m\right\|^{\frac{1}{2}}_{L^{p'}\to L^{p'}} = \left\|T_m\right\|_{L^p\to L^p}, \tag{2.5.13}$$

since $1/2 = (1/2)/p + (1/2)/p'$. Theorem 1.3.4 also gives that

$$\left\|m\right\|_{\mathcal{M}_p} \le \left\|m\right\|_{\mathcal{M}_q}$$

whenever $1 \le q \le p \le 2$. Thus the $\mathcal{M}_p$'s form an increasing family of spaces as p increases from 1 to 2.

Example 2.5.12. The function $m(\xi) = e^{2\pi i\xi\cdot b}$ is an L^p multiplier for all $b \in \mathbf{R}^n$, since the corresponding operator $T_m(f)(x) = f(x+b)$ is bounded on $L^p(\mathbf{R}^n)$. Clearly $\left\|m\right\|_{\mathcal{M}_p} = 1$.

Proposition 2.5.13. *For $1 \le p < \infty$, the normed space $\big(\mathcal{M}_p, \|\cdot\|_{\mathcal{M}_p}\big)$ is a Banach space. Furthermore, $\mathcal{M}_p$ is closed under pointwise multiplication and is a Banach algebra.*

Proof. It suffices to consider the case $1 \le p \le 2$. It is straightforward that if m_1, m_2 are in $\mathcal{M}_p$ and $b \in \mathbf{C}$ then $m_1 + m_2$ and bm_1 are also in $\mathcal{M}_p$. Observe that $m_1 m_2$ is the multiplier that corresponds to the operator $T_{m_1}T_{m_2} = T_{m_1 m_2}$ and thus

$$\left\|m_1 m_2\right\|_{\mathcal{M}_p} = \left\|T_{m_1}T_{m_2}\right\|_{L^p\to L^p} \le \left\|m_1\right\|_{\mathcal{M}_p}\left\|m_2\right\|_{\mathcal{M}_p}.$$

This proves that $\mathcal{M}_p$ is an algebra. To show that $\mathcal{M}_p$ is a complete space, take a Cauchy sequence m_j in $\mathcal{M}_p$. It follows from (2.5.13) that m_j is Cauchy in L^∞, and hence it converges to some bounded function m in the L^∞ norm. We have to show that $m \in \mathcal{M}_p$. Fix $f \in \mathcal{S}$. We have

$$T_{m_j}(f)(x) = \int_{\mathbf{R}^n} \widehat{f}(\xi) m_j(\xi) e^{2\pi i x\cdot\xi}\, d\xi \to \int_{\mathbf{R}^n} \widehat{f}(\xi) m(\xi) e^{2\pi i x\cdot\xi}\, d\xi = T_m(f)(x)$$

a.e. by the Lebesgue dominated convergence theorem. Since $\{m_j\}_j$ is a Cauchy sequence in $\mathcal{M}_p$, it is bounded in $\mathcal{M}_p$, and thus $\sup_j \left\|m_j\right\|_{\mathcal{M}_p} \le C$. Fatou's lemma now implies that

$$\begin{aligned}
\int_{\mathbf{R}^n} |T_m(f)|^p\, dx &= \int_{\mathbf{R}^n} \liminf_{j\to\infty} |T_{m_j}(f)|^p\, dx \\
&\le \liminf_{j\to\infty} \int_{\mathbf{R}^n} |T_{m_j}(f)|^p\, dx \\
&\le \liminf_{j\to\infty} \left\|m_j\right\|^p_{\mathcal{M}_p} \left\|f\right\|^p_{L^p} \\
&\le C^p \left\|f\right\|^p_{L^p},
\end{aligned}$$

which implies that $m \in \mathcal{M}_p$. Incidentally, this argument shows that if $\mu_j \in \mathcal{M}_p$ and $\mu_j \to \mu$ a.e., then μ is in $\mathcal{M}_p$ and satisfies

$$\left\|\mu\right\|_{\mathcal{M}_p} \le \liminf_{j\to\infty} \left\|\mu_j\right\|_{\mathcal{M}_p}.$$

Apply this inequality to $\mu_j = m_k - m_j$ and $\mu = m_k - m$ for some fixed k. Then let $k \to \infty$ and use the fact that m_j is a Cauchy sequence in $\mathscr{M}_p$ to obtain that $m_k \to m$ in $\mathscr{M}_p$. This proves that $\mathscr{M}_p$ is a Banach space. $\square$

The following proposition summarizes some simple properties of multipliers.

Proposition 2.5.14. *For all $m \in \mathscr{M}_p$, $1 \le p < \infty$, $x \in \mathbf{R}^n$, and $h > 0$ we have*

$$\big\|\tau^x(m)\big\|_{\mathscr{M}_p} = \big\|m\big\|_{\mathscr{M}_p}, \tag{2.5.14}$$

$$\big\|\delta^h(m)\big\|_{\mathscr{M}_p} = \big\|m\big\|_{\mathscr{M}_p}, \tag{2.5.15}$$

$$\big\|\widetilde{m}\big\|_{\mathscr{M}_p} = \big\|m\big\|_{\mathscr{M}_p},$$

$$\big\|e^{2\pi i(\cdot)\cdot x}m\big\|_{\mathscr{M}_p} = \big\|m\big\|_{\mathscr{M}_p},$$

$$\big\|m\circ A\big\|_{\mathscr{M}_p} = \big\|m\big\|_{\mathscr{M}_p}, \qquad \textit{A is an orthogonal matrix.}$$

Proof. See Exercise 2.5.2. $\square$

Example 2.5.15. We show that for $-\infty<a<b<\infty$ we have $\big\|\chi_{[a,b]}\big\|_{\mathscr{M}_p} = \big\|\chi_{[0,1]}\big\|_{\mathscr{M}_p}$. Indeed, using (2.5.14) we obtain that $\big\|\chi_{[a,b]}\big\|_{\mathscr{M}_p} = \big\|\chi_{[0,b-a]}\big\|_{\mathscr{M}_p}$, and the latter is equal to $\big\|\chi_{[0,1]}\big\|_{\mathscr{M}_p}$ in view of (2.5.15). The fact that $\big\|\chi_{[0,1]}\big\|_{\mathscr{M}_p} < \infty$ for all $1 < p < \infty$ is shown in Chapter 4.

We continue with the following interesting result.

Theorem 2.5.16. *Suppose that $m(\xi,\eta) \in \mathscr{M}_p(\mathbf{R}^{n+m})$, where $1 < p < \infty$. Then for almost every $\xi \in \mathbf{R}^n$ the function $\eta \mapsto m(\xi,\eta)$ is in $\mathscr{M}_p(\mathbf{R}^m)$, with*

$$\big\|m(\xi,\cdot)\big\|_{\mathscr{M}_p(\mathbf{R}^m)} \le \big\|m\big\|_{\mathscr{M}_p(\mathbf{R}^{n+m})}.$$

Proof. If m is only a measurable function, its restriction to lower-dimensional planes is not defined. To avoid technical difficulties of this sort, we first assume that m is continuous at every point. Fix f_1, g_1 in $\mathscr{S}(\mathbf{R}^n)$ and f_2, g_2 in $\mathscr{S}(\mathbf{R}^m)$. Let

$$M(\xi) = \int_{\mathbf{R}^m} m(\xi,\eta)\widehat{f_2}(\eta)\widehat{g_2}(\eta)\,d\eta, \qquad \xi \in \mathbf{R}^n,$$

and observe that

$$\begin{aligned}
\left|\int_{\mathbf{R}^n} \big(M(\cdot)\widehat{f_1}\big)^\vee g_1\,dx\right| &= \left|\int_{\mathbf{R}^n} M(\xi)\widehat{f_1}(\xi)\widehat{g_1}(\xi)\,d\xi\right| \\
&= \left|\iint_{\mathbf{R}^{n+m}} m(\xi,\eta)\widehat{f_1f_2}(\xi,\eta)\widehat{g_1g_2}(\xi,\eta)\,d\xi\,d\eta\right| \\
&= \left|\iint_{\mathbf{R}^{n+m}} (m\widehat{f_1f_2})^\vee g_1g_2\,d\xi\,d\eta\right| \\
&\le \big\|m\big\|_{\mathscr{M}_p(\mathbf{R}^{n+m})}\big\|f_1\big\|_{L^p}\big\|f_2\big\|_{L^p}\big\|g_1\big\|_{L^{p'}}\big\|g_2\big\|_{L^{p'}}.
\end{aligned}$$

Since by duality we have

$$\|(M(\cdot)\widehat{f_1})^\vee\|_{L^p} = \sup_{\|g_1\|_{L^{p'}}\le 1} \left| \int_{\mathbf{R}^n} (M(\cdot)\widehat{f_1})^\vee g_1\,dx \right|,$$

it follows that $M(\xi)$ is in $\mathscr{M}_p(\mathbf{R}^n)$ with norm

$$\|M\|_{\mathscr{M}_p(\mathbf{R}^n)} \le \|m\|_{\mathscr{M}_p(\mathbf{R}^{n+m})} \|f_2\|_{L^p} \|g_2\|_{L^{p'}}.$$

Since $\|M\|_{L^\infty} \le \|M\|_{\mathscr{M}_p}$ and m is continuous, we obtain that for all $\xi \in \mathbf{R}^n$,

$$\left| \int_{\mathbf{R}^m} (m(\xi,\cdot)\widehat{f_2})^\vee g_2\,dy \right| = |M(\xi)| \le \|m\|_{\mathscr{M}_p(\mathbf{R}^{n+m})} \|f_2\|_{L^p} \|g_2\|_{L^{p'}}, \tag{2.5.16}$$

which of course implies the required conclusion for m continuous. The passage to a general m is achieved via a regularization argument. Define the family of functions $m_\varepsilon(\xi,\eta) = (2\varepsilon)^{-n-m}(m * \chi_{|\xi|\le\varepsilon,|\eta|\le\varepsilon})$. By Exercise 2.5.3 we have that $\|m_\varepsilon\|_{\mathscr{M}_p(\mathbf{R}^{n+m})} \le \|m\|_{\mathscr{M}_p(\mathbf{R}^{n+m})}$, and clearly the m_ε's are continuous functions. From this observation and (2.5.16), it follows that

$$\left| \int_{\mathbf{R}^m} m_\varepsilon(\xi,\eta)\widehat{f_2}(\eta)\widehat{g_2}(\eta)\,d\xi\,d\eta \right| \le \|m\|_{\mathscr{M}_p(\mathbf{R}^{n+m})} \|f_2\|_{L^p} \|g_2\|_{L^{p'}}.$$

Now let $\varepsilon \to 0$ and use the Lebesgue dominated convergence theorem. The conclusion follows. □

Example 2.5.17. (The cone multiplier) On $\mathbf{R}^{n+1}$ define the function

$$m_\lambda(\xi_1,\dots,\xi_{n+1}) = \left(1 - \frac{\xi_1^2+\cdots+\xi_n^2}{\xi_{n+1}^2}\right)_+^\lambda, \qquad \lambda > 0,$$

where the plus sign indicates that $m_\lambda = 0$ if the expression inside the parentheses is negative. The multiplier m_λ is called the *cone multiplier with parameter* λ. If m_λ is in $\mathscr{M}_p(\mathbf{R}^{n+1})$, then the function $b_\lambda(\xi) = (1-|\xi|^2)_+^\lambda$ defined on $\mathbf{R}^n$ is in $\mathscr{M}_p(\mathbf{R}^n)$. Indeed, by Theorem 2.5.16 we have that for some $\xi_{n+1} = h$, $b_\lambda(\xi_1/h,\dots,\xi_n/h)$ is in $\mathscr{M}_p(\mathbf{R}^n)$ and hence so is b_λ by property (2.5.15).

Exercises

2.5.1. Prove that if $f \in L^q(\mathbf{R}^n)$ and $1 \le q < \infty$, then

$$\|\tau^h(f)+f\|_{L^q} \to 2^{1/q}\|f\|_{L^q} \qquad \text{as } |h| \to \infty.$$

2.5.2. Prove Proposition 2.5.14. Also prove that if $\delta_j^{h_j}$ is a dilation operator in the jth variable (for instance $\delta_1^{h_1}(f)(x) = f(h_1x_1, x_2, \dots, x_n)$), then

$$\left\|\delta_1^{h_1}\cdots\delta_n^{h_n}(m)\right\|_{\mathscr{M}_p} = \left\|m\right\|_{\mathscr{M}_p}.$$

2.5.3. Let $m \in \mathscr{M}_p(\mathbf{R}^n)$, $1 \le p < \infty$.
(a) If ψ is a function on $\mathbf{R}^n$ whose inverse Fourier transform is an integrable function, then prove that

$$\left\|\psi m\right\|_{\mathscr{M}_p} \le \left\|\psi^\vee\right\|_{L^1}\left\|m\right\|_{\mathscr{M}_p}.$$

(b) If ψ is in $L^1(\mathbf{R}^n)$, then prove that

$$\left\|\psi * m\right\|_{\mathscr{M}_p} \le \left\|\psi\right\|_{L^1}\left\|m\right\|_{\mathscr{M}_p}.$$

2.5.4. Fix a multi-index γ.
(a) Prove that the map $T(f) = f * \partial^\gamma\delta_0$ maps $\mathscr{S}$ continuously into $\mathscr{S}$.
(b) Prove that when $1/p - 1/q \ne |\gamma|/n$, T does not extend to an element of the space $\mathscr{M}^{p,q}$.

2.5.5. Let $K_\gamma(x) = |x|^{-n+\gamma}$, where $0 < \gamma < n$. Use Theorem 1.4.24 to show that the operator

$$T_\gamma(f) = f * K_\gamma, \qquad f \in \mathscr{S},$$

extends to a bounded operator in $\mathscr{M}^{p,q}$, where $1/p - 1/q = \gamma/n$, $1 < p < q < \infty$. This provides an example of a nontrivial operator in $\mathscr{M}^{p,q}$ when $p < q$.

2.5.6. (a) Use the ideas of the proof of Proposition 2.5.13 to show that if $m_j \in \mathscr{M}_p$, $1 \le p < \infty$, $\left\|m_j\right\|_{\mathscr{M}_p} \le C$ for all $j = 1, 2, \dots$, and $m_j \to m$ a.e., then $m \in \mathscr{M}_p$ and

$$\left\|m\right\|_{\mathscr{M}_p} \le \liminf_{j\to\infty}\left\|m_j\right\|_{\mathscr{M}_p} \le C.$$

(b) Suppose that for some $1 \le p < \infty$, $m_t \in \mathscr{M}_p$ for all $0 < t < \infty$. Prove that

$$\int_0^\infty \left\|m_t\right\|_{\mathscr{M}_p}\frac{dt}{t} < \infty \implies m(\xi) = \int_0^\infty m_t(\xi)\frac{dt}{t} \in \mathscr{M}_p.$$

(c) Use part (a) to prove that if $m \in \mathscr{M}_p$, $1 \le p < \infty$, then $m_0(x) = \lim_{R\to\infty} m(x/R)$ is also in $\mathscr{M}_p$ and satisfies $\left\|m_0\right\|_{\mathscr{M}_p} \le \left\|m\right\|_{\mathscr{M}_p}$.
(d) If $m \in \mathscr{M}_p$ has left and right limits at the origin, then prove that

$$\left\|m\right\|_{\mathscr{M}_p} \ge \max(|m(0+)|, |m(0-)|).$$

2.5.7. Let $1 \le p < \infty$ and suppose that $m \in \mathscr{M}_p(\mathbf{R}^n)$ has no zeros. Prove that the operator $T(f) = (\widehat{f}m^{-1})^\vee$ satisfies $\left\|T(f)\right\|_{L^p} \ge c_p\left\|f\right\|_{L^p}$, where $c_p = \left\|m\right\|_{\mathscr{M}_p}^{-1}$.

2.5.8. (a) Prove that if $m \in L^\infty(\mathbf{R}^n)$ satisfies $m^\vee \ge 0$, then for all $1 \le p < \infty$ we have

$$\|m\|_{\mathcal{M}_p} = \|m^\vee\|_{L^1}\,.$$

(b) (*L. Colzani and E. Laeng*) Let $m_1(\xi) = -1$ for $\xi > 0$ and $m_1(\xi) = 1$ for $\xi < 0$. Let $m_2(\xi) = \min(\xi - 1, 0)$ for $\xi > 0$ and $m_2(\xi) = \max(\xi + 1, 0)$ for $\xi < 0$. Prove that

$$\|m_1\|_{\mathcal{M}_p} = \|m_2\|_{\mathcal{M}_p}$$

for all $1 < p < \infty$.
[*Hint:* Part (a): Use Exercise 1.2.9. Part (b): Use part (a) to show that $\|m_2 m_1^{-1}\|_{\mathcal{M}_p} = 1$. Deduce that $\|m_2\|_{\mathcal{M}_p} \le \|m_1\|_{\mathcal{M}_p}$. For the converse use Exercise 2.5.6(c).]

2.5.9. (*de Leeuw [74]*) Let $1 < p < \infty$ and $0 < A < \infty$. Prove that the following are equivalent:
(a) The operator $f \mapsto \sum_{m\in\mathbf{Z}^n} a_m f(x-m)$ is bounded on $L^p(\mathbf{R}^n)$ with norm A.
(b) The $\mathcal{M}_p$ norm of the function $\sum_{m\in\mathbf{Z}^n} a_m e^{-2\pi i m\cdot x}$ is exactly A.
(c) The operator given by convolution with the sequence $\{a_m\}$ is bounded on $\ell^p(\mathbf{Z}^n)$ with norm A.

2.5.10. (*Jodeit [141]*) Let $m(x)$ in $\mathcal{M}_p(\mathbf{R}^n)$ be supported in $[0,1]^n$. Then the periodic extension of m in $\mathbf{R}^n$,

$$M(x) = \sum_{k\in\mathbf{Z}^n} m(x-k)\,,$$

is also in $\mathcal{M}_p(\mathbf{R}^n)$.

2.5.11. Suppose that u is a $\mathscr{C}^\infty$ function on $\mathbf{R}^n \setminus \{0\}$ that is homogeneous of degree $-n+i\tau$, $\tau \in \mathbf{R}$. Prove that the operator given by convolution with u maps $L^2(\mathbf{R}^n)$ to $L^2(\mathbf{R}^n)$.

2.5.12. (*Hahn [117]*) Let $m_1 \in L^r(\mathbf{R}^n)$ and $m_2 \in L^{r'}(\mathbf{R}^n)$ for some $2 \le r \le \infty$. Prove that $m_1 * m_2 \in \mathcal{M}_p(\mathbf{R}^n)$ when $\frac{1}{p} - \frac{1}{2} = \frac{1}{r}$ and $1 \le p \le 2$.
[*Hint:* Prove that the trilinear operator $(m_1, m_2, f) \mapsto \big((m_1 * m_2)\widehat{f}\,\big)^\vee$ is bounded from $L^2 \times L^2 \times L^1 \to L^1$ and $L^\infty \times L^1 \times L^2 \to L^2$. Apply trilinear complex interpolation (Exercise 1.4.17) to deduce the required conclusion for $1 \le p \le 2$.]

2.6 Oscillatory Integrals

Oscillatory integrals have played an important role in harmonic analysis from its outset. The Fourier transform is the prototype of oscillatory integrals and provides the simplest example of a nontrivial phase, a linear function of the variable of integration. More complicated phases naturally appear in the subject; for instance, Bessel functions provide examples of oscillatory integrals in which the phase is a sinusoidal function.

In this section we take a quick look at oscillatory integrals. We mostly concentrate on one-dimensional results, which already require some significant analysis. We examine only a very simple higher-dimensional situation. Our analysis here is far from adequate.

Definition 2.6.1. An *oscillatory integral* is an expression of the form

$$I(\lambda) = \int_{\mathbf{R}^n} e^{i\lambda\varphi(x)}\psi(x)\,dx, \tag{2.6.1}$$

where λ is a positive real number, φ is a real-valued function on $\mathbf{R}^n$ called the *phase*, and ψ is a complex-valued and smooth integrable function on $\mathbf{R}^n$, which is often taken to have compact support.

2.6.1 Phases with No Critical Points

We begin by studying the simplest possible one-dimensional case. Suppose that φ and ψ are smooth functions on the real line such that supp ψ is a closed interval and

$$\varphi'(x) \neq 0 \qquad \text{for all } x \in \operatorname{supp}\psi.$$

Since φ' has no zeros, it must be either strictly positive or strictly negative everywhere on the support of ψ. It follows that φ is monotonic on the support of ψ and we are allowed to change variables

$$u = \varphi(x)$$

in (2.6.1). Then $dx = (\varphi'(x))^{-1}du = (\varphi^{-1})'(u)\,du$, where φ^{-1} is the inverse function of φ. We transform the integral in (2.6.1) into

$$\int_{\mathbf{R}} e^{i\lambda u}\psi(\varphi^{-1}(u))(\varphi^{-1})'(u)\,du \tag{2.6.2}$$

and we note that the function $\theta(u) = \psi(\varphi^{-1}(u))(\varphi^{-1})'(u)$ is smooth and has compact support on $\mathbf{R}$. We therefore interpret the integral in (2.6.1) as $\widehat{\theta}(-\lambda/2\pi)$, where $\widehat{\theta}$ is the Fourier transform of θ. Since θ is a smooth function with compact support, it follows that the integral in (2.6.2) has rapid decay as $\lambda \to \infty$.

A quick way to see that the expression $\widehat{\theta}(-\lambda/2\pi)$ has decay of order λ^{-N} for all $N > 0$ as λ tends to ∞ is the following. Write

$$e^{i\lambda u} = \frac{1}{(i\lambda)^N}\frac{d^N}{du^N}(e^{i\lambda u})$$

and integrate by parts N times to express the integral in (2.6.2) as

$$\frac{(-1)^N}{(i\lambda)^N}\int_{\mathbf{R}} e^{i\lambda u}\frac{d^N\theta(u)}{du^N}\,du,$$

from which the assertion follows. Hence

$$|I(\lambda)| = |\widehat{\theta}(-\lambda/2\pi)| \le C_N\lambda^{-N}, \tag{2.6.3}$$

where $C_N = \big\|\theta^{(N)}\big\|_{L^1}$, which depends on derivatives of φ and ψ.

We now turn to a higher-dimensional analogue of this situation.

Definition 2.6.2. We say that a point x_0 is a *critical point* of a phase function φ if

$$\nabla\varphi(x_0) = \big(\partial_1\varphi(x_0),\dots,\partial_n\varphi(x_0)\big) = 0.$$

Example 2.6.3. Let $\xi \in \mathbf{R}^n\setminus\{0\}$. Then the phase functions $\varphi_1(x) = x\cdot\xi$, $\varphi_2(x) = e^{x\cdot\xi}$ have no critical points, while the phase function $\varphi_3(x) = |x|^2 - x\cdot\xi$ has one critical point at $x_0 = \frac{1}{2}\xi$.

The next result concerns the behavior of oscillatory integrals whose phase functions have no critical points.

Proposition 2.6.4. *Suppose that ψ is a compactly supported smooth function on $\mathbf{R}^n$ and that φ is a real-valued $\mathscr{C}^1$ function on $\mathbf{R}^n$ that has no critical points on the support of ψ. Then the oscillatory integral*

$$I(\lambda) = \int_{\mathbf{R}^n} e^{i\lambda\varphi(x)}\psi(x)\,dx \tag{2.6.4}$$

obeys a bound of the form $|I(\lambda)| \le C_N\lambda^{-N}$ for all $\lambda \ge 1$ and all $N > 0$, where C_N depends on N and on φ and ψ.

Proof. Since the case $n = 1$ has already been discussed, we concentrate on dimensions $n \ge 2$. For each y in the support of ψ there is a unit vector θ_y such that

$$\theta_y\cdot\nabla\varphi(y) = |\nabla\varphi(y)|.$$

By the continuity of $\nabla\varphi$ there is a small neighborhood $B(y,r_y)$ of y such that for all $x \in B(y,r_y)$ we have

$$\theta_y\cdot\nabla\varphi(x) \ge \frac{1}{2}|\nabla\varphi(y)| > 0.$$

Cover the support of ψ by a finite number of balls $B(y_j,r_{y_j})$, $j = 1,\dots,m$, and pick $c = \min_j \frac{1}{2}|\nabla\varphi(y_j)|$; we have

$$\theta_{y_j}\cdot\nabla\varphi(x) \ge c > 0 \tag{2.6.5}$$

for all $x \in B(y_j,r_{y_j})$ and $j = 1,\dots,m$.

Next we find a smooth partition of unity of $\mathbf{R}^n$ such that each member ζ_k of the partition is supported in some ball $B(y_j,r_{y_j})$ or lies outside the support of ψ. We therefore write

$$I(\lambda) = \sum_k \int_{\mathbf{R}^n} e^{i\lambda\varphi(x)} \psi(x)\zeta_k(x)\,dx, \tag{2.6.6}$$

where the sum contains only a finite number of indices, since only a finite number of the ζ_k's meet the support of ψ. It suffices to show that every term in the sum in (2.6.6) has rapid decay in λ as $\lambda \to \infty$.

To this end, we fix a k and we pick a j such that the support of $\psi\zeta_k$ is contained in some ball $B(y_j, r_{y_j})$. We find unit vectors $\theta_{y_j,2}, \dots, \theta_{y_j,n}$, such that the system $\{\theta_{y_j}, \theta_{y_j,2}, \dots, \theta_{y_j,n}\}$ is an orthonormal basis of $\mathbf{R}^n$. Let e_j be the unit column vector on $\mathbf{R}^n$ whose jth coordinate is one and whose remaining coordinates are zero. We find an orthogonal matrix R such that $Re_1 = \theta_{y_j}$ and we introduce the change of variables $u = y_j + R(x - y_j)$ in the integral

$$I_k(\lambda) = \int_{\mathbf{R}^n} e^{i\lambda\varphi(x)} \psi(x)\zeta_k(x)\,dx.$$

The map $x \mapsto u = (u_1, \dots, u_n)$ is a rotation that fixes y_j and preserves the ball $B(y_j, r_{y_j})$. Defining $\varphi(x) = \varphi^o(u)$, $\psi(x) = \psi^o(u)$, $\zeta_k(x) = \zeta_k^o(u)$, under this new coordinate system we write

$$I_k(\lambda) = \int_K \left\{ \int_{\mathbf{R}} e^{i\lambda\varphi^o(u)} \psi^o(u_1, \dots, u_n)\zeta_k^o(u_1, \dots, u_n)\,du_1 \right\} du_2 \cdots du_n, \tag{2.6.7}$$

where K is a compact subset of $\mathbf{R}^{n-1}$. Since R is an orthogonal matrix, $R^{-1} = R^t$, and the change of variables $x = y_j + R^t(u - y_j)$ implies that

$$\frac{\partial x}{\partial u_1} = \text{first row of } R^t = \text{first column of } R = \theta_{y_j}.$$

Thus for all $x \in B(y_j, r_j)$ we have

$$\frac{\partial \varphi^o(u)}{\partial u_1} = \frac{\partial \varphi(y_j + R^t(u - y_j))}{\partial u_1} = \nabla\varphi(x) \cdot \frac{\partial x}{\partial u_1} = \nabla\varphi(x) \cdot \theta_{y_j} \geq c > 0$$

in view of condition (2.6.5). This lower estimate is valid for all $u \in B(y_j, r_{y_j})$, and therefore the inner integral inside the curly brackets in (2.6.7) is at most $C_N\lambda^{-N}$ by estimate (2.6.3). Integrating over K results in the same conclusion for $I(\lambda)$ defined in (2.6.4). □

2.6.2 *Sublevel Set Estimates and the Van der Corput Lemma*

We discuss a sharp decay estimate for one-dimensional oscillatory integrals. This estimate is obtained as a consequence of delicate size estimates for the Lebesgue measures of the sublevel sets $\{|u| \leq \alpha\}$ for a function u. In what follows, $u^{(k)}$ denotes the kth derivative of a function $u(t)$ defined on $\mathbf{R}$, and $\mathscr{C}^k$ the space of all functions whose kth derivative exists and is continuous.

Lemma 2.6.5. *Let $k \geq 1$ and suppose that $a_0, \dots, a_k$ are distinct real numbers. Let $a = \min(a_j)$ and $b = \max(a_j)$ and let f be a real-valued $\mathscr{C}^{k-1}$ function on $[a,b]$ that is $\mathscr{C}^k$ on (a,b). Then there exists a point y in (a,b) such that*

$$\sum_{m=0}^{k} c_m f(a_m) = f^{(k)}(y),$$

where $c_m = (-1)^k k! \prod_{\substack{\ell=0 \\ \ell \neq m}}^{k} (a_\ell - a_m)^{-1}$.

Proof. Suppose we could find a polynomial $p_k(x) = \sum_{j=0}^{k} b_j x^j$ such that the function

$$\varphi(x) = f(x) - p_k(x) \tag{2.6.8}$$

satisfies $\varphi(a_m) = 0$ for all $0 \leq m \leq k$. Since the a_j are distinct, we apply Rolle's theorem k times to find a point y in (a,b) such that $f^{(k)}(y) = k!\, b_k$.

The existence of a polynomial p_k such that (2.6.8) is satisfied is equivalent to the existence of a solution to the matrix equation

$$\begin{pmatrix} a_0^k & a_0^{k-1} & \dots & a_0 & 1 \\ a_1^k & a_1^{k-1} & \dots & a_1 & 1 \\ \vdots & \vdots & \vdots & \vdots & \vdots \\ a_{k-1}^k & a_{k-1}^{k-1} & \dots & a_{k-1} & 1 \\ a_k^k & a_k^{k-1} & \dots & a_k & 1 \end{pmatrix} \begin{pmatrix} b_k \\ b_{k-1} \\ \vdots \\ b_1 \\ b_0 \end{pmatrix} = \begin{pmatrix} f(a_0) \\ f(a_1) \\ \vdots \\ f(a_{k-1}) \\ f(a_k) \end{pmatrix}.$$

The determinant of the square matrix on the left is called the *Vandermonde determinant* and is equal to

$$\prod_{\ell=0}^{k-1} \prod_{j=\ell+1}^{k} (a_\ell - a_j) \neq 0.$$

Since the a_j are distinct, it follows that the system has a unique solution. Using Cramer's rule, we solve this system to obtain

$$\begin{aligned} b_k &= \sum_{m=0}^{k} (-1)^m f(a_m) \frac{\prod_{\substack{\ell=0 \\ \ell \neq m}}^{k-1} \prod_{\substack{j=\ell+1 \\ j \neq m}}^{k} (a_\ell - a_j)}{\prod_{\ell=0}^{k-1} \prod_{j=\ell+1}^{k} (a_\ell - a_j)} \\ &= \sum_{m=0}^{k} (-1)^m f(a_m) \prod_{\substack{\ell=0 \\ \ell \neq m}}^{k} (a_\ell - a_m)^{-1} (-1)^{k-m}. \end{aligned}$$

The required conclusion now follows with c_m as claimed. □

Lemma 2.6.6. *Let E be a measurable subset of $\mathbf{R}$ with finite nonzero Lebesgue measure and let $k \in \mathbf{Z}^+$. Then there exist $a_0, \dots, a_k$ in E such that for all $\ell = 0, 1, \dots, k$ we have*

$$\prod_{\substack{j=0\\ j\neq \ell}}^{k} |a_j - a_\ell| \geq (|E|/2e)^k. \tag{2.6.9}$$

Proof. Given a measurable set E with finite measure, pick a compact subset E' of E such that $|E \setminus E'| < \delta$, for some $\delta > 0$. For $x \in \mathbf{R}$ define $T(x) = |(-\infty, x) \cap E'|$. Then T enjoys the distance-decreasing property

$$|T(x) - T(y)| \leq |x - y|$$

for all $x, y \in E'$; consequently, by the intermediate value theorem, T is a surjective map from E' to $[0, |E'|]$. Let a_j be points in E' such that $T(a_j) = \frac{j}{k}|E'|$ for $j = 0, 1, \dots, k$. For k an even integer, we have

$$\prod_{\substack{j=0\\ j\neq \ell}}^{k} |a_j - a_\ell| \geq \prod_{\substack{j=0\\ j\neq \ell}}^{k} \left| \frac{j}{k}|E'| - \frac{\ell}{k}|E'| \right| \geq \prod_{\substack{j=0\\ j\neq \frac{k}{2}}}^{k} \left| \frac{j}{k} - \frac{1}{2} \right| |E'|^k = \prod_{r=0}^{\frac{k}{2}-1} \left(\frac{r - \frac{k}{2}}{k} \right)^2 |E'|^k,$$

and it is easily shown that $\big((k/2)!\big)^2 k^{-k} \geq (2e)^{-k}$.

For k an odd integer we have

$$\prod_{\substack{j=0\\ j\neq \ell}}^{k} |a_j - a_\ell| \geq \prod_{\substack{j=0\\ j\neq \ell}}^{k} \left| \frac{j}{k}|E'| - \frac{\ell}{k}|E'| \right| \geq \prod_{\substack{j=0\\ j\neq \frac{k+1}{2}}}^{k} \left| \frac{j}{k} - \frac{k+1}{2k} \right| |E'|^k,$$

while the last product is at least

$$\left\{ \frac{1}{k} \cdot \frac{2}{k} \cdots \frac{\frac{k-1}{2}}{k} \right\}^2 \frac{k+1}{2k} \geq (2e)^{-k}.$$

We have therefore proved (2.6.9) with E' replacing E. Since $|E \setminus E'| < \delta$ and $\delta > 0$ is arbitrarily small, the required conclusion follows. □

The following is the main result of this section.

Proposition 2.6.7. *(a) Let u be a real-valued $\mathscr{C}^k$ function, $k \in \mathbf{Z}^+$, that satisfies $u^{(k)}(t) \geq 1$ for all $t \in \mathbf{R}$. Then the following estimate is valid for all $\alpha > 0$:*

$$\big|\{t \in \mathbf{R} : |u(t)| \leq \alpha\}\big| \leq (2e)((k+1)!)^{\frac{1}{k}} \alpha^{\frac{1}{k}}. \tag{2.6.10}$$

(b) For all $k \geq 2$, for every real-valued $\mathscr{C}^k$ function u on the line that satisfies $u^{(k)}(t) \geq 1$, for any $-\infty < a < b < \infty$, and every $\lambda > 0$, the following is valid:

$$\left| \int_a^b e^{i\lambda u(t)} dt \right| \leq 12k|\lambda|^{-\frac{1}{k}}. \tag{2.6.11}$$

(c) If $k = 1$, $u'(t)$ is monotonic on (a,b), and $u'(t) \geq 1$ for all $t \in (a,b)$, then

$$\left| \int_a^b e^{i\lambda u(t)} dt \right| \leq 3|\lambda|^{-1}. \tag{2.6.12}$$

Proof. Part (a): Let $E = \{t \in \mathbf{R} : |u(t)| \leq \alpha\}$. If $|E|$ is nonzero, then by Lemma 2.6.6 there exist $a_0, a_1, \dots, a_k$ in E such that for all ℓ we have

$$|E|^k \leq (2e)^k \prod_{\substack{j=0 \\ j \neq \ell}}^{k} |a_j - a_\ell|. \tag{2.6.13}$$

Lemma 2.6.5 implies that there exists $y \in \left(\min a_j, \max a_j\right)$ such that

$$u^{(k)}(y) = (-1)^k k! \sum_{m=0}^{k} u(a_m) \prod_{\substack{\ell=0 \\ \ell \neq m}}^{k} (a_\ell - a_m)^{-1}. \tag{2.6.14}$$

Using (2.6.13), we obtain that the expression on the right in (2.6.14) is in absolute value at most

$$(k+1)! \max_{0 \leq j \leq k} |u(a_j)| (2e)^k |E|^{-k} \leq (k+1)! \alpha (2e)^k |E|^{-k},$$

since $a_j \in E$. The bound $u^{(k)}(t) \geq 1$ now implies

$$|E|^k \leq (k+1)! (2e)^k \alpha$$

as claimed. This proves (2.6.10).

Part (b): We now take $k \geq 2$ and we split the interval (a,b) in (2.6.11) into the sets

$$\begin{aligned} R_1 &= \{t \in (a,b) : |u'(t)| \leq \beta\}, \\ R_2 &= \{t \in (a,b) : |u'(t)| > \beta\}, \end{aligned}$$

for some parameter β to be chosen momentarily. The function $v = u'$ satisfies $v^{(k-1)} \geq 1$ and $k - 1 \geq 1$. It follows from part (a) that

$$\left| \int_{R_1} e^{i\lambda u(t)} dt \right| \leq |R_1| \leq 2e (k!)^{\frac{1}{k-1}} \beta^{\frac{1}{k-1}} \leq 6k \beta^{\frac{1}{k-1}}.$$

To obtain the corresponding estimate over R_2, we note that if $u^{(k)} \geq 1$, then the set $\{|u'| > \beta\}$ is the union of at most $2k - 2$ intervals on each of which u' is monotone. Let (c,d) be one of these intervals on which u' is monotone. Then u' has a fixed sign on (c,d) and we have

$$\begin{aligned}
\left|\int_c^d e^{i\lambda u(t)}\,dt\right| &= \left|\int_c^d \left(e^{i\lambda u(t)}\right)' \frac{1}{\lambda u'(t)}\,dt\right| \\
&\le \left|\int_c^d e^{i\lambda u(t)}\left(\frac{1}{\lambda u'(t)}\right)' dt\right| + \frac{1}{|\lambda|}\left|\frac{e^{i\lambda u(d)}}{u'(d)} - \frac{e^{i\lambda u(c)}}{u'(c)}\right| \\
&\le \frac{1}{|\lambda|}\int_c^d \left|\left(\frac{1}{u'(t)}\right)'\right| dt + \frac{2}{|\lambda|\beta} \\
&= \frac{1}{|\lambda|}\left|\int_c^d \left(\frac{1}{u'(t)}\right)' dt\right| + \frac{2}{|\lambda|\beta} \\
&\le \frac{1}{|\lambda|}\left|\frac{1}{u'(d)} - \frac{1}{u'(c)}\right| + \frac{2}{|\lambda|\beta} \le \frac{3}{|\lambda|\beta},
\end{aligned}$$

where we use the monotonicity of $1/u'(t)$ in moving the absolute value from inside the integral to outside. It follows that

$$\left|\int_{R_2} e^{i\lambda u(t)}\,dt\right| \le \frac{6k}{|\lambda|\beta}.$$

Choosing $\beta = |\lambda|^{-(k-1)/k}$ to optimize and adding the corresponding estimates for R_1 and R_2, we deduce the claimed estimate (2.6.11).

Part (c): Repeat the argument in part (b) setting $\beta = 1$ and replacing the interval (c,d) by (a,b). □

Corollary 2.6.8. *Let (a,b), $u(t)$, $\lambda > 0$, and k be as in Proposition 2.6.7. Then for any function ψ on (a,b) with an integrable derivative and $k \ge 2$, we have*

$$\left|\int_a^b e^{i\lambda u(t)}\psi(t)\,dt\right| \le 12k\lambda^{-1/k}\left[|\psi(b)| + \int_a^b |\psi'(s)|\,ds\right].$$

We also have

$$\left|\int_a^b e^{i\lambda u(t)}\psi(t)\,dt\right| \le 3\lambda^{-1}\left[|\psi(b)| + \int_a^b |\psi'(s)|\,ds\right],$$

when $k = 1$ and u' is monotonic on (a,b).

Proof. Set

$$F(x) = \int_a^x e^{i\lambda u(t)}\,dt$$

and use integration by parts to write

$$\int_a^b e^{i\lambda u(t)}\psi(t)\,dt = F(b)\psi(b) - \int_a^b F(t)\psi'(t)\,dt.$$

The conclusion easily follows. □

Example 2.6.9. The *Bessel function* of order m is defined as

$$J_m(r) = \frac{1}{2\pi}\int_0^{2\pi} e^{ir\sin\theta} e^{-im\theta}\, d\theta\,.$$

Here we take both r and m to be real numbers, and we suppose that $m > -\frac{1}{2}$; we refer to Appendix B for an introduction to Bessel functions and their basic properties.

We use Corollary 2.6.8 to calculate the decay of the Bessel function $J_m(r)$ as $r \to \infty$. Set

$$\varphi(\theta) = \sin(\theta)$$

and note that $\varphi'(\theta)$ vanishes only at $\theta = \pi/2$ and $3\pi/2$ inside the interval $[0,2\pi]$ and that $\varphi''(\pi/2) = -1$, while $\varphi''(3\pi/2) = 1$. We now write $1 = \psi_1 + \psi_2 + \psi_3$, where ψ_1 is smooth and compactly supported in a small neighborhood of $\pi/2$, and ψ_2 is smooth and compactly supported in a small neighborhood of $3\pi/2$. For $j = 1,2$, Corollary 2.6.8 yields

$$\left|\int_0^{2\pi} e^{ir\sin(\theta)}\big(\psi_j(\theta)e^{-im\theta}\big)\,d\theta\right| \le C m r^{-1/2}$$

for some constant C, while the corresponding integral containing ψ_3 has arbitrary decay in r in view of estimate (2.6.3) (or Proposition 2.6.4 when $n = 1$).

Exercises

2.6.1. Suppose that u is a $\mathscr{C}^k$ function on the line that satisfies $|u^{(k)}(t)| \ge c_0 > 0$ for some $k \ge 2$ and all $t \in (a,b)$. Prove that for $\lambda > 0$ we have

$$\left|\int_a^b e^{i\lambda u(t)}\,dt\right| \le 12k(\lambda c_0)^{-1/k}$$

and that the same conclusion is valid when $k = 1$, provided u' is monotonic.

2.6.2. Show that if u' is not monotonic in part (c) of Proposition 2.6.7, then the conclusion may fail.
[*Hint:* Let $\varphi(t)$ be a smooth function on the real line that is equal to $10t$ on intervals $[2\pi k + \varepsilon, 2\pi(k+\frac{1}{2}) - \varepsilon]$ and equal to t on intervals $[2\pi(k+\frac{1}{2}) + \varepsilon, 2\pi(k+1) - \varepsilon]$. Show that the imaginary part of the oscillatory integral in question may tend to infinity over the union of several such intervals.]

2.6.3. Prove that the dependence on k of the constant in part (b) of Proposition 2.6.7 is indeed linear.
[*Hint:* Take $u(t) = t^k/k!$ over the interval $(0,k!)$.]

2.6.4. Follow the steps below to give an alternative proof of part (b) of Proposition 2.6.7. Assume that the statement is known for some $k \ge 2$ and some constant $C(k)$

for all intervals $[a,b]$ and all $\mathscr{C}^k$ functions satisfying $u^{(k)} \geq 1$ on $[a,b]$. Let c be the unique point at which the function $u^{(k)}$ attains its minimum in $[a,b]$.
(a) If $u^{(k)}(c) = 0$, then for all $\delta > 0$ we have $u^{(k)}(t) \geq \delta$ in the complement of the interval $(c-\delta, c+\delta)$ and derive the bound

$$\left| \int_a^b e^{i\lambda u(t)}\, dt \right| \leq 2C(k)(\lambda\delta)^{-1/k} + 2\delta .$$

(b) If $u^{(k)}(c) \neq 0$, then we must have $c \in \{a,b\}$. Obtain the bound

$$\left| \int_a^b e^{i\lambda u(t)}\, dt \right| \leq C(k)(\lambda\delta)^{-1/k} + \delta .$$

(c) Choose a suitable δ to optimize and deduce the validity of the statement for $k+1$ with $C(k+1) = 2C(k) + 2 = 5 \cdot 2^k - 2$. (Note that $C(1) = 3$.)

2.6.5. (a) Prove that for some constant C and all $\lambda \in \mathbf{R}$ and $\varepsilon \in (0,1)$ we have

$$\left| \int_{\varepsilon \leq |t| \leq 1} e^{i\lambda t} \frac{dt}{t} \right| \leq C .$$

(b) Prove that for some $C' < \infty$, all $\lambda \in \mathbf{R}$, $k > 0$, and $\varepsilon \in (0,1)$ we have

$$\left| \int_{\varepsilon \leq |t| \leq 1} e^{i\lambda t \pm t^k} \frac{dt}{t} \right| \leq C' .$$

(c) Show that there is a constant C'' such that for any $0 < \varepsilon < N < \infty$, for all ξ_1, ξ_2 in $\mathbf{R}$, and for all integers $k \geq 2$, we have

$$\left| \int_{\varepsilon \leq |s| \leq N} e^{i(\xi_1 s + \xi_2 s^k)} \frac{ds}{s} \right| \leq C'' .$$

[*Hint:* Part (a): For $|\lambda|$ small use the inequality $|e^{i\lambda t} - 1| \leq |\lambda t|$. If $|\lambda|$ is large, split the domains of integration into the regions $|t| \leq |\lambda|^{-1}$ and $|t| \geq |\lambda|^{-1}$ and use integration by parts in the second case. Part (b): Write

$$\frac{e^{i(\lambda t + \pm t^k)} - 1}{t} = e^{i\lambda t} \frac{e^{\pm i t^k} - 1}{t} + \frac{e^{i\lambda t}}{t}$$

and use part (a). Part (c): When $\xi_1 = \xi_2 = 0$ it is trivial. If $\xi_2 = 0$, $\xi_1 \neq 0$, change variables $t = \xi_1 s$ and then split the domain of integration into the sets $|t| \leq 1$ and $|t| \geq 1$. In the interval over the set $|t| \leq 1$ apply part (b) and over the set $|t| \geq 1$ use integration by parts. In the case $\xi_2 \neq 0$, change variables $t = |\xi_2|^{1/k} s$ and split the domain of integration into the sets $|t| \geq 1$ and $|t| \leq 1$. When $|t| \leq 1$ use part (b) and in the case $|t| \geq 1$ use Corollary 2.6.8, noting that $\frac{d^k(\xi_1 |\xi_2|^{-1/k} t \pm t^k)}{dt} = k! \geq 1$.]

2.6.6. (a) Show that for all $a \geq 1$ and $\lambda > 0$ the following is valid:

$$\left| \int_{|t| \le a\lambda} e^{i\lambda \log t} \, dt \right| \le 6a \, .$$

(b) Prove that there is a constant $c > 0$ such that for all $b > \lambda > 10$ we have

$$\left| \int_0^b e^{i\lambda t \log t} \, dt \right| \le \frac{c}{\lambda \log \lambda} \, .$$

[*Hint:* Part (b): Consider the intervals $(0, \delta)$ and $[\delta, b)$ for some δ. Apply Proposition 2.6.7 with $k = 1$ on one of these intervals and with $k = 2$ on the other. Then optimize over δ.]

2.6.7. Show that there is a constant $C < \infty$ such that for all nonintegers $\gamma > 1$ and all $\lambda, b > 1$ we have

$$\left| \int_0^b e^{i\lambda t^\gamma} dt \right| \le \frac{C}{\lambda^\gamma} \, .$$

[*Hint:* On the interval $(0, \delta)$ apply Proposition 2.6.7 with $k = [\gamma] + 1$ and on the interval (δ, b) with $k = [\gamma]$. Then optimize by choosing $\delta = \lambda^{-1/\gamma}$.]

HISTORICAL NOTES

The one-dimensional maximal function originated in the work of Hardy and Littlewood [123]. Its n-dimensional analogue was introduced by Wiener [291], who used Lemma 2.1.5, a variant of the Vitali covering lemma, to derive its L^p boundedness. One may consult the books of de Guzmán [72], [73] for extensions and other variants of such covering lemmas. The actual covering lemma proved by Vitali [285] says that if a family of closed cubes in $\mathbf{R}^n$ has the property that for every point $x \in A \subseteq \mathbf{R}^n$ there exists a sequence of cubes in the family that tends to x, then it is always possible to extract a sequence of pairwise disjoint cubes E_j from the family such that $|A \setminus \bigcup_j E_j| = 0$. We refer to Saks [233] for details and extensions of this theorem.

The class $L \log L$ was introduced by Zygmund to give a sufficient condition on the local integrability of the Hardy–Littlewood maximal operator. The necessity of this condition was observed by Stein [255]. Stein [259] also showed that the $L^p(\mathbf{R}^n)$ norm of the centered Hardy–Littlewood maximal operator $\mathcal{M}$ is bounded above by some dimension-free constant; see also Stein and Strömberg [262]. Analogous results for maximal operators associated with convex bodies are contained in Bourgain [29], Carbery [42], and Müller [204]. The situation for the uncentered maximal operator M is different, since given any $1 < p < \infty$ there exists $C_p > 1$ such that $\|M\|_{L^p(\mathbf{R}^n) \to L^p(\mathbf{R}^n)} \ge C_p^n$ (see Exercise 2.1.8 for a value of such a constant C_p and also the article of Grafakos and Montgomery-Smith [109] for a larger value). The centered maximal function $\mathcal{M}_\mu$ with respect to a general inner regular locally finite positive measure μ on $\mathbf{R}^n$ is bounded on $L^p(\mathbf{R}^n, \mu)$ without the additional hypothesis that the measure is doubling; see Fefferman [93]. The proof of this result requires the following covering lemma, obtained by Besicovitch [23]: Given any family of closed balls whose centers form a bounded subset of $\mathbf{R}^n$, there exists an at most countable subfamily of balls that covers the set of centers and has bounded overlap, i.e., no point in $\mathbf{R}^n$ belongs to more than a finite number (depending on the dimension) of the balls in the subfamily. A similar version of this lemma was obtained independently by Morse [202]. See also Ziemer [300] for an alternative formulation. The uncentered maximal operator M_μ of Exercise 2.1.1 may not be weak type $(1,1)$ if the measure μ is nondoubling, as shown by Sjögren [243]; related positive weak type $(1,1)$ results are contained in the article of Vargas [283]. The precise value of the operator norm of the uncentered Hardy–Littlewood maximal function on $L^p(\mathbf{R})$ was shown by Grafakos and Montgomery-Smith [109] to be the unique positive solution of the equation $(p-1)x^p - px^{p-1} - 1 = 0$. This constant raised to the power n is the operator norm of the strong maximal function M_s on $L^p(\mathbf{R}^n)$ for $1 < p \le \infty$. The best weak type $(1,1)$ constant for the centered Hardy–Littlewood maximal operator was shown by Melas [193] to be the largest root of the quadratic equation $12x^2 - 22x + 5 = 0$. The strong maximal operator M_s is not weak type $(1,1)$, but it satisfies the substitute inequality $d_{M_s(f)}(\alpha) \le C \int_{\mathbf{R}^n} \frac{|f(x)|}{\alpha}(1 + \log^+ \frac{|f(x)|}{\alpha})^{n-1} dx$. This result is due to Jessen, Marcinkiewicz, and Zygmund [140], but a geometric proof of it was obtained by Córdoba and Fefferman [58].

The basic facts about the Fourier transform go back to Fourier [95]. The definition of distributions used here is due to Schwartz [235]. For a concise introduction to the theory of distributions we refer to Hörmander [130] and Yosida [296]. Homogeneous distributions were considered by Riesz [222] in the study of the Cauchy problem in partial differential equations, although some earlier accounts are found in the work of Hadamard. They were later systematically studied by Gelfand and Šilov [100], [101]. References on the uncertainty principle include the articles of Fefferman [90] and Folland and Sitaram [94]. The best possible constant B_p in the Hausdorff–Young inequality $\|\widehat{f}\|_{L^{p'}(\mathbf{R}^n)} \le B_p \|f\|_{L^p(\mathbf{R}^n)}$ when $1 \le p \le 2$ was shown by Beckner [16] to be $B_p = (p^{1/p}(p')^{-1/p'})^{n/2}$. This best constant was previously obtained by Babenko [13] in the case when p' is an even integer.

A nice treatise of the spaces $\mathscr{M}^{p,q}$ is found in Hörmander [129]. This reference also contains Theorem 2.5.6, which is due to him. Theorem 2.5.16 is due to de Leeuw [74], but the proof presented here is taken from Jodeit [142]. De Leeuw's result in Exercise 2.5.9 says that periodic elements of $\mathscr{M}_p(\mathbf{R}^n)$ can be isometrically identified with elements of $\mathscr{M}(\mathbf{T}^n)$, the latter being the space of all multipliers on $\ell^p(\mathbf{Z}^n)$.

Parts (b) and (c) of Proposition 2.6.7 are due to van der Corput [282] and are referred to in the literature as van der Corput's lemma. The refinenment in part (a) was subsequently obtained by Arhipov, Karachuba, and Čubarikov [6]. The treatment of these results in the text is based on the article of Carbery, Christ, and Wright [44], which also investigates higher-dimensional analogues of the theory. Precise asymptotics can be obtained for a variety of oscillatory integrals via the method of stationary phase; see Hörmander [130]. References on oscillatory integrals include the books of Titchmarsh [280], Erdélyi [83], Zygmund [303], [304], Stein [261], and Sogge [248]. The latter provides a treatment of Fourier integral operators.

Chapter 3
Fourier Analysis on the Torus

Principles of Fourier series go back to ancient times. The attempts of the Pythagorean school to explain musical harmony in terms of whole numbers embrace early elements of a trigonometric nature. The theory of epicycles in the *Almagest* of Ptolemy, based on work related to the circles of Appolonius, contains ideas of astronomical periodicities that we would interpret today as harmonic analysis. Early studies of acoustical and optical phenomena, as well as periodic astronomical and geophysical occurrences, provided a stimulus of the physical sciences to the rigorous study of expansions of periodic functions. This study is carefully pursued in this chapter.

The modern theory of Fourier series begins with attempts to solve boundary value problems using trigonometric functions. The work of d'Alembert, Bernoulli, Euler, and Clairaut on the vibrating string led to the belief that it might be possible to represent arbitrary periodic functions as sums of sines and cosines. Fourier announced belief in this possibility in his solution of the problem of heat distribution in spatial bodies (in particular, for the cube $\mathbf{T}^3$) by expanding an arbitrary function of three variables as a triple sine series. Fourier's approach, although heuristic, was appealing and eventually attracted attention. It was carefully studied and further developed by many scientists, but most notably by Laplace and Dirichlet, who were the first to investigate the validity of the representation of a function in terms of its Fourier series. This is the main topic of study in this chapter.

3.1 Fourier Coefficients

We discuss some basic facts of Fourier analysis on the torus $\mathbf{T}^n$. Throughout this chapter, n denotes the dimension, i.e., a fixed positive integer.

L. Grafakos, *Classical Fourier Analysis, Second Edition*,
DOI: 10.1007/978-0-387-09432-8_3,

3.1.1 The n-Torus $\mathbf{T}^n$

The n-torus $\mathbf{T}^n$ is the cube $[0,1]^n$ with opposite sides identified. This means that the points $(x_1,\dots,0,\dots,x_n)$ and $(x_1,\dots,1,\dots,x_n)$ are identified whenever 0 and 1 appear in the same coordinate. A more precise definition can be given as follows: For x,y in $\mathbf{R}^n$, we say that

$$x \equiv y \tag{3.1.1}$$

if $x - y \in \mathbf{Z}^n$. Here $\mathbf{Z}^n$ is the additive subgroup of all points in $\mathbf{R}^n$ with integer coordinates. If (3.1.1) holds, then we write $x = y \pmod{\mathbf{1}}$. It is a simple fact that $\equiv$ is an equivalence relation that partitions $\mathbf{R}^n$ into equivalence classes. The n-torus $\mathbf{T}^n$ is then defined as the set $\mathbf{R}^n/\mathbf{Z}^n$ of all such equivalence classes. When $n = 1$, this set can be geometrically viewed as a circle by bending the line segment $[0,1]$ so that its endpoints are brought together. When $n = 2$, the identification brings together the left and right sides of the unit square $[0,1]^2$ and then the top and bottom sides as well. The resulting figure is a two-dimensional manifold embedded in $\mathbf{R}^3$ that looks like a donut. See Figure 3.1.

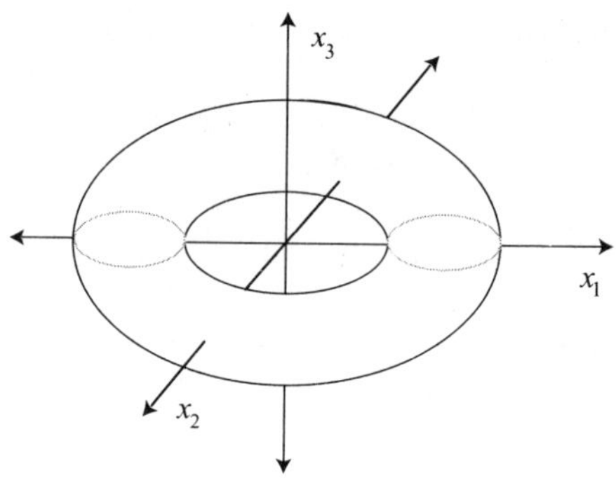

Fig. 3.1 The graph of the two-dimensional torus $\mathbf{T}^2$.

The n-torus is an additive group, and zero is the identity element of the group, which of course coincides with every $e_j = (0,\dots,0,1,0,\dots,0)$. To avoid multiple appearances of the identity element in the group, we often think of the n-torus as the set $[-1/2,1/2]^n$. Since the group $\mathbf{T}^n$ is additive, the inverse of an element $x \in \mathbf{T}^n$ is denoted by $-x$. For example, $-(1/3,1/4) \equiv (2/3,3/4)$ on $\mathbf{T}^2$, or, equivalently, $-(1/3,1/4) = (2/3,3/4) \pmod{\mathbf{1}}$.

The n-torus $\mathbf{T}^n$ can also be thought of as the following subset of $\mathbf{C}^n$,

$$\{(e^{2\pi i x_1},\dots,e^{2\pi i x_n}) \in \mathbf{C}^n \,:\, (x_1,\dots,x_n) \in [0,1]^n\}, \tag{3.1.2}$$

in a way analogous to which the unit interval $[0,1]$ can be thought of as the unit circle in $\mathbf{C}$ once 1 and 0 are identified.

Functions on $\mathbf{T}^n$ are functions f on $\mathbf{R}^n$ that satisfy $f(x+m)=f(x)$ for all $x\in\mathbf{R}^n$ and $m\in\mathbf{Z}^n$. Such functions are called 1-*periodic* in every coordinate. Haar measure on the n-torus is the restriction of n-dimensional Lebesgue measure to the set $\mathbf{T}^n = [0,1]^n$. This measure is still denoted by dx, while the measure of a set $A\subseteq\mathbf{T}^n$ is denoted by $|A|$. Translation invariance of the Lebesgue measure and the periodicity of functions on $\mathbf{T}^n$ imply that for all f on $\mathbf{T}^n$, we have

$$\int_{\mathbf{T}^n} f(x)\,dx = \int_{[-1/2,1/2]^n} f(x)\,dx = \int_{[a_1,1+a_1]\times\cdots\times[a_n,1+a_n]} f(x)\,dx \tag{3.1.3}$$

for any real numbers $a_1,\dots,a_n$. The L^p spaces on $\mathbf{T}^n$ are nested and L^1 contains all L^p spaces for $p\geq 1$.

Elements of $\mathbf{Z}^n$ are denoted by $m=(m_1,\dots,m_n)$. For $m\in\mathbf{Z}^n$, we define the *total size* of m to be the number $|m|=(m_1^2+\cdots+m_n^2)^{1/2}$. Recall that for $x=(x_1,\dots,x_n)$ and $y=(y_1,\dots,y_n)$ in $\mathbf{R}^n$,

$$x\cdot y = x_1y_1+\cdots+x_ny_n$$

denotes the usual dot product. Finally, for $x\in\mathbf{T}^n$, $|x|$ denotes the usual Euclidean norm of x. If we identify $\mathbf{T}^n$ with $[-1/2,1/2]^n$, then $|x|$ can be interpreted as the distance of the element x from the origin, and then we have that $0\leq|x|\leq\sqrt{n}/2$ for all $x\in\mathbf{T}^n$.

3.1.2 Fourier Coefficients

Definition 3.1.1. For a complex-valued function f in $L^1(\mathbf{T}^n)$ and m in $\mathbf{Z}^n$, we define

$$\widehat{f}(m) = \int_{\mathbf{T}^n} f(x)e^{-2\pi i m\cdot x}dx. \tag{3.1.4}$$

We call $\widehat{f}(m)$ the mth *Fourier coefficient* of f. We note that $\widehat{f}(m)$ is not defined for $\xi\in\mathbf{R}^n\setminus\mathbf{Z}^n$, since the function $x\mapsto e^{-2\pi i\xi\cdot x}$ is not 1-periodic in every coordinate and therefore not well defined on $\mathbf{T}^n$.

The *Fourier series* of f at $x\in\mathbf{T}^n$ is the series

$$\sum_{m\in\mathbf{Z}^n}\widehat{f}(m)e^{2\pi i m\cdot x}. \tag{3.1.5}$$

It is not clear at present in which sense and for which $x\in\mathbf{T}^n$ (3.1.5) converges. The study of convergence of Fourier series is the main topic of study in this chapter.

We quickly recall the notation we introduced in Chapter 2. We denote by $\overline{f}$ the complex conjugate of the function f, by $\widetilde{f}$ the function $\widetilde{f}(x)=f(-x)$, and by $\tau^y(f)$ the function $\tau^y(f)(x)=f(x-y)$ for all $y\in\mathbf{T}^n$. We mention some elementary properties of Fourier coefficients.

Proposition 3.1.2. *Let f, g be in $L^1(\mathbf{T}^n)$. Then for all $m,k \in \mathbf{Z}^n$, $\lambda \in \mathbf{C}$, $y \in \mathbf{T}^n$, and all multi-indices α we have*

(1) $\widehat{f+g}(m) = \widehat{f}(m) + \widehat{g}(m)$,

(2) $\widehat{\lambda f}(m) = \lambda \widehat{f}(m)$,

(3) $\widehat{\overline{f}}(m) = \overline{\widehat{f}(-m)}$,

(4) $\widehat{\widetilde{f}}(m) = \widehat{f}(-m)$,

(5) $\widehat{\tau^y(f)}(m) = \widehat{f}(m)e^{-2\pi i m\cdot y}$,

(6) $(e^{2\pi i k(\cdot)}f)^{\widehat{}}(m) = \widehat{f}(m-k)$,

(7) $\widehat{f}(0) = \int_{\mathbf{T}^n} f(x)\,dx$,

(8) $\sup_{m\in\mathbf{Z}^n} |\widehat{f}(m)| \le \|f\|_{L^1(\mathbf{T}^n)}$,

(9) $\widehat{f*g}(m) = \widehat{f}(m)\widehat{g}(m)$,

(10) $\widehat{\partial^\alpha f}(m) = (2\pi i m)^\alpha \widehat{f}(m)$, *whenever* $f \in \mathscr{C}^\alpha$.

Proof. The proof of Proposition 3.1.2 is obvious and is left to the reader. We only sketch the proof of (9). We have

$$\widehat{f*g}(m) = \int_{\mathbf{T}^n}\int_{\mathbf{T}^n} f(x-y)g(y)e^{-2\pi i m\cdot(x-y)}e^{-2\pi i m\cdot y}\,dy\,dx = \widehat{f}(m)\widehat{g}(m),$$

where the interchange of integrals is justified by the absolute convergence of the integrals and Fubini's theorem. □

Remark 3.1.3. The Fourier coefficients have the following property. For a function f_1 on $\mathbf{T}^{n_1}$ and a function f_2 on $\mathbf{T}^{n_2}$, the tensor function

$$(f_1 \otimes f_2)(x_1,x_2) = f_1(x_1)f_2(x_2)$$

is a periodic function on $\mathbf{T}^{n_1+n_2}$ whose Fourier coefficients are

$$\widehat{f_1 \otimes f_2}(m_1,m_2) = \widehat{f_1}(m_1)\widehat{f_2}(m_2), \tag{3.1.6}$$

for all $m_1 \in \mathbf{Z}^{n_1}$ and $m_2 \in \mathbf{Z}^{n_2}$.

Definition 3.1.4. A *trigonometric polynomial* on $\mathbf{T}^n$ is a function of the form

$$P(x) = \sum_{m\in\mathbf{Z}^n} a_m e^{2\pi i m\cdot x}, \tag{3.1.7}$$

where $\{a_m\}_{m\in\mathbf{Z}^n}$ is a finitely supported sequence in $\mathbf{Z}^n$. The *degree* of P is the largest number $|q_1|+\cdots+|q_n|$ such that a_q is nonzero, where $q=(q_1,\dots,q_n)$.

Example 3.1.5. A *trigonometric monomial* is a function of the form

$$P(x)=a\,e^{2\pi i(q_1x_1+\cdots+q_nx_n)}$$

for some $q=(q_1,\dots,q_n)\in\mathbf{Z}^n$ and $a\in\mathbf{C}$. Observe that $\widehat{P}(q)=a$ and $\widehat{P}(m)=0$ for $m\neq q$.

Let $P(x)=\sum_{|m|\le N}a_m e^{2\pi i m\cdot x}$ be a trigonometric polynomial and let f be in $L^1(\mathbf{T}^n)$. Exercise 3.1.1 gives that $(f*P)(x)=\sum_{|m|\le N}a_m\widehat{f}(m)e^{2\pi i m\cdot x}$. This implies that the partial sums $\sum_{|m|\le N}\widehat{f}(m)e^{2\pi i m\cdot x}$ of the Fourier series of f given in (3.1.5) can be obtained by convolving f with the functions

$$D_N(x)=\sum_{|m|\le N}e^{2\pi i m\cdot x}. \tag{3.1.8}$$

These expressions are named after Dirichlet, as the following definition indicates.

3.1.3 The Dirichlet and Fejér Kernels

Definition 3.1.6. Let $0\le R<\infty$. The *square Dirichlet kernel* on $\mathbf{T}^n$ is the function

$$D(n,R)(x)=\sum_{\substack{m\in\mathbf{Z}^n\\ |m_j|\le R}}e^{2\pi i m\cdot x}. \tag{3.1.9}$$

The *circular (or spherical) Dirichlet kernel* on $\mathbf{T}^n$ is the function

$$\widetilde{D}(n,R)(x)=\sum_{\substack{m\in\mathbf{Z}^n\\ |m|\le R}}e^{2\pi i m\cdot x}. \tag{3.1.10}$$

In dimension 1, the function $D(1,R)=\widetilde{D}(1,R)$ (for $R\ge 0$) is called the *Dirichlet kernel* and is denoted by D_R as in (3.1.8). The function D_5 is plotted in Figure 3.2.

Both the square and circular (or spherical) Dirichlet kernels are trigonometric polynomials. The square Dirichlet kernel on $\mathbf{T}^n$ is equal to a product of one-dimensional Dirichlet kernels, that is,

$$D(n,R)(x_1,\dots,x_n)=D_R(x_1)\cdots D_R(x_n). \tag{3.1.11}$$

We have the following two equivalent ways to write the Dirichlet kernel D_N:

$$D_N(x)=\sum_{|m|\le N}e^{2\pi i m\cdot x}=\frac{\sin((2N+1)\pi x)}{\sin(\pi x)},\qquad x\in[0,1]. \tag{3.1.12}$$

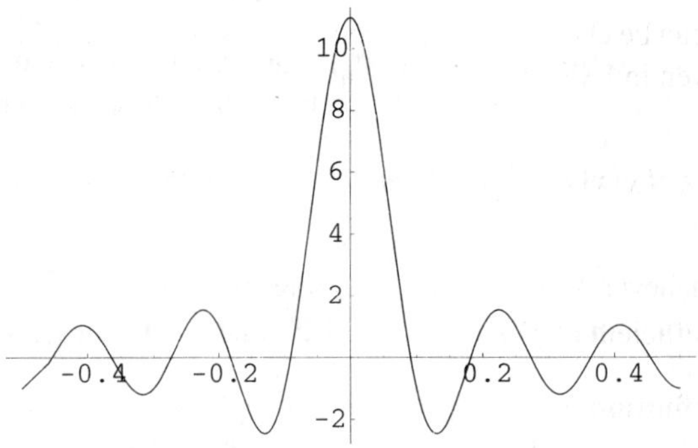

Fig. 3.2 The graph of the Dirichlet kernel D_5 plotted on the interval $[-1/2,1/2]$.

To verify the validity of (3.1.12), sum the geometric series on the left in (3.1.12) to obtain

$$e^{-2\pi iNx}\frac{e^{2\pi i(2N+1)x}-1}{e^{2\pi ix}-1}=\frac{e^{2\pi i(N+1)x}-e^{-2\pi iNx}}{e^{\pi ix}(e^{\pi ix}-e^{-\pi ix})}=\frac{\sin((2N+1)\pi x)}{\sin(\pi x)}.$$

It follows that for $R\in\mathbf{R}^+\cup\{0\}$ we have

$$D_R(x)=\frac{\sin(\pi x(2[R]+1))}{\sin(\pi x)}. \tag{3.1.13}$$

It is reasonable to ask whether the family $\{D_R\}_{R>0}$ forms an approximate identity as $R\to\infty$. Using (3.1.12) we see that each D_R is integrable over $[-1/2,1/2]$ with integral equal to 1. But we can easily obtain from (3.1.12) that for all $\delta>0$ there is a constant $c_\delta>0$ such that

$$\int_{1/2\geq|x|\geq\delta}|D_R(x)|\,dx\geq c_\delta$$

for all $R>0$. Therefore the family $\{D_R\}_{R>0}$ does not satisfy property (iii) in Definition 1.2.15. More important, it follows from Exercise 3.1.8 that $\|D_R\|_{L^1}\approx\log R$ as $R\to\infty$, and therefore property (i) in Definition 1.2.15 also fails for D_R. We conclude that the family $\{D_R\}_{R>0}$ is not an approximate identity on $\mathbf{T}^1$, a fact that significantly complicates the study of Fourier series. It follows immediately that the family $\{D(n,R)\}_{R>0}$ does not form an approximate identity on $\mathbf{T}^n$. The same is true for the family of circular (or spherical) Dirichlet kernels $\{\widetilde{D}(n,R)\}_{R>0}$, although this is harder to prove. It will be a consequence of the results in Section 3.4.

A typical situation encountered in analysis is that the mean of a sequence behaves better than the original sequence. This fact led Cesàro and independently Fejér to consider the arithmetic means of the Dirichlet kernel in dimension 1, that is, the expressions

$$F_N(x)=\frac{1}{N+1}\big[D_0(x)+D_1(x)+D_2(x)+\cdots+D_N(x)\big]. \tag{3.1.14}$$

It can be checked (see Exercise 3.1.3) that (3.1.14) is in fact equal to the Fejér kernel given in Example 1.2.18, that is,

$$F_N(x) = \sum_{j=-N}^{N} \left(1 - \frac{|j|}{N+1}\right) e^{2\pi i j x} = \frac{1}{N+1} \left(\frac{\sin(\pi(N+1)x)}{\sin(\pi x)} \right)^2, \tag{3.1.15}$$

whenever N is a nonnegative integer. Identity (3.1.15) implies that the mth Fourier coefficient of F_N is $\left(1 - \frac{|m|}{N+1}\right)$ if $|m| \le N$ and zero otherwise.

Definition 3.1.7. Let N be a nonnegative integer. The *Fejér kernel* $F(n,N)$ on $\mathbf{T}^n$ is defined as the average of the product of the Dirichlet kernels in each variable, precisely,

$$\begin{aligned} F(n,N)(x_1,\dots,x_n) &= \frac{1}{(N+1)^n} \sum_{k_1=0}^{N} \cdots \sum_{k_n=0}^{N} D_{k_1}(x_1) \cdots D_{k_n}(x_n) \\ &= \prod_{j=1}^{n} \left(\frac{1}{N+1} \sum_{k=0}^{N} D_k(x_j) \right) \\ &= \prod_{j=1}^{n} F_N(x_j) \,. \end{aligned}$$

So $F(n,N)$ is equal to the product of the Fejér kernels in each variable. Note that $F(n,N)$ is a trigonometric polynomial of degree nN.

Remark 3.1.8. Using the first expression for F_N in (3.1.15), we can write

$$F(n,N)(x) = \sum_{\substack{m \in \mathbf{Z}^n \\ |m_j| \le N}} \left(1 - \frac{|m_1|}{N+1}\right) \cdots \left(1 - \frac{|m_n|}{N+1}\right) e^{2\pi i m \cdot x} \tag{3.1.16}$$

for $N \ge 0$ an integer. Observe that $F(n,0)(x) = 1$ for all $x \in \mathbf{T}^n$.

Remark 3.1.9. To verify that the Fejér kernel $F(n,N)$ is an approximate identity on $\mathbf{T}^n$, we use the second expression for $F(1,N)$ in (3.1.15) to obtain

$$F(n,N)(x_1,\dots,x_n) = \frac{1}{(N+1)^n} \prod_{j=1}^{n} \left(\frac{\sin(\pi(N+1)x_j)}{\sin(\pi x_j)} \right)^2. \tag{3.1.17}$$

Properties (i) and (iii) of approximate identities (see Definition 1.2.15) can be proved using the identity (3.1.17), while property (ii) follows from identity (3.1.16). See Exercise 3.1.3 for details.

Having introduced the Fejér kernel, let us see how we can use it to obtain some interesting results.

3.1.4 Reproduction of Functions from Their Fourier Coefficients

Proposition 3.1.10. *The set of trigonometric polynomials is dense in* $L^p(\mathbf{T}^n)$ *for* $1 \le p < \infty$.

Proof. Given f in $L^p(\mathbf{T}^n)$ for $1 \le p < \infty$, consider $f * F(n,N)$. Because of Exercise 3.1.1, $f * F(n,N)$ is also a trigonometric polynomial. In view of Theorem 1.2.19 (1), $f * F(n,N)$ converges to f in L^p as $N \to \infty$. □

Corollary 3.1.11. ***(Weierstrass approximation theorem for trigonometric polynomials)*** *Every continuous function on the torus is a uniform limit of trigonometric polynomials.*

Proof. Since f is continuous on $\mathbf{T}^n$ and $\mathbf{T}^n$ is a compact set, Theorem 1.2.19 (2) gives that $f * F(n,N)$ converges uniformly to f as $N \to \infty$. Since $f * F(n,N)$ is a trigonometric polynomial, we conclude that every continuous function on $\mathbf{T}^n$ can be uniformly approximated by trigonometric polynomials. □

We now define partial sums of Fourier series.

Definition 3.1.12. For $R \ge 0$ the expressions

$$(f * D(n,R))(x) = \sum_{\substack{m\in\mathbf{Z}^n \\ |m_j|\le R}} \widehat{f}(m) e^{2\pi i m\cdot x}$$

are called the *square partial sums of the Fourier series* of f, and the expressions

$$(f * \widetilde{D}(n,R))(x) = \sum_{\substack{m\in\mathbf{Z}^n \\ |m|\le R}} \widehat{f}(m) e^{2\pi i m\cdot x}$$

are called the *circular (or spherical) partial sums* of the Fourier series of f. Similarly, for $N \in \mathbf{Z}^+ \cup \{0\}$ the expressions

$$(f * F(n,N))(x) = \sum_{\substack{m\in\mathbf{Z}^n \\ |m_j|\le N}} \left(1 - \frac{|m_1|}{N+1}\right) \cdots \left(1 - \frac{|m_n|}{N+1}\right) \widehat{f}(m) e^{2\pi i m\cdot x}$$

are called the *square Cesàro means* (or *square Fejér means*) of f. Finally, for $R \ge 0$ the expressions

$$(f * \widetilde{F}(n,R))(x) = \sum_{\substack{m\in\mathbf{Z}^n \\ |m|\le R}} \left(1 - \frac{|m|}{R}\right) \widehat{f}(m) e^{2\pi i m\cdot x}$$

are called the *circular Cesàro means* (or *circular Fejér means*) of f.

Observe that $f * \widetilde{F}(n,R)$ is equal to the average of the expressions $f * \widetilde{D}(n,R)$ from 0 to R in the following sense:

$$(f * \widetilde{F}(n,R))(x) = \frac{1}{R}\int_0^R (\widetilde{D}(n,r) * f)(x)\,dr.$$

This is analogous to the fact that the Fejér kernel F_N is the average of the Dirichlet kernels $D_0, D_1, \dots, D_N$. Also observe that $f * F(n,R)$ can also be defined for $R \ge 0$, but it would be constant on intervals of the form $[a, a+1)$, where $a \in \mathbf{Z}^+$.

A fundamental problem is in what sense the partial sums of the Fourier series converge back to the function as $N \to \infty$. This problem is of central importance in harmonic analysis and is partly investigated in this chapter.

We now ask the question whether the Fourier coefficients uniquely determine the function. The answer is affirmative and simple.

Proposition 3.1.13. *If $f, g \in L^1(\mathbf{T}^n)$ satisfy $\widehat{f}(m) = \widehat{g}(m)$ for all m in $\mathbf{Z}^n$, then $f = g$ a.e.*

Proof. By linearity of the problem, it suffices to assume that $g = 0$. If $\widehat{f}(m) = 0$ for all $m \in \mathbf{Z}^n$, Exercise 3.1.1 implies that $F(n,N) * f = 0$ for all $N \in \mathbf{Z}^+$. The sequence $\{F(n,N)\}_{N \in \mathbf{Z}^+}$ is an approximate identity as $N \to \infty$. Therefore,

$$\big\|f - F(n,N) * f\big\|_{L^1} \to 0$$

as $N \to \infty$; hence $\big\|f\big\|_{L^1} = 0$, from which we conclude that $f = 0$ a.e. □

A useful consequence of the result just proved is the following.

Proposition 3.1.14. ***(Fourier inversion)*** *Suppose that $f \in L^1(\mathbf{T}^n)$ and that*

$$\sum_{m \in \mathbf{Z}^n} |\widehat{f}(m)| < \infty.$$

Then

$$f(x) = \sum_{m \in \mathbf{Z}^n} \widehat{f}(m) e^{2\pi i m \cdot x} \qquad a.e., \tag{3.1.18}$$

and therefore f is almost everywhere equal to a continuous function.

Proof. It is straightforward to check that both functions in (3.1.18) are well defined and have the same Fourier coefficients. Therefore, they must be almost everywhere equal by Proposition 3.1.13. Moreover, the function on the right in (3.1.18) is everywhere continuous. □

We continue with a short discussion of Fourier series of square summable functions.

Let H be a separable Hilbert space with complex inner product $\langle \cdot \,|\, \cdot \rangle$. Recall that a subset E of H is called *orthonormal* if $\langle f \,|\, g \rangle = 0$ for all f, g in E with $f \neq g$, while $\langle f \,|\, f \rangle = 1$ for all f in E. A *complete orthonormal system* is a subset of H having the additional property that the only vector orthogonal to all of its elements is the zero vector. We refer to Rudin [229] for the relevant definitions and theorems and in particular for the proof of the following proposition:

Proposition 3.1.15. *Let H be a separable Hilbert space and let $\{\varphi_k\}_{k\in\mathbf{Z}}$ be an orthonormal system in H. Then the following are equivalent:*
(1) $\{\varphi_k\}_{k\in\mathbf{Z}}$ is a complete orthonormal system.
(2) For every $f\in H$ we have

$$\|f\|_H^2=\sum_{k\in\mathbf{Z}}|\langle f\,|\,\varphi_k\rangle|^2\,.$$

(3) For every $f\in H$ we have

$$f=\lim_{N\to\infty}\sum_{|k|\le N}\langle f\,|\,\varphi_k\rangle\varphi_k\,,$$

where the series converges in H.

Now consider the Hilbert space space $L^2(\mathbf{T}^n)$ with inner product

$$\langle f\,|\,g\rangle=\int_{\mathbf{T}^n}f(t)\overline{g(t)}\,dt\,.$$

Let φ_m be the sequence of functions $\xi\mapsto e^{2\pi i m\cdot\xi}$ indexed by $m\in\mathbf{Z}^n$. The orthonormality of the sequence $\{\varphi_m\}$ is a consequence of the following simple but powerful identity:

$$\int_{[0,1]^n}e^{2\pi i m\cdot x}\overline{e^{2\pi i k\cdot x}}\,dx=\begin{cases}1 & \text{when } m=k,\\ 0 & \text{when } m\neq k.\end{cases}$$

The completeness of the sequence $\{\varphi_m\}$ is also evident. Since $\langle f\,|\,\varphi_m\rangle=\widehat{f}(m)$ for all $f\in L^2(\mathbf{T}^n)$, it follows from Proposition 3.1.13 that if $\langle f\,|\,\varphi_m\rangle=0$ for all $m\in\mathbf{Z}^n$, then $f=0$ a.e.

The next result is a consequence of Proposition 3.1.15.

Proposition 3.1.16. *The following are valid for $f,g\in L^2(\mathbf{T}^n)$:*
(1) ***(Plancherel's identity)***

$$\|f\|_{L^2}^2=\sum_{m\in\mathbf{Z}^n}|\widehat{f}(m)|^2\,.$$

(2) The function $f(t)$ is a.e. equal to the $L^2(\mathbf{T}^n)$ limit of the sequence

$$\lim_{M\to\infty}\sum_{|m|\le M}\widehat{f}(m)e^{2\pi i m\cdot t}\,.$$

(3) ***(Parseval's relation)***

$$\int_{\mathbf{T}^n}f(t)\overline{g(t)}\,dt=\sum_{m\in\mathbf{Z}^n}\widehat{f}(m)\overline{\widehat{g}(m)}\,.$$

(4) The map $f\mapsto\{\widehat{f}(m)\}_{m\in\mathbf{Z}^n}$ is an isometry from $L^2(\mathbf{T}^n)$ onto ℓ^2.

(5) For all $k \in \mathbf{Z}^n$ we have

$$\widehat{fg}(k) = \sum_{m\in\mathbf{Z}^n} \widehat{f}(m)\widehat{g}(k-m) = \sum_{m\in\mathbf{Z}^n} \widehat{f}(k-m)\widehat{g}(m) .$$

Proof. (1) and (2) follow from the corresponding statements in Proposition 3.1.15. Parseval's relation (3) follows from polarization. First replace f by $f+g$ in (1) and expand the squares. We obtain that the real parts of the expressions in (3) are equal. Next replace f by $f+ig$ in (1) and expand the squares. We obtain that the imaginary parts of the expressions in (3) are equal. Thus (3) holds. Next we prove (4). We already know that the map $f \mapsto \{\widehat{f}(m)\}_{m\in\mathbf{Z}^n}$ is an injective isometry. It remains to show that it is onto. Given a square summable sequence $\{a_m\}_{m\in\mathbf{Z}^n}$ of complex numbers, define

$$f_N(t) = \sum_{|m|\le N} a_m e^{2\pi i m\cdot t}.$$

Observe that f_N is a Cauchy sequence in $L^2(\mathbf{T}^n)$ and it therefore converges to some $f \in L^2(\mathbf{T}^n)$. Then we have $\widehat{f}(m) = a_m$ for all $m \in \mathbf{Z}^n$. Finally, (5) is a consequence of (3) and Proposition 3.1.2 (6) and (3). □

3.1.5 The Poisson Summation Formula

We end this section with a useful result that connects Fourier analysis on the torus with Fourier analysis on $\mathbf{R}^n$. Suppose that f is an integrable function on $\mathbf{R}^n$ and let $\widehat{f}$ be its Fourier transform. Restrict $\widehat{f}$ on $\mathbf{Z}^n$ and form the "Fourier series" (assuming that it converges)

$$\sum_{m\in\mathbf{Z}^n} \widehat{f}(m)e^{2\pi i m\cdot x}.$$

What does this series represent? Since the preceding function is 1-periodic in every variable, it follows that it cannot be equal to f, unless it is identically zero. However, it should not come as a surprise that in many cases it is equal to the periodization of f on $\mathbf{R}^n$. More precisely, we have the following.

Theorem 3.1.17. ***(Poisson summation formula)*** *Suppose that $f, \widehat{f} \in L^1(\mathbf{R}^n)$ satisfy*

$$|f(x)| + |\widehat{f}(x)| \le C(1+|x|)^{-n-\delta}$$

for some $C, \delta > 0$. Then f and $\widehat{f}$ are both continuous, and for all $x \in \mathbf{R}^n$ we have

$$\sum_{m\in\mathbf{Z}^n} \widehat{f}(m)e^{2\pi i m\cdot x} = \sum_{m\in\mathbf{Z}^n} f(x+m), \tag{3.1.19}$$

and in particular $\sum_{m\in\mathbf{Z}^n} \widehat{f}(m) = \sum_{m\in\mathbf{Z}^n} f(m).$

Proof. Since $\widehat{f}$ is integrable on $\mathbf{R}^n$, inversion holds and f can be identified with a continuous function. Define a 1-periodic function on $\mathbf{T}^n$ by setting

$$F(x) = \sum_{m\in\mathbf{Z}^n} f(x+m).$$

It is straightforward to verify that $F \in L^1(\mathbf{T}^n)$. The calculation

$$\widehat{F}(m) = \int_{\mathbf{T}^n} F(x)e^{-2\pi i m\cdot x}\,dx = \sum_{k\in\mathbf{Z}^n}\int_{[-\frac{1}{2},\frac{1}{2}]^n-k} f(x)e^{-2\pi i m\cdot x}\,dx = \widehat{f}(m)$$

gives that the sequence of the Fourier coefficients of F coincides with the restriction of the Fourier transform of f on $\mathbf{Z}^n$. Since we have that

$$\sum_{m\in\mathbf{Z}^n} |\widehat{F}(m)| = \sum_{m\in\mathbf{Z}^n} |\widehat{f}(m)| \le C\sum_{m\in\mathbf{Z}^n}\frac{1}{(1+|m|)^{n+\delta}} < \infty,$$

Proposition 3.1.14 implies conclusion (3.1.19). □

Example 3.1.18. We have seen earlier (see Exercise 2.2.11) that the following identity gives the Fourier transform of the Poisson kernel in $\mathbf{R}^n$:

$$(e^{-2\pi|x|})^\wedge(\xi) = \frac{\Gamma(\frac{n+1}{2})}{\pi^{\frac{n+1}{2}}}\frac{1}{(1+|\xi|^2)^{\frac{n+1}{2}}}.$$

The Poisson summation formula yields the identity

$$\frac{\Gamma(\frac{n+1}{2})}{\pi^{\frac{n+1}{2}}}\sum_{k\in\mathbf{Z}^n}\frac{\varepsilon}{(\varepsilon^2+|k+x|^2)^{\frac{n+1}{2}}} = \sum_{k\in\mathbf{Z}^n} e^{-2\pi\varepsilon|k|}e^{-2\pi i k\cdot x}. \tag{3.1.20}$$

It follows that

$$\sum_{k\in\mathbf{Z}^n\setminus\{0\}}\frac{1}{(\varepsilon^2+|k|^2)^{\frac{n+1}{2}}} = \frac{1}{\varepsilon}\left(\frac{\pi^{\frac{n+1}{2}}}{\Gamma(\frac{n+1}{2})}\sum_{k\in\mathbf{Z}^n} e^{-2\pi\varepsilon|k|} - \frac{1}{\varepsilon^n}\right),$$

from which we obtain the identity

$$\sum_{k\in\mathbf{Z}^n\setminus\{0\}}\frac{1}{|k|^{n+1}} = \lim_{\varepsilon\to 0}\frac{1}{\varepsilon}\left(\frac{\pi^{\frac{n+1}{2}}}{\Gamma(\frac{n+1}{2})}\sum_{k\in\mathbf{Z}^n} e^{-2\pi\varepsilon|k|} - \frac{1}{\varepsilon^n}\right). \tag{3.1.21}$$

The limit in (3.1.21) can be calculated easily in dimension 1, since the sum inside the parentheses in (3.1.21) is a geometric series. Carrying out the calculation, we obtain

$$\sum_{k\neq 0}\frac{1}{k^2} = \frac{\pi^2}{3}.$$

Example 3.1.19. Let η and g be as in Example 2.4.9. Let $0 < \operatorname{Re} \alpha < n$ and let $x \in [-\frac{1}{2}, \frac{1}{2})^n$. The Poisson summation formula gives

$$\sum_{m \in \mathbf{Z}^n \setminus \{0\}} \frac{e^{2\pi i m \cdot x}}{|m|^\alpha} = \sum_{m \in \mathbf{Z}^n} \frac{\eta(m) e^{2\pi i m \cdot x}}{|m|^\alpha} = g(x) + \sum_{m \in \mathbf{Z}^n \setminus \{0\}} g(x+m) .$$

It was shown in Example 2.4.9 that $g(\xi)$ decays faster than the reciprocal of any polynomial at infinity and is equal to $\pi^{\alpha-\frac{n}{2}} \Gamma(\frac{n-\alpha}{2}) \Gamma(\frac{\alpha}{2})^{-1} |\xi|^{\alpha-n} + h(\xi)$, where h is a smooth function on $\mathbf{R}^n$. Then, for $x \in [-\frac{1}{2}, \frac{1}{2})^n$, the function

$$\sum_{m \in \mathbf{Z}^n \setminus \{0\}} g(x+m)$$

is also smooth, and we conclude that

$$\sum_{m \in \mathbf{Z}^n \setminus \{0\}} \frac{e^{2\pi i m \cdot x}}{|m|^\alpha} = \frac{\pi^{\alpha-\frac{n}{2}} \Gamma(\frac{n-\alpha}{2})}{\Gamma(\frac{\alpha}{2})} |x|^{\alpha-n} + h_1(x) ,$$

where $h_1(x)$ is a $\mathscr{C}^\infty$ function on $[-\frac{1}{2}, \frac{1}{2})^n$.

For other applications of the Poisson summation formula related to lattice points, see Exercises 3.1.12 and 3.1.13.

Exercises

3.1.1. Let P be a trigonometric polynomial on $\mathbf{T}^n$.
(a) Prove that $P(x) = \sum \widehat{P}(m) e^{2\pi i m \cdot x}$.
(b) Let f be in $L^1(\mathbf{T}^n)$. Prove that $(f * P)(x) = \sum \widehat{P}(m) \widehat{f}(m) e^{2\pi i m \cdot x}$.

3.1.2. On $\mathbf{T}^1$ let P be a trigonometric polynomial of degree $N > 0$. Show that P has at most $2N$ zeros. Construct a trigonometric polynomial with exactly $2N$ zeros.

3.1.3. Prove the identities (3.1.15), (3.1.16), and (3.1.17) about the Fejér kernel $F(n,N)$ on $\mathbf{T}^n$. Deduce from them that the family $\{F_N\}_N$ is an approximate identity as $N \to \infty$.
[*Hint:* Express the functions $\sin^2(\pi x)$ and $\sin^2(\pi(N+1)x)$ in terms of exponentials.]

3.1.4. (*de la Vallée Poussin kernel*). On $\mathbf{T}^1$ define

$$V_N(x) = 2F_{2N+1}(x) - F_N(x) .$$

(a) Show that the sequence V_N is an approximate identity.
(b) Prove that $\widehat{V_N}(m) = 1$ when $|m| \le N+1$, and $\widehat{V_N}(m) = 0$ when $|m| \ge 2N+2$.

3.1.5. (*Hausdorff–Young inequality*) Prove that when $f \in L^p$, $1 \le p \le 2$, the sequence of Fourier coefficients of f is in $l^{p'}$ and

$$\Big(\sum_{m\in\mathbf{Z}^n} |\widehat{f}(m)|^{p'}\Big)^{1/p'} \le \big\|f\big\|_{L^p}.$$

Also observe that 1 is the best constant in the preceding inequality.

3.1.6. Use without proof that there exists a constant $C > 0$ such that for all $t \in \mathbf{R}$ we have

$$\left|\sum_{k=2}^{N} e^{ik\log k} e^{ikt}\right| \le C\sqrt{N}, \qquad N = 2,3,4,\dots,$$

to construct an example of a continuous function g on $\mathbf{T}^1$ with

$$\sum_{m\in\mathbf{Z}} |\widehat{g}(m)|^q = \infty$$

for all $q < 2$. Thus the Hausdorff–Young inequality of Exercise 3.1.5 fails for $p > 2$. [*Hint:* Consider $g(x) = \sum_{k=2}^{\infty} \frac{e^{ik\log k}}{k^{1/2}(\log k)^2} e^{2\pi ikx}$. For a proof of the previous estimate, see Zygmund [303, Theorem (4.7) p. 199].]

3.1.7. The Poisson kernel on $\mathbf{T}^n$ is the function

$$P_{r_1,\dots,r_n}(x) = \sum_{m\in\mathbf{Z}^n} r_1^{|m_1|}\cdots r_n^{|m_n|} e^{2\pi i m\cdot x}$$

and is defined for $0 < r_1,\dots,r_n < 1$. Prove that $P_{r_1,\dots,r_n}$ can be written as

$$P_{r_1,\dots,r_n}(x_1,\dots,x_n) = \prod_{j=1}^{n} \mathrm{Re}\left(\frac{1+r_je^{2\pi ix_j}}{1-r_je^{2\pi ix_j}}\right) = \prod_{j=1}^{n} \frac{1-r_j^2}{1-2r_j\cos(2\pi x_j)+r_j^2},$$

and conclude that $P_{r,\dots,r}(x)$ is an approximate identity as $r \uparrow 1$.

3.1.8. Let $D_N = D(1,N)$ be the Dirichlet kernel on $\mathbf{T}^1$. Prove that

$$\frac{4}{\pi^2}\sum_{k=1}^{N}\frac{1}{k} \le \big\|D_N\big\|_{L^1} \le 2+\frac{\pi}{4}+\frac{4}{\pi^2}\sum_{k=1}^{N}\frac{1}{k}.$$

Conclude that the numbers $\big\|D_N\big\|_{L^1}$ grow logarithmically as $N \to \infty$ and therefore the family $\{D_N\}_N$ is not an approximate identity on $\mathbf{T}^1$. The numbers $\big\|D_N\big\|_{L^1}$, $N = 1,2,\dots$, are called the *Lebesgue constants*.
[*Hint:* Use that $\left|\frac{1}{\sin(\pi x)} - \frac{1}{\pi x}\right| \le \frac{\pi}{4}$ when $|x| \le \frac{1}{2}$.]

3.1.9. Let D_N be the Dirichlet kernel on $\mathbf{T}^1$. Prove that for all $1 < p < \infty$ there exist two constants $C_p, c_p > 0$ such that

$$c_p\,(2N+1)^{1/p'} \le \big\|D_N\big\|_{L^p} \le C_p\,(2N+1)^{1/p'}.$$

[*Hint:* Consider the two closest zeros of D_N near the origin and split the integral into the intervals thus obtained.]

3.1.10. (*S. Bernstein*) Let $P(x)$ be a trigonometric polynomial of degree N on $\mathbf{T}^1$. Prove that $\|P'\|_{L^\infty} \le 4\pi N \|P\|_{L^\infty}$.
[*Hint:* Prove first that $P'(x)/2\pi i N$ is equal to

$$\big((e^{-2\pi i N(\cdot)}P) * F_{N-1}\big)(x)\, e^{2\pi i N x} - \big((e^{2\pi i N(\cdot)}P) * F_{N-1}\big)(x)\, e^{-2\pi i N x}$$

and then take L^∞ norms.]

3.1.11. (*Fejér and F. Riesz*) Let $P(\xi) = \sum_{k=-N}^{N} a_k e^{2\pi i k \xi}$ be a trigonometric polynomial on $\mathbf{T}^1$ of degree N such that $P(\xi) > 0$ for all ξ. Prove that there exists a trigonometric polynomial $Q(\xi)$ of the form $\sum_{k=0}^{N} b_k e^{2\pi i k \xi}$ such that $P(\xi) = |Q(\xi)|^2$.
[*Hint:* Note that N zeros of the polynomial $R(z) = \sum_{k=-N}^{N} a_k z^{k+N}$ lie inside the unit circle and the other N lie outside.]

3.1.12. (*Landau [167]*) Points in $\mathbf{Z}^n$ are called *lattice points*. Follow the following steps to obtain the number of lattice points $N(R)$ inside a closed ball of radius R in $\mathbf{R}^n$. Let B be the closed unit ball in $\mathbf{R}^n$, χ_B its characteristic function, and v_n its volume.
(a) Using the results in Appendices B.6 and B.7, observe that there is a constant C_n such that for all $\xi \in \mathbf{R}^n$ we have

$$|\widehat{\chi_B}(\xi)| \le C_n (1+|\xi|)^{-\frac{n+1}{2}}.$$

(b) For $0 < \varepsilon < \frac{1}{10}$ let $\Phi^\varepsilon = \chi_{(1-\frac{\varepsilon}{2})B} * \zeta_\varepsilon$, where $\zeta_\varepsilon(x)\frac{1}{\varepsilon^n}\zeta(\frac{x}{\varepsilon})$ and ζ is a smooth function that is supported in $|x| \le \frac{1}{2}$ and has integral equal to 1. Also let $\Psi^\varepsilon = \chi_{(1+\frac{\varepsilon}{2})B} * \zeta_\varepsilon$. Prove that

$$\begin{aligned} &\Phi^\varepsilon(x) = 1 \text{ when } |x| \le 1-\varepsilon \quad \text{and} \quad \Phi^\varepsilon(x) = 0 \text{ when } |x| \ge 1, \\ &\Psi^\varepsilon(x) = 1 \text{ when } |x| \le 1 \quad \text{and} \quad \Psi^\varepsilon(x) = 0 \text{ when } |x| \ge 1+\varepsilon, \end{aligned}$$

and also that

$$\big|\widehat{\Phi^\varepsilon}(\xi)\big| + \big|\widehat{\Psi^\varepsilon}(\xi)\big| \le C_{n,N}(1+|\xi|)^{-\frac{n+1}{2}}(1+\varepsilon|\xi|)^{-N}$$

for every $\xi \in \mathbf{R}^n$ and N a large positive number.
(c) Use the result in (b) and the Poisson summation formula to obtain

$$\begin{aligned} \sum_{m\in\mathbf{Z}^n} \chi_B\big(\tfrac{m}{R}\big) \ge \sum_{m\in\mathbf{Z}^n} \Phi^\varepsilon\big(\tfrac{m}{R}\big) &= R^n\widehat{\Phi^\varepsilon}(0) + \sum_{m\in\mathbf{Z}^n\setminus\{0\}} R^n\widehat{\Phi^\varepsilon}(Rm) \\ &\ge v_n(1-\varepsilon)^n - C_{n,N}\sum_{m\in\mathbf{Z}^n\setminus\{0\}} R^n(1+R|m|)^{-\frac{n+1}{2}}(1+\varepsilon R|m|)^{-N}. \end{aligned}$$

Now use $(1-\varepsilon)^n \geq 1-n\varepsilon$ and pick ε such that $\varepsilon R^n = \varepsilon^{-\frac{n-1}{2}}$ to deduce the estimate $N(R) \geq v_n R^n + O(R^{n\frac{n-1}{n+1}})$ as $R \to \infty$. Argue similarly with Ψ^ε to obtain the identity

$$N(R) = v_n R^n + O(R^{n\frac{n-1}{n+1}}),$$

as $R \to \infty$.

3.1.13. (*Minkowski*) Let S be an open convex symmetric set in $\mathbf{R}^n$ and assume that the Fourier transform of its characteristic function satisfies the decay estimate

$$|\widehat{\chi_S}(\xi)| \leq C(1+|\xi|)^{-\frac{n+1}{2}}.$$

(This is the case if the boundary of S has nonzero Gaussian curvature.) Assume that $|S| > 2^n$. Prove that S contains at least one lattice point other than the origin.
[*Hint:* Assume the contrary, set $f = \chi_{\frac{1}{2}S} * \chi_{\frac{1}{2}S}$, and apply the Poisson summation formula to f to prove that $f(0) \geq \widehat{f}(0)$.]

3.2 Decay of Fourier Coefficients

In this section we investigate the interplay between the smoothness of a function and the decay of its Fourier coefficients.

3.2.1 Decay of Fourier Coefficients of Arbitrary Integrable Functions

We begin with the classical result asserting that the Fourier coefficients of any integrable function tend to zero at infinity. One should compare the following proposition with Proposition 2.2.17.

Proposition 3.2.1. ***(Riemann–Lebesgue lemma)*** *Let f be in $L^1(\mathbf{T}^n)$. Then $|\widehat{f}(m)| \to 0$ as $|m| \to \infty$.*

Proof. Given $f \in L^1(\mathbf{T}^n)$ and $\varepsilon > 0$, let P be a trigonometric polynomial such that $\|f-P\|_{L^1} < \varepsilon$. If $|m| > \text{degree}(P)$, then $\widehat{P}(m) = 0$ and thus

$$|\widehat{f}(m)| = |\widehat{f}(m) - \widehat{P}(m)| \leq \|f-P\|_{L^1} < \varepsilon.$$

This proves that $|\widehat{f}(m)| \to 0$ as $|m| \to \infty$. □

Several questions are naturally raised. How fast may the Fourier coefficients of an L^1 function tend to zero? Does additional smoothness of the function imply faster decay of the Fourier coefficients? Can such a decay be quantitatively expressed in terms of the smoothness of the function?

We answer the first question. Fourier coefficients of an L^1 function can tend to zero arbitrarily slowly, that is, more slowly than any given rate of decay.

Theorem 3.2.2. *Let $(d_m)_{m\in\mathbf{Z}^n}$ be a sequence of positive real numbers with $d_m \to 0$ as $|m| \to \infty$. Then there exists a $g \in L^1(\mathbf{T}^n)$ such that $|\widehat{g}(m)| \geq d_m$ for all $m \in \mathbf{Z}^n$. In other words, given any rate of decay, there exists an integrable function on the torus whose Fourier coefficients have slower rate of decay.*

We first prove this theorem when $n = 1$ and then extend it to higher dimensions. We need the following two lemmas.

Lemma 3.2.3. *Given a sequence of positive real numbers $\{a_m\}_{m=0}^{\infty}$ that tends to zero as $m \to \infty$, there exists a sequence $\{c_m\}_{m=0}^{\infty}$ that satisfies*

$$c_m \geq a_m, \qquad c_m \downarrow 0, \quad \text{and} \quad c_{m+2} + c_m \geq 2c_{m+1}$$

for all $m = 0, 1, \ldots$. We call such sequences convex.

Lemma 3.2.4. *Given a convex decreasing sequence $\{c_m\}_{m=0}^{\infty}$ of positive real numbers satisfying $\lim_{m\to\infty} c_m = 0$ and a fixed integer $s \geq 0$, we have that*

$$\sum_{r=0}^{\infty} (r+1)(c_{r+s} + c_{r+s+2} - 2c_{r+s+1}) = c_s. \tag{3.2.1}$$

We first prove Lemma 3.2.3.

Proof. Let $k_0 = 0$ and suppose that $a_m \leq M$ for all $m \geq 0$. Find $k_1 > k_0$ such that for $m \geq k_1$ we have $a_m \leq M/2$. Now find $k_2 > k_1 + \frac{k_1 - k_0}{2}$ such that for $m \geq k_2$ we have $a_m \leq M/4$. Next find $k_3 > k_2 + \frac{k_2 - k_1}{2}$ such that for $m \geq k_3$ we have $a_m \leq M/8$. Continue inductively in this way and construct a subsequence $k_0 < k_1 < k_2 < \cdots$ of the integers such that for $m \geq k_j$ we have $a_m \leq 2^{-j}M$ and $k_{j+1} > k_j + \frac{k_j - k_{j-1}}{2}$ for $j \geq 1$. Join the points $(k_0, 2M)$, (k_1, M), $(k_2, M/2)$, $(k_3, M/4)$, $\ldots$ by straight lines and note that by the choice of the subsequence $\{k_j\}_{j=0}^{\infty}$ the resulting piecewise linear function h is convex on $[0, \infty)$. Define $c_m = h(m)$ and observe that the sequence $\{c_m\}_{m=0}^{\infty}$ satisfies the required properties. See also Exercise 3.2.1 for an alternative proof. □

We now prove Lemma 3.2.4. The proof appears more natural after one has solved Exercise 3.2.3(a).

Proof. We have that

$$\begin{aligned}\sum_{r=0}^{N} &(r+1)(c_{r+s} + c_{r+s+2} - 2c_{r+s+1}) \\ &= c_s - (N+1)(c_{s+N+1} - c_{s+N+2}) - c_{s+N+1}.\end{aligned} \tag{3.2.2}$$

To show that the last expression tends to c_s as $N \to \infty$, we take $M = [\frac{N}{2}]$ and we use convexity $\left(c_{s+M+j} - c_{s+M+j+1} \geq c_{s+M+j+1} - c_{s+M+j+2}\right)$ to obtain

$$\begin{aligned}
c_{s+M+1} - c_{s+N+2} &= c_{s+M+1} - c_{s+M+2} \\
&\quad + c_{s+M+2} - c_{s+M+3} \\
&\quad + \cdots \\
&\quad + c_{s+N+1} - c_{s+N+2} \\
&\geq (N-M+1)(c_{s+N+1} - c_{s+N+2}) \\
&\geq \tfrac{N+1}{2}(c_{s+N+1} - c_{s+N+2}) \geq 0 .
\end{aligned}$$

The preceding calculation implies that $(N+1)(c_{s+N+1} - c_{s+N+2})$ tends to zero as $N \to \infty$ and thus the expression in (3.2.2) converges to c_s as $N \to \infty$. □

We now continue with the proof of Theorem 3.2.2 when $n = 1$.

Proof. We are given a sequence of positive numbers $\{a_m\}_{m\in\mathbf{Z}}$ that converges to zero as $|m| \to \infty$ and we would like to find an integrable function on $\mathbf{T}^1$ with $|\widehat{f}(m)| \geq a_m$ for all $m \in \mathbf{Z}$. Apply Lemma 3.2.3 to the sequence $\{a_m + a_{-m}\}_{m\geq 0}$ to find a convex sequence $\{c_m\}_{m\geq 0}$ that dominates $\{a_m + a_{-m}\}_{m\geq 0}$ and decreases to zero as $m \to \infty$. Extend c_m for $m < 0$ by setting $c_m = c_{|m|}$. Now define

$$f(x) = \sum_{j=0}^{\infty} (j+1)(c_j + c_{j+2} - 2c_{j+1}) F_j(x) , \tag{3.2.3}$$

where F_j is the (one-dimensional) Fejér kernel. The convexity of the sequence c_m and the positivity of the Fejér kernel imply that $f \geq 0$. Lemma 3.2.4 with $s = 0$ gives that

$$\sum_{j=0}^{\infty} (j+1)(c_j + c_{j+2} - 2c_{j+1}) \big\|F_j\big\|_{L^1} = c_0 < \infty , \tag{3.2.4}$$

since $\big\|F_j\big\|_{L^1} = 1$ for all j. Therefore (3.2.3) defines an integrable function f on $\mathbf{T}^1$. We now compute the Fourier coefficients of f. Since the series in (3.2.3) converges in L^1, for $m \in \mathbf{Z}$ we have

$$\begin{aligned}
\widehat{f}(m) &= \sum_{j=0}^{\infty} (j+1)(c_j + c_{j+2} - 2c_{j+1}) \widehat{F_j}(m) \\
&= \sum_{j=|m|}^{\infty} (j+1)(c_j + c_{j+2} - 2c_{j+1}) \left(1 - \frac{|m|}{j+1}\right) \\
&= \sum_{r=0}^{\infty} (r+1)(c_{r+|m|} + c_{r+|m|+2} - 2c_{r+|m|+1}) = c_{|m|} = c_m ,
\end{aligned} \tag{3.2.5}$$

where we used Lemma 3.2.4 with $s = |m|$.

Let us now extend this result on $\mathbf{T}^n$. Let $(d_m)_{m\in\mathbf{Z}^n}$ be a positive sequence with $d_m \to 0$ as $|m| \to \infty$. By Exercise 3.2.2, there exists a positive sequence $(a_j)_{j\in\mathbf{Z}}$ with

$a_{m_1} \cdots a_{m_n} \geq d_{(m_1,\dots,m_n)}$ and $a_j \to 0$ as $|j| \to \infty$. Let

$$g(x_1,\dots,,x_n) = f(x_1)\cdots f(x_n),$$

where f is the function previously constructed when $n = 1$. It can be seen easily using (3.1.6) that $\widehat{g}(m) \geq d_m$. □

3.2.2 Decay of Fourier Coefficients of Smooth Functions

We next study the decay of the Fourier coefficients of functions that possess a certain amount of smoothness. In this section we see that the decay of the Fourier coefficients reflects the smoothness of the function in a rather precise quantitative way. Conversely, if the Fourier coefficients of an integrable function have polynomial decay faster than the dimension, then a certain amount of smoothness can be inferred about the function.

Definition 3.2.5. For $0 \leq \gamma < 1$ define

$$\|f\|_{\dot{\Lambda}_\gamma} = \sup_{x,h\in\mathbf{T}^n} \frac{|f(x+h) - f(x)|}{|h|^\gamma}$$

and

$$\dot{\Lambda}_\gamma(\mathbf{T}^n) = \{f : \mathbf{T}^n \to \mathbf{C} \text{ with } \|f\|_{\dot{\Lambda}_\gamma} < \infty\}.$$

We call $\dot{\Lambda}_\gamma(\mathbf{T}^n)$ the *homogeneous Lipschitz space* of order γ on the torus. Functions f on $\mathbf{T}^n$ with $\|f\|_{\dot{\Lambda}_\gamma} < \infty$ are called *homogeneous Lipschitz* functions of order γ.

Some remarks are in order.

Remark 3.2.6. $\dot{\Lambda}_\gamma(\mathbf{T}^n)$ is called the *homogeneous Lipschitz space* of order γ on $\mathbf{T}^n$, in contrast to the space $\Lambda_\gamma(\mathbf{T}^n)$, which is called the Lipschitz space of order γ. The latter space is defined as

$$\Lambda_\gamma(\mathbf{T}^n) = \left\{f : \mathbf{T}^n \to \mathbf{C} \text{ with } \|f\|_{\Lambda_\gamma} < \infty\right\},$$

where

$$\|f\|_{\Lambda_\gamma} = \|f\|_{L^\infty} + \|f\|_{\dot{\Lambda}_\gamma}.$$

Remark 3.2.7. The positive functional $\|\cdot\|_{\dot{\Lambda}_\gamma}$ satisfies the triangle inequality, but it does not satisfy the property $\|f\|_{\dot{\Lambda}_\gamma} = 0 \implies f = 0$ a.e. required to be a norm. It is therefore a seminorm on $\dot{\Lambda}_\gamma(\mathbf{T}^n)$. However, if we identify functions whose difference is a constant, we form the space of all equivalence classes $\dot{\Lambda}_\gamma(\mathbf{T}^n)/\{\text{constants}\}$ (defined for $0 \leq \gamma < 1$) on which the functional $f \to \|f\|_{\dot{\Lambda}_\gamma}$ is a norm.

Remark 3.2.8. Homogeneous Lipschitz functions of order $\gamma = 0$ are bounded and of order $\gamma \in (0,1)$ are continuous and thus bounded. Therefore, $\dot{\Lambda}_\gamma(\mathbf{T}^n) \subseteq L^\infty(\mathbf{T}^n)$ set-theoretically. However, the norm inequality $\|f\|_{L^\infty} \le C\|f\|_{\dot{\Lambda}_\gamma}$ fails for any constant C independent of all functions f. Take, for example, $f = N + \sin(2\pi x)$ on $\mathbf{T}^1$ with $N \to \infty$ to obtain a counterexample. Nevertheless, under the identification of functions whose difference is a constant, the space $\dot{\Lambda}_\gamma(\mathbf{T}^n)$ embeds in $L^\infty(\mathbf{T}^n)$. To achieve this, fix a point $t_0 \in \mathbf{T}^n$ and define an embedding

$$f \mapsto f - f(t_0)$$

from $\dot{\Lambda}_\gamma(\mathbf{T}^n)$ to $L^\infty(\mathbf{T}^n)$. The kernel of this map is the space of all constant functions on $\mathbf{T}^n$, and thus $\dot{\Lambda}_\gamma/\{\text{constants}\}$ can be identified with a subspace of L^∞ for all $0 \le \gamma < 1$.

The following theorem clearly indicates how the smoothness of a function is reflected by the decay of its Fourier coefficients.

Theorem 3.2.9. *Let $s \in \mathbf{Z}$ with $s \ge 0$.*
(a) Suppose that $\partial^\alpha f$ exist and are integrable for all $|\alpha| \le s$. Then

$$|\widehat{f}(m)| \le \left(\frac{\sqrt{n}}{2\pi}\right)^s \frac{\sup\limits_{|\alpha|=s} |\widehat{\partial^\alpha f}(m)|}{|m|^s}, \qquad m \ne 0, \tag{3.2.6}$$

and thus $|\widehat{f}(m)|(1+|m|^s) \to 0$ as $|m| \to \infty$.
(b) Suppose that $\partial^\alpha f$ exist for all $|\alpha| \le s$ and whenever $|\alpha| = s$, $\partial^\alpha f$ are in $\dot{\Lambda}_\gamma(\mathbf{T}^n)$ for some $0 \le \gamma < 1$. Then

$$|\widehat{f}(m)| \le \frac{(\sqrt{n})^{s+\gamma}}{(2\pi)^s 2^{\gamma+1}} \frac{\sup\limits_{|\alpha|=s} \|\partial^\alpha f\|_{\dot{\Lambda}_\gamma}}{|m|^{s+\gamma}}, \qquad m \ne 0. \tag{3.2.7}$$

Proof. Fix $m \in \mathbf{Z}^n \setminus \{0\}$ and pick a j such that $|m_j| = \sup_{1 \le k \le n} |m_k|$. Then clearly $m_j \ne 0$. Integrating by parts s times with respect to the variable x_j, we obtain

$$\widehat{f}(m) = \int_{\mathbf{T}^n} f(x) e^{-2\pi i x\cdot m}\, dx = (-1)^s \int_{\mathbf{T}^n} (\partial_j^s f)(x) \frac{e^{-2\pi i x\cdot m}}{(-2\pi i m_j)^s}\, dx, \tag{3.2.8}$$

where the boundary terms all vanish because of the periodicity of the integrand. Taking absolute values and using $|m| \le \sqrt{n}\,|m_j|$, we obtain assertion (3.2.6).

We now turn to the second part of the theorem. Let $e_j = (0,\dots,1,\dots,0)$ be the element of the torus $\mathbf{T}^n$ whose jth coordinate is one and all the others are zero. A simple change of variables together with the fact that $e^{\pi i} = -1$ gives that

$$\int_{\mathbf{T}^n} (\partial_j^s f)(x) e^{-2\pi i x\cdot m}\, dx = -\int_{\mathbf{T}^n} (\partial_j^s f)(x - \tfrac{e_j}{2m_j}) e^{-2\pi i x\cdot m}\, dx,$$

which implies that

$$\int_{\mathbf{T}^n} (\partial_j^s f)(x) e^{-2\pi i x\cdot m}\, dx = \frac{1}{2}\int_{\mathbf{T}^n} \left[(\partial_j^s f)(x) - (\partial_j^s f)(x - \tfrac{e_j}{2m_j})\right] e^{-2\pi i x\cdot m}\, dx.$$

Now use the estimate

$$|(\partial_j^s f)(x) - (\partial_j^s f)(x - \tfrac{e_j}{2m_j})| \le \frac{\|\partial_j^s f\|_{\dot{\Lambda}_\gamma}}{(2|m_j|)^\gamma}$$

and identity (3.2.8) to conclude the proof of (3.2.7). □

The following is an immediate consequence.

Corollary 3.2.10. *Let $s \in \mathbf{Z}$ with $s \ge 0$.*
(a) Suppose that $\partial^\alpha f$ exist and are integrable for all $|\alpha| \le s$. Then for some constant $c_{n,s}$ we have

$$|\widehat{f}(m)| \le c_{n,s} \frac{\max\left(\|f\|_{L^1}, \sup_{|\alpha|=s} |\widehat{\partial^\alpha f}(m)|\right)}{(1+|m|)^s}. \tag{3.2.9}$$

(b) Suppose that $\partial^\alpha f$ exist for all $|\alpha| \le s$ and whenever $|\alpha| = s$, $\partial^\alpha f$ are in $\dot{\Lambda}_\gamma(\mathbf{T}^n)$ for some $0 \le \gamma < 1$. Then for some constant $c'_{n,s}$ we have

$$|\widehat{f}(m)| \le c'_{n,s} \frac{\max\left(\|f\|_{L^1}, \sup_{|\alpha|=s} \|\partial^\alpha f\|_{\dot{\Lambda}_\gamma}\right)}{(1+|m|)^{s+\gamma}}. \tag{3.2.10}$$

Remark 3.2.11. The conclusions of Theorem 3.2.9 and Corollary 3.2.10 are also valid when $\gamma = 1$. In this case the spaces $\dot{\Lambda}_\gamma$ should be replaced by the space Lip 1 equipped with the seminorm

$$\|f\|_{\mathrm{Lip}\,1} = \sup_{x,h\in\mathbf{T}^n} \frac{|f(x+h) - f(x)|}{|h|}.$$

There is a slight lack of uniformity in the notation here, since in the theory of Lipschitz spaces the notation $\dot{\Lambda}_1$ is usually reserved for the space with seminorm

$$\|f\|_{\dot{\Lambda}_1} = \sup_{x,h\in\mathbf{T}^n} \frac{|f(x+h) + f(x-h) - 2f(x)|}{|h|}.$$

The following proposition provides a partial converse to Theorem 3.2.9. We denote below by $[[s]]$ the largest integer strictly less than a given real number s.

Proposition 3.2.12. *Let $s > 0$ and suppose that f is an integrable function on the torus with*

$$|\widehat{f}(m)| \le C(1+|m|)^{-s-n}$$

for all $m \in \mathbf{Z}^n$. Then f has partial derivatives of all orders $|\alpha| \le [[s]]$, and for $0 < \gamma < s - [[s]]$, $\partial^\alpha f \in \dot{\Lambda}_\gamma$ for all multi-indices α satisfying $|\alpha| = [[s]]$.

Proof. Since f has an absolutely convergent Fourier series, Proposition 3.1.14 gives that

$$f(x) = \sum_{m\in\mathbf{Z}^n} \widehat{f}(m)e^{2\pi i x\cdot m}, \tag{3.2.11}$$

for almost all $x \in \mathbf{T}^n$.

The series in (3.2.11) can be differentiated with respect to ∂_x^α, where $|\alpha| = [[s]]$, since

$$\sum_{m\in\mathbf{Z}^n} \widehat{f}(m)\partial^\alpha e^{2\pi i x\cdot m} = \sum_{m\in\mathbf{Z}^n} \widehat{f}(m)(2\pi i m)^\alpha e^{2\pi i x\cdot m}$$

and the last series converges absolutely in view of the decay assumptions on the Fourier coefficients of f. Moreover, we have

$$(\partial^\alpha f)(x) = \sum_{m\in\mathbf{Z}^n} \widehat{f}(m)(2\pi i m)^\alpha e^{2\pi i x\cdot m}$$

for all multi-indices $(\alpha_1,\dots,\alpha_n)$ with $|\alpha| = [[s]]$. Now suppose that $0<\gamma<s-[[s]]$. Then

$$\begin{aligned}|(\partial^\alpha f)(x+h)-(\partial^\alpha f)(x)| &= \Big|\sum_{m\in\mathbf{Z}^n} \widehat{f}(m)(2\pi i m)^\alpha e^{2\pi i x\cdot m}\big(e^{2\pi i m\cdot h}-1\big)\Big| \\ &\le 2^{1-\gamma}(2\pi)^s \sum_{m\in\mathbf{Z}^n} |m|^{[[s]]}\frac{|h|^\gamma|m|^\gamma}{(1+|m|)^{n+s}} \\ &= C'_{\gamma,n,s}|h|^\gamma,\end{aligned}$$

where we used that $[[s]]+\gamma-s<0$ to obtain the convergence of the integral and the fact that

$$|e^{2\pi i m\cdot h}-1| \le \min(2, 2\pi|m|\,|h|) \le 2^{1-\gamma}(2\pi)^\gamma|m|^\gamma|h|^\gamma.$$

□

We have seen that if a function on $\mathbf{T}^1$ has an integrable derivative, then its Fourier coefficients tend to zero when divided by $|m|^{-1}$. In this case we say that the Fourier coefficients of f are $o(|m|^{-1})$ as $|m|\to\infty$. We denote by L_1^1 the class of all functions on $\mathbf{T}^1$ whose derivative is also in L^1. Next we introduce a slightly larger class of functions on $\mathbf{T}^1$ whose Fourier coefficients decay like $|m|^{-1}$ as $|m|\to\infty$.

Definition 3.2.13. A measurable function f on $\mathbf{T}^1$ is said to be of *bounded variation* if it is defined everywhere and

$$\mathrm{Var}(f) = \sup\Big\{\sum_{j=1}^M |f(x_j)-f(x_{j-1})| : 0=x_0<x_1<\cdots<x_M=1\Big\}<\infty,$$

where the supremum is taken over all partitions of the interval $[0,1]$. The expression $\mathrm{Var}(f)$ is called the *total variation* of f. The class of functions of bounded variation is denoted by BV.

The following result concerns functions of bounded variation.

Proposition 3.2.14. *If f is in $BV(\mathbf{T}^1)$, then*

$$|\widehat{f}(m)| \leq \frac{\mathrm{Var}(f)}{2\pi|m|}$$

whenever $m \neq 0$.

Proof. If f is a function of bounded variation, then the Lebesgue–Stieltjes integral with respect to f is well defined. Integration by parts gives

$$\widehat{f}(m) = \int_{\mathbf{T}^1} f(x)e^{-2\pi imx}\,dx = \int_{\mathbf{T}^1} \frac{e^{-2\pi imx}}{-2\pi im}\,df\,,$$

where the boundary terms vanish because of periodicity. The conclusion follows from the fact that the norm of the measure df is the total variation of f. □

For the sequences of Fourier coefficients $\{\widehat{f}(m)\}_m$ of functions f in the spaces

$$L_1^1(\mathbf{T}^1) \subseteq BV(\mathbf{T}^1) \subseteq L^\infty(\mathbf{T}^1)\,,$$

we have derived the following rate of decay, respectively,

$$o(|m|^{-1}),\ O(|m|^{-1}),\ o(1)\,,$$

as $|m| \to \infty$.

3.2.3 Functions with Absolutely Summable Fourier Coefficients

Decay for the Fourier coefficients can also be indirectly deduced from a certain knowledge about the summability of these coefficients. The simplest such kind of summability is in the sense of ℓ^1. It is therefore natural to consider the class of functions on the torus whose Fourier coefficients form an absolutely summable series.

Definition 3.2.15. An integrable function f on the torus is said to have an *absolutely convergent* Fourier series if

$$\sum_{m\in\mathbf{Z}^n} |\widehat{f}(m)| < +\infty.$$

We denote by $A(\mathbf{T}^n)$ the space of all integrable functions on the torus $\mathbf{T}^n$ whose Fourier series are absolutely convergent. We then introduce a norm on $A(\mathbf{T}^n)$ by setting

$$\|f\|_{A(\mathbf{T}^n)} = \sum_{m\in\mathbf{Z}^n} |\widehat{f}(m)|\,.$$

It is straightforward that every function in $A(\mathbf{T}^n)$ must be bounded. The following theorem gives us a sufficient condition for a function to be in $A(\mathbf{T}^n)$.

Theorem 3.2.16. *Let s be a nonnegative integer and let $0 \le \alpha < 1$. Assume that f is a function defined on $\mathbf{T}^n$ all of whose partial derivatives of order s lie in the space $\dot{\Lambda}_\alpha$. Suppose that $s+\alpha > n/2$. Then $f \in A(\mathbf{T}^n)$ and*

$$\big\|f\big\|_{A(\mathbf{T}^n)} \le C \sup_{|\beta|=s} \big\|\partial^\beta f\big\|_{\dot{\Lambda}_\alpha},$$

where C depends on n, α, and s.

Proof. For $1 \le j \le n$, let e_j be the element of $\mathbf{R}^n$ with zero entries except for the jth coordinate, which is 1. Let l be a positive integer and let $h_j = 2^{-l-2}e_j$.

Then for a multi-index $m = (m_1, \dots, m_n)$ satisfying $2^l \le |m| \le 2^{l+1}$ and for j in $\{1, \dots, n\}$ chosen such that $|m_j| = \sup_k |m_k|$ we have

$$\frac{|m_j|}{2^l} \ge \frac{|m|}{2^l\sqrt{n}} \ge \frac{1}{\sqrt{n}}.$$

We use the elementary fact that $|t| \le \pi \implies |e^{it}-1| \ge 2|t|/\pi$ to obtain

$$|e^{2\pi i m\cdot h_j} - 1| = |e^{2\pi i m_j 2^{-l-2}} - 1| \ge \frac{2}{\pi}\frac{|2\pi m_j|}{2^{l+2}} = \frac{|m_j|}{2^l} \ge \frac{1}{\sqrt{n}}$$

whenever $\frac{|2\pi m_j|}{2^{l+2}} \le \pi$, which is always true since $\frac{|2\pi m_j|}{2^{l+2}} \le \frac{2\pi 2^{l+1}}{2^{l+2}} \le \pi$.

We now have

$$\begin{aligned}
\Big(\sum_{2^l\le|m|<2^{l+1}} |\widehat{f}(m)|\Big)^2 &\le \Big(\sum_{2^l\le|m|<2^{l+1}} 1^2\Big)\Big(\sum_{2^l\le|m|<2^{l+1}} |\widehat{f}(m)|^2\Big) \\
&\le C_n 2^{ln} \sum_{j=1}^{n} \sum_{\substack{2^l\le|m|<2^{l+1}\\ |m_j|=\sup_k|m_k|}} |\widehat{f}(m)|^2 \\
&\le C_n' 2^{ln} \sum_{j=1}^{n} \sum_{\substack{2^l\le|m|<2^{l+1}\\ |m_j|=\sup_k|m_k|}} |e^{2\pi i m\cdot h_j}-1|^2 |\widehat{f}(m)|^2 \frac{|2\pi m_j|^{2s}}{|2\pi m_j|^{2s}} \\
&\le C_{n,s} 2^{l(n-2s)} \sum_{j=1}^{n} \sum_{m\in\mathbf{Z}^n} |e^{2\pi i m\cdot h_j}-1|^2 |\widehat{\partial_j^s f}(m)|^2 \\
&= C_{n,s} 2^{l(n-2s)} \sum_{j=1}^{n} \big\|\partial_j^s f - \partial_j^s f(\cdot + h_j)\big\|_{L^2}^2 \\
&\le C_{n,s}' 2^{l(n-2s)} (2^{-(l+3)})^{2\alpha} \sup_{|\beta|=s} \big\|\partial^\beta f\big\|_{\dot{\Lambda}_\alpha}^2.
\end{aligned}$$

Taking square roots, summing over all positive integers l, and using that $s+\alpha > n/2$, we obtain the desired conclusion. □

Exercises

3.2.1. Given a sequence $\{a_n\}_{n=0}^{\infty}$ of positive numbers such that $a_n \to 0$ as $n \to \infty$, find a nonnegative integrable function h on $[0,1]$ such that

$$\int_0^1 h(t)t^m\,dt \geq a_m.$$

Use this result to deduce a different proof of Lemma 3.2.3.
[*Hint:* Try $h = e\sum_{k=0}^{\infty}(\sup_{j\geq k} a_j - \sup_{j\geq k+1} a_j)(k+2)\chi_{[\frac{k+1}{k+2},1]}$.]

3.2.2. Prove that given a positive sequence $\{d_m\}_{m\in\mathbf{Z}^n}$ with $d_m \to 0$ as $|m| \to \infty$, there exists a positive sequence $\{a_j\}_{j\in\mathbf{Z}}$ with $a_{m_1}\cdots a_{m_n} \geq d_{(m_1,\dots,m_n)}$ and $a_j \to 0$ as $|j| \to \infty$.

3.2.3. (a) Use the idea of the proof of Lemma 3.2.4 to prove that if a twice continuously differentiable function $f \geq 0$ is defined on $(0,\infty)$ and satisfies $f'(x) \leq 0$ and $f''(x) \geq 0$ for all $x > 0$, then $\lim_{x\to\infty} xf'(x) = 0$.
(b) Suppose that a twice continuously differentiable function g is defined on $(0,\infty)$ and satisfies $g \geq 0$, $g' \leq 0$, and $\int_1^\infty g(x)\,dx < +\infty$. Prove that

$$\lim_{x\to\infty} xg(x) = 0.$$

3.2.4. Prove that for $0 \leq \gamma < \delta < 1$ we have $\|f\|_{\dot{\Lambda}_\gamma} \leq C_{n,\gamma,\delta}\|f\|_{\dot{\Lambda}_\delta}$ for all functions f and thus $\dot{\Lambda}_\delta$ is a subspace of $\dot{\Lambda}_\gamma$.

3.2.5. Prove the inclusions $L_1^1(\mathbf{T}^1) \subseteq BV(\mathbf{T}^1) \subseteq L^\infty(\mathbf{T}^1)$ as follows.
(a) If $f \in L_1^1(\mathbf{T}^1)$, then $\mathrm{Var}(f) \leq \|f'\|_{L^1}$.
(b) If $f \in BV(\mathbf{T}^1)$, then $\|f\|_{L^\infty} \leq \mathrm{Var}(f) + |f(0)|$.

3.2.6. Suppose that f is a differentiable function on $\mathbf{T}^1$ whose derivative f' is in $L^2(\mathbf{T}^1)$. Prove that $f \in A(\mathbf{T}^1)$ and that

$$\|f\|_{A(\mathbf{T}^1)} \leq \|f\|_{L^1} + \frac{1}{2\pi}\Big(\sum_{j\neq 0} j^{-2}\Big)^{1/2}\|f'\|_{L^2}.$$

3.2.7. (a) Prove that the product of two functions in $A(\mathbf{T}^n)$ is also in $A(\mathbf{T}^n)$ and that

$$\|fg\|_{A(\mathbf{T}^n)} \leq \|f\|_{A(\mathbf{T}^n)}\|g\|_{A(\mathbf{T}^n)}.$$

(b) Prove that the convolution of two square integrable functions on $\mathbf{T}^n$ always gives a function in $A(\mathbf{T}^n)$.

3.2.8. Fix $0 < \alpha < 1$ and define f on $\mathbf{T}^1$ by setting

$$f(x) = \sum_{k=0}^{\infty} 2^{-\alpha k} e^{2\pi i 2^k x}.$$

Prove that $f \in \dot{\Lambda}_\alpha$. Conclude that the decay $|\widehat{f}(m)| \le C|m|^{-\alpha}$ is best possible for $f \in \dot{\Lambda}_\alpha$.

3.2.9. Use without proof that there exists a constant $C > 0$ such that

$$\sup_{t\in\mathbf{R}} \left| \sum_{k=2}^{N} e^{ik\log k} e^{ikt} \right| \le C\sqrt{N}, \qquad N = 2,3,4,\dots,$$

to prove that the function

$$g(x) = \sum_{k=2}^{\infty} \frac{e^{ik\log k}}{k} e^{2\pi ikx}$$

is in $\dot{\Lambda}_{1/2}(\mathbf{T}^1)$ but not in $A(\mathbf{T}^1)$. Conclude that the restriction $s > 1/2$ in Theorem 3.2.16 is sharp.

3.2.10. Use a result from functional analysis to show that there exist sequences $\{a_m\}_{m\in\mathbf{Z}^n}$ that tend to zero as $|m| \to \infty$ for which there do not exist functions f in $L^1(\mathbf{T}^n)$ with $\widehat{f}(m) = a_m$ for all m.

3.3 Pointwise Convergence of Fourier Series

In this section we are concerned with the pointwise convergence of the square partial sums and the Fejér means of a function defined on the torus.

3.3.1 Pointwise Convergence of the Fejér Means

We saw in Section 3.1 that the Fejér kernel is an approximate identity. This implies that the Fejér (or Cesàro) means of an L^p function f on $\mathbf{T}^n$ converge to it in L^p for any $1 \le p < \infty$. Moreover, if f is continuous at x_0, then the means $(F(n,N) * f)(x_0)$ converge to $f(x_0)$ as $N \to \infty$ in view of Theorem 1.2.19 (2). Although this is a satisfactory result, it is restrictive, since it applies only to continuous functions. It is natural to ask what happens for more general functions.

Using properties of the Fejér kernel, we obtain the following one-dimensional result regarding the convergence of the Fejér means:

Theorem 3.3.1. *(Fejér) If a function f in $L^1(\mathbf{T}^1)$ has left and right limits at a point x_0, denoted by $f(x_0-)$ and $f(x_0+)$, respectively, then*

$$(F_N * f)(x_0) \to \frac{1}{2}\big(f(x_0+) + f(x_0-)\big) \qquad \text{as} \qquad N \to \infty. \tag{3.3.1}$$

In particular, this is the case for functions of bounded variation.

Proof. Let us identify $\mathbf{T}^1$ with $[-1/2, 1/2]$. Given $\varepsilon > 0$, find $\delta > 0$ ($\delta < 1/2$) such that

$$0 < t < \delta \implies \left| \frac{f(x_0+t)+f(x_0-t)}{2} - \frac{f(x_0+)+f(x_0-)}{2} \right| < \varepsilon. \tag{3.3.2}$$

Using the second expression for F_N in (3.1.15), we can find an $N_0 > 0$ such that for $N \ge N_0$ we have

$$\sup_{t\in[\delta,1/2]} F_N(t) < \varepsilon. \tag{3.3.3}$$

We now have

$$\begin{aligned} (F_N * f)(x_0) - f(x_0+) &= \int_{\mathbf{T}^1} F_N(-t)\big(f(x_0+t) - f(x_0+)\big)\,dt\,, \\ (F_N * f)(x_0) - f(x_0-) &= \int_{\mathbf{T}^1} F_N(t)\big(f(x_0-t) - f(x_0-)\big)\,dt\,. \end{aligned}$$

Averaging these two identities and using that the integrand is even, we obtain

$$\begin{aligned} &(F_N * f)(x_0) - \frac{f(x_0+)+f(x_0-)}{2} \\ &\quad = 2\int_0^{1/2} F_N(t)\left(\frac{f(x_0+t)+f(x_0-t)}{2} - \frac{f(x_0+)+f(x_0-)}{2}\right)dt\,. \end{aligned} \tag{3.3.4}$$

We split the integral in (3.3.4) into two pieces, the integral over $[0,\delta)$ and the integral over $[\delta, 1/2]$. By (3.3.2), the integral over $[0,\delta)$ is controlled by $\varepsilon \int_{\mathbf{T}^1} F_N(t)\,dt = \varepsilon$. Also (3.3.3) gives that for $N \ge N_0$

$$\begin{aligned} &\left| \int_\delta^{1/2} F_N(t)\left(\frac{f(x_0-t)+f(x_0+t)}{2} - \frac{f(x_0-)+f(x_0+)}{2}\right)dt \right| \\ &\qquad \le \frac{\varepsilon}{2}\big(\|f - f(x_0-)\|_{L^1} + \|f - f(x_0+)\|_{L^1}\big) = \varepsilon\, c(f, x_0)\,, \end{aligned}$$

where $c(f,x_0)$ is a constant depending on f and x_0. We have now proved that given $\varepsilon > 0$ there exists an N_0 such that for $N \ge N_0$ the second expression in (3.3.4) is bounded by $2\varepsilon\,(c(f,x_0)+1)$. This proves the required conclusion.

Functions of bounded variation can be written as differences of increasing functions, and since increasing functions have left and right limits everywhere, (3.3.1) holds for these functions. □

We continue with an elementary but very useful proposition. We refer to Exercise 3.3.2 for some of its applications.

Proposition 3.3.2. *(a) Let f be in $L^1(\mathbf{T}^n)$. If x_0 is a point of continuity of f and the square partial sums of the Fourier series of f converge at x_0, then they must converge to $f(x_0)$.*
(b) In dimension 1, if $f(x)$ has left and right limits as $x \to x_0$ and the partial sums of the Fourier series of f converge, then they must converge to $\frac{1}{2}\big(f(x_0+) + f(x_0-)\big)$.

Proof. (a) We observed before that if $f \in L^1(\mathbf{T}^n)$ is continuous at x_0, then

$$(F(n,N) * f)(x_0) \to f(x_0)$$

as $N \to \infty$. If $(D(n,N) * f)(x_0) \to A(x_0)$ as $N \to \infty$, then the arithmetic means of this sequence must converge to the same number as the sequence. Therefore,

$$(F(n,N) * f)(x_0) \to A(x_0)$$

as $N \to \infty$ and thus $A(x_0) = f(x_0)$. Part (b) is proved using the same argument and the result of Theorem 3.3.1. □

3.3.2 Almost Everywhere Convergence of the Fejér Means

We have seen that the Fejér means of a relatively nice function (such as of bounded variation) converge everywhere. What can we say about the Fejér means of a general integrable function? Since the Fejér kernel is a well-behaved approximate identity, the following result should not come as a surprise.

Theorem 3.3.3. *(a) For $f \in L^1(\mathbf{T}^n)$, let*

$$\mathscr{H}(f) = \sup_{N \in \mathbf{Z}^+} |f * F(n,N)|.$$

Then $\mathscr{H}$ maps $L^1(\mathbf{T}^n)$ to $L^{1,\infty}(\mathbf{T}^n)$ and $L^p(\mathbf{T}^n)$ to itself for $1 < p \le \infty$.
(b) For any function $f \in L^1(\mathbf{T}^n)$, we have

$$(F(n,N) * f)(x) \to f(x)$$

as $N \to \infty$ for almost all $x \in \mathbf{T}^n$.

Proof. It is an elementary fact that $|t| \le \frac{\pi}{2} \implies |\sin t| \ge \frac{2}{\pi}|t|$; see Appendix E. Using this fact and the expression (3.1.15) we obtain for all t in $[-\frac{1}{2}, \frac{1}{2}]$,

$$\begin{aligned} |F_N(t)| &= \frac{1}{N+1}\left|\frac{\sin(\pi(N+1)t)}{\sin(\pi t)}\right|^2 \\ &\le \frac{N+1}{4}\left|\frac{\sin(\pi(N+1)t)}{(N+1)t}\right|^2 \\ &\le \frac{N+1}{4}\min\left(\pi^2, \frac{1}{(N+1)^2t^2}\right) \\ &\le \frac{\pi^2}{2}\frac{N+1}{1+(N+1)^2|t|^2}. \end{aligned}$$

For $t \in \mathbf{R}$ let us set $\varphi(t) = (1+|t|^2)^{-1}$ and $\varphi_\varepsilon(t) = \frac{1}{\varepsilon}\varphi(\frac{t}{\varepsilon})$ for $\varepsilon > 0$. For $x = (x_1, \dots, x_n) \in \mathbf{R}^n$ and $\varepsilon > 0$ we also set

$$\Phi(x) = \varphi(x_1)\cdots\varphi(x_n)$$

and $\Phi_\varepsilon(x) = \varepsilon^{-n}\Phi(\varepsilon^{-1}x)$. Then for $|t| \le \frac{1}{2}$ we have $|F_N(t)| \le \frac{\pi^2}{2}\varphi_\varepsilon(t)$ with $\varepsilon = (N+1)^{-1}$, and for $y \in [-\frac{1}{2},\frac{1}{2}]^n$ we have

$$|F(n,N)(y)| \le (\tfrac{\pi^2}{2})^n\,\Phi_\varepsilon(y), \qquad \text{with } \varepsilon = (N+1)^{-1}.$$

Now let f be an integrable function on $\mathbf{T}^n$ and let f_0 denote its periodic extension on $\mathbf{R}^n$. For $x \in [-\frac{1}{2},\frac{1}{2}]^n$ we have

$$\begin{aligned}
\mathscr{H}(f)(x) &\le \sup_{N>0}\left|\int_{\mathbf{T}^n} F(n,N)(y) f(x-y)\,dy\right| \\
&\le (\tfrac{\pi^2}{2})^n \sup_{\varepsilon>0}\int_{[-\frac{1}{2},\frac{1}{2}]^n} |\Phi_\varepsilon(y)|\,|f_0(x-y)|\,dy \\
&\le 5^n \sup_{\varepsilon>0}\int_{\mathbf{R}^n} |\Phi_\varepsilon(y)|\,|(f_0\chi_Q)(x-y)|\,dy \\
&= 5^n \mathscr{G}(f_0\chi_Q)(x),
\end{aligned} \tag{3.3.5}$$

where Q is the cube $[-1,1]^n$ and $\mathscr{G}$ is the operator

$$\mathscr{G}(h) = \sup_{\varepsilon>0} |h| * \Phi_\varepsilon .$$

If we can show that $\mathscr{G}$ maps $L^1(\mathbf{R}^n)$ to $L^{1,\infty}(\mathbf{R}^n)$, the corresponding conclusion for $\mathscr{H}$ on $\mathbf{T}^n$ would follow from the fact $\mathscr{H}(f) \le 5^n\mathscr{G}(f_0\chi_Q)$ proved in (3.3.5) and the sequence of inequalities

$$\begin{aligned}
\big\|\mathscr{H}(f)\big\|_{L^{1,\infty}(\mathbf{T}^n)} &\le 5^n \big\|\mathscr{G}(f_0\chi_Q)\big\|_{L^{1,\infty}(\mathbf{R}^n)} \\
&\le 5^n C \big\|f_0\chi_Q\big\|_{L^1(\mathbf{R}^n)} \\
&= C'\big\|f\big\|_{L^1(\mathbf{T}^n)}.
\end{aligned}$$

Moreover, the L^p conclusion about $\mathscr{H}$ follows from the weak type $(1,1)$ result and the trivial L^∞ inequality, in view of the Marcinkiewicz interpolation theorem (Theorem 1.3.2). The required weak type $(1,1)$ estimate for $\mathscr{G}$ on $\mathbf{R}^n$ is a consequence of Lemma 3.3.4. This completes the proof of the statement in part (a) of the theorem. To prove the statement in part (b) observe that for $f \in \mathscr{C}^\infty(\mathbf{T}^n)$, which is a dense subspace of L^1, we have $F(n,N) * f \to f$ uniformly on $\mathbf{T}^n$ as $N \to \infty$, since the sequence $\{F_N\}_N$ is an approximate identity. Since by part (a), $\mathscr{H}$ maps $L^1(\mathbf{T}^n)$ to $L^{1,\infty}(\mathbf{T}^n)$, Theorem 2.1.14 applies and gives that for all $f \in L^1(\mathbf{T}^n)$, $F(n,N) * f \to f$ a.e. □

We now prove the weak type $(1,1)$ boundedness of $\mathscr{G}$ used earlier.

Lemma 3.3.4. *Let* $\Phi(x_1,\dots,x_n) = (1+|x_1|^2)^{-1}\cdots(1+|x_n|^2)^{-1}$ *and for* $\varepsilon > 0$ *let* $\Phi_\varepsilon(x) = \varepsilon^{-n}\Phi(\varepsilon^{-1}x)$. *Then the maximal operator*

$$\mathscr{G}(f) = \sup_{\varepsilon>0} |f| * \Phi_\varepsilon$$

maps $L^1(\mathbf{R}^n)$ *to* $L^{1,\infty}(\mathbf{R}^n)$.

Proof. Let $I_0 = [-1,1]$ and $I_k = \{t \in \mathbf{R} : 2^{k-1} \le |t| \le 2^k\}$ for $k = 1,2,\dots$. Also, let $\widetilde{I_k}$ be the convex hull of I_k, that is, the interval $[-2^k, 2^k]$. For $a_2,\dots,a_n$ fixed positive numbers, let $M_{a_2,\dots,a_n}$ be the maximal operator obtained by averaging a function on $\mathbf{R}^n$ over all products of closed intervals $J_1 \times \cdots \times J_n$ containing a given point with

$$|J_1| = 2^{a_2}|J_2| = \cdots = 2^{a_n}|J_n|.$$

In view of Exercise 2.1.9(c), we have that $M_{a_2,\dots,a_n}$ maps L^1 to $L^{1,\infty}$ with some constant independent of the a_j's. (This is due to the nice doubling property of this family of rectangles.) For a fixed $\varepsilon > 0$ we need to estimate the expression

$$(\Phi_\varepsilon * |f|)(0) = \int_{\mathbf{R}^n} \frac{|f(-\varepsilon y)|\,dy}{(1+y_1^2)\cdots(1+y_n^2)}.$$

Split $\mathbf{R}^n$ into $n!$ of regions of the form $|y_{j_1}| \ge \cdots \ge |y_{j_n}|$, where $\{j_1,\dots,j_n\}$ is a permutation of the set $\{1,\dots,n\}$. By the symmetry of the problem, let us look at the region $\mathscr{R}$ where $|y_1| \ge \cdots \ge |y_n|$. Then for some constant $C > 0$ we have

$$\int_{\mathscr{R}} \frac{|f(-\varepsilon y)|\,dy}{(1+y_1^2)\cdots(1+y_n^2)} \le C \sum_{k_1=0}^{\infty} \sum_{k_2=0}^{k_1} \cdots \sum_{k_n=0}^{k_{n-1}} 2^{-(2k_1+\cdots+2k_n)} \int_{I_{k_1}} \cdots \int_{I_{k_n}} |f(-\varepsilon y)|\,dy,$$

and the last expression can be trivially controlled by the corresponding expression, where the I_k's are replaced by the $\widetilde{I_k}$'s. This, in turn, is controlled by

$$C' \sum_{k_1=0}^{\infty} \sum_{k_2=0}^{k_1} \cdots \sum_{k_n=0}^{k_{n-1}} 2^{-(k_1+\cdots+k_n)} M_{k_1-k_2,\dots,k_1-k_n}(f)(0). \tag{3.3.6}$$

Now set $s_2 = k_1 - k_2, \dots, s_n = k_1 - k_n$, observe that $s_j \ge 0$, use that

$$2^{-(k_1+\cdots+k_n)} \le 2^{-k_1/2} 2^{-s_2/2n} \cdots 2^{-s_n/2n},$$

and change the indices of summation to estimate the expression in (3.3.6) by

$$C'' \sum_{k_1=0}^{\infty} \sum_{s_2=0}^{\infty} \cdots \sum_{s_n=0}^{\infty} 2^{-k_1/2} 2^{-s_2/2n} \cdots 2^{-s_n/2n} M_{s_2,\dots,s_n}(f)(0).$$

Argue similarly for the remaining regions $|y_{j_1}| \ge \cdots \ge |y_{j_n}|$ and translate to an arbitrary point x to obtain the estimate

$$|(\Phi_\varepsilon * f)(x)| \le C'' n! \sum_{s_2=0}^{\infty} \cdots \sum_{s_n=0}^{\infty} 2^{-s_2/2n} \cdots 2^{-s_n/2n} M_{s_2,\dots,s_n}(f)(x).$$

Now take the supremum over all $\varepsilon > 0$ and use the fact that the maximal functions $M_{s_2,\dots,s_n}$ map L^1 to $L^{1,\infty}$ uniformly in $s_2,\dots,s_n$ as well as the result of Exercise 1.4.10 to obtain the desired conclusion for $\mathscr{G}$. □

3.3.3 Pointwise Divergence of the Dirichlet Means

We now pass to the more difficult question of convergence of the square partial sums of a Fourier series. It is natural to start our investigation with the class of continuous functions. Do the partial sums of the Fourier series of continuous functions converge pointwise? The following simple proposition gives us a certain warning about the behavior of partial sums.

Proposition 3.3.5. ***(duBois Reymond)*** *There exist a continuous function f on $\mathbf{T}^n$ and an $x_0 \in \mathbf{T}^n$ such that the sequence*

$$(D(n,N) * f)(x_0) = \sum_{\substack{m\in\mathbf{Z}^n \\ |m_j|\le N}} \widehat{f}(m)e^{2\pi i x_0\cdot m}$$

satisfies

$$\limsup_{N\to\infty} |(D(n,N) * f)(x_0)| = \infty .$$

In other words, the square partial sums of a continuous function may diverge at a point.

Proof. It suffices to prove the proposition when $n = 1$. The one-dimensional example f can be easily transferred to n dimensions by considering the function $F(x_1,\dots,x_n) = f(x_1)$, which actually diverges on an $(n-1)$-dimensional plane.

We give a functional-analytic proof. For a constructive proof, see Exercise 3.3.6. Let $C(\mathbf{T}^1)$ be the Banach space of all continuous functions on the circle equipped with the L^∞ norm. Consider the continuous linear functionals

$$f \to T_N(f) = (D_N * f)(0)$$

on $C(\mathbf{T}^1)$ for $N = 1,2,\dots$. We show that the norms of the T_N's on $C(\mathbf{T}^1)$ converge to infinity as $N \to \infty$. To see this, given any integer $N \ge 100$, let $\varphi_N(x)$ be a continuous even function on $[-\frac{1}{2},\frac{1}{2}]$ that is bounded by 1 and is equal to the sign of $D_N(x)$ except at small intervals of length $(2N+1)^{-2}$ around the $2N+1$ zeros of D_N. Call the union of all these intervals B_N and set $A_N = [-\frac{1}{2},\frac{1}{2}] \setminus B_N$. Then

$$\int_{B_N} |D_N(x)|\,dx + \left| \int_{B_N} \varphi_N(x)D_N(x)\,dx \right| \le 2|B_N|(2N+1) = 2.$$

Using this estimate we obtain

$$\begin{aligned}\|T_N\|_{C(\mathbf{T}^1)\to\mathbf{C}} &\geq |T_N(\varphi_N)| = \left|\int_{\mathbf{T}^1} D_N(-x)\varphi_N(x)\,dx\right| \\ &\geq \int_{A_N} |D_N(x)|\,dx - \left|\int_{B_N} D_N(x)\varphi_N(x)\,dx\right| \\ &= \int_{\mathbf{T}^1} |D_N(x)|\,dx - \left|\int_{B_N} D_N(x)\varphi_N(x)\,dx\right| - \int_{B_N} |D_N(x)|\,dx \\ &\geq \frac{4}{\pi^2}\sum_{k=1}^{N}\frac{1}{k} - 2.\end{aligned}$$

It follows that the norms of the linear functionals T_N are not uniformly bounded. The uniform boundedness principle now implies the existence of an $f \in C(\mathbf{T}^1)$ and of a sequence $N_j \to \infty$ such that $|T_{N_j}(f)| \to \infty$ as $j \to \infty$. The Fourier series of this f diverges at $x = 0$. □

3.3.4 Pointwise Convergence of the Dirichlet Means

We have seen that continuous functions may have divergent Fourier series. How about Lipschitz continuous functions? As it turns out, there is a more general condition due to Dini that implies convergence for the Fourier series of functions that satisfy a certain integrability condition.

Theorem 3.3.6. *(Dini ($n = 1$), Tonelli ($n \geq 2$)) Let f be an integrable function on $\mathbf{T}^n$ and let $a = (a_1, \dots, a_n) \in \mathbf{T}^n$. If*

$$\int_{|x_1-a_1|\leq\frac{1}{2}} \cdots \int_{|x_n-a_n|\leq\frac{1}{2}} \frac{|f(x)-f(a)|}{|x_1-a_1|\cdots|x_n-a_n|}\,dx < \infty, \tag{3.3.7}$$

then we have $(D(n,N)*f)(a) \to f(a)$.

Proof. Replacing $f(x)$ by $f(x+a) - f(a)$, we may assume that $a = 0$ and $f(a) = 0$. Using identities (3.1.12) and (3.1.11), we can write

$$\begin{aligned}(D(n,N)*f)(0) &= \int_{\mathbf{T}^n} f(-x)\prod_{j=1}^{n}\frac{\sin((2N+1)\pi x_j)}{\sin(\pi x_j)}\,dx \\ &= \int_{\mathbf{T}^n} f(-x)\prod_{j=1}^{n}\left(\frac{\sin(2N\pi x_j)\cos(\pi x_j)}{\sin(\pi x_j)} + \cos(2N\pi x_j)\right)dx.\end{aligned} \tag{3.3.8}$$

Expanding out the product, we obtain a sum of terms each of which contains a factor of $\cos(2N\pi x_j)$ or $\sin(2N\pi x_j)$ and a term of the form

$$f(-x)\prod_{j\in I}\frac{\cos(\pi x_j)}{\sin(\pi x_j)}, \tag{3.3.9}$$

where I is a subset of $\{1,2,\dots,n\}$. The function in (3.3.9) is integrable on $[-\frac{1}{2},\frac{1}{2}]^n$ except possibly in a neighborhood of the origin. But condition (3.3.7) with $a=0$ guarantees that any function of the form (3.3.9) is also integrable in a neighborhood of the origin. It is now a consequence of the Riemann–Lebesgue lemma (Lemma 3.2.1) that the expression in (3.3.8) tends to zero as $N\to\infty$. □

The following are consequences of Dini's test.

Corollary 3.3.7. ***(Riemann's principle of localization)*** *Let $f\in L^1(\mathbf{T}^1)$ and assume that f vanishes on an open interval I. Then $D_N * f$ converges to zero on the ball I.*

Proof. Simply observe that (3.3.7) holds in this case. □

Corollary 3.3.8. *Let $a\in\mathbf{T}^n$ and suppose that $f\in L^1(\mathbf{T}^n)$ satisfies*

$$|f(x)-f(a)|\le C|x_1-a_1|^{\delta_1}\cdots|x_n-a_n|^{\delta_n}$$

for some $C,\delta_j>0$. (When $n=1$, this is saying that f is Lipschitz continuous.) Then the square partial sums $(D(n,N)*f)(a)$ converge to $f(a)$.

Proof. Note that condition (3.3.7) holds. □

Corollary 3.3.9. ***(Dirichlet)*** *If f is defined on $\mathbf{T}^1$ and is a differentiable function at a point a in $\mathbf{T}^1$, then $(D_N*f)(a)\to f(a)$.*

Proof. There exists a $\delta>0$ (say less than $1/2$) such that $|f(x)-f(a)|/|x-a|$ is bounded by $|f'(a)|+1$ for $|x-a|\le\delta$. Also $|f(x)-f(a)|/|x-a|$ is bounded by $|f(x)-f(a)|/\delta$ when $|x-a|>\delta$. It follows that condition (3.3.7) holds. □

Exercises

3.3.1. Identify $\mathbf{T}^1$ with $[-1/2,1/2)$ and fix $0<b<1/2$. Prove the following:

(a) The mth Fourier coefficient of the function x is $i\frac{(-1)^m}{2\pi m}$ when $m\neq 0$ and zero when $m=0$.
(b) The mth Fourier coefficient of the function $\chi_{[-b,b]}$ is $\dfrac{\sin(2\pi bm)}{m\pi}$.
(c) The mth Fourier coefficient of the function $\left(1-\frac{|x|}{b}\right)_+$ is $\dfrac{\sin^2(\pi bm)}{bm^2\pi^2}$.
(d) The mth Fourier coefficient of the function $|x|$ is $-\frac{1}{2m^2\pi^2}+\frac{(-1)^m}{2m^2\pi^2}$ when $m\neq 0$ and $\frac{1}{4}$ when $m=0$.
(e) The mth Fourier coefficient of the function x^2 is $\frac{(-1)^m}{2m^2\pi^2}$ when $m\neq 0$ and $\frac{1}{12}$ when $m=0$.
(f) The mth Fourier coefficient of the function $\cosh(2\pi x)$ is $\frac{(-1)^m}{1+m^2}\frac{\sinh\pi}{\pi}$.
(g) The mth Fourier coefficient of the function $\sinh(2\pi x)$ is $\frac{im(-1)^m}{1+m^2}\frac{\sinh\pi}{\pi}$.

3.3.2. Use Exercise 3.3.1 and Proposition 3.3.2 to prove that

$$\sum_{k\in\mathbf{Z}} \frac{1}{(2k+1)^2} = \frac{\pi^2}{4}, \qquad \sum_{k\in\mathbf{Z}\setminus\{0\}} \frac{1}{k^2} = \frac{\pi^2}{3},$$

$$\sum_{k\in\mathbf{Z}\setminus\{0\}} \frac{(-1)^{k+1}}{k^2} = \frac{\pi^2}{6}, \qquad \sum_{k\in\mathbf{Z}} \frac{(-1)^k}{k^2+1} = \frac{2\pi}{e^{\pi}-e^{-\pi}}.$$

3.3.3. Let $M > N$ be given positive integers.
(a) For $f \in L^1(\mathbf{T}^1)$, prove the following identity:

$$\begin{aligned}(D_N * f)(x) &= \frac{M+1}{M-N}(F_M * f)(x) - \frac{N+1}{M-N}(F_N * f)(x) \\ &\quad - \frac{M+1}{M-N} \sum_{N<|j|\le M} \left(1 - \frac{|j|}{M+1}\right) \widehat{f}(j) e^{2\pi i j x}.\end{aligned}$$

(b) (*G. H. Hardy*) Suppose that a function f on $\mathbf{T}^1$ satisfies the following condition: there exists an $a > 0$ such that for any $\varepsilon > 0$ there is a $k_0 > 0$ such that for all $k \ge k_0$ we have

$$\sum_{k<|m|\le[ak]} |\widehat{f}(m)| < \varepsilon.$$

Use part (a) to prove that if $(F_N * f)(x)$ converges to $A(x)$ as $N \to \infty$, then $(D_N * f)(x)$ also converges to $A(x)$ as $N \to \infty$.

3.3.4. Use Exercise 3.3.3 to prove that if f is a function of bounded variation on $\mathbf{T}^1$, then

$$(D_N * f)(x) \to \frac{1}{2}(f(x+0) + f(x-0))$$

for every $t \in \mathbf{T}^1$. Apply this result to the function $\chi_{[-b,b]}$ of Exercise 3.3.1(b) to obtain that

$$\lim_{N\to\infty} \sum_{m=-N}^{N} \frac{\sin(2\pi b m)}{m\pi} e^{2\pi i b m} = \frac{1}{2} - 2b.$$

3.3.5. (a) Prove that the Riemann–Lebesgue lemma holds uniformly on compact subsets of $L^1(\mathbf{T}^n)$. This means that given any compact subset of $L^1(\mathbf{T}^n)$ and $\varepsilon > 0$ there exists an $N_0 > 0$ such that for $|m| \ge N_0$ we have $|\widehat{f}(m)| \le \varepsilon$ for all $f \in K$.
(b) Use part (a) to prove the following sharpening of the localization theorem. If f vanishes on an open ball B in $\mathbf{T}^n$, then $D(n,N) * f$ converges to zero uniformly on compact subsets of B.

3.3.6. Follow the steps given to obtain a constructive proof of a continuous function whose Fourier series diverges at a point. On $\mathbf{T}^1$ let $g(x) = -2\pi i(x - 1/2)$.
(a) Prove that $\widehat{g}(m) = 1/m$ when $m \ne 0$ and zero otherwise.
(b) Prove that for all nonnegative integers M and N we have

$$\big((e^{2\pi i N(\cdot)}(g*D_N))*D_M\big)(x) = e^{2\pi i N x} \sum_{1\le |r|\le N} \frac{1}{r} e^{2\pi i r x}$$

when $M \ge 2N$ and

$$\big((e^{2\pi i N(\cdot)}(g*D_N))*D_M\big)(x) = e^{2\pi i N x} \sum_{\substack{-N\le r\le M-N\\ r\neq 0}} \frac{1}{r} e^{2\pi i r x}$$

when $M < 2N$. Conclude that there exists a constant $C > 0$ such that for all M, N, and $x \neq 0$ we have

$$\big|\big(e^{2\pi i N(\cdot)}(g*D_N)*D_M\big)(x)\big| \le \frac{C}{|x|}.$$

(c) Show that there exists a constant $C_1 > 0$ such that

$$\sup_{N>0}\sup_{x\in\mathbf{T}^1} \big|(g*D_N)(x)\big| = \sup_{N>0}\sup_{x\in\mathbf{T}^1} \left|\sum_{1\le |r|\le N} \frac{1}{r} e^{2\pi i r x}\right| \le C_1 < \infty.$$

(d) Let $\lambda_k = 1 + e^{e^k}$. Define

$$f(x) = \sum_{k=1}^{\infty} \frac{1}{k^2} e^{2\pi i \lambda_k x}(g*D_{\lambda_k})(x)$$

and prove that f is continuous on $\mathbf{T}^1$ and that its Fourier series converges at every $x \neq 0$, but $\limsup_{M\to\infty} |(f*D_M)(0)| = \infty$.
$\big[$*Hint:* Take $M = e^{e^m}$ with $m \to \infty$.$\big]$

3.4 Divergence of Fourier Series and Bochner–Riesz Summability

We saw in the previous section that the Fourier series of a continuous function may diverge at a point. As expected, the situation can only get worse as the functions get worse. In this section we present an example, due to A. N. Kolmogorov, of an integrable function on $\mathbf{T}^1$ whose Fourier series diverges almost everywhere. Using this example, we may construct integrable functions on $\mathbf{T}^n$ whose square Dirichlet means diverge a.e.; see Exercise 3.4.1.

3.4.1 Motivation for Bochner–Riesz Summability

We now consider an analogous question for the circular Dirichlet means of integrable functions on $\mathbf{T}^n$. In dimension 1 we saw that the Fejér means of integrable functions are better behaved than their Dirichlet means. We investigate whether there is a similar phenomenon in higher dimensions. Recall that the circular (or spherical) partial sums of the Fourier series of f are given by

$$(f * \widetilde{D}(n,R))(x) = \sum_{\substack{m\in\mathbf{Z}^n\\|m|\le R}} \widehat{f}(m)e^{2\pi i m\cdot x},$$

where $R \ge 0$. Taking the averages of these expressions, we obtain

$$\frac{1}{R}\int_0^R (f * \widetilde{D}(n,r))(x)\,dr = \sum_{\substack{m\in\mathbf{Z}^n\\|m|\le R}} \big(1-\tfrac{|m|}{R}\big)\widehat{f}(m)e^{2\pi i m\cdot x},$$

and we call these expressions the *circular Cesàro means* (or *circular Fejér means*) of f. It turns out that the circular Cesàro means of integrable functions on $\mathbf{T}^2$ always converge in L^1, but in dimension 3, this may fail. Theorem 3.4.6 gives an example of an integrable function f on $\mathbf{T}^3$ whose circular Cesàro means diverge a.e. However, we show that this is not the case if the circular Cesàro means of a function f in $L^1(\mathbf{T}^3)$ are replaced by the only slightly different-looking means

$$\sum_{\substack{m\in\mathbf{Z}^n\\|m|\le R}} \left(1-\tfrac{|m|}{R}\right)^{1+\varepsilon} \widehat{f}(m)e^{2\pi i m\cdot x},$$

for some $\varepsilon > 0$. The previous discussion suggests that the preceding expressions behave better as ε increases, but for a fixed ε they get worse as the dimension increases. To study this situation more carefully, we define the family of operators for which the exponent $1+\varepsilon$ is replaced by a general nonnegative index $\alpha \ge 0$.

Definition 3.4.1. Let $\alpha \ge 0$. The *Bochner–Riesz means* of order α of an integrable function f on $\mathbf{T}^n$ are defined as follows:

$$B_R^\alpha(f)(x) = \sum_{\substack{m\in\mathbf{Z}^n\\|m|\le R}} \left(1-\tfrac{|m|^2}{R^2}\right)^{\alpha} \widehat{f}(m)e^{2\pi i m\cdot x}. \tag{3.4.1}$$

This family of operators forms a natural "spherical" analogue of the Cesàro–Fejér sums. It turns out that there is no different behavior of the means if the expression $\big(1-\frac{|m|^2}{R^2}\big)^{\alpha}$ in (3.4.1) is replaced by the expression $\big(1-\frac{|m|}{R}\big)^{\alpha}$. See Exercise 3.6.1, on the equivalence of means generated by these two expressions. The advantage of the quadratic expression in (3.4.1) is that it has an easily computable kernel. Moreover, the appearance of the quadratic term in the definition of the Bochner–Riesz means is responsible for the following reproducing formula:

$$B_R^\alpha(f) = \frac{2\Gamma(\alpha+1)}{\Gamma(\alpha-\beta)\Gamma(\beta+1)}\,\frac{1}{R}\int_0^R \left(1-\frac{r^2}{R^2}\right)^{\alpha-\beta-1}\left(\frac{r^2}{R^2}\right)^{\beta+\frac{1}{2}} B_r^\beta(f)\,dr, \tag{3.4.2}$$

which precisely quantifies the way in which B_R^α is smoother than B_R^β when $\alpha > \beta$. Identity (3.4.2) also says that when $\alpha > \beta$, the operator $B_R^\alpha(f)$ is an average of the operators $B_r^\beta(f)$, $0 < r < R$, with respect to a certain density.

Note that the Bochner–Riesz means of order zero coincide with the circular (or spherical) Dirichlet means, and as we have seen, these converge in $L^2(\mathbf{T}^n)$. We now indicate why the Bochner–Riesz means $B_R^\alpha(f)$ converge to f in $L^1(\mathbf{T}^n)$ as $R \to \infty$ when $\alpha > (n-1)/2$. Consider the function

$$m_\alpha(\xi) = (1 - |\xi|^2)_+^\alpha$$

defined for ξ in $\mathbf{R}^n$. Using an identity proved in Appendix B.5, we have that

$$(m_\alpha)^\vee(x) = K^\alpha(x) = \frac{\Gamma(\alpha+1)}{\pi^\alpha} \frac{J_{\frac{n}{2}+\alpha}(2\pi|x|)}{|x|^{\frac{n}{2}+\alpha}}, \tag{3.4.3}$$

where J_λ is the Bessel function of order λ. The estimates in Appendices B.6 and B.7 yield that if $\alpha > (n-1)/2$, then the function K^α obeys the inequality

$$|K^\alpha(x)| \le C_{n,\alpha}(1+|x|)^{-n-(\alpha-\frac{n-1}{2})}, \tag{3.4.4}$$

and hence it is in $L^1(\mathbf{R}^n)$. Using the Poisson summation formula, we write

$$\begin{aligned} B_R^\alpha(f)(x) &= \sum_{l\in\mathbf{Z}^n} m_\alpha(\tfrac{l}{R})\widehat{f}(l)e^{2\pi i l\cdot x} \\ &= \sum_{l\in\mathbf{Z}^n} (f * (K^\alpha)_{1/R})(x+l) \\ &= (f * (L^\alpha)_{1/R})(x), \end{aligned}$$

where $L^\alpha(x) = \sum_{k\in\mathbf{Z}^n} K^\alpha(x+k)$ and $g_{1/R}(x) = R^n g(Rx)$. Using (3.4.4), we show easily that the function L^α is an integrable 1-periodic function on $\mathbf{T}^n$. Moreover,

$$\int_{\mathbf{T}^n} L^\alpha(t)\,dt = \int_{\mathbf{R}^n} K^\alpha(x)\,dx = m_\alpha(0) = 1\,.$$

This fact suggests that when $\alpha > \frac{n-1}{2}$, the family $\{(L^\alpha)_\varepsilon\}_{\varepsilon>0}$ is an approximate identity on $\mathbf{T}^n$ as $\varepsilon \to 0$. To see this we need only to verify the third property in Definition 1.2.15. For $\delta < \frac{1}{2}$ using (3.4.4) we have

$$\frac{1}{\varepsilon^n}\int_{\frac{1}{2}\ge|x_j|\ge\delta} \left|L^\alpha(x/\varepsilon)\right|dx \le C_{n,\alpha}\varepsilon^{\alpha-\frac{n-1}{2}} \int_{\frac{1}{2}\ge|x_j|\ge\delta} \sum_{\ell\in\mathbf{Z}^n} \frac{1}{|x+\ell|^{n+\alpha-\frac{n-1}{2}}}\,dx \to 0$$

as $\varepsilon \to 0$, since the sum over ℓ converges uniformly in $x \in [-1/2,1/2]^n \setminus [-\delta,\delta]^n$.

Using Theorem 1.2.19, we obtain these conclusions for $\alpha > (n-1)/2$:

(a) For $f \in L^p(\mathbf{T}^n)$, $1 \le p < \infty$, $B_R^\alpha(f)$ converge to f in L^p as $R \to \infty$.
(b) For f continuous on $\mathbf{T}^n$, $B_R^\alpha(f)$ converge to f uniformly as $R \to \infty$.

One may wonder whether there are analogous results for $\alpha \le (n-1)/2$. Theorem 3.4.6 warns that the Bochner–Riesz means may diverge in L^1 when $\alpha = (n-1)/2$. For this reason, the number $\alpha = (n-1)/2$ is referred to as the *critical index*. The question of determining the range of α's for which the Bochner–Riesz means of order α converge in $L^p(\mathbf{T}^n)$ when $1 < p < \infty$ is investigated in Chapter 10.

3.4.2 Divergence of Fourier Series of Integrable Functions

It is natural to start our investigation with the case $n = 1$. We begin with the following important result:

Theorem 3.4.2. *There exists an integrable function on the circle $\mathbf{T}^1$ whose Fourier series diverges almost everywhere.*

Proof. The proof of this theorem is a bit involved, and we need a sequence of lemmas, which we prove first.

Lemma 3.4.3. ***(Kronecker)*** *Suppose that $n \in \mathbf{Z}^+$ and*

$$\{x_1, x_2, \dots, x_n, 1\}$$

is a linearly independent set over the rationals. Then for any $\varepsilon > 0$ and any complex numbers $z_1, z_2, \dots, z_n$ with $|z_j| = 1$, there exists an integer $m \in \mathbf{Z}$ such that

$$|e^{2\pi i m x_j} - z_j| < \varepsilon \qquad \textit{for all} \quad 1 \le j \le n.$$

Proof. Identifying $\mathbf{T}^n$ with the set $\{(e^{2\pi i t_1}, \dots, e^{2\pi i t_n}) : 0 \le t_j \le 1\}$, the required conclusion is a consequence of the fact that for a fixed $x = (x_1, \dots, x_n)$ the set $\{mx : m \in \mathbf{Z}\}$ is dense in $\mathbf{T}^n$. If this were not the case, then there would exist an open set U in $\mathbf{T}^n$ that contains no elements of the set $\{mx : m \in \mathbf{Z}\}$. Pick a smooth, nonzero, and nonnegative function f on $\mathbf{T}^n$ supported in U. Then $f(mx) = 0$ for all $m \in \mathbf{Z}$, but

$$\widehat{f}(0) = \int_{\mathbf{T}^n} f(x)\, dx > 0\,.$$

Then we have

$$\begin{aligned} 0 = \frac{1}{N}\sum_{m=0}^{N-1} f(mx) &= \frac{1}{N}\sum_{m=0}^{N-1}\Big(\sum_{l\in\mathbf{Z}^n} \widehat{f}(l) e^{2\pi i l\cdot mx}\Big) \\ &= \sum_{l\in\mathbf{Z}^n} \widehat{f}(l)\Big(\frac{1}{N}\sum_{m=0}^{N-1} e^{2\pi i m(l\cdot x)}\Big) \\ &= \sum_{l\in\mathbf{Z}^n\setminus\{0\}} \widehat{f}(l)\Big(\frac{1}{N}\frac{e^{2\pi i N(l\cdot x)}-1}{e^{2\pi i(l\cdot x)}-1}\Big) + \widehat{f}(0)\,. \end{aligned}$$

In the last identity we used the fact that $e^{2\pi i(l\cdot x)} \neq 1$, since by assumption the set $\{x_1, x_2, \dots, x_n, 1\}$ is linearly independent over the rationals. But the expression inside the parentheses above is bounded by 1 and tends to 0 as $N \to \infty$. Since

$$|\widehat{f}(l)| \leq C(f,n)(1+|l|)^{-100n},$$

taking limits as $N \to \infty$ and using the Lebesgue dominated convergence theorem, we obtain that

$$\begin{aligned} 0 &= \lim_{N\to\infty} \sum_{l \in \mathbf{Z}^n \setminus \{0\}} \widehat{f}(l) \left(\frac{1}{N} \frac{e^{2\pi i N(l\cdot x)} - 1}{e^{2\pi i(l\cdot x)} - 1} \right) + \widehat{f}(0) \\ &= \widehat{f}(0), \end{aligned}$$

which contradicts our assumption on f. □

Lemma 3.4.4. *Let N be a large positive integer. Then there exists a positive measure μ_N on $\mathbf{T}^1$ with $\mu_N(\mathbf{T}^1) = 1$ such that*

$$\sup_{L\geq 1} \big|(\mu_N * D_L)(x)\big| = \sup_{L\geq 1} \left| \sum_{k=-L}^{L} \widehat{\mu_N}(k) e^{2\pi i k x} \right| \geq c \log N \tag{3.4.5}$$

for almost all $x \in \mathbf{T}^1$ (c is a fixed constant).

Proof. We choose points $0 \leq x_1 < x_2 < \cdots < x_N \leq 1$ such that

$$\frac{1}{2N} \leq |x_{j+1} - x_j| \leq \frac{2}{N}, \qquad 1 \leq j \leq N, \tag{3.4.6}$$

where we defined $x_{N+1} = x_1 + 1$, and such that the set

$$\{x_1, \dots, x_N, 1\}$$

is linearly independent over the rationals. Let

$$E_N = \big\{x \in [0,1] : \{x - x_1, \dots, x - x_N, 1\} \quad \text{is linearly independent over } \mathbf{Q}\big\}$$

and observe that almost all[1] x in $\mathbf{T}^1$ belong to E_N.

Next, we define the probability measure

$$\mu_N = \frac{1}{N} \sum_{j=1}^{N} \delta_{x_j},$$

where δ_{x_j} are Dirac delta masses at the points x_j. For this measure we have

[1] Every x in $[0,1] \setminus \mathbf{Q}[x_1, \dots, x_N]$ belongs to E_N. Here $\mathbf{Q}[x_1, \dots, x_N]$ denotes the field extension of $\mathbf{Q}$ obtained by attaching to it the linearly independent elements $\{x_1, \dots, x_N\}$.

$$\begin{aligned}\left|\sum_{k=-L}^{L}\widehat{\mu_N}(k)e^{2\pi ikx}\right| &= \left|\sum_{k=-L}^{L}\left(\frac{1}{N}\sum_{j=1}^{N}e^{-2\pi ikx_j}\right)e^{2\pi ikx}\right| \\ &= \left|\frac{1}{N}\sum_{j=1}^{N}D_L(x-x_j)\right| \\ &= \left|\frac{1}{N}\sum_{j=1}^{N}\frac{\sin(2\pi(L+\frac{1}{2})(x-x_j))}{\sin(\pi(x-x_j))}\right| \\ &= \left|\frac{1}{N}\sum_{j=1}^{N}\frac{\operatorname{Im}\left[e^{2\pi i(L+\frac{1}{2})(x-x_j)}\right]\operatorname{sgn}\big(\sin(\pi(x-x_j))\big)}{|\sin(\pi(x-x_j))|}\right|,\end{aligned} \tag{3.4.7}$$

where the signum function is defined as $\operatorname{sgn} a = 1$ for $a > 0$, -1 for $a < 0$, and zero if $a = 0$. By Lemma 3.4.3, for all $x \in E_N$ there exists an $L \in \mathbf{Z}^+$ such that

$$\left|e^{2\pi iL(x-x_j)} - ie^{-2\pi i\frac{1}{2}(x-x_j)}\operatorname{sgn}\big(\sin(\pi(x-x_j))\big)\right| < \frac{1}{2},$$

which can be equivalently written as

$$\left|e^{2\pi i(L+\frac{1}{2})(x-x_j)}\operatorname{sgn}\big(\sin(\pi(x-x_j))\big) - i\right| < \frac{1}{2}. \tag{3.4.8}$$

It follows from (3.4.8) that

$$\operatorname{Im}\left[e^{2\pi i(L+\frac{1}{2})(x-x_j)}\right]\operatorname{sgn}\big(\sin(\pi(x-x_j))\big) > \frac{1}{2}.$$

Combining this with the result of the calculation in (3.4.7), we obtain that

$$\left|\sum_{k=-L}^{L}\widehat{\mu_N}(k)e^{2\pi ikx}\right| > \frac{1}{2N}\sum_{j=1}^{N}\frac{1}{|\sin(\pi(x-x_j))|} \geq \frac{1}{2\pi N}\sum_{j=1}^{N}\frac{1}{|x-x_j|}.$$

But for every $x \in [0,1]$ there exists a j_0 such that $x \in [x_{j_0}, x_{j_0+1})$. It follows from (3.4.6) that $|x - x_j| \leq C(|j - j_0| + 1)N^{-1}$ and thus

$$\sum_{j=1}^{N}\frac{1}{|x-x_j|} \geq c'N\log N.$$

Thus for every $x \in E_N$ there exists an $L \in \mathbf{Z}^+$ such that

$$\left|\sum_{k=-L}^{L}\widehat{\mu_N}(k)e^{2\pi ikx}\right| > c\log N,$$

which proves the required conclusion. □

Lemma 3.4.5. *For each $0 < M < \infty$ there exists a trigonometric polynomial g_M and a measurable subset A_M of $\mathbf{T}^1$ with measure $|A_M| > 1 - 2^{-M}$ such that $\|g_M\|_{L^1} = 1$,* *and such that*

$$\inf_{x\in A_M}\sup_{L\geq 1}\big|(D_L*g_M)(x)\big| = \inf_{x\in A_M}\sup_{L\geq 1}\bigg|\sum_{k=-L}^{L}\widehat{g_M}(k)e^{2\pi ikx}\bigg| > 2^M . \tag{3.4.9}$$

Proof. Given an M with $0<M<\infty$, we pick an integer $N(M)$ such that $c\log N(M) > 2^{M+2}$, where c is as in (3.4.5), and we also pick the measure $\mu_{N(M)}$, which satisfies (3.4.5). By Fatou's lemma we have

$$\begin{aligned}1 &= \big|\big\{x\in \mathbf{T}^1 :\ \lim_{L\to\infty}\sup_{1\leq j\leq L}|(D_j*\mu_{N(M)})(x)|\geq 2^{M+1}\big\}\big|\\ &\leq \liminf_{L\to\infty}\big|\big\{x\in \mathbf{T}^1 :\ \sup_{1\leq j\leq L}|(D_j*\mu_{N(M)})(x)|\geq 2^{M+1}\big\}\big| ,\end{aligned}$$

and thus we can find a positive integer $L(M)$ such that the set

$$A_M = \big\{x\in \mathbf{T}^1 :\ \sup_{1\leq j\leq L(M)}|(D_j*\mu_{N(M)})(x)|\geq 2^{M+1}\big\}$$

has measure greater than $1-2^{-M}$. We pick a positive integer $K(M)$ such that

$$\sup_{1\leq j\leq L(M)}\big\|F_{K(M)}*D_j-D_j\big\|_{L^\infty}\leq 1 ,$$

where F_K is the Fejér kernel. This is possible, since the Fejér kernel is an approximate identity and $\{D_j :\ 1\leq j\leq L\}$ is a finite family of continuous functions. Then we define $g_M=\mu_{N(M)}*F_{K(M)}$. Since $\mu_{N(M)}$ is a probability measure, we obtain

$$|(D_j*g_M)(x)-(D_j*\mu_{N(M)})(x)|\leq \big\|D_j*F_{K(M)}-D_j\big\|_{L^\infty}\leq 1$$

for all $x\in[0,1]$ and $1\leq j\leq L$. It follows that for $x\in A_M$ and $1\leq j\leq L$ we have

$$|(D_j*g_M)(x)|\geq |(D_j*\mu_{N(M)})(x)|-1\geq 2^{M+2}-1\geq 2^{M+1} .$$

Therefore, (3.4.9) is satisfied for this g_M and A_M. Since μ_N is a probability measure and $F_{K(M)}$ is nonnegative and has L^1 norm 1, we have that

$$\big\|g_M\big\|_{L^1}=\big\|\mu_{N(M)}*F_{K(M)}\big\|_{L^1}=\big\|\mu_{N(M)}\big\|_{\mathcal{M}}\big\|F_{K(M)}\big\|_{L^1}=1 .$$

□

We now have the tools needed to construct an example of a function whose Fourier series diverges almost everywhere. The example is given as a series of functions each of which has a behavior that worsens as its index becomes bigger. The function we wish to construct is a sum of the form

$$g=\sum_{j=1}^{\infty}\varepsilon_j g_{M_j} , \tag{3.4.10}$$

for a choice of sequences $\varepsilon_j \to 0$ and $M_j \to \infty$, where g_M are as in Lemma 3.4.5.

Let us be specific. We set $\varepsilon_0 = M_0 = d_0 = 1$. Assume that we have defined ε_j, M_j, and d_j for all $0 \le j < N$. We first set

$$\varepsilon_N = 2^{-N}(3d_{N-1})^{-1}. \tag{3.4.11}$$

Then we pick M_N such that

$$\varepsilon_N 2^{M_N} \ge 2^N + d_{N-1} + 1. \tag{3.4.12}$$

Finally, we set

$$d_N = \max_{1\le s\le N} \text{degree}\,(g_{M_s}), \tag{3.4.13}$$

where g_M is the trigonometric polynomial of Lemma 3.4.5. This defines ε_N, M_N, and d_N for a given N, provided these numbers are known for all $j < N$. By induction we define ε_N, M_N, and d_N for all natural numbers N.

We observe that the selections of ε_j and M_j force the inequalities $\varepsilon_j \le 2^{-j}$ and $d_j \le d_{j+1}$ for all $j \ge 0$. Since each g_{M_j} has L^1 norm 1 and $\varepsilon_j \le 2^{-j}$, the function g in (3.4.10) is integrable and has L^1 norm at most 1.

For a given $j \ge 0$ and $x \in A_{M_j}$, by Lemma 3.4.5 there exists an $L \ge 1$ such that $|(D_L * g_{M_j})(x)| > 2^{M_j}$. Set $k = k(x) = \min(L, d_j)$. Then we have

$$|(D_k * g)(x)| \ge \varepsilon_j |(D_k * g_{M_j})(x)| - \sum_{1<s<j} \varepsilon_s |(D_k * g_{M_s})(x)| - \sum_{s>j} \varepsilon_s |(D_k * g_{M_s})(x)|.$$

We make the following observations:

(i) $|(D_k * g_{M_j})(x)| = |(D_L * g_{M_j})(x)| > 2^{M_j}$.

(ii) $|(D_k * g_{M_s})(x)| = |(D_{\min(d_s,k)} * g_{M_s})(x)| \le \big\|D_{\min(d_s,L)}\big\|_{L^\infty} \le 3d_s$, when $s < j$.

(iii) $|(D_k * g_{M_s})(x)| = |(D_{\min(d_s,k)} * g_{M_s})(x)| \le \big\|D_{\min(d_j,L)}\big\|_{L^\infty} \le 3d_j$, when $s > j$.

In these estimates we have used that $k = \min(L, d_j)$, $\big\|D_m\big\|_{L^\infty} \le 2m+1 \le 3m$, and that

$$D_r * g_{M_s} = D_{\min(r,d_s)} * g_{M_s}\,,$$

which follows easily by examining the corresponding Fourier coefficients.

Using the estimates in (i), (ii), and (iii), for this x in A_{M_j} and $k = k(x)$ we obtain

$$|(D_k * g)(x)| \ge \varepsilon_j 2^{M_j} - 3\sum_{1<s<j} \varepsilon_s d_s - 3\sum_{s>j} \varepsilon_s d_j\,. \tag{3.4.14}$$

Our selection of ε_j and M_j now ensures that (3.4.14) is a large number. In fact, we have

$$3\sum_{s>j} \varepsilon_s d_j \le \sum_{s>j} 2^{-s} d_j (d_{s-1})^{-1} \le \sum_{s>j} 2^{-s} \le 1$$

and

$$3\sum_{1<s<j}\varepsilon_s d_s \le 3d_{j-1}\sum_{1<s<j}\varepsilon_s \le d_{j-1}\sum_{1<s<j}2^{-s}(d_{s-1})^{-1}\le d_{j-1}\,.$$

Therefore, the expression in (3.4.14) is at least $\varepsilon_j 2^{M_j} - d_{j-1} - 1 \ge 2^j$. It follows that for every $j \ge 0$ and every $x \in A_{M_j}$ there exists a $k = k(x)$ such that

$$|(D_k * g)(x)| \ge 2^j\,. \tag{3.4.15}$$

We conclude that for every $j \ge 0$ and $x \in A_{M_j}$ we have

$$\sup_{k\ge 1}|(D_k * g)(x)| \ge 2^j\,.$$

Thus, for all x in the set $A = \bigcap_{j=0}^{\infty}\bigcup_{r=j}^{\infty} A_{M_r}$ we have

$$\sup_{k\ge 1}|(D_k * g)(x)| = \infty\,. \tag{3.4.16}$$

But A is a countable intersection of subsets of $\mathbf{T}^1$ of full measure. Therefore, A has measure 1 and the required conclusion follows. □

3.4.3 Divergence of Bochner–Riesz Means of Integrable Functions

We now turn to the corresponding n-dimensional problem for spherical summability of Fourier series. The situation here is quite similar at the critical index $\alpha = \frac{n-1}{2}$.

Theorem 3.4.6. *Let $n > 1$. There exists an integrable function f on $\mathbf{T}^n$ such that*

$$\limsup_{R\to\infty}\left|B_R^{\frac{n-1}{2}}(f)(x)\right| = \limsup_{R\to\infty}\left|\sum_{\substack{m\in\mathbf{Z}^n\\ |m|\le R}}\Big(1-\tfrac{|m|^2}{R^2}\Big)^{\frac{n-1}{2}}\widehat{f}(m)e^{2\pi i m\cdot x}\right| = \infty$$

for almost all $x \in \mathbf{T}^n$. Furthermore, such a function can be constructed such that it is supported in an arbitrarily small given neighborhood of the origin.

Proof. We start by defining the set

$$S = \big\{x \in \mathbf{R}^n : \{|x-m| : m \in \mathbf{Z}^n\} \text{ is linearly independent over } \mathbf{Q}\big\}\,.$$

We show that S has full measure in $\mathbf{R}^n$. Indeed, if $x \in \mathbf{R}^n \setminus S$, then there exist $k \in \mathbf{Z}^+$, $m_1,\dots m_k \in \mathbf{Z}^n$, and $a_{m_1},\dots,a_{m_k}$ nonzero rational numbers such that

$$\sum_{j=1}^{k} a_{m_j}|x-m_j| = 0\,. \tag{3.4.17}$$

Since the function

$$t \to \sum_{j=1}^{k} a_{m_j} |t - m_j|$$

is nonzero and real analytic on $\mathbf{R}^n \setminus \mathbf{Z}^n$, it must vanish only on a set of Lebesgue measure zero. Therefore, there exists a set $A_{m_1,\dots,m_k,a_{m_1},\dots,a_{m_k}}$ of Lebesgue measure zero such that (3.4.17) holds exactly when x is in this set. Then

$$\mathbf{R}^n \setminus S \subseteq \bigcup_{k=1}^{\infty} \bigcup_{m_1,\dots,m_k \in \mathbf{Z}^n} \bigcup_{a_{m_1},\dots,a_{m_k} \in \mathbf{Q}} A_{m_1,\dots,m_k,a_{m_1},\dots,a_{m_k}},$$

from which it follows that $\mathbf{R}^n \setminus S$ has Lebesgue measure zero.

Let us set

$$K_R^{\alpha}(x) = \sum_{|m| \le R} \left(1 - \tfrac{|m|^2}{R^2}\right)^{\alpha} e^{2\pi i m \cdot x}.$$

We need the following lemma regarding K_R^{α}:

Lemma 3.4.7. *For each $x \in S \cap \mathbf{T}^n$, $n \ge 2$, we have*

$$\limsup_{R \to \infty} |K_R^{\frac{n-1}{2}}(x)| = \infty.$$

It is noteworthy to compare the result of this lemma with the analogous one-dimensional statement

$$\limsup_{R \to \infty} |D_R(x)| = \infty$$

for the Dirichlet kernel, which holds exactly when $x = 0$. Thus the uniform ill behavior of the kernel $K_R^{\frac{n-1}{2}}$ reflects in some sense its lack of localization.

Proof. Using (3.4.3) and the Poisson summation formula (Theorem 3.1.17), we obtain the identity

$$K_R^{\alpha}(x) = \frac{\Gamma(\alpha+1)}{\pi^{\alpha}} R^{\frac{n}{2}-\alpha} \sum_{m \in \mathbf{Z}^n} \frac{J_{\frac{n}{2}+\alpha}(2\pi R|x-m|)}{|x-m|^{\frac{n}{2}+\alpha}}, \tag{3.4.18}$$

which is valid for all $x \in \mathbf{T}^n \setminus \mathbf{Z}^n$. Because of the asymptotics in Appendix B.7, the sum (3.4.18) converges for $\alpha > \frac{n-1}{2}$. The same asymptotics imply that for $x \notin \mathbf{Z}^n$ and $R \ge 1$ we have

$$\begin{aligned} J_{\frac{n}{2}+\alpha}(2\pi R|x-m|) &= \frac{e^{2\pi i R|x-m|} e^{-i\frac{\pi}{2}(\frac{n}{2}+\alpha)-i\frac{\pi}{4}} + e^{-2\pi i R|x-m|} e^{i\frac{\pi}{2}(\frac{n}{2}+\alpha)+i\frac{\pi}{4}}}{\pi\sqrt{R|x-m|}} \\ &\quad + O\big((R|x-m|)^{-\frac{3}{2}}\big) \end{aligned}$$

for all $\alpha > 0$. It is not possible to let $\alpha \to \frac{n-1}{2}$ in (3.4.18), since the series on the right of that identity diverges for this value of α. It is a remarkable fact, however, that if we average over R first, we obtain an oscillatory factor that allows us to let $\alpha = \frac{n-1}{2}$ in the previous identity. Now for $x \notin \mathbf{Z}^n$ and $T > 1$ we obtain

$$\begin{aligned}
&\frac{1}{T}\int_1^T K_R^\alpha(x) e^{2\pi i\lambda R}\,dR \\
&\qquad = \frac{\Gamma(\alpha+1)}{\pi^\alpha}\sum_{m\in\mathbf{Z}^n}\frac{e^{-i\frac{\pi}{2}(\frac{n}{2}+\alpha)-i\frac{\pi}{4}}}{|x-m|^{\frac{n+1}{2}+\alpha}}\frac{1}{T}\int_1^T e^{2\pi iR(\lambda+|x-m|)}R^{\frac{n-1}{2}-\alpha}\,dR \\
&\qquad + \frac{\Gamma(\alpha+1)}{\pi^\alpha}\sum_{m\in\mathbf{Z}^n}\frac{e^{i\frac{\pi}{2}(\frac{n}{2}+\alpha)+i\frac{\pi}{4}}}{|x-m|^{\frac{n+1}{2}+\alpha}}\frac{1}{T}\int_1^T e^{2\pi iR(\lambda-|x-m|)}R^{\frac{n-1}{2}-\alpha}\,dR \\
&\qquad + \frac{\Gamma(\alpha+1)}{\pi^\alpha}\sum_{m\in\mathbf{Z}^n} O\Big(\frac{1}{|x-m|^{\frac{n+3}{2}+\alpha}}\Big)\frac{1}{T}\int_1^T R^{\frac{n-3}{2}-\alpha}\,dR.
\end{aligned}$$

We now let $\alpha \to \frac{n-1}{2}$ in the preceding expression. Then we have

$$\begin{aligned}
&\frac{1}{T}\int_1^T K_R^{\frac{n-1}{2}}(x) e^{2\pi i\lambda R}\,dR \qquad\qquad (3.4.19)\\
&\qquad = \frac{\Gamma(\frac{n+1}{2})}{\pi^{\frac{n-1}{2}}}\sum_{m\in\mathbf{Z}^n}\frac{e^{-i\frac{\pi}{2}(\frac{2n-1}{2})-i\frac{\pi}{4}}}{|x-m|^n}\frac{1}{T}\int_1^T e^{2\pi iR(\lambda+|x-m|)}\,dR \\
&\qquad + \frac{\Gamma(\frac{n+1}{2})}{\pi^{\frac{n-1}{2}}}\sum_{m\in\mathbf{Z}^n}\frac{e^{i\frac{\pi}{2}(\frac{2n-1}{2})+i\frac{\pi}{4}}}{|x-m|^n}\frac{1}{T}\int_1^T e^{2\pi iR(\lambda-|x-m|)}\,dR \\
&\qquad + \frac{\Gamma(\frac{n+1}{2})}{\pi^{\frac{n-1}{2}}}\sum_{m\in\mathbf{Z}^n} O\Big(\frac{1}{|x-m|^{n+1}}\Big)\frac{1}{T}\int_1^T \frac{dR}{R},
\end{aligned}$$

and the wonderful fact is that the first two sums converge because of the appearance of the oscillatory factors. See Exercise 3.4.8(c). It follows from the previous identity that if $\lambda > 0$ and $\lambda \neq |x-m_0|$ for any $m_0 \in \mathbf{Z}^n$, then the expression in (3.4.19) converges to zero as $T \to \infty$, while it converges to

$$\frac{\Gamma(\frac{n+1}{2})}{\pi^{\frac{n-1}{2}}}\frac{e^{\pm i(\frac{\pi}{2}(\frac{2n-1}{2})+i\frac{\pi}{4})}}{|x-m_0|^n} = \frac{\Gamma(\frac{n+1}{2})}{\pi^{\frac{n-1}{2}}}\frac{e^{\pm i\frac{\pi n}{2}}}{|x-m_0|^n}$$

if $\lambda = \pm|x-m_0|$ for some $m_0 \in \mathbf{Z}^n$. We now fix $x_0 \in S \cap \mathbf{T}^n$ and we set

$$\Lambda_{x_0} = \{|x_0 - m| : m \in \mathbf{Z}^n\} = \{\lambda_1, \lambda_2, \lambda_3, \dots\},$$

where $0 < \lambda_1 < \lambda_2 < \lambda_3 < \cdots$. Observe that

$$\sum_{j=1}^\infty \frac{1}{\lambda_j^n} = \infty. \qquad (3.4.20)$$

We have shown that

$$\lim_{T\to\infty}\frac{1}{T}\int_1^T K_t^{\frac{n-1}{2}}(x_0)e^{2\pi i\lambda t}\,dt=\begin{cases}\dfrac{\Gamma(\frac{n+1}{2})}{\pi^{\frac{n-1}{2}}}\dfrac{e^{i\frac{\pi n}{2}}}{\lambda_j^n} & \text{if } \lambda=\lambda_j,\\ 0 & \text{if } \lambda\neq\pm\lambda_j,\\ \dfrac{\Gamma(\frac{n+1}{2})}{\pi^{\frac{n-1}{2}}}\dfrac{e^{-i\frac{\pi n}{2}}}{\lambda_j^n} & \text{if } \lambda=-\lambda_j.\end{cases}\tag{3.4.21}$$

Since $x_0\in S\cap\mathbf{T}^n$, the λ_j are linearly independent and thus no expression of the form $\pm\lambda_{j_1}\pm\cdots\pm\lambda_{j_s}$ is equal to any other λ_j. It follows from (3.4.21) that

$$\lim_{T\to\infty}\frac{1}{T}\int_1^T K_t^{\frac{n-1}{2}}(x_0)\prod_{j=1}^N\left[1+\frac{e^{-i\frac{\pi n}{2}}e^{2\pi i\lambda_j t}+e^{i\frac{\pi n}{2}}e^{-2\pi i\lambda_j t}}{2}\right]dt=\frac{\Gamma(\frac{n+1}{2})}{\pi^{\frac{n-1}{2}}}\sum_{j=1}^N\frac{1}{\lambda_j^n}.$$

Suppose that for $x_0\in S\cap\mathbf{T}^n$ we had

$$\sup_{R\geq 1}|K_R^{\frac{n-1}{2}}(x_0)|\leq A_{x_0}<\infty.$$

Then it would follow from the previous identity that

$$\begin{aligned}\frac{\Gamma(\frac{n+1}{2})}{\pi^{\frac{n-1}{2}}}\sum_{j=1}^N\frac{1}{\lambda_j^n}&\leq A_{x_0}\lim_{T\to\infty}\frac{1}{T}\int_1^T\prod_{j=1}^N\left[1+\frac{e^{-i\frac{\pi n}{2}}e^{2\pi i\lambda_j t}+e^{i\frac{\pi n}{2}}e^{-2\pi i\lambda_j t}}{2}\right]dt\\&=A_{x_0},\end{aligned}$$

which contradicts (3.4.20). We deduce that $\sup_{R\geq 1}|K_R^{\frac{n-1}{2}}(x_0)|=\infty$ for every point $x_0\in S\cap\mathbf{T}^n$ and this concludes the proof of Lemma 3.4.7. □

We now proceed with the proof of Theorem 3.4.6. This part of the proof is similar to the proof of Theorem 3.4.2. Lemma 3.4.7 says that the means $B_R^{\frac{n-1}{2}}(\delta_0)(x)$, where δ_0 is the Dirac mass at 0, do not converge for almost all $x\in\mathbf{T}^n$. Our goal is to replace this Dirac mass by a series of integrable functions on $\mathbf{T}^n$ that have a peak at the origin.

Let us fix a nonnegative $\mathscr{C}^\infty$ radial function $\widehat{\Phi}$ on $\mathbf{R}^n$ that is supported in the unit ball $|\xi|\leq 1$ and has integral equal to 1. We now set

$$\varphi_\varepsilon(x)=\sum_{m\in\mathbf{Z}^n}\tfrac{1}{\varepsilon^n}\widehat{\Phi}(\tfrac{x+m}{\varepsilon})=\sum_{m\in\mathbf{Z}^n}\Phi(\varepsilon m)e^{2\pi i m\cdot x},$$

where the identity is valid because of the Poisson summation formula. It follows that the mth Fourier coefficient of φ_ε is $\Phi(\varepsilon m)$. Therefore, we have the estimate

$$\sup_{x\in\mathbf{T}^n}\sup_{R>0}|B_R^{\frac{n-1}{2}}(\varphi_\varepsilon)(x)|\leq\sum_{m\in\mathbf{Z}^n}|\Phi(\varepsilon m)|\leq\sum_{m\in\mathbf{Z}^n}\frac{C_n'}{(1+\varepsilon|m|)^{n+1}}\leq\frac{C_n}{\varepsilon^n}.\tag{3.4.22}$$

For any $j\geq 1$, we construct measurable subsets E_j of $\mathbf{T}^n$ that satisfy $|E_j|\geq 1-\frac{1}{j}$, a sequence of positive numbers $0<R_1<R_2<\cdots$, and two sequences of positive numbers $\varepsilon_j\leq\delta_j$ such that

$$\sup_{R\le R_j}\Big|B_R^{\frac{n-1}{2}}\Big(\sum_{s=1}^{\infty}2^{-s}(\varphi_{\varepsilon_s}-\varphi_{\delta_s})\Big)(x)\Big|\ge j \qquad \text{for } x\in E_j. \tag{3.4.23}$$

We pick $E_1=\emptyset$, $R_1=1$, and $\varepsilon_1=\delta_1=1$. Let $k>1$ and suppose that we have selected E_j, R_j, δ_j, and ε_j for all $1\le j\le k-1$ such that (3.4.23) is satisfied. We construct E_k, R_k, δ_k, and ε_k such that (3.4.23) is satisfied with $j=k$. We begin by choosing δ_k. Let B be a constant such that

$$|\Phi(x)-\Phi(y)|\le B|x-y|$$

for all $x,y\in\mathbf{R}^n$. Pick δ_k small enough that

$$B\delta_k\sum_{|m|\le R_{k-1}}|m|\le 1\,. \tag{3.4.24}$$

Then we let

$$A_k=C_n2^{-k}\delta_k^{-n}+C_n\sum_{j=1}^{k-1}2^{-j}(\varepsilon_j^{-n}+\delta_j^{-n})\,,$$

where C_n is the constant in (3.4.22), and observe that in view of (3.4.22) we have

$$\sup_{x\in\mathbf{T}^n}\sup_{R>0}\Big|B_R^{\frac{n-1}{2}}\Big(-2^{-k}\varphi_{\delta_k}+\sum_{j=1}^{k-1}2^{-j}(\varphi_{\varepsilon_j}-\varphi_{\delta_j})\Big)(x)\Big|\le A_k\,. \tag{3.4.25}$$

Let δ_0 be the Dirac mass at the origin in $\mathbf{T}^n$. Since by Fatou's lemma and Lemma 3.4.7 we have

$$\liminf_{N\to\infty}\Big|\Big\{x\in\mathbf{T}^n:\ \sup_{0<R\le N}\Big|B_R^{\frac{n-1}{2}}(\delta_0)(x)\Big|>A_k+k+2\Big\}\Big|=1\,,$$

there exists an $R_k>R_{k-1}$ such that the set

$$E_k=\Big\{x\in\mathbf{T}^n:\ \sup_{0<R\le R_k}\Big|B_R^{\frac{n-1}{2}}(\delta_0)(x)\Big|>A_k+k+2\Big\}$$

has measure at least $1-\frac{1}{k}$. We now choose $\varepsilon_k\le\delta_k$ such that

$$\sup_{x\in\mathbf{T}^n}\big|B_R^{\frac{n-1}{2}}(\delta_0)(x)-B_R^{\frac{n-1}{2}}(\varphi_{\varepsilon_k})(x)\big|\le\sum_{|m|\le R_k}\Big(1-\tfrac{|m|^2}{R_k^2}\Big)^{\frac{n-1}{2}}|1-\widehat{\varphi_{\varepsilon_k}}(m)|\le 1\,.$$

This is possible, since the preceding expression in the middle tends to zero as $\varepsilon_k\to 0$. Then for $x\in E_k$ we have

$$\inf_{x\in E_k}\sup_{R\le R_k}2^{-k}\big|B_R^{\frac{n-1}{2}}(\varphi_{\varepsilon_k})(x)\big|\ge A_k+k+1\,. \tag{3.4.26}$$

Observe that the construction of δ_k gives the estimate

$$
\begin{aligned}
\sup_{x\in\mathbf{T}^n}\sup_{R\le R_{k-1}} |B_R^{\frac{n-1}{2}}(\varphi_{\varepsilon_k}-\varphi_{\delta_k})(x)| &\le \sum_{|m|\le R_{k-1}} |\Phi(\varepsilon_k m)-\Phi(\delta_k m)| \\
&\le B(\delta_k-\varepsilon_k)\sum_{|m|\le R_{k-1}} |m| \\
&\le B\delta_k \sum_{|m|\le R_{k-1}} |m| \le 1
\end{aligned} \tag{3.4.27}
$$

using (3.4.24). The inductive selection of the parameters can be described schematically as follows:

$$\delta_1, R_1, E_1, \varepsilon_1 \implies \delta_2 \implies A_2 \implies R_2, E_2 \implies \varepsilon_2 \implies \delta_3 \implies \text{etc.}$$

Let us now prove (3.4.23) for $j=k$. Write

$$
\begin{aligned}
B_R^{\frac{n-1}{2}}\Big(\sum_{s=1}^{\infty} 2^{-s}(\varphi_{\varepsilon_s}-\varphi_{\delta_s})\Big)(x) &= B_R^{\frac{n-1}{2}}\Big(-2^{-k}\varphi_{\delta_k}+\sum_{s=1}^{k-1} 2^{-s}(\varphi_{\varepsilon_s}-\varphi_{\delta_s})\Big)(x) \\
&\quad + B_R^{\frac{n-1}{2}}(2^{-k}\varphi_{\varepsilon_k})(x) \\
&\quad + B_R^{\frac{n-1}{2}}\Big(\sum_{s=k+1}^{\infty} 2^{-s}(\varphi_{\varepsilon_s}-\varphi_{\delta_s})\Big)(x).
\end{aligned}
$$

In view of (3.4.25), (3.4.26), and (3.4.27) for all $x \in E_k$, we obtain

$$\sup_{R\le R_{k-1}} \Big| B_R^{\frac{n-1}{2}}\Big(\sum_{s=1}^{\infty} 2^{-s}(\varphi_{\varepsilon_s}-\varphi_{\delta_s})\Big)(x)\Big| \ge k,$$

which clearly implies (3.4.23) (with $j=k$), since $R_k > R_{k-1}$. Setting

$$f = \sum_{s=1}^{\infty} 2^{-s}(\varphi_{\varepsilon_s}-\varphi_{\delta_s}) \in L^1(\mathbf{T}^n)$$

we have now proved that $\sup_{R>0} \big|B_R^{\frac{n-1}{2}}(f)(x)\big| = \infty$ for all x in

$$\bigcap_{k=1}^{\infty}\bigcup_{r=k}^{\infty} E_r.$$

Since the latter set has full measure in $\mathbf{T}^n$, the required conclusion follows.

By taking ε_1 arbitrarily small (instead of picking $\varepsilon_1 = 1$), we force f to be supported in an arbitrarily small neighborhood of the origin. □

The previous argument shows that the Bochner–Riesz means B_R^α are badly behaved on $L^1(\mathbf{T}^n)$ when $\alpha = \frac{n-1}{2}$. It follows that the "rougher" spherical Dirichlet means $\widetilde{D}(n,N) * f$ (which correspond to $\alpha = 0$) are also ill behaved on $L^1(\mathbf{T}^n)$. See Exercise 3.4.5. In Chapter 10 we establish the stronger negative result that the spherical Dirichlet means of L^p functions may also diverge in L^p when $p \neq 2$.

Exercises

3.4.1. Prove that if $f \in L^1(\mathbf{T}^1)$ satisfies $\limsup_{N\to\infty} |(D_N * f)(x)| = \infty$ for almost all $x \in \mathbf{T}^1$, then the function

$$F(x_1,\dots,x_n) = f(x_1)$$

on $\mathbf{T}^n$ satisfies $\limsup_{N\to\infty} |(D(n,N) * F)(x)| = \infty$ for almost all $x \in \mathbf{T}^n$.

3.4.2. (*H. Weyl*) A sequence $\{a_k\}_{k=0}^{\infty}$ with values in $\mathbf{T}^n$ is called *equidistributed* if for every square Q in $\mathbf{T}^n$ we have

$$\lim_{N\to\infty} \frac{\#\{k:\ 0 \le k \le N-1, \quad a_k \in Q\}}{N} = |Q|\,.$$

Show that the following are equivalent:
(a) The sequence $\{a_k\}_{k=0}^{\infty}$ is equidistributed.
(b) For every smooth function f on $\mathbf{T}^n$ we have that

$$\lim_{N\to\infty} \frac{1}{N} \sum_{k=0}^{N-1} f(a_k) = \int_{\mathbf{T}^n} f(x)\,dx\,.$$

(c) For every $m \in \mathbf{Z}^n \setminus \{0\}$ we have

$$\lim_{N\to\infty} \frac{1}{N} \sum_{k=0}^{N-1} e^{2\pi i m\cdot a_k} = 0\,.$$

[*Hint:* Prove that $(a) \implies (b) \implies (c) \implies (b) \implies (a)$. In proving $(a) \iff (b)$, approximate f by step functions. In proving $(c) \implies (b)$, use Fourier inversion.]

3.4.3. Suppose that $x = (x_1,\dots,x_n) \in \mathbf{T}^n$ and $m\cdot x$ is irrational for all $m \in \mathbf{Z}^n \setminus \{0\}$. Use Exercise 3.4.2 to show that the sequence $\{([kx_1],\dots,[kx_n])\}_{k=0}^{\infty}$ is equidistributed. (In dimension 1 the hypothesis is satisfied if x is irrational.)

3.4.4. The beta function is defined in Appendix A.2. Derive the identity

$$t^{\alpha} = \frac{1}{B(\alpha-\beta,\beta+1)} \int_0^t (t-s)^{\alpha-\beta-1} s^{\beta}\,ds$$

and show that the function $K_R^{\alpha}(x) = \sum_{|m|\le R} \big(1 - \frac{|m|^2}{R^2}\big)^{\alpha} e^{2\pi i m\cdot x}$ satisfies (3.4.2). [*Hint:* Take $t = 1 - \frac{|m|^2}{R^2}$ and change variables $s = \frac{r^2-|m|^2}{R^2}$ in the previous beta function identity.]

3.4.5. Use Exercise 3.4.4 to obtain that if for some $x_0 \in \mathbf{T}^n$ we have

$$\limsup_{R\to\infty} |K_R^{\alpha}(x_0)| < \infty,$$

then for all $\beta > \alpha$ we have

$$\sup_{R>0} |K_R^\beta(x_0)| < \infty .$$

Conclude that the circular (spherical) Dirichlet means of the function f constructed in the proof of Theorem 3.4.6 diverge a.e. The same conclusion is true for the Bochner–Riesz means of f of every order $\alpha \le \frac{n-1}{2}$.

3.4.6. For $t \in [0,\infty)$ let

$$N(t) = \#\{m \in \mathbf{Z}^n : \ |m| \le t\} .$$

Let $0 = r_0 < r_1 < r_2 < \cdots$ be the sequence all of numbers r for which there exist $m \in \mathbf{Z}^n$ such that $|m| = r$.
(a) Observe that N is right continuous and constant on intervals of the form $[r_j, r_{j+1})$.
(b) Show that the distributional derivative of N is the measure

$$\mu(t) = \#\{m \in \mathbf{Z}^n : \ |m| = t\} ,$$

defined via the identity $\langle \mu, \varphi \rangle = \sum_{j=0}^{\infty} \#\{m \in \mathbf{Z}^n : \ |m| = r_j\} \, \varphi(r_j)$.

3.4.7. Let $f \in \mathscr{C}^1([0,\infty))$ and $0 \le a < b < \infty$ not equal to any r_j as defined in Exercise 3.4.6. Derive the useful identity

$$\sum_{\substack{m\in \mathbf{Z}^n \\ a\le |m| \le b}} f(|m|) = f(b)N(b) - f(a)N(a) - \int_a^b f'(x)N(x)\, dx .$$

3.4.8. (a) Let $0 < \lambda < \infty$ and fix a transcendental number γ in $(0,1)$. Prove that for $k \in \mathbf{Z}^+$ we have

$$\sum_{\substack{m\in \mathbf{Z}^n \\ k+\gamma\le |m| \le k+1+\gamma}} \frac{e^{i|m|}}{|m|^\lambda} = \frac{-i\,\omega_{n-1}\, e^{i(k+1+\gamma)}}{(k+1+\gamma)^{\lambda-(n-1)}} - \frac{-i\,\omega_{n-1}\, e^{i(k+\gamma)}}{(k+\gamma)^{\lambda-(n-1)}} + O(k^{-\lambda+(n-1)-\frac{n-1}{n+1}}) ,$$

as $k \to \infty$, where ω_{n-1} is the volume of $\mathbf{S}^{n-1}$.

(b) Use part (a) to show that if $\lambda \le n-1$, the limit

$$\lim_{R\to\infty} \sum_{\substack{m\in \mathbf{Z}^n\setminus\{0\} \\ |m|\le R}} \frac{e^{i|m|}}{|m|^\lambda}$$

does not exist.
(c) Show that if $\lambda > n - \frac{n-1}{n+1}$, the following series converges:

$$\sum_{m\in \mathbf{Z}^n\setminus\{0\}} \frac{e^{i|m|}}{|m|^\lambda} ,$$

where the infinite sum is interpreted as the limit in part (b).
[*Hint:* Part (a): One may need Exercises 3.4.7 and 3.1.12. Part (b): Suppose the limit exists and let $\beta_k = \frac{-i\omega_{n-1}e^{i(k+1+\gamma)}}{(k+1+\gamma)^{\lambda-(n-1)}} - \frac{-i\omega_{n-1}e^{i(k+\gamma)}}{(k+\gamma)^{\lambda-(n-1)}}$. If the series $\sum_{m\in\mathbf{Z}^n\setminus\{0\}} \frac{e^{i|m|}}{|m|^\lambda}$ converged, then we would have $\beta_k \to 0$ as $k \to \infty$. But then $k^{\lambda-(n-1)}\beta_k$ would also tend to zero, which gives a contradiction.]

3.4.9. (*Pinsky, Stanton, and Trapa [214]*) Prove that the spherical partial sums of the Fourier series of the characteristic function of the ball $B(0, \frac{1}{2\pi})$ in $\mathbf{T}^n$ diverges at $x = 0$ when $n \geq 3$.
[*Hint:* Use the idea of Exercise 3.4.8 with $\lambda = \frac{n+1}{2}$.]

3.5 The Conjugate Function and Convergence in Norm

In this section we address the following fundamental question: Do Fourier series converge in norm? We begin with some abstract necessary and sufficient conditions that guarantee such a convergence. In one dimension, we are able to reduce matters to the study of the so-called conjugate function on the circle, a sister operator of the Hilbert transform, which is the center of study of the next chapter. In higher dimensions the situation is more complicated, but we are able to give a positive answer in the case of square summability.

3.5.1 Equivalent Formulations of Convergence in Norm

The question we pose is for which $1 \leq p < \infty$ we have

$$\big\|D(n,N) * f - f\big\|_{L^p(\mathbf{T}^n)} \to 0 \qquad \text{as } N \to \infty, \tag{3.5.1}$$

and similarly for the circular Dirichlet kernel $\widetilde{D}(n,N)$. We tackle this question by looking at an equivalent formulation of it.

Theorem 3.5.1. *Fix $1 \leq p < \infty$ and $\{a_m\}$ in $\ell^\infty(\mathbf{Z}^n)$. For each $R \geq 0$, let $\{a_m(R)\}_{m\in\mathbf{Z}^n}$ be a compactly supported sequence (whose support depends on R) that satisfies $\lim_{R\to\infty} a_m(R) = a_m$. For $f \in L^p(\mathbf{T}^n)$ define*

$$S_R(f)(x) = \sum_{m\in\mathbf{Z}^n} a_m(R)\widehat{f}(m)e^{2\pi i m\cdot x}$$

and for $h \in \mathscr{C}^\infty(\mathbf{T}^n)$ define

$$A(h)(x) = \sum_{m\in\mathbf{Z}^n} a_m\widehat{h}(m)e^{2\pi i m\cdot x}.$$

Then for all $f \in L^p(\mathbf{T}^n)$ the sequence $S_R(f)$ converges in L^p as $R \to \infty$ if and only if there exists a constant $K < \infty$ such that

$$\sup_{R\geq 0} \big\|S_R\big\|_{L^p \to L^p} \leq K. \tag{3.5.2}$$

Furthermore, if (3.5.2) holds, then for the same constant K we have

$$\sup_{0\neq h\in\mathscr{C}^\infty} \frac{\big\|A(h)\big\|_{L^p}}{\big\|h\big\|_{L^p}} \leq K, \tag{3.5.3}$$

and then A extends to a bounded operator $\widetilde{A}$ from $L^p(\mathbf{T}^n)$ to itself; moreover, for every $f \in L^p(\mathbf{T}^n)$ we have that $S_R(f) \to \widetilde{A}(f)$ in L^p as $R \to \infty$.

Proof. If $S_R(f)$ converges in L^p, then $\big\|S_R(f)\big\|_{L^p} \leq C_f$ for some constant C_f that depends on f. The uniform boundedness theorem now gives that the operator norms of S_R from L^p to L^p are bounded uniformly in R. This proves (3.5.2).

Conversely, assume (3.5.2). For $h \in \mathscr{C}^\infty(\mathbf{T}^n)$ Fatou's lemma gives

$$\big\|A(h)\big\|_{L^p} = \big\|\lim_{R\to\infty} S_R(h)\big\|_{L^p} \leq \liminf_{R\to\infty} \big\|S_R(h)\big\|_{L^p} \leq K\big\|h\big\|_{L^p};$$

hence (3.5.3) holds. Thus A extends to a bounded operator $\widetilde{A}$ on $L^p(\mathbf{T}^n)$ by density. We show that for all $f \in L^p(\mathbf{T}^n)$ we have $S_R(f) \to \widetilde{A}(f)$ in L^p as $R \to \infty$. Fix f in $L^p(\mathbf{T}^n)$ and let $\varepsilon > 0$ be given. Pick a trigonometric polynomial P satisfying $\big\|f - P\big\|_{L^p} \leq \varepsilon$. Let d be the degree of P. Then

$$\begin{aligned} \big\|S_R(P) - A(P)\big\|_{L^p} &\leq \big\|S_R(P) - A(P)\big\|_{L^\infty} \\ &\leq \sum_{|m_1|+\cdots+|m_n|\leq d} |a_m(R) - a_m|\,|\widehat{P}(m)| \leq \varepsilon, \end{aligned}$$

provided $R > R_0$, since $a_m(R) \to a_m$ for every m with $|m_1| + \cdots + |m_n| \leq d$. Then

$$\begin{aligned} \big\|S_R(f)-\widetilde{A}(f)\big\|_{L^p} &\leq \big\|S_R(f)-S_R(P)\big\|_{L^p} + \big\|S_R(P)-\widetilde{A}(P)\big\|_{L^p} + \big\|\widetilde{A}(P)-\widetilde{A}(f)\big\|_{L^p} \\ &\leq K\varepsilon + \varepsilon + K\varepsilon = (2K+1)\varepsilon \end{aligned}$$

for $R > R_0$. This proves that $S_R(f)$ converges to $\widetilde{A}(f)$ in L^p as $R \to \infty$. □

The most interesting situation arises, of course, when $a_m(R) \to a_m = 1$ for all $m \in \mathbf{Z}^n$. In this case we expect the operators $S_R(f)$ to converge back to f as $R \to \infty$. We should keep in mind the following three examples:

(a) the sequence $a_m(R) = 1$ when $\max_{1\leq j\leq n} |m_j| \leq R$ and zero otherwise, in which case the operator S_R of Theorem 3.5.1 is

$$S_R(f) = f * D(n,R); \tag{3.5.4}$$

(b) the sequence $a_m(R) = 1$ when $|m| \leq R$ and zero otherwise, in which case the S_R of Theorem 3.5.1 is

$$\widetilde{S}_R(f) = f * \widetilde{D}(n,R); \tag{3.5.5}$$

(c) the sequence $a_m(R) = \left(1 - \frac{|m|^2}{R^2}\right)_+^\alpha$, in which case $S_R = B_R^\alpha$.

Corollary 3.5.2. *Let $1 \le p < \infty$ and $\alpha \ge 0$. Let S_R and $\widetilde{S}_R$ be as in (3.5.4) and (3.5.5), respectively, and let B_R^α be the Bochner–Riesz means as defined in (3.4.1). Then*

$$\forall f \in L^p(\mathbf{T}^n),\ D(n,R) * f \to f \quad \text{in } L^p \iff \sup_{R \ge 0} \left\| S_R \right\|_{L^p \to L^p} < \infty.$$

$$\forall f \in L^p(\mathbf{T}^n),\ \widetilde{D}(n,R) * f \to f \quad \text{in } L^p \iff \sup_{R \ge 0} \left\| \widetilde{S}_R \right\|_{L^p \to L^p} < \infty.$$

$$\forall f \in L^p(\mathbf{T}^n),\ B_R^\alpha(f) \to f \quad \text{in } L^p \iff \sup_{R \ge 0} \left\| B_R^\alpha \right\|_{L^p \to L^p} < \infty.$$

Example 3.5.3. We investigate the one-dimensional case in some detail. We take $n = 1$ and we define $a_m(N) = 1$ for all $-N \le m \le N$ and zero otherwise. Then $S_N(f) = \widetilde{S}_N(f) = D_N * f$, where D_N is the Dirichlet kernel. Clearly, the expressions $\left\| S_N \right\|_{L^p \to L^p}$ can be estimated from above by the L^1 norm of D_N, but this estimate is quite rough as it yields a bound that blows up as $N \to \infty$. We later show, via a more delicate argument, that the expressions $\left\| S_N \right\|_{L^p \to L^p}$ are uniformly bounded in N when $1 < p < \infty$.

This reasoning, however, allows us to deduce that for some function $g \in L^1(\mathbf{T}^1)$, $S_N(g)$ may not converge in L^1. This is also a consequence of the proof of Theorem 3.4.2; see (3.4.16). Note that since the Fejér kernel F_M has L^1 norm 1, we have

$$\left\| S_N \right\|_{L^1 \to L^1} \ge \lim_{M \to \infty} \left\| D_N * F_M \right\|_{L^1} = \left\| D_N \right\|_{L^1}.$$

This implies that the expressions $\left\| S_N \right\|_{L^1 \to L^1}$ are not uniformly bounded in N, and therefore Corollary 3.5.2 gives that for some $f \in L^1(\mathbf{T}^1)$, $S_N(f)$ does not converge to f (nor to any other integrable function) in L^1.

Although convergence of the partial sums of Fourier series fail in L^1, it is a consequence of Plancherel's theorem that it holds in L^2. More precisely, if $f \in L^2(\mathbf{T}^n)$, then

$$\left\| \widetilde{D}_N * f - f \right\|_{L^2}^2 = \sum_{|m| > N} |\widehat{f}(m)|^2 \to 0$$

as $N \to \infty$ and the same result is true for D_N. The following question is therefore naturally raised. Does L^p convergence hold for $p \neq 2$? This question was answered in the affirmative by M. Riesz in dimension 1. In higher dimensions a certain interesting dichotomy appears. Although it is a consequence of the one-dimensional result that the square partial sums $D(n,N) * f$ converge to f in $L^p(\mathbf{T}^n)$, this is not the case for the circular partial sums, since there exists $f \in L^p(\mathbf{T}^n)$ such that $\widetilde{D}(n,N) * f$ do not converge in L^p if $1 < p \neq 2 < \infty$. We study this issue in Chapter 10.

We begin the discussion with the one-dimensional situation.

Definition 3.5.4. For $f \in \mathscr{C}^\infty(\mathbf{T}^1)$ define the *conjugate function* $\widetilde{f}$ by

$$\widetilde{f}(x) = -i \sum_{m \in \mathbf{Z}^1} \operatorname{sgn}(m) \widehat{f}(m) e^{2\pi i m x},$$

where $\operatorname{sgn}(m) = 1$ for $m > 0$, -1 for $m < 0$, and 0 for $m = 0$. Also define the *Riesz projections* P_+ and P_- by

$$P_+(f)(x) = \sum_{m=1}^{\infty} \widehat{f}(m) e^{2\pi i m x}, \tag{3.5.6}$$

$$P_-(f)(x) = \sum_{m=-\infty}^{-1} \widehat{f}(m) e^{2\pi i m x}. \tag{3.5.7}$$

Observe that $f = P_+(f) + P_-(f) + \widehat{f}(0)$, while $\widetilde{f} = -iP_+(f) + iP_-(f)$, when f is in $\mathscr{C}^\infty(\mathbf{T}^1)$. The following is a consequence of Theorem 3.5.1.

Proposition 3.5.5. *Let $1 \le p < \infty$. Then the expressions $S_N(f) = D_N * f$ converge to f in $L^p(\mathbf{T}^1)$ as $N \to \infty$ if and only if there exists a constant $C_p > 0$ such that for all smooth f we have $\|\widetilde{f}\|_{L^p(\mathbf{T}^1)} \le C_p \|f\|_{L^p(\mathbf{T}^1)}$.*

Proof. Observe that

$$P_+(f) = \frac{1}{2}(f + i\widetilde{f}) - \frac{1}{2}\widehat{f}(0)$$

and therefore the L^p boundedness of the operator $f \mapsto \widetilde{f}$ is equivalent to that of the operator $f \mapsto P_+(f)$.

Next, note the validity of the identity

$$e^{-2\pi i N x} \sum_{m=0}^{2N} \big(f(\cdot)e^{2\pi i N(\cdot)}\big)\widehat{\ }(m) e^{2\pi i m x} = \sum_{m=-N}^{N} \widehat{f}(m) e^{2\pi i m x}.$$

Since multiplication by exponentials does not affect L^p norms, this identity implies that the norm of the operator $S_N(f) = D_N * f$ from L^p to L^p is equal to that of the operator

$$S_N'(g)(x) = \sum_{m=0}^{2N} \widehat{g}(m) e^{2\pi i m x}$$

from L^p to L^p. Therefore,

$$\sup_{N \ge 0} \|S_N\|_{L^p \to L^p} < \infty \iff \sup_{N \ge 0} \|S_N'\|_{L^p \to L^p} < \infty. \tag{3.5.8}$$

Suppose now that for all $f \in L^p(\mathbf{T}^1)$, $S_N(f) \to f$ in L^p as $N \to \infty$. Corollary 3.5.2 yields $\sup_{N\ge 0} \|S_N\|_{L^p \to L^p} < \infty$ and thus $\sup_{N \ge 0} \|S_N'\|_{L^p \to L^p} < \infty$ by (3.5.8). Theorem 3.5.1 applied to the sequence $a_m(R) = 1$ for $0 \le m \le R$ and $a_m(R) = 0$ otherwise gives that the operator $A(f) = P_+(f) + \widehat{f}(0)$ is bounded on $L^p(\mathbf{T}^1)$. Hence so is P_+.

Conversely, suppose that P_+ extends to a bounded operator from $L^p(\mathbf{T}^1)$ to itself. For all f smooth we can write

$$\begin{aligned} S'_N(f)(x) &= \sum_{m=0}^{\infty} \widehat{f}(m)e^{2\pi imx} - \sum_{m=2N+1}^{\infty} \widehat{f}(m)e^{2\pi imx} \\ &= \sum_{m=0}^{\infty} \widehat{f}(m)e^{2\pi imx} - e^{2\pi i(2N)x} \sum_{m=1}^{\infty} \widehat{f}(m+2N)e^{2\pi imx} \\ &= P_+(f)(x) - e^{2\pi i(2N)x} P_+(e^{-2\pi i(2N)(\cdot)}f) + \widehat{f}(0). \end{aligned}$$

The previous identity implies that

$$\sup_{N\geq 0} \big\|S'_N(f)\big\|_{L^p} \leq \big(2\big\|P_+\big\|_{L^p\to L^p}+1\big)\big\|f\big\|_{L^p} \tag{3.5.9}$$

for all f smooth, and by density for all $f \in L^p(\mathbf{T}^1)$. (Note that S'_N is well defined on $L^p(\mathbf{T}^1)$.) In view of (3.5.8), estimate (3.5.9) also holds for S_N. Theorem 3.5.1 applied again gives that $S_N(f) \to f$ in L^p for all $f \in L^p(\mathbf{T}^1)$. □

3.5.2 The L^p Boundedness of the Conjugate Function

We know now that convergence of Fourier series in L^p is equivalent to the L^p boundedness of the conjugate function or either of the two Riesz projections. It is natural to ask whether these operators are L^p bounded.

Theorem 3.5.6. *Given $1 < p < \infty$, there is a constant $A_p > 0$ such that for all f in $\mathscr{C}^\infty(\mathbf{T}^1)$ we have*

$$\big\|\widetilde{f}\big\|_{L^p} \leq A_p\big\|f\big\|_{L^p}. \tag{3.5.10}$$

Consequently, the Fourier series of L^p functions on the circle converge back to the functions in L^p for $1 < p < \infty$.

Proof. We present a relatively short proof of this theorem due to S. Bochner. Let $f(t)$ be a trigonometric polynomial on $\mathbf{T}^1$ with coefficients c_j. We write

$$f(t) = \sum_{j=-N}^{N} c_j e^{2\pi ijt} = \left[\sum_{j=-N}^{N} \frac{c_j+\overline{c_{-j}}}{2} e^{2\pi ijt}\right] + i\left[\sum_{j=-N}^{N} \frac{c_j-\overline{c_{-j}}}{2i} e^{2\pi ijt}\right]$$

and we note that the expressions inside the square brackets are real-valued trigonometric polynomials. We may therefore assume that f is real-valued and by subtracting a constant we can assume that $\widehat{f}(0) = 0$. Since f is real-valued, we have that $\widehat{f}(-m) = \overline{\widehat{f}(m)}$ for all m, and since $\widehat{f}(0) = 0$, we may write

$$\widetilde{f}(t) = -i\sum_{m>0} \widehat{f}(m)e^{2\pi imt} + i\sum_{m>0} \widehat{f}(-m)e^{-2\pi imt} = 2\mathrm{Re}\left[-i\sum_{m>0} \widehat{f}(m)e^{2\pi imt}\right],$$

which implies that $\widetilde{f}$ is also real-valued (see also Exercise 3.5.4(b)). Therefore the polynomial $f+i\widetilde{f}$ contains only positive frequencies. Thus for $k \in \mathbf{Z}^+$ we have

$$\int_{\mathbf{T}^1} (f(t)+i\widetilde{f}(t))^{2k}\,dt = 0.$$

Expanding the $2k$ power and taking real parts, we obtain

$$\sum_{j=0}^{k} (-1)^{k-j} \binom{2k}{2j} \int_{\mathbf{T}^1} \widetilde{f}(t)^{2k-2j} f(t)^{2j}\,dt = 0,$$

where we used that f is real-valued. Therefore,

$$\begin{aligned} \big\|\widetilde{f}\big\|_{L^{2k}}^{2k} &\le \sum_{j=1}^{k} \binom{2k}{2j} \int_{\mathbf{T}^1} \widetilde{f}(t)^{2k-2j} f(t)^{2j}\,dt \\ &\le \sum_{j=1}^{k} \binom{2k}{2j} \big\|\widetilde{f}\big\|_{L^{2k}}^{2k-2j} \big\|f\big\|_{L^{2k}}^{2j}, \end{aligned}$$

by applying Hölder's inequality with exponents $2k/(2k-2j)$ and $2k/(2j)$ to the jth term of the sum. Dividing the last inequality by $\big\|f\big\|_{L^{2k}}^{2k}$, we obtain

$$R^{2k} \le \sum_{j=1}^{k} \binom{2k}{2j} R^{2k-2j}, \tag{3.5.11}$$

where $R = \big\|\widetilde{f}\big\|_{L^{2k}} / \big\|f\big\|_{L^{2k}}$. It is an elementary fact that if $R>0$ satisfies (3.5.11), then there exists a positive constant C_{2k} such that $R \le C_{2k}$. We conclude that

$$\big\|\widetilde{f}\big\|_{L^p} \le C_p \big\|f\big\|_{L^p} \qquad \text{when } p = 2k. \tag{3.5.12}$$

We can now remove the assumption that $\widehat{f}(0)=0$. Apply (3.5.12) to $f-\widehat{f}(0)$, observe that the conjugate function of a constant is zero, and use the triangle inequality and the fact that $|\widehat{f}(0)| \le \big\|f\big\|_{L^1} \le \big\|f\big\|_{L^p}$ to obtain $\big\|\widetilde{f}\big\|_{L^p} \le 2C_p \big\|f\big\|_{L^p}$ when $p=2k$ and f is a real-valued trigonometric polynomial. Since a general trigonometric polynomial can be written as $P+iQ$, where P and Q are real-valued trigonometric polynomials, we obtain the inequality $\big\|\widetilde{f}\big\|_{L^p} \le 4C_p \big\|f\big\|_{L^p}$ for all trigonometric polynomials f when $p=2k$. Since trigonometric polynomials are dense in L^p, it follows that (3.5.10) holds for all smooth functions when $p=2k$. It also follows that the conjugate function has a bounded extension on $L^p(\mathbf{T}^1)$ when $p=2k$ and in particular, this extension is well defined for simple functions.

Every real number $p \ge 2$ lies in an interval of the form $[2k, 2k+2]$, for some $k \in \mathbf{Z}^+$. Theorem 1.3.4 gives that

$$\big\|\widetilde{f}\big\|_{L^p} \le A_p \big\|f\big\|_{L^p} \tag{3.5.13}$$

for some $A_p > 0$ and all $2 \le p < \infty$ when f is a simple function. By density the same result is valid for all L^p functions when $p \ge 2$. Finally, we observe that the adjoint operator of $f \mapsto \widetilde{f}$ is $f \mapsto -\widetilde{f}$. By duality, estimate (3.5.13) is also valid for $1 < p \le 2$ with constant $A_{p'} = A_p$. □

We extend the preceding result to higher dimensions.

Theorem 3.5.7. *Let $1 < p < \infty$ and $f \in L^p(\mathbf{T}^n)$. Then $D(n,N) * f$ converges to f in L^p as $N \to \infty$.*

Proof. Let us prove this theorem in dimension $n = 2$. The same proof can be adjusted to work in every dimension. In view of Corollary 3.5.2, it suffices to prove that for all f smooth on $\mathbf{T}^2$ we have

$$\sup_{N\ge 0} \int_0^1 \int_0^1 \bigg| \sum_{|m_1|\le N} \sum_{|m_2|\le N} e^{2\pi i(m_1 x_1 + m_2 x_2)} \widehat{f}(m_1, m_2) \bigg|^p dx_1\, dx_2 \le K^{2p} \|f\|^p_{L^p(\mathbf{T}^2)}.$$

For fixed $f \in \mathscr{C}^\infty(\mathbf{T}^2)$, $N \ge 0$, and $x_2 \in [0,1]$, define a trigonometric polynomial g_{N,x_2} on $\mathbf{T}^1$ by setting

$$\sum_{|m_2|\le N} e^{2\pi i m_2 x_2} \widehat{f}(m_1, m_2) = \widehat{g_{N,x_2}}(m_1)$$

for all $m_1 \in \mathbf{Z}$. Then we have

$$(D_N * g_{N,x_2})(x_1) = \sum_{|m_1|\le N} e^{2\pi i m_1 x_1} \widehat{g_{N,x_2}}(m_1)$$

and also

$$g_{N,x_2}(x_1) = \sum_{|m_2|\le N} e^{2\pi i m_2 x_2} \bigg[\sum_{m_1 \in \mathbf{Z}} e^{2\pi i m_1 x_1} \widehat{f}(m_1, m_2) \bigg] = (D_N * f_{x_1})(x_2),$$

where f_{x_1} is the function defined by $f_{x_1}(y) = f(x_1, y)$. We have

$$\begin{aligned} &\int_0^1 \int_0^1 \bigg| \sum_{|m_1|\le N} \sum_{|m_2|\le N} e^{2\pi i(m_1 x_1 + m_2 x_2)} \widehat{f}(m_1, m_2) \bigg|^p dx_1\, dx_2 \\ &\qquad = \int_0^1 \int_0^1 \Big| (D_N * g_{N,x_2})(x_1) \Big|^p dx_1\, dx_2 \\ &\qquad \le K^p \int_0^1 \int_0^1 |g_{N,x_2}(x_1)|^p\, dx_1\, dx_2 \\ &\qquad = K^p \int_0^1 \int_0^1 \Big| (D_N * f_{x_1})(x_2) \Big|^p dx_2\, dx_1 \\ &\qquad \le K^{2p} \int_0^1 \int_0^1 |f_{x_1}(x_2)|^p\, dx_2\, dx_1 \\ &\qquad = K^{2p} \|f\|^p_{L^p(\mathbf{T}^2)}. \end{aligned}$$

We used twice the fact that the one-dimensional partial sums are uniformly bounded in L^p when $1 < p < \infty$, a consequence of Corollary 3.5.2, Proposition 3.5.5, and Theorem 3.5.6. $\square$

Exercises

3.5.1. If $f \in \mathscr{C}^\infty(\mathbf{T}^n)$, then show that $D(n,N) * f$ and $\widetilde{D}(n,N) * f$ converge to f uniformly and in L^p for $1 \le p \le \infty$.

3.5.2. Prove that the norms of the Riesz projections on $L^2(\mathbf{T}^1)$ are at most 1, while the operation of conjugation $f \mapsto \widetilde{f}$ is an isometry on $L^2(\mathbf{T}^1)$.

3.5.3. Let $-\infty \le a_j < b_j \le +\infty$ for $1 \le j \le n$. Consider the rectangular projection operator defined on $\mathscr{C}^\infty(\mathbf{T}^n)$ by

$$P(f)(x) = \sum_{a_j \le m_j \le b_j} \widehat{f}(m) e^{2\pi i(m_1 x_1 + \cdots + m_n x_n)} .$$

Prove that when $1 < p < \infty$, P extends to a bounded operator from $L^p(\mathbf{T}^n)$ to itself with bounds independent of the a_j, b_j.
[*Hint:* Express P in terms of the Riesz projection P_+.]

3.5.4. Let $P_r(t)$ be the Poisson kernel on $\mathbf{T}^1$ as defined in Exercise 3.1.7. For $0 < r < 1$, define the *conjugate Poisson kernel* $Q_r(t)$ on the circle by

$$Q_r(t) = -i \sum_{m=-\infty}^{+\infty} \operatorname{sgn}(m)\, r^{|m|} e^{2\pi i m t} .$$

(a) For $0 < r < 1$, prove the identity

$$Q_r(t) = \frac{2r\sin(2\pi t)}{1 - 2r\cos(2\pi t) + r^2} .$$

(b) Prove that $\widetilde{f}(t) = \lim_{r \to 1}(Q_r * f)(t)$ whenever f is smooth. Conclude that if f is real-valued, then so is $\widetilde{f}$.
(c) Let $f \in L^1(\mathbf{T}^1)$. Prove that the functions $z \mapsto (P_r * f)(t)$ and $z \mapsto (Q_r * f)(t)$ are harmonic functions of $z = re^{2\pi i t}$ in the region $|z| < 1$.
(d) Let $f \in L^1(\mathbf{T}^1)$. Prove that the function

$$z \mapsto (P_r * f)(t) + i(Q_r * f)(t)$$

is analytic in $z = re^{2\pi i t}$ and thus $(P_r * f)(t)$ and $(Q_r * f)(t)$ are conjugate harmonic functions.

3.5.5. Let f be in $\dot{\Lambda}_\alpha(\mathbf{T}^1)$ for some $0 < \alpha < 1$. Prove that the conjugate function $\widetilde{f}$ is well defined and can be written as

$$\begin{aligned}\widetilde{f}(x) &= \lim_{\varepsilon\to 0}\int_{\varepsilon\le|t|\le 1/2} f(x-t)\cot(\pi t)\,dt \\ &= \int_{|t|\le 1/2}\big(f(x-t)-f(x)\big)\cot(\pi t)\,dt.\end{aligned}$$

[*Hint:* Use part (b) of Exercise 3.5.4 and the fact that Q_r has integral zero over the circle to write $(f*Q_r)(x) = \big((f-f(x))*Q_r\big)(x)$, allowing use of the Lebesgue dominated convergence theorem.]

3.5.6. Suppose that f is a real-valued function on $\mathbf{T}^1$ with $|f|\le 1$ and $0\le\lambda<\pi/2$.
(a) Prove that

$$\int_{\mathbf{T}^1} e^{\lambda\widetilde{f}(t)}\,dt \le \frac{1}{\cos(\lambda)}.$$

(b) Conclude that for $0\le\lambda<\pi/2$ we have

$$\int_{\mathbf{T}^1} e^{\lambda|\widetilde{f}(t)|}\,dt \le \frac{2}{\cos(\lambda)}.$$

[*Hint:* Part (a): Consider the analytic function $F(z)$ on the disk $|z|<1$ defined by $F(z) = -i(P_r*f)(\theta)+(Q_r*f)(\theta)$, where $z=re^{2\pi i\theta}$. Then Re $e^{\lambda F(z)}$ is harmonic and its average over the circle $|z|=r$ is equal to its value at the origin, which is $\cos(\lambda f(0))\le 1$. Let $r\uparrow 1$ and use that for $z=e^{2\pi it}$ on the circle we have Re $e^{\lambda F(z)}\ge e^{\lambda\widetilde{f}(t)}\cos(\lambda)$.]

3.5.7. Prove that for $0<\alpha<1$ there is a constant C_α such that

$$\big\|\widetilde{f}\big\|_{\dot{\Lambda}_\alpha(\mathbf{T}^1)} \le C_\alpha\big\|f\big\|_{\dot{\Lambda}_\alpha(\mathbf{T}^1)}.$$

[*Hint:* Using Exercise 3.5.5, for $|h|\le 1/10$ write $\widetilde{f}(x+h)-\widetilde{f}(x)$ as

$$\begin{aligned}&\int_{|t|\le 5|h|}\big(f(x-t)-f(x+h)\big)\cot(\pi(t+h))\,dt \\ &\quad+\int_{|t|\le 5|h|}\big(f(x-t)-f(x)\big)\cot(\pi t)\,dt \\ &\quad+\int_{5|h|\le|t|\le 1/2}\big(f(x-t)-f(x)\big)\big(\cot(\pi(t+h))-\cot(\pi t)\big)\,dt \\ &\quad+\big(f(x)-f(x+h)\big)\int_{5|h|\le|t|\le 1/2}\cot(\pi(t+h))\,dt.\end{aligned}$$

You may use the fact that $\cot(\pi t) = \frac{1}{\pi t}+b(t)$, where $b(t)$ is a bounded function when $|t|\le 1/2$. The case $|h|\ge 1/10$ is easy.]

3.5.8. (a) Show that for M,N positive integers we have

$$(F_M*D_N)(x) = \begin{cases} F_M(x) & \text{for } M\le N, \\ F_N(x)+\frac{M-N}{(M+1)(N+1)}\sum\limits_{|k|\le N}|k|\,e^{2\pi ikx} & \text{for } M>N.\end{cases}$$

(b) Prove that for some constant $c > 0$ we have

$$\int_{\mathbf{T}^1} \left| \sum_{|k| \le N} |k| \, e^{2\pi i k x} \right| dx \ge c N \log N$$

as $N \to \infty$.
$\big[$*Hint:* Part (b): Show that for $x \in [-\frac{1}{2}, \frac{1}{2}]$ we have

$$\sum_{|k| \le N} |k| \, e^{2\pi i k x} = (N+1)(D_N(x) - F_N(x))$$

and use the result of Exercise 3.1.8.$\big]$

3.5.9. Show that the Fourier series of the integrable functions

$$f_1(x) = \sum_{j=0}^{\infty} 2^{-j} F_{2^{2^{2^j}}}(x), \qquad f_2(x) = \sum_{j=1}^{\infty} \frac{1}{j^2} F_{2^{2^j}}(x), \qquad x \in \mathbf{T}^1,$$

do not converge in $L^1(\mathbf{T}^1)$.
$\big[$*Hint:* Let $M_j = 2^{2^{2^j}}$ or $M_j = 2^{2^j}$ depending on the situation. For fixed N let j_N be the least integer j such that $M_j > N$. Then for $j \ge j_N + 1$ we have $M_j \ge M_{j_N}^2 > N^2 \ge 2N + 1$, hence $\frac{M_j - N}{M_j + 1} \ge \frac{1}{2}$. Split the summation indices into the sets $j \ge j_N$ and $j < j_N$. Conclude that $\left\| f_1 * D_N \right\|_{L^1}$ and $\left\| f_2 * D_N \right\|_{L^1}$ tend to infinity as $N \to \infty$ using Exercise 3.5.8.$\big]$

3.5.10. (*Stein [251]*) Note that if $\alpha \ge 0$, then B_R^α are bounded on $L^2(\mathbf{T}^n)$ uniformly in $R > 0$. Show that if $\alpha > \frac{n-1}{2}$, then B_R^α are bounded on $L^1(\mathbf{T}^n)$ uniformly in $R > 0$. Use complex interpolation to prove that for $\alpha > \frac{n-1}{2}\left|\frac{1}{p} - \frac{1}{2}\right|$, the B_R^α are bounded on $L^p(\mathbf{T}^n)$ uniformly in $R > 0$. Compare this problem with Exercise 1.3.5.

3.6 Multipliers, Transference, and Almost Everywhere Convergence

In Chapter 2 we saw that bounded operators from $L^p(\mathbf{R}^n)$ to $L^q(\mathbf{R}^n)$ that commute with translations are given by convolution with tempered distributions on $\mathbf{R}^n$. In particular, when $p = q$, these tempered distributions have bounded Fourier transforms, called Fourier multipliers. Convolution operators that commute with translations can also be defined on the torus. These lead to Fourier multipliers on the torus.

3.6.1 Multipliers on the Torus

In analogy with the nonperiodic case, we could identify convolution operators on $\mathbf{T}^n$ with appropriate distributions on the torus; see Exercise 3.6.2 for an introduction to this topic. However, it is simpler to avoid this point of view and consider the multipliers directly, bypassing the discussion of distributions on the torus. The reason for this is the following theorem.

Theorem 3.6.1. *Suppose that T is a linear operator that commutes with translations and maps $L^p(\mathbf{T}^n)$ to $L^q(\mathbf{T}^n)$ for some $1 \le p,q \le \infty$. Then there exists a bounded sequence $\{a_m\}_{m\in\mathbf{Z}^n}$ such that*

$$T(f)(x) = \sum_{m\in\mathbf{Z}^n} a_m \widehat{f}(m) e^{2\pi i m\cdot x} \tag{3.6.1}$$

for all $f \in \mathscr{C}^\infty(\mathbf{T}^n)$. Moreover, we have $\big\|\{a_m\}\big\|_{\ell^\infty} \le \big\|T\big\|_{L^p\to L^q}$.

Proof. Consider the functions $e_m(x) = e^{2\pi i m\cdot x}$ defined on $\mathbf{T}^n$ for m in $\mathbf{Z}^n$. Since T is translation invariant for all $h \in \mathbf{T}^n$, we have

$$T(e_m)(x-h) = T(\tau^h(e_m))(x) = e^{-2\pi i m\cdot h} T(e_m)(x)$$

for every $x \in F_h$, where F_h is a set of full measure on $\mathbf{T}^n$. For $x \in \mathbf{T}^n$ define $D(x) = |\{h \in \mathbf{T}^n : x \in F_h\}|$. Then $D(x) \le 1$ for all x and by Fubini's theorem D has integral 1 on $\mathbf{T}^n$. Therefore there exists an $x_0 \in \mathbf{T}^n$ such that $D(x_0) = 1$. It follows that for almost all $h \in \mathbf{T}^n$ (i.e., for all h in the set $\{h \in \mathbf{T}^n : x_0 \in F_h\}$) we have $T(e_m)(x_0 - h) = e^{-2\pi i m\cdot h} T(e_m)(x_0)$. Replacing $x_0 - h$ by x, we obtain

$$T(e_m)(x) = e^{2\pi i m\cdot x}\big(e^{-2\pi i m\cdot x_0} T(e_m)(x_0)\big) = a_m e_m(x) \tag{3.6.2}$$

for almost all $x \in \mathbf{T}^n$, where we set $a_m = e^{-2\pi i m\cdot x_0} T(e_m)(x_0)$, for $m \in \mathbf{Z}^n$. Taking L^q norms in (3.6.2), we deduce $|a_m| = \big\|T(e_m)\big\|_{L^q} \le \big\|T\big\|_{L^p\to L^q}$, and thus a_m is bounded. Moreover, since $T(e_m) = a_m e_m$ for all m in $\mathbf{Z}^n$, it follows that (3.6.1) holds for all trigonometric polynomials. By density this extends to all $f \in \mathscr{C}^\infty(\mathbf{T}^n)$ and the theorem is proved. □

Definition 3.6.2. Let $1 \le p,q \le \infty$. We call a bounded sequence $\{a_m\}_{m\in\mathbf{Z}^n}$ an (L^p,L^q) *multiplier* if the corresponding operator given by (3.6.1) maps $L^p(\mathbf{T}^n)$ to $L^q(\mathbf{T}^n)$. If $p = q$, (L^p,L^p) multipliers are called simply L^p multipliers. When $1 \le p < \infty$, the space of all L^p multipliers on $\mathbf{T}^n$ is denoted by $\mathscr{M}_p(\mathbf{Z}^n)$. This notation follows the convention that $\mathscr{M}_p(\widehat{G})$ denote the space of L^p multipliers on $L^p(G)$, where G is a locally compact group and $\widehat{G}$ is its dual group. The norm of an element $\{a_m\}$ in $\mathscr{M}_p(\mathbf{Z}^n)$ is the norm of the operator T given by (3.6.1) from $L^p(\mathbf{T}^n)$ to itself. This norm is denoted by $\big\|\{a_m\}\big\|_{\mathscr{M}_p}$.

We now examine some special cases. We begin with the case $p = q = 2$. As expected, it turns out that $\mathscr{M}_2(\mathbf{Z}^n) = \ell^\infty(\mathbf{Z}^n)$.

Theorem 3.6.3. *A linear operator T that commutes with translations maps $L^2(\mathbf{T}^n)$ to itself if and only if there exists a sequence $\{a_m\}_{m\in\mathbf{Z}^n}$ in ℓ^∞ such that*

$$T(f)(x) = \sum_{m\in\mathbf{Z}^n} a_m \widehat{f}(m) e^{2\pi i m\cdot x} \tag{3.6.3}$$

for all $f \in \mathscr{C}^\infty(\mathbf{T}^n)$. Moreover, in this case we have $\big\|T\big\|_{L^2\to L^2} = \big\|\{a_m\}\big\|_{\ell^\infty}$.

Proof. The existence of such a sequence is guaranteed by Theorem 3.6.1, which also gives $\big\|\{a_m\}\big\|_{\ell^\infty} \le \big\|T\big\|_{L^2\to L^2}$. Conversely, any operator given by the form (3.6.3) satisfies

$$\big\|T(f)\big\|_{L^2}^2 = \sum_{m\in\mathbf{Z}^n} |a_m \widehat{f}(m)|^2 \le \big\|\{a_m\}\big\|_{\ell^\infty}^2 \sum_{m\in\mathbf{Z}^n} |\widehat{f}(m)|^2 ,$$

and thus $\big\|T\big\|_{L^2\to L^2} \le \big\|\{a_m\}\big\|_{\ell^\infty}$. □

We continue with the case $p = q = 1$. Recall the definition of a finite Borel measure on $\mathbf{T}^n$. Given such a measure μ, its Fourier coefficients are defined by

$$\widehat{\mu}(m) = \int_{\mathbf{T}^n} e^{-2\pi i x\cdot m}\, d\mu(x), \qquad m \in \mathbf{Z}^n .$$

Clearly all the Fourier coefficients of the measure μ are bounded by the total variation $\big\|\mu\big\|$ of μ. See Exercise 3.6.3 for basic properties of Fourier transforms of distributions on the torus.

Theorem 3.6.4. *A linear operator T that commutes with translations maps $L^1(\mathbf{T}^n)$ to itself if and only if there exists a finite Borel measure μ on the torus such that*

$$T(f)(x) = \sum_{m\in\mathbf{Z}^n} \widehat{\mu}(m) \widehat{f}(m) e^{2\pi i m\cdot x} \tag{3.6.4}$$

for all $f \in \mathscr{C}^\infty(\mathbf{T}^n)$. Moreover, in this case we have $\big\|T\big\|_{L^1\to L^1} = \big\|\mu\big\|$. In other words, $\mathscr{M}_1(\mathbf{Z}^n)$ is the set of all sequences given by Fourier coefficients of finite Borel measures on $\mathbf{T}^n$.

Proof. Fix $f \in L^1(\mathbf{T}^n)$. If (3.6.4) is valid, then $\widehat{T(f)}(m) = \widehat{f}(m)\widehat{\mu}(m)$ for all $m \in \mathbf{Z}^n$. But Exercise 3.6.3 gives that $\widehat{f*\mu}(m) = \widehat{f}(m)\widehat{\mu}(m)$ for all $m \in \mathbf{Z}^n$; therefore, the integrable functions $f*\mu$ and $T(f)$ have the same Fourier coefficients and they must be equal. Thus $T(f) = f*\mu$, which implies that T is bounded on L^1 and $\big\|T(f)\big\|_{L^1} \le \big\|\mu\big\|\big\|f\big\|_{L^1}$.

To prove the converse direction, we suppose that T commutes with translations and maps $L^1(\mathbf{T}^n)$ to itself. We recall the following identity obtained in (3.1.20):

$$P_\varepsilon(x) = \sum_{m\in\mathbf{Z}^n} e^{-2\pi|m|\varepsilon} e^{2\pi i m\cdot x} = \frac{\Gamma(\frac{n+1}{2})}{\pi^{\frac{n+1}{2}}} \sum_{m\in\mathbf{Z}^n} \frac{\varepsilon^{-n}}{(1+|\frac{x+m}{\varepsilon}|^2)^{\frac{n+1}{2}}} \ge 0 \tag{3.6.5}$$

for all $x \in \mathbf{T}^n$. Integrating the second series in (3.6.5) over $[-1/2,1/2]^n$, expressing the result as an integral over $\mathbf{R}^n$, and using the fact that the Poisson kernel on $\mathbf{R}^n$ has integral one (cf. Example 2.1.13), we conclude that $\|P_\varepsilon\|_{L^1(\mathbf{T}^n)} = 1$. It follows that

$$\|T(P_\varepsilon)\|_{L^1(\mathbf{T}^n)} \le \|T\|_{L^1\to L^1}$$

for all $\varepsilon > 0$. The Banach–Alaoglu theorem gives that there exist a sequence $\varepsilon_j \downarrow 0$ and a finite Borel measure μ on $\mathbf{T}^n$ such that $T(P_{\varepsilon_j})$ tends to μ weakly as $j \to \infty$. This means that for all continuous functions g on $\mathbf{T}^n$ we have

$$\lim_{j\to\infty} \int_{\mathbf{T}^n} g(x) T(P_{\varepsilon_j})(x)\, dx = \int_{\mathbf{T}^n} g(x)\, d\mu(x). \tag{3.6.6}$$

It follows from (3.6.6) that for all g continuous on $\mathbf{T}^n$ we have

$$\left| \int_{\mathbf{T}^n} g(x)\, d\mu(x) \right| \le \sup_j \|T(P_{\varepsilon_j})\|_{L^1} \|g\|_{L^\infty} \le \|T\|_{L^1\to L^1} \|g\|_{L^\infty}.$$

Since by the Riesz representation theorem we have that the norm of the linear functional

$$g \mapsto \int_{\mathbf{T}^n} g(x)\, d\mu(x)$$

on $C(\mathbf{T}^n)$ is $\|\mu\|$, it follows that

$$\|\mu\| \le \|T\|_{L^1\to L^1}. \tag{3.6.7}$$

It remains to prove that T has the form given in (3.6.4). By Theorem 3.6.1 we have that there exists a bounded sequence $\{a_m\}$ on $\mathbf{Z}^n$ such that (3.6.1) is satisfied. Taking $g(x) = e^{-2\pi i k\cdot x}$ in (3.6.6) and using the representation for T in (3.6.1), we obtain

$$\widehat{\mu}(k) = \int_{\mathbf{T}^n} e^{-2\pi i k\cdot x}\, d\mu(x) = \lim_{j\to\infty} \int_{\mathbf{T}^n} e^{-2\pi i k\cdot x} \sum_{m\in\mathbf{Z}^n} a_m e^{-2\pi\varepsilon_j|m|} e^{2\pi i m\cdot x}\, dx = a_k.$$

This proves assertion (3.6.4). It follows from (3.6.4) that $T(f) = f * \mu$ and thus $\|T\|_{L^1\to L^1} \le \|\mu\|$. This fact combined with (3.6.7) gives $\|T\|_{L^1\to L^1} = \|\mu\|$. □

Remark 3.6.5. It is not hard to see that most basic properties of the space $\mathscr{M}_p(\mathbf{R}^n)$ of L^p Fourier multipliers on $\mathbf{R}^n$ are also valid for $\mathscr{M}_p(\mathbf{Z}^n)$. In particular, $\mathscr{M}_p(\mathbf{Z}^n)$ is a closed subspace of ℓ^∞ and thus a Banach space itself. Moreover, sums, scalar multiples, and products of elements of $\mathscr{M}_p(\mathbf{Z}^n)$ are also in $\mathscr{M}_p(\mathbf{Z}^n)$, which makes this space a Banach algebra. As in the nonperiodic case, we also have $\mathscr{M}_p(\mathbf{Z}^n) = \mathscr{M}_{p'}(\mathbf{Z}^n)$ when $1 < p < \infty$.

3.6.2 Transference of Multipliers

It is clear by now that multipliers on $L^1(\mathbf{T}^n)$ and $L^1(\mathbf{R}^n)$ are very similar, and the same is true for $L^2(\mathbf{T}^n)$ and $L^2(\mathbf{R}^n)$. These similarities became obvious when we characterized L^1 and L^2 multipliers on both $\mathbf{R}^n$ and $\mathbf{T}^n$. So far, there is no known nontrivial characterization of $\mathscr{M}_p(\mathbf{R}^n)$, but we might ask whether this space is related to $\mathscr{M}_p(\mathbf{Z}^n)$. There are several connections of this type and there are general ways to produce multipliers on the torus from multipliers on $\mathbf{R}^n$ and vice versa. General methods of this sort are called transference of multipliers.

We begin with a useful definition.

Definition 3.6.6. Let $t_0 \in \mathbf{R}^n$. A bounded function b on $\mathbf{R}^n$ is called *regulated at the point* t_0 if

$$\lim_{\varepsilon\to 0} \frac{1}{\varepsilon^n} \int_{|t|\le\varepsilon} \big(b(t_0-t) - b(t_0)\big)\, dt = 0. \tag{3.6.8}$$

The function b is called *regulated* if it is regulated at every $t_0 \in \mathbf{R}^n$.

Condition (3.6.8) says that the point t_0 is a Lebesgue point of b. This is certainly the case if the function b is continuous at $t_0 \in \mathbf{R}^n$. If $b(t_0) = 0$, condition (3.6.8) also holds when $b(t_0-t) = -b(t_0+t)$ whenever $|t| \le \varepsilon$ for some $\varepsilon > 0$.

The first transference result we discuss is the following.

Theorem 3.6.7. *Suppose that b is a regulated function that lies in $\mathscr{M}_p(\mathbf{R}^n)$ for some $1 \le p < \infty$. Then the sequence $\{b(m)\}_{m\in\mathbf{Z}^n}$ is in $\mathscr{M}_p(\mathbf{Z}^n)$ and moreover,*

$$\big\|\{b(m)\}\big\|_{\mathscr{M}_p(\mathbf{Z}^n)} \le \big\|b\big\|_{\mathscr{M}_p(\mathbf{R}^n)}.$$

Also, for all $R > 0$, the sequences $\{b(m/R)\}_{m\in\mathbf{Z}^n}$ are in $\mathscr{M}_p(\mathbf{Z}^n)$ and we have

$$\sup_{R>0} \big\|\{b(m/R)\}\big\|_{\mathscr{M}_p(\mathbf{Z}^n)} \le \big\|b\big\|_{\mathscr{M}_p(\mathbf{R}^n)}.$$

The second conclusion of the theorem is a consequence of the first conclusion ($R = 1$), since the functions $b(\xi/R)$ and $b(\xi)$ have the same norm in $\mathscr{M}_p(\mathbf{R}^n)$. Before we begin the proof, we state the following lemma, which we derive after the proof of Theorem 3.6.7.

Lemma 3.6.8. *Let T be the operator on $\mathbf{R}^n$ whose multiplier is $b(\xi)$, and let S be the operator on $\mathbf{T}^n$ whose multiplier is the sequence $\{b(m)\}_{m\in\mathbf{Z}^n}$. Assume that $b(\xi)$ is regulated at every point $m \in \mathbf{Z}^n$. Suppose that P and Q are trigonometric polynomials on $\mathbf{T}^n$ and let $L_\varepsilon(x) = e^{-\pi\varepsilon|x|^2}$ for $x \in \mathbf{R}^n$ and $\varepsilon > 0$. Then the following identity is valid whenever $\alpha, \beta > 0$ and $\alpha + \beta = 1$:*

$$\lim_{\varepsilon\to 0} \varepsilon^{\frac{n}{2}} \int_{\mathbf{R}^n} T(PL_{\varepsilon\alpha})(x)\overline{Q(x)L_{\varepsilon\beta}(x)}\, dx = \int_{\mathbf{T}^n} S(P)(x)\overline{Q(x)}\, dx. \tag{3.6.9}$$

Proof. We give the proof of Theorem 3.6.7. The case $p = 1$ can be proved easily using Theorems 2.5.8, 3.6.4, and Exercise 3.6.4 and is left to the reader. Let us therefore consider the case $1 < p < \infty$. We are assuming that T maps $L^p(\mathbf{R}^n)$ to itself and we need to show that S maps $L^p(\mathbf{T}^n)$ to itself. We prove this using duality. For P and Q trigonometric polynomials, using Lemma 3.6.8, we have

$$
\begin{aligned}
&\left|\int_{\mathbf{T}^n} S(P)(x)\overline{Q(x)}\,dx\right| \\
&= \left|\lim_{\varepsilon\to 0} \varepsilon^{\frac{n}{2}} \int_{\mathbf{R}^n} T(PL_{\varepsilon/p})(x)\overline{Q(x)L_{\varepsilon/p'}(x)}\,dx\right| \\
&\le \big\|T\big\|_{L^p\to L^p} \limsup_{\varepsilon\to 0} \varepsilon^{\frac{n}{2}} \big\|PL_{\varepsilon/p}\big\|_{L^p(\mathbf{R}^n)} \big\|QL_{\varepsilon/p'}\big\|_{L^{p'}(\mathbf{R}^n)} \\
&= \big\|T\big\|_{L^p\to L^p} \limsup_{\varepsilon\to 0} \left(\varepsilon^{\frac{n}{2}} \int_{\mathbf{R}^n} |P(x)|^p e^{-\varepsilon\pi|x|^2}\,dx\right)^{\frac{1}{p}} \left(\varepsilon^{\frac{n}{2}} \int_{\mathbf{R}^n} |Q(x)|^{p'} e^{-\varepsilon\pi|x|^2}\,dx\right)^{\frac{1}{p'}} \\
&= \big\|T\big\|_{L^p\to L^p} \left(\int_{\mathbf{T}^n} |P(x)|^p\,dx\right)^{\frac{1}{p}} \left(\int_{\mathbf{T}^n} |Q(x)|^{p'}\,dx\right)^{\frac{1}{p'}},
\end{aligned}
$$

provided for all continuous (periodic) functions g on $\mathbf{T}^n$ we have that

$$
\lim_{\varepsilon\to 0} \varepsilon^{\frac{n}{2}} \int_{\mathbf{R}^n} g(x) e^{-\varepsilon\pi|x|^2}\,dx = \int_{\mathbf{T}^n} g(x)\,dx. \tag{3.6.10}
$$

Assuming (3.6.10) for the moment, we take the supremum over all trigonometric polynomials Q on $\mathbf{T}^n$ with $L^{p'}$ norm at most 1 to obtain that S maps $L^p(\mathbf{T}^n)$ to itself with norm at most $\big\|T\big\|_{L^p\to L^p}$, yielding the required conclusion.

We now prove (3.6.10). Use the Poisson summation formula to write the left-hand side of (3.6.10) as

$$
\begin{aligned}
\varepsilon^{\frac{n}{2}} \sum_{k\in\mathbf{Z}^n} \int_{\mathbf{T}^n} g(x-k) e^{-\varepsilon\pi|x-k|^2}\,dx &= \int_{\mathbf{T}^n} g(x) \varepsilon^{\frac{n}{2}} \sum_{k\in\mathbf{Z}^n} e^{-\varepsilon\pi|x-k|^2}\,dx \\
&= \int_{\mathbf{T}^n} g(x) \sum_{k\in\mathbf{Z}^n} e^{-\pi|k|^2/\varepsilon} e^{2\pi i x\cdot k}\,dx \\
&= \int_{\mathbf{T}^n} g(x)\,dx + A_\varepsilon ,
\end{aligned}
$$

where

$$
|A_\varepsilon| \le \big\|g\big\|_{L^\infty} \sum_{|k|\ge 1} e^{-\pi|k|^2/\varepsilon} \to 0
$$

as $\varepsilon \to 0$. This completes the proof of Theorem 3.6.7. □

We now turn to the proof of Lemma 3.6.8.

Proof. It suffices to prove the required assertion for $P(x) = e^{2\pi i m\cdot x}$ and $Q(x) = e^{2\pi i k\cdot x}$, $k,m \in \mathbf{Z}^n$, since the general case follows from this case by linearity. In view of Parseval's relation (Proposition 3.1.16 (3)), we have

$$
\int_{\mathbf{T}^n} S(P)(x)\overline{Q(x)}\,dx = \sum_{r\in\mathbf{Z}^n} b(r)\widehat{P}(r)\overline{\widehat{Q}(r)} = \begin{cases} b(m) & \text{when } k = m, \\ 0 & \text{when } k \ne m. \end{cases} \tag{3.6.11}
$$

On the other hand, using the identity in Theorem 2.2.14 (3), we obtain

$$\varepsilon^{\frac{n}{2}} \int_{\mathbf{R}^n} T(PL_{\varepsilon\alpha})(x)\overline{Q(x)L_{\varepsilon\beta}(x)}\,dx$$
$$= \varepsilon^{\frac{n}{2}} \int_{\mathbf{R}^n} b(\xi)\widehat{PL_{\varepsilon\alpha}}(\xi)\overline{\widehat{QL_{\varepsilon\beta}}(\xi)}\,d\xi$$
$$= \varepsilon^{\frac{n}{2}} \int_{\mathbf{R}^n} b(\xi)(\varepsilon\alpha)^{-\frac{n}{2}} e^{-\pi\frac{|\xi-m|^2}{\varepsilon\alpha}} (\varepsilon\beta)^{-\frac{n}{2}} e^{-\pi\frac{|\xi-k|^2}{\varepsilon\beta}}\,d\xi$$
$$= (\varepsilon\alpha\beta)^{-\frac{n}{2}} \int_{\mathbf{R}^n} b(\xi) e^{-\pi\frac{|\xi-m|^2}{\varepsilon\alpha}} e^{-\pi\frac{|\xi-k|^2}{\varepsilon\beta}}\,d\xi\,. \tag{3.6.12}$$

Now if $m = k$, since $\alpha + \beta = 1$, the expression in (3.6.12) is equal to

$$(\varepsilon\alpha\beta)^{-\frac{n}{2}} \int_{\mathbf{R}^n} b(\xi) e^{-\pi\frac{|\xi-m|^2}{\varepsilon\alpha\beta}}\,d\xi\,, \tag{3.6.13}$$

which tends to $b(m)$ if b is continuous at m, since the family $\varepsilon^{-\frac{n}{2}} e^{-\pi\frac{|\xi|^2}{\varepsilon}}$ is an approximate identity as $\varepsilon \to 0$. If b is not continuous at m but satisfies condition (3.6.8) with $t_0 = m$, then still the expression in (3.6.13) tends to $b(m)$ as $\varepsilon \to 0$ in view of the result of Exercise 3.6.6.

We now consider the case $m \neq k$ in (3.6.12). If $|m-k| \geq 1$, then every ξ in $\mathbf{R}^n$ must satisfy either $|\xi - m| \geq 1/2$ or $|\xi - k| \geq 1/2$. Therefore, the expression in (3.6.12) is controlled by

$$(\varepsilon\alpha\beta)^{-\frac{n}{2}} \left(\int_{|\xi-m|\geq\frac{1}{2}} b(\xi) e^{-\frac{\pi}{4\varepsilon\alpha}} e^{-\pi\frac{|\xi-k|^2}{\varepsilon\beta}}\,d\xi + \int_{|\xi-k|\geq\frac{1}{2}} b(\xi) e^{-\frac{\pi}{4\varepsilon\beta}} e^{-\pi\frac{|\xi-m|^2}{\varepsilon\alpha}}\,d\xi \right),$$

which is in turn controlled by

$$\|b\|_{L^\infty} \left(\alpha^{-\frac{n}{2}} e^{-\frac{\pi}{4\varepsilon\alpha}} + \beta^{-\frac{n}{2}} e^{-\frac{\pi}{4\varepsilon\beta}} \right),$$

which tends to zero as $\varepsilon \to 0$. This proves that the expression in (3.6.11) is equal to the limit of the expression in (3.6.12) as $\varepsilon \to 0$. This completes the proof of Lemma 3.6.8 □

We now obtain a converse of Theorem 3.6.7. If $b(\xi)$ is a bounded function on $\mathbf{R}^n$ and the sequence $\{b(m)\}_{m\in\mathbf{Z}^n}$ is in $\mathscr{M}_p(\mathbf{Z}^n)$, then we cannot necessarily obtain that b is in $\mathscr{M}_p(\mathbf{R}^n)$, since such a conclusion would have to depend on all the values of b and not on the values of b on a set of measure zero such as the integer lattice. However, a converse can be formulated if we assume that for all $R > 0$, the sequences $\{b(Rm)\}_{m\in\mathbf{Z}^n}$ are in $\mathscr{M}_p(\mathbf{Z}^n)$ uniformly in R. Then we obtain that $b(R\xi)$ is in $\mathscr{M}_p(\mathbf{R}^n)$ uniformly in $R > 0$, which is equivalent to saying that $b \in \mathscr{M}_p(\mathbf{R}^n)$, since dilations of multipliers on $\mathbf{R}^n$ do not affect their norms (see Proposition 2.5.14). These remarks can be precisely expressed in the following theorem.

Theorem 3.6.9. *Suppose that $b(\xi)$ is a bounded Riemann integrable function on $\mathbf{R}^n$ and that the sequences $\{b(\frac{m}{R})\}_{m\in\mathbf{Z}^n}$ are in $\mathscr{M}_p(\mathbf{Z}^n)$ uniformly in $R > 0$ for some $1 \leq p < \infty$. Then b is in $\mathscr{M}_p(\mathbf{R}^n)$ and we have*

$$\|b\|_{\mathscr{M}_p(\mathbf{R}^n)} \le \sup_{R>0} \|\{b(\tfrac{m}{R})\}_{m\in\mathbf{Z}^n}\|_{\mathscr{M}_p(\mathbf{Z}^n)}. \tag{3.6.14}$$

Proof. Suppose that f and g are smooth functions with compact support on $\mathbf{R}^n$. Then there is an $R_0 > 0$ such that for $R \ge R_0$, the functions $x \mapsto f(Rx)$ and $x \mapsto g(Rx)$ are supported in $[-1/2, 1/2]^n$. We define periodic functions

$$F_R(x) = \sum_{k\in\mathbf{Z}^n} f(R(x-k)) \quad \text{and} \quad G_R(x) = \sum_{k\in\mathbf{Z}^n} g(R(x-k))$$

on $\mathbf{T}^n$. Observe that the mth Fourier coefficient of F_R is $\widehat{F_R}(m) = R^{-n}\widehat{f}(m/R)$ and that of G_R is $\widehat{G_R}(m) = R^{-n}\widehat{g}(m/R)$. Set

$$C_b = \sup_{R>0} \|\{b(m/R)\}_{m\in\mathbf{Z}^n}\|_{\mathscr{M}_p(\mathbf{Z}^n)}.$$

Now for $R \ge R_0$ we have

$$\Big|\sum_{m\in\mathbf{Z}^n} b(m/R)\widehat{f}(m/R)\overline{\widehat{g}(m/R)}\,\text{Volume}\,\big([\tfrac{m}{R}, \tfrac{m+1}{R}]^n\big)\Big| \tag{3.6.15}$$

$$\begin{aligned}
&= \Big|R^n \sum_{m\in\mathbf{Z}^n} b(m/R)\widehat{F_R}(m)\overline{\widehat{G_R}(m)}\Big| \\
&= \Big|R^n \int_{\mathbf{T}^n} \Big(\sum_{m\in\mathbf{Z}^n} b(m/R)\widehat{F_R}(m)e^{2\pi i m\cdot x}\Big)\overline{G_R(x)}\,dx\Big| \\
&\le R^n \|\{b(m/R)\}_m\|_{\mathscr{M}_p(\mathbf{Z}^n)} \|F_R\|_{L^p(\mathbf{T}^n)} \|G_R\|_{L^{p'}(\mathbf{T}^n)} \\
&< C_b R^n \|F_R\|_{L^p(\mathbf{R}^n)} \|G_R\|_{L^{p'}(\mathbf{R}^n)} \\
&= C_b \|f\|_{L^p(\mathbf{R}^n)} \|g\|_{L^{p'}(\mathbf{R}^n)}.
\end{aligned} \tag{3.6.16}$$

Since the function b is continuous and bounded, it is Riemann integrable over $\mathbf{R}^n$ and the same is true for the function $b(\xi)\widehat{f}(\xi)\overline{\widehat{g}(\xi)}$. The expressions in (3.6.15) tend to

$$\Big|\int_{\mathbf{R}^n} b(\xi)\widehat{f}(\xi)\overline{\widehat{g}(\xi)}\,d\xi\Big|$$

as $R \to \infty$ by the definition of the Riemann integral. We deduce that

$$\int_{\mathbf{R}^n} b(\xi)\widehat{f}(\xi)\overline{\widehat{g}(\xi)}\,d\xi = \int_{\mathbf{R}^n} (b\widehat{f})^\vee(x)\,\overline{g(x)}\,dx$$

is controlled in absolute value by the expression in (3.6.16). This implies the conclusion of the theorem. □

3.6.3 Applications of Transference

Having established two main transference theorems, we turn to an application.

Corollary 3.6.10. *Let* $1 \le p < \infty$, $f \in L^p(\mathbf{T}^n)$, *and* $\alpha \ge 0$. *Then*

(a) $\big\|D(n,R) * f - f\big\|_{L^p(\mathbf{T}^n)} \to 0$ *as* $R \to \infty$ *if and only if* $\chi_{[-1,1]^n} \in \mathscr{M}_p(\mathbf{R}^n)$.
(b) $\big\|\widetilde{D}(n,R) * f - f\big\|_{L^p(\mathbf{T}^n)} \to 0$ *as* $R \to \infty$ *if and only if* $\chi_{B(0,1)} \in \mathscr{M}_p(\mathbf{R}^n)$.
(c) $\big\|B_R^\alpha(f) - f\big\|_{L^p(\mathbf{T}^n)} \to 0$ *as* $R \to \infty$ *if and only if* $(1-|\xi|^2)_+^\alpha \in \mathscr{M}_p(\mathbf{R}^n)$.

Proof. First observe that in view of Corollary 3.5.2, the first statements in (a), (b), and (c) are equivalent to the statements

$$\sup_{R>0} \big\|D(n,R) * f\big\|_{L^p(\mathbf{T}^n)} \le C_p \big\|f\big\|_{L^p(\mathbf{T}^n)},$$
$$\sup_{R>0} \big\|\widetilde{D}(n,R) * f\big\|_{L^p(\mathbf{T}^n)} \le C_p \big\|f\big\|_{L^p(\mathbf{T}^n)},$$
$$\sup_{R>0} \big\|B_R^\alpha(f)\big\|_{L^p(\mathbf{T}^n)} \le C_p \big\|f\big\|_{L^p(\mathbf{T}^n)},$$

for some constant $0 < C_p < \infty$ and all f in $L^p(\mathbf{T}^n)$. Now define

$$\widetilde{\chi}_{B(0,1)}(x) = \begin{cases} 1 & \text{when } |x| < 1, \\ 1/2 & \text{when } |x| = 1, \\ 0 & \text{when } |x| > 1, \end{cases}$$

and

$$\widetilde{\chi}_{[-1,1]^n}(x_1,\dots,x_n) = \begin{cases} 1 & \text{when all } |x_j| < 1, \\ 1/2 & \text{when some but not all } |x_j| = 1, \\ 1/2^n & \text{when all } |x_j| = 1, \\ 0 & \text{when some } |x_j| > 1. \end{cases}$$

It is not difficult to see that the functions $\widetilde{\chi}_{B(0,1)}$ and $\widetilde{\chi}_{[-1,1]^n}$ are regulated and Riemann integrable; see Exercise 3.6.7. The function $(1-|\xi|^2)_+^\alpha$ is continuous and therefore it is both regulated and Riemann integrable. Theorems 3.6.7 and 3.6.9 imply that the uniform (in $R > 0$) boundedness of the operators $D(n,R)$, $\widetilde{D}(n,R)$, and B_R^α on $L^p(\mathbf{T}^n)$ is equivalent to the statements that the functions $\widetilde{\chi}_{B(0,1)}$, $\widetilde{\chi}_{[-1,1]^n}$, and $(1-|\xi|^2)_+^\alpha$ are in $\mathscr{M}_p(\mathbf{R}^n)$, respectively. Since $\widetilde{\chi}_{[-1,1]^n} = \chi_{[-1,1]^n}$ a.e. and $\widetilde{\chi}_{B(0,1)} = \chi_{B(0,1)}$ a.e., the required conclusion follows. □

3.6.4 Transference of Maximal Multipliers

We now prove a theorem concerning maximal multipliers analogous to Theorems 3.6.7 and 3.6.9. This enables us to reduce problems related to almost everywhere convergence of Fourier series on the torus to problems of boundedness of maximal operators on $\mathbf{R}^n$.

Let b be a bounded, regulated, and Riemann integrable function defined on all of $\mathbf{R}^n$. Suppose that $b \in \mathscr{M}_p(\mathbf{R}^n)$ for some $1 < p < \infty$. For $R > 0$, we introduce multiplier operators

$$S_{b,R}(F)(x) = \sum_{m\in\mathbf{Z}^n} \widehat{F}(m)b(m/R)e^{2\pi i m\cdot x}, \tag{3.6.17}$$

$$T_{b,R}(f)(x) = \int_{\mathbf{R}^n} \widehat{f}(\xi)b(\xi/R)e^{2\pi i\xi\cdot x}\,d\xi, \tag{3.6.18}$$

initially defined for smooth functions with compact support f on $\mathbf{R}^n$ and smooth functions F on $\mathbf{T}^n$.

In view of Theorems 3.6.7 and 3.6.9, $S_{b,R}$ admits a bounded extension on $L^p(\mathbf{T}^n)$ and $T_{b,R}$ admits a bounded extension on $L^p(\mathbf{R}^n)$. These extensions are denoted in the same way. We introduce maximal operators

$$M_b(F)(x) = \sup_{R>0}\big|S_{b,R}(F)(x)\big|, \tag{3.6.19}$$

$$N_b(f)(x) = \sup_{R>0}\big|T_{b,R}(f)(x),\big| \tag{3.6.20}$$

and we have the following result concerning them:

Theorem 3.6.11. *Suppose that $b(\xi)$ is a bounded, regulated, and Riemann integrable function defined for all $\xi \in \mathbf{R}^n$. Let $1 < p < \infty$, and suppose that b lies in $\mathscr{M}_p(\mathbf{R}^n)$. Let M_b and N_b be as in (3.6.19) and (3.6.20). Then the following assertions are equivalent for some finite constant C_p:*

$$\big\|M_b(F)\big\|_{L^p(\mathbf{T}^n)} \le C_p\big\|F\big\|_{L^p(\mathbf{T}^n)}, \qquad F\in L^p(\mathbf{T}^n), \tag{3.6.21}$$

$$\big\|N_b(f)\big\|_{L^p(\mathbf{R}^n)} \le C_p\big\|f\big\|_{L^p(\mathbf{R}^n)}, \qquad f\in L^p(\mathbf{R}^n). \tag{3.6.22}$$

Proof. Using Exercise 3.6.9, it suffices to prove the required equivalences for the maximal operators

$$M_b^{\mathscr{F}}(F)(x) = \sup_{t_1<\cdots<t_k}\big|S_{b,t_j}(F)(x)\big|,$$

$$N_b^{\mathscr{F}}(f)(x) = \sup_{t_1<\cdots<t_k}\big|T_{b,t_j}(f)(x)\big|,$$

uniformly in the choice of the finite subset

$$\mathscr{F} = \{t_1,\dots,t_k\}$$

of $\mathbf{R}^+$. Then $M_b^{\mathscr{F}}$ may be viewed as an operator defined on $L^p(\mathbf{T}^n)$ and taking values in $L^p(\mathbf{T}^n, l^\infty(\mathscr{F}))$, and $N_b^{\mathscr{F}}$ defined on $L^p(\mathbf{R}^n)$ with values in $L^p(\mathbf{R}^n, l^\infty(\mathscr{F}))$. Using this reduction and duality, estimates (3.6.21) and (3.6.22) are equivalent to the pair of inequalities

$$\Big\| \sum_{j=1}^{k} \sum_{m\in\mathbf{Z}^n} \widehat{F_j}(m) b(m/t_j) e^{2\pi i m\cdot x} \Big\|_{L^{p'}(\mathbf{T}^n)} \le C_p \Big\| \sum_{j=1}^{k} |F_j| \Big\|_{L^{p'}(\mathbf{T}^n)}, \tag{3.6.23}$$

$$\Big\| \sum_{j=1}^{k} \int_{\mathbf{R}^n} \widehat{f_j}(\xi) b(\xi/t_j) e^{2\pi i \xi\cdot x} \Big\|_{L^{p'}(\mathbf{R}^n)} \le C_p \Big\| \sum_{j=1}^{k} |f_j| \Big\|_{L^{p'}(\mathbf{R}^n)}, \tag{3.6.24}$$

where f_j are functions on $\mathbf{R}^n$, and F_j are functions on $\mathbf{T}^n$. In proving the equivalence of (3.6.23) and (3.6.24), by density, we may work with smooth functions with compact support f_j and trigonometric polynomials F_j. Suppose that (3.6.23) holds and let $f_1,\dots,f_k,g$ be smooth functions with compact support on $\mathbf{R}^n$. Then for $R\ge R_0$ we have that the functions $F_{j,R}(x)=f_j(Rx)$ and $G_R(x)=g(Rx)$ are supported in $[-1/2,1/2]^n$ and they can be viewed as functions on $\mathbf{T}^n$ once they are periodized. As before, the mth Fourier coefficient of $F_{j,R}$ is $R^{-n}\widehat{f_j}(m/R)$ and that of G_R is $R^{-n}\widehat{g}(m/R)$. Set

$$C_b = \sup_{R>0} \big\| \{b(m/R)\}_{m\in\mathbf{Z}^n} \big\|_{\mathscr{M}_p(\mathbf{Z}^n)}.$$

As in the proof of Theorem 3.6.9, for $R\ge R_0$ we have

$$\Big| \sum_{j=1}^{k} \sum_{m\in\mathbf{Z}^n} b(m/Rt_j) \widehat{f_j}(m/R) \overline{\widehat{g}(m/R)} \,\text{Volume}\left(\left[\tfrac{m}{R},\tfrac{m+1}{R}\right]^n\right) \Big| \tag{3.6.25}$$

$$\begin{aligned}
&= \Big| R^n \int_{\mathbf{T}^n} \Big(\sum_{j=1}^{k} \sum_{m\in\mathbf{Z}^n} b(m/R) \widehat{F_{j,R}}(m) e^{2\pi i m\cdot x} \Big) \overline{G_R(x)}\, dx \Big| \\
&\le C_b R^n \Big\| \sum_{j=1}^{k} |F_{j,R}| \Big\|_{L^{p'}(\mathbf{T}^n)} \big\| G_R \big\|_{L^p(\mathbf{T}^n)} \\
&\le C_b R^n \Big\| \sum_{j=1}^{k} |F_{j,R}| \Big\|_{L^{p'}(\mathbf{R}^n)} \big\| G_R \big\|_{L^p(\mathbf{R}^n)} \\
&= C_b \Big\| \sum_{j=1}^{k} |f_j| \Big\|_{L^{p'}(\mathbf{R}^n)} \big\| g \big\|_{L^p(\mathbf{R}^n)}.
\end{aligned}$$

Set $\delta^{t_j^{-1}}(b)(\xi)=b(\xi/t_j)$. Using that b is Riemann integrable, and realizing the limit of the partial sums in (3.6.25) when $R\to\infty$ as a Riemann integral, we obtain

$$\Big| \int_{\mathbf{R}^n} \sum_{j=1}^{k} (\delta^{t_j^{-1}}(b)\widehat{f_j})^\vee(x) \overline{g(x)}\, dx \Big| \le C_p \Big\| \sum_{j=1}^{k} |f_j| \Big\|_{L^{p'}(\mathbf{R}^n)} \big\| g \big\|_{L^p(\mathbf{R}^n)}.$$

Taking the supremum over all smooth functions with compact support g whose L^p norm is at most 1, we deduce (3.6.24).

We now turn to the converse. Assume that (3.6.24) holds. Let $P_1,\dots,P_k$ and Q be trigonometric polynomials on $\mathbf{T}^n$. Set $L_\varepsilon(x)=e^{-\pi\varepsilon|x|^2}$. Since b is regulated at every point in $\mathbf{R}^n$, Lemma 3.6.8 gives

$$\left|\int_{\mathbf{T}^n}\Big(\sum_{j=1}^{k}\sum_{m\in\mathbf{Z}^n}\widehat{P_j}(m)b(m/Rt_j)e^{2\pi i m\cdot x}\Big)\overline{Q(x)}\,dx\right|$$

$$=\left|\lim_{\varepsilon\to 0}\varepsilon^{\frac{n}{2}}\int_{\mathbf{R}^n}\Big(\sum_{j=1}^{k}\int_{\mathbf{R}^n}P_j(\xi)L_{\varepsilon/p'}(\xi)b(\xi/Rt_j)e^{2\pi i\xi\cdot x}d\xi\Big)\overline{Q(x)L_{\varepsilon/p}(x)}\,dx\right|$$

$$\le C_p\limsup_{\varepsilon\to 0}\left[\varepsilon^{\frac{n}{2}}\Big\|\sum_{j=1}^{k}|P_jL_{\varepsilon/p'}|\Big\|_{L^{p'}(\mathbf{R}^n)}\big\|QL_{\varepsilon/p'}\big\|_{L^p(\mathbf{R}^n)}\right]$$

$$=C_p\limsup_{\varepsilon\to 0}\left[\varepsilon^{\frac{n}{2p'}}\Big\|\sum_{j=1}^{k}|P_j|L_{\varepsilon/p'}\Big\|_{L^{p'}(\mathbf{R}^n)}\Big(\varepsilon^{\frac{n}{2}}\int_{\mathbf{R}^n}|Q(x)|^pe^{-\varepsilon\pi|x|^2}dx\Big)^{\frac{1}{p}}\right]$$

$$=C_p\Big\|\sum_{j=1}^{k}|P_j|\Big\|_{L^{p'}(\mathbf{R}^n)}\big\|Q\big\|_{L^p(\mathbf{T}^n)},$$

where we used (3.6.10) in the last equality. Taking the supremum over all trigonometric polynomials Q with L^p norm 1, we obtain (3.6.23), and this completes the proof of the theorem. □

Remark 3.6.12. Under the hypotheses of Theorem 3.6.11, the following two inequalities are also equivalent:

$$\big\|M_b(F)\big\|_{L^{p,\infty}(\mathbf{T}^n)}\le C_p\big\|F\big\|_{L^p(\mathbf{T}^n)},\qquad F\in L^p(\mathbf{T}^n),\tag{3.6.26}$$

$$\big\|N_b(f)\big\|_{L^{p,\infty}(\mathbf{R}^n)}\le C_p\big\|f\big\|_{L^p(\mathbf{R}^n)},\qquad f\in L^p(\mathbf{R}^n).\tag{3.6.27}$$

Indeed, Exercise 1.4.12 gives that the pair of inequalities (3.6.26) and (3.6.27) is equivalent to the pair of inequalities

$$\Big\|\sum_{j=1}^{k}\sum_{m\in\mathbf{Z}^n}\widehat{F_j}(m)b(m/t_j)e^{2\pi i m\cdot x}\Big\|_{L^{p'}(\mathbf{T}^n)}\le C_p\Big\|\sum_{j=1}^{k}|F_j|\Big\|_{L^{p',1}(\mathbf{T}^n)},\tag{3.6.28}$$

$$\Big\|\sum_{j=1}^{k}\int_{\mathbf{R}^n}\widehat{f_j}(\xi)b(\xi/t_j)e^{2\pi i\xi\cdot x}\Big\|_{L^{p'}(\mathbf{R}^n)}\le C_p\Big\|\sum_{j=1}^{k}|f_j|\Big\|_{L^{p',1}(\mathbf{R}^n)},\tag{3.6.29}$$

where $L^{p',1}$ is the Lorentz space.

Now (3.6.29) follows from (3.6.28) in exactly the same way that (3.6.24) follows from (3.6.23). Conversely, assuming (3.6.29), in order to prove (3.6.28) it will suffice to know that

$$\lim_{\varepsilon\to 0}\varepsilon^{\frac{n}{2p'}}\Big\|\sum_{j=1}^{k}|P_j|L_{\varepsilon/p'}\Big\|_{L^{p',1}(\mathbf{R}^n)}=\Big\|\sum_{j=1}^{k}|P_j|\Big\|_{L^{p',1}(\mathbf{T}^n)}.\tag{3.6.30}$$

For this we refer to Exercise 3.6.8.

3.6.5 Transference and Almost Everywhere Convergence

Next we consider the issue of almost everywhere convergence of Fourier series of functions on $\mathbf{T}^1$. The following results are valid.

Theorem 3.6.13. *There exists a constant $C > 0$ such that for any function F in $L^2(\mathbf{T}^1)$ we have*

$$\big\| \sup_{N>0} |F * D_N| \big\|_{L^{2,\infty}} \le C \big\| F \big\|_{L^2}.$$

Consequently, for any function $F \in L^2(\mathbf{T}^1)$, we have

$$\lim_{N\to\infty} \sum_{|m|\le N} \widehat{F}(m) e^{2\pi i m x} = F(x)$$

for almost every $x \in [0,1]$.

This theorem can be extended to L^p functions on $\mathbf{T}^1$ for $1 < p < \infty$.

Theorem 3.6.14. *For every $1 < p < \infty$ there exists a finite constant C_p such that for all $F \in L^p(\mathbf{T}^1)$ we have*

$$\Big\| \sup_{N>0} |F * D_N| \Big\|_{L^p} \le C_p \big\| F \big\|_{L^p}.$$

Consequently, for any $F \in L^p(\mathbf{T}^1)$, we have

$$\lim_{N\to\infty} \sum_{|m|\le N} \widehat{F}(m) e^{2\pi i m x} = F(x)$$

for almost every $x \in [0,1]$.

The proofs of Theorems 3.6.13 and 3.6.14 are lengthy and involved. They are consequences of Theorems 11.1.1 and 11.2.1, respectively. We discuss the relationship between the aforementioned pairs of theorems.

Consider the following function defined on $\mathbf{R}$:

$$b(x)(x) = \begin{cases} 1 & \text{when } |x| < 1, \\ 1/2 & \text{when } |x| = 1, \\ 0 & \text{when } |x| > 1. \end{cases} \tag{3.6.31}$$

Then b is easily seen to be regulated. Let $\{D_R\}_{R>0}$ be the family of Dirichlet kernels as defined in (3.1.13). Since $D_R = D_{R+\varepsilon}$ whenever $0 < \varepsilon < 1$, for all $F \in L^p(\mathbf{T}^1)$, $1 < p < \infty$, we have

$$\sup_{R>0} \big\| F * D_R \big\|_{L^p(\mathbf{T}^1)} = \sup_{N\in\mathbf{Z}^+} \big\| F * D_N \big\|_{L^p(\mathbf{T}^1)} \le C_p \big\| F \big\|_{L^p(\mathbf{T}^1)}, \tag{3.6.32}$$

where the last estimate follows from Theorem 3.5.1, Proposition 3.5.5, and Theorem 3.5.6. (The constant C_p naturally depends only on p.)

Let $S_{b,R}$ be as in (3.6.17). For an integrable function F on $\mathbf{T}^1$ we have

$$S_{b,R}(F) = F * D_R + Q_R(F),$$

where

$$Q_R(F)(x) = \begin{cases} \frac{1}{2}\widehat{F}(R)e^{2\pi ixR} + \frac{1}{2}\widehat{F}(-R)e^{-2\pi ixR} & \text{when } R \in \mathbf{Z}^+, \\ 0 & \text{when } R \in \mathbf{R}^+ \setminus \mathbf{Z}^+. \end{cases}$$

Since Q_R is bounded on $L^p(\mathbf{T}^1)$ with norm 1, using (3.6.32) we conclude that

$$\sup_{R>0} \big\|S_{b,R}(F)\big\|_{L^p(\mathbf{T}^1)} \le (C_p + 1)\big\|F\big\|_{L^p(\mathbf{T}^1)}$$

for all F in $L^p(\mathbf{T}^1)$. Appealing to Theorem 3.6.9, we deduce that the function b defined in (3.6.31) lies in $\mathscr{M}_p(\mathbf{R})$ (i.e., it is an L^p Fourier multiplier).

Next we discuss the boundedness of the corresponding maximal multipliers. If M_b is as in (3.6.19), then

$$M_b(F)(x) = \sup_{R>0} \big|(F * D_R)(x) + Q_R(F)(x)\big|,$$

whenever F is a function on $\mathbf{T}^1$ and

$$N_b(f)(x) = \sup_{R>0} \left| \int_{\mathbf{R}} \widehat{f}(\xi) b(\xi/R) e^{2\pi ix\xi}\, d\xi \right| = \sup_{R>0} \left| \int_{-R}^{+R} \widehat{f}(\xi) e^{2\pi ix\xi}\, d\xi \right|$$

for f in $L^p(\mathbf{R})$, $1 < p < \infty$. Both integrals may not be absolutely convergent for all $f \in L^p(\mathbf{R})$, but they should be interpreted as the quantity $T_{b,R}(f)(x)$, which of course coincides with them for nice f. (The operator $T_{b,R}$ is defined in (3.6.18).)

Since the sublinear operator $F \to \sup_{R>0} |Q_R(F)(x)|$ is clearly bounded on $L^p(\mathbf{T}^1)$, it follows from Theorem 3.6.11 that the boundedness of the maximal operator M_b on $L^p(\mathbf{T}^1)$ is equivalent to that of the maximal operator N_b on $L^p(\mathbf{R})$. (The operator N_b is defined in (3.6.20) and is associated with the function b in (3.6.31).)

The maximal operator N_b is called the *Carleson operator* and is denoted by $\mathscr{C}$. Then

$$\mathscr{C}(f)(x) = \sup_{R>0} \left| \int_{-R}^{+R} \widehat{f}(\xi) e^{2\pi ix\xi}\, d\xi \right|.$$

The boundedness of this operator on $L^p(\mathbf{R})$ is obtained in Chapter 11.

The extension of Theorem 3.6.14 to higher dimensions is a rather straightforward consequence of the one-dimensional result.

Theorem 3.6.15. *For every $1 < p < \infty$, there exists a finite constant $C_{p,n}$ such that for all $f \in L^p(\mathbf{T}^n)$ we have*

$$\Big\| \sup_{N>0} |D(n,N) * f| \Big\|_{L^p(\mathbf{T}^n)} \le C_{p,n} \big\|f\big\|_{L^p(\mathbf{T}^n)} \tag{3.6.33}$$

and consequently

$$\lim_{N\to\infty} \sum_{\substack{m\in\mathbf{Z}^n\\ |m_j|\le N}} \widehat{f}(m)e^{2\pi i m\cdot x} = f(x)$$

for almost every $x \in \mathbf{T}^n$.

Proof. We prove Theorem 3.6.15 when $n = 2$. Fix a p with $1 < p < \infty$. Since the Riesz projection P_+ is bounded on $L^p(\mathbf{T}^1)$, transference gives that the function $\chi_{(0,\infty)}$ is in $\mathscr{M}_p(\mathbf{R})$. It follows that the characteristic function of any half-space of the form $x_j > 0$ in $\mathbf{R}^2$ is in $\mathscr{M}_p(\mathbf{R}^2)$. (These functions have to be suitably defined on the line $x_j = 0$ to be regulated.) Since rotations of multipliers do not affect their norms, it follows that the characteristic function of the half-space $x_2 > x_1$ is in $\mathscr{M}_p(\mathbf{R}^2)$. The product of two multipliers is a multiplier; thus the characteristic function of the truncated cone $|x_1| \le |x_2| \le L$ is also in $\mathscr{M}_p(\mathbf{R}^2)$. Transference gives that the sequence $\{a_{m_1,m_2}\}_{m_1,m_2}$ defined by $a_{m_1,m_2} = 1$ when $|m_1| \le |m_2| \le L$ and zero otherwise is in $\mathscr{M}_p(\mathbf{Z}^2)$ with norm independent of $L > 0$. This means that for some constant B_p we have the following inequality for all f in $L^p(\mathbf{T}^2)$:

$$\int_{\mathbf{T}^2} \Bigg| \sum_{\substack{m_2\in\mathbf{Z}\\ |m_2|\le L}} \sum_{\substack{m_1\in\mathbf{Z}\\ |m_1|\le|m_2|}} \widehat{f}(m_1,m_2)e^{2\pi i(m_1x_1+m_2x_2)}\Bigg|^p dx_2\,dx_1 \le B_p^p \|f\|^p_{L^p(\mathbf{T}^n)}, \tag{3.6.34}$$

where the constant B_p is independent of $L > 0$. Now let $1 < p < \infty$ and suppose that $f \in L^p(\mathbf{T}^2)$. For fixed $x_1 \in \mathbf{T}^1$ define a function f_{x_1} on $\mathbf{T}^1$ as follows:

$$f_{x_1}(x_2) = \sum_{m_2\in\mathbf{Z}\ \|m_2|\le L} \sum_{\substack{m_1\in\mathbf{Z}\\ |m_1|\le|m_2|}} \left[\widehat{f}(m_1,m_2)e^{2\pi i m_1x_1}\right] e^{2\pi i m_2 x_2}.$$

Then $f_{x_1} \in L^p(\mathbf{T}^1)$ and its Fourier coefficients are zero for $|m_2| > L$ and equal to

$$\widehat{f_{x_1}}(m_2) = \sum_{|m_1|\le|m_2|} \widehat{f}(m_1,m_2)e^{2\pi i m_1 x_1}$$

for $|m_2| \le L$. We now have

$$\begin{aligned}
&\int_{\mathbf{T}^1}\int_{\mathbf{T}^1} \sup_{0<N\le L} \Bigg|\sum_{|m_1|\le N}\sum_{|m_2|\le N} \widehat{f}(m_1,m_2)e^{2\pi i m_1x_1}e^{2\pi i m_2x_2}\Bigg|^p dx_2\,dx_1\\
&\le 2\int_{\mathbf{T}^1}\int_{\mathbf{T}^1} \sup_{0<N\le L} \Bigg|\sum_{|m_2|\le N}\Bigg[\sum_{|m_1|\le|m_2|} \widehat{f}(m_1,m_2)e^{2\pi i m_1x_1}\Bigg]e^{2\pi i m_2x_2}\Bigg|^p dx_2\,dx_1\\
&= 2\int_{\mathbf{T}^1}\int_{\mathbf{T}^1} \sup_{0<N\le L} |(D_N * f_{x_1})(x_2)|^p\,dx_2\,dx_1\\
&\le 2C_p^p \int_{\mathbf{T}^1}\int_{\mathbf{T}^1} |f_{x_1}(x_2)|^p\,dx_2\,dx_1\\
&\le 2C_p^p B_p^p \|f\|^p_{L^p(\mathbf{T}^2)},
\end{aligned}$$

where we used Theorem 3.6.14 in the penultimate inequality and estimate (3.6.34) in the last inequality. Since the last estimate we obtained is independent of $L > 0$, letting $L \to \infty$ and applying Fatou's lemma, we obtain the conclusion (3.6.33) for $n = 2$. The case of a general dimension $n \geq 3$ presents no additional difficulties. $\square$

Exercises

3.6.1. Let $\alpha \geq 0$. Prove that the function $(1 - |\xi|^2)_+^\alpha$ is in $\mathscr{M}_p(\mathbf{R}^n)$ if and only if the function $(1 - |\xi|)_+^\alpha$ is in $\mathscr{M}_p(\mathbf{R}^n)$.
[*Hint:* Use that smooth functions with compact support lie in $\mathscr{M}_p$.]

3.6.2. The purpose of this exercise is to introduce distributions on the torus. The set of test functions on the torus is $\mathscr{C}^\infty(\mathbf{T}^n)$ equipped with the following topology. Given f_j, f in $\mathscr{C}^\infty(\mathbf{T}^n)$, we say that $f_j \to f$ in $\mathscr{C}^\infty(\mathbf{T}^n)$ if

$$\left\|\partial^\alpha f_j - \partial^\alpha f\right\|_{L^\infty(\mathbf{T}^n)} \to 0 \quad \text{as } j \to \infty,\ \forall\, \alpha.$$

Under this notion of convergence, $\mathscr{C}^\infty(\mathbf{T}^n)$ is a topological vector space with topology induced by the family of seminorms $\rho_\alpha(\varphi) = \sup_{x \in \mathbf{T}^n} |(\partial^\alpha f)(x)|$, where α ranges over all multi-indices. The dual space of $\mathscr{C}^\infty(\mathbf{T}^n)$ under this topology is the set of all distributions on $\mathbf{T}^n$ and is denoted by $\mathscr{D}'(\mathbf{T}^n)$. The definition implies that for u_j and u in $\mathscr{D}'(\mathbf{T}^n)$ we have $u_j \to u$ in $\mathscr{D}'(\mathbf{T}^n)$ if and only if

$$\left\langle u_j, f\right\rangle \to \left\langle u, f\right\rangle \quad \text{as } j \to \infty \text{ for all } f \in \mathscr{C}^\infty(\mathbf{T}^n).$$

The following operations can be defined on elements of $\mathscr{D}'(\mathbf{T}^n)$: differentiation (as in Definition 2.3.6), translation and reflection (as in Definition 2.3.11), convolution with a $\mathscr{C}^\infty$ function (as in Definition 2.3.13), multiplication by a $\mathscr{C}^\infty$ function (as in Definition 2.3.15), the support of a distribution (as in Definition 2.3.16). Use the same ideas as in $\mathbf{R}^n$ to prove the following:
(a) Prove that if $u \in \mathscr{D}'(\mathbf{T}^n)$ and $f \in \mathscr{C}^\infty(\mathbf{T}^n)$, then $f * u$ is the $\mathscr{C}^\infty$ function $x \mapsto \left\langle u, \tau^x(\widetilde{f})\right\rangle$.
(b) In contrast to the situation in $\mathbf{R}^n$, the convolution of two distributions on the torus can be defined. For $u, v \in \mathscr{D}'(\mathbf{T}^n)$ and $f \in \mathscr{C}^\infty(\mathbf{T}^n)$ define

$$\left\langle u * v, f\right\rangle = \left\langle u, f * \widetilde{v}\right\rangle.$$

Check that convolution of distributions on $\mathscr{D}'(\mathbf{T}^n)$ is associative, commutative, and distributive.
(c) Prove the analogue of Proposition 2.3.23: $\mathscr{C}^\infty(\mathbf{T}^n)$ is dense in $\mathscr{D}'(\mathbf{T}^n)$.

3.6.3. For $u \in \mathscr{D}'(\mathbf{T}^n)$ and $m \in \mathbf{Z}^n$ define the Fourier coefficient $\widehat{u}(m)$ by

$$\widehat{u}(m) = u(e^{-2\pi i m \cdot (\cdot)}) = \left\langle u, e^{-2\pi i m \cdot (\cdot)}\right\rangle.$$

Prove properties (1), (2), (4), (5), (6), (8), (9), (11), and (12) of Proposition 2.3.22 regarding the Fourier coefficients of distributions on the circle. Moreover, prove that for any u, v in $\mathscr{D}'(\mathbf{T}^n)$ we have $(u*v)\widehat{\ }(m) = \widehat{u}(m)\widehat{v}(m)$. In particular, this is valid for finite Borel measures.

3.6.4. Let μ be a finite Borel measure on $\mathbf{R}^n$ and let ν be the periodization of μ, that is, ν is a measure on $\mathbf{T}^n$ defined by

$$\nu(A) = \sum_{m\in\mathbf{Z}^n} \mu(A+m)$$

for all measurable subsets A of $\mathbf{T}^n$. Prove that the restriction of the Fourier transform of μ on $\mathbf{Z}^n$ coincides with the sequence of the Fourier coefficients of the measure ν.

3.6.5. Let T be an operator that commutes with translations and maps $L^p(\mathbf{T}^n)$ to $L^q(\mathbf{T}^n)$ for some $1 \le p,q \le \infty$. Prove that there exists a distribution u on $\mathbf{T}^n$ such that $T(f) = f*u$.

3.6.6. (*G. Weiss*) Suppose that the function b on $\mathbf{R}^n$ is regulated at the point x_0 in the sense that

$$\lim_{\varepsilon\to 0}\frac{1}{\varepsilon^n}\int_{|t|\le\varepsilon}\big(b(x_0-t)-b(x_0)\big)\,dt = 0.$$

Let $K_\varepsilon(x) = \varepsilon^{-n}e^{-\pi|x/\varepsilon|^2}$ for $\varepsilon>0$. Prove that $(b*K_\varepsilon)(x_0)\to b(x_0)$ as $\varepsilon\to 0$.
[*Hint:* Prove that for all $\delta>0$ we have

$$\begin{aligned}\big|(b*K_\varepsilon)(x_0)-b(x_0)\big| \le{}& 2\big\|b\big\|_{L^\infty}\int_{|y|\ge\delta/\varepsilon} e^{-\pi|y|^2}\,dy\\ &+|F_{x_0}(\delta)|\,\frac{\delta^n}{\varepsilon^n}\,e^{-\pi\delta^2/\varepsilon^2}\\ &+2\pi\sup_{0<r\le\delta}|F_{x_0}(r)|\int_0^{\frac{\delta}{\varepsilon}} r^{n+1}e^{-\pi r^2}\,dr,\end{aligned}$$

where $F_{x_0}(\delta) = \frac{1}{\delta^n}\int_{|t|\le\delta}\big(b(x_0-t)-b(x_0)\big)\,dt.$]

3.6.7. Let v_n be the volume of the unit ball in $\mathbf{R}^n$ and $e_1 = (1,0,\dots,0)$. Prove that

$$\lim_{\varepsilon\to 0}\frac{1}{v_n\varepsilon^n}\int_{|x-e_1|\le\varepsilon}\chi_{|x|\le 1}\,dx \to \frac{1}{2}.$$

Conclude that the function

$$\widetilde{\chi}_{B(0,1)}(x) = \begin{cases} 1 & \text{when } |x|<1,\\ 1/2 & \text{when } |x|=1,\\ 0 & \text{when } |x|>1\end{cases}$$

is regulated.

3.6.8. Let $L_\varepsilon(x) = e^{-\pi\varepsilon|x|^2}$ for $\varepsilon > 0$. Given a continuous function g on $\mathbf{T}^n$, prove that

$$\lim_{\varepsilon\to 0} \varepsilon^{\frac{n}{2q}} \big\| g L_{\varepsilon/q} \big\|_{L^{q,1}(\mathbf{R}^n)} = \big\| g \big\|_{L^{q,1}(\mathbf{T}^n)}$$

for all $1 < q < \infty$.

3.6.9. Suppose that $\{f_t\}_{t\in\mathbf{R}}$ is a family of measurable functions on a measure space X that satisfies

$$\Big\| \sup_{t\in F} |f_t| \Big\|_{L^p} \le b < \infty$$

for every finite subset F of $\mathbf{R}$. Prove that for any t there is a measurable function $\widetilde{f_t}$ on X that is a.e. equal to f_t such that

$$\Big\| \sup_{t\in\mathbf{R}} |\widetilde{f_t}| \Big\|_{L^p} \le b\,.$$

$\big[$*Hint:* Let $a = \sup_F \big\| \sup_{t\in F} |f_t| \big\|_{L^p} \le b$, where the supremum is taken over all finite subsets F of $\mathbf{R}$. Pick a sequence of sets F_n such that $\big\| \sup_{t\in F_n} |f_t| \big\|_{L^p} \to a$ as $n\to\infty$. Let $g = \sup_n \sup_{t\in F_n} |f_t|$ and note that $\big\|g\big\|_{L^p} = a$. Then for any $s \in \mathbf{R}$ we have

$$\Big\| \sup\big(|f_s|,\ \sup_{1\le k\le n} \sup_{t\in F_k} |f_t|\big) \Big\|_{L^p} \le a\,.$$

This implies $\big\| \max(|f_s|, g) \big\|_{L^p} \le a = \big\|g\big\|_{L^p}$, so that $|f_s| \le g$ a.e. for all $s\in\mathbf{R}$. This means that g is an a.e. upper bound for all f_t.$\big]$

3.6.10. (*E. Prestini*) Let $p \ge 2$ and $k > 0$. Show that for $f \in L^p(\mathbf{T}^2)$ we have that

$$\sum_{\substack{|m_1|\le N\\ |m_2|\le N^k}} \widehat{f}(m_1,m_2) e^{2\pi i(m_1x_1+m_2x_2)} \to f(x_1,x_2)$$

for almost all (x_1,x_2) in $\mathbf{T}^2$.
$\big[$*Hint:* It suffices to take $p=2$. Use the splitting $\widehat{f}(m_1,m_2) = \widehat{f}(m_1,m_2)\chi_{|m_2|\le|m_1|^k} + \widehat{f}(m_1,m_2)\chi_{|m_2|>|m_1|^k}$ and apply the idea of the proof of Theorem 3.6.15.$\big]$

3.7 Lacunary Series

In this section we take a quick look at lacunary series. These series provide examples of functions that possess some remarkable properties.

3.7.1 Definition and Basic Properties of Lacunary Series

We begin by defining lacunary sequences.

Definition 3.7.1. A sequence of positive integers $\{\lambda_k\}_{k=1}^{\infty}$ is called *lacunary* if there exists a constant $A > 1$ such that $\lambda_{k+1} \geq A\lambda_k$ for all $k \in \mathbf{Z}^+$.

Examples of lacunary sequences are provided by exponential sequences, such as $\lambda_k = 2^k, 3^k, 4^k, \ldots$. Observe that polynomial sequences such as $\lambda_k = 1 + k^2$ are not lacunary. Note that lacunary sequences tend to infinity as $k \to \infty$.

We begin with the following result.

Proposition 3.7.2. *Let λ_k be a lacunary sequence and let f be an integrable function on the circle that is differentiable at a point and has Fourier coefficients*

$$\widehat{f}(m) = \begin{cases} a_m & \text{when } m = \lambda_k\,, \\ 0 & \text{when } m \neq \lambda_k\,. \end{cases} \tag{3.7.1}$$

Then we have

$$\lim_{k \to +\infty} \widehat{f}(\lambda_k)\lambda_k = 0\,.$$

Proof. Applying translation, we may assume that the point at which f is differentiable is the origin. Replacing f by

$$g(t) = f(t) - f(0)\cos(t) - f'(0)\sin(t) = f(t) - f(0)\frac{e^{it} + e^{-it}}{2} - f'(0)\frac{e^{it} - e^{-it}}{2i}$$

we may assume that $f(0) = f'(0) = 0$. (We have $\widehat{g}(m) = \widehat{f}(m)$ for $|m| \geq 2$ and thus the final conclusion for f is equivalent to that for g.)

Using condition and (3.7.1), we obtain that

$$1 \leq |m - \lambda_k| < \min(A - 1, 1 - A^{-1})\lambda_k \implies \widehat{f}(m) = 0\,. \tag{3.7.2}$$

Given $\varepsilon > 0$, pick a positive integer k_0 such that if $[\min(A-1, 1-A^{-1})\lambda_{k_0}] = 2N_0$, then $N_0^{-2} < \varepsilon$, and

$$\sup_{|x| < N_0^{-\frac{1}{4}}} \left| \frac{f(x)}{x} \right| < \varepsilon\,. \tag{3.7.3}$$

The expression in (3.7.3) can be made arbitrarily small, since f is differentiable at the origin. Now take an integer k with $k \geq k_0$ and set $2N = [\min(A - 1, 1 - A^{-1})\lambda_k]$, which is of course at least $2N_0$. Using (3.7.2), we obtain that for any trigonometric polynomial K_N of degree $2N$ with $\widehat{K_N}(0) = 1$ we have

$$\widehat{f}(\lambda_k) = \int_{|x| \leq \frac{1}{2}} f(x) K_N(x) e^{-2\pi i \lambda_k x}\, dx\,. \tag{3.7.4}$$

We take $K_N = (F_N/\|F_N\|_{L^2})^2$, where F_N is the Fejér kernel. Using (3.1.15), we obtain first the identity

$$\|F_N\|_{L^2}^2 = \sum_{j=-N}^{N}\left(1-\frac{|j|}{N+1}\right)^2 = 1+\frac{1}{3}\frac{N(2N+1)}{N+1} > \frac{N}{3} \tag{3.7.5}$$

and also the estimate

$$F_N(x)^2 \le \left(\frac{1}{N+1}\frac{1}{4x^2}\right)^2, \tag{3.7.6}$$

which is valid for $|x| \le 1/2$. In view of (3.7.5) and (3.7.6), we have the estimate

$$K_N(x) \le \frac{3}{16}\frac{1}{N^3}\frac{1}{x^4}. \tag{3.7.7}$$

We now use (3.7.4) to obtain

$$\lambda_k \widehat{f}(\lambda_k) = \lambda_k \int_{|x|\le\frac{1}{2}} f(x)K_N(x)e^{-2\pi i\lambda_k x}\,dx = I_k^1 + I_k^2 + I_k^3,$$

where

$$\begin{aligned}
I_k^1 &= \lambda_k \int_{|x|\le N^{-1}} f(x)K_N(x)e^{-2\pi i\lambda_k x}\,dx,\\
I_k^2 &= \lambda_k \int_{N^{-1}<|x|\le N^{-\frac{1}{4}}} f(x)K_N(x)e^{-2\pi i\lambda_k x}\,dx,\\
I_k^3 &= \lambda_k \int_{N^{-\frac{1}{4}}<|x|\le\frac{1}{2}} f(x)K_N(x)e^{-2\pi i\lambda_k x}\,dx.
\end{aligned}$$

Since $\|K_N\|_{L^1} = 1$, it follows that

$$|I_k^1| \le \frac{\lambda_k}{N}\sup_{|x|<N^{-1}}\left|\frac{f(x)}{x}\right| \le \frac{(2N+1)\,\varepsilon}{\min(A-1,1-A^{-1})N},$$

which can be made arbitrarily small if ε is small. Also, using (3.7.7), we obtain

$$|I_k^2| \le \frac{3\lambda_k}{16N^3}\sup_{|x|<N^{-\frac{1}{4}}}\left|\frac{f(x)}{x}\right|\int_{N^{-1}<|x|\le N^{-\frac{1}{4}}}\frac{dx}{x^3} \le \frac{3\lambda_k}{16N}\sup_{|x|<N^{-\frac{1}{4}}}\left|\frac{f(x)}{x}\right|,$$

which, as observed, is bounded by a constant multiple of ε. Finally, using again (3.7.7), we obtain

$$|I_k^3| \le \frac{3}{16N^3}\frac{1}{N^{-\frac{1}{4}}}\int_{N^{-\frac{1}{4}}<|x|\le\frac{1}{2}}|f(x)|\,dx \le \frac{3}{16N^2}\|f\|_{L^1} < \frac{3\varepsilon}{16}\|f\|_{L^1}.$$

It follows that for all $k \ge k_0$ we have

$$|\lambda_k \widehat{f}(\lambda_k)| \le |I_k^1| + |I_k^2| + |I_k^3| \le C(f)\,\varepsilon$$

for some fixed constant $C(f)$. This proves the required conclusion. □

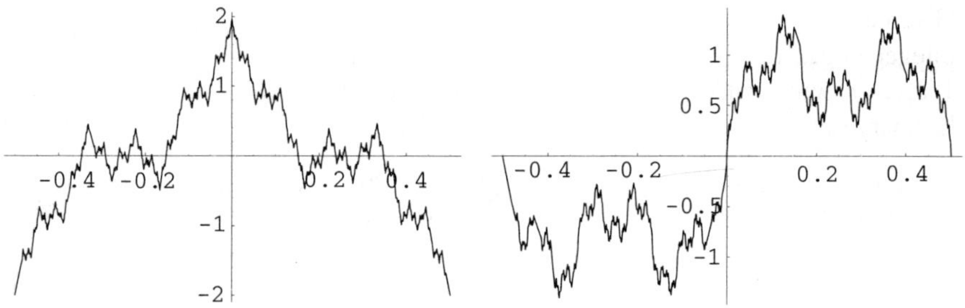

Fig. 3.3 The graph of the real and imaginary parts of the function $f(t) = \sum_{k=0}^{\infty} 2^{-k} e^{2\pi i 3^k t}$.

Corollary 3.7.3. *(**Weierstrass**) There exists a continuous function on the circle that is nowhere differentiable.*

Proof. Consider the 1-periodic function

$$f(t) = \sum_{k=0}^{\infty} 2^{-k} e^{2\pi i 3^k t}.$$

Since this series converges absolutely and uniformly, f is a continuous function. If f were differentiable at a point, then by Proposition 3.7.2 we would have that $3^k \widehat{f}(3^k)$ tends to zero as $k \to \infty$. Since $\widehat{f}(3^k) = 2^{-k}$ for $k \ge 0$, this is not the case. Therefore, f is nowhere differentiable. The real and imaginary parts of this function are displayed in Figure 3.3. □

3.7.2 Equivalence of L^p Norms of Lacunary Series

We now turn to one of the most important properties of lacunary series, equivalence of their norms. It is a remarkable result that lacunary Fourier series have comparable L^p norms for $1 \le p < \infty$. More precisely, we have the following theorem:

Theorem 3.7.4. *Let $0 < \lambda_1 < \lambda_2 < \lambda_3 < \cdots$ be a lacunary sequence with constant $A > 1$. Set $\Lambda = \{\lambda_k : k \in \mathbf{Z}^+\}$. Then for all $1 < p < \infty$ there exists a constant $C_p(A)$ such that for all $f \in L^1(\mathbf{T}^1)$ with $\widehat{f}(k) = 0$ when $k \in \mathbf{Z} \setminus \Lambda$ we have*

$$\big\|f\big\|_{L^p(\mathbf{T}^1)} \le C_p(A) \big\|f\big\|_{L^1(\mathbf{T}^1)}. \tag{3.7.8}$$

Note that the converse inequality to (3.7.8) is trivial. Therefore, L^p norms of lacunary Fourier series are all equivalent for $1 \le p < \infty$.

Proof. We suppose first that $f \in L^2(\mathbf{T}^1)$ and we define

$$f_N(x) = \sum_{j=1}^{N} \widehat{f}(\lambda_j) e^{2\pi i \lambda_j x}. \tag{3.7.9}$$

Given a $2 \le p < \infty$, we pick an integer m with $2m > p$ and we also pick a positive integer r such that $A^r > m$. Then we can write f_N as a sum of r functions φ_s, $s = 1, 2, \dots, r$, where each φ_s has Fourier coefficients that vanish except possibly on the lacunary set

$$\{\lambda_{kr+s} : k \in \mathbf{Z}^+ \cup \{0\}\} = \{\mu_1, \mu_2, \mu_3, \dots\}.$$

It is a simple fact that the sequence $\{\mu_k\}_k$ is lacunary with constant A^r. Then we have

$$\int_0^1 |\varphi_s(x)|^{2m}\, dx = \sum_{\substack{1 \le j_1, \dots, j_m, k_1, \dots, k_m \le N \\ \mu_{j_1} + \cdots + \mu_{j_m} = \mu_{k_1} + \cdots + \mu_{k_m}}} \widehat{\varphi_s}(\mu_{j_1}) \cdots \widehat{\varphi_s}(\mu_{j_m}) \overline{\widehat{\varphi_s}(\mu_{k_1})} \cdots \overline{\widehat{\varphi_s}(\mu_{k_m})}.$$

We claim that if $\mu_{j_1} + \cdots + \mu_{j_m} = \mu_{k_1} + \cdots + \mu_{k_m}$, then

$$\max(\mu_{j_1}, \dots, \mu_{j_m}) = \max(\mu_{k_1}, \dots, \mu_{k_m}).$$

Indeed, if $\max(\mu_{j_1}, \dots, \mu_{j_m}) > \max(\mu_{k_1}, \dots, \mu_{k_m})$, then

$$\max(\mu_{j_1}, \dots, \mu_{j_m}) \le \mu_{k_1} + \cdots + \mu_{k_m} \le m \max(\mu_{k_1}, \dots, \mu_{k_m}).$$

But since

$$A^r \max(\mu_{k_1}, \dots, \mu_{k_m}) \le \max(\mu_{j_1}, \dots, \mu_{j_m}),$$

it would follow that $A^r \le m$, which contradicts our choice of r. Likewise, we eliminate the case $\max(\mu_{j_1}, \dots, \mu_{j_m}) < \max(\mu_{k_1}, \dots, \mu_{k_m})$. We conclude that these numbers are equal. We can now continue the same reasoning by induction to conclude that if $\mu_{j_1} + \cdots + \mu_{j_m} = \mu_{k_1} + \cdots + \mu_{k_m}$, then

$$\{\mu_{k_1}, \dots, \mu_{k_m}\} = \{\mu_{j_1}, \dots, \mu_{j_m}\}.$$

Using this fact in the evaluation of the previous multiple sum, we obtain

$$\int_0^1 |\varphi_s(x)|^{2m}\, dx = \sum_{j_1=1}^{N} \cdots \sum_{j_m=1}^{N} |\widehat{\varphi_s}(\mu_{j_1})|^2 \cdots |\widehat{\varphi_s}(\mu_{j_m})|^2 = \left(\|\varphi_s\|_{L^2}^2\right)^m,$$

which implies that $\|\varphi_s\|_{L^{2m}} = \|\varphi_s\|_{L^2}$ for all $s \in \{1, 2, \dots, r\}$. Thus we have

$$\|f_N\|_{L^p} \le \|f_N\|_{L^{2m}} \le \sqrt{r} \Big(\sum_{s=1}^{r} \|\varphi_s\|_{L^{2m}}^2 \Big)^{\frac{1}{2}} = \sqrt{r} \Big(\sum_{s=1}^{r} \|\varphi_s\|_{L^2}^2 \Big)^{\frac{1}{2}} = \sqrt{r}\, \|f_N\|_{L^2},$$

since the functions φ_s are orthogonal on L^2. Since r can be chosen to be $[\log_A m]+1$ and m can be taken to be $[\frac{p}{2}]+1$, we have now established the inequality

$$\big\|f_N\big\|_{L^p(\mathbf{T}^1)} \le C_p(A)\big\|f_N\big\|_{L^2(\mathbf{T}^1)}, \qquad p \ge 2, \tag{3.7.10}$$

with $C_p(A) = \sqrt{1+\left[\log_A\left(\left[\frac{p}{2}\right]+1\right)\right]}$ for all f_N that have the form (3.7.9). To extend (3.7.10) to all $f \in L^2(\mathbf{T}^1)$, we observe that $f_N \to f$ in L^2 and some subsequence of them f_{N_j} tends to f a.e. Then Fatou's lemma gives

$$\begin{aligned}
\int_0^1 |f(x)|^p\,dx &= \int_0^1 \liminf_{j\to\infty} |f_{N_j}(x)|^p\,dx \\
&\le \liminf_{j\to\infty} \int_0^1 |f_{N_j}(x)|^p\,dx \\
&\le C_p(A)^p \liminf_{j\to\infty} \big\|f_{N_j}\big\|_{L^2}^p \\
&= C_p(A)^p \big\|f\big\|_{L^2}^p,
\end{aligned}$$

which proves (3.7.10) for all $f \in L^2$. We now turn our attention to (3.7.8) in the case $1 < p < 2$. By interpolation we obtain for $1 < p < 2$

$$\big\|f\big\|_{L^2} \le \big\|f\big\|_{L^4}^{\frac{2}{3}} \big\|f\big\|_{L^1}^{\frac{1}{3}} \le \left([\log_A 3]+1\right)^{\frac{1}{2}\cdot\frac{2}{3}} \big\|f\big\|_{L^2}^{\frac{2}{3}} \big\|f\big\|_{L^1}^{\frac{1}{3}}.$$

This implies that for $1 < p < 2$ we have

$$\big\|f\big\|_{L^p(\mathbf{T}^1)} \le \big\|f\big\|_{L^2(\mathbf{T}^1)} \le \left([\log_A 3]+1\right)\big\|f\big\|_{L^1(\mathbf{T}^1)}.$$

Combining this with (3.7.10), which now holds for all $f \in L^2$, yields (3.7.8). □

Theorem 3.7.4 describes the equivalence of the L^p norms of lacunary Fourier series for $p < \infty$. The question that remains is whether there is a similar characterization of the L^∞ norms of lacunary Fourier series. Such a characterization is given in Theorem 3.7.6. Before we state and prove this theorem, we need a classical tool, referred to as a Riesz product.

Definition 3.7.5. A *Riesz product* is a function of the form

$$P_N(x) = \prod_{j=1}^{N} \left(1 + a_j \cos(2\pi\lambda_j x + 2\pi\gamma_j)\right), \tag{3.7.11}$$

where N is a positive integer, $\lambda_1 < \lambda_2 < \cdots < \lambda_N$ is a lacunary sequence of integers, a_j are complex numbers, and $\gamma_j \in [0,1]$.

We make a few observations about Riesz products. A simple calculation gives that if $P_{N,j}(x) = 1 + a_j \cos(2\pi\lambda_j x + 2\pi\gamma_j)$, then

$$\widehat{P_{N,j}}(m) = \begin{cases} 1 & \text{when } m = 0, \\ \frac{1}{2}a_j e^{2\pi i \gamma_j} & \text{when } m = \lambda_j, \\ \frac{1}{2}a_j e^{-2\pi i \gamma_j} & \text{when } m = -\lambda_j, \\ 0 & \text{when } m \notin \{0, \lambda_j, -\lambda_j\}. \end{cases} \tag{3.7.12}$$

Assume that the constant A associated with the lacunary sequence $\lambda_1 < \lambda_2 < \cdots < \lambda_N$ is at least 3. Then each integer m has at most one representation as a sum

$$m = \varepsilon_1 \lambda_1 + \cdots + \varepsilon_N \lambda_N,$$

where $\varepsilon_j \in \{-1, 1, 0\}$. See Exercise 3.7.1. We now calculate the Fourier coefficients of the Riesz product defined in (3.7.11). For a fixed integer b, let us denote by δ_b the sequence of integers that is equal to 1 at b and zero otherwise. Then, using (3.7.12), we obtain that

$$\widehat{P_{N,j}} = \delta_0 + \tfrac{1}{2}a_j e^{2\pi i \gamma_j}\delta_{\lambda_j} + \tfrac{1}{2}a_j e^{-2\pi i \gamma_j}\delta_{-\lambda_j},$$

and thus $\widehat{P_N}$ is the N-fold convolution of these functions. Using that $\delta_a * \delta_b = \delta_{a+b}$, we obtain

$$\widehat{P_N}(m) = \begin{cases} 1 & \text{when } m = 0, \\ \prod_{j=1}^N \frac{1}{2}a_j e^{2\pi i \varepsilon_j \gamma_j} & \text{when } m = \sum_{j=1}^N \varepsilon_j \lambda_j \text{ and } \sum_{j=1}^N |\varepsilon_j| > 0, \\ 0 & \text{otherwise.} \end{cases}$$

It follows that $\widehat{P_N}(\lambda_j) = \frac{1}{2}a_j e^{2\pi i \gamma_j}$ for $1 \le j \le N$ and that $\widehat{P_N}(\lambda_j) = 0$ for $j \ge N+1$, since each λ_j can be written uniquely as a sum of λ_k's as $0 \cdot \lambda_1 + \cdots + 0 \cdot \lambda_{j-1} + 1 \cdot \lambda_j$. See Exercise 3.7.1.

We recall the space $A(\mathbf{T}^1)$ of all functions with absolutely summable Fourier coefficients with norm the ℓ^1 norm of the coefficients.

Theorem 3.7.6. *Let $0 < \lambda_1 < \lambda_2 < \lambda_3 < \cdots$ be a lacunary sequence of integers with constant $A > 1$. Set $\Lambda = \{\lambda_k : k \in \mathbf{Z}^+\}$. Then there exists a constant $C(A)$ such that for all $f \in L^\infty(\mathbf{T}^1)$ with $\widehat{f}(k) = 0$ when $k \in \mathbf{Z} \setminus \Lambda$ we have*

$$\big\|f\big\|_{A(\mathbf{T}^1)} = \sum_{k \in \Lambda} |\widehat{f}(k)| \le C(A)\big\|f\big\|_{L^\infty(\mathbf{T}^1)}. \tag{3.7.13}$$

Proof. Let us assume first that $A \ge 3$. Also fix $f \in L^\infty(\mathbf{T}^1)$. We consider the Riesz product

$$P_N(x) = \prod_{j=1}^N \big(1 + \cos(2\pi\lambda_j x + 2\pi\gamma_j)\big),$$

where γ_j is defined via the identity $|\widehat{f}(\lambda_j)| = e^{2\pi i \gamma_j}\overline{\widehat{f}(\lambda_j)}$. Then $P_N \ge 0$ and since $\widehat{P_N}(0) = 1$, it follows that $\big\|P_N\big\|_{L^1} = 1$. By Parseval's relation we obtain

$$\left|\sum_{m\in\mathbf{Z}} \widehat{P_N}(m)\overline{\widehat{f}(m)}\right| = \left|\int_0^1 P_N(x)\overline{f(x)}\,dx\right| \le \|f\|_{L^\infty}, \tag{3.7.14}$$

and the sum in (3.7.14) is finite, since the Fourier coefficients of $\widehat{P_N}$ form a finitely supported sequence. But $\widehat{f}(m) = 0$ for $m \notin \Lambda$, while $\widehat{P_N}(\lambda_j) = \frac{1}{2}e^{2\pi i\gamma_j}$ for $1 \le j \le N$. Moreover, $\widehat{P_N}(\lambda_j) = 0$ for $j \ge N+1$, as observed earlier. Thus (3.7.14) reduces to

$$\frac{1}{2}\sum_{j=1}^{N} |\widehat{f}(\lambda_j)| = \left|\sum_{j=1}^{N} \frac{1}{2} e^{2\pi i\gamma_j}\overline{\widehat{f}(\lambda_j)}\right| \le \|f\|_{L^\infty}.$$

Letting $N \to \infty$, we obtain that $\sum_{j=1}^{\infty} |\widehat{f}(\lambda_j)| \le 2\|f\|_{L^\infty}$, which proves (3.7.13) when $A \ge 3$.

To prove the theorem for $1 < A < 3$, we pick a positive integer r with $A^r \ge 3$ (take $r = [\log_A 3] + 1$). We now consider the sequences

$$\{\lambda_{kr+s}\}_k\,, \qquad k \in \mathbf{Z}^+ \cup \{0\},$$

and we observe that each such sequence is lacunary with constant A^r. The preceding construction gives

$$\sum_{j=1}^{\infty} |\widehat{f}(\lambda_{jr+s})| \le 2\|f\|_{L^\infty}.$$

Summing over s in the set $\{1,2,\dots,r\}$, we obtain the required conclusion with $C(A) = 2r = 2[\log_A 3] + 2$. □

It follows from Theorem 3.7.6 that if $\Lambda = \{\lambda_k : k \in \mathbf{Z}^+\}$ is a lacunary set and f is a bounded function on the circle that satisfies $\widehat{f}(k) = 0$ when $k \in \mathbf{Z}\setminus\Lambda$, then we have

$$f(x) = \sum_{k\in\Lambda} \widehat{f}(k)e^{2\pi ikx}.$$

This is a consequence of the inversion result in Proposition 3.1.14.

Given a subset Λ of the integers, we denote by C_Λ the space of all continuous functions on $\mathbf{T}^1$ such that

$$m \in \mathbf{Z}\setminus\Lambda \implies \widehat{f}(m) = 0. \tag{3.7.15}$$

It is straightforward that C_Λ is a closed subspace of all bounded functions on the circle $\mathbf{T}^1$ with the standard L^∞ norm.

Definition 3.7.7. A set of integers Λ is called a *Sidon set* if every function in C_Λ has an absolutely convergent Fourier series.

Example 3.7.8. Every lacunary set is a Sidon set. Indeed, if f satisfies (3.7.15), then Theorem 3.7.6 gives that

$$\sum_{m\in\Lambda} |\widehat{f}(m)| \le C(A)\|f\|_{L^\infty};$$

hence f has an absolutely convergent Fourier series.

Example 3.7.9. There exist subsets of $\mathbf{R}$ that are not Sidon. For example, $\mathbf{Z}\setminus\{0\}$ is not a Sidon set. See Exercise 3.7.2.

Exercises

3.7.1. Suppose that $0<\lambda_1<\lambda_2<\cdots<\lambda_N$ is a lacunary sequence of integers with lacunarity constant $A\geq 3$. Prove that for every integer m there exists at most one N-tuple $(\varepsilon_1,\dots,\varepsilon_N)$ with each $\varepsilon_j\in\{-1,1,0\}$ such that

$$m=\varepsilon_1\lambda_1+\cdots+\varepsilon_N\lambda_N\,.$$

[*Hint:* Suppose there exist two such N-tuples. Pick the largest k such that the coefficients of λ_k are different.]

3.7.2. Consider the 1-periodic continuous function $h(t)=\cos(2\pi t)$. Then we have $\widehat{h}(0)=0$, but show that $\sum_k|\widehat{h}(k)|=\infty$. Thus $\mathbf{Z}\setminus\{0\}$ is not a Sidon set.

3.7.3. Suppose that $0<\lambda_1<\lambda_2<\cdots$ is a lacunary sequence and let f be a bounded function on the circle that satisfies $\widehat{f}(m)=0$ whenever $m\in\mathbf{Z}\setminus\{\lambda_1,\lambda_2,\dots\}$. Suppose also that

$$\sup_{t\neq 0}\frac{|f(t)-f(0)|}{|t|^{\alpha}}=B<\infty$$

for some $0<\alpha<1$.
(a) Prove that there is a constant C such that $|\widehat{f}(\lambda_k)|\leq CB\lambda_k^{-\alpha}$ for all $k\geq 1$.
(b) Prove that $f\in\dot{\Lambda}_\alpha(\mathbf{T}^1)$.
[*Hint:* Let $2N=[\min(A-1,1-A^{-1})\lambda_k]$ and let K_N be as in the proof of Proposition 3.7.2. Write

$$\begin{aligned}\widehat{f}(\lambda_k) &= \int_{|x|\leq N^{-1}}(f(x)-f(0))e^{-2\pi i\lambda_k x}K_N(x)\,dx\\ &\quad+\int_{N^{-1}\leq|x|\leq\frac12}(f(x)-f(0))e^{-2\pi i\lambda_k x}K_N(x)\,dx\,.\end{aligned}$$

Use that $\big\|K_N\big\|_{L^1}=1$ and also the estimate (3.7.6). Part (b): Use the estimate in part (a).]

3.7.4. Let f be an integrable function on the circle whose Fourier coefficients vanish outside a lacunary set $\Lambda=\{\lambda_1,\lambda_2,\lambda_3,\dots\}$. Suppose that f vanishes identically in a small neighborhood of the origin. Show that f is in $\mathscr{C}^\infty(\mathbf{T}^1)$.
[*Hint:* Let $2N=[\min(A-1,1-A^{-1})\lambda_k]$ and let K_N be as in the proof of Proposition 3.7.2. Write

$$\widehat{f}(\lambda_k)=\int_{|x|\leq\frac12}f(x)e^{-2\pi i\lambda_k x}K_N(x)\,dx$$

and use estimate (3.7.6) to obtain that f is in $\mathscr{C}^2$. Continue by induction.]

3.7.5. Let $1 < a,b < \infty$. Consider the 1-periodic function

$$f(x) = \sum_{k=0}^{\infty} a^{-k} e^{2\pi i b^k x}.$$

Prove that the following statements are equivalent:
(a) f is differentiable at a point.
(b) $b < a$.
(c) f is differentiable everywhere.

3.7.6. Let Λ be a subset of the integers such that for any sequence of complex numbers $\{d_\lambda\}_{\lambda\in\Lambda}$ with $|d_\lambda| = 1$ there is a finite Borel measure μ on $\mathbf{T}^1$ such that

$$|\widehat{\mu}(\lambda) - d_\lambda| < \frac{1}{2}$$

for all $\lambda \in \Lambda$. Show that Λ is a Sidon set.

3.7.7. Let $\Lambda \subseteq \mathbf{Z}^+$. Suppose that there is a constant $A < \infty$ such that for any $n \in \Lambda \cup \{0\}$ the number of elements in the set

$$\Big\{(\varepsilon_1,\dots,\varepsilon_m) \in \{-1,1\}^m : \, n = \sum_{j=1}^{m} \varepsilon_j n_j, \, n_1 < \dots < n_m, \, n_j \in \Lambda\Big\}$$

is at most A^m. Show that Λ is a Sidon set.
[*Hint:* Construct a suitable measure μ and use Exercise 3.7.6.]

3.7.8. Show that the set

$$\{3^{2^{m+2}} + 3^{2^m + k}, \, 0 \le k \le 2^{m-1}, \, m \ge 1\}$$

is a Sidon set.
[*Hint:* Use Exercise 3.7.7.]

HISTORICAL NOTES

Trigonometric series in one dimension were first considered in the study of the vibrating string problem and are implicitly contained in the work of d'Alembert, D. Bernoulli, Clairaut, and Euler. The analogous problem for vibrating higher-dimensional bodies naturally suggested the use of multiple trigonometric series. However, it was the work of Fourier on steady-state heat conduction that incited the subsequent systematic development of such series. Fourier announced his results in 1811, although his classical book *Théorie de la chaleur* was published in 1822. This book contains several examples of heuristic use of trigonometric expansions and motivated other mathematicians to carefully study such expansions.

The fact that the Fourier series of a continuous function can diverge was first observed by DuBois Reymond in 1876. The Riemann–Lebesgue lemma was first proved by Riemann in his memoir on trigonometric series (appeared between 1850 and 1860). It carries Lebesgue's name today because Lebesgue later extended it to his notion of integral. The rebuilding of the theory of Fourier series based on Lebesgue's integral was mainly achieved by de la Vallée-Poussin and Fatou.

Theorem 3.2.16 was obtained by S. Bernstein in dimension $n = 1$. Higher-dimensional analogues of the Hardy–Littlewood series of Exercise 3.2.9 were studied by Wainger [287]. These series can be used to produce examples indicating that the restriction $s > \alpha + n/2$ in Bernstein's theorem is sharp even in higher dimensions. Part (b) of Theorem 3.3.3 is due to Lebesgue when $n = 1$ and Marcinkiewicz and Zygmund [190] when $n = 2$. Marcinkiewicz and Zygmund's proof also extends to higher dimensions. The proof given here is based on Lemma 3.3.4 proved by Stein [260] in a different context. The proof of Lemma 3.3.4 presented here was suggested by T. Tao.

The development of the complex methods in the study of Fourier series was pioneered by the Russian school, especially Luzin and his students Kolmogorov, Menshov, and Privalov. The existence of an integrable function on $\mathbf{T}^1$ whose Fourier series diverges almost everywhere (Theorem 3.4.2) is due to Kolmogorov [156]. An example of an integrable function whose Fourier series diverges everywhere was also produced by Kolmogorov [159] three years later. Localization of the Bochner–Riesz means at the critical exponent $\alpha = \frac{n-1}{2}$ fails for L^1 functions on $\mathbf{T}^n$ (see Bochner [25]) but holds for functions f such that $|f|\log^+|f|$ is integrable over $\mathbf{T}^n$ (see Stein [252]). The latter article also contains the L^p boundedness of the maximal Bochner–Riesz operator $\sup_{R>0}|B_R^\alpha(f)|$ for $1<p<\infty$ when $\alpha > |\frac{1}{p}-\frac{1}{2}|$. Theorem 3.4.6 is also due to Stein [254]. The technique that involves the points for which the set $\{|x-m| : m \in \mathbf{Z}^n\}$ is linearly independent over the rationals was introduced by Bochner [25].

The boundedness of the conjugate function on the circle (Theorem 3.5.6) and hence the L^p convergence of one-dimensional Fourier series was announced by Riesz in [219], but its proof appeared a little later in [220]. Luzin's conjecture [182] on almost everywhere convergence of the Fourier series of continuous functions was announced in 1913 and settled by Carleson [45] in 1965 for the more general class of square summable functions (Theorem 3.6.13). Carleson's theorem was later extended by Hunt [135] for the class of L^p functions for all $1 < p < \infty$ (Theorem 3.6.14). Sjölin [245] sharpened this result by showing that the Fourier series of functions f with $|f|(\log^+|f|)(\log^+\log^+|f|)$ integrable over $\mathbf{T}^1$ converge almost everywhere. Antonov [3] improved Sjölin's result by extending it to functions f with $|f|(\log^+|f|)(\log^+\log^+\log^+|f|)$ integrable over $\mathbf{T}^1$. One should also consult the related results of Soria [250] and Arias de Reyna [7]. The book [8] of Arias de Reyna contains a historically motivated comprehensive study of topics related to the Carleson–Hunt theorem. Counterexamples due to Konyagin [161] show that Fourier series of functions f with $|f|(\log^+|f|)^{\frac{1}{2}}(\log^+\log^+|f|)^{-\frac{1}{2}-\varepsilon}$ integrable over $\mathbf{T}^1$ may diverge when $\varepsilon > 0$. Examples of continuous functions whose Fourier series diverge exactly on given sets of measure zero are given in Katznelson [152] and Kahane and Katznelson [145].

The extension of the Carleson–Hunt theorem to higher dimensions for square summability of Fourier series (Theorem 3.6.15) is a rather straightforward consequence of the one-dimensional result and was independently obtained by Fefferman [88], Sjölin [245], and Tevzadze [277]. An example showing that the circular partial sums of a Fourier series may not converge in $L^p(\mathbf{T}^n)$ for $n \ge 2$ and $p \neq 2$ was obtained by Fefferman [89]. This example also shows that there exist L^p functions on $\mathbf{T}^n$ for $n \ge 2$ whose circular partial sums do not converge almost everywhere when $1 \le p < 2$. Indeed, if the opposite happened, then the maximal operator $f \to \sup_{N\ge 0}|\widetilde{D}(n,N) * f|$ would have to be finite a.e. for all $f \in L^p(\mathbf{T}^n)$, and by Stein's theorem [254] it would have to be of weak type (p,p) for some $1 < p < 2$. But this would contradict Fefferman's counterexample on L^{p_1} for some $p < p_1 < 2$. On the other hand, almost everywhere is valid for the square partial sums of functions f with $|f|(\log^+|f|)^n(\log^+\log^+\log^+|f|)$ integrable over $\mathbf{T}^n$, as shown by Antonov [4]; see also Sjölin and Soria [247].

Transference of regulated multipliers originated in the article of de Leeuw [74]. The methods of transference in Section 3.6 were beautifully placed into the framework of a general theory by

Coifman and Weiss [55]. Transference of maximal multipliers (Theorem 3.6.11) was first obtained by Kenig and Tomas [154] and later elaborated by Asmar, Berkson, and Gillespie [11], [12].

The main references for trigonometric series are the books of Bary [15] and Zygmund [303], [304]. Other references for one-dimensional Fourier series include the books of Edwards [82], Dym and McKean [81], Katznelson [153], Körner [162], and the first eight chapters in Torchinsky [281]. The reader may also consult the book of Krantz [163] for a historical introduction to the subject of Fourier series.

A classical treatment of multiple Fourier series can be found in the last chapter of Bochner's book [26] and in parts of his other book [27]. Other references include the last chapter in Zygmund [304], the books of Yanushauskas [295] (in Russian) and Zhizhiashvili [299], the last chapter in Stein and Weiss [265], and the article of Alimov, Ashurov, and Pulatov in [2]. A brief survey article on the subject was written by Ash [10]. More extensive expositions were written by Shapiro [240], Igari [136], and Zhizhiashvili [298]. A short note on the history of Fourier series was written by Zygmund [305].

Chapter 4
Singular Integrals of Convolution Type

In this chapter we take up the one of the fundamental topics covered in this book, that of singular integrals. This topic is motivated by its intimate connection with some of the most important problems in Fourier analysis, such as convergence of Fourier series. As we have seen, the L^p boundedness of the conjugate function on the circle is equivalent to the L^p convergence of Fourier series of L^p functions. And since the Hilbert transform on the line is just a version of the conjugate function, it plays the same role in the convergence of Fourier integrals on the line as the conjugate function does on the circle.

The Hilbert transform is the prototype of all singular integrals, and a careful study of it provides the insight and inspiration for subsequent development of the subject. Historically, the theory of the Hilbert transform depended on techniques of complex analysis. With the development of the Calderón–Zygmund school, real-variable methods slowly replaced complex analysis, and this led to the introduction of singular integrals in other areas of mathematics. Singular integrals are nowadays intimately connected with partial differential equations, operator theory, several complex variables, and other fields. There are two kinds of singular integral operators: those of convolution type and those of nonconvolution type. In this chapter we study singular integrals of convolution type.

4.1 The Hilbert Transform and the Riesz Transforms

We begin the investigation of singular integrals with a careful study of the Hilbert transform. This study provides a great model for the development of the theory of singular integrals, presented in the remaining sections and in Chapter 8.

L. Grafakos, *Classical Fourier Analysis, Second Edition*,
DOI: 10.1007/978-0-387-09432-8_4, © Springer Science+Business Media, LLC 2008

4.1.1 Definition and Basic Properties of the Hilbert Transform

There are several equivalent ways to introduce the Hilbert transform; in this exposition we first define it as a convolution operator with a certain principal value distribution, but we later discuss other equivalent definitions.

We begin by defining a distribution W_0 in $\mathscr{S}'(\mathbf{R})$ as follows:

$$\langle W_0, \varphi \rangle = \frac{1}{\pi} \lim_{\varepsilon \to 0} \int_{\varepsilon \le |x| \le 1} \frac{\varphi(x)}{x}\, dx + \frac{1}{\pi} \int_{|x| \ge 1} \frac{\varphi(x)}{x}\, dx, \tag{4.1.1}$$

for φ in $\mathscr{S}(\mathbf{R})$. The function $1/x$ integrated over $[-1,-\varepsilon] \bigcup [\varepsilon, 1]$ has mean value zero, and we may replace $\varphi(x)$ by $\varphi(x)-\varphi(0)$ in the first integral in (4.1.1). Since $(\varphi(x)-\varphi(0))x^{-1}$ is controlled by $\|\varphi'\|_{L^\infty}$, it follows that the limit in (4.1.1) exists. To see that W_0 is indeed in $\mathscr{S}'(\mathbf{R})$, we go an extra step in the previous reasoning and obtain the estimate

$$\big|\langle W_0, \varphi \rangle\big| \le \frac{2}{\pi} \|\varphi'\|_{L^\infty} + \frac{2}{\pi} \sup_{x \in \mathbf{R}} |x\varphi(x)|. \tag{4.1.2}$$

This guarantees that $W_0 \in \mathscr{S}'(\mathbf{R})$.

Definition 4.1.1. The *truncated Hilbert transform* of $f \in \mathscr{S}(\mathbf{R})$ (at height ε) is defined by

$$H^{(\varepsilon)}(f)(x) = \frac{1}{\pi} \int_{|y| \ge \varepsilon} \frac{f(x-y)}{y}\, dy = \frac{1}{\pi} \int_{|x-y| \ge \varepsilon} \frac{f(y)}{x-y}\, dy. \tag{4.1.3}$$

The *Hilbert transform* of $f \in \mathscr{S}(\mathbf{R})$ is defined by

$$H(f)(x) = (W_0 * f)(x) = \lim_{\varepsilon \to 0} H^{(\varepsilon)}(f)(x). \tag{4.1.4}$$

The integral

$$\int_{-\infty}^{+\infty} \frac{f(x-y)}{y}\, dy$$

does not converge absolutely but is defined as a limit of the absolutely convergent integrals

$$\int_{|y| \ge \varepsilon} \frac{f(x-y)}{y}\, dy,$$

as $\varepsilon \to 0$. Such limits are called *principal value integrals* and are denoted by the letters p.v. Using this notation, the Hilbert transform is

$$H(f)(x) = \frac{1}{\pi} \text{p.v.} \int_{-\infty}^{+\infty} \frac{f(x-y)}{y}\, dy = \frac{1}{\pi} \text{p.v.} \int_{-\infty}^{+\infty} \frac{f(y)}{x-y}\, dy. \tag{4.1.5}$$

Remark 4.1.2. Note that for given $x \in \mathbf{R}$, $H(f)(x)$ is defined for all integrable functions f on $\mathbf{R}$ that satisfy a Hölder condition near the point x, that is,

$$|f(x) - f(y)| \le C_x |x-y|^{\varepsilon_x}$$

for some $C_x > 0$ and $\varepsilon_x > 0$ whenever $|y-x| < \delta_x$. Indeed, suppose that this is the case. Then we write

$$\begin{aligned} H^{(\varepsilon)}(f)(x) &= \frac{1}{\pi} \int\limits_{\varepsilon<|x-y|<\delta_x} \frac{f(y)}{x-y}\, dy + \frac{1}{\pi} \int\limits_{|x-y|\ge\delta_x} \frac{f(y)}{x-y}\, dy \\ &= \frac{1}{\pi} \int\limits_{\varepsilon<|x-y|<\delta_x} \frac{f(y)-f(x)}{x-y}\, dy + \frac{1}{\pi} \int\limits_{|x-y|\ge\delta_x} \frac{f(y)}{x-y}\, dy . \end{aligned}$$

Both integrals converge absolutely; hence the limit of $H^{(\varepsilon)}(f)(x)$ exists as $\varepsilon \to 0$. Therefore, the Hilbert transform of a piecewise smooth integrable function is well defined at all points of Hölder–Lipschitz continuity of the function. On the other hand, observe that $H^{(\varepsilon)}(f)$ is well defined for all $f \in L^p$, $1 \le p < \infty$, which follows from Hölder's inequality, since $1/x$ is integrable to the power p' on the set $|x| \ge \varepsilon$.

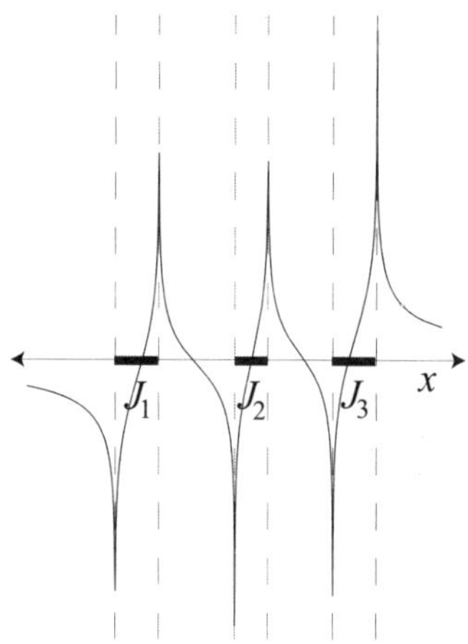

Fig. 4.1 The graph of the function $H(\chi_E)$ when E is a union of three disjoint intervals $J_1 \cup J_2 \cup J_3$.

Example 4.1.3. Consider the characteristic function $\chi_{[a,b]}$ of an interval $[a,b]$. It is a simple calculation to show that

$$H(\chi_{[a,b]})(x) = \frac{1}{\pi} \log \frac{|x-a|}{|x-b|} . \tag{4.1.6}$$

Let us verify this identity. Pick $\varepsilon < \min(|x-a|, |x-b|)$. To show (4.1.6) consider the three cases $0 < x-b$, $x-a < 0$, and $x-b < 0 < x-a$. In the first two cases, (4.1.6) follows immediately. In the third case we have

$$H(\chi_{[a,b]})(x) = \frac{1}{\pi} \lim_{\varepsilon \to 0} \left(\log \frac{|x-a|}{\varepsilon} + \log \frac{\varepsilon}{|x-b|} \right), \tag{4.1.7}$$

which gives (4.1.6). It is crucial to observe how the cancellation of the odd kernel $1/x$ is manifested in (4.1.7). Note that $H(\chi_{[a,b]})(x)$ blows up logarithmically in x near the points a and b and decays like $|x|^{-1}$ as $x \to \infty$. See Figure 4.1.

Example 4.1.4. Let $\log^+ x = \log x$ when $x \geq 1$ and zero otherwise. Observe that the calculation in the previous example actually gives

$$H^{(\varepsilon)}(\chi_{[a,b]})(x) = \begin{cases} \dfrac{1}{\pi} \log^+ \dfrac{|x-a|}{\max(\varepsilon, |x-b|)} & \text{when } x > b, \\[2ex] -\dfrac{1}{\pi} \log^+ \dfrac{|x-b|}{\max(\varepsilon, |x-a|)} & \text{when } x < a, \\[2ex] \dfrac{1}{\pi} \log^+ \dfrac{|x-a|}{\varepsilon} - \dfrac{1}{\pi} \log^+ \dfrac{|x-b|}{\varepsilon} & \text{when } a < x < b. \end{cases}$$

We now give an alternative characterization of the Hilbert transform using the Fourier transform. To achieve this we need to compute the Fourier transform of the distribution W_0 defined in (4.1.1). Fix a Schwartz function φ on $\mathbf{R}$. Then

$$\begin{aligned} \langle \widehat{W_0}, \varphi \rangle = \langle W_0, \widehat{\varphi} \rangle &= \frac{1}{\pi} \lim_{\varepsilon \to 0} \int_{|\xi| \geq \varepsilon} \widehat{\varphi}(\xi) \frac{d\xi}{\xi} \\ &= \frac{1}{\pi} \lim_{\varepsilon \to 0} \int_{\frac{1}{\varepsilon} \geq |\xi| \geq \varepsilon} \int_{\mathbf{R}} \varphi(x) e^{-2\pi i x \xi} \, dx \, \frac{d\xi}{\xi} \\ &= \lim_{\varepsilon \to 0} \int_{\mathbf{R}} \varphi(x) \left[\frac{1}{\pi} \int_{\frac{1}{\varepsilon} \geq |\xi| \geq \varepsilon} e^{-2\pi i x \xi} \frac{d\xi}{\xi} \right] dx \\ &= \lim_{\varepsilon \to 0} \int_{\mathbf{R}} \varphi(x) \left[\frac{-i}{\pi} \int_{\frac{1}{\varepsilon} \geq |\xi| \geq \varepsilon} \sin(2\pi x \xi) \frac{d\xi}{\xi} \right] dx. \end{aligned} \tag{4.1.8}$$

Now use the results (a) and (b) of Exercise 4.1.1 to deduce that the expressions inside the square brackets in (4.1.8) are uniformly bounded by 8 and converge as $\varepsilon \to 0$ to

$$\lim_{\varepsilon \to 0} \int_{\frac{1}{\varepsilon} \geq |\xi| \geq \varepsilon} \sin(2\pi x \xi) \frac{d\xi}{\xi} = \pi \operatorname{sgn} x = \begin{cases} \pi & \text{when } x > 0, \\ 0 & \text{when } x = 0, \\ -\pi & \text{when } x < 0. \end{cases} \tag{4.1.9}$$

The Lebesgue dominated convergence theorem and these facts allow us to pass the limit inside the integral in (4.1.8) to obtain that

$$\langle \widehat{W_0}, \varphi \rangle = \int_{\mathbf{R}} \varphi(x) (-i \operatorname{sgn}(x)) \, dx. \tag{4.1.10}$$

This implies that

$$\widehat{W_0}(\xi) = -i\,\mathrm{sgn}\,\xi\,. \tag{4.1.11}$$

In particular, identity (4.1.11) says that $\widehat{W_0}$ is a (bounded) function.

We now use identity (4.1.11) to write

$$H(f)(x) = \big(\widehat{f}(\xi)(-i\,\mathrm{sgn}\,\xi)\big)^{\vee}(x)\,. \tag{4.1.12}$$

This formula can be used to give an alternative definition of the Hilbert transform. An immediate consequence of (4.1.12) is that

$$\big\|H(f)\big\|_{L^2} = \big\|f\big\|_{L^2}\,, \tag{4.1.13}$$

that is, H is an isometry on $L^2(\mathbf{R})$. Moreover, H satisfies

$$H^2 = HH = -I\,, \tag{4.1.14}$$

where I is the identity operator. Equation (4.1.14) is a simple consequence of the fact that $(-i\,\mathrm{sgn}\,\xi)^2 = -1$. The adjoint operator H^* of H is uniquely defined via the identity

$$\big\langle f\,|\,H(g)\big\rangle = \int_{\mathbf{R}} f\,\overline{H(g)}\,dx = \int_{\mathbf{R}} H^*(f)\,\overline{g}\,dx = \big\langle H^*(f)\,|\,g\big\rangle\,,$$

and we can easily obtain that H^* has multiplier $\overline{-i\,\mathrm{sgn}\,\xi} = i\,\mathrm{sgn}\,\xi$. We conclude that $H^* = -H$. Likewise, we obtain $H^t = -H$.

4.1.2 Connections with Analytic Functions

We now investigate connections of the Hilbert transform with the Poisson kernel. Recall the definition of the Poisson kernel P_y given in Example 1.2.17. Then for $f \in L^p(\mathbf{R})$, $1 \le p < \infty$, we have

$$(P_y * f)(x) = \frac{y}{\pi}\int_{-\infty}^{+\infty} \frac{f(t)}{(x-t)^2 + y^2}\,dt\,, \tag{4.1.15}$$

and the integral in (4.1.15) converges absolutely by Hölder's inequality, since the function $t \mapsto ((x-t)^2 + y^2)^{-1}$ is in $L^{p'}(\mathbf{R})$ whenever $y > 0$.

Let $\mathrm{Re}\,z$ and $\mathrm{Im}\,z$ denote the real and imaginary parts of a complex number z. Observe that

$$(P_y * f)(x) = \mathrm{Re}\left(\frac{i}{\pi}\int_{-\infty}^{+\infty} \frac{f(t)}{x-t+iy}\,dt\right) = \mathrm{Re}\left(\frac{i}{\pi}\int_{-\infty}^{+\infty} \frac{f(t)}{z-t}\,dt\right),$$

where $z = x + iy$. The function

$$F_f(z) = \frac{i}{\pi}\int_{-\infty}^{+\infty} \frac{f(t)}{z-t}\,dt$$

defined on

$$\mathbf{R}^2_+ = \{z = x+iy : y > 0\}$$

is analytic, since its $\partial/\partial\bar{z}$ derivative is zero. The real part of $F_f(x+iy)$ is $(P_y * f)(x)$. The imaginary part of $F_f(x+iy)$ is

$$\operatorname{Im}\left(\frac{i}{\pi}\int_{-\infty}^{+\infty} \frac{f(t)}{x-t+iy}\,dt\right) = \frac{1}{\pi}\int_{-\infty}^{+\infty} \frac{f(t)(x-t)}{(x-t)^2+y^2}\,dt = (f * Q_y)(x),$$

where Q_y is called the *conjugate Poisson kernel* and is given by

$$Q_y(x) = \frac{1}{\pi}\frac{x}{x^2+y^2}\,. \tag{4.1.16}$$

The function $u_f + iv_f$ is analytic and thus $u_f(x+iy) = (f * P_y)(x)$ and $v_f(x+iy) = (f * Q_y)(x)$ are *conjugate harmonic functions*. Since the family P_y, $y > 0$, is an approximate identity, it follows from Theorem 1.2.19 that $P_y * f \to f$ in $L^p(\mathbf{R})$ as $y \to 0$. The following question therefore arises: What is the limit of $f * Q_y$ as $y \to 0$? As we show next, this limit has to be $H(f)$.

Theorem 4.1.5. *Let $1 \le p < \infty$. For any $f \in L^p(\mathbf{R})$ we have*

$$f * Q_\varepsilon - H^{(\varepsilon)}(f) \to 0 \tag{4.1.17}$$

in L^p and almost everywhere as $\varepsilon \to 0$.

Proof. We see that

$$(Q_\varepsilon * f)(x) - \frac{1}{\pi}\int_{|t|\ge\varepsilon} \frac{f(x-t)}{t}\,dt = \frac{1}{\pi}(f * \psi_\varepsilon)(x),$$

where $\psi_\varepsilon(x) = \varepsilon^{-1}\psi(\varepsilon^{-1}x)$ and

$$\psi(t) = \begin{cases} \frac{t}{t^2+1} - \frac{1}{t} & \text{when } |t| \ge 1, \\ \frac{t}{t^2+1} & \text{when } |t| < 1. \end{cases} \tag{4.1.18}$$

Note that ψ has integral zero. Furthermore, the integrable function

$$\Psi(t) = \begin{cases} \frac{1}{t^2+1} & \text{when } |t| \ge 1, \\ 1 & \text{when } |t| < 1, \end{cases} \tag{4.1.19}$$

is a radially decreasing majorant of ψ. It follows from Theorem 1.2.21 and Corollary 2.1.19 (with $a = 0$) that $f * \psi_\varepsilon \to 0$ in L^p and almost everywhere as $\varepsilon \to 0$. □

Remark 4.1.6. For $f \in \mathscr{S}(\mathbf{R})$ we know that $\lim_{\varepsilon\to 0} H^{(\varepsilon)}(f) = H(f)$, and we therefore conclude from (4.1.17) that

$$Q_\varepsilon * f \to H(f) \qquad \text{a.e.}$$

as $\varepsilon \to 0$. This convergence is also valid for general functions $f \in L^p(\mathbf{R})$. This is a consequence of the result of Theorem 4.1.12, estimate (4.1.31), and Theorem 2.1.14.

4.1.3 L^p Boundedness of the Hilbert Transform

As a consequence of the result in Exercise 4.1.4 and of the fact that

$$x \le \tfrac{1}{2}(e^x - e^{-x}),$$

we obtain that

$$|\{x : |H(\chi_E)(x)| > \alpha\}| \le \frac{2}{\pi}\frac{|E|}{\alpha}, \qquad \alpha > 0, \tag{4.1.20}$$

for all subsets E of the real line of finite measure. Theorem 1.4.19 with $p_0 = q_0 = 1$ and $p_1 = q_1 = 2$ now implies that H is bounded on L^p for $1 < p < 2$. Duality gives that $H^* = -H$ is bounded on L^p for $2 < p < \infty$ and hence so is H.

We give another proof of the boundedness of the Hilbert transform H on $L^p(\mathbf{R})$, which has the advantage that it gives the best possible constant in the resulting norm inequality when p is a power of 2.

Theorem 4.1.7. *For all $1 < p < \infty$, there exists a positive constant C_p such that*

$$\|H(f)\|_{L^p} \le C_p \|f\|_{L^p}$$

for all f in $\mathscr{S}(\mathbf{R})$. Moreover, the constant C_p satisfies $C_p \le 2p$ for $2 \le p < \infty$ and $C_p \le 2p/(p-1)$ for $1 < p \le 2$. Therefore, the Hilbert transform H admits an extension to a bounded operator on $L^p(\mathbf{R})$ when $1 < p < \infty$.

Proof. The proof we give is based on the interesting identity

$$H(f)^2 = f^2 + 2H(fH(f)), \tag{4.1.21}$$

valid whenever f is a real-valued Schwartz function. Before we prove (4.1.21), we discuss its origin. The function $f + iH(f)$ has a holomorphic extension on $\mathbf{R}^2_+$ and therefore so does its square

$$(f + iH(f))^2 = f^2 - H(f)^2 + i2fH(f).$$

Then $f^2 - H(f)^2$ has a harmonic extension u on the upper half-space whose conjugate harmonic function v must have boundary values $H(f^2 - H(f)^2)$. Thus $H(f^2 - H(f)^2) = 2fH(f)$, which implies (4.1.21) as $H^2 = -I$.

To give an alternative proof of (4.1.21) we take Fourier transforms. Let

$$m(\xi) = -i \operatorname{sgn} \xi$$

be the symbol of the Hilbert transform. We have

$$\begin{aligned}
&\widehat{f^2}(\xi)+2[H(fH(f))]\widehat{\ }(\xi)\\
&\quad=(\widehat{f}*\widehat{f})(\xi)+2m(\xi)(\widehat{f}*\widehat{H(f)})(\xi)\\
&\quad=\int_{\mathbf{R}}\widehat{f}(\eta)\widehat{f}(\xi-\eta)\,d\eta+2m(\xi)\int_{\mathbf{R}}\widehat{f}(\eta)\widehat{f}(\xi-\eta)m(\eta)\,d\eta \qquad (4.1.22)\\
&\quad=\int_{\mathbf{R}}\widehat{f}(\eta)\widehat{f}(\xi-\eta)\,d\eta+2m(\xi)\int_{\mathbf{R}}\widehat{f}(\eta)\widehat{f}(\xi-\eta)m(\xi-\eta)\,d\eta. \qquad (4.1.23)
\end{aligned}$$

Averaging (4.1.22) and (4.1.23) we obtain

$$\widehat{f^2}(\xi)+2[H(fH(f))]\widehat{\ }(\xi)=\int_{\mathbf{R}}\widehat{f}(\eta)\widehat{f}(\xi-\eta)\big[1+m(\xi)\big(m(\eta)+m(\xi-\eta)\big)\big]\,d\eta\,.$$

But the last displayed expression is equal to

$$\int_{\mathbf{R}}\widehat{f}(\eta)\widehat{f}(\xi-\eta)m(\eta)m(\xi-\eta)\,d\eta=(\widehat{H(f)}*\widehat{H(f)})(\xi)$$

in view of the identity

$$m(\eta)m(\xi-\eta)=1+m(\xi)m(\eta)+m(\xi)m(\xi-\eta),$$

which is valid for the function $m(\xi)=-i\operatorname{sgn}\xi$.

Having established (4.1.21), we can easily obtain L^p bounds for H when $p=2^k$ is a power of 2. We already know that H is bounded on L^p with norm one when $p=2^k$ and $k=1$. Suppose that H is bounded on L^p with bound c_p for $p=2^k$ for some $k\in\mathbf{Z}^+$. Then

$$\begin{aligned}
\big\|H(f)\big\|_{L^{2p}}=\big\|H(f)^2\big\|_{L^p}^{\frac12}&\le\big(\big\|f^2\big\|_{L^p}+\big\|2H(fH(f))\big\|_{L^p}\big)^{\frac12}\\
&\le\big(\big\|f\big\|_{L^{2p}}^2+2c_p\big\|fH(f)\big\|_{L^p}\big)^{\frac12}\\
&\le\big(\big\|f\big\|_{L^{2p}}^2+2c_p\big\|f\big\|_{L^{2p}}\big\|H(f)\big\|_{L^{2p}}\big)^{\frac12}.
\end{aligned}$$

We obtain that

$$\left(\frac{\|H(f)\|_{L^{2p}}}{\|f\|_{L^{2p}}}\right)^2-2c_p\frac{\|H(f)\|_{L^{2p}}}{\|f\|_{L^{2p}}}-1\le 0.$$

If follows that

$$\frac{\|H(f)\|_{L^{2p}}}{\|f\|_{L^{2p}}}\le c_p+\sqrt{c_p^2+1}\,,$$

and from this we conclude that H is bounded on L^{2p} with bound

$$c_{2p}\le c_p+\sqrt{c_p^2+1}\,. \qquad (4.1.24)$$

This completes the induction. We have proved that H maps L^p to L^p when $p = 2^k$, $k = 1, 2, \dots$. Interpolation now gives that H maps L^p to L^p for all $p \geq 2$. Since $H^* = -H$, duality gives that H is also bounded on L^p for $1 < p \leq 2$.

The previous proof of the boundedness of the Hilbert transform provides us with some useful information about the norm of this operator on $L^p(\mathbf{R})$. Let us begin with the identity

$$\cot\frac{x}{2} = \cot x + \sqrt{1 + \cot^2 x},$$

valid for $0 < x < \frac{\pi}{2}$. If $c_p \leq \cot\frac{\pi}{2p}$, then (4.1.24) gives that

$$c_{2p} \leq c_p + \sqrt{c_p^2 + 1} \leq \cot\frac{\pi}{2p} + \sqrt{1 + \cot^2\frac{\pi}{2p}} = \cot\frac{\pi}{2 \cdot 2p},$$

and since $1 = \cot\frac{\pi}{4} = \cot\frac{\pi}{2 \cdot 2}$, we obtain by induction that the numbers $\cot\frac{\pi}{2p}$ are indeed bounds for the norm of H on L^p when $p = 2^k$, $k = 1, 2, \dots$. Duality now gives that the numbers $\cot\frac{\pi}{2p'} = \tan\frac{\pi}{2p}$ are bounds for the norm of H on L^p when $p = \frac{2^k}{2^k - 1}$, $k = 1, 2, \dots$. These bounds allow us to derive good estimates for the norm $\|H\|_{L^p \to L^p}$ as $p \to 1$ and $p \to \infty$. Indeed, since $\cot\frac{\pi}{2p} \leq p$ when $p \geq 2$, the Riesz–Thorin interpolation theorem gives that $\|H\|_{L^p \to L^p} \leq 2p$ for $2 \leq p < \infty$ and by duality $\|H\|_{L^p \to L^p} \leq \frac{2p}{p-1}$ for $1 < p \leq 2$. This completes the proof of this theorem. It is worth comparing this proof with the one given in Theorem 3.5.6. □

Remark 4.1.8. The numbers $\cot\frac{\pi}{2p}$ for $2 \leq p < \infty$ and $\tan\frac{\pi}{2p}$ for $1 < p \leq 2$ are indeed equal to the norms of the Hilbert transform H on $L^p(\mathbf{R})$. This requires a more delicate argument; see Exercise 4.1.12.

Remark 4.1.9. We may wonder what happens when $p = 1$ or $p = \infty$. The Hilbert transform of $\chi_{[a,b]}$ computed in Example 4.1.3 is easily seen to be unbounded and not integrable, since it behaves like $1/|x|$ as $x \to \infty$. This behavior near infinity suggests that the Hilbert transform may map L^1 to $L^{1,\infty}$. This is indeed the case, but this will not be shown until Section 4.3.

We now introduce the maximal Hilbert transform.

Definition 4.1.10. The *maximal Hilbert transform* is the operator

$$H^{(*)}(f)(x) = \sup_{\varepsilon > 0} \left| H^{(\varepsilon)}(f)(x) \right| \tag{4.1.25}$$

defined for all f in L^p, $1 \leq p < \infty$. For such f, $H^{(\varepsilon)}(f)$ is well defined as a convergent integral by Hölder's inequality. Hence $H^{(*)}(f)$ makes sense for $f \in L^p(\mathbf{R})$, although for some values of x, $H^{(*)}(f)(x)$ may be infinite.

Example 4.1.11. Using the result of Example 4.1.4, we obtain that

$$H^{(*)}(\chi_{[a,b]})(x) = \frac{1}{\pi} \left| \log\frac{|x-a|}{|x-b|} \right|. \tag{4.1.26}$$

We see that in general, $H^{(*)}(f)(x) \neq |H(f)(x)|$ by taking f to be the characteristic function of the union of two disjoint closed intervals.

The definition of H yields that $H^{(\varepsilon)}(f)$ converges pointwise to $H(f)$ whenever f is a smooth function with compact support. If we have the estimate $\big\|H^{(*)}(f)\big\|_{L^p} \le C_p\big\|f\big\|_{L^p}$ for $f \in L^p(\mathbf{R})$, Theorem 2.1.14 yields that $H^{(\varepsilon)}(f)$ converges to $H(f)$ a.e. as $\varepsilon \to 0$ for any $f \in L^p$. This almost everywhere limit provides a way to describe $H(f)$ for general $f \in L^p(\mathbf{R})$. Note that Theorem 4.1.7 implies only that H has a (unique) bounded extension on L^p, but it does not provide a way to describe $H(f)$ when f is a general L^p function.

The next theorem is a simple consequence of these ideas.

Theorem 4.1.12. *There exists a constant C such that for all $1 < p < \infty$ we have*

$$\big\|H^{(*)}(f)\big\|_{L^p} \le C\max\big(p,(p-1)^{-1}\big)\big\|f\big\|_{L^p}\,. \tag{4.1.27}$$

Moreover, for all f in $L^p(\mathbf{R})$, $H^{(\varepsilon)}(f)$ converges to $H(f)$ a.e. and in L^p.

Proof. The following proof yields the slightly weaker bound $C\max\big(p,(p-1)^{-2}\big)$. Another proof of this theorem with the asserted bound in (4.1.27) is given in Theorem 8.2.3.

Recall the kernels P_ε and Q_ε defined in (4.1.15) and (4.1.16). Fix $1 < p < \infty$ and suppose momentarily that

$$f * Q_\varepsilon = H(f) * P_\varepsilon\,, \qquad \varepsilon > 0\,, \tag{4.1.28}$$

holds whenever f is an L^p function. Then we have

$$H^{(\varepsilon)}(f) = H^{(\varepsilon)}(f) - f * Q_\varepsilon + H(f) * P_\varepsilon\,. \tag{4.1.29}$$

Using the identity

$$H^{(\varepsilon)}(f)(x) - (f * Q_\varepsilon)(x) = -\frac{1}{\pi}\int_{\mathbf{R}} f(x-t)\,\psi_\varepsilon(t)\,dt\,, \tag{4.1.30}$$

where ψ is as in (4.1.18), and applying Corollary 2.1.12, we obtain the estimate

$$\sup_{\varepsilon>0}\big|H^{(\varepsilon)}(f)(x) - (f * Q_\varepsilon)(x)\big| \le \frac{1}{\pi}\big\|\Psi\big\|_{L^1} M(f)(x)\,, \tag{4.1.31}$$

where Ψ is as in (4.1.19) and M is the Hardy–Littlewood maximal function. In view of (4.1.29) and (4.1.31), we obtain for $f \in L^p(\mathbf{R}^n)$ that

$$|H^{(*)}(f)(x)| \le \big\|\Psi\big\|_{L^1} M(f)(x) + M(H(f))(x)\,. \tag{4.1.32}$$

It follows immediately from (4.1.32) that $H^{(*)}$ is L^p bounded with norm at most $C\max\big(p,(p-1)^{-2}\big)$.

It suffices therefore to establish (4.1.28). In the proof of (4.1.28), we might as well assume that f is a Schwartz function. Taking Fourier transforms, we see that (4.1.28) is a consequence of the identity

$$\big((-i\,\mathrm{sgn}\,\xi)e^{-2\pi|\xi|}\big)^{\vee}(x) = \frac{1}{\pi}\frac{x}{x^2+1}. \tag{4.1.33}$$

Writing the inverse Fourier transform as an integral from $-\infty$ to $+\infty$ and then changing this to an integral from 0 to ∞, we obtain that (4.1.33) is equivalent to the identity

$$-i\int_0^\infty e^{-2\pi\xi}\big[e^{2\pi i x\xi} - e^{-2\pi i x\xi}\big]\,d\xi = \frac{1}{\pi}\frac{x}{x^2+1},$$

which can be easily checked using integration by parts twice.

The statement in the theorem about the almost everywhere convergence of $H^{(\varepsilon)}(f)$ to $H(f)$ is a consequence of (4.1.27), of the fact that the alleged convergence holds for Schwartz functions, and of Theorem 2.1.14. Finally, the L^p convergence follows from the almost everywhere convergence and the Lebesgue dominated convergence theorem in view of the validity of (4.1.32). □

4.1.4 The Riesz Transforms

We now study an n-dimensional analogue of the Hilbert transform. It turns out that there exist n operators in $\mathbf{R}^n$, called the Riesz transforms, with properties analogous to those of the Hilbert transform on $\mathbf{R}$.

To define the Riesz transforms, we first introduce tempered distributions W_j on $\mathbf{R}^n$, for $1 \le j \le n$, as follows. For $\varphi \in \mathscr{S}(\mathbf{R}^n)$, let

$$\langle W_j, \varphi\rangle = \frac{\Gamma(\frac{n+1}{2})}{\pi^{\frac{n+1}{2}}}\lim_{\varepsilon\to 0}\int_{|y|\ge\varepsilon}\frac{y_j}{|y|^{n+1}}\varphi(y)\,dy.$$

One should check that indeed $W_j \in \mathscr{S}'(\mathbf{R}^n)$. Observe that the normalization of W_j is similar to that of the Poisson kernel.

Definition 4.1.13. For $1 \le j \le n$, the jth *Riesz transform* of f is given by convolution with the distribution W_j, that is,

$$R_j(f)(x) = (f * W_j)(x) = \frac{\Gamma(\frac{n+1}{2})}{\pi^{\frac{n+1}{2}}}\,\mathrm{p.v.}\int_{\mathbf{R}^n}\frac{x_j - y_j}{|x-y|^{n+1}}f(y)\,dy, \tag{4.1.34}$$

for all $f \in \mathscr{S}(\mathbf{R}^n)$. Definition 4.1.13 makes sense for any integrable function f that has the property that for all x there exist $C_x > 0$, $\varepsilon_x > 0$, and $\delta_x > 0$ such that for y satisfying $|y-x| < \delta_x$ we have $|f(x) - f(y)| \le C_x|x-y|^{\varepsilon_x}$. The principal value integral in (4.1.34) is as in Definition 4.1.1.

We now give a characterization of R_j using the Fourier transform. For this we need to compute the Fourier transform of W_j.

Proposition 4.1.14. *The jth Riesz transform R_j is given on the Fourier transform side by multiplication by the function $-i\xi_j/|\xi|$. That is, for any f in $\mathscr{S}(\mathbf{R}^n)$ we have*

$$R_j(f)(x) = \Big(-\frac{i\xi_j}{|\xi|}\widehat{f}(\xi)\Big)^{\vee}(x). \tag{4.1.35}$$

Proof. The proof is essentially a reprise of the corresponding proof for the Hilbert transform, but it involves a few technical difficulties. Fix a Schwartz function φ on $\mathbf{R}^n$. Then for $1 \le j \le n$ we have

$$\begin{aligned}
\big\langle \widehat{W_j}, \varphi \big\rangle &= \big\langle W_j, \widehat{\varphi} \big\rangle \qquad (4.1.36)\\
&= \frac{\Gamma(\frac{n+1}{2})}{\pi^{\frac{n+1}{2}}} \lim_{\varepsilon\to 0} \int_{|\xi|\ge\varepsilon} \widehat{\varphi}(\xi)\frac{\xi_j}{|\xi|^{n+1}}\,d\xi\\
&= \frac{\Gamma(\frac{n+1}{2})}{\pi^{\frac{n+1}{2}}} \lim_{\varepsilon\to 0} \int_{\frac{1}{\varepsilon}\ge|\xi|\ge\varepsilon} \int_{\mathbf{R}^n} \varphi(x)e^{-2\pi i x\cdot\xi}\,dx\,\frac{\xi_j}{|\xi|^{n+1}}\,d\xi\\
&= \lim_{\varepsilon\to 0}\int_{\mathbf{R}^n}\varphi(x)\left[\frac{\Gamma(\frac{n+1}{2})}{\pi^{\frac{n+1}{2}}}\int_{\frac{1}{\varepsilon}\ge|\xi|\ge\varepsilon} e^{-2\pi i x\cdot\xi}\frac{\xi_j}{|\xi|^{n+1}}\,d\xi\right]dx\\
&= \lim_{\varepsilon\to 0}\int_{\mathbf{R}^n}\varphi(x)\left[\frac{\Gamma(\frac{n+1}{2})}{\pi^{\frac{n+1}{2}}}\int_{\mathbf{S}^{n-1}}\int_{\varepsilon\le r\le\frac{1}{\varepsilon}} e^{-2\pi i r x\cdot\theta}\frac{r}{r^{n+1}}r^{n-1}\,dr\,\theta_j\,d\theta\right]dx\\
&= \int_{\mathbf{R}^n}\varphi(x)\left[-i\frac{\Gamma(\frac{n+1}{2})}{\pi^{\frac{n+1}{2}}}\int_{\mathbf{S}^{n-1}}\int_0^\infty \sin(2\pi r x\cdot\theta)\frac{dr}{r}\,\theta_j\,d\theta\right]dx\\
&= \int_{\mathbf{R}^n}\varphi(x)\left[-i\frac{\pi}{2}\frac{\Gamma(\frac{n+1}{2})}{\pi^{\frac{n+1}{2}}}\int_{\mathbf{S}^{n-1}}\operatorname{sgn}(x\cdot\theta)\,\theta_j\,d\theta\right]dx\\
&= \int_{\mathbf{R}^n} -i\varphi(x)\frac{x_j}{|x|}\,dx,
\end{aligned}$$

where in the penultimate equality we used the identity $\int_0^\infty \frac{\sin t}{t}\,dt = \frac{\pi}{2}$, for which we refer to Exercise 4.1.1, while in the last equality we used the identity

$$-i\frac{\pi}{2}\frac{\Gamma(\frac{n+1}{2})}{\pi^{\frac{n+1}{2}}}\int_{\mathbf{S}^{n-1}}\operatorname{sgn}(x\cdot\theta)\theta_j\,d\theta = -i\frac{x_j}{|x|}, \tag{4.1.37}$$

which needs to be established. The passage of the limit inside the integral in the previous calculation is a consequence of the Lebesgue dominated convergence theorem, which is justified from the fact that

$$\left|\int_\varepsilon^{1/\varepsilon}\frac{\sin(2\pi r\theta)}{r}\,dr\right| \le 4 \tag{4.1.38}$$

for all $\varepsilon > 0$. For a proof of (4.1.38) we again refer to Exercise 4.1.1. □

It remains to establish (4.1.37). Let us recall that $O(n)$ is the set of all orthogonal $n \times n$ matrices with real entries. An invertible matrix A is called orthogonal if its transpose A^t is equal to its inverse A^{-1}, that is, $AA^t = A^tA = I$.

Lemma 4.1.15. *The following identity is valid for all* $\xi \in \mathbf{R}^n \setminus \{0\}$*:*

$$\int_{\mathbf{S}^{n-1}} \operatorname{sgn}(\xi \cdot \theta)\, \theta_j \, d\theta = \frac{2\pi^{\frac{n-1}{2}}}{\Gamma(\frac{n+1}{2})} \frac{\xi_j}{|\xi|}. \tag{4.1.39}$$

Therefore (4.1.37) *holds.*

Proof. We begin with the identity

$$\int_{\mathbf{S}^{n-1}} \operatorname{sgn}(\theta_k)\, \theta_j \, d\theta = \begin{cases} 0 & \text{if } k \neq j, \\ \displaystyle\int_{\mathbf{S}^{n-1}} |\theta_j| \, d\theta & \text{if } k = j, \end{cases} \tag{4.1.40}$$

which can be proved by noting that for $k \neq j$, $\operatorname{sgn}(\theta_k)$ has a constant sign on the hemispheres $\theta_k > 0$ and $\theta_k < 0$, on either of which the function $\theta \mapsto \theta_j$ has integral zero.

It suffices to prove (4.1.39) for a unit vector ξ. Given $\xi \in \mathbf{S}^{n-1}$, pick an orthogonal $n \times n$ matrix $A = (a_{kl})_{k,l}$ such that $Ae_j = \xi$. Then the jth column of the matrix A is the vector $(\xi_1, \xi_2, \dots, \xi_n)^t$. We have

$$\begin{aligned} \int_{\mathbf{S}^{n-1}} \operatorname{sgn}(\xi \cdot \theta)\, \theta_j \, d\theta &= \int_{\mathbf{S}^{n-1}} \operatorname{sgn}(Ae_j \cdot \theta)\, \theta_j \, d\theta \\ &= \int_{\mathbf{S}^{n-1}} \operatorname{sgn}(e_j \cdot A^t\theta)\, (AA^t\theta)_j \, d\theta \\ &= \int_{\mathbf{S}^{n-1}} \operatorname{sgn}(e_j \cdot \theta)\, (A\theta)_j \, d\theta \\ &= \int_{\mathbf{S}^{n-1}} \operatorname{sgn}(\theta_j)\, (a_{j1}\theta_1 + \dots + \xi_j\theta_j + \dots + a_{jn}\theta_n) \, d\theta \\ &= \xi_j \int_{\mathbf{S}^{n-1}} \operatorname{sgn}(\theta_j)\, \theta_j \, d\theta + \sum_{1 \le m \neq j \le n} 0 \\ &= \frac{\xi_j}{|\xi|} \int_{\mathbf{S}^{n-1}} |\theta_j| \, d\theta \,. \end{aligned}$$

Next, for all $j \in \{1, 2, \dots, n\}$, we compute the value of the integral

$$\int_{\mathbf{S}^{n-1}} |\theta_j| \, d\theta = \int_{\mathbf{S}^{n-1}} |\theta_1| \, d\theta \,,$$

which is obviously independent of j by symmetry. In view of the result of Appendix D.2, we write

$$
\begin{aligned}
\int_{\mathbf{S}^{n-1}} |\theta_1|\,d\theta &= \int_{-1}^{1} |s| \int_{\sqrt{1-s^2}\,\mathbf{S}^{n-2}} d\varphi \frac{ds}{(1-s^2)^{\frac{1}{2}}} \\
&= \omega_{n-2} \int_{-1}^{1} |s|(1-s^2)^{\frac{n-3}{2}}\,ds \\
&= \omega_{n-2} \int_{0}^{1} u^{\frac{n-3}{2}}\,du \\
&= \frac{2\omega_{n-2}}{n-1} \\
&= \frac{2\pi^{\frac{n-1}{2}}}{\Gamma(\frac{n-1}{2})\frac{n-1}{2}} \\
&= \frac{2\pi^{\frac{n-1}{2}}}{\Gamma(\frac{n+1}{2})},
\end{aligned}
$$

having used the expression for ω_{n-2} in Appendix A.3. This proves (4.1.39). The proof of the lemma and hence that of Proposition 4.1.14 is complete. □

Proposition 4.1.16. *The Riesz transforms satisfy*

$$
-I = \sum_{j=1}^{n} R_j^2, \tag{4.1.41}
$$

where I is the identity operator.

Proof. Use the Fourier transform and the identity $\sum_{j=1}^{n}(-i\xi_j/|\xi|)^2 = -1$ to obtain that $\sum_{j=1}^{n} R_j^2(f) = -f$ for any f in the Schwartz class. □

Next we discuss a use of the Riesz transforms to partial differential equations.

Example 4.1.17. Suppose that f is a given Schwartz function on $\mathbf{R}^n$ and that u is a distribution that solves *Laplace's equation*

$$
\Delta(u) = f.
$$

Then we can express all second-order derivatives of u in terms of the Riesz transforms of f. First we note that

$$
(-4\pi^2|\xi|^2)\widehat{u}(\xi) = \widehat{f}(\xi).
$$

It follows that for all $1 \le j,k \le n$ we have

$$
\begin{aligned}
\partial_j\partial_k u &= \big[(2\pi i\xi_j)(2\pi i\xi_k)\widehat{u}(\xi)\big]^{\vee} \\
&= \Big[(2\pi i\xi_j)(2\pi i\xi_k)\frac{\widehat{f}(\xi)}{-4\pi^2|\xi|^2}\Big]^{\vee} \\
&= -R_jR_k(f),
\end{aligned}
$$

and in particular, we conclude that $\partial_j\partial_k u$ are functions.

Thus the Riesz transforms provide an explicit way to recover second-order derivatives in terms of the Laplacian. Such representations are useful in controlling quantitative expressions (such as norms) of second-order derivatives in terms of the corresponding expressions for the Laplacian. For instance, this is the case with the L^p norm; the L^p boundedness of the Riesz transforms is one of the main results of the next section. We refer to Exercises 4.2.9 and 4.2.10 for similar applications.

Exercises

4.1.1. (a) Show that for all $0 < a < b < \infty$ we have

$$\left|\int_a^b \frac{\sin x}{x}\,dx\right| \le 4.$$

(b) For $a > 0$ define

$$I(a) = \int_0^\infty \frac{\sin x}{x} e^{-ax}\,dx$$

and show that $I(a)$ is continuous at zero. Differentiate in a and look at the behavior of $I(a)$ as $a \to \infty$ to obtain the identity

$$I(a) = \frac{\pi}{2} - \arctan(a).$$

Deduce that $I(0) = \frac{\pi}{2}$ and also derive the following identity used in (4.1.9):

$$\int_{-\infty}^{+\infty} \frac{\sin(bx)}{x}\,dx = \pi \operatorname{sgn}(b).$$

(c) Argue as in part (b) to prove for $a \ge 0$ the identity

$$\int_0^\infty \frac{1-\cos x}{x^2} e^{-ax}\,dx = \frac{\pi}{2} - \arctan(a) + a\log\frac{a}{\sqrt{1+a^2}}.$$

[*Hint:* Part (a): Consider the cases $b \le 1$, $a \le 1 \le b$, $1 \le a$. When $a \ge 1$, integrate by parts.]

4.1.2. (a) Let φ be a compactly supported $\mathscr{C}^{m+1}$ function on $\mathbf{R}$ for some m in $\mathbf{Z}^+ \bigcup \{0\}$. Prove that if $\varphi^{(m)}$ is the mth derivative of φ, then

$$|H(\varphi^{(m)})(x)| \le C_{m,\varphi}\,(1+|x|)^{-m-1}$$

for some $C_{m,\varphi} > 0$.

(b) Let φ be a compactly supported $\mathscr{C}^{m+1}$ function on $\mathbf{R}^n$ for some $m \in \mathbf{Z}^+$. Show that

$$|R_j(\partial^\alpha \varphi)(x)| \le C_{n,m,\varphi}\,(1+|x|)^{-n-m}$$

for some $C_{n,m,\varphi} > 0$ and all multi-indices α with $|\alpha| = m$.
(c) Let I be an interval on the line and assume that a function h is equal to 1 on the left half of I, is equal to -1 on the right half of I, and vanishes outside I. Prove that for $x \notin \frac{3}{2}I$ we have

$$|H(h)(x)| \le 4|I|^2|x-\text{center}(I)|^{-2}.$$

[*Hint:* Use that when $|t| \le \frac{1}{2}$ we have $\log(1+t) = t + R_1(t)$, where $|R_1(t)| \le 2|t|^2$.]

4.1.3. (a) In view of identity (4.1.12) one may define $H(f)$ as an element of $\mathscr{S}'(\mathbf{R})$ for bounded functions f on the line whose Fourier transform vanishes in a neighborhood of the origin. Using this interpretation, prove that

$$\begin{aligned} H(e^{ix}) &= -ie^{ix}, \\ H(\cos x) &= \sin x, \\ H(\sin x) &= -\cos x, \\ H(\sin(\pi x)/\pi x) &= (\cos(\pi x) - 1)/\pi x. \end{aligned}$$

(b) Show that the operators given by convolution with the smooth function $\sin(t)/t$ and the distribution p.v. $\cos(t)/t$ are bounded on $L^p(\mathbf{R})$ whenever $1 < p < \infty$.

4.1.4. (*Stein and Weiss [264]*) Show that the distribution function of the Hilbert transform of a characteristic function of a measurable subset E of $\mathbf{R}$ of finite measure is

$$d_{H(\chi_E)}(\alpha) = \frac{4|E|}{e^{\pi\alpha} - e^{-\pi\alpha}}, \qquad \alpha > 0.$$

[*Hint:* First take $E = \bigcup_{j=1}^N (a_j, b_j)$, where $b_j < a_{j+1}$. Show that the equation $H(\chi_E)(x) = \pi\alpha$ has exactly one root ρ_j in each open interval (a_j, b_j) for $1 \le j \le N$ and exactly one root r_j in each interval (b_j, a_{j+1}) for $1 \le j \le N$, $(a_{N+1} = \infty)$. Then $|\{x \in \mathbf{R} : H(\chi_E)(x) > \pi\alpha\}| = \sum_{j=1}^N r_j - \sum_{j=1}^N \rho_j$, and this can be expressed in terms of $\sum_{j=1}^N a_j$ and $\sum_{j=1}^N b_j$. Argue similarly for the set $\{x \in \mathbf{R} : H(\chi_E)(x) < -\pi\alpha\}$. For a general measurable set E, find sets E_n such that each E_n is a finite union of intervals and that $\chi_{E_n} \to \chi_E$ in L^2. Then $H(\chi_{E_n}) \to H(\chi_E)$ in measure; thus $H(\chi_{E_{n_k}}) \to H(\chi_E)$ a.e. for some subsequence n_k. The Lebesgue dominated convergence theorem gives $d_{H(\chi_{E_{n_k}})} \to d_{H(\chi_E)}$. See Figure 4.1.]

4.1.5. Let $1 \le p < \infty$. Suppose that there exists a constant $C > 0$ such that for all $f \in \mathscr{S}(\mathbf{R})$ with L^p norm one we have

$$|\{x : |H(f)(x)| > 1\}| \le C.$$

Using only this inequality, prove that H maps $L^p(\mathbf{R})$ to $L^{p,\infty}(\mathbf{R})$. Here H is the Hilbert transform. State properties for a general operator such that the same conclusion is valid.
[*Hint:* Try functions of the form $\lambda^{-1/p} f(\lambda^{-1}x)$.]

4.1.6. Let φ be in $\mathscr{S}(\mathbf{R})$. Prove that

$$\lim_{N\to\infty} \text{p.v.} \int_{\mathbf{R}} \frac{e^{2\pi i N x}}{x} \varphi(x)\,dx = \varphi(0)\pi i,$$
$$\lim_{N\to-\infty} \text{p.v.} \int_{\mathbf{R}} \frac{e^{2\pi i N x}}{x} \varphi(x)\,dx = -\varphi(0)\pi i.$$

4.1.7. Let T_α, $\alpha \in \mathbf{R}$, be the operator given by convolution with the distribution whose Fourier transform is the function

$$u_\alpha(\xi) = e^{-\pi i \alpha \operatorname{sgn}\xi}.$$

(a) Show that the T_α's are isometries on $L^2(\mathbf{R})$ that satisfy

$$(T_\alpha)^{-1} = T_{2-\alpha}.$$

(b) Express T_α in terms of the identity operator and the Hilbert transform.

4.1.8. Let $Q_y^{(j)}$ be the jth *conjugate Poisson kernel* of P_y defined by

$$Q_y^{(j)}(x) = \frac{\pi^{\frac{n+1}{2}}}{\Gamma(\frac{n+1}{2})} \frac{x_j}{(|x|^2+y^2)^{\frac{n+1}{2}}}.$$

Prove that $(Q_y^{(j)})^{\widehat{}}(\xi) = -i\xi_j e^{-2\pi|\xi|}/|\xi|$. Conclude that $R_j(P_y) = Q_y^{(j)}$ and that for f in $L^2(\mathbf{R}^n)$ we have $R_j(f) * P_y = f * Q_y^{(j)}$. These results are analogous to the statements $\widehat{Q_y}(\xi) = -i\operatorname{sgn}(\xi)\widehat{P_y}(\xi)$, $H(P_y) = Q_y$, and $H(f) * P_y = f * Q_y$.

4.1.9. Let $f_0, f_1, \dots, f_n$ all belong to $L^2(\mathbf{R}^n)$ and let $u_j = P_y * f_j$ be their corresponding Poisson integrals for $0 \le j \le n$. Show that a necessary and sufficient condition for

$$f_j = R_j(f_0), \qquad j = 1, \dots, n,$$

is that the following system of generalized Cauchy-Riemann equations holds:

$$\sum_{j=0}^{n} \frac{\partial u_j}{\partial x_j} = 0,$$
$$\frac{\partial u_j}{\partial x_k} = \frac{\partial u_k}{\partial x_j}, \qquad j \neq k, \qquad x_0 = y.$$

4.1.10. Prove the distributional identity

$$\partial_j |x|^{-n+1} = (1-n)\text{p.v.} \frac{x_j}{|x|^{n+1}}.$$

Then take Fourier transforms of both sides and use Theorem 2.4.6 to obtain another proof of Proposition 4.1.14.

4.1.11. (a) Prove that if T is a bounded linear operator on $L^2(\mathbf{R})$ that commutes with translations and dilations and anticommutes with the reflection $f(x) \mapsto \widetilde{f}(x) = f(-x)$, then T is a constant multiple of the Hilbert transform.
(b) Prove that if T is a bounded operator on $L^2(\mathbf{R})$ that commutes with translations and dilations and vanishes when applied to functions whose Fourier transform is supported in $[0,\infty)$, then T is a constant multiple of the operator $f \mapsto \big(\widehat{f}\chi_{(-\infty,0]}\big)^\vee$.

4.1.12. (*Pichorides [213]*) Fix $1 < p \le 2$.
(a) Show that the function $(x,y) \mapsto \text{Re}\,(|x|+iy)^p$ is subharmonic on $\mathbf{R}^2$.
(b) Prove that for f in $\mathscr{C}_0^\infty(\mathbf{R})$ we have

$$\int_{\mathbf{R}} \text{Re}\,(|f(x)|+iH(f)(x))^p\,dx \ge 0.$$

(c) Prove that for all a and b reals we have

$$|b|^p \le \Big(\tan\frac{\pi}{2p}\Big)^p |a|^p - D_p \,\text{Re}\,(|a|+ib)^p$$

for some $D_p > 0$. Then use part (b) to conclude that

$$\big\|H\big\|_{L^p\to L^p} \le \tan\frac{\pi}{2p}.$$

(d) To deduce that this constant is sharp, take $\pi/2p' < \gamma < \pi/2p$ and let $f_\gamma(x) = (x+1)^{-1}|x+1|^{2\gamma/\pi}|x-1|^{-2\gamma/\pi}\cos\gamma$. Then

$$H(f_\gamma)(x) = \begin{cases} \frac{1}{x+1}\frac{|x+1|}{|x-1|}^{2\gamma/\pi}\sin\gamma & \text{when } |x|>1, \\ -\frac{1}{x+1}\frac{|x+1|}{|x-1|}^{2\gamma/\pi}\sin\gamma & \text{when } |x|<1. \end{cases}$$

[*Hint:* Part (b): Let C_R be the circle of radius R centered at $(0,R)$ in $\mathbf{R}^2$. Use that the integral of the subharmonic function

$$(x,y) \mapsto \text{Re}\,(|(P_y * f)(x)| + i(Q_y * f)(x))^p$$

over C_R is at least $2\pi R\,\text{Re}\,(|(P_R * f)(0)| + i(Q_R * f)(0))^p$ and let $R \to \infty$. Part (d): This is best seen by considering the restriction of the analytic function

$$F(z) = (z+1)^{-1}\Big(\frac{iz+i}{z-1}\Big)^{2\gamma/\pi}$$

on $\mathbf{R}\times\{0\}$.]

4.2 Homogeneous Singular Integrals and the Method of Rotations

So far we have introduced the Hilbert and the Riesz transforms and we have derived the L^p boundedness of the former. The boundedness properties of the Riesz transforms on L^p spaces are consequences of the results discussed in this section.

4.2.1 Homogeneous Singular and Maximal Singular Integrals

We introduce singular integral operators on $\mathbf{R}^n$ that appropriately generalize the Riesz transforms on $\mathbf{R}^n$. Here is the setup. We fix Ω to be an integrable function of the unit sphere $\mathbf{S}^{n-1}$ with mean value zero. Observe that the kernel

$$K_\Omega(x) = \frac{\Omega(x/|x|)}{|x|^n}, \qquad x \neq 0, \tag{4.2.1}$$

is homogeneous of degree $-n$ just like the functions $x_j/|x|^{n+1}$. Since K_Ω is not in $L^1(\mathbf{R}^n)$, convolution with K_Ω cannot be defined as an operation on Schwartz functions on $\mathbf{R}^n$. For this reason we introduce a distribution W_Ω in $\mathscr{S}'(\mathbf{R}^n)$ by setting

$$\langle W_\Omega, \varphi \rangle = \lim_{\varepsilon \to 0} \int_{|x| \geq \varepsilon} K_\Omega(x)\varphi(x)\,dx = \lim_{\varepsilon \to 0} \int_{\varepsilon \leq |x| \leq \varepsilon^{-1}} K_\Omega(x)\varphi(x)\,dx \tag{4.2.2}$$

for $\varphi \in \mathscr{S}(\mathbf{R}^n)$. Using the fact that Ω has mean value zero, we can easily see that W_Ω is a well defined tempered distribution on $\mathbf{R}^n$. Indeed, since K_Ω has integral zero over all annuli centered at the origin, we have

$$\begin{aligned}
\big|\langle W_\Omega, \varphi \rangle\big| &= \left| \lim_{\varepsilon \to 0} \int_{\varepsilon \leq |x| \leq 1} \frac{\Omega(x/|x|)}{|x|^n}(\varphi(x) - \varphi(0))\,dx + \int_{|x| \geq 1} \frac{\Omega(x/|x|)}{|x|^n}\varphi(x)\,dx \right| \\
&\leq \big\|\nabla\varphi\big\|_{L^\infty} \int_{|x| \leq 1} \frac{|\Omega(x/|x|)|}{|x|^{n-1}}\,dx + \sup_{y \in \mathbf{R}^n} |y|\,|\varphi(y)| \int_{|x| \geq 1} \frac{|\Omega(x/|x|)|}{|x|^{n+1}}\,dx \\
&\leq C_1 \big\|\nabla\varphi\big\|_{L^\infty} \big\|\Omega\big\|_{L^1} + C_2 \sum_{|\alpha| \leq 1} \big\|\varphi(x)x^\alpha\big\|_{L^\infty} \big\|\Omega\big\|_{L^1},
\end{aligned}$$

for suitable C_1 and C_2, where we used (2.2.2) in the last estimate. Note that the distribution W_Ω coincides with the function K_Ω on $\mathbf{R}^n \setminus \{0\}$.

The Hilbert transform and the Riesz transforms are examples of these general operators T_Ω. For instance, the function $\Omega(\theta) = \frac{\theta}{\pi|\theta|} = \frac{1}{\pi}\operatorname{sgn}\theta$ defined on the unit sphere $\mathbf{S}^0 = \{-1, 1\} \subseteq \mathbf{R}$ gives rise to the Hilbert transform, while the function

$$\Omega(\theta) = \frac{\Gamma(\frac{n+1}{2})}{\pi^{\frac{n+1}{2}}} \frac{\theta_j}{|\theta|}$$

defined on $\mathbf{S}^{n-1} \subseteq \mathbf{R}^n$ gives rise to the jth Riesz transform.

Definition 4.2.1. Let Ω be integrable on the sphere $\mathbf{S}^{n-1}$ with mean value zero. For $0 < \varepsilon < N$ and $f \in \bigcup_{1 \le p < \infty} L^p(\mathbf{R}^n)$ we define the *truncated singular integral*

$$T_\Omega^{(\varepsilon,N)}(f)(x) = \int_{\varepsilon \le |y| \le N} f(x-y)\,\frac{\Omega(y/|y|)}{|y|^n}\,dy. \tag{4.2.3}$$

Note that for $f \in L^p(\mathbf{R}^n)$ we have

$$\big\|T_\Omega^{(\varepsilon,N)}(f)\big\|_{L^p} \le \big\|\Omega\big\|_{L^1} \log(N/\varepsilon)\,\big\|f\big\|_{L^p(\mathbf{R}^n)},$$

which implies that (4.2.3) is finite a.e. and therefore well defined. We denote by T_Ω the singular integral operator whose kernel is the distribution W_Ω, that is,

$$T_\Omega(f)(x) = (f * W_\Omega)(x) = \lim_{\substack{\varepsilon \to 0 \\ N \to \infty}} T_\Omega^{(\varepsilon,N)}(f)(x),$$

defined for $f \in \mathscr{S}(\mathbf{R}^n)$. The associated *maximal singular integral* is defined by

$$T_\Omega^{(**)}(f) = \sup_{0<N<\infty}\sup_{0<\varepsilon<N} \big|T_\Omega^{(\varepsilon,N)}(f)\big|. \tag{4.2.4}$$

We note that if Ω is bounded, there is no need to use the upper truncations in the definition of $T_\Omega^{(\varepsilon,N)}$ given in (4.2.3). In this case the maximal singular integrals could be defined as

$$T_\Omega^{(*)}(f) = \sup_{\varepsilon>0} \big|T_\Omega^{(\varepsilon)}(f)\big|, \tag{4.2.5}$$

where for $f \in \bigcup_{1 \le p < \infty} L^p(\mathbf{R})$, $\varepsilon > 0$, and $x \in \mathbf{R}^n$, $T_\Omega^{(\varepsilon)}(f)(x)$ is defined in terms of the absolutely convergent integral

$$T_\Omega^{(\varepsilon)}(f)(x) = \int_{|y| \ge \varepsilon} f(x-y)\frac{\Omega(y/|y|)}{|y|^n}\,dy.$$

To examine the relationship between $T_\Omega^{(*)}$ and $T_\Omega^{(**)}$ for $\Omega \in L^\infty(\mathbf{S}^{n-1})$, notice that

$$\left|\int_{\varepsilon \le |y| \le N} f(x-y)\frac{\Omega(y/|y|)}{|y|^n}\,dy\right| \le \sup_{0<N<\infty} \big|T_\Omega^{(\varepsilon,N)}(f)(x)\big|. \tag{4.2.6}$$

Then for $f \in L^p(\mathbf{R}^n)$, $1 \le p < \infty$, we let $N \to \infty$ on the left in (4.2.6) and we note that the limit exists in view of the absolute convergence of the integral. Then we take the supremum over $\varepsilon > 0$ to deduce that $T_\Omega^{(*)}$ is pointwise bounded by $T_\Omega^{(**)}$. Since $T_\Omega^{(\varepsilon,N)} = T_\Omega^{(\varepsilon)} - T_\Omega^{(N)}$, it also follows that $T_\Omega^{(**)} \le 2T_\Omega^{(*)}$; thus $T_\Omega^{(*)}$ and $T_\Omega^{(**)}$ are pointwise comparable when Ω lies in $L^\infty(\mathbf{S}^{n-1})$. This is the case with the Hilbert transform, that is, $H^{(**)}$ is comparable to $H^{(*)}$; likewise with the Riesz transforms.

A certain class of multipliers can be realized as singular integral operators of the kind discussed. Recall from Proposition 2.4.7 that if m is homogeneous of degree 0 and infinitely differentiable on the sphere, then $m^\vee$ is given by

$$m^\vee = c\,\delta_0 + W_\Omega,$$

for some complex constant c and some smooth Ω on $\mathbf{S}^{n-1}$ with mean value zero. Therefore, all convolution operators whose multipliers are homogeneous of degree zero smooth functions on $\mathbf{S}^{n-1}$ can be realized as a constant multiple of the identity plus an operator of the form T_Ω.

Example 4.2.2. Let $P(\xi) = \sum_{|\alpha|=k} b_\alpha \xi^\alpha$ be a homogeneous polynomial of degree k in $\mathbf{R}^n$ that vanishes only at the origin. Let α be a multi-index of order k. Then the function

$$m(\xi) = \frac{\xi^\alpha}{P(\xi)} \tag{4.2.7}$$

is infinitely differentiable on the sphere and homogeneous of degree zero. The operator given by multiplication on the Fourier transform by $m(\xi)$ is a constant multiple of the identity plus an operator given by convolution with a distribution of the form W_Ω for some Ω in $\mathscr{C}^\infty(\mathbf{S}^{n-1})$ with mean value zero. In this section we establish the L^p boundedness of such operators when Ω has appropriate smoothness on the sphere. This, in particular, implies that $m(\xi)$ defined by (4.2.7) lies in the space $\mathscr{M}_p(\mathbf{R}^n)$, defined in Section 2.5, for $1 < p < \infty$.

4.2.2 L^2 Boundedness of Homogeneous Singular Integrals

Next we would like to compute the Fourier transform of W_Ω. This provides information as to whether the operator given by convolution with K_Ω is L^2 bounded. We have the following result.

Proposition 4.2.3. *Let $n \geq 2$ and $\Omega \in L^1(\mathbf{S}^{n-1})$ have mean value zero. Then the Fourier transform of W_Ω is a (finite a.e.) function given by the formula*

$$\widehat{W_\Omega}(\xi) = \int_{\mathbf{S}^{n-1}} \Omega(\theta)\left(\log\frac{1}{|\xi\cdot\theta|} - \frac{i\pi}{2}\,\mathrm{sgn}\,(\xi\cdot\theta)\right)d\theta\,. \tag{4.2.8}$$

Remark 4.2.4. We need to show that the function of ξ on the right in (4.2.8) is well defined and finite for almost all ξ in $\mathbf{R}^n$. Write $\xi = |\xi|\xi'$ where $\xi' \in \mathbf{S}^{n-1}$ and decompose $\log\frac{1}{|\xi\cdot\theta|}$ as $\log\frac{1}{|\xi|} + \log\frac{1}{|\xi'\cdot\theta|}$. Since Ω has mean value zero, the term $\log\frac{1}{|\xi|}$ multiplied by $\Omega(\theta)$ vanishes when integrated over the sphere.

We need to show that

$$\int_{\mathbf{S}^{n-1}} |\Omega(\theta)|\log\frac{1}{|\xi'\cdot\theta|}\,d\theta < \infty \tag{4.2.9}$$

for almost all $\xi' \in \mathbf{S}^{n-1}$. Integrate (4.2.9) over $\xi' \in \mathbf{S}^{n-1}$ and apply Fubini's theorem to obtain

$$\begin{aligned}
\int_{\mathbf{S}^{n-1}} |\Omega(\theta)| \int_{\mathbf{S}^{n-1}} \log \frac{1}{|\xi' \cdot \theta|} \, d\xi' \, d\theta
&= \int_{\mathbf{S}^{n-1}} |\Omega(\theta)| \int_{\mathbf{S}^{n-1}} \log \frac{1}{|\xi_1|} \, d\xi \, d\theta \\
&= \omega_{n-2} \int_{\mathbf{S}^{n-1}} |\Omega(\theta)| \int_{-1}^{+1} \left(\log \frac{1}{|s|} \right) (1-s^2)^{\frac{n-3}{2}} \, ds \, d\theta \\
&= C_n \|\Omega\|_{L^1(\mathbf{S}^{n-1})} < \infty ,
\end{aligned}$$

since we are assuming that $n \geq 2$. (The second-to-last identity follows from the identity in Appendix D.2.) We conclude that (4.2.9) holds for almost all $\xi' \in \mathbf{S}^{n-1}$.

Since the function of ξ on the right in (4.2.8) is homogeneous of degree zero, it follows that it is a locally integrable function on $\mathbf{R}^n$.

Before we return to the proof of Proposition 4.2.3, we discuss the following lemma:

Lemma 4.2.5. *Let a be a nonzero real number. Then for $0 < \varepsilon < N < \infty$ we have*

$$\lim_{\substack{\varepsilon \to 0 \\ N \to \infty}} \int_{\varepsilon}^{N} \frac{\cos(ra) - \cos(r)}{r} \, dr = \log \frac{1}{|a|} , \tag{4.2.10}$$

$$\left| \int_{\varepsilon}^{N} \frac{\cos(ra) - \cos(r)}{r} \, dr \right| \leq 2 \left| \log \frac{1}{|a|} \right| \qquad \textit{for all } N > \varepsilon > 0, \tag{4.2.11}$$

$$\lim_{\substack{\varepsilon \to 0 \\ N \to \infty}} \int_{\varepsilon}^{N} \frac{e^{-ira} - \cos(r)}{r} \, dr = \log \frac{1}{|a|} - i \frac{\pi}{2} \operatorname{sgn} a , \tag{4.2.12}$$

$$\left| \int_{\varepsilon}^{N} \frac{e^{-ira} - \cos(r)}{r} \, dr \right| \leq 2 \left| \log \frac{1}{|a|} \right| + 4 \qquad \textit{for all } N > \varepsilon > 0. \tag{4.2.13}$$

Proof. We first prove (4.2.10) and (4.2.11). By the fundamental theorem of calculus we can write

$$\begin{aligned}
\int_{\varepsilon}^{N} \frac{\cos(ra) - \cos(r)}{r} \, dr &= \int_{\varepsilon}^{N} \frac{\cos(r|a|) - \cos(r)}{r} \, dr \\
&= - \int_{\varepsilon}^{N} \int_{1}^{|a|} \sin(tr) \, dt \, dr \\
&= - \int_{1}^{|a|} \int_{\varepsilon}^{N} \sin(tr) \, dr \, dt \\
&= - \int_{1}^{|a|} \frac{\cos(\varepsilon t)}{t} \, dt + \int_{N}^{N|a|} \frac{\cos(t)}{t} \, dt ,
\end{aligned}$$

and from this expression, we clearly obtain (4.2.11). But the first integral of the same expression converges to $-\log|a|$ as $\varepsilon \to 0$ while the second integral converges to zero as $N \to \infty$ by an integration by parts. This proves (4.2.10).

To prove (4.2.12) and (4.2.13) we need to know that the expressions

$$\left| \int_{\varepsilon}^{N} \frac{\sin(ra)}{r}\, dr \right| = \left| \int_{\varepsilon|a|}^{N|a|} \frac{\sin(r)}{r}\, dr \right| \tag{4.2.14}$$

tend to $\frac{\pi}{2}$ as $\varepsilon \to 0$ and $N \to \infty$ and are bounded by 4. Both statements follow from Exercise 4.1.1. □

Let us now prove Proposition 4.2.3.

Proof. Let us set $\xi' = \xi/|\xi|$. We have the following:

$$\begin{aligned}
\langle \widehat{W_\Omega}, \varphi \rangle &= \langle W_\Omega, \widehat{\varphi} \rangle \\
&= \lim_{\varepsilon \to 0} \int_{|x| \geq \varepsilon} \frac{\Omega(x/|x|)}{|x|^n} \widehat{\varphi}(x)\, dx \\
&= \lim_{\substack{\varepsilon \to 0 \\ N \to \infty}} \int_{\mathbf{R}^n} \varphi(\xi) \int_{\varepsilon \leq |x| \leq N} \frac{\Omega(x/|x|)}{|x|^n} e^{-2\pi i x \cdot \xi}\, dx\, d\xi \\
&= \lim_{\substack{\varepsilon \to 0 \\ N \to \infty}} \int_{\mathbf{R}^n} \varphi(\xi) \int_{\mathbf{S}^{n-1}} \Omega(\theta) \int_{\varepsilon \leq r \leq N} e^{-2\pi i r \theta \cdot \xi} \frac{dr}{r}\, d\theta\, d\xi \\
&= \lim_{\substack{\varepsilon \to 0 \\ N \to \infty}} \int_{\mathbf{R}^n} \varphi(\xi) \int_{\mathbf{S}^{n-1}} \Omega(\theta) \int_{\varepsilon \leq r \leq N} \left(e^{-2\pi r |\xi| i \theta \cdot \xi'} - \cos(2\pi r |\xi|) \right) \frac{dr}{r}\, d\theta\, d\xi \\
&= \lim_{\substack{\varepsilon \to 0 \\ N \to \infty}} \int_{\mathbf{R}^n} \varphi(\xi) \int_{\mathbf{S}^{n-1}} \Omega(\theta) \int_{\frac{\varepsilon}{2\pi|\xi|} \leq r \leq \frac{N}{2\pi|\xi|}} \frac{e^{-ir\theta \cdot \xi'} - \cos(r)}{r}\, dr\, d\theta\, d\xi \\
&= \int_{\mathbf{R}^n} \varphi(\xi) \int_{\mathbf{S}^{n-1}} \Omega(\theta) \left(\log \frac{1}{|\xi' \cdot \theta|} - \frac{i\pi}{2} \operatorname{sgn}(\xi \cdot \theta) \right) d\theta\, d\xi,
\end{aligned}$$

by the Lebesgue dominated convergence theorem, Lemma 4.2.5, and Remark 4.2.4. We were able to subtract $\cos(2\pi r|\xi|)$ from the r integral in the previous calculation, since Ω has mean value zero over the sphere. Also, the use of the dominated convergence theorem is justified from the fact that the function

$$(\theta, \xi) \mapsto |\Omega(\theta)|\, |\varphi(\xi)| \left(\log \frac{1}{|\xi' \cdot \theta|} + 4 \right)$$

lies in $L^1(\mathbf{S}^{n-1} \times \mathbf{R}^n)$. □

Corollary 4.2.6. *Let $\Omega \in L^1(\mathbf{S}^{n-1})$ have mean value zero. Then for almost all ξ' in $\mathbf{S}^{n-1}$ the integral*

$$\int_{\mathbf{S}^{n-1}} \Omega(\theta) \log \frac{1}{|\xi' \cdot \theta|}\, d\theta \tag{4.2.15}$$

converges absolutely. Moreover, the associated operator T_Ω maps $L^2(\mathbf{R}^n)$ to itself if and only if

$$\operatorname*{ess.sup}_{\xi' \in \mathbf{S}^{n-1}} \left| \int_{\mathbf{S}^{n-1}} \Omega(\theta) \log \frac{1}{|\xi' \cdot \theta|} d\theta \right| < \infty. \tag{4.2.16}$$

Proof. To obtain the absolute convergence of the integral in (4.2.15) we integrate over $\xi' \in \mathbf{S}^{n-1}$ and we apply Fubini's theorem. The assertion concerning the boundedness of T_Ω on L^2 is an immediate consequence of Proposition 4.2.3 and Theorem 2.5.10. □

There exist functions Ω in $L^1(\mathbf{S}^{n-1})$ with mean value zero such that the expressions in (4.2.16) are equal to infinity; consequently, not all such Ω give rise to bounded operators on $L^2(\mathbf{R}^n)$. Observe, however, that for Ω odd (i.e., $\Omega(-\theta) = -\Omega(\theta)$ for all $\theta \in \mathbf{S}^{n-1}$), (4.2.16) trivially holds, since $\log \frac{1}{|\xi \cdot \theta|}$ is even and its product against an odd function must have integral zero over $\mathbf{S}^{n-1}$. We conclude that singular integrals T_Ω with odd Ω are always L^2 bounded.

4.2.3 The Method of Rotations

Having settled the issue of L^2 boundedness for singular integrals of the form T_Ω with Ω odd, we turn our attention to their L^p boundedness. A simple procedure called the method of rotations plays a crucial role in the study of operators T_Ω when Ω is an odd function.

Theorem 4.2.7. *If Ω is odd and integrable over $\mathbf{S}^{n-1}$, then T_Ω and $T_\Omega^{(*)}$ are L^p bounded for all $1 < p < \infty$. More precisely, T_Ω initially defined on Schwartz functions has a bounded extension on $L^p(\mathbf{R}^n)$ (which is also denoted by T_Ω).*

Proof. We introduce the directional Hilbert transforms. Fix a unit vector θ in $\mathbf{R}^n$. For a Schwartz function f on $\mathbf{R}^n$ let

$$\mathscr{H}_\theta(f)(x) = \frac{1}{\pi} \text{p.v.} \int_{-\infty}^{+\infty} f(x - t\theta) \frac{dt}{t}.$$

We call $\mathscr{H}_\theta(f)$ the *directional Hilbert transform* of f in the direction θ. Let e_j be the usual unit vectors in $\mathbf{S}^{n-1}$. Then $\mathscr{H}_{e_1}$ is simply obtained by applying the Hilbert transform in the first variable followed by the identity operator in the remaining variables. Clearly, $\mathscr{H}_{e_1}$ is bounded on $L^p(\mathbf{R}^n)$ with norm equal to that of the Hilbert transform on $L^p(\mathbf{R})$. Next observe that the following identity is valid for all matrices $A \in O(n)$:

$$\mathscr{H}_{A(e_1)}(f)(x) = \mathscr{H}_{e_1}(f \circ A)(A^{-1}x). \tag{4.2.17}$$

This implies that the L^p boundedness of $\mathscr{H}_\theta$ can be reduced to that of $\mathscr{H}_{e_1}$. We conclude that $\mathscr{H}_\theta$ is L^p bounded for $1 < p < \infty$ with norm bounded by the norm of the Hilbert transform on $L^p(\mathbf{R})$ for every $\theta \in \mathbf{S}^{n-1}$.

Likewise, we define the *directional maximal Hilbert transforms*. For a function f in $\bigcup_{1 \le p < \infty} L^p(\mathbf{R}^n)$ and $0 < \varepsilon < N < \infty$ we let

$$\mathscr{H}_\theta^{(\varepsilon,N)}(f)(x) = \frac{1}{\pi} \int_{\varepsilon \le |t| \le N} f(x - t\theta) \frac{dt}{t},$$

$$\mathscr{H}_\theta^{(**)}(f)(x) = \sup_{0<\varepsilon<N<\infty} \left| \mathscr{H}_\theta^{(\varepsilon,N)}(f)(x) \right|.$$

We observe that for any fixed $0 < \varepsilon < N < \infty$ and $f \in L^p(\mathbf{R}^n)$, $\mathscr{H}_\theta^{(\varepsilon,N)}(f)$ is well defined almost everywhere. Indeed, by Minkowski's integral inequality we obtain

$$\left\| \mathscr{H}_\theta^{(\varepsilon,N)}(f) \right\|_{L^p(\mathbf{R}^n)} \le \frac{2}{\pi} \|f\|_{L^p(\mathbf{R}^n)} \log \frac{N}{\varepsilon} < \infty,$$

which implies that $\mathscr{H}_\theta^{(\varepsilon,N)}(f)(x)$ is finite for almost all $x \in \mathbf{R}^n$. Thus $\mathscr{H}_\theta^{(**)}(f)$ is well defined a.e. for f in $\bigcup_{1\le p<\infty} L^p(\mathbf{R}^n)$.

Identity (4.2.17) is also valid for $\mathscr{H}_\theta^{(\varepsilon,N)}$ and $\mathscr{H}_\theta^{(**)}$. Consequently, $\mathscr{H}_\theta^{(**)}$ is bounded on $L^p(\mathbf{R}^n)$ for $1 < p < \infty$ with norm at most that of $H^{(**)}$ on $L^p(\mathbf{R})$.

Next we realize a general singular integral T_Ω with Ω odd as an average of the directional Hilbert transforms $\mathscr{H}_\theta$. We start with f in $\bigcup_{1\le p<\infty} L^p(\mathbf{R}^n)$ and the following identities:

$$\begin{aligned}
\int_{\varepsilon\le|y|\le N} \frac{\Omega(y/|y|)}{|y|^n} f(x-y)\,dy &= +\int_{\mathbf{S}^{n-1}} \Omega(\theta) \int_{r=\varepsilon}^{N} f(x-r\theta)\frac{dr}{r}\,d\theta \\
&= -\int_{\mathbf{S}^{n-1}} \Omega(\theta) \int_{r=\varepsilon}^{N} f(x+r\theta)\frac{dr}{r}\,d\theta,
\end{aligned}$$

where the first follows by switching to polar coordinates and the second one is a consequence of the first one and the fact that Ω is odd via the change variables $\theta \mapsto -\theta$. Averaging the two identities, we obtain

$$\begin{aligned}
&\int_{\varepsilon\le|y|\le N} \frac{\Omega(y/|y|)}{|y|^n} f(x-y)\,dy \\
&\qquad = \frac{1}{2}\int_{\mathbf{S}^{n-1}} \Omega(\theta) \int_{r=\varepsilon}^{N} \frac{f(x-r\theta)-f(x+r\theta)}{r}\,dr\,d\theta \qquad (4.2.18)\\
&\qquad = \frac{\pi}{2}\int_{\mathbf{S}^{n-1}} \Omega(\theta)\,\mathscr{H}_\theta^{(\varepsilon,N)}(f)(x)\,d\theta.
\end{aligned}$$

It follows from the identity in (4.2.18) that

$$\int_{\varepsilon\le|y|\le N} \frac{\Omega(y/|y|)}{|y|^n} f(x-y)\,dy = \frac{\pi}{2}\int_{\mathbf{S}^{n-1}} \Omega(\theta)\,\mathscr{H}_\theta^{(\varepsilon,N)}(f)(x)\,d\theta, \qquad (4.2.19)$$

from which we conclude that

$$T_\Omega^{(**)}(f)(x) \le \frac{\pi}{2}\int_{\mathbf{S}^{n-1}} |\Omega(\theta)|\,\mathscr{H}_\theta^{(**)}(f)(x)\,d\theta. \qquad (4.2.20)$$

Using the Lebesgue dominated convergence theorem, we see that for f in $\mathscr{S}(\mathbf{R}^n)$, we can pass the limits as $\varepsilon \to 0$ and $N \to \infty$ inside the integral in (4.2.19), concluding that

$$T_\Omega(f)(x) = \frac{\pi}{2} \int_{\mathbf{S}^{n-1}} \Omega(\theta)\, \mathscr{H}_\theta(f)(x)\, d\theta\,, \tag{4.2.21}$$

for $f \in \mathscr{S}(\mathbf{R}^n)$. The L^p boundedness of T_Ω and $T_\Omega^{(**)}$ for Ω odd are then trivial consequences of (4.2.21) and (4.2.20) via Minkowski's integral inequality. □

Corollary 4.2.8. *The Riesz transforms R_j and the maximal Riesz transforms $R_j^{(*)}$ are bounded on $L^p(\mathbf{R}^n)$ for $1 < p < \infty$.*

Proof. The Riesz transforms have odd kernels. □

Remark 4.2.9. It follows from the proof of Theorem 4.2.7 and from Theorems 4.1.7 and 4.1.12 that whenever Ω is an odd function on $\mathbf{S}^{n-1}$, we have

$$\big\|T_\Omega\big\|_{L^p\to L^p} \le \big\|\Omega\big\|_{L^1} \begin{cases} a\,p & \text{when } p \ge 2, \\ a\,(p-1)^{-1} & \text{when } 1 < p \le 2, \end{cases}$$

$$\big\|T_\Omega^{(**)}\big\|_{L^p\to L^p} \le \big\|\Omega\big\|_{L^1} \begin{cases} a\,p & \text{when } p \ge 2, \\ a\,(p-1)^{-1} & \text{when } 1 < p \le 2, \end{cases}$$

for some $a > 0$ independent of p and the dimension.

4.2.4 Singular Integrals with Even Kernels

Since a general integrable function Ω on $\mathbf{S}^{n-1}$ with mean value zero can be written as a sum of an odd and an even function, it suffices to study singular integral operators T_Ω with even kernels. For the rest of this section, fix an integrable even function Ω on $\mathbf{S}^{n-1}$ with mean value zero. The following idea is fundamental in the study of such singular integrals. Proposition 4.1.16 implies that

$$T_\Omega = -\sum_{j=1}^{n} R_j R_j T_\Omega\,. \tag{4.2.22}$$

If $R_j T_\Omega$ were another singular integral operator of the form T_{Ω_j} for some odd Ω_j, then the boundedness of T_Ω would follow from that of T_{Ω_j} via the identity (4.2.22) and Theorem 4.2.7. It turns out that $R_j T_\Omega$ does have an odd kernel, but it may not be integrable on $\mathbf{S}^{n-1}$ unless Ω itself possesses an additional amount of integrability. The amount of extra integrability needed is logarithmic, more precisely of this sort:

$$c_\Omega = \int_{\mathbf{S}^{n-1}} |\Omega(\theta)| \log^+ |\Omega(\theta)|\, d\theta < \infty\,. \tag{4.2.23}$$

Observe that

$$\|\Omega\|_{L^1} \le c_\Omega + 2\omega_{n-1} \le C_n (c_\Omega + 1),$$

which says that the norm $\|\Omega\|_{L^1}$ is always controlled by a dimensional constant multiple of $c_\Omega + 1$. The following theorem is the main result of this section.

Theorem 4.2.10. *Let $n \ge 2$ and let Ω be an even integrable function on $\mathbf{S}^{n-1}$ with mean value zero that satisfies (4.2.23). Then the corresponding singular integral T_Ω is bounded on $L^p(\mathbf{R}^n)$, $1 < p < \infty$, with norm at most a dimensional constant multiple of the quantity* $\max\big((p-1)^{-2}, p^2\big)(c_\Omega + 1)$.

If the operator T_Ω in Theorem 4.2.10 is weak type $(1,1)$, then the estimate on the L^p operator norm of T_Ω can be improved to $\|T_\Omega\|_{L^p \to L^p} \le C_n (p-1)^{-1}$ as $p \to 1$. This is indeed the case; see the historical comments at the end of this chapter.

Proof. Let W_Ω be the distributional kernel of T_Ω. Using Proposition 4.2.3 and the fact that Ω is an even function, we obtain the formula

$$\widehat{W_\Omega}(\xi) = \int_{\mathbf{S}^{n-1}} \Omega(\theta) \log \frac{1}{|\xi \cdot \theta|}\, d\theta, \tag{4.2.24}$$

which implies that $\widehat{W_\Omega}$ is itself an even function. Now, using Exercise 4.2.3 and condition (4.2.23), we conclude that $\widehat{W_\Omega}$ is a bounded function. Therefore, T_Ω is L^2 bounded. To obtain the L^p boundedness of T_Ω, we use the idea mentioned earlier involving the Riesz transforms. In view of (4.1.41), we have that

$$T_\Omega = -\sum_{j=1}^{n} R_j T_j, \tag{4.2.25}$$

where $T_j = R_j T_\Omega$. Equality (4.2.25) makes sense as an operator identity on $L^2(\mathbf{R}^n)$, since T_Ω and each R_j are well defined and bounded on $L^2(\mathbf{R}^n)$.

The kernel of the operator T_j is the inverse Fourier transform of the distribution $-i\frac{\xi_j}{|\xi|}\widehat{W_\Omega}(\xi)$, which we denote by K_j. At this point we know only that K_j is a tempered distribution whose Fourier transform is the function $-i\frac{\xi_j}{|\xi|}\widehat{W_\Omega}(\xi)$. Our first goal is to show that K_j coincides with an integrable function on an annulus. To prove this assertion we write

$$W_\Omega = W_\Omega^0 + W_\Omega^1 + W_\Omega^\infty,$$

where W_Ω^0 is a distribution and $W_\Omega^1, W_\Omega^\infty$ are functions defined by

$$\begin{aligned}
\langle W_\Omega^0, \varphi \rangle &= \lim_{\varepsilon \to 0} \int_{\varepsilon < |x| \le \frac{1}{2}} \frac{\Omega(x/|x|)}{|x|^n} \varphi(x)\, dx, \\
W_\Omega^1(x) &= \frac{\Omega(x/|x|)}{|x|^n} \chi_{\frac{1}{2} \le |x| \le 2}, \\
W_\Omega^\infty(x) &= \frac{\Omega(x/|x|)}{|x|^n} \chi_{2 < |x|}.
\end{aligned}$$

We now fix a $j \in \{1,2,\dots,n\}$ and we write

$$K_j = K_j^0 + K_j^1 + K_j^\infty ,$$

where

$$\begin{aligned} K_j^0 &= \big(-i\tfrac{\xi_j}{|\xi|}\widehat{W_\Omega^0}(\xi)\big)^\vee ,\\ K_j^1 &= \big(-i\tfrac{\xi_j}{|\xi|}\widehat{W_\Omega^1}(\xi)\big)^\vee ,\\ K_j^\infty &= \big(-i\tfrac{\xi_j}{|\xi|}\widehat{W_\Omega^\infty}(\xi)\big)^\vee . \end{aligned}$$

Define the annulus

$$A = \{x \in \mathbf{R}^n :\ 2/3 < |x| < 3/2\}.$$

For $x \in A$, the convolution of W_Ω^0 with the kernel of the Riesz transform R_j can be written as the convergent integral inside the absolute value:

$$\begin{aligned} &\left|\frac{\Gamma(\frac{n+1}{2})}{\pi^{\frac{n+1}{2}}} \lim_{\varepsilon\to 0}\int_{\varepsilon<|y|<\frac12} \frac{x_j-y_j}{|x-y|^{n+1}}\,\frac{\Omega(y/|y|)}{|y|^n}\,dy\right| \qquad (4.2.26)\\ &\quad= \frac{\Gamma(\frac{n+1}{2})}{\pi^{\frac{n+1}{2}}}\left|\int_{|y|<\frac12}\left(\frac{x_j-y_j}{|x-y|^{n+1}}-\frac{x_j}{|x|^{n+1}}\right)\frac{\Omega(y/|y|)}{|y|^n}\,dy\right|\\ &\quad\le \int_{|y|\le\frac12} C_n|y|\,\frac{|\Omega(y/|y|)|}{|y|^n}\,dy\\ &\quad= C_n'\big\|\Omega\big\|_{L^1}, \end{aligned}$$

where we used the fact that $\Omega(y/|y|)|y|^{-n}$ has integral zero over annuli of the form $\varepsilon < |y| < \frac12$, the mean value theorem applied to the function $x_j|x|^{-(n+1)}$, and the fact that $|x-y| \ge 1/6$ for x in the annulus A. We conclude that on A, K_j^0 coincides with the bounded function inside the absolute value in (4.2.26).

Likewise, for $x \in A$ we have

$$\begin{aligned} &\frac{\Gamma(\frac{n+1}{2})}{\pi^{\frac{n+1}{2}}}\left|\int_{|y|>2} \frac{x_j-y_j}{|x-y|^{n+1}}\,\frac{\Omega(y/|y|)}{|y|^n}\,dy\right| \qquad (4.2.27)\\ &\quad\le \frac{\Gamma(\frac{n+1}{2})}{\pi^{\frac{n+1}{2}}}\int_{|y|>2}\frac{1}{|x-y|^n}\,\frac{|\Omega(y/|y|)|}{|y|^n}\,dy\\ &\quad\le \frac{\Gamma(\frac{n+1}{2})}{\pi^{\frac{n+1}{2}}}\int_{|y|>2}\frac{4^n}{|y|^{2n}}|\Omega(y/|y|)|\,dy\\ &\quad= C\big\|\Omega\big\|_{L^1}, \end{aligned}$$

from which it follows that on the annulus A, K_j^∞ coincides with the bounded function inside the absolute value in (4.2.27).

Now observe that condition (4.2.23) gives that the function W_Ω^1 satisfies

$$\begin{aligned}\int_{|x|\le 2} |W_\Omega^1(x)| \log^+ |W_\Omega^1(x)|\,dx &\\ \le \int_{1/2}^{2}\int_{\mathbf{S}^{n-1}} \frac{|\Omega(\theta)|}{r^n} \log^+[2^n|\Omega(\theta)|]\,d\theta\, r^{n-1}\frac{dr}{r} &\\ \le (\log 4)\left[n(\log 2)\|\Omega\|_{L^1} + c_\Omega\right] < \infty . &\end{aligned}$$

Since the Riesz transform R_j maps L^p to L^p with norm at most $4(p-1)^{-1}$ for $1<p<2$, it follows from Exercise 1.3.7 that $K_j^1 = R_j(W_\Omega^1)$ is integrable over the ball $|x|\le 3/2$ and moreover, it satisfies

$$\begin{aligned}\int_A |K_j^1(x)|\,dx &\le C_n\left[\int_{|x|\le 2} |W_\Omega^1(x)| \log^+ |W_\Omega^1(x)|\,dx + 1\right]\\ &\le C_n'(c_\Omega + 1).\end{aligned}$$

We have proved that K_j is a distribution that coincides with an integrable function on the annulus A. Furthermore, since $\widehat{K_j}$ is homogeneous of degree zero, we have that K_j is a homogeneous distribution of degree $-n$ (Exercise 2.3.9). This means that for all test functions φ and all $\lambda>0$ we have

$$\left\langle K_j, \delta^\lambda(\varphi)\right\rangle = \left\langle K_j, \varphi\right\rangle .$$

But then for φ supported in the annulus $3/4<|x|<4/3$ and for λ in $(8/9, 9/8)$ we have that $\delta^\lambda(\varphi)$ is supported in A and thus

$$\int K_j(x)\varphi(\lambda x)\,dx = \left\langle K_j, \delta^\lambda(\varphi)\right\rangle = \left\langle K_j, \varphi\right\rangle = \int \lambda^{-n} K_j(\lambda^{-1}x)\varphi(x)\,dx .$$

From this we conclude that $K_j(x) = \lambda^{-n}K_j(\lambda^{-1}x)$ for $3/4<|x|<4/3$ and $8/9<\lambda<9/8$. Thus for $8/9<|x|<9/8$ we have

$$K_j(x) = |x|^{-n}K_j(x/|x|) = |x|^{-n}\Omega_j(x/|x|), \tag{4.2.28}$$

where we defined Ω_j to be the restriction of K_j over $\mathbf{S}^{n-1}$. The integrability of K_j over the annulus $8/9<|x|<9/8$ implies the integrability (and hence finiteness a.e.) of Ω_j over $\mathbf{S}^{n-1}$ via (4.2.28).

Pick a nonnegative, radial, smooth, and nonzero function ψ on $\mathbf{R}^n$ supported in $8/9<|x|<9/8$. Let $e_1 = (1,0,\dots,0)$. Switching to polar coordinates, we obtain

$$\begin{aligned}\left\langle K_j, \psi\right\rangle &= \int_{\mathbf{R}^n} \frac{\Omega_j(x/|x|)}{|x|^n}\psi(x)\,dx = \left(\int_{8/9}^{9/8} \psi(re_1)\frac{dr}{r}\right)\int_{\mathbf{S}^{n-1}} \Omega_j(\theta)\,d\theta,\\ \left\langle K_j, \psi\right\rangle &= \left\langle \widehat{K_j}, \widehat{\psi}\right\rangle = \int_{\mathbf{R}^n} \frac{-i\xi_j}{|\xi|}\widehat{W_\Omega}(\xi)\widehat{\psi}(\xi)\,d\xi = c_\psi'\int_{\mathbf{S}^{n-1}} \frac{-i\theta_j}{|\theta|}\widehat{W_\Omega}(\theta)\,d\theta = 0,\end{aligned}$$

since by (4.2.24), $\frac{-i\xi_j}{|\xi|}\widehat{W_\Omega}(\xi)$ is an odd function. We conclude that Ω_j has mean value zero over $\mathbf{S}^{n-1}$. We are now in a position to define the distribution W_{Ω_j}. We claim that

$$K_j = W_{\Omega_j}. \tag{4.2.29}$$

To establish the claim, we use (4.2.28) to obtain that the homogeneous distributions K_j and W_{Ω_j} agree on the open set $8/9 < |x| < 9/8$, and thus they must agree everywhere on $\mathbf{R}^n \setminus \{0\}$ (check that $\langle K_j, \varphi\rangle = \langle W_{\Omega_j}, \varphi\rangle$ for all $\varphi \in \mathscr{C}_0^\infty(\mathbf{R}^n \setminus \{0\})$ by dilating and translating their support). Therefore, $K_j - W_{\Omega_j}$ is supported at the origin, and since it is homogeneous of degree $-n$, it must be equal to $b\delta_0$, a constant multiple of the Dirac mass. But $\widehat{K_j}$ is an odd function and hence K_j is also odd. It follows that W_{Ω_j} is an odd function on $\mathbf{R}^n \setminus \{0\}$, which implies that Ω_j is an odd function. Defining odd distributions in the natural way, we obtain that $K_j - W_{\Omega_j}$ is an odd distribution, and thus the previous multiple of the Dirac mass must be an odd distribution. But if $b\delta_0$ is odd, then $b = 0$. We conclude that for each j there exists an odd integrable function Ω_j on $\mathbf{S}^{n-1}$ with $\|\Omega_j\|_{L^1}$ controlled by a constant multiple of $c_\Omega + 1$ such that (4.2.29) holds.

Then we use (4.2.25) and (4.2.29) to write

$$T_\Omega = -\sum_{j=1}^{n} R_j T_{\Omega_j},$$

and appealing to the boundedness of each T_{Ω_j} (Theorem 4.2.7) and to that of the Riesz transforms, we obtain the required L^p boundedness for T_Ω. □

We note that Theorem 4.2.10 holds for all $\Omega \in L^1(\mathbf{S}^{n-1})$ that satisfy (4.2.23), not necessarily even Ω. Simply write $\Omega = \Omega_e + \Omega_o$, where Ω_e is even and Ω_o is odd, and check that condition (4.2.23) holds for Ω_e.

4.2.5 Maximal Singular Integrals with Even Kernels

We have the corresponding theorem for maximal singular integrals.

Theorem 4.2.11. *Let Ω be an even integrable function on $\mathbf{S}^{n-1}$ with mean value zero that satisfies (4.2.23). Then the corresponding maximal singular integral $T_\Omega^{(**)}$, defined in (4.2.4), is bounded on $L^p(\mathbf{R}^n)$ for $1 < p < \infty$ with norm at most a dimensional constant multiple of $\max(p^2, (p-1)^{-2})(c_\Omega + 1)$.*

Proof. For $f \in L^1_{\mathrm{loc}}(\mathbf{R}^n)$, define the maximal function of f in the direction θ by setting

$$M_\theta(f)(x) = \sup_{a>0} \frac{1}{2a} \int_{|r|\le a} |f(x - r\theta)|\, dr. \tag{4.2.30}$$

In view of Exercise 4.2.6(a) we have that M_θ is bounded on $L^p(\mathbf{R}^n)$ with norm at most $3\,p\,(p-1)^{-1}$.

Fix Φ a smooth radial function such that $\Phi(x) = 0$ for $|x| < 1/4$, $\Phi(x) = 1$ for $|x| > 3/4$, and $0 \le \Phi(x) \le 1$ for all x in $\mathbf{R}^n$. For $f \in L^p(\mathbf{R}^n)$ and $0 < \varepsilon < N < \infty$ we introduce the smoothly truncated singular integral

$$\widetilde{T}_{\Omega}^{(\varepsilon,N)}(f)(x) = \int_{\mathbf{R}^n} \frac{\Omega\left(\frac{x-y}{|x-y|}\right)}{|x-y|^n} \left(\Phi\left(\tfrac{x-y}{\varepsilon}\right) - \Phi\left(\tfrac{x-y}{N}\right)\right) f(y)\, dy$$

and the corresponding maximal singular integral operator

$$\widetilde{T}_{\Omega}^{(**)}(f) = \sup_{0<N<\infty} \sup_{0<\varepsilon<N} |\widetilde{T}_{\Omega}^{(\varepsilon,N)}(f)|\,.$$

For f in $L^p(\mathbf{R}^n)$ (for some $1 < p < \infty$), we have

$$\begin{aligned}
&\sup_{0<\varepsilon<N<\infty} \left|\widetilde{T}_{\Omega}^{(\varepsilon,N)}(f)(x) - T_{\Omega}^{(\varepsilon,N)}(f)(x)\right| \\
&= \sup_{0<\varepsilon<N<\infty} \Bigg| \int_{\frac{\varepsilon}{4}\le|y|\le\varepsilon} \frac{\Omega\left(\frac{y}{|y|}\right)}{|y|^n} \Phi\left(\tfrac{y}{\varepsilon}\right) f(x-y)\, dy \\
&\qquad\qquad - \int_{\frac{N}{4}\le|y|\le N} \frac{\Omega\left(\frac{y}{|y|}\right)}{|y|^n} \Phi\left(\tfrac{y}{N}\right) f(x-y)\, dy \Bigg| \\
&\le \sup_{0<\varepsilon<N<\infty} \left[\int_{\frac{\varepsilon}{4}\le|y|\le\varepsilon} \frac{|\Omega\left(\frac{y}{|y|}\right)|}{|y|^n} |f(x-y)|\, dy + \int_{\frac{N}{4}\le|y|\le N} \frac{|\Omega\left(\frac{y}{|y|}\right)|}{|y|^n} |f(x-y)|\, dy \right] \\
&\le \sup_{0<\varepsilon<N<\infty} \int_{\mathbf{S}^{n-1}} |\Omega(\theta)| \left[\frac{4}{\varepsilon} \int_{\frac{\varepsilon}{4}}^{\varepsilon} |f(x-r\theta)|\, dr + \frac{4}{N} \int_{\frac{N}{4}}^{N} |f(x-r\theta)|\, dr \right] d\theta \\
&\le 16 \int_{\mathbf{S}^{n-1}} |\Omega(\theta)|\, M_\theta(f)(x)\, d\theta\,.
\end{aligned}$$

Using the result of Exercise 4.2.6(a) we conclude that

$$\big\|\widetilde{T}_{\Omega}^{(**)}(f) - T_{\Omega}^{(**)}(f)\big\|_{L^p} \le 96\, \big\|\Omega\big\|_{L^1} \max(p, (p-1)^{-1})\, \big\|f\big\|_{L^p}\,.$$

This implies that it suffices to obtain the required L^p bound for the smoothly truncated maximal singular integral operator $\widetilde{T}_{\Omega}^{(**)}$.

Let K_j, Ω_j, and T_j be as in the previous theorem, and let F_j be the Riesz transform of the function $\Omega(x/|x|)\Phi(x)|x|^{-n}$. Let $f \in L^p(\mathbf{R}^n)$. A calculation yields the identity

$$\begin{aligned}
\widetilde{T}_{\Omega}^{(\varepsilon,N)}(f)(x) &= \int_{\mathbf{R}^n} \left[\frac{1}{\varepsilon^n} \frac{\Omega(\frac{y}{\varepsilon}/|\frac{y}{\varepsilon}|)}{|\frac{y}{\varepsilon}|^n} \Phi(\tfrac{y}{\varepsilon}) - \frac{1}{N^n} \frac{\Omega(\frac{y}{N}/|\frac{y}{N}|)}{|\frac{y}{N}|^n} \Phi(\tfrac{y}{N}) \right] f(x-y)\, dy \\
&= -\Big(\sum_{j=1}^{n} \left[\tfrac{1}{\varepsilon^n} F_j\left(\tfrac{\cdot}{\varepsilon}\right) - \tfrac{1}{N^n} F_j\left(\tfrac{\cdot}{N}\right) \right] * R_j(f) \Big)(x)\,,
\end{aligned}$$

where in the last step we used Proposition 4.1.16. Therefore we may write

$$
\begin{aligned}
-\widetilde{T}_\Omega^{(\varepsilon,N)}(f)(x) &= \sum_{j=1}^{n}\int_{\mathbf{R}^n}\Big[\tfrac{1}{\varepsilon^n}F_j\big(\tfrac{x-y}{\varepsilon}\big)-\tfrac{1}{N^n}F_j\big(\tfrac{x-y}{N}\big)\Big]R_j(f)(y)\,dy \\
&= A_1^{(\varepsilon,N)}(f)(x)+A_2^{(\varepsilon,N)}(f)(x)+A_3^{(\varepsilon,N)}(f)(x)\,,
\end{aligned}
\tag{4.2.31}
$$

where

$$
\begin{aligned}
A_1^{(\varepsilon,N)}(f)(x) &= \sum_{j=1}^{n}\frac{1}{\varepsilon^n}\int_{|x-y|\le\varepsilon}F_j\big(\tfrac{x-y}{\varepsilon}\big)R_j(f)(y)\,dy \\
&\qquad -\sum_{j=1}^{n}\frac{1}{N^n}\int_{|x-y|\le N}F_j\big(\tfrac{x-y}{N}\big)R_j(f)(y)\,dy\,, \\
A_2^{(\varepsilon,N)}(f)(x) &= \sum_{j=1}^{n}\int_{\mathbf{R}^n}\Big[\tfrac{1}{\varepsilon^n}\chi_{|x-y|>\varepsilon}\big\{F_j\big(\tfrac{x-y}{\varepsilon}\big)-K_j\big(\tfrac{x-y}{\varepsilon}\big)\big\} \\
&\qquad -\tfrac{1}{N^n}\chi_{|x-y|>N}\big\{F_j\big(\tfrac{x-y}{N}\big)-K_j\big(\tfrac{x-y}{N}\big)\big\}\Big]R_j(f)(y)\,dy\,, \\
A_3^{(\varepsilon,N)}(f)(x) &= \sum_{j=1}^{n}\int_{\mathbf{R}^n}\Big[\tfrac{1}{\varepsilon^n}\chi_{|x-y|>\varepsilon}K_j\big(\tfrac{x-y}{\varepsilon}\big)-\tfrac{1}{N^n}\chi_{|x-y|>N}K_j\big(\tfrac{x-y}{N}\big)\Big]R_j(f)(y)\,dy\,.
\end{aligned}
$$

It follows from the definitions of F_j and K_j that

$$
\begin{aligned}
F_j(z)-K_j(z) &= \frac{\Gamma(\frac{n+1}{2})}{\pi^{\frac{n+1}{2}}}\lim_{\varepsilon\to0}\int_{\varepsilon\le|y|}\frac{\Omega(y/|y|)}{|y|^n}(\Phi(y)-1)\frac{z_j-y_j}{|z-y|^{n+1}}\,dy \\
&= \frac{\Gamma(\frac{n+1}{2})}{\pi^{\frac{n+1}{2}}}\int_{|y|\le\frac34}\frac{\Omega(y/|y|)}{|y|^n}(\Phi(y)-1)\Big\{\frac{z_j-y_j}{|z-y|^{n+1}}-\frac{z_j}{|z|^{n+1}}\Big\}\,dy
\end{aligned}
$$

whenever $|z|\ge1$. But using the mean value theorem, the last expression is easily seen to be bounded by

$$
C_n\int_{|y|\le\frac34}\frac{\Omega(y/|y|)}{|y|^n}\frac{|y|}{|z|^{n+1}}\,dy=C_n'\big\|\Omega\big\|_{L^1}|z|^{-(n+1)}\,,
$$

whenever $|z|\ge1$. Using this estimate, we obtain that the jth term in $A_2^{(\varepsilon,N)}(f)(x)$ is bounded by

$$
C_n\frac{\|\Omega\|_{L^1}}{\varepsilon^n}\int\limits_{|x-y|>\varepsilon}\frac{|R_j(f)(y)|\,dy}{(|x-y|/\varepsilon)^{n+1}}\le C_n\frac{2\|\Omega\|_{L^1}}{2^{-n}\varepsilon^n}\int\limits_{\mathbf{R}^n}\frac{|R_j(f)(y)|\,dy}{\big(1+\frac{|x-y|}{\varepsilon}\big)^{n+1}}\,.
$$

It follows that for functions f in L^p we have

$$
\sup_{0<\varepsilon<N<\infty}|A_2^{(\varepsilon,N)}(f)|\le C_n\big\|\Omega\big\|_{L^1}M(R_j(f))\,,
$$

in view of Theorem 2.1.10. (M here is the Hardy–Littlewood maximal operator.) By Theorem 2.1.6, M maps $L^p(\mathbf{R}^n)$ to itself with norm bounded by a dimensional constant multiple of $\max(1,(p-1)^{-1})$. Since by Remark 4.2.9 the norm $\|R_j\|_{L^p\to L^p}$ is controlled by a dimensional constant multiple of $\max(p,(p-1)^{-1})$, it follows that

$$\Big\| \sup_{0<\varepsilon<N<\infty} |A_2^{(\varepsilon,N)}(f)|\Big\|_{L^p} \le C_n \big\|\Omega\big\|_{L^1} \max(p,(p-1)^{-1})\big\|f\big\|_{L^p}. \tag{4.2.32}$$

Next, recall that in the proof of Theorem 4.2.10 we showed that

$$K_j(x) = \frac{\Omega_j(x/|x|)}{|x|^n},$$

where Ω_j are integrable functions on $\mathbf{S}^{n-1}$ that satisfy

$$\big\|\Omega_j\big\|_{L^1} \le C_n(c_\Omega + 1). \tag{4.2.33}$$

Consequently, for functions f in $L^p(\mathbf{R}^n)$ we have

$$\sup_{0<\varepsilon<N<\infty} |A_3^{(\varepsilon,N)}(f)| \le 2\sum_{j=1}^{n} T_{\Omega_j}^{(**)}(R_j(f)),$$

and by Remark 4.2.9 this last expression has L^p norm at most a dimensional constant multiple of $\big\|\Omega_j\big\|_{L^1}\max(p,(p-1)^{-1})\big\|R_j(f)\big\|_{L^p}$. It follows that

$$\Big\| \sup_{0<\varepsilon<N<\infty} |A_3^{(\varepsilon,N)}(f)|\Big\|_{L^p} \le C_n \max(p^2,(p-1)^{-2})(c_\Omega+1)\, \big\|f\big\|_{L^p}. \tag{4.2.34}$$

Finally, we turn our attention to the term $A_1^{(\varepsilon,N)}(f)$. To prove the required estimate, we first show that there exist nonnegative homogeneous of degree zero functions G_j on $\mathbf{R}^n$ that satisfy

$$|F_j(x)| \le G_j(x) \qquad \text{when } |x|\le 1 \tag{4.2.35}$$

and

$$\int_{\mathbf{S}^{n-1}} |G_j(\theta)|\,d\theta \le C_n(c_\Omega+1). \tag{4.2.36}$$

To prove (4.2.35), first note that if $|x|\le 1/8$, then

$$\begin{aligned} |F_j(x)| &= \frac{\Gamma(\frac{n+1}{2})}{\pi^{\frac{n+1}{2}}}\bigg| \int_{\mathbf{R}^n} \frac{\Omega(y/|y|)}{|y|^n}\Phi(y)\frac{x_j-y_j}{|x-y|^{n+1}}\,dy\bigg| \\ &\le C_n \int_{|y|\ge\frac14} \frac{|\Omega(y/|y|)|}{|y|^{2n}}\,dy \\ &\le C_n' \big\|\Omega\big\|_{L^1}. \end{aligned}$$

We now fix an x satisfying $1/8 \le |x| \le 1$ and we write

$$\begin{aligned}|F_j(x)| &\le \Phi(x)|K_j(x)|+|F_j(x)-\Phi(x)K_j(x)|\\&\le |K_j(x)|+\frac{\Gamma(\frac{n+1}{2})}{\pi^{\frac{n+1}{2}}}\left|\lim_{\varepsilon\to 0}\int_{|y|>\varepsilon}\frac{x_j-y_j}{|x-y|^{n+1}}\big(\Phi(y)-\Phi(x)\big)\frac{\Omega(y/|y|)}{|y|^n}dy\right|\\&=|K_j(x)|+\frac{\Gamma(\frac{n+1}{2})}{\pi^{\frac{n+1}{2}}}\big(P_1(x)+P_2(x)+P_3(x)\big),\end{aligned}$$

where

$$\begin{aligned}P_1(x)&=\left|\int_{|y|\le\frac{1}{16}}\left(\frac{x_j-y_j}{|x-y|^{n+1}}-\frac{x_j}{|x|^{n+1}}\right)\big(\Phi(y)-\Phi(x)\big)\frac{\Omega(y/|y|)}{|y|^n}\,dy\right|,\\P_2(x)&=\left|\int_{\frac{1}{16}\le|y|\le 2}\frac{x_j-y_j}{|x-y|^{n+1}}\big(\Phi(y)-\Phi(x)\big)\frac{\Omega(y/|y|)}{|y|^n}\,dy\right|,\\P_3(x)&=\left|\int_{|y|\ge 2}\frac{x_j-y_j}{|x-y|^{n+1}}\big(\Phi(y)-\Phi(x)\big)\frac{\Omega(y/|y|)}{|y|^n}\,dy\right|.\end{aligned}$$

But since $1/8\le|x|\le 1$, we see that

$$P_1(x)\le C_n\int_{|y|\le\frac{1}{16}}\frac{|y|}{|x|^{n+1}}\frac{|\Omega(y/|y|)|}{|y|^n}\,dy\le C_n'\|\Omega\|_{L^1}$$

and that

$$P_3(x)\le C_n\int_{|y|\ge 2}\frac{|\Omega(y/|y|)|}{|y|^{2n}}\,dy\le C_n'\|\Omega\|_{L^1}.$$

For $P_2(x)$ we use the estimate $|\Phi(y)-\Phi(x)|\le C|x-y|$ to obtain

$$\begin{aligned}P_2(x)&\le\int_{\frac{1}{16}\le|y|\le 2}\frac{C}{|x-y|^{n-1}}\frac{|\Omega(y/|y|)|}{|y|^n}\,dy\\&\le 4C\int_{\frac{1}{16}\le|y|\le 2}\frac{|\Omega(y/|y|)|}{|x-y|^{n-1}|y|^{n-\frac{1}{2}}}\,dy\\&\le 4C\int_{\mathbf{R}^n}\frac{|\Omega(y/|y|)|}{|x-y|^{n-1}|y|^{n-\frac{1}{2}}}\,dy.\end{aligned}$$

Recall that $K_j(x)=\Omega_j(x/|x|)|x|^{-n}$. We now set

$$G_j(x)=C_n\left(\|\Omega\|_{L^1}+\left|\Omega_j\left(\frac{x}{|x|}\right)\right|+|x|^{n-\frac{3}{2}}\int_{\mathbf{R}^n}\frac{|\Omega(y/|y|)|\,dy}{|x-y|^{n-1}|y|^{n-\frac{1}{2}}}\right)\tag{4.2.37}$$

and we observe that G_j is a homogeneous of degree zero function, it satisfies (4.2.35), and it is integrable over the annulus $\frac{1}{2}\le|x|\le 2$. To verify the last assertion, we split up the double integral

$$I=\int_{\frac{1}{2}\le|x|\le 2}\int_{\mathbf{R}^n}\frac{|\Omega(y/|y|)|\,dy}{|x-y|^{n-1}|y|^{n-\frac{1}{2}}}\,dx$$

into the pieces $1/4 \le |y| \le 4$, $|y| > 4$, and $|y| < 1/4$. The part of I where $1/4 \le |y| \le 4$ is pointwise bounded by a constant multiple of

$$\int_{\frac{1}{4}\le|y|\le4} \left|\Omega\left(\frac{y}{|y|}\right)\right| \int_{\frac{1}{2}\le|x|\le2} \frac{dx}{|y-x|^{n-1}}\,dy \le \int_{\frac{1}{4}\le|y|\le4} \left|\Omega\left(\frac{y}{|y|}\right)\right| \int_{|x-y|\le6} \frac{dx}{|y-x|^{n-1}}\,dy,$$

which is pointwise controlled by a constant multiple of $\|\Omega\|_{L^1}$. In the part of I where $|y| > 4$ we use that $|x-y|^{-n+1} \le (|y|/2)^{-n+1}$ to obtain rapid decay in y and hence a bound by a constant multiple of $\|\Omega\|_{L^1}$. Finally, in the part of I where $|y| < 1/4$ we use that $|x-y|^{-n+1} \le (1/4)^{-n+1}$, and then we also obtain a similar bound. It follows from (4.2.37) and (4.2.33) that

$$\int_{\frac{1}{2}\le|x|\le2} |G_j(x)|\,dx \le C_n\big(\|\Omega\|_{L^1} + \|\Omega_j\|_{L^1} + \|\Omega\|_{L^1}\big) \le C_n(c_\Omega + 1).$$

Since G_j is homogeneous of degree zero, we deduce (4.2.36).

To complete the proof, we argue as follows:

$$\begin{aligned}
&\sup_{0<\varepsilon<N<\infty} |A_1^{(\varepsilon,N)}(f)(x)| \\
&\qquad \le 2\sup_{\varepsilon>0} \sum_{j=1}^n \frac{1}{\varepsilon^n} \int_{|z|\le\varepsilon} |F_j(z)|\,|R_j(f)(x-z)|\,dz \\
&\qquad \le 2\sup_{\varepsilon>0} \sum_{j=1}^n \frac{1}{\varepsilon^n} \int_{r=0}^{\varepsilon} \int_{\mathbf{S}^{n-1}} |F_j(r\theta)|\,|R_j(f)(x-r\theta)|\,r^{n-1}\,d\theta\,dr \\
&\qquad \le 2\sum_{j=1}^n \int_{\mathbf{S}^{n-1}} |G_j(\theta)| \left\{ \sup_{\varepsilon>0} \frac{1}{\varepsilon^n} \int_{r=0}^{\varepsilon} |R_j(f)(x-r\theta)|\,r^{n-1}\,dr \right\} d\theta \\
&\qquad \le 4\sum_{j=1}^n \int_{\mathbf{S}^{n-1}} |G_j(\theta)| M_\theta(R_j(f))(x)\,d\theta\,.
\end{aligned}$$

Using (4.2.36) together with the L^p boundedness of the Riesz transforms and of M_θ we obtain

$$\Big\| \sup_{0<\varepsilon<N<\infty} |A_1^{(\varepsilon,N)}(f)| \Big\|_{L^p} \le C_n \max(p,(p-1)^{-2})(c_\Omega+1)\|f\|_{L^p}\,. \tag{4.2.38}$$

Combining (4.2.38), (4.2.32), and (4.2.34), we obtain the required conclusion. □

The following corollary is a consequence of Theorem 4.2.11.

Corollary 4.2.12. *Let Ω be as in Theorem 4.2.11. Then for $1 < p < \infty$ and f in $L^p(\mathbf{R}^n)$ the functions $T_\Omega^{(\varepsilon,N)}(f)$ converge to $T_\Omega(f)$ in L^p and almost everywhere as $\varepsilon \to 0$ and $N \to \infty$.*

Proof. The a.e. convergence is a consequence of Theorem 2.1.14. The L^p convergence is a consequence of the Lebesgue dominated convergence theorem since for $f \in L^p(\mathbf{R}^n)$ we have that $|T_\Omega^{(\varepsilon,N)}(f)| \le T_\Omega^{(**)}(f)$ and $T_\Omega^{(**)}(f)$ is in $L^p(\mathbf{R}^n)$. □

Exercises

4.2.1. Show that the directional Hilbert transform $\mathscr{H}_\theta$ is given by convolution with the distribution w_θ in $\mathscr{S}'(\mathbf{R}^n)$ defined by

$$\langle w_\theta, \varphi \rangle = \frac{1}{\pi} \text{p.v.} \int_{-\infty}^{+\infty} \frac{\varphi(t\theta)}{t} \, dt.$$

Compute the Fourier transform of w_θ and prove that $\mathscr{H}_\theta$ maps $L^1(\mathbf{R}^n)$ to $L^{1,\infty}(\mathbf{R}^n)$.

4.2.2. Extend the definitions of W_Ω and T_Ω to $\Omega = d\mu$ a finite signed Borel measure on $\mathbf{S}^{n-1}$ with mean value zero. Compute the Fourier transform of such W_Ω and find a necessary and sufficient condition on measures $\Omega = d\mu$ so that T_Ω is L^2 bounded.

4.2.3. Use the inequality $AB \le A \log A + e^B$ for $A \ge 1$ and $B > 0$ to prove that if Ω satisfies (4.2.23) then it must satisfy (4.2.16). Conclude that if $|\Omega| \log^+ |\Omega|$ is in $L^1(\mathbf{S}^{n-1})$, then T_Ω is L^2 bounded.
[*Hint:* Use that $\int_{\mathbf{S}^{n-1}} |\xi \cdot \theta|^{-\alpha} \, d\theta$ converges when $\alpha < 1$. See Appendix D.3.]

4.2.4. Let Ω be a nonzero integrable function on $\mathbf{S}^{n-1}$ with mean value zero. Let $f \ge 0$ be nonzero and integrable over $\mathbf{R}^n$. Prove that $T_\Omega(f)$ in not in $L^1(\mathbf{R}^n)$.
[*Hint:* Show that $\widehat{T_\Omega(f)}$ cannot be continuous at zero.]

4.2.5. Use the idea of the boundedness of $\mathscr{H}_\theta$ to show that M_θ maps $L^p(\mathbf{R}^n)$ to itself with the same norm as the norm of the centered Hardy–Littlewood maximal operator on $L^p(\mathbf{R})$.

4.2.6. (a) Let $\theta \in \mathbf{S}^{n-1}$. Use an identity similar to (4.2.17) to show that the maximal operators

$$\sup_{a>0} \frac{1}{a} \int_0^a |f(x - r\theta)| \, dr, \qquad \sup_{a>0} \frac{1}{2a} \int_{-a}^{+a} |f(x - r\theta)| \, dr$$

are $L^p(\mathbf{R}^n)$ bounded for $1 < p < \infty$ with norm at most $3\,p\,(p-1)^{-1}$.
(b) For $\Omega \in L^1(\mathbf{S}^{n-1})$ and f locally integrable on $\mathbf{R}^n$, define

$$M_\Omega(f)(x) = \sup_{R>0} \frac{1}{v_n R^n} \int_{|y| \le R} |\Omega(y/|y|)| \, |f(x-y)| \, dy.$$

Apply the method of rotations to prove that M_Ω maps $L^p(\mathbf{R}^n)$ to itself for $1 < p < \infty$.

4.2.7. Let $\Omega(x, \theta)$ be a function on $\mathbf{R}^n \times \mathbf{S}^{n-1}$ satisfying
(a) $\Omega(x, -\theta) = -\Omega(x, \theta)$ for all x and θ.
(b) $\displaystyle\int_{\mathbf{S}^{n-1}} \Omega(x, \theta) \, d\theta = 0$ for all $x \in \mathbf{R}^n$.
(c) $\sup_x |\Omega(x, \theta)|$ is in $L^1(\mathbf{S}^{n-1})$.
Use the method of rotations to prove that

$$T_\Omega(f)(x) = \text{p.v.} \int_{\mathbf{R}^n} \frac{\Omega(x, y/|y|)}{|y|^n} f(x-y)\, dy$$

is bounded on $L^p(\mathbf{R}^n)$ for $1 < p < \infty$.

4.2.8. Let $\Omega \in L^1(\mathbf{S}^{n-1})$ have mean value zero. Prove that if T_Ω maps $L^p(\mathbf{R}^n)$ to $L^q(\mathbf{R}^n)$, then $p = q$.
[*Hint:* Use dilations.]

4.2.9. Prove that for all $1 < p < \infty$ there exists a constant $A_p > 0$ such that for every complex-valued $\mathscr{C}^2(\mathbf{R}^2)$ function f with compact support we have the bound

$$\big\|\partial_{x_1} f\big\|_{L^p} + \big\|\partial_{x_2} f\big\|_{L^p} \le A_p \big\|\partial_{x_1} f + i\partial_{x_2} f\big\|_{L^p}.$$

4.2.10. (a) Let $\Delta = \sum_{j=1}^n \partial_{x_j}^2$ be the usual Laplacian on $\mathbf{R}^n$. Prove that for all $1 < p < \infty$ there exists a constant $A_p > 0$ such that for all $\mathscr{C}^2$ functions f with compact support we have the bound

$$\big\|\partial_{x_j}\partial_{x_k} f\big\|_{L^p} \le A_p \big\|\Delta(f)\big\|_{L^p}.$$

(b) Let $\Delta^m = \overbrace{\Delta \circ \cdots \circ \Delta}^{m \text{ times}}$. Show that for any $1 < p < \infty$ there exists a $C_p > 0$ such that for all f of class $\mathscr{C}^{2m}$ with compact support and all differential monomials ∂_x^α of order $|\alpha| = 2m$ we have

$$\big\|\partial_x^\alpha f\big\|_{L^p} \le C_p \big\|\Delta^m(f)\big\|_{L^p}.$$

4.2.11. Use the same idea as in Lemma 4.2.5 to show that if f is continuous on $[0,\infty)$, differentiable in $(0,\infty)$, and satisfies

$$\lim_{N\to\infty} \int_N^{Na} \frac{f(u)}{u}\, du = 0$$

for all $a > 0$, then

$$\lim_{\substack{\varepsilon\to 0\\ N\to\infty}} \int_\varepsilon^N \frac{f(at) - f(t)}{t}\, dt = f(0) \log\frac{1}{a}.$$

4.2.12. Let Ω_o be an odd integrable function on $\mathbf{S}^{n-1}$ and Ω_e an even function on $\mathbf{S}^{n-1}$ that satisfies (4.2.23). Let f be a function supported in a ball B in $\mathbf{R}^n$. Prove that
(a) If $|f|\log^+|f|$ is integrable over a ball B, then $T_{\Omega_o}(f)$ and $T_{\Omega_o}^{(**)}(f)$ are integrable over B.
(b) If $|f|(\log^+|f|)^2$ is integrable over a ball B, then $T_{\Omega_e}(f)$ and $T_{\Omega_e}^{(**)}(f)$ are integrable over B.
[*Hint:* Use Exercise 1.3.7.]

4.2.13. (*Sjögren and Soria [244]*) Let Ω be integrable on $\mathbf{S}^{n-1}$ with mean value zero. Use Jensen's inequality to show that for some $C > 0$ and every radial function $f \in L^2(\mathbf{R}^n)$ we have

$$\|T_\Omega(f)\|_{L^2} \le C\|f\|_{L^2}.$$

This inequality subsumes that T_Ω is well defined for f radial.

4.3 The Calderón–Zygmund Decomposition and Singular Integrals

The behavior of singular integral operators on $L^1(\mathbf{R}^n)$ is a more subtle issue than that on L^p for $1 < p < \infty$. It turns out that singular integrals are not bounded from L^1 to L^1. See Example 4.1.3 and also Exercise 4.2.4. In this section we see that singular integrals map L^1 into the larger space $L^{1,\infty}$. This result strengthens their L^p boundedness.

4.3.1 The Calderón–Zygmund Decomposition

To make some advances in the theory of singular integrals, we need to introduce the Calderón–Zygmund decomposition. This is a powerful stopping-time construction that has many other interesting applications. We have already encountered an example of a stopping-time argument in Section 2.1.

Recall that a dyadic cube in $\mathbf{R}^n$ is the set

$$[2^k m_1, 2^k(m_1+1)) \times \cdots \times [2^k m_n, 2^k(m_n+1)),$$

where $k, m_1, \dots, m_n \in \mathbf{Z}$. Two dyadic cubes are either disjoint or related by inclusion.

Theorem 4.3.1. *Let $f \in L^1(\mathbf{R}^n)$ and $\alpha > 0$. Then there exist functions g and b on $\mathbf{R}^n$ such that*

(1) $f = g + b$.

(2) $\|g\|_{L^1} \le \|f\|_{L^1}$ *and* $\|g\|_{L^\infty} \le 2^n\alpha$.

(3) $b = \sum_j b_j$, *where each b_j is supported in a dyadic cube Q_j. Furthermore, the cubes Q_k and Q_j are disjoint when $j \ne k$.*

(4) $\displaystyle\int_{Q_j} b_j(x)\,dx = 0$.

(5) $\|b_j\|_{L^1} \le 2^{n+1}\alpha|Q_j|$.

(6) $\sum_j |Q_j| \le \alpha^{-1}\|f\|_{L^1}$.

Remark 4.3.2. This decomposition is called the *Calderón–Zygmund decomposition* of f at height α. The function g is called the *good function* of the decomposition, since it is both integrable and bounded; hence the letter g. The function b is called the *bad function*, since it contains the singular part of f (hence the letter b), but it is carefully chosen to have mean value zero. It follows from (1) and (2) that the bad function b is integrable and satisfies

$$\|b\|_{L^1} \leq \|f\|_{L^1} + \|g\|_{L^1} \leq 2\|f\|_{L^1}.$$

By (2) the good function is integrable and bounded; hence it lies in all the L^p spaces for $1 \leq p \leq \infty$. More specifically, we have the following estimate:

$$\|g\|_{L^p} \leq \|g\|_{L^1}^{\frac{1}{p}} \|g\|_{L^\infty}^{1-\frac{1}{p}} \leq \|f\|_{L^1}^{\frac{1}{p}} (2^n\alpha)^{1-\frac{1}{p}} = 2^{\frac{n}{p'}} \alpha^{\frac{1}{p'}} \|f\|_{L^1}^{\frac{1}{p}}. \tag{4.3.1}$$

Proof. Decompose $\mathbf{R}^n$ into a mesh of disjoint dyadic cubes of the same size such that

$$|Q| \geq \frac{1}{\alpha}\|f\|_{L^1}$$

for every cube Q in the mesh. Call these cubes of zero generation. Subdivide each cube of zero generation into 2^n congruent cubes by bisecting each of its sides. We now have a new mesh of dyadic cubes, which we call of generation one. Select a cube Q of generation one if

$$\frac{1}{|Q|}\int_Q |f(x)|\,dx > \alpha. \tag{4.3.2}$$

Let $S^{(1)}$ be the set of all selected cubes of generation one. Now subdivide each nonselected cube of generation one into 2^n congruent subcubes by bisecting each side and call these cubes of generation two. Then select all cubes Q of generation two if (4.3.2) holds. Let $S^{(2)}$ be the set of all selected cubes of generation two. Repeat this procedure indefinitely.

The set of all selected cubes $\bigcup_{m=1}^{\infty} S^{(m)}$ is countable and is exactly the set of the cubes Q_j proclaimed in the proposition. Note that in some instances this set may be empty, in which case $b = 0$ and $g = f$. Let us observe that the selected cubes are disjoint, for otherwise some Q_k would be a proper subset of some Q_j, which is impossible since the selected cube Q_j was never subdivided. Now define

$$b_j = \left(f - \frac{1}{|Q_j|}\int_{Q_j} f\,dx\right)\chi_{Q_j},$$

$b = \sum_j b_j$, and $g = f - b$.

For a selected cube Q_j there exists a unique nonselected cube Q' with twice its side length that contains Q_j. Let us call this cube the parent of Q_j. Since the parent Q' of Q_j was not selected, we have $|Q'|^{-1}\int_{Q'}|f|\,dx \leq \alpha$. Then

$$\frac{1}{|Q_j|}\int_{Q_j}|f(x)|\,dx \le \frac{1}{|Q_j|}\int_{Q'}|f(x)|\,dx = \frac{2^n}{|Q'|}\int_{Q'}|f(x)|\,dx \le 2^n\alpha.$$

Consequently,

$$\int_{Q_j}|b_j|\,dx \le \int_{Q_j}|f|\,dx + |Q_j|\left|\frac{1}{|Q_j|}\int_{Q_j} f\,dx\right| \le 2\int_{Q_j}|f|\,dx \le 2^{n+1}\alpha|Q_j|,$$

which proves (5). To prove (6), simply observe that

$$\sum_j |Q_j| \le \frac{1}{\alpha}\sum_j \int_{Q_j}|f|\,dx = \frac{1}{\alpha}\int_{\bigcup_j Q_j}|f|\,dx \le \frac{1}{\alpha}\|f\|_{L^1}.$$

Next we need to obtain the estimates concerning g. We obviously have

$$g = \begin{cases} f & \text{on } \mathbf{R}^n \setminus \bigcup_j Q_j, \\ \frac{1}{|Q_j|}\displaystyle\int_{Q_j} f\,dx & \text{on } Q_j. \end{cases} \tag{4.3.3}$$

On the cube Q_j, g is equal to the constant $|Q_j|^{-1}\int_{Q_j} f\,dx$, and this is bounded by $2^n\alpha$. It suffices to show that g is bounded outside the union of the Q_j's. Indeed, for each $x \in \mathbf{R}^n \setminus \bigcup_j Q_j$ and for each $k = 0,1,2,\dots$ there exists a unique nonselected dyadic cube $Q_x^{(k)}$ of generation k that contains x. Then for each $k \ge 0$, we have

$$\left|\frac{1}{|Q_x^{(k)}|}\int_{Q_x^{(k)}} f(y)\,dy\right| \le \frac{1}{|Q_x^{(k)}|}\int_{Q_x^{(k)}}|f(y)|\,dy \le \alpha.$$

The intersection of the closures of the cubes $Q_x^{(k)}$ is the singleton $\{x\}$. Using a version of Corollary 2.1.16 where the balls are replaced with cubes, we deduce that for almost all $x \in \mathbf{R}^n \setminus \bigcup_j Q_j$ we have

$$f(x) = \lim_{k\to\infty}\frac{1}{|Q_x^{(k)}|}\int_{Q_x^{(k)}} f(y)\,dy.$$

Since these averages are at most α, we conclude that $|f| \le \alpha$ a.e. on $\mathbf{R}^n \setminus \bigcup_j Q_j$, hence $|g| \le \alpha$ a.e. on this set. Finally, it follows from (4.3.3) that $\|g\|_{L^1} \le \|f\|_{L^1}$. This finishes the proof of the theorem. □

We now apply the Calderón–Zygmund decomposition to obtain weak type $(1,1)$ bounds for a wide class of singular integral operators that includes the operators T_Ω we studied in the previous section.

4.3.2 General Singular Integrals

Let K be a measurable function defined on $\mathbf{R}^n \setminus \{0\}$ that satisfies the size condition

$$\sup_{R>0} \int_{R\le|x|\le 2R} |K(x)|\,dx = A_1 < \infty. \tag{4.3.4}$$

This condition is less restrictive than the standard size estimate

$$\sup_{x\in\mathbf{R}^n} |x|^n |K(x)| < \infty, \tag{4.3.5}$$

but it is strong enough to capture size properties of kernels $K(x) = \Omega(x/|x|)/|x|^n$, where $\Omega \in L^1(\mathbf{S}^{n-1})$. We also note that condition (4.3.4) is equivalent to

$$\sup_{R>0} \frac{1}{R} \int_{|x|\le R} |K(x)|\,|x|\,dx < \infty. \tag{4.3.6}$$

See Exercise 4.3.1.

The size condition (4.3.4) is sufficient to make K a tempered distribution away from the origin. Indeed, for $\varphi \in \mathscr{S}(\mathbf{R}^n)$ we have

$$\begin{aligned}\int_{|x|\ge 1} |K(x)\varphi(x)|\,dx &\le \sum_{m=0}^{\infty} \int_{2^{m+1}\ge|x|\ge 2^m} \frac{|K(x)|(1+|x|)^N|\varphi(x)|}{(1+2^m)^N}\,dx \\ &\le \sum_{m=0}^{\infty} \frac{A_1}{(1+2^m)^N} \sup_{x\in\mathbf{R}^n} (1+|x|)^N|\varphi(x)|,\end{aligned}$$

and the latter is controlled by a finite sum of Schwartz seminorms of φ.

We are interested in tempered distributions W on $\mathbf{R}^n$ that extend the function K defined on $\mathbf{R}^n \setminus \{0\}$ and that have the form

$$W(\varphi) = \lim_{j\to\infty} \int_{|x|\ge\delta_j} K(x)\varphi(x)\,dx, \qquad \varphi \in \mathscr{S}(\mathbf{R}^n), \tag{4.3.7}$$

for some sequence $\delta_j \downarrow 0$ as $j \to \infty$. It is not hard to see that there exists a tempered distribution W satisfying (4.3.7) for all $\varphi \in \mathscr{S}(\mathbf{R}^n)$ if and only if

$$\lim_{j\to\infty} \int_{1\ge|x|\ge\delta_j} K(x)\,dx = L \tag{4.3.8}$$

exists. See Exercise 4.3.2. If such a distribution W exists it may not be unique, since it depends on the choice of the sequence δ_j. Two different sequences tending to zero may give two different tempered distributions W of the form (4.3.7), both coinciding with the function K on $\mathbf{R}^n \setminus \{0\}$. See Example 4.4.2 and Remark 4.4.3. Furthermore, not all functions K on $\mathbf{R}^n \setminus \{0\}$ give rise to distributions W defined by (4.3.7); take, for example, $K(x) = |x|^{-n}$.

If condition (4.3.8) is satisfied, we can define

$$W(\varphi) = \lim_{j\to\infty} \int_{j\geq|x|\geq\delta_j} K(x)\varphi(x)\,dx \tag{4.3.9}$$

and the limit exists as $j \to \infty$ for all $\varphi \in \mathscr{S}(\mathbf{R}^n)$ and is equal to

$$W(\varphi) = \int_{|x|\leq 1} K(x)(\varphi(x) - \varphi(0))\,dx + \varphi(0)L + \int_{|x|\geq 1} K(x)\varphi(x)\,dx.$$

Moreover, the previous calculations show that W is an element of $\mathscr{S}'(\mathbf{R}^n)$.

Next we assume that the given function K on $\mathbf{R}^n \setminus \{0\}$ satisfies a certain smoothness condition. There are three kinds of smoothness conditions that we encounter: first, the *gradient condition*

$$|\nabla K(x)| \leq A_2|x|^{-n-1}, \qquad x \neq 0; \tag{4.3.10}$$

next, the weaker *Lipschitz condition*,

$$|K(x-y) - K(x)| \leq A_2 \frac{|y|^\delta}{|x|^{n+\delta}}, \qquad \text{whenever } |x| \geq 2|y|; \tag{4.3.11}$$

and finally the even weaker smoothness condition

$$\sup_{y\neq 0} \int_{|x|\geq 2|y|} |K(x-y) - K(x)|\,dx = A_2, \tag{4.3.12}$$

for some $A_2 < \infty$. One should verify that (4.3.12) is a weaker condition than (4.3.11), which in turn is weaker than (4.3.10). Condition (4.3.12) is often referred to as *Hörmander's condition.*

4.3.3 L^r Boundedness Implies Weak Type $(1,1)$ Boundedness

This next theorem provides a very classical application of the Calderón–Zygmund decomposition.

Theorem 4.3.3. *Assume that K is defined on $\mathbf{R}^n \setminus \{0\}$ and satisfies (4.3.12) for some $A_2 < \infty$. Let $W \in \mathscr{S}'(\mathbf{R}^n)$ be as in (4.3.7) coinciding with K on $\mathbf{R}^n \setminus \{0\}$. Suppose that the operator T given by convolution with W maps $L^r(\mathbf{R}^n)$ to itself with norm B for some $1 < r \leq \infty$. Then T has an extension that maps $L^1(\mathbf{R}^n)$ to $L^{1,\infty}(\mathbf{R}^n)$ with norm*

$$\big\|T\big\|_{L^1\to L^{1,\infty}} \leq C_n(A_2 + B), \tag{4.3.13}$$

and T also extends to a bounded operator from $L^p(\mathbf{R}^n)$ to itself for $1 < p < \infty$ with norm

$$\big\|T\big\|_{L^p\to L^p} \leq C_n' \max\left(p, (p-1)^{-1}\right)(A_2 + B), \tag{4.3.14}$$

where C_n, C_n' are constants that depend on the dimension but not on r or p.

Proof. We first explain the idea of the proof. We write $f = g + b$; hence

$$T(f) = T(g) + T(b).$$

The function $T(g)$ is in L^r and thus it satisfies a weak type L^r estimate. The bad part of f is a sum of functions with mean value zero. Cancellation is used to subtract a suitable term from every piece of the bad function that allows us to use Hörmander's condition (4.3.12). Let us proceed with the details. We work out the case $r < \infty$ and we refer to Exercise 4.3.7 for the case $r = \infty$.

Fix $\alpha > 0$ and let f be in $L^1(\mathbf{R}^n)$. We assume that f is in the Schwartz class since $T(f)$ may not be a priori defined for $f \in L^1(\mathbf{R}^n)$. Once (4.3.13) is obtained for f in $\mathscr{S}(\mathbf{R}^n)$, a density argument gives that T admits an extension on L^1 that also satisfies (4.3.13). Apply the Calderón–Zygmund decomposition to f at height $\gamma\alpha$, where γ is a positive constant to be chosen later. That is, write the function f as the sum

$$f = g + b,$$

where conditions (1)–(6) of Theorem 4.3.1 are satisfied with the constant α replaced by $\gamma\alpha$. We denote by $\ell(Q)$ the side length of a cube Q. Let Q_j^* be the unique cube with sides parallel to the axes having the same center as Q_j and having side length

$$\ell(Q_j^*) = 2\sqrt{n}\,\ell(Q_j).$$

We have

$$\begin{aligned}
&\big|\{x \in \mathbf{R}^n : |T(f)(x)| > \alpha\}\big| \\
&\quad\le \Big|\Big\{x \in \mathbf{R}^n : |T(g)(x)| > \frac{\alpha}{2}\Big\}\Big| + \Big|\Big\{x \in \mathbf{R}^n : |T(b)(x)| > \frac{\alpha}{2}\Big\}\Big| \\
&\quad\le \frac{2^r}{\alpha^r}\|T(g)\|_{L^r}^r + \Big|\bigcup_j Q_j^*\Big| + \Big|\Big\{x \notin \bigcup_j Q_j^* : |T(b)(x)| > \frac{\alpha}{2}\Big\}\Big| \\
&\quad\le \frac{2^r}{\alpha^r}B^r\|g\|_{L^r}^r + \sum_j |Q_j^*| + \frac{2}{\alpha}\int_{(\bigcup_j Q_j^*)^c} |T(b)(x)|\,dx \\
&\quad\le \frac{2^r}{\alpha^r}2^{\frac{nr}{r'}}B^r(\gamma\alpha)^{\frac{r}{r'}}\|f\|_{L^1} + (2\sqrt{n})^n\frac{\|f\|_{L^1}}{\gamma\alpha} + \frac{2}{\alpha}\sum_j\int_{(Q_j^*)^c}|T(b_j)(x)|\,dx \\
&\quad\le \left(\frac{(2^{n+1}B\gamma)^r}{2^n\gamma} + \frac{(2\sqrt{n})^n}{\gamma}\right)\frac{\|f\|_{L^1}}{\alpha} + \frac{2}{\alpha}\sum_j\int_{(Q_j^*)^c}|T(b_j)(x)|\,dx\,.
\end{aligned}$$

It suffices to show that the last sum is bounded by some constant multiple of the L^1 norm of f. It is here where we use the fact that b_j has mean value zero and Hörmander's condition (4.3.12).

Let y_j be the center of the cube Q_j. We have

$$
\begin{aligned}
\sum_j \int_{(Q_j^*)^c} |T(b_j)(x)|\,dx &= \sum_j \int_{(Q_j^*)^c} \left| \int b_j(y) K(x-y)\,dy \right| dx \\
&= \sum_j \int_{(Q_j^*)^c} \left| \int_{Q_j} b_j(y) \big(K(x-y) - K(x-y_j)\big)\,dy \right| dx \\
&\le \sum_j \int_{Q_j} |b_j(y)| \int_{(Q_j^*)^c} |K(x-y) - K(x-y_j)|\,dx\,dy \\
&= \sum_j \int_{Q_j} |b_j(y)| \int_{-y_j+(Q_j^*)^c} |K(x-(y-y_j)) - K(x)|\,dx\,dy \\
&\le \sum_j \int_{Q_j} |b_j(y)| \int_{|x|\ge 2|y-y_j|} |K(x-(y-y_j)) - K(x)|\,dx\,dy \\
&\le A_2 \sum_j \|b_j\|_{L^1} \\
&\le A_2 2^{n+1} \|f\|_{L^1},
\end{aligned}
$$

where we used the fact that if $x \in -y_j + (Q_j^*)^c$ then $|x| \ge \frac{1}{2}\ell(Q_j^*) = \sqrt{n}\,\ell(Q_j)$. But since $y - y_j \in -y_j + Q_j$, we have $|y - y_j| \le \frac{\sqrt{n}}{2}\ell(Q_j)$, thus $|x| \ge 2|y - y_j|$. Here we used the fact that the diameter of a cube is equal to $\sqrt{n}$ times its side length. See Figure 4.2.

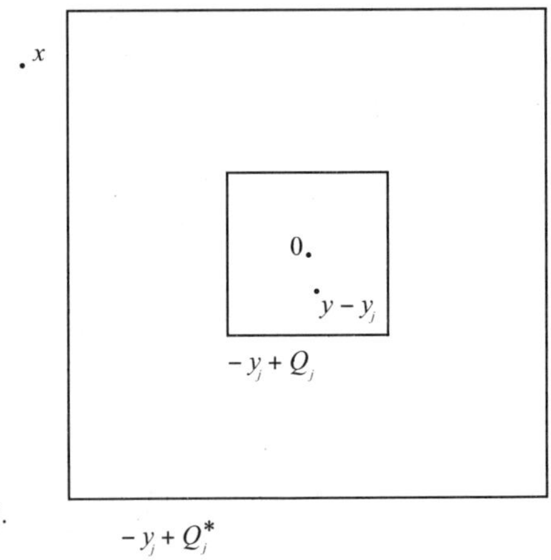

Fig. 4.2 The cubes $-y_j + Q_j$ and $-y_j + Q_j^*$.

Choosing $\gamma = 2^{-(n+1)}B^{-1}$, we deduce the weak type $(1,1)$ estimate (4.3.13) for T with $C_n = 2 + 2^{n+1}(2\sqrt{n})^n + 2^{n+2}$.

In view of Exercise 1.3.2, we have that T maps L^p to L^p with bound at most $C_n'(A_2 + B)(p-1)^{-1/p}$ whenever $1 < p < r$. This proves (4.3.14) for $1 < p < r$. To obtain a similar conclusion for $r < p < \infty$ we use duality. Notice that the adjoint operator T^* of T, defined by

$$\langle T(f) \,|\, g \rangle = \langle f \,|\, T^*(g) \rangle ,$$

has a kernel that coincides with the function $K^*(x) = \overline{K(-x)}$ on $\mathbf{R}^n \setminus \{0\}$. Next we notice that since K satisfies (4.3.12), then so does K^* and with the same bound. Therefore, T^*, which maps $L^{r'}$ to $L^{r'}$, has a kernel that satisfies Hörmander's condition. It must therefore map L^1 to $L^{1,\infty}$ and $L^{p'}$ to $L^{p'}$ for $1 < p' < r'$ with norm at most $C_n'(A_2 + B)(p' - 1)^{-1}$, by the argument just shown. It follows that T maps L^p to L^p with norm at most $C_n'(A_2 + B)p$ for $r < p < \infty$, and this proves (4.3.14). □

4.3.4 Discussion on Maximal Singular Integrals

In this subsection we introduce maximal singular integrals and we derive their boundedness under certain smoothness conditions on the kernels, assuming boundedness of the associated linear operator.

Suppose that K is a kernel on $\mathbf{R}^n \setminus \{0\}$ that satisfies the size condition

$$|K(x)| \le A_1 |x|^{-n} \tag{4.3.15}$$

for $x \neq 0$. Then for any $\varepsilon > 0$ the function $K^{(\varepsilon)}(x) = |x|^{-n}\chi_{|x|\ge\varepsilon}$ lies in $L^{p'}(\mathbf{R}^n)$ (with norm $c_{p,n}\varepsilon^{-n/p}$) for all $1 \le p < \infty$. Consequently, by Hölder's inequality, the integral

$$(f * K^{(\varepsilon)})(x) = \int_{|y|\ge\varepsilon} f(x-y)K(y)\,dy$$

converges absolutely for all $x \in \mathbf{R}^n$ and all $f \in L^p(\mathbf{R}^n)$, when $1 \le p < \infty$.

Let $f \in \bigcup_{1\le p<\infty} L^p(\mathbf{R}^n)$. We define the truncated singular integrals $T^{(\varepsilon)}(f)$ associated with the kernel K by setting

$$T^{(\varepsilon)}(f) = f * K^{(\varepsilon)};$$

we also define the *maximal truncated singular integral operator* associated with K by setting

$$T^{(*)}(f) = \sup_{\varepsilon>0} |(f * K^{(\varepsilon)})| = \sup_{\varepsilon>0} |T^{(\varepsilon)}(f)|.$$

This operator is well defined, but possibly infinite, for certain points in $\mathbf{R}^n$.

We now consider the situation in which the kernel K satisfies an integrability condition over concentric annuli centered at the origin, a condition that is certainly a weaker condition than (4.3.15). Precisely, suppose that K is a function on $\mathbf{R}^n \setminus \{0\}$ for which there is a constant $A_1 < \infty$ such that

$$\sup_{R>0} \int_{R\le|x|\le 2R} |K(x)|\,dx \le A_1 < \infty. \tag{4.3.16}$$

Such kernels may not be integrable to the power $p' > 1$ over the region $|x| \geq \varepsilon$. For this reason, it is not possible to define $T^{(\varepsilon)}$ as an absolutely convergent integral. To overcome this difficulty, we consider double truncations. We define the doubly truncated kernel $K^{(\varepsilon,N)}$ by setting

$$K^{(\varepsilon,N)}(x) = K(x)\chi_{\varepsilon\leq|x|\leq N}(x). \tag{4.3.17}$$

A repeated application of (4.3.16) yields that

$$\int |K^{(\varepsilon,N)}(x)|\,dx \leq A_1\Big(\Big[\log_2\frac{N}{\varepsilon}\Big]+1\Big),$$

which implies that $K^{(\varepsilon,N)}$ is integrable over concentric annuli centered at the origin. Next, we define the doubly truncated singular integrals $T^{(\varepsilon,N)}$ by setting

$$T^{(\varepsilon,N)}(f) = f * K^{(\varepsilon,N)},$$

and we observe that these operators are well defined when f in L^p, for $1 \leq p \leq \infty$. Indeed, Theorem 1.2.10 yields that

$$\big\|T^{(\varepsilon,N)}(f)\big\|_{L^p} \leq \big\|f\big\|_{L^p}\int |K^{(\varepsilon,N)}(x)|\,dx < \infty$$

for functions f in L^p, $1 \leq p \leq \infty$. Consequently, for almost every $x \in \mathbf{R}^n$ we have

$$|T^{(\varepsilon,N)}(f)(x)| < \infty.$$

For functions in $\bigcup_{1\leq p\leq\infty} L^p(\mathbf{R}^n)$ we define the *doubly truncated maximal singular integral operator* $T^{(**)}$ associated with K by setting

$$T^{(**)}(f) = \sup_{0<\varepsilon<N<\infty} |T^{(\varepsilon,N)}(f)|. \tag{4.3.18}$$

For such functions and for almost all $x \in \mathbf{R}^n$, $T^{(**)}(f)(x)$ is well defined, but potentially infinite.

One observation is that under condition (4.3.16), one can also define $T^{(*)}(g)$ for general integrable functions g with compact support. In this case, say that the ball $B(0,R)$ contains the support of g. Let $x \in B(0,M)$ and $N = M+R$. Then $|T^{(\varepsilon)}(g)(x)| \leq |g| * |K^{(\varepsilon,N)}|(x)$, which is finite a.e. as the convolution of two L^1 functions; consequently, the integral defining $T^{(\varepsilon)}(g)(x)$ converges absolutely for all $x \in B(0,R)$. Since $R > 0$ is arbitrary, $T^{(\varepsilon)}(g)(x)$ is defined and finite for almost all $x \in \mathbf{R}^n$.

Obviously $T^{(*)}$ and $T^{(**)}$ are related. If K satisfies condition (4.3.15), then

$$\left|\int_{\varepsilon\leq|y|} f(x-y)K(y)\,dy\right| \leq \sup_{N>0}\left|\int_{\varepsilon\leq|y|\leq N} f(x-y)K(y)\,dy\right|,$$

which implies that

$$T^{(*)}(f) \le T^{(**)}(f)$$

for all $f \in \bigcup_{1\le p<\infty} L^p$. Also, $T^{(\varepsilon,N)}(f) = T^{(\varepsilon)}(f) - T^{(N)}(f)$; hence

$$T^{(**)}(f) \le 2T^{(*)}(f).$$

Therefore, for kernels satisfying (4.3.15), $T^{(**)}$ and $T^{(*)}$ are comparable and the boudnedness properties of $T^{(**)}$ and $T^{(*)}$ are equivalent

Theorem 4.3.4. (Cotlar's inequality.) *Let $0 < A_1, A_2, A_3 < \infty$ and suppose that K is defined on $\mathbf{R}^n \setminus \{0\}$ and satisfies the size condition,*

$$|K(x)|\,dx \le A_1|x|^{-n}, \qquad x \neq 0, \tag{4.3.19}$$

the smoothness condition

$$|K(x-y) - K(x)| \le A_2|y|^{\delta}\,|x|^{-n-\delta}, \tag{4.3.20}$$

whenever $|x| \ge 2|y| > 0$, and the cancellation condition

$$\sup_{0<r<R<\infty} \left| \int_{r<|x|<R} K(x)\,dx \right| \le A_3. \tag{4.3.21}$$

Let W be any tempered distribution on $\mathbf{R}^n$ that coincides with K on $\mathbf{R}^n \setminus \{0\}$ and let T be the operator given by convolution with W. Then there is a constant $C_{n,\delta}$ such that the following inequality is valid:

$$T^{(*)}(f) \le M(T(f)) + C_{n,\delta}\,(A_1 + A_2 + A_3)\,M(f), \tag{4.3.22}$$

for all $f \in L^p$, $1 < p < \infty$, where M is the Hardy–Littlewood maximal operator. Thus the L^p boundedness of $T^{(*)}$ for $1 < p < \infty$ can be deduced from that of T.

Proof. Let φ be a radially decreasing smooth function with integral 1 supported in the ball $B(0,1/2)$. For a function g and $\varepsilon > 0$ we use the notation $g_\varepsilon(x) = \varepsilon^{-n} g(\varepsilon^{-1}x)$. For a distribution W we define W_ε analogously, i.e. as the unique distribution with the property $\langle W_\varepsilon, \psi\rangle = \varepsilon^{-n}\langle W, \psi_{\varepsilon^{-1}}\rangle$. We begin by observing that $K_{\varepsilon^{-1}}(x) = \varepsilon^n K(\varepsilon x)$ satisfies (4.3.19), (4.3.20), and (4.3.21) uniformly in $\varepsilon > 0$.

Set, as before, $K^{(\varepsilon)}(x) = K(x)\chi_{|x|\ge\varepsilon}$. Fix $f \in L^p(\mathbf{R}^n)$ for some $1 < p < \infty$. Obviously we have

$$f * K^{(\varepsilon)} = f * \big((K_{\varepsilon^{-1}})^{(1)}\big)_\varepsilon = f * W * \varphi_\varepsilon + f * \big((K_{\varepsilon^{-1}})^{(1)} - W_{\varepsilon^{-1}} * \varphi\big)_\varepsilon. \tag{4.3.23}$$

Next we prove the following estimate for all $\varepsilon > 0$:

$$\big|\big((K_{\varepsilon^{-1}})^{(1)} - W_{\varepsilon^{-1}} * \varphi\big)(x)\big| \le C(A_1 + A_2)(1+|x|)^{-n-\delta} \tag{4.3.24}$$

for all $x \in \mathbf{R}^n$. Indeed, for $|x| \ge 1$ we express the left-hand side in (4.3.24) as

$$\left|\int_{\mathbf{R}^n} \big(K_{\varepsilon^{-1}}(x) - K_{\varepsilon^{-1}}(x-y)\big)\varphi(y)\,dy\right|.$$

Since φ is supported in $|y| \le 1/2$, we have $|x| \ge 2|y|$, and condition (4.3.20) yields that the expression on the left-hand side of (4.3.24) is bounded by

$$\frac{A_2}{|x|^{n+\delta}} \int_{\mathbf{R}^n} |y|^{\delta} |\varphi(y)|\,dy \le c\,\frac{A_2}{(1+|x|)^{n+\delta}}\,,$$

which proves (4.3.24) in the case $|x| \ge 1$. When $|x| < 1$, the left-hand side of (4.3.24) can be written as

$$(W_{\varepsilon^{-1}} * \varphi)(x) = \lim_{\delta_j \to 0} \int_{|x-y| \ge \delta_j} K_{\varepsilon^{-1}}(x-y)\varphi(y)\,dy \tag{4.3.25}$$

for some sequence $\delta_j \downarrow 0$; see the discussion in Section 4.3.2. The expression in (4.3.25) is equal to

$$I_1 + I_2 + I_3\,,$$

where

$$\begin{aligned}
I_1 &= \int_{|x-y| \ge \frac{1}{8}} K_{\varepsilon^{-1}}(x-y)\varphi(y)\,dy\,,\\
I_2 &= \int_{|x-y| \le \frac{1}{8}} K_{\varepsilon^{-1}}(x-y)\big(\varphi(y) - \varphi(x)\big)\,dy\,,\\
I_3 &= \varphi(x) \lim_{\delta_j \to 0} \int_{|x-y| \ge \delta_j} K_{\varepsilon^{-1}}(x-y)\,dy\,.
\end{aligned}$$

In I_1 we have $1/8 \le |x-y| \le 1 + 1/2 = 3/2$; hence I_1 is bounded by a multiple of A_1. Since $|\varphi(x) - \varphi(y)| \le c|x-y|$, the same is valid for I_2. Finally, I_3 is bounded by a multiple of A_3. Combining these facts yields the proof of (4.3.24) in the case $|x| < 1$ as well.

Use Corollary 2.1.12 to deduce that

$$\sup_{\varepsilon > 0} \left| f * \big((K_{\varepsilon^{-1}})^{(1)} - K_{\varepsilon^{-1}} * \varphi\big)_{\varepsilon} \right| \le c\,(A_1 + A_2 + A_3) M(f)\,.$$

Finally, take the supremum over $\varepsilon > 0$ in (4.3.23) and use (4.3.24) and Corollary 2.1.12 one more time to deduce the estimate

$$T^{(*)}(f) \le M(f * W) + C\,(A_1 + A_2 + A_3) M(f)\,,$$

where C depends on n and δ, thus concluding the proof of (4.3.22). □

4.3.5 Boundedness for Maximal Singular Integrals Implies Weak Type $(1,1)$ Boundedness

We now state and prove a result analogous to that in Theorem 4.3.3 for maximal singular integrals.

Theorem 4.3.5. *Let $K(x)$ be function on $\mathbf{R}^n \setminus \{0\}$ satisfying (4.3.4) with constant $A_1 < \infty$ and Hörmander's condition (4.3.12) with constant $A_2 < \infty$. Suppose that the operator $T^{(**)}$ as defined in (4.3.18) maps $L^2(\mathbf{R}^n)$ to itself with norm B. Then $T^{(**)}$ maps $L^1(\mathbf{R}^n)$ to $L^{1,\infty}(\mathbf{R}^n)$ with norm*

$$\big\|T^{(**)}\big\|_{L^1\to L^{1,\infty}} \le C_n(A_1+A_2+B),$$

where C_n is some dimensional constant.

Proof. The proof of this theorem is only a little more involved than the proof of Theorem 4.3.3. We fix an $L^1(\mathbf{R}^n)$ function f. We apply the Calderón–Zygmund decomposition of f at height $\gamma\alpha$ for some $\gamma,\alpha>0$. We then write $f=g+b$, where $b=\sum_j b_j$ and each b_j is supported in some cube Q_j. We define Q_j^* as the cube with the same center as Q_j and with sides parallel to the sides of Q_j having length $\ell(Q_j^*)=5\sqrt{n}\,\ell(Q_j)$. This is only a minor change compared with the definition of Q_j in Theorem 4.3.3. The main change in the proof is in the treatment of the term

$$\Big|\Big\{x\in\Big(\bigcup_j Q_j^*\Big)^c:\ |T^{(**)}(b)(x)|>\frac{\alpha}{2}\Big\}\Big|. \tag{4.3.26}$$

We show that for all $\gamma\le(2^{n+5}A_1)^{-1}$ we have

$$\Big|\Big\{x\in\Big(\bigcup_j Q_j^*\Big)^c:\ |T^{(**)}(b)(x)|>\frac{\alpha}{2}\Big\}\Big|\le 2^{n+8}A_2\frac{\|f\|_{L^1}}{\alpha}. \tag{4.3.27}$$

Let us conclude the proof of the theorem assuming for the moment the validity of (4.3.27). As in the proof of Theorem 4.3.3, we can show that

$$\Big|\Big\{x\in\mathbf{R}^n:\ |T^{(**)}(g)(x)|>\frac{\alpha}{2}\Big\}\Big|+\Big|\bigcup_j Q_j^*\Big|\le\Big(2^{n+2}B^2\gamma+\frac{(5\sqrt{n})^n}{\gamma}\Big)\frac{\|f\|_{L^1}}{\alpha}.$$

Combining this estimate with (4.3.27) and choosing

$$\gamma=(2^{n+5}(A_1+A_2+B))^{-1},$$

we obtain the required estimate

$$\big|\{x\in\mathbf{R}^n:\ |T^{(**)}(f)(x)|>\alpha\}\big|\le C_n(A_1+A_2+B)\frac{\|f\|_{L^1}}{\alpha}$$

with $C_n=2^{-3}+(5\sqrt{n})^n2^{n+5}+2^{n+8}$.

It remains to prove (4.3.27). This estimate will be a consequence of the fact that for $x \in \left(\bigcup_j Q_j^*\right)^c$ we have the key inequality

$$T^{(**)}(b)(x) \le 4E_1(x) + 2^{n+2}\alpha\gamma E_2(x) + 2^{n+3}\alpha\gamma A_1\,, \tag{4.3.28}$$

where

$$\begin{aligned} E_1(x) &= \sum_j \int_{Q_j} |K(x-y) - K(x-y_j)|\,|b_j(y)|\,dy\,, \\ E_2(x) &= \sum_j \int_{Q_j} |K(x-y) - K(x-y_j)|\,dy\,, \end{aligned}$$

and y_j is the center of Q_j.

If we had (4.3.28), then we could easily derive (4.3.27). Indeed, fix a γ satisfying $\gamma \le (2^{n+5}A_1)^{-1}$. Then we have $2^{n+3}\alpha\gamma A_1 < \frac{\alpha}{3}$, and using (4.3.28), we obtain

$$\begin{aligned} &\left|\left\{x \in \Big(\bigcup_j Q_j^*\Big)^c : |T^{(**)}(b)(x)| > \frac{\alpha}{2}\right\}\right| \\ &\qquad \le \left|\left\{x \in \Big(\bigcup_j Q_j^*\Big)^c : 4E_1(x) > \frac{\alpha}{12}\right\}\right| \\ &\qquad\qquad + \left|\left\{x \in \Big(\bigcup_j Q_j^*\Big)^c : 2^{n+2}\alpha\gamma E_2(x) > \frac{\alpha}{12}\right\}\right| \\ &\qquad \le \frac{48}{\alpha}\int_{(\bigcup_j Q_j^*)^c} E_1(x)\,dx + 2^{n+6}\gamma \int_{(\bigcup_j Q_j^*)^c} E_2(x)\,dx\,, \end{aligned} \tag{4.3.29}$$

since $\frac{\alpha}{2} = \frac{\alpha}{3} + \frac{\alpha}{12} + \frac{\alpha}{12}$. We have

$$\begin{aligned} &\int_{(\bigcup_j Q_j^*)^c} E_1(x)\,dx \\ &\qquad \le \sum_j \int_{Q_j} |b_j(y)| \int_{(Q_j^*)^c} |(K(x-y) - K(x-y_j)|\,dx\,dy \\ &\qquad \le \sum_j \int_{Q_j} |b_j(y)| \int_{|x-y_j| \ge 2|y-y_j|} |K(x-y) - K(x-y_j)|\,dx\,dy \\ &\qquad \le A_2 \sum_j \int_{Q_j} |b_j(y)|\,dy = A_2 \sum_j \|b_j\|_{L^1} \le A_2 2^{n+1}\|f\|_{L^1}\,, \end{aligned} \tag{4.3.30}$$

where we used the fact that if $x \in (Q_j^*)^c$, then $|x-y_j| \ge \frac{1}{2}\ell(Q_j^*) = \frac{5}{2}\sqrt{n}\,\ell(Q_j)$. But since $|y-y_j| \le \frac{\sqrt{n}}{2}\ell(Q_j)$, this implies that $|x-y_j| \ge 2|y-y_j|$. Here we used the fact that the diameter of a cube is equal to $\sqrt{n}$ times its side length. Likewise, we obtain that

$$\int_{(\bigcup_j Q_j^*)^c} E_2(x)\,dx \le A_2 \sum_j |Q_j| \le A_2 \frac{\|f\|_{L^1}}{\alpha\gamma}\,. \tag{4.3.31}$$

Combining (4.3.30) and (4.3.31) with (4.3.29) yields (4.3.27).

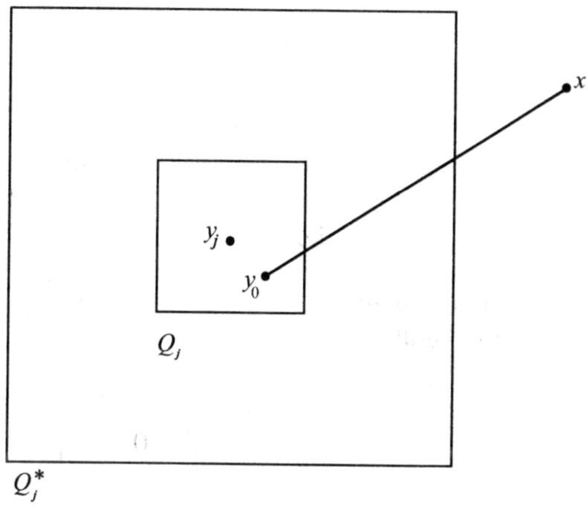

Fig. 4.3 The cubes Q_j and Q_j^*.

Therefore, the main task of the proof is to prove (4.3.28). Since $b = \sum_j b_j$, to estimate $T^{(**)}(b)$, it suffices to estimate each $|T^{(\varepsilon,N)}(b_j)|$ uniformly in ε and N. To achieve this we use the estimate

$$|T^{(\varepsilon,N)}(b_j)| \le |T^{(\varepsilon)}(b_j)| + |T^{(N)}(b_j)|, \tag{4.3.32}$$

noting that the truncated singular integrals $T^{(\varepsilon)}(b_j)$ are well defined. Indeed, say x lies in a compact set K_0. Pick M such that $K_0 - Q_j$ is contained in a ball $B(0,M)$. Then

$$|T^{(\varepsilon)}(b_j)(x)| \le |b_j| * |K^{(\varepsilon,M)}|(x),$$

which is finite a.e. as the convolution of two L^1 functions; thus the integral defining $T^{(\varepsilon)}(b_j)(x)$ converges absolutely and the expression $T^{(\varepsilon)}(b_j)(x)$ is well defined for almost all x.

We work with $T^{(\varepsilon)}$ and we note that $T^{(N)}$ can be treated similarly. Fix $x \notin \bigcup_j Q_j^*$ and $\varepsilon > 0$ and define

$$\begin{aligned} J_1(x,\varepsilon) &= \{j : \forall y \in Q_j \text{ we have } |x-y| < \varepsilon\}, \\ J_2(x,\varepsilon) &= \{j : \forall y \in Q_j \text{ we have } |x-y| > \varepsilon\}, \\ J_3(x,\varepsilon) &= \{j : \exists y \in Q_j \text{ we have } |x-y| = \varepsilon\}. \end{aligned}$$

Note that

$$T^{(\varepsilon)}(b_j)(x) = 0$$

whenever $x \notin \bigcup_j Q_j^*$ and $j \in J_1(x,\varepsilon)$. Also note that

$$K^{(\varepsilon)}(x-y) = K(x-y)$$

whenever $x \notin \bigcup_j Q_j^*$, $j \in J_2(x,\varepsilon)$ and $y \in Q_j$. Therefore,

$$\sup_{\varepsilon>0} |T^{(\varepsilon)}(b)(x)| \le \sup_{\varepsilon>0} \Big| \sum_{j\in J_2(x,\varepsilon)} T(b_j)(x) \Big| + \sup_{\varepsilon>0} \Big| \sum_{j\in J_3(x,\varepsilon)} T(b_j \chi_{|x-\cdot|\ge\varepsilon})(x) \Big| ,$$

but since

$$\sup_{\varepsilon>0} \Big| \sum_{j\in J_2(x,\varepsilon)} T(b_j)(x) \Big| \le \sum_j |T(b_j)(x)| \le E_1(x) , \tag{4.3.33}$$

it suffices to estimate the term

$$\sup_{\varepsilon>0} \Big| \sum_{j\in J_3(x,\varepsilon)} T(b_j \chi_{|x-\cdot|\ge\varepsilon})(x) \Big| .$$

We now make some geometric observations; see Figure 4.3. Fix $\varepsilon > 0$ and a cube Q_j with $j \in J_3(x,\varepsilon)$; recall that x lies in $(\bigcup_j Q_j^*)^c$. Then we have

$$\varepsilon \ge \frac{1}{2}\big(\ell(Q_j^*) - \ell(Q_j)\big) = \frac{1}{2}(5\sqrt{n}-1)\ell(Q_j) \ge 2\sqrt{n}\,\ell(Q_j) . \tag{4.3.34}$$

Since $j \in J_3(x,\varepsilon)$, there exists a $y_0 \in Q_j$ with

$$|x - y_0| = \varepsilon .$$

Using (4.3.34), we obtain that for any $y \in Q_j$ we have

$$\begin{aligned} \frac{\varepsilon}{2} \le \varepsilon - \sqrt{n}\,\ell(Q_j) \le |x-y_0| - |y-y_0| \le |x-y| ,\\ |x-y| \le |x-y_0| + |y-y_0| \le \varepsilon + \sqrt{n}\,\ell(Q_j) \le \frac{3\varepsilon}{2} . \end{aligned}$$

Therefore, we have proved that

$$\bigcup_{j\in J_3(x,\varepsilon)} Q_j \subseteq B(x, \tfrac{3\varepsilon}{2}) \setminus B(x, \tfrac{\varepsilon}{2}) .$$

Letting

$$c_j(\varepsilon) = \frac{1}{|Q_j|} \int_{Q_j} b_j(y) \chi_{|x-y|\ge\varepsilon}(y)\, dy ,$$

we note that in view of property (5) of the Calderón–Zygmund decomposition (Theorem 4.3.1), the estimate $|c_j(\varepsilon)| \le 2^{n+1}\alpha\gamma$ holds. Then

$$
\begin{aligned}
&\sup_{\varepsilon>0}\bigg|\sum_{j\in J_3(x,\varepsilon)}\int_{Q_j} K(x-y)b_j(y)\chi_{|x-y|\geq\varepsilon}(y)\,dy\bigg| \\
&\qquad\leq \sup_{\varepsilon>0}\bigg|\sum_{j\in J_3(x,\varepsilon)}\int_{Q_j} K(x-y)\big(b_j(y)\chi_{|x-y|\geq\varepsilon}(y)-c_j(\varepsilon)\big)\,dy\bigg| \\
&\qquad\qquad\qquad\qquad + \sup_{\varepsilon>0}\bigg|\sum_{j\in J_3(x,\varepsilon)} c_j(\varepsilon)\int_{Q_j} K(x-y)\,dy\bigg| \\
&\qquad\leq \sup_{\varepsilon>0}\bigg|\sum_{j\in J_3(x,\varepsilon)}\int_{Q_j} \big(K(x-y)-K(x-y_j)\big)\big(b_j(y)\chi_{|x-y|\geq\varepsilon}(y)-c_j(\varepsilon)\big)\,dy\bigg| \\
&\qquad\qquad\qquad\qquad + 2^{n+1}\alpha\gamma\sup_{\varepsilon>0}\int_{B(x,\frac{3\varepsilon}{2})\setminus B(x,\frac{\varepsilon}{2})}|K(x-y)|\,dy \\
&\qquad\leq \sum_j\int_{Q_j}|K(x-y)-K(x-y_j)|\big(|b_j(y)|+2^{n+1}\alpha\gamma\big)\,dy \\
&\qquad\qquad\qquad\qquad + 2^{n+1}\alpha\gamma\sup_{\varepsilon>0}\int_{\frac{\varepsilon}{2}\leq|x-y|\leq\frac{3\varepsilon}{2}}|K(x-y)|\,dy \\
&\qquad\leq E_1(x)+2^{n+1}\alpha\gamma E_2(x)+2^{n+1}\alpha\gamma(2A_1)\,.
\end{aligned}
$$

The last estimate, together with (4.3.33), with (4.3.32), and with the analogous estimate for $\sup_{N>0}|T^{(N)}(b_j)(x)|$ (which is similarly obtained), yields (4.3.28). □

The value of the previous theorem lies in the following: Since we know that for some sequences $\varepsilon_j\downarrow 0$, $N_j\uparrow\infty$ the pointwise limit $T^{(\varepsilon_j,N_j)}(f)$ exists a.e. for all f in a dense subclass of L^1, then Theorem 4.3.5 allows us to deduce that $T^{(\varepsilon_j,N_j)}(f)$ exists a.e. for all f in $L^1(\mathbf{R}^n)$.

If the singular integrals have kernels of the form $\Omega(x/|x|)|x|^{-n}$ with Ω in L^∞, such as the Hilbert transform and the Riesz transforms, then the upper truncations are not needed for K in (4.3.17). In this case

$$T_\Omega^{(\varepsilon)}(f)(x)=\int_{|y|\geq\varepsilon} f(x-y)\frac{\Omega(y/|y|)}{|y|^n}\,dy$$

is well defined for $f\in\bigcup_{1\leq p<\infty}L^p(\mathbf{R}^n)$ by Hölder's inequality and is equal to

$$\lim_{N\to\infty}\int_{\varepsilon\leq|y|\leq N} f(x-y)\frac{\Omega(y/|y|)}{|y|^n}\,dy\,.$$

Corollary 4.3.6. *The maximal Hilbert transform $H^{(*)}$ and the maximal Riesz transforms $R_j^{(*)}$ are weak type $(1,1)$. Secondly, $\lim_{\varepsilon\to0}H^{(\varepsilon)}(f)$ and $\lim_{\varepsilon\to0}R_j^{(\varepsilon)}(g)$ exist a.e. for all $f\in L^1(\mathbf{R})$ and $g\in L^1(\mathbf{R}^n)$, as $\varepsilon\to0$.*

Proof. Since the kernels $1/x$ on $\mathbf{R}$ and $x_j/|x|^n$ on $\mathbf{R}^n$ satisfy (4.3.10), the first statement in the corollary is an immediate consequence of Theorem 4.3.5. The second statement follows from Theorem 2.1.14 and Corollary 4.2.8, since these limits exist for Schwartz functions. □

Corollary 4.3.7. *Under the hypotheses of Theorem 4.3.5, $T^{(**)}$ maps $L^p(\mathbf{R}^n)$ to itself for $1<p<2$ with norm*

$$\|T^{(**)}\|_{L^p\to L^p} \le \frac{C_n(A_1+A_2+B)}{p-1},$$

where C_n is some dimensional constant.

Exercises

4.3.1. Let A_1 be defined in (4.3.4). Prove that

$$\frac{1}{2}A_1 \le \sup_{R>0}\frac{1}{R}\int_{|x|\le R}|K(x)|\,|x|\,dx \le 2A_1;$$

thus the expressions in (4.3.6) and (4.3.4) are equivalent.

4.3.2. Suppose that K is a locally integrable function on $\mathbf{R}^n\setminus\{0\}$ that satisfies (4.3.4). Suppose that $\delta_j\downarrow 0$. Prove that the principal value operation

$$W(\varphi)=\lim_{j\to\infty}\int_{\delta_j\le|x|\le 1}K(x)\varphi(x)\,dx$$

defines a distribution in $\mathscr{S}'(\mathbf{R}^n)$ if and only if the following limit exists:

$$\lim_{j\to\infty}\int_{\delta_j\le|x|\le 1}K(x)\,dx.$$

4.3.3. Suppose that a function K on $\mathbf{R}^n\setminus\{0\}$ satisfies condition (4.3.4) with constant A_1 and condition (4.3.12) with constant A_2.
(a) Show that the functions $K(x)\chi_{|x|\ge\varepsilon}$ also satisfy condition (4.3.12) uniformly in $\varepsilon>0$ with constant A_1+A_2.
(b) Obtain the same conclusion for the upper truncations $K(x)\chi_{|x|\le N}$.
(c) Deduce a similar conclusion for the double truncations $K^{(\varepsilon,N)}(x)=K(x)\chi_{\varepsilon\le|x|\le N}$.

4.3.4. Modify the proof of Theorem 4.3.5 to prove that if $T^{(**)}$ maps L^r to L^r for some $1<r<\infty$, and K satisfies condition (4.3.12), then $T^{(**)}$ maps L^1 to $L^{1,\infty}$.

4.3.5. Assume that T is a linear operator acting on measurable functions on $\mathbf{R}^n$ such that whenever a function f is supported in a cube Q, then $T(f)$ is supported in a fixed multiple of Q.
(a) Suppose that T maps L^p to itself for some $1<p<\infty$ with norm B. Prove that T extends to a bounded operator from L^1 to $L^{1,\infty}$ with norm a constant multiple of B.
(b) Suppose that T maps L^p to L^q for some $1<q<p<\infty$ with norm B. Prove that T extends to a bounded operator from L^1 to $L^{s,\infty}$ with norm a multiple of B, where

$$\frac{1}{p'} + \frac{1}{q} = \frac{1}{s}.$$

4.3.6. (a) Prove that the good function g in the Calderón–Zygmund decomposition of $f = g + b$ at height α lies in the Lorentz space $L^{q,1}$ for $1 \le q < \infty$. Moreover, for some dimensional constant C_n we have $\|g\|_{L^{q,1}} \le C_n \alpha^{1/q'} \|f\|_{L^1}$.
(b) Using this result, prove the following generalization of Theorem 4.3.3: If T maps $L^{q,1}$ to $L^{q,\infty}$ with norm B for some $1 < q < \infty$, then T is weak type $(1,1)$ with norm at most a multiple of $A_2 + B$.
(c) When $1 < q < \infty$, use the results of Exercise 1.1.12 and Exercise 1.4.7 to prove that if

$$|\{x : \, |T(\chi_E)(x)| > \alpha\}| \le B \frac{|E|}{\alpha^q}$$

for all subsets E of $\mathbf{R}^n$ with finite measure, then T is weak type $(1,1)$ with norm at most a multiple of $A_2 + B$.

4.3.7. Let K satisfy (4.3.12) for some $A_2 > 0$, let $W \in \mathscr{S}'(\mathbf{R}^n)$ be an extension of K on $\mathbf{R}^n$ as in (4.3.7), and let T be the operator given by convolution with W. Obtain the case $r = \infty$ in Theorem 4.3.3. Precisely, prove that if T maps $L^\infty(\mathbf{R}^n)$ to itself with constant B, then T has an extension on $L^1 + L^\infty$ that satisfies

$$\|T\|_{L^1 \to L^{1,\infty}} \le C_n' (A_2 + B),$$

and for $1 < p < \infty$ it satisfies

$$\|T\|_{L^p \to L^p} \le C_n \frac{1}{(p-1)^{1/p}} (A_2 + B),$$

where C_n, C_n' are constants that depend only on the dimension.
[*Hint:* Apply the Calderón–Zygmund decomposition $f = g + b$ at height $\alpha\gamma$, where $\gamma = (2^{n+1}B)^{-1}$. Since $|g| \le 2^n \alpha\gamma$, observe that

$$|\{x : \, |T(f)(x)| > \alpha\}| \le |\{x : \, |T(b)(x)| > \alpha/2\}|.$$

For the interpolation use the result of Exercise 1.3.2.]

4.3.8. (*Calderón–Zygmund decomposition on L^q*) Fix a function $f \in L^q(\mathbf{R}^n)$ for some $1 \le q < \infty$ and let $\alpha > 0$. Then there exist functions g and b on $\mathbf{R}^n$ such that

(1) $f = g + b$.

(2) $\|g\|_{L^q} \le \|f\|_{L^q}$ and $\|g\|_{L^\infty} \le 2^{\frac{n}{q}} \alpha$.

(3) $b = \sum_j b_j$, where each b_j is supported in a cube Q_j. Furthermore, the cubes Q_k and Q_j have disjoint interiors when $j \ne k$.

(4) $\|b_j\|_{L^q}^q \le 2^{n+q} \alpha^q |Q_j|$.

(5) $\int_{Q_j} b_j(x)\, dx = 0$.

(6) $\sum_j |Q_j| \le \alpha^{-q} \|f\|_{L^q}^q$.

(7) $\|b\|_{L^q} \le 2^{\frac{n+q}{q}} \|f\|_{L^q}$ and $\|b\|_{L^1} \le 2\alpha^{1-q} \|f\|_{L^q}^q$.

[*Hint:* Imitate the basic idea of the proof of Theorem 4.3.1, but select a cube Q if $\left(\frac{1}{|Q|}\int_Q |f(x)|^q\,dx\right)^{1/q} > \alpha$. Define g and b as in the proof of Theorem 4.3.1.]

4.3.9. Let $f \in L^1(\mathbf{R}^n)$. Then for any $\alpha > 0$, prove that there exist disjoint cubes Q_j in $\mathbf{R}^n$ such that the set $E_\alpha = \{x \in \mathbf{R}^n : M_c(f)(x) > \alpha\}$ is contained in $\bigcup_j 3Q_j$ and $\frac{\alpha}{4^n} < \frac{1}{|Q_j|}\int_{Q_j} |f(t)|\,dt \le \frac{\alpha}{2^n}$.
[*Hint:* For given $\alpha > 0$, select all maximal dyadic cubes $Q_j(\alpha)$ such that the average of f over them is bigger than α. Given $x \in E_\alpha$, pick a cube R that contains x such that the average of $|f|$ over R is bigger than α and find a dyadic cube Q such that $2^{-n}|Q| < |R| \le |Q|$ and that $\int_{R\cap Q} |f|\,dx > 2^{-n}\alpha|R|$. Conclude that Q is contained in some $Q_k(4^{-n}\alpha)$ and thus R is contained in $3Q_k(4^{-n}\alpha)$. The collection of all $Q_j = Q_j(4^{-n}\alpha)$ is the required one.]

4.3.10. Let $K(x)$ be a function on $\mathbf{R}^n \setminus \{0\}$ that satisfies $|K(x)| \le A|x|^{-n}$. Let $\eta(x)$ be a smooth function that is equal to 1 when $|x| \ge 2$ and vanishes when $|x| \le 1$. For $f \in L^p$, $1 \le p < \infty$, define truncated singular integral operators

$$T^{(\varepsilon)}(f)(x) = \int_{|y|\ge\varepsilon} K(y) f(x-y)\,dy,$$

$$T_\eta^{(\varepsilon)}(f)(x) = \int_{\mathbf{R}^n} \eta(y/\varepsilon) K(y) f(x-y)\,dy.$$

Show that the truncated maximal singular integral $T^{(*)}(f) = \sup_{\varepsilon>0} |T^{(\varepsilon)}(f)|$ is L^p bounded for $1 < p < \infty$ if and only if the smoothly truncated maximal singular integral $T_\eta^{(*)}(f) = \sup_{\varepsilon>0} |T_\eta^{(\varepsilon)}(f)|$ is L^p bounded. Formulate an analogous statement for $p = 1$.

4.3.11. (M. Mastylo) Let $1 \le p < \infty$. Suppose that T_ε are linear operators defined on $L^p(\mathbf{R}^n)$ such that for all $f \in L^p(\mathbf{R}^n)$ we have $|T_\varepsilon(f)| \le A\varepsilon^{-a}\|f\|_{L^p}$ for some $0 < a, A < \infty$. Also suppose that there is a constant $C < \infty$ such that the maximal operator $T_*(f) = \sup_{\varepsilon>0} |T_\varepsilon(f)|$ satisfies $\|T_*(h)\|_{L^p} \le C\|h\|_{L^p}$ for all $h \in \mathscr{S}(\mathbf{R}^n)$. Prove that the same inequality is valid for all $f \in L^p(\mathbf{R}^n)$.
[*Hint:* For a fixed $\delta > 0$ define $S_\delta(f) = \sup_{\varepsilon>\delta} |T_\varepsilon(f)|$, which is a subadditive functional on $L^p(\mathbf{R}^n)$. For a fixed $f_0 \in L^p(\mathbf{R}^n)$ define a linear space $X_0 = \{\lambda f_0 : \lambda \in \mathbf{C}\}$ and a linear functional T_0 on X_0 by setting $T_0(\lambda f_0) = \lambda S_\delta(f_0)$. By the Hahn–Banach theorem there is an extension $\widetilde{T}_0$ of T_0 that satisfies $|\widetilde{T}_0(f)| \le S_\delta(f)$ for all $f \in L^p(\mathbf{R}^n)$. Since S_δ is L^p is bounded on Schwartz functions with norm at most C, then so is $\widetilde{T}_0$. But $\widetilde{T}_0$ is linear and by density it is bounded on $L^p(\mathbf{R}^n)$ with norm at most C; consequently, $\|S_\delta(f_0)\|_{L^p} = \|T_0(f_0)\|_{L^p} = \|\widetilde{T}_0(f_0)\|_{L^p} \le C\|f_0\|_{L^p}$. The required conclusion for T_* follows by Fatou's lemma.]

4.4 Sufficient Conditions for L^p Boundedness

We have used the Calderón–Zygmund decomposition to prove weak type $(1,1)$ boundedness for singular integral and maximal singular integral operators, assuming that these operators are already L^2 bounded. It is therefore natural to ask for sufficient conditions that imply L^2 boundedness for such operators. Precisely, what are sufficient conditions on functions K on $\mathbf{R}^n \setminus \{0\}$ so that the corresponding singular and maximal singular integral operators associated with K are L^2 bounded? We saw in Section 4.2 that if K has the special form $K(x) = \Omega(x/|x|)/|x|^n$ for some $\Omega \in L^1(\mathbf{S}^{n-1})$ with mean value zero, then condition (4.2.16) is necessary and sufficient for the L^2 boundedness of T, while the L^2 boundedness of $T^{(*)}$ requires the stronger smoothness condition (4.2.23).

For the general K considered in this section (for which the corresponding operator does not necessarily commute with dilations), we only give some sufficient conditions for L^2 boundedness of T and $T^{(**)}$.

Throughout this section K denotes a locally integrable function on $\mathbf{R}^n \setminus \{0\}$ that satisfies the "size" condition

$$\sup_{R>0} \int_{R\le|x|\le 2R} |K(x)|\,dx = A_1 < \infty, \tag{4.4.1}$$

the "smoothness" condition

$$\sup_{y\neq 0} \int_{|x|\ge 2|y|} |K(x-y) - K(x)|\,dx = A_2 < \infty, \tag{4.4.2}$$

and the "cancellation" condition

$$\sup_{0<R_1<R_2<\infty} \left| \int_{R_1<|x|<R_2} K(x)\,dx \right| = A_3 < \infty, \tag{4.4.3}$$

for some $A_1, A_2, A_3 > 0$. As mentioned earlier, condition (4.4.2) is often referred to as Hörmander's condition. In this section we show that these three conditions give rise to convolution operators that are bounded on L^p.

4.4.1 Sufficient Conditions for L^p Boundedness of Singular Integrals

We first note that under conditions (4.4.1), (4.4.2), and (4.4.3), there exists a tempered distribution W that coincides with K on $\mathbf{R}^n \setminus \{0\}$. Indeed, condition (4.4.3) implies that there exists a sequence $\delta_j \downarrow 0$ such that

$$\lim_{j\to\infty} \int_{\delta_j<|x|\le 1} K(x)\,dx = L$$

exists. Using (4.3.8), we conclude that there exists such a tempered distribution W. Note that we must have $|L| \le A_3$.

We observe that the difference of two distributions W and W' that coincide with K on $\mathbf{R}^n \setminus \{0\}$ must be supported at the origin.

Theorem 4.4.1. *Assume that K satisfies (4.4.1), (4.4.2), and (4.4.3), and let W be a tempered distribution that coincides with K on $\mathbf{R}^n \setminus \{0\}$. Then we have*

$$\sup_{0<\varepsilon<N<\infty} \sup_{\xi \neq 0} |(K\chi_{\varepsilon<|\cdot|<N})\widehat{\ }(\xi)| \le 15(A_1+A_2+A_3). \tag{4.4.4}$$

Thus the operator given by convolution with W maps $L^2(\mathbf{R}^n)$ to itself with norm at most $15(A_1+A_2+A_3)$. Consequently, it also maps $L^1(\mathbf{R}^n)$ to $L^{1,\infty}(\mathbf{R}^n)$ with bound at most a dimensional constant multiple of $A_1+A_2+A_3$ and $L^p(\mathbf{R}^n)$ to itself with bound at most $C_n \max(p,(p-1)^{-1})(A_1+A_2+A_3)$, for some dimensional constant C_n, whenever $1<p<\infty$.

Proof. Let us set $K^{(\varepsilon,N)}(x) = K(x)\chi_{\varepsilon<|x|<N}$. If we prove (4.4.4), then for all f in $\mathscr{S}(\mathbf{R}^n)$ we will have the estimate

$$\big\|f * K^{(\delta_j,j)}\big\|_{L^2} \le 15(A_1+A_2+A_3)\big\|f\big\|_{L^2}$$

uniformly in j. Using this, (4.3.9), and Fatou's lemma, we obtain that

$$\big\|f * W\big\|_{L^2} \le 15(A_1+A_2+A_3)\big\|f\big\|_{L^2},$$

thus proving the second conclusion of the theorem.

Let us now fix a ξ with $\varepsilon < |\xi|^{-1} < N$ and prove (4.4.4). Write $\widehat{K^{(\varepsilon,N)}}(\xi) = I_1(\xi)+I_2(\xi)$, where

$$\begin{aligned} I_1(\xi) &= \int_{\varepsilon<|x|<|\xi|^{-1}} K(x)e^{-2\pi i x\cdot\xi}\,dx, \\ I_2(\xi) &= \int_{|\xi|^{-1}<|x|<N} K(x)e^{-2\pi i x\cdot\xi}\,dx. \end{aligned}$$

We now have

$$I_1(\xi) = \int_{\varepsilon<|x|<|\xi|^{-1}} K(x)\,dx + \int_{\varepsilon<|x|<|\xi|^{-1}} K(x)\,(e^{-2\pi i x\cdot\xi}-1)\,dx. \tag{4.4.5}$$

It follows that

$$|I_1(\xi)| \le A_3 + 2\pi|\xi|\int_{|x|<|\xi|^{-1}} |x|\,|K(x)|\,dx \le A_3 + 2\pi(2A_1)$$

uniformly in ε. Let us now examine $I_2(\xi)$. Let $z = \frac{\xi}{2|\xi|^2}$ so that $e^{2\pi i z\cdot\xi} = -1$ and $2|z| = |\xi|^{-1}$. By changing variables $x = x' - z$, rewrite I_2 as

$$I_2(\xi) = -\int_{|\xi|^{-1}<|x'-z|<N} K(x'-z)\, e^{-2\pi i x'\cdot\xi}\, dx';$$

hence averaging gives

$$I_2(\xi) = \frac{1}{2}\int_{|\xi|^{-1}<|x|<N} K(x) e^{-2\pi i x\cdot\xi} dx - \frac{1}{2}\int_{|\xi|^{-1}<|x-z|<N} K(x-z)\, e^{-2\pi i x\cdot\xi}\, dx.$$

Now use that

$$\int_A F\,dx - \int_B G\,dx = \int_B (F-G)\,dx + \int_{A\setminus B} F\,dx - \int_{B\setminus A} F\,dx \tag{4.4.6}$$

to write $I_2(\xi) = J_1(\xi) + J_2(\xi) + J_3(\xi) + J_4(\xi) + J_5(\xi)$, where

$$J_1(\xi) = +\frac{1}{2} \int\limits_{|\xi|^{-1}<|x-z|<N} \big(K(x) - K(x-z)\big) e^{-2\pi i x\cdot\xi}\, dx, \tag{4.4.7}$$

$$J_2(\xi) = +\frac{1}{2} \int\limits_{\substack{|\xi|^{-1}<|x|<N \\ |x-z|\le|\xi|^{-1}}} K(x)\, e^{-2\pi i x\cdot\xi}\, dx, \tag{4.4.8}$$

$$J_3(\xi) = +\frac{1}{2} \int\limits_{\substack{|\xi|^{-1}<|x|<N \\ |x-z|\ge N}} K(x)\, e^{-2\pi i x\cdot\xi}\, dx, \tag{4.4.9}$$

$$J_4(\xi) = -\frac{1}{2} \int\limits_{\substack{|\xi|^{-1}<|x-z|<N \\ |x|\le|\xi|^{-1}}} K(x)\, e^{-2\pi i x\cdot\xi}\, dx, \tag{4.4.10}$$

$$J_5(\xi) = -\frac{1}{2} \int\limits_{\substack{|\xi|^{-1}<|x-z|<N \\ |x|\ge N}} K(x)\, e^{-2\pi i x\cdot\xi}\, dx. \tag{4.4.11}$$

Since $2|z| = |\xi|^{-1}$, $J_1(\xi)$ is bounded in absolute value by $\frac{1}{2}A_2$, in view of (4.4.2).

Next observe that $|\xi|^{-1} \le |x| \le \frac{3}{2}|\xi|^{-1}$ in (4.4.8), while $\frac{1}{2}|\xi|^{-1} \le |x| \le |\xi|^{-1}$ in (4.4.10); hence both of these terms are bounded by $\frac{1}{2}A_1$. Finally, we have $\frac{1}{2}N < |x| < N$ in (4.4.9) (since $|x| > N - \frac{1}{2}|\xi|^{-1}$), and similarly we have $N \le |x| < \frac{3}{2}N$ in (4.4.11). Thus both J_3 and J_5 are bounded above by $\frac{1}{2}A_1$.

We are left to consider the cases $\varepsilon < N \le |\xi|^{-1}$ and $|\xi|^{-1} \le \varepsilon < N$. In the first case we estimate

$$\int_{\varepsilon<|x|<N} K(x) e^{-2\pi i x\cdot\xi}\, dx$$

by adapting the previous argument for the term I_1, while in the second case we run the argument used for the term I_2 to complete the proof. $\square$

4.4.2 An Example

We now give an example of a distribution that satisfies conditions (4.4.1), (4.4.2), and (4.4.3).

Example 4.4.2. Let τ be a nonzero real number and let $K(x) = \frac{1}{|x|^{n+i\tau}}$ defined for $x \in \mathbf{R}^n \setminus \{0\}$. For a sequence $\delta_k \downarrow 0$ and φ a Schwartz function on $\mathbf{R}^n$, define

$$\langle W, \varphi \rangle = \lim_{k\to\infty} \int_{\delta_k \le |x|} \varphi(x) \frac{dx}{|x|^{n+i\tau}}, \tag{4.4.12}$$

whenever the limit exists. We claim that for some choices of sequences δ_k, W is a well defined tempered distribution on $\mathbf{R}^n$. Take, for example, $\delta_k = e^{-2\pi k/\tau}$. For this sequence δ_k, observe that

$$\int_{\delta_k \le |x| \le 1} \frac{1}{|x|^{n+i\tau}} \, dx = \omega_{n-1} \frac{1 - (e^{-2\pi k/\tau})^{-i\tau}}{-i\tau} = 0,$$

and thus

$$\langle W, \varphi \rangle = \int_{|x|\le 1} (\varphi(x) - \varphi(0)) \frac{dx}{|x|^{n+i\tau}} + \int_{|x|\ge 1} \varphi(x) \frac{dx}{|x|^{n+i\tau}}, \tag{4.4.13}$$

which implies that $W \in \mathscr{S}'(\mathbf{R}^n)$, since

$$\big|\langle W, \varphi \rangle\big| \le C\big[\|\nabla\varphi\|_{L^\infty} + \big\| |x|\, \varphi(x) \big\|_{L^\infty}\big].$$

If φ is supported in $\mathbf{R}^n \setminus \{0\}$, then

$$\langle W, \varphi \rangle = \int K(x)\varphi(x)\, dx.$$

Therefore W coincides with the function K away from the origin. Moreover, (4.4.1) and (4.4.2) are clearly satisfied for K, while (4.4.3) is also satisfied, since

$$\left| \int_{R_1<|x|<R_2} \frac{1}{|x|^{n+i\tau}} \, dx \right| = \omega_{n-1} \left| \frac{R_1^{-i\tau} - R_2^{-i\tau}}{-i\tau} \right| \le \frac{2\omega_{n-1}}{|\tau|}.$$

Remark 4.4.3. It is important to emphasize that the limit in (4.4.12) *may not* exist for all sequences $\delta_k \to 0$. For example, the limit in (4.4.12) does not exist if $\delta_k = e^{-\pi k/\tau}$. Moreover, for a different choice of a sequence δ_k for which the limit in (4.4.12) exists (for example, $\delta_k = e^{-\pi(2k+1)/\tau}$), we obtain a different distribution W_1 that coincides with the function $K(x) = |x|^{-n-i\tau}$.

We discuss a point of caution. We can directly check that the distributions W defined by (4.4.12) are not homogeneous distributions of degree $-n-i\tau$. In fact, the only homogeneous distribution of degree $-n-i\tau$ that coincides with the function $|x|^{-n-i\tau}$ away from zero is a multiple of the distribution $u_{-n-i\tau}$, where u_z is defined in (2.4.6). Let us investigate the relationship between $u_{-n-i\tau}$ and W defined in (4.4.13). Recall that (2.4.7) gives

$$\begin{aligned}\langle u_{-n-i\tau},\varphi\rangle &= \int_{|x|\ge 1}\varphi(x)\frac{\pi^{-i\frac{\tau}{2}}}{\Gamma(-i\frac{\tau}{2})}|x|^{-n-i\tau}\,dx\\ &\quad+\int_{|x|\le 1}(\varphi(x)-\varphi(0))\frac{\pi^{-i\frac{\tau}{2}}}{\Gamma(-i\frac{\tau}{2})}|x|^{-n-i\tau}\,dx+\frac{\omega_{n-1}\pi^{-i\frac{\tau}{2}}}{-i\tau\Gamma(-i\frac{\tau}{2})}\varphi(0)\,.\end{aligned}$$

Using (4.4.13), we conclude that $u_{-n-i\tau}-c_1W=c_2\delta_0$ for suitable nonzero constants c_1 and c_2. Since the Dirac mass at the origin is not a homogeneous distribution of degree $-n-i\tau$, it follows that neither is W.

Since $\widehat{u_{-n-i\tau}}=u_{i\tau}=c_3|\xi|^{i\tau}$, the identity $u_{-n-i\tau}-c_1W=c_2\delta_0$ can be used to obtain a formula for the Fourier transform of W and thus produce a different proof that convolution with W is a bounded operator on $L^2(\mathbf{R}^n)$.

4.4.3 Necessity of the Cancellation Condition

Although conditions (4.4.1), (4.4.2), and (4.4.3) are sufficient for L^2 boundedness, they are not necessary. However, (4.4.3) is also necessary. We have the following:

Proposition 4.4.4. *Suppose that K is a function on $\mathbf{R}^n\setminus\{0\}$ that satisfies (4.4.1). Let W be a tempered distribution on $\mathbf{R}^n$ extending K given by (4.3.7). If the operator $T(f)=f*W$ maps $L^2(\mathbf{R}^n)$ to itself (equivalently if $\widehat{W}$ is an L^∞ function), then the function K must satisfy (4.4.3).*

Proof. Pick a radial $\mathscr{C}^\infty$ function φ supported in the ball $|x|\le 2$ with $0\le\varphi\le 1$, and $\varphi(x)=1$ when $|x|\le 1$. For $R>0$ let $\varphi^R(x)=\varphi(x/R)$. Fourier inversion for distributions gives the second equality,

$$(W*\varphi^R)(0)=\langle W,\varphi^R\rangle=\langle\widehat{W},\widehat{\varphi^R}\rangle=\int_{\mathbf{R}^n}\widehat{W}(\xi)R^n\widehat{\varphi}(R\xi)\,d\xi\,,$$

and the preceding identity implies that

$$|(W*\varphi^R)(0)|\le\big\|\widehat{W}\big\|_{L^\infty}\big\|\widehat{\varphi}\big\|_{L^1}=\big\|T\big\|_{L^2\to L^2}\big\|\widehat{\varphi}\big\|_{L^1}$$

uniformly in $R>0$. Fix $0<R_1<R_2<\infty$. If $R_2\le 2R_1$, we have

$$\left|\int_{R_1<|x|<R_2}K(x)\,dx\right|\le\int_{R_1<|x|<2R_1}|K(x)|\,dx\le A_1\,,$$

which implies the required conclusion. We may therefore assume that $2R_1<R_2$. Since the part of the integral in (4.4.3) over the set $R_1<|x|<2R_1$ is controlled by

A_1, it suffices to control the integral of $K(x)$ over the set $2R_1 < |x| < R_2$. Since the function $\varphi^{R_2} - \varphi^{R_1}$ is supported away from the origin, the action of the distribution W on it can be written as integration against the function K. We have

$$\int_{\mathbf{R}^n} K(x)(\varphi^{R_2}(x) - \varphi^{R_1}(x))\,dx$$
$$= \int\limits_{2R_1<|x|<R_2} K(x)\,dx + \int\limits_{R_1<|x|<2R_1} K(x)(1-\varphi^{R_1}(x))\,dx + \int\limits_{R_2<|x|<2R_2} K(x)\varphi^{R_2}(x)\,dx.$$

The sum of the last two integrals is bounded by $3A_1$ (since $0 \le \varphi \le 1$), while the first integral is equal to

$$(W * \varphi^{R_2})(0) - (W * \varphi^{R_1})(0)$$

and is therefore bounded by $2\|T\|_{L^2\to L^2}\|\widehat{\varphi}\|_{L^1}$. We conclude that the function K must satisfy (4.4.3) with constant

$$A_3 \le 3A_1 + 2\|\widehat{\varphi}\|_{L^1}\|T\|_{L^2\to L^2} \le c\big(A_1 + \|T\|_{L^2\to L^2}\big).$$

□

4.4.4 Sufficient Conditions for L^p Boundedness of Maximal Singular Integrals

We now discuss the analogous result to Theorem 4.4.1 for the maximal singular integral operator $T^{(**)}$.

Theorem 4.4.5. *Suppose that K satisfies (4.4.1), (4.4.2), and (4.4.3) and let $T^{(**)}$ be as in* (4.3.18). *Then $T^{(**)}$ is bounded on $L^p(\mathbf{R}^n)$, $1 < p < \infty$, with norm*

$$\|T^{(**)}\|_{L^p\to L^p} \le C_n \max(p,(p-1)^{-1})(A_1 + A_2 + A_3),$$

where C_n is a dimensional constant.

Proof. We first define an operator T associated with K that satisfies (4.4.1), (4.4.2), and (4.4.3). Because of condition (4.4.3), there exists a sequence $\delta_j \downarrow 0$ such that

$$\lim_{j\to\infty} \int_{\delta_j<|x|\le 1} K(x)\,dx$$

exists. Therefore, for $\varphi \in \mathscr{S}(\mathbf{R}^n)$ we can define a tempered distribution

$$\langle W, \varphi\rangle = \lim_{j\to\infty} \int_{\delta_j\le|x|\le j} K(x)\varphi(x)\,dx$$

and an operator T given by $T(f) = f * W$ for $f \in \mathscr{S}(\mathbf{R}^n)$. In view of Theorems 4.4.1 and 4.3.3, T admits an L^p bounded extension ($1 < p < \infty$) with

$$\|T\|_{L^p \to L^p} \le c_n \max(p, (p-1)^{-1})(A_1 + A_2 + A_3) \tag{4.4.14}$$

and is weak type $(1,1)$. This extension is still denoted by T.

Fix $1 < p < \infty$ and $f \in L^p(\mathbf{R}^n) \cap L^\infty(\mathbf{R}^n)$ with compact support. We have

$$\begin{aligned}
&T^{(\varepsilon,N)}(f)(x) \\
&= \int_{\varepsilon \le |x-y| < N} K(x-y) f(y)\,dy = T^{(\varepsilon)}(f)(x) - T^{(N)}(f)(x) \\
&= \int_{\varepsilon \le |x-y|} K(x-y) f(y)\,dy - \int_{N \le |x-y|} K(x-y) f(y)\,dy \\
&= \int_{\varepsilon \le |x-y|} (K(x-y) - K(z_1 - y)) f(y)\,dy + \int_{\varepsilon \le |x-y|} K(z_1 - y) f(y)\,dy \\
&\qquad - \int_{N \le |x-y|} (K(x-y) - K(z_2 - y)) f(y)\,dy - \int_{N \le |x-y|} K(z_2 - y) f(y)\,dy \\
&= \int_{\varepsilon \le |x-y|} (K(x-y) - K(z_1 - y)) f(y)\,dy + T(f)(z_1) - T(f\chi_{|x-\cdot|<\varepsilon})(z_1) \\
&\qquad - \int_{N \le |x-y|} (K(x-y) - K(z_2 - y)) f(y)\,dy - T(f)(z_2) + T(f\chi_{|x-\cdot|<N})(z_2),
\end{aligned}$$

where z_1 and z_2 are arbitrary points in $\mathbf{R}^n$ that satisfy $|z_1 - x| \le \frac{\varepsilon}{2}$ and $|z_2 - x| \le \frac{N}{2}$. We used that f has compact support in order to be able to write $T^{(\varepsilon)}(f)(x)$ and $T^{(N)}(f)(x)$ as convergent integrals for almost every x.

At this point we take absolute values, average over $|z_1 - x| \le \frac{\varepsilon}{2}$ and $|z_2 - x| \le \frac{N}{2}$, and we apply Hölder's inequality in two terms. We obtain the estimate

$$\begin{aligned}
&|T^{(\varepsilon,N)}(f)(x)| \\
&\le \frac{1}{v_n}\left(\frac{2}{\varepsilon}\right)^n \int_{|z_1-x| \le \frac{\varepsilon}{2}} \int_{|x-y| \ge \varepsilon} |K(x-y) - K(z_1-y)|\,|f(y)|\,dy\,dz_1 \\
&\quad + \frac{1}{v_n}\left(\frac{2}{\varepsilon}\right)^n \int_{|z_1-x| \le \frac{\varepsilon}{2}} |T(f)(z_1)|\,dz_1 \\
&\quad + \left(\frac{1}{v_n}\left(\frac{2}{\varepsilon}\right)^n \int_{|z_1-x| \le \frac{\varepsilon}{2}} |T(f\chi_{|x-\cdot|<\varepsilon})(z_1)|^p\,dz_1\right)^{\frac{1}{p}} \\
&\quad + \frac{1}{v_n}\left(\frac{2}{N}\right)^n \int_{|z_2-x| \le \frac{N}{2}} \int_{|x-y| \ge N} |K(x-y) - K(z_2-y)|\,|f(y)|\,dy\,dz_2 \\
&\quad + \frac{1}{v_n}\left(\frac{2}{N}\right)^n \int_{|z_2-x| \le \frac{N}{2}} |T(f)(z_2)|\,dz_2 \\
&\quad + \left(\frac{1}{v_n}\left(\frac{2}{N}\right)^n \int_{|z_2-x| \le \frac{N}{2}} |T(f\chi_{|x-\cdot|<N})(z_2)|^p\,dz_2\right)^{\frac{1}{p}},
\end{aligned}$$

where v_n is the volume of the unit ball in $\mathbf{R}^n$. Applying condition (4.4.2) and estimate (4.4.14), we obtain for f in $L^p(\mathbf{R}^n)\cap L^\infty(\mathbf{R}^n)$ with compact support that

$$\begin{aligned}
&|T^{(\varepsilon,N)}(f)(x)|\\
&\le \frac{1}{v_n}\Big(\frac{2}{\varepsilon}\Big)^n\int_{|z_1-x|\le\frac{\varepsilon}{2}}|T(f)(z_1)|\,dz_1+\frac{1}{v_n}\Big(\frac{2}{N}\Big)^n\int_{|z_2-x|\le\frac{N}{2}}|T(f)(z_2)|\,dz_2\\
&\quad+c_n\Big(\sum_{j=1}^3 A_j\Big)\max(p,(p-1)^{-1})\Big(\frac{1}{v_n}\Big(\frac{2}{\varepsilon}\Big)^n\int_{|z_1-x|\le\varepsilon}|f(z_1)|^p\,dz_1\Big)^{\frac{1}{p}}\\
&\quad+c_n\Big(\sum_{j=1}^3 A_j\Big)\max(p,(p-1)^{-1})\Big(\frac{1}{v_n}\Big(\frac{2}{N}\Big)^n\int_{|z_2-x|\le N}|f(z_2)|^p\,dz_2\Big)^{\frac{1}{p}}\\
&\quad+2A_2\big\|f\big\|_{L^\infty}.
\end{aligned}$$

We now use density to remove the compact support condition on f and obtain the last displayed estimate for all functions f in $L^p(\mathbf{R}^n)\cap L^\infty(\mathbf{R}^n)$. Taking the supremum over all $0<\varepsilon<N$ and over all $N>0$, we deduce that for all f in $L^p(\mathbf{R}^n)\cap L^\infty(\mathbf{R}^n)$ we have the estimate

$$T^{(**)}(f)(x)\le 2A_2\big\|f\big\|_{L^\infty}+S_p(f)(x), \tag{4.4.15}$$

where S_p is the sublinear operator defined by

$$S_p(f)(x)=2M(T(f))(x)+3^{n+1}c_n\Big(\sum_{j=1}^3 A_j\Big)\max(p,(p-1)^{-1})(M(|f|^p)(x))^{\frac{1}{p}},$$

and M is the Hardy–Littlewood maximal operator.

Recalling that M maps L^1 to $L^{1,\infty}$ with bound at most 3^n and also L^p to $L^{p,\infty}$ with bound at most $2\cdot 3^{n/p}$ for $1<p<\infty$ (Exercise 2.1.4), we conclude that S_p maps $L^p(\mathbf{R}^n)$ to $L^{p,\infty}(\mathbf{R}^n)$ with norm at most

$$\big\|S_p\big\|_{L^p\to L^{p,\infty}}\le\widetilde{c}_n(A_1+A_2+A_3)\max(p,(p-1)^{-1}), \tag{4.4.16}$$

where $\widetilde{c}_n$ is another dimensional constant.

Now write $f=f_\alpha+f^\alpha$, where

$$f_\alpha=f\chi_{|f|\le\alpha/(16A_2)}\qquad\text{and}\qquad f^\alpha=f\chi_{|f|>\alpha/(16A_2)}.$$

The function f_α is in $L^\infty\cap L^p$ and f^α is in $L^1\cap L^p$. Moreover, we see that

$$\big\|f^\alpha\big\|_{L^1}\le(16A_2/\alpha)^{p-1}\big\|f\big\|_{L^p}^p. \tag{4.4.17}$$

Apply the Calderón–Zygmund decomposition (Theorem 4.3.1) to the function f^α at height $\alpha\gamma$ to write $f^\alpha=g^\alpha+b^\alpha$, where g^α is the good function and b^α is the bad function of this decomposition. Using (4.3.1), we obtain

$$\big\|g^\alpha\big\|_{L^p} \le 2^{n/p'}(\alpha\gamma)^{1/p'}\big\|f^\alpha\big\|_{L^1}^{1/p} \le 2^{(n+4)/p'}(A_2\gamma)^{1/p'}\big\|f\big\|_{L^p}. \tag{4.4.18}$$

We now use (4.4.15) to get

$$|\{x\in\mathbf{R}^n:\, T^{(**)}(f)(x)>\alpha\}| \le b_1+b_2+b_3, \tag{4.4.19}$$

where

$$\begin{aligned} b_1 &= |\{x\in\mathbf{R}^n:\, 2A_2\big\|f_\alpha\big\|_{L^\infty}+S_p(f_\alpha)(x)>\alpha/4\}|,\\ b_2 &= |\{x\in\mathbf{R}^n:\, 2A_2\big\|g^\alpha\big\|_{L^\infty}+S_p(g^\alpha)(x)>\alpha/4\}|,\\ b_3 &= |\{x\in\mathbf{R}^n:\, T^{(**)}(b^\alpha)(x)>\alpha/2\}|. \end{aligned}$$

Observe that $2A_2\big\|f_\alpha\big\|_{L^\infty}\le\alpha/8$. Selecting $\gamma=2^{-n-5}(A_1+A_2)^{-1}$ and using property (2) in Theorem 4.3.1, we obtain

$$2A_2\big\|g^\alpha\big\|_{L^\infty}\le A_2 2^{n+1}\alpha\gamma\le\alpha 2^{-4}<\frac{\alpha}{8}$$

and therefore

$$\begin{aligned} b_1 &\le |\{x\in\mathbf{R}^n:\, S_p(f_\alpha)(x)>\alpha/8\}|,\\ b_2 &\le |\{x\in\mathbf{R}^n:\, S_p(g^\alpha)(x)>\alpha/8\}|. \end{aligned} \tag{4.4.20}$$

Since $\gamma\le(2^{n+5}A_1)^{-1}$, it follows from (4.3.27) that

$$b_3\le\Big|\bigcup_j Q_j^*\Big|+2^{n+8}A_2\frac{\big\|f^\alpha\big\|_{L^1}}{\alpha}\le\left(\frac{(5\sqrt{n})^n}{\gamma}+2^{n+8}A_2\right)\frac{\big\|f^\alpha\big\|_{L^1}}{\alpha},$$

and using (4.4.17), we obtain

$$b_3\le C_n(A_1+A_2)^p\alpha^{-p}\big\|f\big\|_{L^p}^p.$$

Using Chebyshev's inequality in (4.4.20) and (4.4.16), we finally obtain that

$$b_1+b_2\le(8/\alpha)^p\,(\widetilde{c}_n)^p(A_1+A_2+A_3)^p\max(p,(p-1)^{-1})^p\big(\big\|f\big\|_{L^p}^p+\big\|g^\alpha\big\|_{L^p}^p\big).$$

Combining the estimates for b_1, b_2, and b_3 and using (4.4.18), we deduce

$$\big\|T^{(**)}(f)\big\|_{L^{p,\infty}}\le C_n(A_1+A_2+A_3)\max(p,(p-1)^{-1})\big\|f\big\|_{L^p(\mathbf{R}^n)}. \tag{4.4.21}$$

Finally, we need to obtain a similar estimate to (4.4.21), in which the weak L^p norm on the left is replaced by the L^p norm. This is a consequence of Theorem 1.3.2 via interpolation between the estimates $L^{\frac{p+1}{2}}\to L^{\frac{p+1}{2},\infty}$ and $L^{2p}\to L^{2p,\infty}$ for $2<p<\infty$ and between the estimates $L^{2p}\to L^{2p,\infty}$ and $L^1\to L^{1,\infty}$ for $1<p<2$. The latter estimate follows from Theorem 4.3.5. See also Corollary 4.3.7. □

Exercises

4.4.1. Suppose that T is a convolution operator that is L^2 bounded. Suppose that $f \in L^1(\mathbf{R}^n) \cap L^2(\mathbf{R}^n)$ has vanishing integral and that $T(f)$ is integrable. Prove that $T(f)$ also has vanishing integral.

4.4.2. Let K satisfy (4.4.1), (4.4.2), and (4.4.3) and let $W \in \mathscr{S}'$ be an extension of K on $\mathbf{R}^n$. Let f be a Schwartz function on $\mathbf{R}^n$ with mean value zero. Prove that the function $f * W$ is in $L^1(\mathbf{R}^n)$.

4.4.3. Suppose K is a function on $\mathbf{R}^n \setminus \{0\}$ that satisfies (4.4.1), (4.4.2), and (4.4.3). Let $K^{(\varepsilon,N)}(x) = K(x)\chi_{\varepsilon<|x|<N}$ for $0 < \varepsilon < N < \infty$ and let $T^{(\varepsilon,N)}$ be the operator given by convolution with $K^{(\varepsilon,N)}$. Use Theorem 4.4.5 to prove that $T^{(\varepsilon,N)}(f)$ converges to $T(f)$ in $L^p(\mathbf{R}^n)$ and almost everywhere whenever $1 < p < \infty$ and $f \in L^p(\mathbf{R}^n)$ as $\varepsilon \to 0$ and $N \to \infty$.

4.4.4. (a) Prove that for all $x, y \in \mathbf{R}^n$ that satisfy $|x| \geq 2|y|$ we have

$$\left| \frac{x-y}{|x-y|} - \frac{x}{|x|} \right| \leq 2\frac{|y|}{|x|}.$$

(b) Let Ω be an integrable function with mean value zero on the sphere $\mathbf{S}^{n-1}$. Suppose that Ω satisfies a *Lipschitz (Hölder) condition* of order $0 < \alpha < 1$ on $\mathbf{S}^{n-1}$. This means that

$$|\Omega(\theta_1) - \Omega(\theta_2)| \leq B_0 |\theta_1 - \theta_2|^\alpha$$

for all $\theta_1, \theta_2 \in \mathbf{S}^{n-1}$. Prove that $K(x) = \Omega(x/|x|)/|x|^n$ satisfies Hörmander's condition with constant at most a multiple of $B_0 + \left\|\Omega\right\|_{L^\infty}$.

4.4.5. Let Ω be an L^1 function on $\mathbf{S}^{n-1}$ with mean value zero.
(a) Let $\omega_\infty(t) = \sup\{|\Omega(\theta_1) - \Omega(\theta_2)| : \theta_1, \theta_2 \in \mathbf{S}^{n-1}, \ |\theta_1 - \theta_2| \leq t\}$ and suppose that the following *Dini condition* holds:

$$\int_0^1 \omega_\infty(t) \frac{dt}{t} < \infty.$$

Prove that the function $K(x) = \Omega(x/|x|)|x|^{-n}$ satisfies Hörmander's condition.
(b) (*A. Calderón and A. Zygmund*) For $A \in O(n)$, let

$$\left\|A\right\| = \sup\{|\theta - A(\theta)| : \theta \in \mathbf{S}^{n-1}\}.$$

Suppose that Ω satisfies the more general Dini-type condition

$$\int_0^1 \omega_1(t) \frac{dt}{t} < \infty,$$

where

$$\omega_1(t) = \sup_{\substack{A \in O(n) \\ \|A\| \le t}} \int_{\mathbf{S}^{n-1}} |\Omega(A(\theta)) - \Omega(\theta)| \, d\theta \, .$$

Prove the same conclusion as in part (a).

[*Hint:* Part (b): Use the result in part (a) of Exercise 4.4.4 and switch to polar coordinates.]

4.5 Vector-Valued Inequalities

Certain nonlinear expressions that appear in Fourier analysis, such as maximal functions and square functions, can be viewed as linear quantities taking values in some Banach space. This point of view provides the motivation for a systematic study of Banach-valued operators. Let us illustrate this line of thinking via an example. Let T be a linear operator acting on L^p of some measure space (X,μ) and taking values in the set of measurable functions of another measure space (Y,ν). The seemingly nonlinear inequality

$$\Big\| \Big(\sum_j |T(f_j)|^2 \Big)^{\frac{1}{2}} \Big\|_{L^p} \le C_p \Big\| \Big(\sum_j |f_j|^2 \Big)^{\frac{1}{2}} \Big\|_{L^p} \tag{4.5.1}$$

can be transformed to a linear one with only a slight change of view. Let us denote by $L^p(X,\ell^2)$ the Banach space of all sequences $\{f_j\}_j$ of measurable functions on X that satisfy

$$\big\| \{f_j\}_j \big\|_{L^p(X,\ell^2)} = \bigg(\int_X \Big(\sum_j |f_j|^2 \Big)^{\frac{p}{2}} d\mu \bigg)^{\frac{1}{p}} < \infty . \tag{4.5.2}$$

Define a linear operator acting on such sequences by setting

$$\vec{T}(\{f_j\}_j) = \{T(f_j)\}_j. \tag{4.5.3}$$

Then (4.5.1) is equivalent to the inequality

$$\big\| \vec{T}(\{f_j\}_j) \big\|_{L^p(Y,\ell^2)} \le C_p \big\| \{f_j\}_j \big\|_{L^p(X,\ell^2)} , \tag{4.5.4}$$

in which $\vec{T}$ is thought of as a linear operator acting on the L^p space of ℓ^2-valued functions on X. This is the basic idea of vector-valued inequalities. A nonlinear inequality such as (4.5.1) can be viewed as a linear norm estimate for an operator acting and taking values in suitable Banach spaces.

4.5.1 ℓ^2-Valued Extensions of Linear Operators

The following result is classical and fundamental in the subject of vector-valued inequalities.

Theorem 4.5.1. *Let $0 < p,q < \infty$ and let (X,μ) and (Y,ν) be two measure spaces. The following are valid:*
(a) Suppose that T is a bounded linear operator from $L^p(X)$ to $L^q(Y)$ with norm A. Then T has an ℓ^2-valued extension, that is, for all complex-valued functions f_j in $L^p(X)$ we have

$$\Big\|\Big(\sum_j |T(f_j)|^2\Big)^{\frac{1}{2}}\Big\|_{L^q} \le C_{p,q} A \Big\|\Big(\sum_j |f_j|^2\Big)^{\frac{1}{2}}\Big\|_{L^p} \tag{4.5.5}$$

for some constant $C_{p,q}$ that depends only on p and q. Moreover, the constant $C_{p,q}$ satisfies $C_{p,q} = 1$ if $p \le q$.
(b) Suppose that T is a bounded linear operator from $L^p(X)$ to $L^{q,\infty}(Y)$ with norm A. Then T has an ℓ^2-valued extension, that is,

$$\Big\|\Big(\sum_j |T(f_j)|^2\Big)^{\frac{1}{2}}\Big\|_{L^{q,\infty}} \le D_{p,q} A \Big\|\Big(\sum_j |f_j|^2\Big)^{\frac{1}{2}}\Big\|_{L^p} \tag{4.5.6}$$

for some constant $D_{p,q}$ that depends only on p and q.

To prove this theorem, we need the following identities.

Lemma 4.5.2. *For any $0 < r < \infty$, define constants*

$$A_r = \left(\frac{\Gamma(\frac{r+1}{2})}{\pi^{\frac{r+1}{2}}}\right)^{\frac{1}{r}} \quad \text{and} \quad B_r = \left(\frac{\Gamma(\frac{r}{2}+1)}{\pi^{\frac{r}{2}}}\right)^{\frac{1}{r}}. \tag{4.5.7}$$

Then for any $\lambda_1, \lambda_2, \dots, \lambda_n \in \mathbf{R}$ we have

$$\left(\int_{\mathbf{R}^n} |\lambda_1 x_1 + \cdots + \lambda_n x_n|^r e^{-\pi|x|^2}\, dx\right)^{\frac{1}{r}} = A_r (\lambda_1^2 + \cdots + \lambda_n^2)^{\frac{1}{2}}, \tag{4.5.8}$$

and for all $w_1, w_2, \dots, w_n \in \mathbf{C}$ we have

$$\left(\int_{\mathbf{C}^n} |w_1 z_1 + \cdots + w_n z_n|^r e^{-\pi|z|^2}\, dz\right)^{\frac{1}{r}} = B_r (|w_1|^2 + \cdots + |w_n|^2)^{\frac{1}{2}}. \tag{4.5.9}$$

Proof. Dividing both sides of (4.5.8) by $(\lambda_1^2 + \cdots + \lambda_n^2)^{\frac{1}{2}}$, we reduce things to the situation in which $\lambda_1^2 + \cdots + \lambda_n^2 = 1$. Let $e_1 = (1,0,\dots,0)^t$ be the standard basis column unit vector on $\mathbf{R}^n$ and find an orthogonal $n \times n$ matrix $A \in O(n)$ (orthogonal means a real matrix satisfying $A^t = A^{-1}$) such that $A^{-1}e_1 = (\lambda_1, \dots, \lambda_n)^t$. Then the first coordinate of Ax is

$$(Ax)_1 = Ax \cdot e_1 = x \cdot A^t e_1 = x \cdot A^{-1} e_1 = \lambda_1 x_1 + \cdots + \lambda_n x_n .$$

Now change variables $y = Ax$ in the integral in (4.5.8) and use the fact that $|Ax| = |x|$ to obtain

$$\begin{aligned}\left(\int_{\mathbf{R}^n} |\lambda_1 x_1 + \cdots + \lambda_n x_n|^r e^{-\pi|x|^2} dx\right)^{\frac{1}{r}} &= \left(\int_{\mathbf{R}^n} |y_1|^r e^{-\pi|y|^2} dy\right)^{\frac{1}{r}} \\ &= \left(2\int_0^\infty t^r e^{-\pi t^2} dt\right)^{\frac{1}{r}} \\ &= \left(\int_0^\infty s^{\frac{r-1}{2}} e^{-\pi s} ds\right)^{\frac{1}{r}} \\ &= \left(\frac{\Gamma(\frac{r+1}{2})}{\pi^{\frac{r+1}{2}}}\right)^{\frac{1}{r}} \\ &= A_r ,\end{aligned}$$

which proves (4.5.8).

The proof of (4.5.9) is almost identical. We normalize by assuming that

$$|w_1|^2 + \cdots + |w_n|^2 = 1 ,$$

and we let ε_1 be the column vector of $\mathbf{C}^n$ having 1 in the first entry and zero elsewhere. We find a unitary $n \times n$ matrix $\mathscr{A}$ such that $\mathscr{A}^{-1}\varepsilon_1 = (\overline{w_1}, \dots, \overline{w_n})^t$. Unitary means $\mathscr{A}^{-1} = \mathscr{A}^*$, where $\mathscr{A}^*$ is the conjugate transpose matrix of $\mathscr{A}$, i.e., the matrix whose entries are the complex conjugates of $\mathscr{A}^t$ and that satisfies $u \cdot \overline{\mathscr{A} v} = \mathscr{A}^* u \cdot v$ for all $u, v \in \mathbf{C}^n$. Then $(\mathscr{A} z)_1 = w_1 z_1 + \cdots + w_n z_n$ and also $|\mathscr{A} z| = |z|$; therefore, changing variables $\zeta = \mathscr{A} z$ in the integral in (4.5.9), we can rewrite that integral as

$$\begin{aligned}\left(\int_{\mathbf{C}^n} |\zeta_1|^r e^{-\pi|\zeta|^2} d\zeta\right)^{\frac{1}{r}} &= \left(\int_{\mathbf{C}} |\zeta_1|^r e^{-\pi|\zeta_1|^2} d\zeta_1\right)^{\frac{1}{r}} \\ &= \left(2\pi \int_0^\infty t^r e^{-\pi t^2} t\, dt\right)^{\frac{1}{r}} \\ &= \left(\pi \int_0^\infty s^{\frac{r}{2}} e^{-\pi s} ds\right)^{\frac{1}{r}} \\ &= B_r .\end{aligned}$$

□

Let us now continue with the proof of Theorem 4.5.1.

Proof. If T maps real-valued functions to real-valued functions, then we may use conclusion (4.5.8) of Lemma 4.5.2. In general, T maps complex-valued functions to complex-valued functions, and we use conclusion (4.5.9).

Part (a): Assume first that $q \le p$ and let B_r be as in (4.5.7). We may assume that the sequence $\{f_j\}_j$ is indexed by $\mathbf{Z}^+$. Use successively identity (4.5.9), the boundedness of T, Hölder's inequality with exponents p/q and $(p/q)'$ with respect to the measure $e^{-\pi|z|^2}dz$, and identity (4.5.9) again to deduce for $n \in zp$

$$\begin{aligned}
\Big\|\Big(\sum_{j=1}^{n}|T(f_j)|^2\Big)^{\frac12}\Big\|_{L^q(Y)}^q &= (B_q)^{-q}\int_Y\int_{\mathbf{C}^n}|z_1T(f_1)+\cdots+z_nT(f_n)|^q e^{-\pi|z|^2}\,dz\,d\nu\\
&= (B_q)^{-q}\int_{\mathbf{C}^n}\int_Y|T(z_1f_1+\cdots+z_nf_n)|^q\,d\nu\, e^{-\pi|z|^2}\,dz\\
&\le (B_q)^{-q}A^q\int_{\mathbf{C}^n}\Big(\int_X|z_1f_1+\cdots+z_nf_n|^p\,d\mu\Big)^{\frac{q}{p}} e^{-\pi|z|^2}\,dz\\
&\le (B_q)^{-q}A^q\Big(\int_{\mathbf{C}^n}\int_X|z_1f_1+\cdots+z_nf_n|^p\,d\mu\, e^{-\pi|z|^2}\,dz\Big)^{\frac{q}{p}}\\
&= (B_q)^{-q}A^q\Big(B_p^p\int_X\Big(\sum_{j=1}^{n}|f_j|^2\Big)^{\frac{p}{2}}\,d\mu\Big)^{\frac{q}{p}}\\
&= (B_pB_q^{-1})^qA^q\Big\|\Big(\sum_{j=1}^{n}|f_j|^2\Big)^{\frac12}\Big\|_{L^p(X)}^q.
\end{aligned}$$

Now, letting $n \to \infty$ in the previous inequality, we obtain the required conclusion with $C_{p,q} = B_pB_q^{-1}$. Note that $C_{p,q} = 1$ if $p = q$.

We now turn to the case $q > p$. Using similar reasoning, we obtain

$$\begin{aligned}
\Big\|\Big(\sum_{j=1}^{n}|T(f_j)|^2\Big)^{\frac12}\Big\|_{L^q(Y)}^q &= (B_q)^{-q}\int_Y\int_{\mathbf{C}^n}|z_1T(f_1)+\cdots+z_nT(f_n)|^q e^{-\pi|z|^2}\,dz\,d\nu\\
&= (B_q)^{-q}\int_{\mathbf{C}^n}\int_Y|T(z_1f_1+\cdots+z_nf_n)|^q\,d\nu\, e^{-\pi|z|^2}\,dz\\
&\le (AB_q^{-1})^q\int_{\mathbf{C}^n}\Big(\int_X|z_1f_1+\cdots+z_nf_n|^p\,d\mu\Big)^{\frac{q}{p}} e^{-\pi|z|^2}\,dz\\
&= (AB_q^{-1})^q\Big\|\int_X|z_1f_1+\cdots+z_nf_n|^p\,d\mu\Big\|_{L^{\frac{q}{p}}(\mathbf{C}^n,e^{-\pi|z|^2}dz)}^{q/p}\\
&\le (AB_q^{-1})^q\Big\{\int_X\Big\||z_1f_1+\cdots+z_nf_n|^p\Big\|_{L^{\frac{q}{p}}(\mathbf{C}^n,e^{-\pi|z|^2}dz)}\,d\mu\Big\}^{\frac{q}{p}}\\
&= (AB_q^{-1})^q\Big\{\int_X\Big(\int_{\mathbf{C}^n}|z_1f_1+\cdots+z_nf_n|^q e^{-\pi|z|^2}dz\Big)^{\frac{p}{q}}\,d\mu\Big\}^{\frac{q}{p}}\\
&= (AB_q^{-1})^q\Big\{\int_X(B_q)^p\Big(\sum_{j=1}^{n}|f_j|^2\Big)^{\frac{p}{2}}\,d\mu\Big\}^{\frac{q}{p}}\\
&= A^q\Big\|\Big(\sum_{j=1}^{n}|f_j|^2\Big)^{\frac12}\Big\|_{L^p(X)}^q.
\end{aligned}$$

Note that we made use of Minkowski's integral inequality (Exercise 1.1.6) in the last inequality.

Part (b): Inequality (4.5.6) will be a consequence of (4.5.5) and of the following result of Exercise 1.4.3 (see also Exercise 1.1.12):

$$\big\|g\big\|_{L^{q,\infty}} \le \sup_{0<\nu(E)<\infty} \nu(E)^{\frac{1}{q}-\frac{1}{r}}\Big(\int_E |g|^r\,d\nu\Big)^{\frac{1}{r}} \le \Big(\frac{q}{q-r}\Big)^{\frac{1}{r}}\big\|g\big\|_{L^{q,\infty}}, \tag{4.5.10}$$

where $0<r<q$ and the supremum is taken over all subsets E of Y of finite measure. Using (4.5.10), we obtain

$$\begin{aligned}
&\Big\|\Big(\sum_j |T(f_j)|^2\Big)^{\frac{1}{2}}\Big\|_{L^{q,\infty}(Y)} \\
&\le \sup_{0<\nu(E)<\infty} \nu(E)^{\frac{1}{q}-\frac{1}{r}}\Big(\int_E\Big(\sum_j |T(f_j)|^2\Big)^{\frac{r}{2}}d\nu\Big)^{\frac{1}{r}} \\
&= \sup_{0<\nu(E)<\infty} \nu(E)^{\frac{1}{q}-\frac{1}{r}}\Big(\int_Y\Big(\sum_j |\chi_E T(f_j)|^2\Big)^{\frac{r}{2}}d\nu\Big)^{\frac{1}{r}} \\
&\le \sup_{0<\nu(E)<\infty} \nu(E)^{\frac{1}{q}-\frac{1}{r}}\big\|T_E\big\|_{L^p\to L^r} C_{p,r}\Big(\int_X\Big(\sum_j |f_j|^2\Big)^{\frac{p}{2}}d\mu\Big)^{\frac{1}{p}},
\end{aligned} \tag{4.5.11}$$

where T_E is defined by $T_E(f)=\chi_E T(f)$. Since for any function f in $L^p(X)$ we have

$$\nu(E)^{\frac{1}{q}-\frac{1}{r}}\big\|T_E(f)\big\|_{L^r} \le \Big(\frac{q}{q-r}\Big)^{\frac{1}{r}}\big\|T(f)\big\|_{L^{q,\infty}} \le \Big(\frac{q}{q-r}\Big)^{\frac{1}{r}} A\big\|f\big\|_{L^p},$$

it follows that for any measurable set E of finite measure the estimate

$$\nu(E)^{\frac{1}{q}-\frac{1}{r}}\big\|T_E\big\|_{L^p\to L^r} \le \Big(\frac{q}{q-r}\Big)^{\frac{1}{r}} A \tag{4.5.12}$$

is valid. Inserting (4.5.12) in (4.5.11), we obtain the required conclusion. □

4.5.2 Applications and ℓ^r-Valued Extensions of Linear Operators

Here is an application of Theorem 4.5.1:

Example 4.5.3. On the real line consider the intervals $I_j=[b_j,\infty)$ for $j\in\mathbf{Z}$. Let T_j be the operator given by multiplication on the Fourier transform by the characteristic function of I_j. Then we have the following two inequalities:

$$\Big\|\Big(\sum_{j\in\mathbf{Z}}|T_j(f_j)|^2\Big)^{\frac{1}{2}}\Big\|_{L^p}\le C_p\Big\|\Big(\sum_{j\in\mathbf{Z}}|f_j|^2\Big)^{\frac{1}{2}}\Big\|_{L^p}, \tag{4.5.13}$$

$$\Big\|\Big(\sum_{j\in\mathbf{Z}}|T_j(f_j)|^2\Big)^{\frac{1}{2}}\Big\|_{L^{1,\infty}}\le C\Big\|\Big(\sum_{j\in\mathbf{Z}}|f_j|^2\Big)^{\frac{1}{2}}\Big\|_{L^1}, \tag{4.5.14}$$

for $1<p<\infty$. To prove these, first observe that the operator $T=\frac{1}{2}(I+iH)$ is given on the Fourier transform by multiplication by the characteristic function of the half-axis $[0,\infty)$ [precisely, the Fourier multiplier of T is equal to 1 on the set $(0,\infty)$ and $1/2$ at the origin; this function is almost everywhere equal to the characteristic function of the half-axis $[0,\infty)$]. Moreover, each T_j is given by

$$T_j(f)(x)=e^{2\pi i b_j x}T(e^{-2\pi i b_j(\cdot)}f)(x)$$

and thus with $g_j(x)=e^{-2\pi i b_j x}f(x)$, (4.5.13) and (4.5.14) can be written respectively as

$$\Big\|\Big(\sum_{j\in\mathbf{Z}}|T(g_j)|^2\Big)^{\frac{1}{2}}\Big\|_{L^p}\le C_p\Big\|\Big(\sum_{j\in\mathbf{Z}}|g_j|^2\Big)^{\frac{1}{2}}\Big\|_{L^p},$$

$$\Big\|\Big(\sum_{j\in\mathbf{Z}}|T(g_j)|^2\Big)^{\frac{1}{2}}\Big\|_{L^{1,\infty}}\le C\Big\|\Big(\sum_{j\in\mathbf{Z}}|g_j|^2\Big)^{\frac{1}{2}}\Big\|_{L^1}.$$

Theorem 4.5.1 gives that both of the previous estimates are valid by in view of the boundedness of $T=\frac{1}{2}(I+iH)$ from L^p to L^p and from $L^1\to L^{1,\infty}$. For a slight generalization and an extension to higher dimensions, see Exercise 4.6.1.

We have now seen that bounded operators from L^p to L^q (or to $L^{q,\infty}$) always admit ℓ^2-valued extensions. It is natural to ask whether they also admit ℓ^r-valued extensions for some $r\neq 2$. For some values of r we may answer this question. Here is a straightforward corollary of Theorem 4.5.1.

Corollary 4.5.4. *Suppose that T is a linear bounded operator from $L^p(X)$ to $L^p(Y)$ with norm A for some $1\le p<\infty$. Let r be a number between p and 2. Then we have*

$$\Big\|\Big(\sum_j|T(f_j)|^r\Big)^{\frac{1}{r}}\Big\|_{L^p}\le A\Big\|\Big(\sum_j|f_j|^r\Big)^{\frac{1}{r}}\Big\|_{L^p}. \tag{4.5.15}$$

Proof. The endpoint case $r=2$ is a consequence of Theorem 4.5.1, while the endpoint case $r=p$ is trivial. Interpolation (see Exercise 4.5.2) gives the required conclusion for r between p and 2. □

We note that Exercise 4.5.2 and Corollary 4.5.4 are also valid for indices less than 1.

Example 4.5.5. The result of Corollary 4.5.4 may fail if r does not lie in the interval with endpoints p and 2. Let us take, for example, $1<p<2$ and consider an $r<p$. Take $X=Y=\mathbf{R}$ and define a linear operator T by setting

$$T(f)(x) = \widehat{f}(x)\chi_{[0,1]}(x).$$

Then T is L^p bounded, since $\|T(f)\|_{L^p} \le \|T(f)\|_{L^{p'}} \le \|f\|_{L^p}$. Now take $f_j = \chi_{[j-1,j]}$ for $j = 1,\dots,N$. A simple calculation gives

$$\Big(\sum_{j=1}^{N} |T(f_j)(x)|^r\Big)^{\frac{1}{r}} = N^{\frac{1}{r}} \left| \frac{e^{-2\pi i x} - 1}{-2\pi i x} \chi_{[0,1]}(x) \right| ,$$

while

$$\Big(\sum_{j=1}^{N} |f_j|^r\Big)^{\frac{1}{r}} = \chi_{[0,N]}\,.$$

It follows that $N^{1/r} \le CN^{1/p}$ for all $N > 1$, and hence (4.5.15) cannot hold if $p > r$.

We have now seen that ℓ^r-valued extensions for $r \neq 2$ may fail in general. But do they fail for some specific operators of interest in Fourier analysis? For instance, is the inequality

$$\Big\|\Big(\sum_{j\in\mathbf{Z}} |H(f_j)|^r\Big)^{\frac{1}{r}}\Big\|_{L^p} \le C_{p,r}\Big\|\Big(\sum_{j\in\mathbf{Z}} |f_j|^r\Big)^{\frac{1}{r}}\Big\|_{L^p} \tag{4.5.16}$$

true for the Hilbert transform H whenever $1 < p,r < \infty$? The answer to this question is affirmative. Inequality (4.5.16) is indeed valid and was first proved using complex function theory. In the next section we plan to study inequalities such as (4.5.16) for general singular integrals using the Calderón–Zygmund theory of the previous section applied to the context of Banach-valued functions.

4.5.3 General Banach-Valued Extensions

We now set up the background required to state the main results of this section. Although the Banach spaces of most interest to us are ℓ^r for $1 \le r \le \infty$, we introduce the basic notions we need in general.

Let $\mathscr{B}$ be a Banach space over the field of complex numbers with norm $\|\ \|_{\mathscr{B}}$, and let $\mathscr{B}^*$ be its dual (with norm $\|\ \|_{\mathscr{B}^*}$). A function F defined on a σ-finite measure space (X,μ) and taking values in $\mathscr{B}$ is called $\mathscr{B}$-measurable if there exists a measurable subset X_0 of X such that $\mu(X \setminus X_0) = 0$, $F[X_0]$ is contained in some separable subspace $\mathscr{B}_0$ of $\mathscr{B}$, and for every $u^* \in \mathscr{B}^*$ the complex-valued map

$$x \mapsto \big\langle u^*, F(x)\big\rangle$$

is measurable. A consequence of this definition is that the positive function $x \mapsto \|F(x)\|_{\mathscr{B}}$ on X is measurable; to see this, use the relevant result in Yosida [296, p. 131].

For $0 < p \le \infty$, denote by $L^p(X,\mathscr{B})$ the space of all $\mathscr{B}$-measurable functions F on X satisfying

$$\left(\int_X \|F(x)\|_{\mathscr{B}}^p \, d\mu(x)\right)^{\frac{1}{p}} < \infty, \tag{4.5.17}$$

with the obvious modification when $p = \infty$. Similarly define $L^{p,\infty}(X,\mathscr{B})$ as the space of all $\mathscr{B}$-measurable functions F on X satisfying

$$\Big\| \|F(\cdot)\|_{\mathscr{B}} \Big\|_{L^{p,\infty}(X)} < \infty. \tag{4.5.18}$$

Then $L^p(X,\mathscr{B})$ (respectively, $L^{p,\infty}(X,\mathscr{B})$) is called the L^p (respectively, $L^{p,\infty}$) space of functions on X with values in $\mathscr{B}$. Similarly, we can define other Lorentz spaces of $\mathscr{B}$-valued functions. The quantity in (4.5.17) (respectively, in (4.5.18)) is the *norm* of F in $L^p(X,\mathscr{B})$ (respectively, in $L^{p,\infty}(X,\mathscr{B})$).

We denote by $L^p(X)$ the space $L^p(X,\mathbf{C})$. Let $L^p(X) \otimes \mathscr{B}$ be the set of all finite linear combinations of elements of $\mathscr{B}$ with coefficients in $L^p(X)$, that is, elements of the form

$$F = f_1 u_1 + \cdots + f_m u_m, \tag{4.5.19}$$

where $f_j \in L^p(X)$, $u_j \in \mathscr{B}$, and $m \in \mathbf{Z}^+$.

If F is an element of $L^1 \otimes \mathscr{B}$ given as in (4.5.19), we define its integral (which is an element of $\mathscr{B}$) by setting

$$\int_X F(x)\,d\mu(x) = \sum_{j=1}^m \left(\int_X f_j(x)\,d\mu(x)\right) u_j.$$

Observe that for every $F \in L^1 \otimes \mathscr{B}$ we have

$$\begin{aligned}
\Big\| \int_X F(x)\,d\mu(x) \Big\|_{\mathscr{B}} &= \sup_{\|u^*\|_{\mathscr{B}^*} \le 1} \Big| \Big\langle u^*, \sum_{j=1}^m \Big(\int_X f_j\,d\mu\Big) u_j \Big\rangle \Big| \\
&= \sup_{\|u^*\|_{\mathscr{B}^*} \le 1} \Big| \int_X \langle u^*, \sum_{j=1}^m f_j u_j \rangle \, d\mu \Big| \\
&\le \int_X \sup_{\|u^*\|_{\mathscr{B}^*} \le 1} \big| \langle u^*, \sum_{j=1}^m f_j u_j \rangle \big| \, d\mu \\
&= \big\| F \big\|_{L^1(X,\mathscr{B})}.
\end{aligned}$$

Thus the linear operator

$$F \mapsto I_F = \int_X F(x)\,d\mu(x)$$

is bounded from $L^1(X) \otimes \mathscr{B}$ into $\mathscr{B}$. Since every element of $L^1(X,\mathscr{B})$ is a (norm) limit of a sequence of elements in $L^1(X) \otimes \mathscr{B}$, by continuity, the operator $F \mapsto I_F$ has a unique extension on $L^1(X,\mathscr{B})$ that we call the *Bochner integral* of F and denote by

$$\int_X F(x)\,d\mu(x).$$

It is not difficult to show that the Bochner integral of F is the only element of $\mathscr{B}$ that satisfies

$$\left\langle u^*, \int_X F(x)\,d\mu(x)\right\rangle = \int_X \langle u^*, F(x)\rangle\,d\mu(x)$$

for all $u^* \in \mathscr{B}^*$.

Proposition 4.5.6. *Let $\mathscr{B}$ be a Banach space. Then the set of functions of the form $\sum_{j=1}^m \chi_{E_j} u_j$, where $u_j \in \mathscr{B}$, $\{E_j\}_{j=1}^m$ are pairwise disjoint subsets of $\mathbf{R}^n$ with finite measure, is dense in $L^p(\mathbf{R}^n, \mathscr{B})$ whenever $0 < p < \infty$. For $p = \infty$, the set of functions of the form $\sum_{j=1}^\infty \chi_{E_j} u_j$, where $u_j \in \mathscr{B}$ and $\{E_j\}_{j=1}^\infty$ is a partition of $\mathbf{R}^n$, is dense in $L^\infty(\mathbf{R}^n, \mathscr{B})$.*

Proof. If $F \in L^p(\mathbf{R}^n, \mathscr{B})$ for $0 < p \le \infty$, then F is $\mathscr{B}$-measurable; thus there exists $K_0 \subset \mathbf{R}^n$ satisfying $|\mathbf{R}^n \setminus K_0| = 0$ and $F[K_0] \subset \mathscr{B}_0$, where $\mathscr{B}_0$ is some separable subspace of $\mathscr{B}$. Choose a countable dense sequence $\{u_j\}_{j=1}^\infty$ of $\mathscr{B}_0$.

First assume that $p < \infty$. For any $\varepsilon > 0$, there exists a bounded subset K_1 of K_0 such that

$$\int_{\mathbf{R}^n\setminus K_1} \|F(x)\|_{\mathscr{B}}^p\,dx < \frac{\varepsilon^p}{3}.$$

Setting $\widetilde{B}(u_j,\varepsilon) = \{u \in \mathscr{B}_0 : \|u-u_j\|_{\mathscr{B}} < \varepsilon(3|K_1|)^{-\frac{1}{p}}\}$, we have $\mathscr{B}_0 \subset \bigcup_{j=1}^\infty \widetilde{B}(u_j,\varepsilon)$. Let $A_1 = \widetilde{B}(u_1,\varepsilon)$ and $A_j = \widetilde{B}(u_j,\varepsilon) \setminus (\bigcup_{i-1}^{j-1} \widetilde{B}(u_i,\varepsilon))$ for $j \ge 2$. It is easily seen that $\{A_j\}_{j=1}^\infty$ are pairwise disjoint and $\bigcup_{j=1}^\infty A_j = \bigcup_{j=1}^\infty \widetilde{B}(u_j,\varepsilon)$. Set $\widetilde{A}_j = A_j \cap F[K_1]$ and $E_j = F^{-1}[\widetilde{A}_j]$. Then $K_1 = \bigcup_{j=1}^\infty E_j$ and $\{E_j\}_{j=1}^\infty$ are pairwise disjoint. Since $|K_1| = \sum_{j=1}^\infty |E_j| < \infty$, it follows that $|E_j| < \infty$ and also that for some $m \in \mathbf{Z}^+$,

$$\int_{\bigcup_{j=m+1}^\infty E_j} \|F(x)\|_{\mathscr{B}}^p\,dx < \frac{\varepsilon^p}{3}. \tag{4.5.20}$$

Moreover, one can easily verify that $\sum_{j=1}^m \chi_{E_j} u_j$ is $\mathscr{B}$-measurable. Notice that $\|F(x) - u_j\|_{\mathscr{B}} < \varepsilon(3|K_1|)^{-1/p}$ for any $x \in E_j$ and $j \in \{1,\dots,m\}$. This fact combined with (4.5.20) and the mutual disjointness of $\{E_j\}_{j=1}^m$ yields that

$$\begin{aligned}
\int_{\mathbf{R}^n} \left\| F(x) - \sum_{j=1}^m \chi_{E_j}(x) u_j \right\|_{\mathscr{B}}^p dx &= \int_{\mathbf{R}^n\setminus K_1} \|F(x)\|_{\mathscr{B}}^p\,dx + \int_{\bigcup_{j=m+1}^\infty E_j} \|F(x)\|_{\mathscr{B}}^p\,dx \\
&\quad + \int_{\bigcup_{j=1}^m E_j} \left\| \sum_{j=1}^m \chi_{E_j}(x)[F(x) - u_j] \right\|_{\mathscr{B}}^p dx \\
&< \frac{\varepsilon^p}{3} + \frac{\varepsilon^p}{3} + \frac{\varepsilon^p}{3} = \varepsilon^p.
\end{aligned}$$

Now consider the case $p = \infty$. Obviously we have $\mathscr{B}_0 \subset \bigcup_{j=1}^\infty B(u_j,\varepsilon)$, where $B(u_j,\varepsilon) = \{u \in \mathscr{B}_0 : \|u - u_j\|_{\mathscr{B}} < \varepsilon\}$. Let $A_1 = B(u_1,\varepsilon)$ and for $j \ge 2$ define sets $A_j = B(u_j,\varepsilon) \setminus (\bigcup_{i=1}^{j-1} B(u_i,\varepsilon))$. Let $E_j = F^{-1}[A_j]$ for $j \ge 1$ and $E_0 = \mathbf{R}^n \setminus (\bigcup_{j=1}^\infty E_j)$. As in the proof of the case $p < \infty$, we have that $\{E_j\}_{j=0}^\infty$ are pairwise disjoint and

$K_0 \subset \bigcup_{j=0}^{\infty} E_j$. Notice that $\sum_{j=0}^{\infty} \chi_{E_j} u_j$ is $\mathscr{B}$-measurable. Since $\|F(x) - u_j\|_{\mathscr{B}} < \varepsilon$ for any $x \in E_j$ and $j \geq 0$, we have

$$\left\| F - \sum_{j=0}^{\infty} \chi_{E_j} u_j \right\|_{L^\infty(\mathbf{R}^n, \mathscr{B})} = \left\| \sum_{j=0}^{\infty} \chi_{E_j} (F - u_j) \right\|_{L^\infty(\mathbf{R}^n, \mathscr{B})} < \varepsilon,$$

which completes the proof in the case $p = \infty$ as well. □

Proposition 4.5.7. *Let $\mathscr{B}$ be a Banach space.*
(a) For any $F \in L^p(\mathbf{R}^n, \mathscr{B})$ with $1 \leq p \leq \infty$ we have

$$\|F\|_{L^p(\mathbf{R}^n, \mathscr{B})} = \sup_{\|G\|_{L^{p'}(\mathbf{R}^n, \mathscr{B}^*)} \leq 1} \left| \int_{\mathbf{R}^n} \langle G(x), F(x) \rangle \, dx \right|.$$

(b) The space $L^p(\mathbf{R}^n, \mathscr{B})$ isometrically embeds in $(L^{p'}(\mathbf{R}^n, \mathscr{B}^*))^*$ when $1 \leq p \leq \infty$.

Proof. Obviously (b) is a consequence of (a), thus we concentrate on (a). Hölder's inequality yields that the right-hand side of (a) is controlled by its left-hand side. It remains to establish the reverse inequality.

For $F \in L^p(\mathbf{R}^n, \mathscr{B})$ and $\varepsilon > 0$, by Proposition 4.5.6, there is $F_\varepsilon(x) = \sum_{j=1}^{m} \chi_{E_j}(x) u_j$ with $m \in \mathbf{Z}^+$ or $m = \infty$ (when $p = \infty$) such that $\|F_\varepsilon - F\|_{L^p(\mathbf{R}^n, \mathscr{B})} < \varepsilon/2$, where $\{E_j\}_{j=1}^m$ are pairwise disjoint subsets of $\mathbf{R}^n$ and $u_j \in \mathscr{B}$. Since $F_\varepsilon \in L^p(\mathbf{R}^n, \mathscr{B})$, we choose a nonnegative function h satisfying $\|h\|_{L^{p'}(\mathbf{R}^n)} \leq 1$ such that

$$\|F_\varepsilon\|_{L^p(\mathbf{R}^n, \mathscr{B})} = \left(\int_{\mathbf{R}^n} \|F_\varepsilon(x)\|_{\mathscr{B}}^p \, dx \right)^{1/p} < \int_{\mathbf{R}^n} h(x) \|F_\varepsilon(x)\|_{\mathscr{B}} \, dx + \frac{\varepsilon}{4}. \tag{4.5.21}$$

When $1 \leq p < \infty$, we can further choose $h \in L^{p'}(\mathbf{R}^n)$ to be a function with bounded support, which ensures that it is integrable. For given $u_j \in \mathscr{B}$, there exists $u_j^* \in \mathscr{B}^*$ satisfying $\|u_j^*\|_{\mathscr{B}^*} = 1$ and

$$\|u_j\|_{\mathscr{B}} < \langle u_j^*, u_j \rangle + \frac{\varepsilon}{4(\|h\|_{L^1(\mathbf{R}^n)} + 1)}. \tag{4.5.22}$$

Set $G(x) = \sum_{j=1}^{m} h(x) \chi_{E_j}(x) u_j^*$. Clearly G is $\mathscr{B}^*$-measurable and $\|G\|_{L^{p'}(\mathbf{R}^n, \mathscr{B}^*)} \leq 1$. It follows from (4.5.21) and (4.5.22) that

$$\begin{aligned}
\int_{\mathbf{R}^n} \langle G(x), F_\varepsilon(x) \rangle \, dx &= \int_{\mathbf{R}^n} h(x) \sum_{j=1}^{m} \chi_{E_j}(x) \langle u_j^*, u_j \rangle \, dx \\
&\geq \int_{\mathbf{R}^n} h(x) \sum_{j=1}^{m} \left(\|u_j\|_{\mathscr{B}} - \frac{\varepsilon}{4(\|h\|_{L^1(\mathbf{R}^n)} + 1)} \right) \chi_{E_j}(x) \, dx \\
&\geq \|F_\varepsilon\|_{L^p(\mathbf{R}^n, \mathscr{B})} - \frac{\varepsilon}{2}.
\end{aligned}$$

Hence, for any $\varepsilon > 0$, we have

$$\|F\|_{L^p(\mathbf{R}^n,\mathscr{B})} \leq \sup_{\|G\|_{L^{p'}(\mathbf{R}^n,\mathscr{B}^*)}\leq 1} \left| \int_{\mathbf{R}^n} \langle G(x), F(x)\rangle \, dx \right| + \varepsilon.$$

Letting $\varepsilon \to 0$ implies the desired inequality, which completes the proof. □

Definition 4.5.8. Let T be a linear operator that maps $L^p(\mathbf{R}^n)$ to $L^q(\mathbf{R}^n)$ (respectively, $L^p(\mathbf{R}^n)$ to $L^{q,\infty}(\mathbf{R}^n)$) for some $0 < p,q \leq \infty$. We define another operator $\vec{T}$ acting on $L^p \otimes \mathscr{B}$ by setting

$$\vec{T}\Big(\sum_{j=1}^m f_j u_j\Big) = \sum_{j=1}^m T(f_j)\, u_j.$$

If $\vec{T}$ happens to have a bounded extension from $L^p(\mathbf{R}^n,\mathscr{B})$ to $L^q(\mathbf{R}^n,\mathscr{B})$ (respectively from $L^p(\mathbf{R}^n,\mathscr{B})$ to $L^{q,\infty}(\mathbf{R}^n,\mathscr{B})$), then we say that T has a bounded *$\mathscr{B}$-valued extension*. In this case we also denote by $\vec{T}$ the $\mathscr{B}$-valued extension of T.

Example 4.5.9. Let $\mathscr{B} = \ell^r$ for some $1 \leq r < \infty$. Then a measurable function $F : X \to \mathscr{B}$ is just a sequence $\{f_j\}_j$ of measurable functions $f_j : X \to \mathbf{C}$. The space $L^p(X,\ell^r)$ consists of all measurable complex-valued sequences $\{f_j\}_j$ on X that satisfy

$$\big\|\{f_j\}_j\big\|_{L^p(X,\ell^r)} = \Big\|\Big(\sum_j |f_j|^r\Big)^{\frac{1}{r}}\Big\|_{L^p(X)} < \infty.$$

The space $L^p(X)\otimes\ell^r$ is the set of all finite sums

$$\sum_{j=1}^m (a_{j1},a_{j2},a_{j3},\dots)\, g_j\,,$$

where $g_j \in L^p(X)$ and $(a_{j1},a_{j2},a_{j3},\dots) \in \ell^r$, $j = 1,\dots,m$. This is certainly a subspace of $L^p(X,\ell^r)$. Now given $(f_1,f_2,\dots) \in L^p(X,\ell^r)$, let $F_m = e_1 f_1 + \cdots + e_m f_m$, where e_j is the infinite sequence with zeros everywhere except at the jth entry, where it has 1. Then $F_m \in L^p(X)\otimes\ell^r$ and approximates f in the norm of $L^p(X,\ell^r)$. This shows the density of $L^p(X)\otimes\ell^r$ in $L^p(X,\ell^r)$.

If T is a linear operator bounded from $L^p(X)$ to $L^q(Y)$, then $\vec{T}$ is defined by

$$\vec{T}(\{f_j\}_j) = \{T(f_j)\}_j.$$

According to Definition 4.5.8, T has a bounded ℓ^r-extension if and only if the inequality

$$\Big\|\Big(\sum_j |T(f_j)|^r\Big)^{\frac{1}{r}}\Big\|_{L^q} \leq C\Big\|\Big(\sum_j |f_j|^r\Big)^{\frac{1}{r}}\Big\|_{L^p}$$

is valid.

A linear operator T acting on measurable functions is called *positive* if it satisfies $f \geq 0 \implies T(f) \geq 0$. It is straightforward to verify that positive operators satisfy

$$\begin{aligned} &f \le g \implies T(f) \le T(g), \\ &|T(f)| \le T(|f|), \\ &\sup_j |T(f_j)| \le T\big(\sup_j |f_j|\big), \end{aligned} \tag{4.5.23}$$

for all f, g, f_j measurable functions. We have the following result regarding vector-valued extensions of positive operators:

Proposition 4.5.10. *Let $0 < p, q \le \infty$ and (X, μ), (Y, ν) be two measure spaces. Let T be a positive linear operator mapping $L^p(X)$ to $L^q(Y)$ (respectively, to $L^{q,\infty}(Y)$) with norm A. Let $\mathscr{B}$ be a Banach space. Then T has a $\mathscr{B}$-valued extension $\vec{T}$ that maps $L^p(X, \mathscr{B})$ to $L^q(Y, \mathscr{B})$ (respectively, to $L^{q,\infty}(Y, \mathscr{B})$) with the same norm.*

Proof. Let us first understand this theorem when $\mathscr{B} = \ell^r$ for $1 \le r \le \infty$. The two endpoint cases $r = 1$ and $r = \infty$ can be checked easily using the properties in (4.5.23). For instance, for $r = 1$ we have

$$\Big\|\sum_j |T(f_j)|\Big\|_{L^q} \le \Big\|\sum_j T(|f_j|)\Big\|_{L^q} = \Big\|T\Big(\sum_j |f_j|\Big)\Big\|_{L^q} \le A\Big\|\sum_j |f_j|\Big\|_{L^p},$$

while for $r = \infty$ we have

$$\Big\|\sup_j |T(f_j)|\Big\|_{L^q} \le \Big\|T(\sup_j |f_j|)\Big\|_{L^q} \le A\,\big\|\sup_j |f_j|\big\|_{L^p}.$$

The required inequality for $1 < r < \infty$,

$$\Big\|\Big(\sum_j |T(f_j)|^r\Big)^{\frac{1}{r}}\Big\|_{L^q} \le A\Big\|\Big(\sum_j |f_j|^r\Big)^{\frac{1}{r}}\Big\|_{L^p},$$

follows from the Riesz–Thorin interpolation theorem (see Exercise 4.5.2).

The result for a general Banach space $\mathscr{B}$ can be proved using the following inequality:

$$\|\vec{T}(F)(x)\|_{\mathscr{B}} \le T\Big(\|F\|_{\mathscr{B}}\Big)(x), \qquad x \in X, \tag{4.5.24}$$

by simply taking L^q norms. To prove (4.5.24), let us take $F = \sum_{j=1}^n f_j u_j$. Then

$$\begin{aligned} \|\vec{T}(F)(x)\|_{\mathscr{B}} = \Big\|\sum_{j=1}^n T(f_j)(x) u_j\Big\|_{\mathscr{B}} &= \sup_{\|u^*\|_{\mathscr{B}^*} \le 1} \Big|\Big\langle u^*, \sum_{j=1}^n T(f_j)(x) u_j\Big\rangle\Big| \\ &= \sup_{\|u^*\|_{\mathscr{B}^*} \le 1} \Big|T\Big(\sum_{j=1}^n f_j \langle u^*, u_j\rangle\Big)(x)\Big| \\ &\le T\Big(\sup_{\|u^*\|_{\mathscr{B}^*} \le 1} \Big|\Big\langle u^*, \sum_{j=1}^n f_j u_j\Big\rangle\Big|\Big)(x) \\ &= T\Big(\Big\|\sum_{j=1}^n f_j u_j\Big\|_{\mathscr{B}}\Big)(x) = T\Big(\|F\|_{\mathscr{B}}\Big)(x), \end{aligned}$$

where the inequality makes use of the fact that T is a positive operator. □

We end this section with a simple extension of Theorem 4.5.1.

Proposition 4.5.11. *Let $\mathscr{H}$ be a Hilbert space and let $0 < p < \infty$. Then every bounded linear operator T from $L^p(\mathbf{R}^n)$ to $L^p(\mathbf{R}^n)$ has an $\mathscr{H}$-valued extension. In particular, for all measurable families of functions $\{f_t\}_{t\in\mathbf{R}^d}$ and for all positive measures μ on $\mathbf{R}^d$ the following estimate is valid:*

$$\Big\|\Big(\int_{\mathbf{R}^d}|T(f_t)|^2\,d\mu(t)\Big)^{\frac{1}{2}}\Big\|_{L^p(\mathbf{R}^n)}\le\|T\|_{L^p\to L^p}\Big\|\Big(\int_{\mathbf{R}^d}|f_t|^2\,d\mu(t)\Big)^{\frac{1}{2}}\Big\|_{L^p(\mathbf{R}^n)}.$$

Proof. If the Hilbert space $\mathscr{H}$ is finite-dimensional, then it is isometrically isomorphic to $\ell^2(\{1,2,\dots,N\})$ for some positive integer N. If $\mathscr{H}$ is infinite-dimensional and separable, then it is isometrically isomorphic to $\ell^2(\mathbf{Z})$. By Theorem 4.5.1, the linear operator T has an ℓ^2-valued extension, and in view of the isometry with $\mathscr{H}$, it must also have an $\mathscr{H}$-valued extension. If the Hilbert space $\mathscr{H}$ is not separable, we obtain a vector-valued extension of T for all separable subspaces of $\mathscr{H}$ with norm independent of the subspace. □

Exercises

4.5.1. Let $\mathscr{B}$ be a Banach space. Prove that
(a) for any $G\in L^{p'}(\mathbf{R}^n,\mathscr{B}^*)$, $1\le p\le\infty$, one has

$$\|G\|_{L^{p'}(\mathbf{R}^n,\mathscr{B}^*)}=\sup_{\|F\|_{L^p(\mathbf{R}^n,\mathscr{B})}\le 1}\left|\int_{\mathbf{R}^n}\langle G(x),F(x)\rangle\,dx\right|;$$

(b) the space $L^{p'}(\mathbf{R}^n,\mathscr{B}^*)$ isometrically embeds in $(L^p(\mathbf{R}^n,\mathscr{B}))^*$ when $1\le p\le\infty$.

4.5.2. Prove the following version of the Riesz–Thorin interpolation theorem. Let $1\le p_0,q_0,,p_1,q_1,r_0,s_0,r_1,s_1\le\infty$ and $0<\theta<1$ satisfy

$$\frac{1-\theta}{p_0}+\frac{\theta}{p_1}=\frac{1}{p},\qquad \frac{1-\theta}{q_0}+\frac{\theta}{q_1}=\frac{1}{q},$$
$$\frac{1-\theta}{r_0}+\frac{\theta}{r_1}=\frac{1}{r},\qquad \frac{1-\theta}{s_0}+\frac{\theta}{s_1}=\frac{1}{s}.$$

Suppose that $\vec{T}$ is a linear operator that maps $L^{p_0}(\mathbf{R}^n,\ell^{r_0})$ to $L^{q_0}(\mathbf{R}^n,\ell^{s_0})$ with norm A_0 and $L^{p_1}(\mathbf{R}^n,\ell^{r_1})$ to $L^{q_1}(\mathbf{R}^n,\ell^{s_2})$ with norm A_1. Prove that $\vec{T}$ maps $L^p(\mathbf{R}^n,\ell^r)$ to $L^q(\mathbf{R}^n,\ell^s)$ with norm at most $A_0^{1-\theta}A_1^{\theta}$.

4.5.3. (a) Prove the following version of the Marcinkiewicz interpolation theorem. Let $0<p_0<p<p_1\le\infty$ and $0<\theta<1$ satisfy

$$\frac{1-\theta}{p_0}+\frac{\theta}{p_1}=\frac{1}{p}.$$

Suppose that $\vec{T}$ is a sublinear operator, that is, it satisfies

$$\|\vec{T}(F+G)\|_{\mathscr{B}_2}\le\|\vec{T}(F)\|_{\mathscr{B}_2}+\|\vec{T}(G)\|_{\mathscr{B}_2},$$

for all F and G. Assume that $\vec{T}$ maps $L^{p_0}(\mathbf{R}^n,\mathscr{B}_1)$ to $L^{p_0,\infty}(\mathbf{R}^n,\mathscr{B}_2)$ with norm A_0 and $L^{p_1}(\mathbf{R}^n,\mathscr{B}_1)$ to $L^{p_1,\infty}(\mathbf{R}^n,\mathscr{B}_2)$ with norm A_1. Show that $\vec{T}$ maps $L^p(\mathbf{R}^n,\mathscr{B}_1)$ to $L^p(\mathbf{R}^n,\mathscr{B}_2)$ with norm at most $2\big(\frac{p}{p-p_0}+\frac{p}{p_1-p}\big)^{\frac{1}{p}}A_0^{1-\theta}A_1^{\theta}$.
(b) Let $p_0=1$. If $\vec{T}$ is linear and maps $L^1(\mathbf{R}^n,\mathscr{B}_1)$ to $L^{1,\infty}(\mathbf{R}^n,\mathscr{B}_2)$ with norm A_0 and $L^{p_1}(\mathbf{R}^n,\mathscr{B}_1)$ to $L^{p_1}(\mathbf{R}^n,\mathscr{B}_2)$ with norm A_1, show that the constant in part (a) can be improved to $8(p-1)^{-1/p}A_0^{1-\theta}A_1^{\theta}$; see also Exercise 1.3.2.

4.5.4. Suppose that all $x\in\mathbf{R}^n$, $K(x)$ is a bounded linear operator from $\mathscr{B}_1$ to $\mathscr{B}_2$ and let $\vec{T}(F)(x)=\int_{\mathbf{R}^n}K(x-y)F(y)\,dy$ be the vector-valued operator given by convolution with K.
(a) Suppose that K satisfies

$$\int_{\mathbf{R}^n}\|K(x)\|_{\mathscr{B}_1\to\mathscr{B}_2}\,dx=C<\infty.$$

Prove that the operator $\vec{T}(F)$ maps $L^p(\mathbf{R}^n,\mathscr{B}_1)$ to $L^p(\mathbf{R}^n,\mathscr{B}_2)$ with norm at most C for $1\le p\le\infty$.
(b) (*Young's inequality*) Suppose that K satisfies

$$\left(\int_{\mathbf{R}^n}\|K(x)\|_{\mathscr{B}_1\to\mathscr{B}_2}^s\,dx\right)^{1/s}=C<\infty.$$

Prove that $\vec{T}(F)$ maps $L^p(\mathbf{R}^n,\mathscr{B}_1)$ to $L^q(\mathbf{R}^n,\mathscr{B}_2)$ with norm at most C whenever $1\le p,q,s\le\infty$ and $1/q+1=1/s+1/p$.
(c) (*Young's inequality for weak type spaces*) Suppose that K satisfies

$$\Big\|\|K(\cdot)\|_{\mathscr{B}_1\to\mathscr{B}_2}\Big\|_{L^{s,\infty}}<\infty.$$

Prove that $\vec{T}(F)$ maps $L^p(\mathbf{R}^n,\mathscr{B}_1)$ to $L^q(\mathbf{R}^n,\mathscr{B}_2)$ whenever $1\le p<\infty$, $1<p,s<\infty$, and $1/q+1=1/s+1/p$.

4.5.5. Prove the following (slight) generalization of the Exercise 4.5.4 when $p=1$. Suppose that K satisfies

$$\int_{\mathbf{R}^n}\|K(x)u\|_{\mathscr{B}_2}\,dx\le C\|u\|_{\mathscr{B}_1}$$

for all $u \in \mathscr{B}_1$. Then $\vec{T}$ maps $L^1(\mathbf{R}^n, \mathscr{B}_1)$ to $L^1(\mathbf{R}^n, \mathscr{B}_2)$ with norm at most C. Show, however, that the preceding condition is not strong enough to imply L^p boundedness for $\vec{T}$ for $1 < p < \infty$.

4.5.6. Use the inequality for the Rademacher functions in Appendix C.2 instead of Lemma 4.5.2 to prove part (a) of Theorem 4.5.1 in the special case $p = q$.

4.5.7. Prove the following extension of Theorem 4.4.1. If T is a bounded linear operator from L^p to the Lorentz space $L^{q,s}$, then it has an ℓ^2-valued extension. Here $0 < p, q, s \le \infty$.

4.5.8. Let $T_j(f)(x) = f(x - j)$ and $f_j(x) = \chi_{[-j, 1-j]}$ for $j = 1, 2, \dots, N$. Use these functions and operators to show that the inequality

$$\Big\| \Big(\sum_j |T_j(f_j)|^2 \Big)^{\frac{1}{2}} \Big\|_{L^p} \le C_p \Big\| \Big(\sum_j |f_j|^2 \Big)^{\frac{1}{2}} \Big\|_{L^p}$$

may be false in general although the linear operators T_j are uniformly bounded from $L^p(\mathbf{R})$ to $L^p(\mathbf{R})$.

4.5.9. Suppose that T is a linear operator that takes real-valued functions to real-valued functions. Prove that

$$\sup_{\substack{f \text{ real-valued} \\ f \neq 0}} \frac{\|T(f)\|_{L^p}}{\|f\|_{L^p}} = \sup_{\substack{f \text{ complex-valued} \\ f \neq 0}} \frac{\|T(f)\|_{L^p}}{\|f\|_{L^p}}.$$

[*Hint:* Use Theorem 4.5.1 (a) with $p = q$.]

4.6 Vector-Valued Singular Integrals

We now discuss some results about vector-valued singular integrals. By this we mean singular integral operators taking values in Banach spaces. At this point we restrict our attention to the situation in which $X = Y = \mathbf{R}^n$.

4.6.1 Banach-Valued Singular Integral Operators

We consider a kernel $\vec{K}$ defined on $\mathbf{R}^n \setminus \{0\}$ that takes values in the space $L(\mathscr{B}_1, \mathscr{B}_2)$ of all bounded linear operators from $\mathscr{B}_1$ to $\mathscr{B}_2$. In other words, for all $x \in \mathbf{R}^n \setminus \{0\}$, $\vec{K}(x)$ is a bounded linear operator from $\mathscr{B}_1$ to $\mathscr{B}_2$, whose norm we denote by $\|\vec{K}(x)\|_{\mathscr{B}_1 \to \mathscr{B}_2}$. We assume that $\vec{K}(x)$ is $L(\mathscr{B}_1, \mathscr{B}_2)$-measurable and locally integrable away from the origin, so that the integral

$$\vec{T}(F)(x) = \int_{\mathbf{R}^n} \vec{K}(x-y)F(y)\,dy \tag{4.6.1}$$

is well defined as an element of $\mathscr{B}_2$ for all $F \in L^\infty(\mathbf{R}^n, \mathscr{B}_1)$ with compact support when x lies outside the support of F.

We assume that the kernel $\vec{K}$ satisfies Hörmander's condition,

$$\int_{|x|\geq 2|y|} \big\|\vec{K}(x-y) - \vec{K}(x)\big\|_{\mathscr{B}_1\to\mathscr{B}_2}\,dx \leq A < \infty, \qquad y \in \mathbf{R}^n \setminus \{0\}, \tag{4.6.2}$$

which is a certain form of regularity familiar to us from the scalar case.

The following vector-valued extension of Theorem 4.3.3 is the main result of this section.

Theorem 4.6.1. *Let $\mathscr{B}_1$ and $\mathscr{B}_2$ be Banach spaces. Suppose that $\vec{T}$ given by (4.6.1) is a bounded linear operator from $L^r(\mathbf{R}^n, \mathscr{B}_1)$ to $L^r(\mathbf{R}^n, \mathscr{B}_2)$ with norm $B = B(r)$ for some $1 < r \leq \infty$. Assume that $\vec{K}$ satisfies Hörmander's condition (4.6.2) for some $A > 0$. Then $\vec{T}$ has well defined extensions on $L^p(\mathbf{R}^n, \mathscr{B}_1)$ for all $1 \leq p < \infty$. Moreover, there exist dimensional constants C_n and C_n' such that*

$$\big\|\vec{T}(F)\big\|_{L^{1,\infty}(\mathbf{R}^n,\mathscr{B}_2)} \leq C_n'(A+B)\big\|F\big\|_{L^1(\mathbf{R}^n,\mathscr{B}_1)} \tag{4.6.3}$$

for all F in $L^1(\mathbf{R}^n, \mathscr{B}_1)$ and

$$\big\|\vec{T}(F)\big\|_{L^p(\mathbf{R}^n,\mathscr{B}_2)} \leq C_n \max\big(p, (p-1)^{-1}\big)(A+B)\big\|F\big\|_{L^p(\mathbf{R}^n,\mathscr{B}_1)} \tag{4.6.4}$$

whenever $1 < p < \infty$ and F is in $L^p(\mathbf{R}^n, \mathscr{B}_1)$.

Proof. We prove the weak type estimate (4.6.3) by applying the Calderón–Zygmund decomposition just as in the scalar case to the function $x \mapsto \big\|F(x)\big\|_{\mathscr{B}_1}$ defined on $\mathbf{R}^n$. The proof of Theorem 4.3.3 is directly applicable here, and an identical repetition of the arguments given in the scalar case with suitable norms replacing absolute values yields (4.6.3).

Next we interpolate between the estimates $\vec{T}: L^1(\mathbf{R}^n, \mathscr{B}_1) \to L^{1,\infty}(\mathbf{R}^n, \mathscr{B}_2)$ and $\vec{T}: L^r(\mathbf{R}^n, \mathscr{B}_1) \to L^r(\mathbf{R}^n, \mathscr{B}_2)$. Using Exercise 4.5.3, we obtain for $1 < p < r$,

$$\big\|\vec{T}(F)\big\|_{L^p(\mathbf{R}^n,\mathscr{B}_2)} \leq C_n \max(1, (p-1)^{-1})(A+B)\big\|F\big\|_{L^p(\mathbf{R}^n,\mathscr{B}_1)}, \tag{4.6.5}$$

where C_n is independent of r, p, $\mathscr{B}_1$, and $\mathscr{B}_2$ (and depends only on n).

We obtain (4.6.4) for $p > r$ via duality. Since $\vec{K}(x)$ is an operator from $\mathscr{B}_1$ to $\mathscr{B}_2$, its adjoint $\vec{K}^*(x)$ is an operator from $\mathscr{B}_2^*$ to $\mathscr{B}_1^*$. Let $\vec{T}^*$ be the Banach-valued operator with kernel $\vec{K}^*$. Obviously $\vec{K}^*(x-y) - \vec{K}^*(x)$ and $\vec{K}(x-y) - \vec{K}(x)$ have the same norm. Therefore, Hörmander's condition (4.6.2) also holds for $\vec{T}^*$, since it can be written as

$$\sup_{y\in\mathbf{R}^n\setminus\{0\}} \int_{|x|\geq 2|y|} \big\|\vec{K}^*(x-y) - \vec{K}^*(x)\big\|_{\mathscr{B}_2^*\to\mathscr{B}_1^*}\,dx = A < \infty. \tag{4.6.6}$$

The assumption on $\vec{T}$ gives that $\vec{T}^*$ is bounded from $L^{r'}(\mathbf{R}^n, \mathscr{B}_2^*)$ to $L^{r'}(\mathbf{R}^n, \mathscr{B}_1^*)$. Indeed, to see this, we fix $F \in L^{r'}(\mathbf{R}^n, \mathscr{B}_2^*)$ and use Exercise 4.5.1(a). We have

$$\begin{aligned}
\big\|\vec{T}^*(F)\big\|_{L^{r'}(\mathbf{R}^n, \mathscr{B}_1^*)} &= \sup_{\|G\|_{L^r(\mathbf{R}^n, \mathscr{B}_1)} \le 1} \left| \int_{\mathbf{R}^n} \langle \vec{T}^*(F)(x), G(x) \rangle \, dx \right| \\
&= \sup_{\|G\|_{L^r(\mathbf{R}^n, \mathscr{B}_1)} \le 1} \left| \int_{\mathbf{R}^n} \langle F(x), \vec{T}(G)(x) \rangle \, dx \right| \\
&\le \sup_{\|G\|_{L^r(\mathbf{R}^n, \mathscr{B}_1)} \le 1} \int_{\mathbf{R}^n} \big\|F(x)\big\|_{\mathscr{B}_2^*} \big\|\vec{T}(G)(x)\big\|_{\mathscr{B}_2} \, dx \\
&\le \sup_{\|G\|_{L^r(\mathbf{R}^n, \mathscr{B}_1)} \le 1} \big\|F\big\|_{L^{r'}(\mathbf{R}^n, \mathscr{B}_2^*)} \big\|\vec{T}(G)\big\|_{L^r(\mathbf{R}^n, \mathscr{B}_2)} \\
&\le B \big\|F\big\|_{L^{r'}(\mathbf{R}^n, \mathscr{B}_2^*)} .
\end{aligned}$$

Combining these facts, we obtain that (4.6.3) holds for $\vec{T}^*$, that is,

$$\big\|\vec{T}^*(F)\big\|_{L^{1,\infty}(\mathbf{R}^n, \mathscr{B}_1^*)} \le C_n (A+B) \big\|F\big\|_{L^1(\mathbf{R}^n, \mathscr{B}_2^*)} .$$

Consequently, we obtain by interpolation for $1 < p' < r'$ the estimate

$$\big\|\vec{T}^*(F)\big\|_{L^{p'}(\mathbf{R}^n, \mathscr{B}_1^*)} \le C_n \max(1, p-1)(A+B) \big\|F\big\|_{L^{p'}(\mathbf{R}^n, \mathscr{B}_2^*)} , \tag{4.6.7}$$

since $(p'-1)^{-1} = p-1$.

We now fix $r < p < \infty$. Let F lie in some dense subspace of $L^p(\mathbf{R}^n, \mathscr{B}_1)$, such that $\big\|\vec{T}(F)\big\|_{L^p(\mathbf{R}^n, \mathscr{B}_2)} < \infty$. We use Proposition 4.5.7(a) to write

$$\begin{aligned}
\big\|\vec{T}(F)\big\|_{L^p(\mathbf{R}^n, \mathscr{B}_2)} &\le \sup_{\|G\|_{L^{p'}(\mathbf{R}^n, \mathscr{B}_2^*)} \le 1} \left| \int_{\mathbf{R}^n} \langle G(x), \vec{T}(F)(x) \rangle \, dx \right| \\
&= \sup_{\|G\|_{L^{p'}(\mathbf{R}^n, \mathscr{B}_2^*)} \le 1} \left| \int_{\mathbf{R}^n} \langle \vec{T}^*(G)(x), F(x) \rangle \, dx \right| \\
&\le \sup_{\|G\|_{L^{p'}(\mathbf{R}^n, \mathscr{B}_2^*)} \le 1} \big\|\vec{T}^*(G)\big\|_{L^{p'}(\mathbf{R}^n, \mathscr{B}_1^*)} \big\|F\big\|_{L^p(\mathbf{R}^n, \mathscr{B}_1)} \\
&\le C_n \max(1, p)(A+B) \big\|F\big\|_{L^p(\mathbf{R}^n, \mathscr{B}_1)} \\
&= C_n \max(1, p)(A+B) \big\|F\big\|_{L^p(\mathbf{R}^n, \mathscr{B}_1)} ,
\end{aligned}$$

where we used (4.6.7). Since F lies in some dense subspace of $L^p(\mathbf{R}^n, \mathscr{B}_1)$, the required conclusion follows. This combined with (4.6.5) implies the required conclusion whenever $1 < r < \infty$. Observe that the case $r = \infty$ is easier and requires only Exercise 4.3.7 adapted to the Banach-valued setting. □

4.6.2 Applications

We proceed with some applications. An important consequence of Theorem 4.6.1 is the following:

Corollary 4.6.2. *Let W_j be a sequence of tempered distributions on $\mathbf{R}^n$ whose Fourier transforms are uniformly bounded functions (i.e., $|\widehat{W_j}| \le B$ for some B). Suppose that each W_j coincides with some locally integrable function K_j on $\mathbf{R}^n \setminus \{0\}$ that satisfies*

$$\int_{|x|\ge 2|y|} \sup_j |K_j(x-y) - K_j(x)|\, dx \le A, \qquad y \in \mathbf{R}^n \setminus \{0\}. \tag{4.6.8}$$

Then there are constants $C_n, C_n' > 0$ such that for all $1 < p, r < \infty$ we have

$$\Big\| \Big(\sum_j |W_j * f_j|^r \Big)^{\frac{1}{r}} \Big\|_{L^{1,\infty}} \le C_n' \max(r,(r-1)^{-1})(A+B) \Big\| \Big(\sum_j |f_j|^r \Big)^{\frac{1}{r}} \Big\|_{L^1},$$

$$\Big\| \Big(\sum_j |W_j * f_j|^r \Big)^{\frac{1}{r}} \Big\|_{L^p} \le C_n c(p,r)(A+B) \Big\| \Big(\sum_j |f_j|^r \Big)^{\frac{1}{r}} \Big\|_{L^p},$$

where $c(p,r) = \max(p,(p-1)^{-1}) \max(r,(r-1)^{-1})$.

Proof. Let T_j be the operator given by convolution with the distribution W_j. It follows from Theorem 4.3.3 that the T_j's are of weak type $(1,1)$ and also bounded on L^r with bounds at most a dimensional constant multiple of $\max(r,(r-1)^{-1})(A+B)$, uniformly in j. Naturally, set $\mathscr{B}_1 = \mathscr{B}_2 = \ell^r$ and define

$$\vec{T}(\{f_j\}_j) = \{W_j * f_j\}_j$$

for $\{f_j\}_j \in L^r(\mathbf{R}^n, \ell^r)$. Summing gives that $\vec{T}$ maps $L^r(\mathbf{R}^n, \ell^r)$ to itself with norm at most a dimensional constant multiple of $\max(r,(r-1)^{-1})(A+B)$.

The operator $\vec{T}$ has the form

$$\vec{T}(F)(x) = \int_{\mathbf{R}^n} \vec{K}(x-y) F(y)\, dy$$

for $F \in L^r(\mathbf{R}^n, \ell^r)$ with compact support and $x \notin \operatorname{support}(F)$, where $\vec{K}(x)$ in $L(\ell^r, \ell^r)$ is the following operator:

$$\vec{K}(x)(\{t_j\}_j) = \{K_j(x) t_j\}_j, \qquad \{t_j\}_j \in \ell^r.$$

Clearly,

$$\big\| \vec{K}(x-y) - \vec{K}(x) \big\|_{\ell^r \to \ell^r} \le \sup_j |K_j(x-y) - K_j(x)|,$$

and therefore Hörmander's condition holds for $\vec{K}$ as a consequence of (4.6.8). The desired conclusion follows from Theorem 4.6.1. □

If all the W_j's are equal, we obtain the following corollary, which contains in particular the result (4.5.16) mentioned earlier.

Corollary 4.6.3. *Let W be an element of $\mathscr{S}'(\mathbf{R}^n)$ whose Fourier transform is a function bounded in absolute value by some $B > 0$. Suppose that W coincides with some locally integrable function K on $\mathbf{R}^n \setminus \{0\}$ that satisfies Hörmander's condition:*

$$\int_{|x|\geq 2|y|} |K(x-y) - K(x)|\,dx \leq A, \qquad y \in \mathbf{R}^n \setminus \{0\}. \tag{4.6.9}$$

Let T be the operator given by convolution with W. Then there exist constants $C_n, C_n' > 0$ such that for all $1 < p, r < \infty$ we have that

$$\Big\|\Big(\sum_j |T(f_j)|^r\Big)^{\frac{1}{r}}\Big\|_{L^{1,\infty}} \leq C_n' \max(r,(r-1)^{-1})(A+B)\Big\|\Big(\sum_j |f_j|^r\Big)^{\frac{1}{r}}\Big\|_{L^1},$$

$$\Big\|\Big(\sum_j |T(f_j)|^r\Big)^{\frac{1}{r}}\Big\|_{L^p} \leq C_n c(p,r)\,(A+B)\Big\|\Big(\sum_j |f_j|^r\Big)^{\frac{1}{r}}\Big\|_{L^p},$$

where $c(p,r) = \max(p,(p-1)^{-1})\max(r,(r-1)^{-1})$. In particular, these inequalities are valid for the Hilbert transform and the Riesz transforms.

Interestingly enough, we can use the very statement of Theorem 4.6.1 to obtain its corresponding vector-valued version.

Proposition 4.6.4. *Let let $1 < p, r < \infty$ and let $\mathscr{B}_1$ and $\mathscr{B}_2$ be two Banach spaces. Suppose that $\vec{T}$ given by (4.6.1) is a bounded linear operator from $L^r(\mathbf{R}^n, \mathscr{B}_1)$ to $L^r(\mathbf{R}^n, \mathscr{B}_2)$ with norm $B = B(r)$. Also assume that for all $x \in \mathbf{R}^n \setminus \{0\}$, $\vec{K}(x)$ is a bounded linear operator from $\mathscr{B}_1$ to $\mathscr{B}_2$ that satisfies Hörmander's condition (4.6.2) for some $A > 0$. Then there exist positive constants C_n, C_n' such that for all $\mathscr{B}_1$-valued functions F_j we have*

$$\Big\|\Big(\sum_j \|\vec{T}(F_j)\|_{\mathscr{B}_2}^r\Big)^{\frac{1}{r}}\Big\|_{L^{1,\infty}(\mathbf{R}^n)} \leq C_n'(A+B)\Big\|\Big(\sum_j \|F_j\|_{\mathscr{B}_1}^r\Big)^{\frac{1}{r}}\Big\|_{L^1(\mathbf{R}^n)},$$

$$\Big\|\Big(\sum_j \|\vec{T}(F_j)\|_{\mathscr{B}_2}^r\Big)^{\frac{1}{r}}\Big\|_{L^p(\mathbf{R}^n)} \leq C_n(A+B)c(p)\Big\|\Big(\sum_j \|F_j\|_{\mathscr{B}_1}^r\Big)^{\frac{1}{r}}\Big\|_{L^p(\mathbf{R}^n)},$$

where $c(p) = \max(p,(p-1)^{-1})$.

Proof. Let us denote by $\ell^r(\mathscr{B}_1)$ the Banach space of all $\mathscr{B}_1$-valued sequences $\{t_j\}_j$ that satisfy

$$\big\|\{t_j\}_j\big\|_{\ell^r(\mathscr{B}_1)} = \Big(\sum_j \|t_j\|_{\mathscr{B}_1}^r\Big)^{\frac{1}{r}} < \infty.$$

Now consider the operator $\vec{S}$ defined by

$$\vec{S}(\{F_j\}_j) = \{\vec{T}(F_j)\}_j.$$

It is obvious that $\vec{S}$ maps $L^r(\mathbf{R}^n, \ell^r(\mathscr{B}_1))$ to $L^r(\mathbf{R}^n, \ell^r(\mathscr{B}_2))$ with norm at most B. Moreover, $\vec{S}$ has kernel $\widetilde{K}(x) \in L(\ell^r(\mathscr{B}_1), \ell^r(\mathscr{B}_2))$ given by

$$\widetilde{K}(x)(\{t_j\}_j) = \{\vec{K}(x)t_j\}_j,$$

where $\vec{K}$ is the kernel of $\vec{T}$. It is not hard to see that the operator norms of $\vec{K}$ and $\widetilde{K}$ coincide and therefore

$$\big\|\vec{K}(x-y) - \vec{K}(x)\big\|_{\mathscr{B}_1 \to \mathscr{B}_2} = \big\|\widetilde{K}(x-y) - \widetilde{K}(x)\big\|_{\ell^r(\mathscr{B}_1) \to \ell^r(\mathscr{B}_2)}.$$

We conclude that $\widetilde{K}$ satisfies the hypotheses of Theorem 4.6.1. The conclusions of Theorem 4.6.1 for $\vec{S}$ are the desired inequalities for $\vec{T}$. □

4.6.3 Vector-Valued Estimates for Maximal Functions

Next, we discuss applications of vector-valued inequalities to some nonlinear operators. We fix an integrable function Φ on $\mathbf{R}^n$ and for $t > 0$ define $\Phi_t(x) = t^{-n}\Phi(t^{-1}x)$. We suppose that Φ satisfies the following *regularity* condition:

$$\int_{|x| \ge 2|y|} \sup_{t>0} |\Phi_t(x-y) - \Phi_t(x)|\,dx = A_\Phi < \infty, \qquad y \in \mathbf{R}^n \setminus \{0\}. \tag{4.6.10}$$

We consider the maximal operator

$$M_\Phi(f)(x) = \sup_{t>0} |(f * \Phi_t)(x)|$$

defined for f in $L^1 + L^\infty$. We are interested in obtaining L^p estimates for M_Φ. It is reasonable to start with $p = \infty$, which yields the easiest of all the L^p estimates for M_Φ, the trivial estimate

$$\big\|M_\Phi(f)\big\|_{L^\infty} \le \big\|\Phi\big\|_{L^1} \big\|f\big\|_{L^\infty}. \tag{4.6.11}$$

We think of M_Φ as a linear operator taking values in a Banach space. Indeed, it is natural to set

$$\mathscr{B}_1 = \mathbf{C} \qquad \text{and} \qquad \mathscr{B}_2 = L^\infty(\mathbf{R}^+)$$

and view M_Φ as the linear operator $f \mapsto \{f * \Phi_\delta\}_{\delta>0}$ that maps $\mathscr{B}_1$-valued functions to $\mathscr{B}_2$-valued functions.

To do this precisely, we define a $\mathscr{B}_2$-valued kernel

$$\vec{K}_\Phi(x) = \{\Phi_\delta(x)\}_{\delta \in \mathbf{R}^+}$$

and a $\mathscr{B}_2$ -valued linear operator

$$\vec{M}_\Phi(f) = f * \vec{K}_\Phi = \{f * \Phi_\delta\}_{\delta \in \mathbf{R}^+}$$

acting on complex-valued functions on $\mathbf{R}^n$. We know that $\vec{M}_\Phi$ maps $L^\infty(\mathbf{R}^n,\mathscr{B}_1) = L^\infty(\mathbf{R}^n)$ to $L^\infty(\mathbf{R}^n,\mathscr{B}_2)$ with norm at most $\|\Phi\|_{L^1}$. Clearly (4.6.10) implies condition (4.6.2) for the kernel $\vec{K}_\Phi$. Applying Theorem 4.6.1, we obtain for $1<p<\infty$,

$$\big\|\vec{M}_\Phi(f)\big\|_{L^p(\mathbf{R}^n,\mathscr{B}_2)} \le C_n \max(p,(p-1)^{-1})\big(A_\Phi+\|\Phi\|_{L^1}\big)\|f\|_{L^p(\mathbf{R}^n)}, \tag{4.6.12}$$

which can be immediately improved to

$$\big\|\vec{M}_\Phi(f)\big\|_{L^r(\mathbf{R}^n,\mathscr{B}_2)} \le C_n \max(1,(r-1)^{-1})\big(A_\Phi+\|\Phi\|_{L^1}\big)\|f\|_{L^r(\mathbf{R}^n)} \tag{4.6.13}$$

via interpolation with estimate (4.6.11) for all $1<r<\infty$.

Next we use estimate (4.6.13) to obtain vector-valued estimates for the sublinear operator M_Φ.

Corollary 4.6.5. *Let Φ be an integrable function on $\mathbf{R}^n$ that satisfies (4.6.10). Then there exist dimensional constants C_n and C_n' such that for all $1<p,r<\infty$ the following vector-valued inequalities are valid:*

$$\Big\|\Big(\sum_j |M_\Phi(f_j)|^r\Big)^{\frac{1}{r}}\Big\|_{L^{1,\infty}} \le C_n' c(r)\big(A_\Phi+\|\Phi\|_{L^1}\big)\Big\|\Big(\sum_j |f_j|^r\Big)^{\frac{1}{r}}\Big\|_{L^1}, \tag{4.6.14}$$

where $c(r) = 1+(r-1)^{-1}$, and

$$\Big\|\Big(\sum_j |M_\Phi(f_j)|^r\Big)^{\frac{1}{r}}\Big\|_{L^p} \le C_n c(p,r)\big(A_\Phi+\|\Phi\|_{L^1}\big)\Big\|\Big(\sum_j |f_j|^r\Big)^{\frac{1}{r}}\Big\|_{L^p}, \tag{4.6.15}$$

where $c(p,r) = \big(1+(r-1)^{-1}\big)\big(p+(p-1)^{-1}\big)$.

Proof. We set $\mathscr{B}_1 = \mathbf{C}$ and $\mathscr{B}_2 = L^\infty(\mathbf{R}^+)$. We use estimate (4.6.13) as a starting point in Proposition 4.6.4, which immediately yields the required conclusions (4.6.14) and (4.6.15). □

Similar estimates hold for the Hardy–Littlewood maximal operator.

Theorem 4.6.6. *For $1<p,r<\infty$ the Hardy–Littlewood maximal function M satisfies the vector-valued inequalities*

$$\Big\|\Big(\sum_j |M(f_j)|^r\Big)^{\frac{1}{r}}\Big\|_{L^{1,\infty}} \le C_n'\big(1+(r-1)^{-1}\big)\Big\|\Big(\sum_j |f_j|^r\Big)^{\frac{1}{r}}\Big\|_{L^1}, \tag{4.6.16}$$

$$\Big\|\Big(\sum_j |M(f_j)|^r\Big)^{\frac{1}{r}}\Big\|_{L^p} \le C_n c(p,r)\Big\|\Big(\sum_j |f_j|^r\Big)^{\frac{1}{r}}\Big\|_{L^p}, \tag{4.6.17}$$

where $c(p,r) = \big(1+(r-1)^{-1}\big)\big(p+(p-1)^{-1}\big)$.

Proof. Let us fix a positive radial symmetrically decreasing Schwartz function Φ on $\mathbf{R}^n$ that satisfies $\Phi(x)\ge 1$ when $|x|\le 1$. Then the Hardy–Littlewood maximal function $M(f)$ is pointwise controlled by a constant multiple of the function $M_\Phi(|f|)$. In view of Corollary 4.6.5, it suffices to check that for such a Φ, (4.6.10) holds. First observe that in view of the decreasing character of Φ, we have

$$\sup_j |f| * \Phi_{2^j} \le M_\Phi(|f|) \le 2^n \sup_j |f| * \Phi_{2^j},$$

and for this reason we choose to work with the easier dyadic maximal operator

$$M_\Phi^d(f) = \sup_j |f * \Phi_{2^j}|.$$

We observe the validity of the simple inequalties

$$2^{-n} M(f) \le \mathcal{M}(f) \le M_\Phi(|f|) \le 2^n M_\Phi^d(|f|). \tag{4.6.18}$$

If we can show that

$$\sup_{y\in\mathbf{R}^n\setminus\{0\}} \int_{|x|\ge 2|y|} \sup_{j\in\mathbf{Z}} |\Phi_{2^j}(x-y) - \Phi_{2^j}(x)|\,dx = C_n < \infty, \tag{4.6.19}$$

then (4.6.14) and (4.6.15) are satisfied with M_Φ^d replacing M_Φ. We therefore turn our attention to (4.6.19). We have

$$\begin{aligned}
&\int_{|x|\ge 2|y|} \sup_{j\in\mathbf{Z}} |\Phi_{2^j}(x-y) - \Phi_{2^j}(x)|\,dx \\
&\quad \le \sum_{j\in\mathbf{Z}} \int_{|x|\ge 2|y|} |\Phi_{2^j}(x-y) - \Phi_{2^j}(x)|\,dx \\
&\quad \le \sum_{2^j>|y|} \int_{|x|\ge 2|y|} \frac{|y|\,|\nabla\Phi(\frac{x-\theta y}{2^j})|}{2^{(n+1)j}}\,dx + \sum_{2^j\le|y|} \int_{|x|\ge 2|y|} (|\Phi_{2^j}(x-y)| + |\Phi_{2^j}(x)|)\,dx \\
&\quad \le \sum_{2^j>|y|} \int_{|x|\ge 2|y|} \frac{|y|}{2^{(n+1)j}} \frac{C_N\,dx}{(1+|2^{-j}(x-\theta y)|)^N} + 2\sum_{2^j\le|y|} \int_{|x|\ge|y|} |\Phi_{2^j}(x)|\,dx \\
&\quad \le \sum_{2^j>|y|} \int_{|x|\ge 2|y|} \frac{|y|}{2^{(n+1)j}} \frac{C_N}{(1+|2^{-j-1}x|)^N}\,dx + 2\sum_{2^j\le|y|} \int_{|x|\ge 2^{-j}|y|} |\Phi(x)|\,dx \\
&\quad \le \sum_{2^j>|y|} \int_{|x|\ge 2^{-j}|y|} \frac{|y|}{2^j} \frac{C_N}{(1+|x|)^N}\,dx + 2\sum_{2^j\le|y|} C_N(2^{-j}|y|)^{-N} \\
&\quad \le C_N \sum_{2^j>|y|} \frac{|y|}{2^j} + C_N = 2C_N,
\end{aligned}$$

where $C_N > 0$ depends on $N > n$, $\theta \in [0,1]$, and $|x-\theta y| \ge |x|/2$ when $|x| \ge 2|y|$.

Now apply (4.6.14) and (4.6.15) to M_Φ^d and use (4.6.18) to obtain the desired vector-valued inequalities. □

Remark 4.6.7. Observe that (4.6.16) and (4.6.17) also hold for $r = \infty$. These endpoint estimates can be proved directly by observing that

$$\sup_j M(f_j) \le M(\sup_j |f_j|).$$

The same is true for estimates (4.6.14) and (4.6.15). Finally, estimates (4.6.17) and (4.6.15) also hold for $p = \infty$.

Exercises

4.6.1. (a) For all $j \in \mathbf{Z}$, let I_j be an interval in $\mathbf{R}$ and let T_j be the operator given on the Fourier transform by multiplication by the characteristic function of I_j. Prove that there exists a constant $C > 0$ such that for all $1 < p, r < \infty$ and for all integrable functions f_j on $\mathbf{R}$ we have

$$\Big\|\Big(\sum_j |T_j(f_j)|^r\Big)^{\frac{1}{r}}\Big\|_{L^p} \le C c(p,r) \Big\|\Big(\sum_j |f_j|^r\Big)^{\frac{1}{r}}\Big\|_{L^p},$$

$$\Big\|\Big(\sum_j |T_j(f_j)|^r\Big)^{\frac{1}{r}}\Big\|_{L^{1,\infty}} \le C \max\big(r,(r-1)^{-1}\big) \Big\|\Big(\sum_j |f_j|^r\Big)^{\frac{1}{r}}\Big\|_{L^1},$$

where $c(p,r) = \max\big(r,(r-1)^{-1}\big)\max\big(p,(p-1)^{-1}\big)$.
(b) Let R_j be arbitrary rectangles on $\mathbf{R}^n$ with sides parallel to the axes and let S_j be the operators given on the Fourier transform by multiplication by the characteristic functions of R_j. Prove that there exists a dimensional constant $C_n < \infty$ such that for all indices $1 < p, r < \infty$ and for all functions f_j in $L^p(\mathbf{R}^n)$ we have

$$\Big\|\Big(\sum_j |S_j(f_j)|^r\Big)^{\frac{1}{r}}\Big\|_{L^p} \le C_n c(p,r)^n \Big\|\Big(\sum_j |f_j|^r\Big)^{\frac{1}{r}}\Big\|_{L^p},$$

where $c(p,r)$ is as in part (a).
[*Hint:* Use Theorem 4.5.1 and the fact that the operator whose multiplier is $\chi_{(a,b)}$ is equal to $\frac{i}{2}\big(M^a H M^{-a} - M^b H M^{-b}\big)$, where $M^a(f)(x) = f(x)e^{2\pi i a x}$ and H is the Hilbert transform.]

4.6.2. For every $t \in \mathbf{R}^d$, let $R(t)$ be a rectangle with sides parallel to the axes in $\mathbf{R}^n$ such that the map $t \mapsto R(t)$ is measurable. Then there is a constant $C_n > 0$ such that for all $1 < p < \infty$, for all σ-finite measures μ on $\mathbf{R}^d$, and for all families of measurable functions f_t on $\mathbf{R}^n$ we have

$$\Big\|\Big(\int_{\mathbf{R}^d} |(\widehat{f_t}\chi_{R(t)})^\vee|^2 \, d\mu(t)\Big)^{\frac{1}{2}}\Big\|_{L^p} \le C_n c(p)^n \Big\|\Big(\int_{\mathbf{R}^d} |f_t|^2 \, d\mu(t)\Big)^{\frac{1}{2}}\Big\|_{L^p},$$

where $c(p) = \max(p,(p-1)^{-1})$.
[*Hint:* Reduce this estimate to Proposition 4.5.11. Observe that when $d = 1$, this provides a continuous version of the result of Exercise 4.6.1.]

4.6.3. (a) Let Φ be a radially decreasing function on $\mathbf{R}^n$ that satisfies

$$\int_{\mathbf{R}^n} |\Phi(x-y)-\Phi(x)|\,dx \le \eta(y), \qquad \int_{|x|\ge R} |\Phi(x)|\,dx \le \eta(R^{-1}),$$

for all $R>1$, where η is an increasing function with $\eta(0)=0$ such that

$$\int_0^1 \frac{\eta(t)}{t}\,dt < \infty.$$

Prove that (4.6.19) holds.
[*Hint:* Modify the calculation in the proof of Theorem 4.6.6.]
(b) Use Theorem 4.6.1 with $r=\infty$ to conclude that the maximal function $f \mapsto \sup_{j\in\mathbf{Z}} |f*\Phi_{2^j}|$ maps $L^p(\mathbf{R}^n)$ to itself for $1<p\le\infty$.

4.6.4. (a) On $\mathbf{R}$, take $f_j=\chi_{[2^{j-1},2^j]}$ to prove that inequality (4.6.17) fails when $p=\infty$ and $1<r<\infty$.
(b) Again on $\mathbf{R}$, take $N>2$ and $f_j=\chi_{[\frac{j-1}{N},\frac{j}{N}]}$ for $j=1,2,\dots,N$ to prove that (4.6.17) fails when $1<p<\infty$ and $r=1$.

4.6.5. Prove that the vector-valued inequality

$$\Big\|\Big(\sum_j |K*f_j|^q\Big)^{\frac{1}{q}}\Big\|_{L^p} \le C_{p,q}\Big\|\Big(\sum_j |f_j|^q\Big)^{\frac{1}{q}}\Big\|_{L^p}$$

may fail in general when $q<1$ when the operator $f\mapsto f*K$ is L^p bounded.
[*Hint:* Take $K=\chi_{[-1,1]}$ and $f_j=\chi_{[\frac{j-1}{N},\frac{j}{N}]}$ for $1\le j\le N$.]

4.6.6. Let $\{Q_j\}_j$ be a countable collection of cubes in $\mathbf{R}^n$ with disjoint interiors. Let c_j be the center of the cube Q_j and d_j its diameter. For $\varepsilon>0$, define the *Marcinkiewicz function* associated with the family $\{Q_j\}_j$ as follows:

$$M_\varepsilon(x)=\sum_j \frac{d_j^{n+\varepsilon}}{|x-c_j|^{n+\varepsilon}+d_j^{n+\varepsilon}}.$$

Prove that for some constants $C_{n,\varepsilon,p}$ and $C_{n,\varepsilon}$ one has

$$\|M_\varepsilon\|_{L^p} \le C_{n,\varepsilon,p}\Big(\sum_j |Q_j|\Big)^{\frac{1}{p}}, \qquad p>\frac{n}{n+\varepsilon},$$

$$\|M_\varepsilon\|_{L^{\frac{n}{n+\varepsilon},\infty}} \le C_{n,\varepsilon}\Big(\sum_j |Q_j|\Big)^{\frac{n+\varepsilon}{n}},$$

and consequently

$$\int_{\mathbf{R}^n} M_\varepsilon(x)\,dx \le C_{n,\varepsilon}\sum_j |Q_j|.$$

[*Hint:* Verify that

$$\frac{d_j^{n+\varepsilon}}{|x-c_j|^{n+\varepsilon}+d_j^{n+\varepsilon}} \le CM(\chi_{Q_j})(x)^{\frac{n+\varepsilon}{n}}$$

and use Corollary 4.6.5.]

HISTORICAL NOTES

The L^p boundedness of the conjugate function on the circle was announced in 1924 by Riesz [219], but its first proof appeared three years later in [221]. In view of the identification of the Hilbert transform with the conjugate function, the L^p boundedness of the Hilbert transform is also attributed to M. Riesz. Riesz's proof was first given for $p = 2k$, $k \in \mathbf{Z}^+$, via an argument similar to that in the proof of Theorem 3.5.6. For $p \neq 2k$ this proof relied on interpolation and was completed with the simultaneous publication of Riesz's article on interpolation of bilinear forms [220]. The weak type $(1,1)$ property of the Hilbert transform is due to Kolmogorov [158]. Additional proofs of the boundedness of the Hilbert transform have been obtained by Stein [267], Loomis [177], and Calderón [33]. The proof of Theorem 4.1.7, based on identity (4.1.21), is a refinement of a proof given by Cotlar [60].

The norm of the conjugate function on $L^p(\mathbf{T}^1)$, and consequently that of the Hilbert transform on $L^p(\mathbf{R})$, was shown by Gohberg and Krupnik [102] to be $\cot(\pi/2p)$ when p is a power of 2. Duality gives that this norm is $\tan(\pi/2p)$ for $1 < p \le 2$ whenever p' is a power of 2. Pichorides [213] extended this result to all $1 < p < \infty$ by refining Calderón's proof of Riesz's theorem. This result was also independently obtained by B. Cole (unpublished). The direct and simplified proof for the Hilbert transform given in Exercise 4.1.12 is in Grafakos [103]. The norm of the operators $\frac{1}{2}(I \pm iH)$ for real-valued functions was found to be $\frac{1}{2}\big[\min(\cos(\pi/2p), \sin(\pi/2p))\big]^{-1}$ by Verbitsky [284] and later independently by Essén [84]. The norm of the same operators for complex-valued functions was shown to be equal to $[\sin(\pi/p)]^{-1}$ by Hollenbeck and Verbitsky [128]. The best constant in the weak type $(1,1)$ estimate for the Hilbert transform is equal to $(1+\frac{1}{3^2}+\frac{1}{5^2}+\cdots)(1-\frac{1}{3^2}+\frac{1}{5^2}-\cdots)^{-1}$ as shown by Davis [71] using Brownian motion; an alternative proof was later obtained by Baernstein [14]. Iwaniec and Martin [138] showed that the norms of the Riesz transforms on $L^p(\mathbf{R}^n)$ coincide with that of the Hilbert transform on $L^p(\mathbf{R})$.

Operators of the kind T_Ω as well as the stopping-time decomposition of Theorem 4.3.1 were introduced by Calderón and Zygmund [37]. In the same article, Calderón and Zygmund used this decomposition to prove Theorem 4.3.3 for operators of the form T_Ω when Ω satisfies a certain weak smoothness condition. The more general condition (4.3.12) first appeared in Hörmander's article [129]. A more flexible condition sufficient to yield weak type $(1,1)$ bounds is contained in the article of Duong and McIntosh [80]. Theorems 4.2.10 and 4.2.11 are also due to Calderón and Zygmund [39]. The latter article contains the method of rotations. Algebras of operators of the form T_Ω were studied in [40]. For more information on algebras of singular integrals see the article of Calderón [36]. Theorem 4.4.1 is due to Benedek, Calderón, and Panzone [18], while Example 4.4.2 is taken from Muckenhoupt [203]. Theorem 4.4.5 is due to Riviere [223]. A weaker version of this theorem, applicable for smoother singular integrals such as the maximal Hilbert transform, was obtained by Cotlar [60] (Theorem 4.3.4). Improvements of the main inequality in Theorem 4.3.4 for homogeneous singular integrals were obtained by Mateu and Verdera [191] and Mateu, Orobitg, and Verdera [192]. For a general overview of singular integrals and their applications, one may consult the expository article of Calderón [35].

Part (a) of Theorem 4.5.1 is due to Marcinkiewicz and Zygmund [189], although the case $p = q$ was proved earlier by Paley [209] with a larger constant. The values of r for which a general linear operator of weak or strong type (p,q) admits bounded ℓ^r extensions are described in Rubio de Francia and Torrea [227]. The L^p and weak L^p spaces in Theorem 4.5.1 can be replaced by general Banach lattices, as shown by Krivine [164] using Grothendieck's inequality. Hilbert-space-valued estimates for singular integrals were obtained by Benedek, Calderón, and Panzone [18]. Other operator-valued singular integral operators were studied by Rubio de Francia, Ruiz, and Torrea [228]. Banach-valued singular integrals are studied in great detail in the book of García-Cuerva and Rubio de Francia [98], which provides an excellent presentation of the subject. The ℓ^r-valued estimates (4.5.16) for the Hilbert transform were first obtained by Boas and Bochner [24]. The corresponding vector-valued estimates for the Hardy–Littlewood maximal function in Theorem 4.6.6 are due to Fefferman and Stein [91]. Conditions of the form (4.6.10) have been applied to several situations and can be traced in Zo [301].

The sharpness of the logarithmic condition (4.2.23) was indicated by Weiss and Zygmund [289], who constructed an example of an integrable function Ω with vanishing integral on $\mathbf{S}^1$ satisfying $\int_{\mathbf{S}^{n-1}} |\Omega(\theta)|\log^+|\Omega(\theta)|\big(\log(2+\log(2+|\Omega(\theta)|))\big)^{-\delta}\,d\theta = \infty$ for all $\delta > 0$ and of a continuous function in $L^p(\mathbf{R}^2)$ for all $1 < p < \infty$ such that $\limsup_{\varepsilon\to 0}|T_\Omega^{(\varepsilon)}(f)(x)| = \infty$ for almost all $x \in \mathbf{R}^2$. The proofs of Theorems 4.2.10 and 4.2.11 can be modified to give that if Ω is in the Hardy space H^1 of $\mathbf{S}^{n-1}$, then T_Ω and $T_\Omega^{(*)}$ map L^p to L^p for $1 < p < \infty$. For T_Ω this fact was proved by Connett [57] and independently by Ricci and Weiss [216]; for $T_\Omega^{(*)}$ this was proved by Fan and Pan [86] and independently by Grafakos and Stefanov [110]. The latter authors [111] also obtained that the logarithmic condition $\text{ess.sup}_{|\xi|=1}\int_{\mathbf{S}^{n-1}}|\Omega(\theta)|(\log\frac{1}{|\xi\cdot\theta|})^{1+\alpha}\,d\theta < \infty$, $\alpha > 0$, implies L^p boundedness for T_Ω and $T_\Omega^{(*)}$ for some $p \neq 2$. See also Fan, Guo, and Pan [85] as well as Ryabogin and Rubin [231] for extensions. Examples of functions Ω for which T_Ω maps L^p to L^p for a certain range of p's but not for other ranges of p's is given in Grafakos, Honzík, and Ryabogin [104].

The relatively weak condition $|\Omega|\log^+|\Omega| \in L^1(\mathbf{S}^{n-1})$ also implies weak type $(1,1)$ boundedness for operators T_Ω. This was obtained by Seeger [239] and later extended by Tao [273] to situations in which there is no Fourier transform structure. Earlier partial results are in Christ and Rubio de Francia [52] and in the simultaneous work of Hofmann [127], both inspired by the work of Christ [50]. Soria and Sjögren [244] showed that for arbitrary Ω in $L^1(\mathbf{S}^{n-1})$, T_Ω is weak type $(1,1)$ when restricted to radial functions. Examples due to Christ (published in [110]) indicate that even for bounded functions Ω on $\mathbf{S}^{n-1}$, T_Ω may not map the endpoint Hardy space $H^1(\mathbf{R}^n)$ to $L^1(\mathbf{R}^n)$. However, Tao and Seeger [275] have showed that T_Ω always maps the Hardy space $H^1(\mathbf{R}^n)$ to the Lorentz space $L^{1,2}(\mathbf{R}^n)$ when $|\Omega|(\log^+|\Omega|)^2$ is integrable over $\mathbf{S}^{n-1}$. This result is sharp in the sense that for such Ω, T_Ω may not map $H^1(\mathbf{R}^n)$ to $L^{1,q}(\mathbf{R}^n)$ when $q < 2$ in general. If T_Ω maps $H^1(\mathbf{R}^n)$ to itself, Daly and Phillips [67] (in dimension $n = 2$) and Daly [66] (in dimensions $n \geq 3$) showed that Ω must lie in the Hardy space $H^1(\mathbf{S}^{n-1})$. There are also results concerning the singular maximal operator $M_\Omega(f)(x) = \sup_{r>0}\frac{1}{v_n r^n}\int_{|y|\leq r}|f(x-y)|\,|\Omega(y)|\,dy$, where Ω is an integrable function on $\mathbf{S}^{n-1}$ of not necessarily vanishing integral. Such operators were studied by Fefferman [92], Christ [50], and Hudson [132]. An excellent treatment of several kinds of singular integral operators with rough kernels is contained in the book of Lu, Ding, and Yan [181].

Chapter 5
Littlewood–Paley Theory and Multipliers

In this chapter we are concerned with orthogonality properties of the Fourier transform. This orthogonality is easily understood on L^2, but at this point it is not clear how it manifests itself on other spaces. Square functions introduce a way to express and quantify orthogonality of the Fourier transform on L^p and other function spaces. The introduction of square functions in this setting was pioneered by Littlewood and Paley, and the theory that subsequently developed is named after them. The extent to which Littlewood–Paley theory characterizes function spaces is remarkable. This topic is investigated Chapter 6.

Historically, Littlewood–Paley theory first appeared in the context of one-dimensional Fourier series and depended on complex function theory. With the development of real-variable methods, the whole theory became independent of complex methods and was extended to $\mathbf{R}^n$. This is the approach that we follow in this chapter. It turns out that the Littlewood–Paley theory is intimately related to the Calderón–Zygmund theory introduced in the previous chapter. This connection is deep and far-reaching, and its central feature is that one is able to derive the main results of one theory from the other.

The thrust and power of the Littlewood–Paley theory become apparent in some of the applications we discuss in this chapter. Such applications include the derivation of certain multiplier theorems, that is, theorems that yield sufficient conditions for bounded functions to be L^p multipliers. As a consequence of Littlewood–Paley theory we also prove that the lacunary partial Fourier integrals $\int_{|\xi|\le 2^N} \widehat{f}(\xi)e^{2\pi i x\cdot\xi}\,d\xi$ converge almost everywhere to an L^p function f on $\mathbf{R}^n$.

5.1 Littlewood–Paley Theory

We begin by examining more closely what we mean by orthogonality of the Fourier transform. If the functions f_j defined on $\mathbf{R}^n$ have Fourier transforms $\widehat{f_j}$ supported in disjoint sets, then they are *orthogonal* in the sense that

L. Grafakos, *Classical Fourier Analysis, Second Edition*,
DOI: 10.1007/978-0-387-09432-8_5, © Springer Science+Business Media, LLC 2008

$$\Big\|\sum_j f_j\Big\|_{L^2}^2 = \sum_j \|f_j\|_{L^2}^2. \tag{5.1.1}$$

Unfortunately, when 2 is replaced by some $p \neq 2$ in (5.1.1), the previous quantities may not even be comparable, as we show in Examples 5.1.8 and 5.1.9. The Littlewood–Paley theorem provides a substitute inequality to (5.1.1) expressing the fact that certain orthogonality considerations are also valid in $L^p(\mathbf{R}^n)$.

5.1.1 The Littlewood–Paley Theorem

The orthogonality we are searching for is best seen in the context of one-dimensional Fourier series (which was the setting in which Littlewood and Paley formulated their result). The primary observation is that the exponential $e^{2\pi i 2^k x}$ oscillates half as much as $e^{2\pi i 2^{k+1}x}$ and is therefore nearly constant in each period of the latter. This observation was instrumental in the proof of Theorem 3.7.4, which implied in particular that for all $1 < p < \infty$ we have

$$\Big\|\sum_{k=1}^N a_k e^{2\pi i 2^k x}\Big\|_{L^p[0,1]} \approx \Big(\sum_{k=1}^N |a_k|^2\Big)^{\frac{1}{2}}. \tag{5.1.2}$$

In other words, we can calculate the L^p norm of $\sum_{k=1}^N a_k e^{2\pi i 2^k x}$ in almost a precise fashion to obtain (modulo multiplicative constants) the same answer as in the L^2 case. Similar calculations are valid for more general blocks of exponentials in the dyadic range $\{2^k+1,\dots,2^{k+1}-1\}$, since the exponentials in each such block behave independently from those in each previous block. In particular, the L^p integrability of a function on $\mathbf{T}^1$ is not affected by the randomization of the sign of its Fourier coefficients in the previous dyadic blocks. This is the intuition behind the Littlewood–Paley theorem.

Motivated by this discussion, we introduce the Littlewood–Paley operators in the continuous setting.

Definition 5.1.1. Let Ψ be an integrable function on $\mathbf{R}^n$ and $j \in \mathbf{Z}$. We define the *Littlewood–Paley operator* Δ_j associated with Ψ by

$$\Delta_j(f) = f * \Psi_{2^{-j}},$$

where $\Psi_{2^{-j}}(x) = 2^{jn}\Psi(2^j x)$ for all x in $\mathbf{R}^n$. Thus we have

$$\widehat{\Psi_{2^{-j}}}(\xi) = \widehat{\Psi}(2^{-j}\xi)$$

for all ξ in $\mathbf{R}^n$. We note that whenever Ψ is a Schwartz function and f is a tempered distribution, the quantity $\Delta_j(f)$ is a well defined function.

These operators depend on the choice of the function Ψ; in most applications we choose Ψ to be a smooth function with compactly supported Fourier transform.

Observe that if $\widehat{\Psi}$ is supported in some annulus $0 < c_1 < |\xi| < c_2 < \infty$, then the Fourier transform of Δ_j is supported in the annulus $c_1 2^j < |\xi| < c_2 2^j$; in other words, it is localized near the frequency $|\xi| \approx 2^j$. Thus the purpose of Δ_j is to isolate the part of frequency of a function concentrated near $|\xi| \approx 2^j$.

The *square function* associated with the Littlewood–Paley operators Δ_j is defined as

$$f \to \Big(\sum_{j\in\mathbf{Z}} |\Delta_j(f)|^2\Big)^{\frac{1}{2}}.$$

It turns out that this quadratic expression captures crucial orthogonality information about the function f. Precisely, we have the following theorem.

Theorem 5.1.2. ***(Littlewood–Paley theorem)*** *Suppose that Ψ is an integrable $\mathscr{C}^1$ function on $\mathbf{R}^n$ with mean value zero that satisfies*

$$|\Psi(x)| + |\nabla\Psi(x)| \le B(1+|x|)^{-n-1}. \tag{5.1.3}$$

Then there exists a constant $C_n < \infty$ such that for all $1 < p < \infty$ and all f in $L^p(\mathbf{R}^n)$ we have

$$\Big\|\Big(\sum_{j\in\mathbf{Z}} |\Delta_j(f)|^2\Big)^{\frac{1}{2}}\Big\|_{L^p(\mathbf{R}^n)} \le C_n B \max\big(p,(p-1)^{-1}\big)\|f\|_{L^p(\mathbf{R}^n)}. \tag{5.1.4}$$

There also exists a $C_n' < \infty$ such that for all f in $L^1(\mathbf{R}^n)$ we have

$$\Big\|\Big(\sum_{j\in\mathbf{Z}} |\Delta_j(f)|^2\Big)^{\frac{1}{2}}\Big\|_{L^{1,\infty}(\mathbf{R}^n)} \le C_n' B\|f\|_{L^1(\mathbf{R}^n)}. \tag{5.1.5}$$

Conversely, let Ψ be a Schwartz function such that either $\widehat{\Psi}(0) = 0$ and

$$\sum_{j\in\mathbf{Z}} |\widehat{\Psi}(2^{-j}\xi)|^2 = 1, \qquad \xi \in \mathbf{R}^n \setminus \{0\}, \tag{5.1.6}$$

or $\widehat{\Psi}$ is compactly supported away from the origin and

$$\sum_{j\in\mathbf{Z}} \widehat{\Psi}(2^{-j}\xi) = 1, \qquad \xi \in \mathbf{R}^n \setminus \{0\}. \tag{5.1.7}$$

Then given a tempered distribution f such that the function $\big(\sum_{j\in\mathbf{Z}} |\Delta_j(f)|^2\big)^{\frac{1}{2}}$ is in $L^p(\mathbf{R}^n)$ for some $1 < p < \infty$, there exists a unique polynomial Q such that the tempered distribution $f - Q$ coincides with an L^p function, and we have

$$\|f-Q\|_{L^p(\mathbf{R}^n)} \le C B \max\big(p,(p-1)^{-1}\big)\Big\|\Big(\sum_{j\in\mathbf{Z}} |\Delta_j(f)|^2\Big)^{\frac{1}{2}}\Big\|_{L^p(\mathbf{R}^n)} \tag{5.1.8}$$

for some constant $C = C_{n,\Psi}$.

Proof. We first prove (5.1.4) when $p = 2$. Using Plancherel's theorem, we see that (5.1.4) is a consequence of the inequality

$$\sum_j |\widehat{\Psi}(2^{-j}\xi)|^2 \le C_n B^2 \tag{5.1.9}$$

for some $C_n < \infty$. Because of (5.1.3), Fourier inversion holds for Ψ. Furthermore, Ψ has mean value zero and we may write

$$\widehat{\Psi}(\xi) = \int_{\mathbf{R}^n} e^{-2\pi i x\cdot\xi}\Psi(x)\,dx = \int_{\mathbf{R}^n}(e^{-2\pi i x\cdot\xi} - 1)\Psi(x)\,dx, \tag{5.1.10}$$

from which we obtain the estimate

$$|\widehat{\Psi}(\xi)| \le \sqrt{4\pi|\xi|}\int_{\mathbf{R}^n}|x|^{\frac{1}{2}}|\Psi(x)|\,dx \le C_n B|\xi|^{\frac{1}{2}}. \tag{5.1.11}$$

For $\xi = (\xi_1,\dots,\xi_n) \ne 0$, let j be such that $|\xi_j| \ge |\xi_k|$ for all $k \in \{1,\dots,n\}$. Integrate by parts with respect to ∂_j in (5.1.10) to obtain

$$\widehat{\Psi}(\xi) = -\int_{\mathbf{R}^n}(-2\pi i\xi_j)^{-1}e^{-2\pi i x\cdot\xi}(\partial_j\Psi)(x)\,dx,$$

from which we deduce the estimate

$$|\widehat{\Psi}(\xi)| \le \sqrt{n}|\xi|^{-1}\int_{\mathbf{R}^n}|\nabla\Psi(x)|\,dx \le C_n B|\xi|^{-1}. \tag{5.1.12}$$

We now break up the sum in (5.1.9) into the parts where $2^{-j}|\xi| \le 1$ and $2^{-j}|\xi| \ge 1$ and use (5.1.11) and (5.1.12), respectively, to obtain (5.1.9). (See also Exercise 5.1.2.) This proves (5.1.4) when $p = 2$.

We now turn our attention to the case $p \ne 2$ in (5.1.4). We view (5.1.4) and (5.1.5) as vector-valued inequalities in the spirit of Section 4.5. Define an operator $\vec{T}$ acting on functions on $\mathbf{R}^n$ as follows:

$$\vec{T}(f)(x) = \{\Delta_j(f)(x)\}_j.$$

The inequalities (5.1.4) and (5.1.5) we wish to prove say simply that $\vec{T}$ is a bounded operator from $L^p(\mathbf{R}^n,\mathbf{C})$ to $L^p(\mathbf{R}^n,\ell^2)$ and from $L^1(\mathbf{R}^n,\mathbf{C})$ to $L^{1,\infty}(\mathbf{R}^n,\ell^2)$. We just proved that this statement is true when $p = 2$, and therefore the first hypothesis of Theorem 4.6.1 is satisfied. We now observe that the operator $\vec{T}$ can be written in the form

$$\vec{T}(f)(x) = \left\{\int_{\mathbf{R}^n}\Psi_{2^{-j}}(x-y)f(y)\,dy\right\}_j = \int_{\mathbf{R}^n}\vec{K}(x-y)(f(y))\,dy,$$

where for each $x \in \mathbf{R}^n$, $\vec{K}(x)$ is a bounded linear operator from $\mathbf{C}$ to ℓ^2 given by

$$\vec{K}(x)(a) = \{\Psi_{2^{-j}}(x)a\}_j. \tag{5.1.13}$$

We clearly have that $\|\vec{K}(x)\|_{\mathbf{C}\to\ell^2} = \big(\sum_j |\Psi_{2^{-j}}(x)|^2\big)^{\frac{1}{2}}$, and to be able to apply Theorem 4.6.1 we need to know that

$$\int_{|x|\geq 2|y|} \|\vec{K}(x-y)-\vec{K}(x)\|_{\mathbf{C}\to\ell^2}\,dx \leq C_n B, \qquad y\neq 0. \tag{5.1.14}$$

Since Ψ is a $\mathscr{C}^1$ function, for $|x|\geq 2|y|$ we have

$$\begin{aligned}
&|\Psi_{2^{-j}}(x-y)-\Psi_{2^{-j}}(x)| \\
&\quad\leq 2^{(n+1)j}|\nabla\Psi(2^j(x-\theta y))|\,|y| && \text{for some } \theta\in[0,1], \\
&\quad\leq B2^{(n+1)j}\big(1+2^j|x-\theta y|\big)^{-(n+1)}|y| \\
&\quad\leq B2^{(n+1)j}\big(1+2^{j-1}|x|\big)^{-(n+1)}|y| && \text{since } |x-\theta y|\geq \tfrac{1}{2}|x|.
\end{aligned} \tag{5.1.15}$$

This estimate implies that

$$|\Psi_{2^{-j}}(x-y)-\Psi_{2^{-j}}(x)| \leq B2^{(n+1)j}|y|. \tag{5.1.16}$$

We also have that

$$\begin{aligned}
&|\Psi_{2^{-j}}(x-y)-\Psi_{2^{-j}}(x)| \\
&\qquad\leq 2^{nj}|\Psi(2^j(x-y))|+2^{jn}|\Psi(2^jx)| \\
&\qquad\leq B2^{nj}\big(1+2^j|x|\big)^{-(n+1)}+B2^{jn}\big(1+2^{j-1}|x|\big)^{-(n+1)} \\
&\qquad\leq 2B2^{nj}\big(1+2^{j-1}|x|\big)^{-(n+1)}.
\end{aligned} \tag{5.1.17}$$

Taking the geometric mean of (5.1.15) and (5.1.17), we obtain

$$|\Psi_{2^{-j}}(x-y)-\Psi_{2^{-j}}(x)| \leq 2B|y|^{\frac{1}{2}}2^{(n+\frac{1}{2})j}\big(1+2^{j-1}|x|\big)^{-(n+1)}. \tag{5.1.18}$$

We now use estimate (5.1.16) when $2^j<\frac{2}{|x|}$ and (5.1.18) when $2^j\geq\frac{2}{|x|}$. We obtain

$$\begin{aligned}
\|\vec{K}(x-y)-\vec{K}(x)\|_{\mathbf{C}\to\ell^2} &= \Big(\sum_{j\in\mathbf{Z}}|\Psi_{2^{-j}}(x-y)-\Psi_{2^{-j}}(x)|^2\Big)^{\frac{1}{2}} \\
&\leq \sum_{j\in\mathbf{Z}}\big|\Psi_{2^{-j}}(x-y)-\Psi_{2^{-j}}(x)\big| \\
&\leq 2B\Big(|y|\sum_{2^j<\frac{2}{|x|}}2^{(n+1)j}+|y|^{\frac{1}{2}}\sum_{2^j\geq\frac{2}{|x|}}2^{(n+\frac{1}{2})j}(2^{(j-1)}|x|)^{-(n+1)}\Big) \\
&\leq C_nB\big(|y||x|^{-n-1}+|y|^{\frac{1}{2}}|x|^{-n-\frac{1}{2}}\big)
\end{aligned}$$

whenever $|x|\geq 2|y|$. Using this bound, we easily deduce (5.1.14) by integrating over the region $|x|\geq 2|y|$. Finally, using Theorem 4.6.1 we conclude the proofs of (5.1.4) and (5.1.5), which establishes one direction of the theorem.

We now turn to the converse direction. Let Δ_j^* be the adjoint operator of Δ_j given by $\widehat{\Delta_j^* f} = \widehat{f}\,\overline{\widehat{\Psi_{2^{-j}}}}$. Let f be in $\mathscr{S}'(\mathbf{R}^n)$. Then the series $\sum_{j\in\mathbf{Z}} \Delta_j^*\Delta_j(f)$ converges in $\mathscr{S}'(\mathbf{R}^n)$. To see this, it suffices to show that the sequence of partial sums $u_N = \sum_{|j|<N} \Delta_j^*\Delta_j(f)$ converges in $\mathscr{S}'$. This means that if we test this sequence against a Schwartz function g, then it is a Cauchy sequence and hence it converges as $N \to \infty$. But an easy argument using duality and the Cauchy–Schwarz and Hölder's inequalities shows that for $M > N$ we have

$$|\langle u_N, g\rangle - \langle u_M, g\rangle| \le \Big\|\Big(\sum_j |\Delta_j(f)|^2\Big)^{\frac{1}{2}}\Big\|_{L^p}\Big\|\Big(\sum_{N\le|j|\le M} |\Delta_j(g)|^2\Big)^{\frac{1}{2}}\Big\|_{L^{p'}},$$

and this can be made small by picking $M > N \ge N_0(g)$. Since the sequence $\langle u_N, g\rangle$ is Cauchy, it converges to some $\Lambda(g)$. Now it remains to show that the map $g \mapsto \Lambda(g)$ is a tempered distribution. Obviously $\Lambda(g)$ is a linear functional. Also,

$$\begin{aligned}|\Lambda(g)| &\le \Big\|\Big(\sum_j |\Delta_j(f)|^2\Big)^{\frac{1}{2}}\Big\|_{L^p}\Big\|\Big(\sum_j |\Delta_j(g)|^2\Big)^{\frac{1}{2}}\Big\|_{L^{p'}}\\ &\le C_{p'}\Big\|\Big(\sum_j |\Delta_j(f)|^2\Big)^{\frac{1}{2}}\Big\|_{L^p}\big\|g\big\|_{L^{p'}},\end{aligned}$$

and since $\|g\|_{L^{p'}}$ is controlled by a finite number of Schwartz seminorms of g, it follows that Λ is in $\mathscr{S}'$. The distribution Λ is the limit of the series $\sum_j \Delta_j^*\Delta_j$.

Under hypothesis (5.1.6), the Fourier transform of the tempered distribution $f - \sum_{j\in\mathbf{Z}} \Delta_j^*\Delta_j(f)$ is supported at the origin. This implies that there exists a polynomial Q such that $f - Q = \sum_{j\in\mathbf{Z}} \Delta_j^*\Delta_j(f)$. Now let g be a Schwartz function. We have

$$\begin{aligned}|\langle f-Q, \overline{g}\rangle| &= \Big|\Big\langle \sum_{j\in\mathbf{Z}} \Delta_j^*\Delta_j(f), \overline{g}\Big\rangle\Big|\\ &= \Big|\sum_{j\in\mathbf{Z}} \langle \Delta_j^*\Delta_j(f), \overline{g}\rangle\Big|\\ &= \Big|\sum_{j\in\mathbf{Z}} \langle \Delta_j(f), \overline{\Delta_j(g)}\rangle\Big|\\ &= \Big|\int_{\mathbf{R}^n} \sum_{j\in\mathbf{Z}} \Delta_j(f)\,\overline{\Delta_j(g)}\,dx\Big|\\ &\le \int_{\mathbf{R}^n} \Big(\sum_{j\in\mathbf{Z}} |\Delta_j(f)|^2\Big)^{\frac{1}{2}}\Big(\sum_{j\in\mathbf{Z}} |\Delta_j(g)|^2\Big)^{\frac{1}{2}}dx\\ &\le \Big\|\Big(\sum_{j\in\mathbf{Z}} |\Delta_j(f)|^2\Big)^{\frac{1}{2}}\Big\|_{L^p}\Big\|\Big(\sum_{j\in\mathbf{Z}} |\Delta_j(g)|^2\Big)^{\frac{1}{2}}\Big\|_{L^{p'}}\\ &\le \Big\|\Big(\sum_{j\in\mathbf{Z}} |\Delta_j(f)|^2\Big)^{\frac{1}{2}}\Big\|_{L^p} C_n B \max\big(p', (p'-1)^{-1}\big)\big\|g\big\|_{L^{p'}},\end{aligned} \tag{5.1.19}$$

having used the definition of the adjoint (Section 2.5.2), the Cauchy–Schwarz inequality, Hölder's inequality, and (5.1.4). Taking the supremum over all g in $L^{p'}$ with norm at most one, we obtain that the tempered distribution $f-Q$ is a bounded linear functional on $L^{p'}$. By the Riesz representation theorem, $f-Q$ coincides with an L^p function whose norm satisfies the estimate

$$\big\|f-Q\big\|_{L^p} \le C_n B \max\big(p,(p-1)^{-1}\big)\Big\|\Big(\sum_{j\in\mathbf{Z}}|\Delta_j(f)|^2\Big)^{\frac{1}{2}}\Big\|_{L^p}.$$

We now show uniqueness. If Q_1 is another polynomial, with $f-Q_1\in L^p$, then $Q-Q_1$ must be an L^p function; but the only polynomial that lies in L^p is the zero polynomial. This completes the proof of the converse of the theorem under hypothesis (5.1.6).

To obtain the same conclusion under the hypothesis (5.1.7) we argue in a similar way but we leave the details as an exercise. (One may adapt the argument in the proof of Corollary 5.1.7 to this setting.) □

Remark 5.1.3. We make some observations. If $\widehat{\Psi}$ is real-valued, then the operators Δ_j are self-adjoint. Indeed,

$$\int_{\mathbf{R}^n}\Delta_j(f)\,\overline{g}\,dx=\int_{\mathbf{R}^n}\widehat{f}\,\widehat{\Psi_{2^{-j}}}\,\overline{\widehat{g}}\,d\xi=\int_{\mathbf{R}^n}\widehat{f}\,\overline{\widehat{\Psi_{2^{-j}}}\,\widehat{g}}\,d\xi=\int_{\mathbf{R}^n}f\,\overline{\Delta_j(g)}\,dx.$$

Moreover, if Ψ is a radial function, we see that the operators Δ_j are self-transpose, that is, they satisfy

$$\int_{\mathbf{R}^n}\Delta_j(f)\,g\,dx=\int_{\mathbf{R}^n}f\,\Delta_j(g)\,dx.$$

Assume now that Ψ is both radial and has a real-valued Fourier transform. Suppose also that Ψ satisfies (5.1.3) and that it has mean value zero. Then the inequality

$$\Big\|\sum_{j\in\mathbf{Z}}\Delta_j(f_j)\Big\|_{L^p} \le C_n B \max\big(p,(p-1)^{-1}\big)\Big\|\Big(\sum_{j\in\mathbf{Z}}|f_j|^2\Big)^{\frac{1}{2}}\Big\|_{L^p} \tag{5.1.20}$$

is true for sequences of functions $\{f_j\}_j$. To see this we use duality. Let $\vec{T}(f)=\{\Delta_j(f)\}_j$. Then $\vec{T}^*(\{g_j\}_j)=\sum_j\Delta_j(g_j)$. Inequality (5.1.4) says that the operator $\vec{T}$ maps $L^p(\mathbf{R}^n,\mathbf{C})$ to $L^p(\mathbf{R}^n,\ell^2)$, and its dual statement is that $\vec{T}^*$ maps $L^{p'}(\mathbf{R}^n,\ell^2)$ to $L^{p'}(\mathbf{R}^n,\mathbf{C})$. This is exactly the statement in (5.1.20) if p is replaced by p'. Since p is any number in $(1,\infty)$, (5.1.20) is proved.

5.1.2 Vector-Valued Analogues

We now obtain a vector-valued extension of Theorem 5.1.2. We have the following.

Proposition 5.1.4. *Let Ψ be an integrable $\mathscr{C}^1$ function on $\mathbf{R}^n$ with mean value zero that satisfies (5.1.3) and let Δ_j be the Littlewood–Paley operator associated with Ψ. Then there exists a constant $C_n < \infty$ such that for all $1 < p, r < \infty$ and all sequences of L^p functions f_j we have*

$$\Big\|\Big(\sum_{j\in\mathbf{Z}}\Big(\sum_{k\in\mathbf{Z}}|\Delta_k(f_j)|^2\Big)^{\frac{r}{2}}\Big)^{\frac{1}{r}}\Big\|_{L^p(\mathbf{R}^n)} \le C_n B\widetilde{C}_{p,r}\Big\|\Big(\sum_{j\in\mathbf{Z}}|f_j|^r\Big)^{\frac{1}{r}}\Big\|_{L^p(\mathbf{R}^n)},$$

where $\widetilde{C}_{p,r} = \max(p,(p-1)^{-1})\max(r,(r-1)^{-1})$. *Moreover, for some $C_n' > 0$ and all sequences of L^1 functions f_j we have*

$$\Big\|\Big(\sum_{j\in\mathbf{Z}}\Big(\sum_{k\in\mathbf{Z}}|\Delta_k(f_j)|^2\Big)^{\frac{r}{2}}\Big)^{\frac{1}{r}}\Big\|_{L^{1,\infty}(\mathbf{R}^n)} \le C_n' B\max(r,(r-1)^{-1})\Big\|\Big(\sum_{j\in\mathbf{Z}}|f_j|^r\Big)^{\frac{1}{r}}\Big\|_{L^1(\mathbf{R}^n)}.$$

In particular,

$$\Big\|\Big(\sum_{j\in\mathbf{Z}}|\Delta_j(f_j)|^r\Big)^{\frac{1}{r}}\Big\|_{L^p(\mathbf{R}^n)} \le C_n B\widetilde{C}_{p,r}\Big\|\Big(\sum_{j\in\mathbf{Z}}|f_j|^r\Big)^{\frac{1}{r}}\Big\|_{L^p(\mathbf{R}^n)}. \tag{5.1.21}$$

Proof. As in the proof of Theorem 5.1.2, we introduce Banach spaces $\mathscr{B}_1 = \mathbf{C}$ and $\mathscr{B}_2 = \ell^2$ and for $f \in L^p(\mathbf{R}^n)$ define an operator

$$\vec{T}(f) = \{\Delta_k(f)\}_{k\in\mathbf{Z}}.$$

In the proof of Theorem 5.1.2 we showed that $\vec{T}$ has a kernel $\vec{K}$ that satisfies condition (5.1.14). Furthermore, $\vec{T}$ obviously maps $L^r(\mathbf{R}^n,\mathbf{C})$ to $L^r(\mathbf{R}^n,\ell^r)$. Applying Proposition 4.6.4, we obtain the first two statements of the proposition. Restricting to $k = j$ yields (5.1.21). □

5.1.3 L^p Estimates for Square Functions Associated with Dyadic Sums

Let us pick a Schwartz function Ψ whose Fourier transform is compactly supported in the annulus $2^{-1} \le |\xi| \le 2^2$ such that (5.1.6) is satisfied. (Clearly (5.1.6) has no chance of being satisfied if $\widehat{\Psi}$ is supported only in the annulus $1 \le |\xi| \le 2$.) The Littlewood–Paley operation $f \mapsto \Delta_j(f)$ represents the smoothly truncated frequency localization of a function f near the dyadic annulus $|\xi| \approx 2^j$. Theorem 5.1.2 says that the square function formed by these localizations has L^p norm comparable to that of the original function. In other words, this square function characterizes the L^p norm of a function. This is the main feature of Littlewood–Paley theory.

One may ask whether Theorem 5.1.2 still holds if the Littlewood–Paley operators Δ_j are replaced by their nonsmooth versions

$$f \mapsto \big(\chi_{2^j \le |\xi| < 2^{j+1}} \widehat{f}(\xi)\big)^\vee(x). \tag{5.1.22}$$

This question has a surprising answer that already signals that there may be some fundamental differences between one-dimensional and higher-dimensional Fourier analysis. The square function formed by the operators in (5.1.22) can be used to characterize $L^p(\mathbf{R})$ in the same way Δ_j did, but not $L^p(\mathbf{R}^n)$ when $n > 1$ and $p \neq 2$. The problem lies in the fact that the characteristic function of the unit disk is not an L^p multiplier on $\mathbf{R}^n$ when $n \ge 2$ unless $p = 2$; this fact is discussed in detail in Section 10.1. The one-dimensional result we alluded to earlier is the following.

For $j \in \mathbf{Z}$ we introduce the one-dimensional operator

$$\Delta_j^\flat(f)(x) = (\widehat{f}\chi_{I_j})^\vee(x), \tag{5.1.23}$$

where

$$I_j = [2^j, 2^{j+1}) \cup (-2^{j+1}, -2^j],$$

and $\Delta_j^\flat$ is a version of the operator Δ_j in which the characteristic function of the set $2^j \le |\xi| < 2^{j+1}$ replaces the function $\widehat{\Psi}(2^{-j}\xi)$.

Theorem 5.1.5. *There exists a constant C_1 such that for all $1 < p < \infty$ and all f in $L^p(\mathbf{R})$ we have*

$$\frac{\|f\|_{L^p}}{C_1(p+\frac{1}{p-1})^2} \le \Big\|\Big(\sum_{j\in\mathbf{Z}} |\Delta_j^\flat(f)|^2\Big)^{\frac12}\Big\|_{L^p} \le C_1(p+\tfrac{1}{p-1})^2\|f\|_{L^p}. \tag{5.1.24}$$

Proof. Pick a Schwartz function ψ on the line whose Fourier transform is supported in the set $2^{-1} \le |\xi| \le 2^2$ and is equal to 1 on the set $1 \le |\xi| \le 2$. Let Δ_j be the Littlewood–Paley operator associated with ψ. Observe that $\Delta_j\Delta_j^\flat = \Delta_j^\flat\Delta_j = \Delta_j^\flat$, since $\widehat{\psi}$ is equal to one on the support of $\Delta_j^\flat(f)\widehat{\ }$. We now use Exercise 4.6.1(a) to obtain

$$\begin{aligned}
\Big\|\Big(\sum_{j\in\mathbf{Z}} |\Delta_j^\flat(f)|^2\Big)^{\frac12}\Big\|_{L^p} &= \Big\|\Big(\sum_{j\in\mathbf{Z}} |\Delta_j^\flat\Delta_j(f)|^2\Big)^{\frac12}\Big\|_{L^p} \\
&\le C\max(p,(p-1)^{-1})\Big\|\Big(\sum_{j\in\mathbf{Z}} |\Delta_j(f)|^2\Big)^{\frac12}\Big\|_{L^p} \\
&\le CB\max(p,(p-1)^{-1})^2\|f\|_{L^p},
\end{aligned}$$

where the last inequality follows from Theorem 5.1.2. The reverse inequality for $1 < p < \infty$ follows just like the reverse inequality (5.1.8) of Theorem 5.1.2 by simply replacing the Δ_j's by the $\Delta_j^\flat$'s and setting the polynomial Q equal to zero. (There is no need to use the Riesz representation theorem here, just the fact that the L^p norm of f can be realized as the supremum of expressions $|\langle f, g\rangle|$ where g has $L^{p'}$ norm at most 1.) □

There is a higher-dimensional version of Theorem 5.1.5 with dyadic rectangles replacing the dyadic intervals. As has already been pointed out, the higher-dimensional version with dyadic annuli replacing the dyadic intervals is false.

Let us introduce some notation. For $j \in \mathbf{Z}$, we denote by I_j the dyadic set $[2^j, 2^{j+1}) \bigcup (-2^{j+1}, -2^j]$ as in the statement of Theorem 5.1.5. For $j_1, \dots, j_n \in \mathbf{Z}$ define a dyadic rectangle

$$R_{j_1,\dots,j_n} = I_{j_1} \times \cdots \times I_{j_n}$$

in $\mathbf{R}^n$. Actually $R_{j_1,\dots,j_n}$ is not a rectangle but a union of 2^n rectangles; with some abuse of language we still call it a rectangle. For notational convenience we write

$$R_j = R_{j_1,\dots,j_n}, \qquad \text{where } j = (j_1, \dots, j_n) \in \mathbf{Z}^n.$$

Observe that for different $j, j' \in \mathbf{Z}^n$ the rectangles R_j and $R_{j'}$ have disjoint interiors and that the union of all the R_j's is equal to $\mathbf{R}^n \setminus \{0\}$. In other words, the family of R_j's, where $j \in \mathbf{Z}^n$, forms a *tiling* of $\mathbf{R}^n$, which we call the *dyadic decomposition* of $\mathbf{R}^n$. We now introduce operators

$$\Delta_j^\flat(f)(x) = (\widehat{f}\chi_{R_j})^\vee(x), \tag{5.1.25}$$

and we have the following n-dimensional extension of Theorem 5.1.5.

Theorem 5.1.6. *For a Schwartz function ψ on the line with integral zero we define the operator*

$$\Delta_j(f)(x) = \big(\widehat{\psi}(2^{-j_1}\xi_1)\cdots\widehat{\psi}(2^{-j_n}\xi_n)\widehat{f}(\xi)\big)^\vee(x), \tag{5.1.26}$$

where $j = (j_1, \dots, j_n) \in \mathbf{Z}^n$. Then there is a dimensional constant C_n such that

$$\Big\|\Big(\sum_{j\in\mathbf{Z}^n}|\Delta_j(f)|^2\Big)^{\frac{1}{2}}\Big\|_{L^p} \le C_n(p+\tfrac{1}{p-1})^n\|f\|_{L^p}. \tag{5.1.27}$$

Let $\Delta_j^\flat$ be the operators defined in (5.1.25). Then there exists a positive constant C_n such that for all $1 < p < \infty$ and all $f \in L^p(\mathbf{R}^n)$ we have

$$\frac{\|f\|_{L^p}}{C_n(p+\frac{1}{p-1})^{2n}} \le \Big\|\Big(\sum_{j\in\mathbf{Z}^n}|\Delta_j^\flat(f)|^2\Big)^{\frac{1}{2}}\Big\|_{L^p} \le C_n(p+\tfrac{1}{p-1})^{2n}\|f\|_{L^p}. \tag{5.1.28}$$

Proof. We first prove (5.1.27). Note that if $j = (j_1, \dots, j_n) \in \mathbf{Z}^n$, then the operator Δ_j is equal to

$$\Delta_j(f) = \Delta_{j_1}^{(j_1)} \cdots \Delta_{j_n}^{(j_n)}(f),$$

where the $\Delta_{j_r}^{(j_r)}$ are one-dimensional operators given on the Fourier transform by multiplication by $\widehat{\psi}(2^{-j_r}\xi_r)$, with the remaining variables fixed. Inequality in (5.1.27) is a consequence of the one-dimensional case. For instance, we discuss the case $n = 2$. Using Proposition 5.1.4, we obtain

$$
\begin{aligned}
&\Big\| \Big(\sum_{j\in\mathbf{Z}^2} |\Delta_j(f)|^2 \Big)^{\frac{1}{2}} \Big\|_{L^p(\mathbf{R}^2)}^p \\
&\qquad = \int_{\mathbf{R}} \bigg[\int_{\mathbf{R}} \Big(\sum_{j_1\in\mathbf{Z}} \sum_{j_2\in\mathbf{Z}} |\Delta_{j_1}^{(1)} \Delta_{j_2}^{(2)}(f)(x_1,x_2)|^2 \Big)^{\frac{p}{2}} dx_1 \bigg] dx_2 \\
&\qquad \le C^p \max(p,(p-1)^{-1})^p \int_{\mathbf{R}} \bigg[\int_{\mathbf{R}} \Big(\sum_{j_2\in\mathbf{Z}} |\Delta_{j_2}^{(2)}(f)(x_1,x_2)|^2 \Big)^{\frac{p}{2}} dx_1 \bigg] dx_2 \\
&\qquad = C^p \max(p,(p-1)^{-1})^p \int_{\mathbf{R}} \bigg[\int_{\mathbf{R}} \Big(\sum_{j_2\in\mathbf{Z}} |\Delta_{j_2}^{(2)}(f)(x_1,x_2)|^2 \Big)^{\frac{p}{2}} dx_2 \bigg] dx_1 \\
&\qquad \le C^{2p} \max(p,(p-1)^{-1})^{2p} \int_{\mathbf{R}} \bigg[\int_{\mathbf{R}} |f(x_1,x_2)|^p \, dx_2 \bigg] dx_1 \\
&\qquad = C^{2p} \max(p,(p-1)^{-1})^{2p} \|f\|_{L^p(\mathbf{R}^2)}^p ,
\end{aligned}
$$

where we also used Theorem 5.1.2 in the calculation. Higher-dimensional versions of this estimate may easily be obtained by induction.

We now turn to the upper inequality in (5.1.28). We pick a Schwartz function ψ whose Fourier transform is supported in the union $[-4,-1/2]\bigcup[1/2,4]$ and is equal to 1 on $[-2,-1]\bigcup[2,-4]$. Then we clearly have

$$
\Delta_j^\flat = \Delta_j^\flat \Delta_j ,
$$

since $\widehat{\psi}(2^{-j_1}\xi_1)\cdots\widehat{\psi}(2^{-j_n}\xi_n)$ is equal to 1 on the rectangle R_j. We now use Exercise 4.6.1(b) and estimate (5.1.27) to obtain

$$
\begin{aligned}
\Big\| \Big(\sum_{j\in\mathbf{Z}^n} |\Delta_j^\flat(f)|^2 \Big)^{\frac{1}{2}} \Big\|_{L^p} &= \Big\| \Big(\sum_{j\in\mathbf{Z}^n} |\Delta_j^\flat \Delta_j(f)|^2 \Big)^{\frac{1}{2}} \Big\|_{L^p} \\
&\le C \max(p,(p-1)^{-1})^n \Big\| \Big(\sum_{j\in\mathbf{Z}^n} |\Delta_j(f)|^2 \Big)^{\frac{1}{2}} \Big\|_{L^p} \\
&\le CB \max(p,(p-1)^{-1})^{2n} \|f\|_{L^p} .
\end{aligned}
$$

The lower inequality in (5.1.28) for $1 < p < \infty$ is proved like inequality (5.1.8) in Theorem 5.1.2. The fundamental ingredient in the proof is that $f = \sum_{j\in\mathbf{Z}^n} \Delta_j^\flat \Delta_j^\flat(f)$ for all Schwartz functions f, where the sum is interpreted as the L^2-limit of the sequence of partial sums. Thus the series converges in $\mathscr{S}'$, and pairing with a Schwartz function $\overline{g}$, we obtain the lower inequality in (5.1.28) for Schwartz functions, by applying the steps in (5.1.19) (with $Q = 0$). To prove the lower inequality in (5.1.28) for a general function $f \in L^p(\mathbf{R}^n)$ we approximate an L^p function by a sequence of Schwartz functions in the L^p norm. Then both sides of the lower inequality in (5.1.28) for the approximating sequence converge to the corresponding sides of the lower inequality in (5.1.28) for f; the convergence of the sequence of L^p norms of the square functions requires the upper inequality in (5.1.28) that was previously established. This concludes the proof of the theorem. □

Next we observe that if the Schwartz function ψ is suitably chosen, then the reverse inequality in estimate (5.1.27) also holds. More precisely, suppose $\widehat{\psi}(\xi)$ is an even smooth real-valued function supported in the set $\frac{9}{10} \le |\xi| \le \frac{21}{10}$ in $\mathbf{R}$ that satisfies

$$\sum_{j\in\mathbf{Z}} \widehat{\psi}(2^{-j}\xi) = 1, \qquad \xi \in \mathbf{R}\setminus\{0\}; \tag{5.1.29}$$

then we have the following.

Corollary 5.1.7. *Suppose that ψ satisfies (5.1.29) and let Δ_j be as in (5.1.26). Let f be an L^p function on $\mathbf{R}^n$ such that the function $\big(\sum_{j\in\mathbf{Z}^n} |\Delta_j(f)|^2\big)^{\frac{1}{2}}$ is in $L^p(\mathbf{R}^n)$. Then there is a constant C_n that depends only on the dimension and ψ such that the lower estimate*

$$\frac{\|f\|_{L^p}}{C_n(p+\frac{1}{p-1})^n} \le \Big\|\Big(\sum_{j\in\mathbf{Z}^n} |\Delta_j(f)|^2\Big)^{\frac{1}{2}}\Big\|_{L^p} \tag{5.1.30}$$

holds.

Proof. If we had $\sum_{j\in\mathbf{Z}} |\widehat{\psi}(2^{-j}\xi)|^2 = 1$ instead of (5.1.29), then we could apply the method used in the lower estimate of Theorem 5.1.2 to obtain the required conclusion. In this case we provide another argument that is very similar in spirit.

We first prove (5.1.30) for Schwartz functions f. Then the series $\sum_{j\in\mathbf{Z}^n} \Delta_j(f)$ converges in L^2 (and hence in $\mathscr{S}'$) to f. Now let g be another Schwartz function. We express the inner product $\langle f, \overline{g}\rangle$ as the action of the distribution $\sum_{j\in\mathbf{Z}^n} \Delta_j(f)$ on the test function $\overline{g}$:

$$\begin{aligned}
|\langle f,\overline{g}\rangle| &= \Big|\Big\langle \sum_{j\in\mathbf{Z}^n} \Delta_j(f),\overline{g}\Big\rangle\Big| \\
&= \Big|\sum_{j\in\mathbf{Z}^n} \langle \Delta_j(f),\overline{g}\rangle\Big| \\
&= \Big|\sum_{j\in\mathbf{Z}^n} \sum_{k_r\in\{j_r-1,j_r,j_r+1\}} \langle \Delta_j(f),\overline{\Delta_k(g)}\rangle\Big| \\
&\le \int_{\mathbf{R}^n} \sum_{j\in\mathbf{Z}^n} \sum_{k_r\in\{j_r-1,j_r,j_r+1\}} |\Delta_j(f)|\,|\Delta_k(g)|\,dx \\
&\le 3^n \int_{\mathbf{R}^n} \Big(\sum_{j\in\mathbf{Z}^n} |\Delta_j(f)|^2\Big)^{\frac{1}{2}} \Big(\sum_{k\in\mathbf{Z}^n} |\Delta_k(g)|^2\Big)^{\frac{1}{2}} dx \\
&\le 3^n \Big\|\Big(\sum_{j\in\mathbf{Z}^n} |\Delta_j(f)|^2\Big)^{\frac{1}{2}}\Big\|_{L^p} \Big\|\Big(\sum_{k\in\mathbf{Z}^n} |\Delta_k(g)|^2\Big)^{\frac{1}{2}}\Big\|_{L^{p'}} \\
&\le C_n^{-1} \max\big(p',(p'-1)^{-1}\big)^n \|g\|_{L^{p'}} \Big\|\Big(\sum_{j\in\mathbf{Z}^n} |\Delta_j(f)|^2\Big)^{\frac{1}{2}}\Big\|_{L^p},
\end{aligned}$$

where we used the fact that $\Delta_j(f)$ and $\Delta_k(g)$ are orthogonal operators unless every coordinate of k is within 1 unit of the corresponding coordinate of j; this is an easy consequence of the support properties of $\widehat{\psi}$. We now take the supremum over all g in $L^{p'}$ with norm at most 1, to obtain (5.1.30) for Schwartz functions f.

To extend this estimate to general L^p functions f, we use the density argument described in the last paragraph in the proof of Theorem 5.1.6. □

5.1.4 Lack of Orthogonality on L^p

We discuss two examples indicating why (5.1.1) cannot hold if the exponent 2 is replaced by some other exponent $q \neq 2$. More precisely, we show that if the functions f_j have Fourier transforms supported in disjoint sets, then the inequality

$$\Big\|\sum_j f_j\Big\|_{L^p}^p \le C_p \sum_j \|f_j\|_{L^p}^p \tag{5.1.31}$$

cannot hold if $p > 2$, and similarly, the inequality

$$\sum_j \|f_j\|_{L^p}^p \le C_p \Big\|\sum_j f_j\Big\|_{L^p}^p \tag{5.1.32}$$

cannot hold if $p < 2$. In both (5.1.31) and (5.1.32) the constants C_p are supposed to be independent of the functions f_j.

Example 5.1.8. Pick a Schwartz function ζ whose Fourier transform is positive and supported in the interval $|\xi| \le 1/4$. Let N be a large integer and let $f_j(x) = e^{2\pi i jx}\zeta(x)$. Then $\widehat{f_j}(\xi) = \widehat{\zeta}(\xi - j)$ and the $\widehat{f_j}$'s have disjoint Fourier transforms. We obviously have

$$\sum_{j=0}^{N} \|f_j\|_{L^p}^p = (N+1)\|\zeta\|_{L^p}^p.$$

On the other hand, we have the estimate

$$\begin{aligned}
\Big\|\sum_{j=0}^{N} f_j\Big\|_{L^p}^p &= \int_{\mathbf{R}} \Big|\frac{e^{2\pi i(N+1)x}-1}{e^{2\pi ix}-1}\Big|^p |\zeta(x)|^p\,dx \\
&\ge c\int_{|x|<\frac{1}{10}(N+1)^{-1}} \frac{(N+1)^p|x|^p}{|x|^p}|\zeta(x)|^p\,dx \\
&= C_\zeta (N+1)^{p-1},
\end{aligned}$$

since ζ does not vanish in a neighborhood of zero. We conclude that (5.1.31) cannot hold for this choice of f_j's for $p > 2$.

Example 5.1.9. We now indicate why (5.1.32) cannot hold for $p < 2$. We pick a smooth function Ψ on the line whose Fourier transform $\widehat{\Psi}$ is supported in $\left[\frac{7}{8}, \frac{17}{8}\right]$, is nonnegative, is equal to 1 on $\left[\frac{9}{8}, \frac{15}{8}\right]$, and satisfies

$$\sum_{j\in\mathbf{Z}} \widehat{\Psi}(2^{-j}\xi)^2 = 1, \qquad \xi > 0.$$

Extend $\widehat{\Psi}$ to be an even function on the whole line and let Δ_j be the Littlewood–Paley operator associated with Ψ. Also pick a nonzero Schwartz function φ on the real line whose Fourier transform is nonnegative and supported in the set $\left[\frac{11}{8}, \frac{13}{8}\right]$. Fix N a large positive integer and let

$$f_j(x) = e^{2\pi i \frac{12}{8} 2^j x}\varphi(x), \tag{5.1.33}$$

for $j = 1, 2, \dots, N$. Then the function $\widehat{f_j}(\xi) = \widehat{\varphi}(\xi - \frac{12}{8}2^j)$ is supported in the set $\left[\frac{11}{8} + \frac{12}{8}2^j, \frac{13}{8} + \frac{12}{8}2^j\right]$, which is contained in $\left[\frac{9}{8}2^j, \frac{15}{8}2^j\right]$ for $j \geq 3$. In other words, $\widehat{\Psi}(2^{-j}\xi)$ is equal to 1 on the support of $\widehat{f_j}$. This implies that

$$\Delta_j(f_j) = f_j \qquad \text{for} \qquad j \geq 3.$$

This observation combined with (5.1.20) gives for $N \geq 3$,

$$\Big\|\sum_{j=3}^N f_j\Big\|_{L^p} = \Big\|\sum_{j=3}^N \Delta_j(f_j)\Big\|_{L^p} \leq C_p\Big\|\Big(\sum_{j=3}^N |f_j|^2\Big)^{\frac{1}{2}}\Big\|_{L^p} = C_p\big\|\varphi\big\|_{L^p}(N-2)^{\frac{1}{2}},$$

where $1 < p < \infty$. On the other hand, (5.1.33) trivially yields that

$$\Big(\sum_{j=3}^N \|f_j\|_{L^p}^p\Big)^{\frac{1}{p}} = \big\|\varphi\big\|_{L^p}(N-2)^{\frac{1}{p}}.$$

Letting $N \to \infty$ we see that (5.1.32) cannot hold for $p < 2$ even when the f_j's have Fourier transforms supported in disjoint sets.

Example 5.1.10. A similar idea illustrates the necessity of the ℓ^2 norm in (5.1.4). To see this, let Ψ and Δ_j be as in Example 5.1.9. Let us fix $1 < p < \infty$ and $q < 2$. We show that the inequality

$$\Big\|\Big(\sum_{j\in\mathbf{Z}} |\Delta_j(f)|^q\Big)^{\frac{1}{q}}\Big\|_{L^p} \leq C_{p,q}\big\|f\big\|_{L^p} \tag{5.1.34}$$

cannot hold. Take $f = \sum_{j=3}^N f_j$, where the f_j are as in (5.1.33) and $N \geq 3$. Then the left-hand side of (5.1.34) is bounded from below by $\big\|\varphi\big\|_{L^p}(N-2)^{1/q}$, while the right-hand side is bounded above by $\big\|\varphi\big\|_{L^p}(N-2)^{1/2}$. Letting $N \to \infty$, we deduce that (5.1.34) is impossible when $q < 2$.

Example 5.1.11. For $1 < p < \infty$ and $2 < q < \infty$, the inequality

$$\big\|g\big\|_{L^p} \le C_{p,q} \Big\| \Big(\sum_{j\in\mathbf{Z}} |\Delta_j(g)|^q \Big)^{\frac{1}{q}} \Big\|_{L^p} \tag{5.1.35}$$

cannot hold even under assumption (5.1.6) on Ψ. Let Δ_j be as in Example 5.1.9. Let us suppose that (5.1.35) did hold for some $q > 2$ for these Δ_j's. Then the self-adjointness of the Δ_j's and duality would give

$$\begin{aligned}
&\Big\| \Big(\sum_{k\in\mathbf{Z}} |\Delta_k(g)|^{q'} \Big)^{\frac{1}{q'}} \Big\|_{L^{p'}} \\
&= \sup_{\big\| \|\{h_k\}_k\|_{\ell^q} \big\|_{L^p} \le 1} \Big| \int_{\mathbf{R}^n} \sum_{k\in\mathbf{Z}} \Delta_k(g)\, \overline{h_k}\, dx \Big| \\
&\le \big\|g\big\|_{L^{p'}} \sup_{\big\| \|\{h_k\}_k\|_{\ell^q} \big\|_{L^p} \le 1} \Big\| \sum_{k\in\mathbf{Z}} \overline{\Delta_k(h_k)} \Big\|_{L^p} \\
&\le C \big\|g\big\|_{L^{p'}} \sup_{\big\| \|\{h_k\}_k\|_{\ell^q} \big\|_{L^p} \le 1} \Big\| \Big(\sum_{j\in\mathbf{Z}} \Big| \Delta_j \Big(\sum_{k\in\mathbf{Z}} \Delta_k(h_k) \Big) \Big|^q \Big)^{\frac{1}{q}} \Big\|_{L^p} \qquad \text{by (5.1.35)} \\
&\le C' \big\|g\big\|_{L^{p'}} \sup_{\big\| \|\{h_k\}_k\|_{\ell^q} \big\|_{L^p} \le 1} \Big\{ \sum_{l=-1}^{1} \Big\| \Big(\sum_{j\in\mathbf{Z}} |\Delta_j \Delta_{j+l}(h_j)|^q \Big)^{\frac{1}{q}} \Big\|_{L^p} \Big\} \\
&\le C'' \big\|g\big\|_{L^{p'}} \sup_{\big\| \|\{h_k\}_k\|_{\ell^q} \big\|_{L^p} \le 1} \Big\| \Big(\sum_{j\in\mathbf{Z}} |h_j|^q \Big)^{\frac{1}{q}} \Big\|_{L^p} = C'' \big\|g\big\|_{L^{p'}},
\end{aligned}$$

where the next-to-last inequality follows from (5.1.21) applied twice, while the one before that follows from support considerations. But since $q' < 2$, this exactly proves (5.1.34), previously shown to be false, a contradiction.

We conclude that if both assertions (5.1.4) and (5.1.8) of Theorem 5.1.2 were to hold, then the ℓ^2 norm inside the L^p norm could not be replaced by an ℓ^q norm for some $q \neq 2$. Exercise 5.1.6 indicates the crucial use of the fact that ℓ^2 is a Hilbert space in the converse inequality (5.1.8) of Theorem 5.1.2.

Exercises

5.1.1. Construct a Schwartz function Ψ that satisfies (5.1.6) and whose Fourier transform is supported in the annulus $\frac{8}{9} \le |x| \le \frac{9}{4}$.
$\big[$*Hint:* Set $\widehat{\Psi}(\xi) = \eta(\xi)\big(\sum_{k\in\mathbf{Z}} \eta(2^{-k}\xi)\big)^{-1}$, where η is a suitable smooth bump.$\big]$

5.1.2. Suppose that a function Ψ satisfies $|\widehat{\Psi}(\xi)| \le B \min(|\xi|^{\delta}, |\xi|^{-\varepsilon})$ for some $\varepsilon, \delta > 0$. Show that for some dimensional constant $C_n < \infty$ we have

$$\sum_{j\in\mathbf{Z}} |\widehat{\Psi}(2^{-j}\xi)| \le C_n B.$$

5.1.3. Let Ψ be an integrable function on $\mathbf{R}^n$ with mean value zero that satisfies

$$|\Psi(x)| \le B|x|^{-n-\varepsilon}, \qquad \int_{\mathbf{R}^n} |\Psi(x-y)-\Psi(x)|\,dx \le B|y|^{\varepsilon},$$

for some $B, \varepsilon > 0$ and all $y \neq 0$.
(a) Prove that $\big|\widehat{\Psi}(\xi)\big| \le B\min\big(|\xi|^{\frac{\varepsilon}{2}}, |\xi|^{-\varepsilon}\big)$ and conclude that (5.1.4) holds for $p = 2$.

(b) Prove that if $\vec{K}$ is defined by (5.1.13), then (5.1.14) holds and therefore deduce the validity of (5.1.4) and (5.1.5).
[*Hint:* Part (a): Write

$$\widehat{\Psi}(\xi) = \int_{\mathbf{R}^n} e^{-2\pi i x\cdot\xi}\Psi(x)\,dx = -\int_{\mathbf{R}^n} e^{-2\pi i x\cdot\xi}\Psi(x-y)\,dx,$$

where $y = \frac{1}{2}\frac{\xi}{|\xi|^2}$ when $|\xi| \ge 1$. For $|\xi| \le 1$ use the mean value property of Ψ. Part (b): Split the sum

$$\sum_{j\in\mathbf{Z}} \int_{|x|\ge 2|y|} \big|\Psi_{2^{-j}}(x-y) - \Psi_{2^{-j}}(x)\big|\,dx$$

into the parts $\sum_{2^j \le |y|^{-1}}$ and $\sum_{2^j > |y|^{-1}}$.]

5.1.4. Under the hypotheses of Theorem 5.1.2, prove the following continuous versions of its conclusions: Show that there exist constants C_n, C_n' such that for all $1 < p < \infty$ and for all $f \in L^p(\mathbf{R}^n)$ we have

$$\Big\| \Big(\int_0^\infty |f * \Psi_t|^2 \frac{dt}{t} \Big)^{\frac{1}{2}} \Big\|_{L^p(\mathbf{R}^n)} \le C_n B \max(p, (p-1)^{-1}) \|f\|_{L^p(\mathbf{R}^n)}$$

and also for all $f \in L^1(\mathbf{R}^n)$ we have

$$\Big\| \Big(\int_0^\infty |f * \Psi_t|^2 \frac{dt}{t} \Big)^{\frac{1}{2}} \Big\|_{L^{1,\infty}} \le C_n' B \|f\|_{L^1}.$$

Under the additional hypothesis that $\int_0^\infty |\widehat{\varphi}(t\xi)|^2 \frac{dt}{t} > 0$, prove the validity of the converse inequality

$$\|f\|_{L^p(\mathbf{R}^n)} \le C_n B \max(p, (p-1)^{-1}) \Big\| \Big(\int_0^\infty |f * \Psi_t|^2 \frac{dt}{t} \Big)^{\frac{1}{2}} \Big\|_{L^p(\mathbf{R}^n)}$$

for all $f \in L^p(\mathbf{R}^n)$.

5.1.5. Prove the following generalization of Theorem 5.1.2. Suppose that $\{K_j\}_j$ is a sequence of tempered distributions on $\mathbf{R}^n$ that coincide with locally integrable functions away from the origin that satisfy

$$\sup_{y\in\mathbf{R}^n\setminus\{0\}}\int_{|x|\geq 2|y|}\Big(\sum_j |K_j(x-y)-K_j(x)|^2\Big)^{\frac{1}{2}}\,dx \leq A<\infty.$$

If the Fourier transforms of K_j are functions satisfying

$$\sum_{j\in\mathbf{Z}}|\widehat{K_j}(\xi)|^2 \leq B^2,$$

then the operator

$$f\to\Big(\sum_{j\in\mathbf{Z}}|K_j*f|^2\Big)^{\frac{1}{2}}$$

maps $L^p(\mathbf{R}^n)$ to itself and is weak type $(1,1)$.

5.1.6. Suppose that $\mathscr{H}$ is a Hilbert space with inner product $\langle\cdot,\cdot\rangle_{\mathscr{H}}$ and that an operator $T: L^2(\mathbf{R}^n)\to L^2(\mathbf{R}^n,\mathscr{H})$ is a multiple of an isometry, that is,

$$\big\|T(f)\big\|_{L^2(\mathbf{R}^n,\mathscr{H})}=A\big\|f\big\|_{L^2(\mathbf{R}^n)}$$

for all f. Then the inequality $\big\|T(f)\big\|_{L^p(\mathbf{R}^n,\mathscr{H})}\leq C_p\big\|f\big\|_{L^p(\mathbf{R}^n)}$ for all $f\in L^p(\mathbf{R}^n)$ and some $p\in(1,\infty)$ implies

$$\big\|f\big\|_{L^{p'}(\mathbf{R}^n)}\leq C_{p'}A^{-2}\big\|T(f)\big\|_{L^{p'}(\mathbf{R}^n,\mathscr{H})}$$

for all f in $L^{p'}(\mathbf{R}^n)$.
[*Hint:* Use the inner product structure and polarization to obtain

$$A^2\bigg|\int_{\mathbf{R}^n}f(x)\overline{g(x)}\,dx\bigg|=\bigg|\int_{\mathbf{R}^n}\Big\langle T(f)(x),T(g)(x)\Big\rangle_{\mathscr{H}}\,dx\bigg|$$

and then argue as in the proof of inequality (5.1.8).]

5.1.7. Suppose that $\{m_j\}_{j\in\mathbf{Z}}$ is a sequence of bounded functions supported in the intervals $[2^j,2^{j+1}]$. Let $T_j(f)=(\widehat{f}m_j)^\vee$ be the corresponding operators. Assume that for all sequences of functions $\{f_j\}_j$ the vector-valued inequality

$$\bigg\|\Big(\sum_j|T_j(f_j)|^2\Big)^{\frac{1}{2}}\bigg\|_{L^p}\leq A_p\bigg\|\Big(\sum_j|f_j|^2\Big)^{\frac{1}{2}}\bigg\|_{L^p}$$

is valid for some $1<p<\infty$. Prove there is a $C_p>0$ such that for all finite subsets S of $\mathbf{Z}$ we have

$$\bigg\|\sum_{j\in S}m_j\bigg\|_{\mathscr{M}_p}\leq C_pA_p.$$

[*Hint:* Use that $\big\langle\sum_{j\in S}T_j(f),g\big\rangle=\sum_{j\in S}\big\langle\Delta_j^\flat T_j(f),\Delta_j^\flat(g)\big\rangle$.]

5.1.8. Let m be a bounded function on $\mathbf{R}^n$ that is supported in the annulus $1 \le |\xi| \le 2$ and define $T_j(f) = \big(\widehat{f}(\xi)m(2^{-j}\xi)\big)^\vee$. Suppose that the square function $f \mapsto \big(\sum_{j\in\mathbf{Z}} |T_j(f)|^2\big)^{1/2}$ is bounded on $L^p(\mathbf{R}^n)$ for some $1 < p < \infty$. Show that for every finite subset S of the integers we have

$$\Big\| \sum_{j\in S} T_j(f) \Big\|_{\mathscr{M}_p} \le C_{p,n} \|f\|_{L^p}$$

for some constant $C_{p,n}$ independent of S.

5.1.9. Fix a nonzero Schwartz function h on the line whose Fourier transform is supported in the interval $\big[-\frac{1}{8}, \frac{1}{8}\big]$. For $\{a_j\}$ a sequence of numbers, set

$$f(x) = \sum_{j=1}^{\infty} a_j e^{2\pi i 2^j x} h(x).$$

Prove that for all $1 < p < \infty$ there exists a constant C_p such that

$$\|f\|_{L^p(\mathbf{R})} \le C_p \big(\sum_j |a_j|^2\big)^{\frac{1}{2}} \|h\|_{L^p}.$$

$\big[$*Hint:* Write $f = \sum_{j=1}^{\infty} \Delta_j(a_j e^{2\pi i 2^j(\cdot)} h)$, where Δ_j is given by convolution with $\varphi_{2^{-j}}$ for some φ whose Fourier transform is supported in the interval $\big[\frac{6}{8}, \frac{10}{8}\big]$ and is equal to 1 on $\big[\frac{7}{8}, \frac{9}{8}\big]$. Then use (5.1.20).$\big]$

5.1.10. Let Ψ be a Schwartz function whose Fourier transform is supported in the annulus $\frac{1}{2} \le |\xi| \le 2$ and that satisfies (5.1.7). Define a Schwartz function Φ by setting

$$\widehat{\Phi}(\xi) = \begin{cases} \sum_{j\le 0} \widehat{\Psi}(2^{-j}\xi) & \text{when } \xi \ne 0, \\ 1 & \text{when } \xi = 0. \end{cases}$$

Let S_0 be the operator given by convolution with Φ.
(a) Prove that for all $f \in \mathscr{S}'(\mathbf{R}^n)$ we have

$$S_0(f) + \sum_{j=1}^{N} \Delta_j(f) \to f$$

in $\mathscr{S}'(\mathbf{R}^n)$.
(a) Prove that for all $f \in \mathscr{S}'(\mathbf{R}^n)/\mathscr{P}$ we have

$$\sum_{j=-N}^{N} \Delta_j(f) \to f$$

in $\mathscr{S}'(\mathbf{R}^n)/\mathscr{P}$.
$\big[$*Hint:* Use Exercise 2.3.12.$\big]$

5.1.11. Let Δ_j and S_0 be as in Exercise 5.1.10. Then for $1 < p < \infty$ we have

$$\|f\|_{L^p} \approx \|S_0(f)\|_{L^p} + \Big\|\Big(\sum_{j=1}^{\infty} |\Delta_j(f)|^2\Big)^{\frac{1}{2}}\Big\|_{L^p},$$

with the following interpretations: for L^p functions f, the right-hand side is controlled by a multiple of the left-hand side; a tempered distribution f with finite right-hand side can be identified with a function whose L^p norm is controlled by a multiple of this quantity.
[*Hint:* Use Theorem 5.1.2 (do not re-prove it), together with the identity $S_0 + \sum_{j=1}^{\infty} \Delta_j = I$, which holds in $\mathscr{S}'(\mathbf{R}^n)$ by Exercise 5.1.10.]

5.2 Two Multiplier Theorems

We now return to the spaces $\mathscr{M}_p$ introduced in Section 2.5. We seek sufficient conditions on L^∞ functions defined on $\mathbf{R}^n$ to be elements of $\mathscr{M}_p$. In this section we are concerned with two fundamental theorems that provide such sufficient conditions. These are the Marcinkiewicz and the Hörmander–Mihlin multiplier theorems. Both multiplier theorems are consequences of the Littlewood–Paley theory discussed in the previous section.

Using the dyadic decomposition of $\mathbf{R}^n$, we can write any L^∞ function m as the sum

$$m = \sum_{j\in\mathbf{Z}^n} m\chi_{R_j},$$

where $j = (j_1, \dots, j_n)$, $R_j = I_{j_1} \times \cdots \times I_{j_n}$, and $I_k = [2^k, 2^{k+1}) \bigcup (-2^{k+1}, -2^k]$. For $j \in \mathbf{Z}^n$ we set $m_j = m\chi_{R_j}$. A consequence of the ideas developed so far is the following characterization of $\mathscr{M}_p(\mathbf{R}^n)$ in terms of a vector-valued inequality.

Proposition 5.2.1. *Let $m \in L^\infty(\mathbf{R}^n)$ and let $m_j = m\chi_{R_j}$. Then m lies in $\mathscr{M}_p(\mathbf{R}^n)$, that is, for some c_p we have*

$$\|(\widehat{f}m)^\vee\|_{L^p} \le c_p\|f\|_{L^p}, \qquad f \in L^p(\mathbf{R}^n),$$

if and only if for some $C_p > 0$ we have

$$\Big\|\Big(\sum_{j\in\mathbf{Z}^n} |(\widehat{f_j}m_j)^\vee|^2\Big)^{\frac{1}{2}}\Big\|_{L^p} \le C_p\Big\|\Big(\sum_{j\in\mathbf{Z}^n} |f_j|^2\Big)^{\frac{1}{2}}\Big\|_{L^p} \tag{5.2.1}$$

for all sequences of functions f_j in $L^p(\mathbf{R}^n)$.

Proof. Suppose that $m \in \mathscr{M}_p(\mathbf{R}^n)$. Exercise 4.6.1 gives the first inequality below

$$\Big\|\Big(\sum_{j\in\mathbf{Z}^n} |(\chi_{R_j}m\widehat{f_j})^\vee|^2\Big)^{\frac{1}{2}}\Big\|_{L^p} \le C_p\Big\|\Big(\sum_{j\in\mathbf{Z}^n} |(m\widehat{f_j})^\vee|^2\Big)^{\frac{1}{2}}\Big\|_{L^p} \le C_p\Big\|\Big(\sum_{j\in\mathbf{Z}^n} |f_j|^2\Big)^{\frac{1}{2}}\Big\|_{L^p},$$

while the second inequality follows from Theorem 4.5.1. (Observe that when $p = q$ in Theorem 4.5.1, then $C_{p,q} = 1$.) Conversely, suppose that (5.2.1) holds for all sequences of functions f_j. Fix a function f and apply (5.2.1) to the sequence $(\widehat{f}\chi_{R_j})^\vee$, where R_j is the dyadic rectangle indexed by $j = (j_1, \dots, j_n) \in \mathbf{Z}^n$. We obtain

$$\Big\|\Big(\sum_{j\in\mathbf{Z}^n} |(\widehat{f}m\chi_{R_j})^\vee|^2\Big)^{\frac{1}{2}}\Big\|_{L^p} \le C_p \Big\|\Big(\sum_{j\in\mathbf{Z}^n} |(\widehat{f}\chi_{R_j})^\vee|^2\Big)^{\frac{1}{2}}\Big\|_{L^p}.$$

Using Theorem 5.1.6, we obtain that the previous inequality is equivalent to the inequality

$$\|(\widehat{f}m)^\vee\|_{L^p} \le c_p \|f\|_{L^p},$$

which implies that $m \in \mathscr{M}_p(\mathbf{R}^n)$. □

5.2.1 *The Marcinkiewicz Multiplier Theorem on* R

Proposition 5.2.1 suggests that the behavior of m on each dyadic rectangle R_j should play a crucial role in determining whether m is an L^p multiplier. The Marcinkiewicz multiplier theorem provides such sufficient conditions on m restricted to any dyadic rectangle R_j. Before stating this theorem, we illustrate its main idea via the following example. Suppose that m is a bounded function that vanishes near $-\infty$, that is differentiable at every point, and whose derivative is integrable. Then we may write

$$m(\xi) = \int_{-\infty}^{\xi} m'(t)\,dt = \int_{-\infty}^{+\infty} \chi_{[t,\infty)}(\xi)m'(t)\,dt,$$

from which it follows that for a Schwartz function f we have

$$(\widehat{f}m)^\vee = \int_{\mathbf{R}} (\widehat{f}\chi_{[t,\infty)})^\vee m'(t)\,dt.$$

Since the operators $f \mapsto (\widehat{f}\chi_{[t,\infty)})^\vee$ map $L^p(\mathbf{R})$ to itself independently of t, it follows that

$$\|(\widehat{f}m)^\vee\|_{L^p} \le C_p \|m'\|_{L^1} \|f\|_{L^p},$$

thus yielding that m is in $\mathscr{M}_p(\mathbf{R})$. The next multiplier theorem is an improvement of this result and is based on the Littlewood–Paley theorem. We begin with the one-dimensional case, which already captures the main ideas.

Theorem 5.2.2. ***(Marcinkiewicz multiplier theorem)*** *Let $m : \mathbf{R} \to \mathbf{R}$ be a bounded function that is $\mathscr{C}^1$ in every dyadic set $(2^j, 2^{j+1}) \cup (-2^{j+1}, -2^j)$ for $j \in \mathbf{Z}$. Assume that the derivative m' of m satisfies*

$$\sup_j \left[\int_{-2^{j+1}}^{-2^j} |m'(\xi)|\,d\xi + \int_{2^j}^{2^{j+1}} |m'(\xi)|\,d\xi\right] \le A < \infty. \tag{5.2.2}$$

Then for all $1 < p < \infty$ *we have that* $m \in \mathscr{M}_p(\mathbf{R})$ *and for some* $C > 0$ *we have*

$$\big\|m\big\|_{\mathscr{M}_p(\mathbf{R})} \le C \max\big(p,(p-1)^{-1}\big)^6\big(\big\|m\big\|_{L^\infty} + A\big). \tag{5.2.3}$$

Proof. Since the function m has an integrable derivative on $(2^j,2^{j+1})$, it has bounded variation in this interval and hence it is a difference of two increasing functions. Therefore, m has left and right limits at the points 2^j and 2^{j+1}, and by redefining m at these points we may assume that m is right continuous at the points 2^j and left continuous at the points -2^j.

Set $I_j = [2^j,2^{j+1})\bigcup(-2^{j+1},-2^j]$ and $I_j^+ = [2^j,2^{j+1})$ whenever $j \in \mathbf{Z}$. Given an interval I in $\mathbf{R}$, we introduce an operator Δ_I defined by $\Delta_I(f) = (\widehat{f}\chi_I)^\vee$. With this notation $\Delta_{I_j^+}(f)$ is "half" of the operator $\Delta_j^\flat$ introduced in the previous section. Given m as in the statement of the theorem, we write $m(\xi) = m_+(\xi) + m_-(\xi)$, where $m_+(\xi) = m(\xi)\chi_{\xi\ge 0}$ and $m_-(\xi) = m(\xi)\chi_{\xi<0}$. We show that both m_+ and m_- are L^p multipliers. Since m' is integrable over all intervals of the form $[2^j,\xi]$ when $2^j \le \xi < 2^{j+1}$, the fundamental theorem of calculus gives

$$m(\xi) = m(2^j) + \int_{2^j}^{\xi} m'(t)\,dt, \qquad \text{for } 2^j \le \xi < 2^{j+1},$$

from which it follows that for a Schwartz function f on the real line we have

$$m(\xi)\widehat{f}(\xi)\chi_{I_j^+}(\xi) = m(2^j)\widehat{f}(\xi)\chi_{I_j^+}(\xi) + \int_{2^j}^{2^{j+1}} \widehat{f}(\xi)\chi_{[t,\infty)}(\xi)\chi_{I_j^+}(\xi)\,m'(t)\,dt\,.$$

We therefore obtain the identity

$$(\widehat{f}\chi_{I_j}m_+)^\vee = (\widehat{f}m\chi_{I_j^+})^\vee = m(2^j)\Delta_{I_j^+}(f) + \int_{2^j}^{2^{j+1}} \Delta_{[t,\infty)}\Delta_{I_j^+}(f)\,m'(t)\,dt\,,$$

which implies that

$$|(\widehat{f}\chi_{I_j}m_+)^\vee| \le \big\|m\big\|_{L^\infty}|\Delta_{I_j^+}(f)| + A^{\frac12}\bigg(\int_{2^j}^{2^{j+1}} \big|\Delta_{[t,\infty)}\Delta_{I_j^+}(f)\big|^2\,|m'(t)|\,dt\bigg)^{\frac12},$$

using the hypothesis (5.2.2). Taking $\ell^2(\mathbf{Z})$ norms we obtain

$$\Big(\sum_{j\in\mathbf{Z}}|(\widehat{f}\chi_{I_j}m_+)^\vee|^2\Big)^{\frac12} \le \big\|m\big\|_{L^\infty}\Big(\sum_{j\in\mathbf{Z}}|\Delta_{I_j^+}(f)|^2\Big)^{\frac12}$$
$$+A^{\frac12}\bigg(\int_0^\infty \big|\Delta_{[t,\infty)}\Delta^\flat_{[\log_2 t]}(f)\big|^2\,|m'(t)|\,dt\bigg)^{\frac12}.$$

Exercise 4.6.2 gives

$$A^{\frac{1}{2}}\Big\|\Big(\int_0^\infty |\Delta_{[t,\infty)}\Delta^\flat_{[\log_2 t]}(f)|^2|m'(t)|\,dt\Big)^{\frac{1}{2}}\Big\|_{L^p}$$
$$\leq C\max(p,(p-1)^{-1})A^{\frac{1}{2}}\Big\|\Big(\int_0^\infty |\Delta^\flat_{[\log_2 t]}(f)|^2|m'(t)|\,dt\Big)^{\frac{1}{2}}\Big\|_{L^p},$$

while the hypothesis on m' implies the inequality

$$\Big\|\Big(\sum_{j\in\mathbf{Z}}|\Delta_{I_j^+}(f)|^2\int_{I_j^+}|m'(t)|\,dt\Big)^{\frac{1}{2}}\Big\|_{L^p}\leq A^{\frac{1}{2}}\Big\|\Big(\sum_j|\Delta_{I_j^+}(f)|^2\Big)^{\frac{1}{2}}\Big\|_{L^p}.$$

Using Theorem 5.1.5 we obtain that

$$\Big\|\Big(\sum_j|\Delta_{I_j^+}(f)|^2\Big)^{\frac{1}{2}}\Big\|_{L^p}\leq C'\max(p,(p-1)^{-1})^2\|(\widehat{f}\chi_{(0,\infty)})^\vee\|_{L^p},$$

and the latter is at most a constant multiple of $\max(p,(p-1)^{-1})^3\|f\|_{L^p}$. Putting things together we deduce that

$$\Big\|\Big(\sum_j|(\widehat{f}\chi_{I_j}m_+)^\vee|^2\Big)^{\frac{1}{2}}\Big\|_{L^p}\leq C''\max(p,(p-1)^{-1})^4(A+\|m\|_{L^\infty})\|f\|_{L^p},\tag{5.2.4}$$

from which we obtain the estimate

$$\|(\widehat{f}m_+)^\vee\|_{L^p}\leq C\max(p,(p-1)^{-1})^6(A+\|m\|_{L^\infty})\|f\|_{L^p},$$

using the lower estimate of Theorem 5.1.5. This proves (5.2.3) for m_+. A similar argument also works for m_-, and this concludes the proof by summing the corresponding estimates for m_+ and m_-. □

We remark that the same proof applies under the more general assumption that m is a function of bounded variation on every interval $[2^j,2^{j+1}]$ and $[-2^{j+1},-2^j]$. In this case the measure $|m'(t)|\,dt$ should be replaced by the total variation $|dm(t)|$ of the Lebesgue–Stieltjes measure $dm(t)$.

Example 5.2.3. Any bounded function that is constant on dyadic intervals is an L^p multiplier. Also, the function

$$m(\xi)=|\xi|2^{-[\log_2|\xi|]}$$

is an L^p multiplier on $\mathbf{R}$ for $1<p<\infty$.

5.2.2 The Marcinkiewicz Multiplier Theorem on $\mathbf{R}^n$

We now extend this theorem on $\mathbf{R}^n$. As usual we denote the coordinates of a point $\xi \in \mathbf{R}^n$ by $(\xi_1, \dots, \xi_n)$. We recall the notation $I_j = (-2^{j+1}, -2^j] \bigcup [2^j, 2^{j+1})$ and $R_j = I_{j_1} \times \cdots \times I_{j_n}$ whenever $j = (j_1, \dots, j_n) \in \mathbf{Z}^n$.

Theorem 5.2.4. *Let m be a bounded function on $\mathbf{R}^n$ that is $\mathscr{C}^n$ in all regions R_j (i.e., $\partial^\alpha m$ are continuous up to the boundary of R_j for all $|\alpha| \le n$). Assume that there is a constant A such that for all $k \in \{1, \dots, n\}$, all $j_1, \dots, j_k \in \{1, 2, \dots, n\}$, all $l_{j_1}, \dots, l_{j_k} \in \mathbf{Z}$, and all $\xi_s \in I_{l_s}$ for $s \in \{1, \dots, n\} \setminus \{j_1, \dots, j_k\}$ we have*

$$\int_{I_{l_{j_1}}} \cdots \int_{I_{l_{j_k}}} \big|(\partial_{j_1} \cdots \partial_{j_k} m)(\xi_1, \dots, \xi_n)\big| \, d\xi_{j_k} \cdots d\xi_{j_1} \le A < \infty . \tag{5.2.5}$$

Then m is in $\mathscr{M}_p(\mathbf{R}^n)$ whenever $1 < p < \infty$ and there is a constant $C_n < \infty$ such that

$$\big\|m\big\|_{\mathscr{M}_p(\mathbf{R}^n)} \le C_n \big(A + \big\|m\big\|_{L^\infty}\big) \max\big(p, (p-1)^{-1}\big)^{6n} . \tag{5.2.6}$$

Proof. We prove this theorem only in dimension $n = 2$, since the general case presents no substantial differences but only some notational inconvenience. We decompose the given function m as

$$m(\xi) = m_{++}(\xi) + m_{-+}(\xi) + m_{+-}(\xi) + m_{--}(\xi),$$

where each of the last four terms is supported in one of the four quadrants. For instance, the function $m_{+-}(\xi_1, \xi_2)$ is supported in the quadrant $\xi_1 \ge 0$ and $\xi_2 < 0$. As in the one-dimensional case, we work with each of these pieces separately. By symmetry we choose to work with m_{++} in the following argument.

Using the fundamental theorem of calculus, we obtain the following simple identity, valid for $2^{j_1} \le \xi_1 < 2^{j_1+1}$ and $2^{j_2} \le \xi_2 < 2^{j_2+1}$:

$$\begin{aligned} m(\xi_1, \xi_2) &= m(2^{j_1}, 2^{j_2}) + \int_{2^{j_1}}^{\xi_1} (\partial_1 m)(t_1, 2^{j_2}) \, dt_1 \\ &\quad + \int_{2^{j_2}}^{\xi_2} (\partial_2 m)(2^{j_1}, t_2) \, dt_2 \\ &\quad + \int_{2^{j_1}}^{\xi_1} \int_{2^{j_2}}^{\xi_2} (\partial_1 \partial_2 m)(t_1, t_2) \, dt_2 \, dt_1 . \end{aligned} \tag{5.2.7}$$

We introduce operators $\Delta_I^{(r)}$, $r \in \{1, 2\}$, acting in the rth variable (with the other variable remaining fixed) given by multiplication on the Fourier transform side by the characteristic function of the interval I. Likewise, we introduce operators $\Delta_j^{\flat(r)}$, $r \in \{1, 2\}$ (also acting in the rth variable), given by multiplication on the Fourier transform side by the characteristic function of the set $(-2^{j+1}, -2^j] \bigcup [2^j, 2^{j+1})$. For notational convenience, for a given Schwartz function f we write

$$f_{++} = \big(\widehat{f}\chi_{(0,\infty)^2}\big)^\vee,$$

and likewise we define f_{+-}, f_{-+}, and f_{--}.

Multiplying both sides of (5.2.7) by the function $\widehat{f}\chi_{R_j}\chi_{(0,\infty)^2}$ and taking inverse Fourier transforms yields

$$\begin{aligned}
(\widehat{f}\chi_{R_j}m_{++})^\vee &= m(2^{j_1},2^{j_2})\Delta_{j_1}^{\flat(1)}\Delta_{j_2}^{\flat(2)}(f_{++}) \\
&+ \int_{2^{j_1}}^{2^{j_1+1}} \Delta_{j_2}^{\flat(2)}\Delta_{[t_1,\infty)}^{(1)}\Delta_{j_1}^{\flat(1)}(f_{++})\,(\partial_1 m)(t_1,2^{j_2})\,dt_1 \\
&+ \int_{2^{j_2}}^{2^{j_2+1}} \Delta_{j_1}^{\flat(1)}\Delta_{[t_2,\infty)}^{(2)}\Delta_{j_2}^{\flat(2)}(f_{++})\,(\partial_2 m)(2^{j_1},t_2)\,dt_2 \\
&+ \int_{2^{j_1}}^{2^{j_1+1}}\int_{2^{j_2}}^{2^{j_2+1}} \Delta_{[t_1,\infty)}^{(1)}\Delta_{j_1}^{\flat(1)}\Delta_{[t_2,\infty)}^{(2)}\Delta_{j_2}^{\flat(2)}(f_{++})\,(\partial_1\partial_2 m)(t_1,t_2)\,dt_2\,dt_1 .
\end{aligned} \tag{5.2.8}$$

We apply the Cauchy–Schwarz inequality in the last three terms of (5.2.8) with respect to the measures $|(\partial_1 m)(t_1,2^{j_2})|dt_1$, $|(\partial_2 m)(2^{j_1},t_2)|dt_2$, $|(\partial_1\partial_2 m)(t_1,t_2)|\,dt_2dt_1$ and we use hypothesis (5.2.5) to deduce

$$\begin{aligned}
&\big|(\widehat{f}\chi_{R_j}m_{++})^\vee\big| \le \|m\|_{L^\infty}\big|\Delta_{j_1}^{\flat(1)}\Delta_{j_2}^{\flat(2)}(f_{++})\big| \\
&+A^{\frac12}\bigg(\int_{2^{j_1}}^{2^{j_1+1}} \big|\Delta_{j_2}^{\flat(2)}\Delta_{[t_1,\infty)}^{(1)}\Delta_{j_1}^{\flat(1)}(f_{++})\big|^2\,|(\partial_1 m)(t_1,2^{j_2})|\,dt_1\bigg)^{\frac12} \\
&+A^{\frac12}\bigg(\int_{2^{j_2}}^{2^{j_2+1}} \big|\Delta_{j_1}^{\flat(1)}\Delta_{[t_2,\infty)}^{(2)}\Delta_{j_2}^{\flat(2)}(f_{++})\big|^2\,|(\partial_2 m)(2^{j_1},t_2)|\,dt_2\bigg)^{\frac12} \\
&+A^{\frac12}\bigg(\int_{2^{j_1}}^{2^{j_1+1}}\int_{2^{j_2}}^{2^{j_2+1}} \big|\Delta_{[t_1,\infty)}^{(1)}\Delta_{j_1}^{\flat(1)}\Delta_{[t_2,\infty)}^{(2)}\Delta_{j_2}^{\flat(2)}(f_{++})\big|^2\,|(\partial_1\partial_2 m)(t_1,t_2)|\,dt_2\,dt_1\bigg)^{\frac12}.
\end{aligned}$$

Both sides of the preceding inequality are sequences indexed by $j\in\mathbf{Z}^2$. We apply $\ell^2(\mathbf{Z}^2)$ norms and use Minkowski's inequality to deduce the pointwise estimate

$$\begin{aligned}
&\Big(\sum_{j\in\mathbf{Z}^2}\big|(\widehat{f}\chi_{R_j}m_{++})^\vee\big|^2\Big)^{\frac12} \le \|m\|_{L^\infty}\Big(\sum_{j\in\mathbf{Z}^2}\big|\Delta_j^\flat(f_{++})\big|^2\Big)^{\frac12} \\
&\quad+A^{\frac12}\bigg(\int_0^\infty\!\!\int_0^\infty \big|\Delta_{[t_1,\infty)}^{(1)}\Delta_{[\log_2 t_2]}^{\flat(2)}\Delta_{[\log_2 t_1]}^{\flat(1)}(f_{++})\big|^2\,\big|(\partial_1 m)(t_1,2^{[\log_2 t_2]})\big|\,dt_1d\nu(t_2)\bigg)^{\frac12} \\
&\quad+A^{\frac12}\bigg(\int_0^\infty\!\!\int_0^\infty \big|\Delta_{[t_2,\infty)}^{(2)}\Delta_{[\log_2 t_1]}^{\flat(1)}\Delta_{[\log_2 t_2]}^{\flat(2)}(f_{++})\big|^2\,\big|(\partial_2 m)(2^{[\log_2 t_1]},t_2)\big|\,d\nu(t_1)dt_2\bigg)^{\frac12} \\
&\quad+A^{\frac12}\bigg(\int_0^\infty\!\!\int_0^\infty \big|\Delta_{[t_1,\infty)}^{(1)}\Delta_{[t_2,\infty)}^{(2)}\Delta_{[\log_2 t_1]}^{\flat(1)}\Delta_{[\log_2 t_2]}^{\flat(2)}(f_{++})\big|^2\big|(\partial_1\partial_2 m)(t_1,t_2)\big|\,dt_1dt_2\bigg)^{\frac12},
\end{aligned}$$

where ν is the counting measure $\sum_{j\in\mathbf{Z}}\delta_{2^j}$ defined by $\nu(A)=\#\{j\in\mathbf{Z}:\ 2^j\in A\}$ for subsets A of $(0,\infty)$. We now take $L^p(\mathbf{R}^2)$ norms and we estimate separately the contribution of each of the four terms on the right side. Using Exercise 4.6.2 we obtain

$$\Big\|\Big(\sum_{j\in\mathbf{Z}^2}|(\widehat{f}\chi_{R_j}m_{++})^\vee|^2\Big)^{\frac12}\Big\|_{L^p} \le \|m\|_{L^\infty}\Big\|\Big(\sum_{j\in\mathbf{Z}^2}|\Delta_j^\flat(f_{++})|^2\Big)^{\frac12}\Big\|_{L^p}$$

$$+ C_2A^{\frac12}\max\big(p,(p-1)^{-1}\big)^2$$

$$\times\Bigg\{\Big\|\Big(\int_0^\infty\!\!\int_0^\infty|\Delta^{\flat(2)}_{[\log_2 t_2]}\Delta^{\flat(1)}_{[\log_2 t_1]}(f_{++})|^2\,|(\partial_1 m)(t_1,2^{[\log_2 t_2]})|\,dt_1\,d\nu(t_2)\Big)^{\frac12}\Big\|_{L^p}$$

$$+\Big\|\Big(\int_0^\infty\!\!\int_0^\infty|\Delta^{\flat(1)}_{[\log_2 t_1]}\Delta^{\flat(2)}_{[\log_2 t_2]}(f_{++})|^2\,|(\partial_2 m)(2^{[\log_2 t_1]},t_2)|\,d\nu(t_1)\,dt_2\Big)^{\frac12}\Big\|_{L^p}$$

$$+\Big\|\Big(\int_0^\infty\!\!\int_0^\infty|\Delta^{\flat(1)}_{[\log_2 t_1]}\Delta^{\flat(2)}_{[\log_2 t_2]}(f_{++})|^2|(\partial_1\partial_2 m)(t_1,t_2)|\,dt_1dt_2\Big)^{\frac12}\Big\|_{L^p}\Bigg\}.$$

But the functions $(t_1,t_2)\mapsto\Delta^{\flat(1)}_{[\log_2 t_1]}\Delta^{\flat(2)}_{[\log_2 t_2]}(f_{++})$ are constant on products of intervals of the form $[2^{j_1},2^{j_1+1})\times[2^{j_2},2^{j_2+1})$; hence using hypothesis (5.2.5) again we deduce the estimate

$$\Big\|\Big(\sum_{j\in\mathbf{Z}^2}|(\widehat{f}\chi_{R_j}m_{++})^\vee|^2\Big)^{\frac12}\Big\|_{L^p(\mathbf{R}^2)}$$

$$\le C_2\big(\|m\|_{L^\infty}+A\big)\max\big(p,(p-1)^{-1}\big)^2\Big\|\Big(\sum_{j\in\mathbf{Z}^2}|\Delta_j^\flat(f_{++})|^2\Big)^{\frac12}\Big\|_{L^p(\mathbf{R}^2)}$$

$$\le C_2\big(\|m\|_{L^\infty}+A\big)\max\big(p,(p-1)^{-1}\big)^4\big\|(\widehat{f}\chi_{(0,\infty)^2})^\vee\big\|_{L^p(\mathbf{R}^2)}$$

$$\le C_2\big(\|m\|_{L^\infty}+A\big)\max\big(p,(p-1)^{-1}\big)^6\|f\|_{L^p(\mathbf{R}^2)},$$

where the penultimate estimate follows from Theorem 5.1.6 and the last estimate by the boundedness of the Hilbert transform (Theorem 4.1.7). We now appeal to the lower estimate of Theorem 5.1.6, which yields the required estimate for m_{++}. A similar argument also works for the remaining parts of m, and this concludes the proof of (5.2.6).

The analogous estimate on $\mathbf{R}^n$ is

$$\Big\|\Big(\sum_{j\in\mathbf{Z}^n}|(\widehat{f}\chi_{R_j}m_{++})^\vee|^2\Big)^{\frac12}\Big\|_{L^p(\mathbf{R}^n)}\le C_n\big(\|m\|_{L^\infty}+A\big)\max\big(p,(p-1)^{-1}\big)^{3n}\|f\|_{L^p(\mathbf{R}^n)}$$

for some dimensional constant $C_n<\infty$ and is obtained in a similar fashion. □

We now give a condition that implies (5.2.5) and is well suited for a variety of applications.

Corollary 5.2.5. *Let m be a bounded function defined away from the coordinate axes on $\mathbf{R}^n$ that is $\mathscr{C}^n$ in that region. Assume furthermore that for all $k\in\{1,\dots,n\}$,* *all* $j_1,\dots,j_k \in \{1,2,\dots,n\}$, *and all* $\xi_r \in \mathbf{R}$ *for* $r \notin \{j_1,\dots,j_k\}$ *we have*

$$\big|(\partial_{j_1}\cdots\partial_{j_k}m)(\xi_1,\dots,\xi_n)\big| \le A\,|\xi_{j_1}|^{-1}\cdots|\xi_{j_k}|^{-1}. \tag{5.2.9}$$

Then m *satisfies (5.2.6).*

Proof. Simply observe that condition (5.2.9) implies (5.2.5). □

Example 5.2.6. The following are examples of functions that satisfy the hypotheses of Corollary 5.2.5:

$$m_1(\xi) = \frac{\xi_1}{\xi_1 + i(\xi_2^2+\cdots+\xi_n^2)},$$

$$m_2(\xi) = \frac{|\xi_1|^{\alpha_1}\cdots|\xi_n|^{\alpha_n}}{(\xi_1^2+\xi_2^2+\cdots+\xi_n^2)^{\alpha/2}},$$

where $\alpha_1+\alpha_2+\cdots+\alpha_n=\alpha$, $\alpha_j>0$,

$$m_3(\xi) = \frac{\xi_2\xi_3^2}{i\xi_1+\xi_2^2+\xi_3^4}.$$

The functions m_1 and m_2 are defined on $\mathbf{R}^n$ and m_3 on $\mathbf{R}^3$.

The previous examples and many other examples that satisfy the hypothesis (5.2.9) of Corollary 5.2.5 are invariant under a set of dilations in the following sense: suppose that there exist $k_1,\dots,k_n \in \mathbf{R}^+$ and $s\in\mathbf{R}$ such that the smooth function m on $\mathbf{R}^n\setminus\{0\}$ satisfies

$$m(\lambda^{k_1}\xi_1,\dots,\lambda^{k_n}\xi_n) = \lambda^{is}m(\xi_1,\dots,\xi_n)$$

for all $\xi_1,\dots,\xi_n\in\mathbf{R}$ and $\lambda>0$. Then m satisfies condition (5.2.9). Indeed, differentiation gives

$$\lambda^{\alpha_1k_1+\cdots+\alpha_nk_n}\partial^\alpha m(\lambda^{k_1}\xi_1,\dots,\lambda^{k_n}\xi_n) = \lambda^{is}\partial^\alpha m(\xi_1,\dots,\xi_n)$$

for every multi-index $\alpha=(\alpha_1,\dots,\alpha_n)$. Now for every $\xi\in\mathbf{R}^n\setminus\{0\}$ pick the unique $\lambda_\xi>0$ such that $(\lambda_\xi^{k_1}\xi_1,\dots,\lambda_\xi^{k_n}\xi_n)\in\mathbf{S}^{n-1}$. Then $\lambda_\xi^{k_j\alpha_j}\le|\xi_j|^{-\alpha_j}$, and it follows that

$$|\partial^\alpha m(\xi_1,\dots,\xi_n)| \le \Big[\sup_{\mathbf{S}^{n-1}}|\partial^\alpha m|\Big]\lambda_\xi^{\alpha_1k_1+\cdots+\alpha_nk_n} \le C_\alpha|\xi_1|^{-\alpha_1}\cdots|\xi_n|^{-\alpha_n}.$$

5.2.3 The Hörmander–Mihlin Multiplier Theorem on $\mathbf{R}^n$

We now discuss a second multiplier theorem.

Theorem 5.2.7. *Let $m(\xi)$ be a complex-valued bounded function on $\mathbf{R}^n\setminus\{0\}$ that satisfies either*
(a) Mihlin's condition

$$|\partial_\xi^\alpha m(\xi)| \le A|\xi|^{-|\alpha|} \tag{5.2.10}$$

for all multi-indices $|\alpha| \le [\frac{n}{2}]+1$,
or (b) Hörmander's condition

$$\sup_{R>0} R^{-n+2|\alpha|}\int_{R<|\xi|<2R} |\partial_\xi^\alpha m(\xi)|^2\,d\xi \le A^2 < \infty \tag{5.2.11}$$

for all multi-indices $|\alpha| \le [n/2]+1$.
Then for all $1<p<\infty$, m lies in $\mathscr{M}_p(\mathbf{R}^n)$ and the following estimate is valid:

$$\big\|m\big\|_{\mathscr{M}_p} \le C_n \max(p,(p-1)^{-1})\big(A+\big\|m\big\|_{L^\infty}\big). \tag{5.2.12}$$

Moreover, the operator $f\mapsto(\widehat{f}m)^\vee$ maps $L^1(\mathbf{R}^n)$ to $L^{1,\infty}(\mathbf{R}^n)$ with norm at most a dimensional constant multiple of $A+\big\|m\big\|_{L^\infty}$.

Proof. First we observe that condition (5.2.11) is a generalization of (5.2.10) and therefore it suffices to assume (5.2.11).

Since m is a bounded function, the operator given by convolution with $W=m^\vee$ is bounded on $L^2(\mathbf{R}^n)$. To prove that this operator maps $L^1(\mathbf{R}^n)$ to $L^{1,\infty}(\mathbf{R}^n)$, it suffices to prove that the distribution W coincides with a function K on $\mathbf{R}^n\setminus\{0\}$ that satisfies Hörmander's condition.

Let $\widehat{\zeta}$ be a smooth function supported in the annulus $\frac{1}{2}\le|\xi|\le 2$ such that

$$\sum_{j\in\mathbf{Z}}\widehat{\zeta}(2^{-j}\xi)=1, \qquad \text{when } \xi\neq 0.$$

Set $m_j(\xi)=m(\xi)\widehat{\zeta}(2^{-j}\xi)$ for $j\in\mathbf{Z}$ and $K_j=m_j^\vee$. We begin by observing that $\sum_{-N}^N K_j$ converges to W in $\mathscr{S}'(\mathbf{R}^n)$. Indeed, for all $\varphi\in\mathscr{S}(\mathbf{R}^n)$ we have

$$\Big\langle\sum_{j=-N}^N K_j,\varphi\Big\rangle=\Big\langle\sum_{j=-N}^N m_j,\varphi^\vee\Big\rangle\to\big\langle m,\varphi^\vee\big\rangle=\big\langle W,\varphi\big\rangle.$$

We set $n_0=[\frac{n}{2}]+1$. We claim that there is a constant $\widetilde{C}_n$ such that

$$\sup_{j\in\mathbf{Z}}\int_{\mathbf{R}^n}|K_j(x)|\,(1+2^j|x|)^{\frac14}\,dx \le \widetilde{C}_n A, \tag{5.2.13}$$

$$\sup_{j\in\mathbf{Z}} 2^{-j}\int_{\mathbf{R}^n}|\nabla K_j(x)|\,(1+2^j|x|)^{\frac14}\,dx \le \widetilde{C}_n A. \tag{5.2.14}$$

To prove (5.2.13) we multiply and divide the integrand in (5.2.13) by the expression $(1+2^j|x|)^{n_0}$. Applying the Cauchy–Schwarz inequality to $|K_j(x)|\,(1+2^j|x|)^{n_0}$ and $(1+2^j|x|)^{-n_0+\frac14}$, we control the integral in (5.2.13) by the product

$$\left(\int_{\mathbf{R}^n} |K_j(x)|^2 (1+2^j|x|)^{2n_0}\,dx\right)^{\frac{1}{2}} \left(\int_{\mathbf{R}^n} (1+2^j|x|)^{-2n_0+\frac{1}{2}}\,dx\right)^{\frac{1}{2}}. \tag{5.2.15}$$

We now note that $-2n_0+\frac{1}{2} < -n$, and hence the second expression in (5.2.15) is equal to a constant multiple of $2^{-jn/2}$. To estimate the first integral in (5.2.15) we use the simple fact that

$$(1+2^j|x|)^{n_0} \le C(n) \sum_{|\gamma|\le n_0} |(2^j x)^\gamma|.$$

We now have that the expression inside the supremum in (5.2.13) is controlled by

$$C'(n)2^{-jn/2} \sum_{|\gamma|\le n_0} \left(\int_{\mathbf{R}^n} |K_j(x)|^2 2^{2j|\gamma|} |x^\gamma|^2\,dx\right)^{\frac{1}{2}}, \tag{5.2.16}$$

which, by Plancherel's theorem, is equal to

$$2^{-jn/2} \sum_{|\gamma|\le n_0} C_\gamma 2^{j|\gamma|} \left(\int_{\mathbf{R}^n} |(\partial^\gamma m_j)(\xi)|^2\,d\xi\right)^{\frac{1}{2}} \tag{5.2.17}$$

for some constants C_γ.

For multi-indices $\delta = (\delta_1,\dots,\delta_n)$ and $\gamma = (\gamma_1,\dots,\gamma_n)$ we introduce the notation $\delta \le \gamma$ to mean $\delta_j \le \gamma_j$ for all $j = 1,\dots,n$. For any $|\gamma| \le n_0$ we use Leibniz's rule to obtain

$$\begin{aligned}
\int_{\mathbf{R}^n} |(\partial^\gamma m_j)(\xi)|^2\,d\xi &\le \sum_{\delta\le\gamma} C_\delta \int_{\mathbf{R}^n} \left|2^{-j|\gamma-\delta|}(\partial_\xi^{\gamma-\delta}\widehat{\zeta})(2^{-j}\xi)(\partial_\xi^\delta m)(\xi)\right|^2 d\xi \\
&\le \sum_{\delta\le\gamma} C_\delta 2^{-2j|\gamma|} 2^{2j|\delta|} \int_{2^{j-1}\le|\xi|\le 2^{j+1}} \left|(\partial_\xi^\delta m)(\xi)\right|^2 d\xi \\
&\le \sum_{\delta\le\gamma} C_\delta 2^{-2j|\gamma|} 2^{2j|\delta|} 2A^2 2^{jn} 2^{-2j|\delta|} \\
&= \widetilde{C}_n A^2 2^{jn} 2^{-2j|\gamma|},
\end{aligned}$$

which is all we need to obtain (5.2.13). To obtain (5.2.14) we repeat the same argument for every derivative $\partial_r K_j$. Since the Fourier transform of $(\partial_r K_j)(x)x^\gamma$ is equal to a constant multiple of $\partial^\gamma\big(\xi_r m(\xi)\widehat{\zeta}(2^{-j}\xi)\big)$, we observe that the extra factor 2^{-j} in (5.2.14) can be combined with ξ_r to write $2^{-j}\partial^\gamma\big(\xi_r m(\xi)\widehat{\zeta}(2^{-j}\xi)\big)$ as $\partial^\gamma\big(m(\xi)\widehat{\zeta}_r(2^{-j}\xi)\big)$, where $\widehat{\zeta}_r(\xi) = \xi_r\widehat{\zeta}(\xi)$. The previous calculation with $\widehat{\zeta}_r$ replacing $\widehat{\zeta}$ can then be used to complete the proof of (5.2.14).

We now show that for all $x \ne 0$, the series $\sum_{j\in\mathbf{Z}} K_j(x)$ converges to a function, which we denote by $K(x)$. Indeed, it is trivial to see that for all $x \in \mathbf{R}^n$ we have

$$|K_j(x)| \le c_n 2^{jn} \|m\|_{L^\infty},$$

which shows that the function $\sum_{j\le 0}|K_j(x)|$ is bounded. Moreover, as a consequence of (5.2.13) we have that

$$(1+2^j\delta)^{\frac{1}{4}}\int_{|x|\ge\delta}|K_j(x)|\,dx\le\widetilde{C}_nA\,,$$

for any $\delta>0$, which implies that the function $\sum_{j>0}|K_j(x)|$ is integrable away from the origin and thus finite almost everywhere. We conclude that the series $\sum_{j\in\mathbf{Z}}K_j(x)$ represents a well defined function $K(x)$ away from the origin that coincides with the distribution $W=m^{\vee}$.

We now prove that the function $K=\sum_{j\in\mathbf{Z}}K_j$ (defined on $\mathbf{R}^n\setminus\{0\}$) satisfies Hörmander's condition. It suffices to prove that for all $y\neq 0$ we have

$$\sum_{j\in\mathbf{Z}}\int_{|x|\ge 2|y|}|K_j(x-y)-K_j(x)|\,dx\le 2C_n'A\,. \tag{5.2.18}$$

Fix a $y\in\mathbf{R}^n\setminus\{0\}$ and pick a $k\in\mathbf{Z}$ such that $2^{-k}\le|y|\le 2^{-k+1}$. The part of the sum in (5.2.18) where $j>k$ is bounded by

$$\begin{aligned}\sum_{j>k}\int_{|x|\ge 2|y|}|K_j(x-y)|+|K_j(x)|\,dx &\le 2\sum_{j>k}\int_{|x|\ge|y|}|K_j(x)|\,dx\\ &\le 2\sum_{j>k}\int_{|x|\ge|y|}|K_j(x)|\frac{(1+2^j|x|)^{\frac{1}{4}}}{(1+2^j|x|)^{\frac{1}{4}}}\,dx\\ &\le\sum_{j>k}\frac{2\widetilde{C}_nA}{(1+2^j|y|)^{\frac{1}{4}}}\\ &\le\sum_{j>k}\frac{2\widetilde{C}_nA}{(1+2^j2^{-k})^{\frac{1}{4}}}=C_n'A\,,\end{aligned}$$

where we used (5.2.13). The part of the sum in (5.2.18) where $j\le k$ is bounded by

$$\begin{aligned}&\sum_{j\le k}\int_{|x|\ge 2|y|}|K_j(x-y)-K_j(x)|\,dx\\ &\qquad\le\sum_{j\le k}\int_{|x|\ge 2|y|}\int_0^1|-y\cdot\nabla K_j(x-\theta y)|\,d\theta\,dx\\ &\qquad\le\int_0^1\sum_{j\le k}2^{-k+1}\int_{\mathbf{R}^n}|\nabla K_j(x-\theta y)|(1+2^j|x-\theta y|)^{\frac{1}{4}}\,dx\,d\theta\\ &\qquad\le\int_0^1\sum_{j\le k}2^{-k+1}\widetilde{C}_nA2^j\,d\theta\le C_n'A\,,\end{aligned}$$

using (5.2.14). Hörmander's condition is satisfied for K, and we appeal to Theorem 4.3.3 to complete the proof of (5.2.12). □

Corollary 5.2.8. *Let $\{m_\ell\}_{\ell\in\mathbf{Z}}$ be bounded functions on $\mathbf{R}^n$ whose L^∞ norms are uniformly controlled by a constant A. Suppose that*

$$\sup_{R>0} R^{-n+2|\alpha|} \sum_{\ell\in\mathbf{Z}} \int_{R<|\xi|<2R} |\partial_\xi^\alpha m_\ell(\xi)|^2 \, d\xi \le A^2 \qquad \text{for all } |\alpha| \le [\tfrac{n}{2}]+1.$$

Then for some $C_n<\infty$ and for all functions f_k we have

$$\Big\| \Big(\sum_{\ell\in\mathbf{Z}} |(\widehat{f_\ell} m_\ell)^\vee|^2 \Big)^{\frac{1}{2}} \Big\|_{L^p} \le C_n (p+(p-1)^{-1}) A \Big\| \Big(\sum_{\ell\in\mathbf{Z}} |f_\ell|^2 \Big)^{\frac{1}{2}} \Big\|_{L^p}. \tag{5.2.19}$$

Proof. Write each $K^\ell = m_\ell^\vee = \sum_j K_j^\ell$ as in the proof of Theorem 5.2.7. Using the hypothesis, we can prove that

$$\sup_{j\in\mathbf{Z}} \int_{\mathbf{R}^n} \sum_\ell |K_j^\ell(x)| \, (1+2^j|x|)^{\frac{1}{4}} \, dx \le \widetilde{C}_n A, \tag{5.2.20}$$

$$\sup_{j\in\mathbf{Z}} 2^{-j} \int_{\mathbf{R}^n} \sum_\ell |\nabla K_j^\ell(x)| \, (1+2^j|x|)^{\frac{1}{4}} \, dx \le \widetilde{C}_n A. \tag{5.2.21}$$

These estimates are proved just like those in (5.2.13) and (5.2.14) with the extra summation on ℓ carried through. Using (5.2.20) and (5.2.21), we now derive that

$$\int_{|x|\ge 2|y|} \sup_\ell |K^\ell(x-y) - K^\ell(x)| \, dx \le \sum_j \int_{|x|\ge 2|y|} \sum_\ell |K_j^\ell(x-y) - K_j^\ell(x)| \, dx < \infty$$

as we did in the proof of Theorem 5.2.7. This is Hörmander condition needed in this setting, which allows the use of Corollary 4.6.2. The proof of (5.2.19) follows. □

Example 5.2.9. Suppose that τ is a real number. Then the function $|\xi|^{i\tau}$ is in $\mathscr{M}_p$ for all $1<p<\infty$, since condition (5.2.10) is satisfied.

We end this section by comparing Theorems 5.2.2/5.2.4 and 5.2.7. It is obvious that in dimension $n=1$, Theorem 5.2.2 is stronger than Theorem 5.2.7. But in higher dimensions neither theorem includes the other. Condition (5.2.10) for all $|\alpha|\le n$ is less restrictive than condition (5.2.9). Thus for functions that are $\mathscr{C}^n$ away from the origin and satisfy condition (5.2.10) for all $|\alpha|\le n$, it is better to use Theorem 5.2.4. However, in Theorem 5.2.7 the function m is assumed only to be $\mathscr{C}^{[n/2]+1}$, requiring almost half the amount of differentiability required by condition (5.2.9).

It should be noted that both theorems have their shortcomings. In particular, they are not L^p sensitive, i.e., delicate enough to detect whether m is bounded on some L^p but not on some other L^q.

Exercises

5.2.1. Let $1 \le k < n$. Use the same idea as in the proof of Proposition 5.2.1 to prove that $m \in \mathscr{M}_p(\mathbf{R}^n)$ if and only if (5.2.1) is satisfied with m_j replaced by

$$m_j(\xi) = m(\xi)\psi(2^{-j_1}\xi_1)\dots\psi(2^{-j_k}\xi_k),$$

where $\psi(\xi)$ is a smooth compactly supported function equal to 1 on the interval $[2^{-1},4]$ that satisfies

$$\sum_{j\in\mathbf{Z}} \psi(2^{-j}\xi) = 1, \qquad \xi \neq 0.$$

5.2.2. (*Calderón reproducing formula*) Let Ψ and Φ be radial Schwartz functions whose Fourier transforms are real-valued and compactly supported away from the origin and satisfy

$$\sum_{j\in\mathbf{Z}} \widehat{\Psi}(2^{-j}\xi)\widehat{\Phi}(2^{-j}\xi) = 1$$

for all $\xi \neq 0$. Prove that for every function f in $\mathscr{S}_\infty(\mathbf{R}^n)$ we have

$$\sum_{j\in\mathbf{Z}} f * \Psi_{2^{-j}} * \Phi_{2^{-j}} = f,$$

where the series converges in $\mathscr{S}_\infty(\mathbf{R}^n)$. Conclude that the identity

$$\sum_{j\in\mathbf{Z}} \Delta_j^\Psi \Delta_j^\Phi = I$$

holds in the sense of $\mathscr{S}'(\mathbf{R}^n)/\mathscr{P}$. Here Δ_j^Ψ is the operator given by convolution with $\Psi_{2^{-j}}$, and Δ_j^Φ is defined likewise.

5.2.3. Consider the differential operators

$$\begin{aligned} L_1 &= \partial_1 - \partial_2^2 + \partial_3^4, \\ L_2 &= \partial_1 + \partial_2^2 + \partial_3^2. \end{aligned}$$

Prove that for every $1 < p < \infty$ there exists a constant $C_p < \infty$ such that for all Schwartz functions f on $\mathbf{R}^3$ we have

$$\begin{aligned} \big\|\partial_2\partial_3^2 f\big\|_{L^p} &\le C_p\big\|L_1(f)\big\|_{L^p}, \\ \big\|\partial_1 f\big\|_{L^p} &\le C_p\big\|L_2(f)\big\|_{L^p}. \end{aligned}$$

[*Hint:* Use Corollary 5.2.5. What is the relevance of multipliers m_1 and m_3 in Example 5.2.6?]

5.2.4. (a) Suppose that $m(\xi)$ is real-valued and satisfies $|\partial^\alpha m(\xi)| \le C_\alpha|\xi|^{-|\alpha|}$ for all multi-indices α satisfying $|\alpha| \le [\frac{n}{2}]+1$ and all $\xi \in \mathbf{R}^n \setminus \{0\}$. Prove that $e^{im(\xi)}$ is

in $\mathscr{M}_p(\mathbf{R}^n)$ for $1 < p < \infty$.
(b) Suppose that $m(\xi)$ is real-valued and satisfies (5.2.9). Prove that $e^{im(\xi)}$ is in $\mathscr{M}_p(\mathbf{R}^n)$ for $1 < p < \infty$.

5.2.5. Suppose that $\varphi(\xi)$ is a smooth function on $\mathbf{R}^n$ that vanishes in a neighborhood of the origin and is equal to 1 in a neighborhood of infinity. Prove that the function $e^{i\xi_j|\xi|^{-1}}\varphi(\xi)$ is in $\mathscr{M}_p(\mathbf{R}^n)$ for $1 < p < \infty$.

5.2.6. Let $\tau, \tau_1, \dots, \tau_n$ be real numbers and $\rho_1, \dots, \rho_n$ be even natural numbers. Prove that the following functions are L^p multipliers on $\mathbf{R}^n$ for $1 < p < \infty$:

$$\begin{aligned} &|\xi_1|^{i\tau_1} \cdots |\xi_n|^{i\tau_n}, \\ &(|\xi_1|^{\rho_1} + \cdots + |\xi_n|^{\rho_n})^{i\tau}, \\ &(|\xi_1|^{-\rho_1} + |\xi_2|^{-\rho_2})^{i\tau}. \end{aligned}$$

5.2.7. Let $\widehat{\zeta}(\xi)$ be a smooth function on the line that is supported in a compact set that does not contain the origin and let a_j be a bounded sequence of complex numbers. Prove that the function

$$m(\xi) = \sum_{j\in\mathbf{Z}} a_j \widehat{\zeta}(2^{-j}\xi)$$

is in $\mathscr{M}_p(\mathbf{R})$ for all $1 < p < \infty$.

5.2.8. Let ζ be as in Exercise 5.2.7 and let $\Delta_j^\zeta(f) = \big(\widehat{f}(\xi)\widehat{\zeta}(2^{-j}\xi)\big)^\vee$. Show that the operator

$$f \to \sup_{N>0} \Big| \sum_{j<N} \Delta_j^\zeta(f) \Big|$$

is bounded on $L^p(\mathbf{R})$ when $1 < p < \infty$.
[*Hint:* Pick a Schwartz function φ satisfying $\sum_{j\in\mathbf{Z}} \widehat{\varphi}(2^{-j}\xi) = 1$ on $\mathbf{R}^n \setminus \{0\}$ with $\widehat{\varphi}$ compactly supported. Then $\Delta_k^\varphi \Delta_j^\zeta = 0$ if $|j-k| < c_0$ and we have

$$\sum_{j<N} \Delta_j^\zeta = \sum_{k<N+c_0} \Delta_k^\varphi \sum_{j<N} \Delta_j^\zeta = \sum_{k<N+c_0} \Delta_k^\varphi \sum_j \Delta_j^\zeta - \sum_{k<N+c_0} \Delta_k^\varphi \sum_{j\ge N} \Delta_j^\zeta,$$

which is a finite sum plus a term controlled by a multiple of the operator

$$f \mapsto M\Big(\sum_{j\in\mathbf{Z}} \Delta_j^\zeta(f)\Big),$$

where M is the Hardy–Littlewood maximal function.]

5.2.9. Let Ψ be a Schwartz function whose Fourier transform is real-valued, supported in a compact set that does not contain the origin, and satisfies

$$\sum_{j\in\mathbf{Z}} \widehat{\Psi}(2^{-j}\xi) = 1 \qquad \text{when } \xi \neq 0.$$

Let Δ_j be the Littlewood–Paley operator associated with Ψ. Prove that

$$\Big\| \sum_{|j|<N} \Delta_j(g) - g \Big\|_{L^p} \to 0$$

as $N \to \infty$ for all functions $g \in \mathscr{S}(\mathbf{R}^n)$. Deduce that Schwartz functions whose Fourier transforms have compact supports that do not contain the origin are dense in $L^p(\mathbf{R}^n)$ for $1 < p < \infty$.
[*Hint:* Use the result of Exercise 5.2.8 and the Lebesgue dominated convergerce theorem.]

5.3 Applications of Littlewood–Paley Theory

We now turn our attention to some important applications of Littlewood–Paley theory. We are interested in obtaining bounds for singular and maximal operators. These bounds are obtained by controlling the corresponding operators by quadratic expressions.

5.3.1 Estimates for Maximal Operators

One way to control the maximal operator $\sup_k |T_k(f)|$ is by introducing a good averaging function φ and using the majorization

$$\begin{aligned} \sup_k |T_k(f)| &\le \sup_k |T_k(f) - f * \varphi_{2^{-k}}| + \sup_k |f * \varphi_{2^{-k}}| \\ &\le \Big(\sum_k |T_k(f) - f * \varphi_{2^{-k}}|^2 \Big)^{\frac{1}{2}} + C_\varphi M(f) \end{aligned} \tag{5.3.1}$$

for some constant C_φ depending on φ. We apply this idea to prove the following theorem.

Theorem 5.3.1. *Let m be a bounded function on $\mathbf{R}^n$ that is $\mathscr{C}^1$ in a neighborhood of the origin and satisfies $m(0) = 1$ and $|m(\xi)| \le C|\xi|^{-\varepsilon}$ for some $C, \varepsilon > 0$ and all $\xi \neq 0$. For each $k \in \mathbf{Z}$ define $T_k(f)(x) = (\widehat{f}(\xi)m(2^{-k}\xi))^\vee(x)$. Then there is a constant C_n such that for all L^2 functions f on $\mathbf{R}^n$ we have*

$$\Big\| \sup_{k\in\mathbf{Z}} |T_k(f)| \Big\|_{L^2} \le C_n \big\| f \big\|_{L^2}. \tag{5.3.2}$$

Proof. Select a Schwartz function φ such that $\widehat{\varphi}(0) = 1$. Then there are positive constants C_1 and C_2 such that $|m(\xi) - \widehat{\varphi}(\xi)| \le C_1|\xi|^{-\varepsilon}$ for $|\xi|$ away from zero and $|m(\xi) - \widehat{\varphi}(\xi)| \le C_2|\xi|$ for $|\xi|$ near zero. These two inequalities imply that

$$\sum_k |m(2^{-k}\xi) - \widehat{\varphi}(2^{-k}\xi)|^2 \le C_3 < \infty,$$

from which the L^2 boundedness of the operator

$$f \mapsto \Big(\sum_k |T_k(f) - f * \varphi_{2^{-k}}|^2\Big)^{1/2}$$

follows easily. Using estimate (5.3.1) and the well-known L^2 estimate for the Hardy–Littlewood maximal function, we obtain (5.3.2). □

If $m(\xi)$ is the characteristic function of a rectangle with sides parallel to the axes, this result can be extended to L^p.

Theorem 5.3.2. *Let $1 < p < \infty$ and let U be the characteristic function of a product of open intervals in $\mathbf{R}^n$ that contain the origin. For each $k \in \mathbf{Z}$ define $T_k(f)(x) = (\widehat{f}(\xi)\chi_U(2^{-k}\xi))^\vee(x)$. Then there is a constant $C_{p,n}$ such that for all L^p functions f on $\mathbf{R}^n$ we have*

$$\big\| \sup_{k\in\mathbf{Z}} |T_k(f)| \big\|_{L^p(\mathbf{R}^n)} \le C_{p,n} \big\| f \big\|_{L^p(\mathbf{R}^n)}.$$

Proof. Let us fix an open annulus A whose interior contains the boundary of U and take a smooth function with compact support $\widehat{\psi}$ that vanishes in a neighborhood of zero and a neighborhood of infinity and is equal to 1 on the annulus A. Then the function $\widehat{\varphi} = (1-\widehat{\psi})\chi_U$ is Schwartz. Since $\chi_U = \chi_U\widehat{\psi} + \widehat{\varphi}$, it follows that for all $f \in L^p(\mathbf{R}^n)$ we have

$$T_k(f) = T_k(f) - f * \varphi_{2^{-k}} + f * \varphi_{2^{-k}} = T_k(f * \psi_{2^{-k}}) + f * \varphi_{2^{-k}}.$$

Taking the supremum over k and using Corollary 2.1.12 we obtain

$$\sup_{k\in\mathbf{Z}} |T_k(f)| \le \Big(\sum_k |T_k(f) - f * \varphi_{2^{-k}}|^2\Big)^{1/2} + C_\varphi M(f). \tag{5.3.3}$$

The operator $T_k(f) - f * \varphi_{2^{-k}}$ is given by multiplication on the Fourier transform side by the multiplier

$$\chi_U(2^{-k}\xi) - \widehat{\varphi}(2^{-k}\xi) = \chi_U(2^{-k}\xi)\widehat{\psi}(2^{-k}\xi) = \chi_{2^kU}(\xi)\widehat{\psi}(2^{-k}\xi).$$

Since $\{2^kU\}_{k\in\mathbf{Z}}$ is a measurable family of rectangles with sides parallel to the axes, Exercise 4.6.1(b) yields the following inequality:

$$\Big\| \Big(\sum_{k\in\mathbf{Z}} |T_k(f) - f * \varphi_{2^{-k}}|^2\Big)^{\frac{1}{2}} \Big\|_{L^p} \le C_{p,n} \Big\| \Big(\sum_{k\in\mathbf{Z}} |f * \psi_{2^{-k}}|^2\Big)^{\frac{1}{2}} \Big\|_{L^p}. \tag{5.3.4}$$

Since $f * \psi_{2^{-k}} = \Delta_j^\psi(f)$, estimate (5.1.4) of Theorem 5.1.2 yields that the expression on the right in (5.3.4) is controlled by a multiple of $\big\|f\big\|_{L^p}$. Taking L^p norms in (5.3.3) and using the L^p estimate for the square function yields the required conclusion. □

The following lacunary version of the Carleson–Hunt theorem is yet another indication of the powerful techniques of Littlewood–Paley theory.

Corollary 5.3.3. *(a) Let f be in $L^2(\mathbf{R}^n)$ and let Ω be an open set that contains the origin in $\mathbf{R}^n$. Then*

$$\lim_{k\to\infty}\int_{2^k\Omega}\widehat{f}(\xi)e^{2\pi i x\cdot\xi}\,d\xi = f(x)$$

for almost all $x \in \mathbf{R}^n$.
(b) Let f be in $L^p(\mathbf{R}^n)$ for some $1 < p < \infty$. Then

$$\lim_{k\to\infty}\int_{\substack{|\xi_1|<2^k\\ \cdots\\ |\xi_n|<2^k}}\widehat{f}(\xi)e^{2\pi i x\cdot\xi}\,d\xi = f(x)$$

for almost all $x \in \mathbf{R}^n$.

Proof. Both limits exist everywhere for functions f in the Schwartz class. To obtain almost everywhere convergence for general f in L^p we appeal to Theorem 2.1.14. The required control of the corresponding maximal operator is a consequence of Theorem 5.3.1 in case (a) and Theorem 5.3.2 in case (b). □

5.3.2 Estimates for Singular Integrals with Rough Kernels

We now turn to another application of the Littlewood–Paley theory involving singular integrals.

Theorem 5.3.4. *Suppose that μ is a finite Borel measure on $\mathbf{R}^n$ with compact support that satisfies $|\widehat{\mu}(\xi)| \le B\min\left(|\xi|^{-b}, |\xi|^{b}\right)$ for some $b > 0$ and all $\xi \neq 0$. Define measures μ_j by setting $\widehat{\mu_j}(\xi) = \widehat{\mu}(2^{-j}\xi)$. Then the operator*

$$T_\mu(f)(x) = \sum_{j\in\mathbf{Z}}(f * \mu_j)(x)$$

is bounded on $L^p(\mathbf{R}^n)$ for all $1 < p < \infty$.

Proof. It is natural to begin with the L^2 boundedness of T_μ. The estimate on $\widehat{\mu}$ implies that

$$\sum_{j\in\mathbf{Z}}|\widehat{\mu}(2^{-j}\xi)| \le \sum_{j\in\mathbf{Z}}B\min\left(|2^{-j}\xi|^{b}, |2^{-j}\xi|^{-b}\right) \le C_b B < \infty. \tag{5.3.5}$$

The L^2 boundedness of T_μ is an immediate consequence of (5.3.5).

We now turn to the L^p boundedness of T_μ for $1 < p < \infty$. We fix a radial Schwartz function ψ whose Fourier transform is supported in the annulus $\frac{1}{2} < |\xi| < 2$ that satisfies

$$\sum_{j\in\mathbf{Z}}\widehat{\psi}(2^{-j}\xi) = 1 \tag{5.3.6}$$

whenever $\xi \neq 0$. We let $\psi_{2^{-k}}(x) = 2^{kn}\psi(2^k x)$, so that $\widehat{\psi_{2^{-k}}}(\xi) = \widehat{\psi}(2^{-k}\xi)$, and we observe that the identity

$$\mu_j = \sum_{k\in\mathbf{Z}} \mu_j * \psi_{2^{-j-k}}$$

is valid by taking Fourier transforms and using (5.3.6). We now define operators S_k by setting

$$S_k(f) = \sum_{j\in\mathbf{Z}} \mu_j * \psi_{2^{-j-k}} * f = \sum_{j\in\mathbf{Z}} (\mu * \psi_{2^{-k}})_{2^{-j}} * f.$$

Then for nice f we have that

$$T_\mu(f) = \sum_{j\in\mathbf{Z}} \mu_j * f = \sum_{j\in\mathbf{Z}}\sum_{k\in\mathbf{Z}} \mu_j * \psi_{2^{-j-k}} * f = \sum_{k\in\mathbf{Z}} S_k(f).$$

It suffices therefore to obtain L^p boundedness for the sum of the S_k's. We begin by investigating the L^2 boundedness of each S_k. Since the product $\widehat{\psi_{2^{-j-k}}}\,\widehat{\psi_{2^{-j'-k}}}$ is nonzero only when $j' \in \{j-1, j, j+1\}$, it follows that

$$\begin{aligned}
\big\|S_k(f)\big\|_{L^2}^2 &\le \sum_{j\in\mathbf{Z}}\sum_{j'\in\mathbf{Z}} \int_{\mathbf{R}^n} |\widehat{\mu_j}(\xi)\overline{\widehat{\mu_{j'}}(\xi)}\,\widehat{\psi}(2^{-j-k}\xi)\overline{\widehat{\psi}(2^{-j'-k}\xi)}|\,|\widehat{f}(\xi)|^2\,d\xi \\
&\le C_1 \sum_{j\in\mathbf{Z}}\sum_{j'=j-1}^{j+1} \int_{|\xi|\approx 2^{j+k}} |\widehat{\mu_j}(\xi)\overline{\widehat{\mu_{j'}}(\xi)}|\,|\widehat{f}(\xi)|^2\,d\xi \\
&\le C_2 \sum_{j\in\mathbf{Z}} \int_{|\xi|\approx 2^{j+k}} B^2 \min(|2^{-j}\xi|^b, |2^{-j}\xi|^{-b})^2 |\widehat{f}(\xi)|^2\,d\xi \\
&\le C_3^2 B^2 2^{-2|k|b} \sum_{j\in\mathbf{Z}} \int_{|\xi|\approx 2^{j+k}} |\widehat{f}(\xi)|^2\,d\xi \\
&= C_3^2 B^2\, 2^{-2|k|b} \big\|f\big\|_{L^2}^2 .
\end{aligned}$$

We have therefore obtained that for all $k \in \mathbf{Z}$ and $f \in \mathscr{S}(\mathbf{R}^n)$ we have

$$\big\|S_k(f)\big\|_{L^2} \le C_3\, B\, 2^{-b|k|} \big\|f\big\|_{L^2}. \tag{5.3.7}$$

Next we show that the kernel of each S_k satisfies Hörmander's condition with constant at most a multiple of $(1+|k|)$. Fix $y \neq 0$. Then

$$\begin{aligned}
&\int_{|x|\ge 2|y|} \bigg| \sum_{j\in\mathbf{Z}} \Big((\mu * \psi_{2^{-k}})_{2^{-j}}(x-y) - (\mu * \psi_{2^{-k}})_{2^{-j}}(x) \Big) \bigg|\,dx \\
&\qquad \le \sum_{j\in\mathbf{Z}} \int_{|x|\ge 2|y|} 2^{jn} \big| (\mu * \psi_{2^{-k}})(2^j x - 2^j y) - (\mu * \psi_{2^{-k}})(2^j x) \big|\,dx \\
&\qquad = \sum_{j\in\mathbf{Z}} I_{j,k}(y),
\end{aligned}$$

where

$$I_{j,k}(y) = \int_{|x|\ge 2^{j+1}|y|} \big|(\mu * \psi_{2^{-k}})(x-2^j y) - (\mu * \psi_{2^{-k}})(x)\big|\,dx.$$

We observe that $I_{j,k}(y) \le C_4\big\|\mu\big\|_{\mathscr{M}}$. To obtain a more delicate estimate for $I_{j,k}(y)$ we argue as follows:

$$\begin{aligned}
I_{j,k}(y) &\le \int\limits_{|x|\ge 2^{j+1}|y|} \int_{\mathbf{R}^n} \big|\psi_{2^{-k}}(x-2^j y - z) - \psi_{2^{-k}}(x-z)\big|\,d\mu(z)\,dx \\
&= \int_{\mathbf{R}^n} 2^{kn} \int\limits_{|x|\ge 2^{j+1}|y|} \big|\psi(2^k x - 2^k z - 2^{j+k} y) - \psi(2^k x - 2^k z)\big|\,dx\,d\mu(z) \\
&\le C_5 \int\limits_{|x|\ge 2^{j+1}|y|} \int_{\mathbf{R}^n} 2^{kn} 2^{j+k}|y|\big|\nabla\psi(2^k x - 2^k z - \theta)\big|\,d\mu(z)\,dx \\
&\le C_6 2^{j+k} \int_{\mathbf{R}^n} \int\limits_{|x|\ge 2^{j+1}|y|} 2^{kn}|y|\big(1+|2^k x - 2^k z - \theta|\big)^{-n-2}\,dx\,d\mu(z) \\
&= C_6 2^{j+k}|y| \int_{\mathbf{R}^n} \int\limits_{|x|\ge 2^{j+k+1}|y|} \big(1+|x - 2^k z - \theta|\big)^{-n-2}\,dx\,d\mu(z),
\end{aligned}$$

where $|\theta| \le 2^{j+k}|y|$. Note that θ depends on j,k, and y. From this and from $I_{j,k}(y) \le C_4\big\|\mu\big\|_{\mathscr{M}}$ we obtain

$$I_{j,k}(y) \le C_7\,\big\|\mu\big\|_{\mathscr{M}} \min\big(1, 2^{j+k}|y|\big), \tag{5.3.8}$$

which is valid for all j,k, and $y \neq 0$. To estimate the last double integral even more delicately, we consider the following two cases: $|x| \ge 2^{k+2}|z|$ and $|x| < 2^{k+2}|z|$. In the first case we have $|x - 2^k z - \theta| \ge \frac{1}{4}|x|$, given the fact that $|x| \ge 2^{j+k+1}|y|$. In the second case we have that $|x| \le 2^{k+2}R$, where $B(0,R)$ contains the support of μ. Applying these observations in the last double integral, we obtain the following estimate:

$$\begin{aligned}
I_{j,k}(y) &\le C_8 2^{j+k}|y| \int_{\mathbf{R}^n} \Bigg[\int\limits_{\substack{|x|\ge 2^{j+k+1}|y| \\ |x|\ge 2^{k+2}|z|}} \frac{dx}{\big(1+\frac{1}{4}|x|\big)^{n+2}} + \int\limits_{\substack{|x|\ge 2^{j+k+1}|y| \\ |x|<2^{k+2}R}} dx \Bigg]\,d\mu(z) \\
&\le C_9 2^{j+k}|y|\big\|\mu\big\|_{\mathscr{M}} \left[\frac{1}{(2^{j+k}|y|)^2} + 0\right] \\
&= C_9 (2^{j+k}|y|)^{-1}\big\|\mu\big\|_{\mathscr{M}},
\end{aligned}$$

provided $2^j|y| \ge 2R$. Combining this estimate with (5.3.8), we obtain

$$I_{j,k}(y) \le C_{10}\big\|\mu\big\|_{\mathscr{M}} \begin{cases} \min\big(1, 2^{j+k}|y|\big) & \text{for all } j,k \text{ and } y, \\ (2^{j+k}|y|)^{-1} & \text{when } 2^j|y| \ge 2R. \end{cases} \tag{5.3.9}$$

We now estimate $\sum_j I_{j,k}(y)$. When $2^k \geq (2R)^{-1}$ we use (5.3.9) to obtain

$$\sum_j I_{j,k}(y) \leq C_{10}\|\mu\|_{\mathscr{M}}\Big[\sum_{2^j\leq\frac{1}{2^k|y|}} 2^{j+k}|y| + \sum_{\frac{1}{2^k|y|}\leq 2^j\leq\frac{2R}{|y|}} 1 + \sum_{2^j\geq\frac{2R}{|y|}} (2^{j+k}|y|)^{-1}\Big]$$
$$\leq C_{11}\|\mu\|_{\mathscr{M}}(|\log R| + |k|).$$

Also when $2^k < (2R)^{-1}$ we again use (5.3.9) to obtain

$$\sum_j I_{j,k}(y) \leq C_{10}\|\mu\|_{\mathscr{M}}\Big[\sum_{2^j\leq\frac{1}{2^k|y|}} 2^{j+k}|y| + \sum_{2^j\geq\frac{1}{2^k|y|}} (2^{j+k}|y|)^{-1}\Big] \leq C_{12}\|\mu\|_{\mathscr{M}},$$

since in the second sum $2^j|y| \geq 2^{-k} > 2R$, which justifies use of the corresponding estimate in (5.3.9). This gives

$$\sum_j I_{j,k}(y) \leq C_{13}\|\mu\|_{\mathscr{M}}(1+|k|), \tag{5.3.10}$$

where the constant C_{13} depends on the dimension and on R. We now use estimates (5.3.7) and (5.3.10) and Theorem 4.3.3 to obtain that each S_k maps $L^1(\mathbf{R}^n)$ to $L^{1,\infty}(\mathbf{R}^n)$ with constant at most

$$C_n(2^{-b|k|}+1+|k|)\|\mu\|_{\mathscr{M}} \leq C_n(2+|k|)\|\mu\|_{\mathscr{M}}.$$

It follows from the Marcinkiewicz interpolation theorem (Theorem 1.3.2) that S_k maps $L^p(\mathbf{R}^n)$ to itself for $1<p<2$ with bound at most $C_{p,n}2^{-b|k|\theta_p}(1+|k|)^{1-\theta_p}$, when $\frac{1}{p} = \frac{\theta_p}{2} + 1 - \theta_p$. Summing over all $k \in \mathbf{Z}$, we obtain that T_μ maps $L^p(\mathbf{R}^n)$ to itself for $1<p<2$. The boundedness of T_μ for $p>2$ follows by duality. □

An immediate consequence of the previous result is the following.

Corollary 5.3.5. *Let μ_j be as in the previous theorem. Then the square function*

$$G(f) = \Big(\sum_{j\in\mathbf{Z}} |\mu_j * f|^2\Big)^{\frac{1}{2}} \tag{5.3.11}$$

maps $L^p(\mathbf{R}^n)$ to itself whenever $1<p<\infty$.

Proof. To obtain the boundedness of the square function in (5.3.11) we use the Rademacher functions $r_j(t)$, introduced in Appendix C.1, reindexed so that their index set is the set of all integers (not the set of nonnegative integers). For each t we introduce the operators

$$T_\mu^t(f) = \sum_{j\in\mathbf{Z}} r_j(t)(f*\mu_j).$$

Next we observe that for each t in $[0,1]$ the operators T_μ^t map $L^p(\mathbf{R}^n)$ to itself with the same constant as the operator T_μ, which is in particular independent of t. Using that the square function in (5.3.11) raised to the power p is controlled by a multiple of the quantity

$$\int_0^1 \Big| \sum_{j\in\mathbf{Z}} r_j(t)(f*\mu_j)\Big|^p dt\,,$$

a fact stated in Appendix C.2, we obtain the required conclusion by integrating over $\mathbf{R}^n$. □

5.3.3 An Almost Orthogonality Principle on L^p

Suppose that T_j are multiplier operators given by $T_j(f) = (\widehat{f}m_j)^\vee$, for some multipliers m_j. If the functions m_j have disjoint supports and they are bounded uniformly in j, then the operator

$$T = \sum_j T_j$$

is bounded on L^2. The following theorem gives an L^p analogue of this result.

Theorem 5.3.6. *Suppose that $1 < p \le 2 \le q < \infty$. Let m_j be Schwartz functions supported in the annuli $2^{j-1} \le |\xi| \le 2^{j+1}$ and let $T_j(f) = (\widehat{f}m_j)^\vee$. Suppose that the T_j's are uniformly bounded operators from $L^p(\mathbf{R}^n)$ to $L^q(\mathbf{R}^n)$, i.e.,*

$$\sup_j \big\|T_j\big\|_{L^p\to L^q} = A < \infty\,.$$

Then for each $f \in L^p(\mathbf{R}^n)$, the series

$$T(f) = \sum_j T_j(f)$$

converges in the L^q norm and there exists a constant $C_{p,q,n} < \infty$ such that

$$\big\|T\big\|_{L^p\to L^q} \le C_{p,q,n}A. \tag{5.3.12}$$

Proof. Fix a radial Schwartz function φ whose Fourier transform $\widehat{\varphi}$ is real, equal to one on the annulus $\frac{1}{2} \le |\xi| \le 2$, and vanishes outside the annulus $\frac{1}{4} \le |\xi| \le 4$. We set $\varphi_{2^{-j}}(x) = 2^{jn}\varphi(2^j x)$, so that $\widehat{\varphi_{2^{-j}}}$ is equal to 1 on the support of each m_j. Setting $\Delta_j(f) = f*\varphi_{2^{-j}}$, we observe that

$$T_j = \Delta_j T_j \Delta_j$$

for all $j \in \mathbf{Z}$. For a positive integer N we set

$$T^N = \sum_{|j|\le N} \Delta_j T_j \Delta_j\,.$$

Fix $f \in L^p(\mathbf{R}^n)$. Clearly for every N, $T^N(f)$ is in $L^q(\mathbf{R}^n)$. Using (5.1.20) we obtain

$$
\begin{aligned}
\big\|T^N(f)\big\|_{L^q} &= \Big\| \sum_{|j|\le N} \Delta_j T_j \Delta_j(f)\Big\|_{L^q} \\
&\le C'_q \Big\| \Big(\sum_{j\in\mathbf{Z}} |T_j\Delta_j(f)|^2\Big)^{\frac12}\Big\|_{L^q} \\
&= C'_q \Big\| \sum_{j\in\mathbf{Z}} |T_j\Delta_j(f)|^2 \Big\|_{L^{q/2}}^{\frac12} \\
&\le C'_q \Big(\sum_{j\in\mathbf{Z}} \Big\| |T_j\Delta_j(f)|^2 \Big\|_{L^{q/2}}\Big)^{\frac12} \\
&= C'_q \Big(\sum_{j\in\mathbf{Z}} \big\| T_j\Delta_j(f)\big\|_{L^q}^2 \Big)^{\frac12},
\end{aligned}
$$

where we used Minkowski's inequality, since $q/2 \ge 1$. Using the uniform boundedness of the T_j's from L^p to L^q, we deduce that

$$
\begin{aligned}
C'_q \Big(\sum_{j\in\mathbf{Z}} \big\| T_j\Delta_j(f)\big\|_{L^q}^2 \Big)^{\frac12} &\le C'_q A \Big(\sum_{j\in\mathbf{Z}} \big\| \Delta_j(f)\big\|_{L^p}^2 \Big)^{\frac12} \\
&= C'_q A \Big(\sum_{j\in\mathbf{Z}} \big\| |\Delta_j(f)|^2\big\|_{L^{p/2}} \Big)^{\frac12} \\
&\le C'_q A \Big(\Big\| \sum_{j\in\mathbf{Z}} |\Delta_j(f)|^2\Big\|_{L^{p/2}} \Big)^{\frac12} \\
&= C'_q A \Big\| \Big(\sum_{j\in\mathbf{Z}} |\Delta_j(f)|^2\Big)^{\frac12} \Big\|_{L^p} \\
&\le C'_q C_p A \big\| f \big\|_{L^p(\mathbf{R}^n)},
\end{aligned}
$$

where we used the result of Exercise 1.1.5(b), since $p \le 2$, and Theorem 5.1.2. We conclude that the operators T^N are uniformly bounded from $L^p(\mathbf{R}^n)$ to $L^q(\mathbf{R}^n)$.

If $\widehat{h}$ is compactly supported in a subset of $\mathbf{R}^n \setminus \{0\}$, then the sequence $T^N(h)$ becomes independent of N for N large enough and hence it is Cauchy in L^q. But in view of Exercise 5.2.9, the set of all such h is dense in $L^p(\mathbf{R}^n)$. Combining these two results with the uniform boundedness of the T^N's from L^p to L^q, a simple $\frac{\varepsilon}{3}$ argument gives that for all $f \in L^p$ the sequence $T^N(f)$ is Cauchy in L^q. Therefore, for all $f \in L^p$ the sequence $\{T^N(f)\}_N$ converges in L^q to some $T(f)$. Fatou's lemma gives

$$
\big\|T(f)\big\|_{L^q} \le C'_q C_p A \big\|f\big\|_{L^p},
$$

which proves (5.3.12). □

Exercises

5.3.1. (*The g-function*) Let $P_t(x) = \frac{\Gamma(\frac{n+1}{2})}{\pi^{\frac{n+1}{2}}} \frac{t}{(t^2+|x|^2)^{\frac{n+1}{2}}}$ be the Poisson kernel.
(a) Use Exercise 5.1.4 with $\Psi(x) = \frac{\partial}{\partial t} P_t(x)\big|_{t=1}$ to obtain that the operator

$$f \to \left(\int_0^\infty t \left| \tfrac{\partial}{\partial t}(P_t * f)(x) \right|^2 dt \right)^{1/2}$$

is L^p bounded for $1 < p < \infty$.
(b) Use Exercise 5.1.3 with $\Psi(x) = \partial_k P_1(x)$ to obtain that the operator

$$f \to \left(\int_0^\infty t |\partial_k (P_t * f)(x)|^2 dt \right)^{1/2}$$

is L^p bounded for $1 < p < \infty$.
(c) Conclude that the g-function

$$g(f)(x) = \left(\int_0^\infty t |\nabla_{x,t} (P_t * f)(x)|^2 dt \right)^{1/2}$$

is L^p bounded for $1 < p < \infty$.

5.3.2. Suppose that μ is a finite Borel measure on $\mathbf{R}^n$ with compact support that satisfies $\widehat{\mu}(0) = 0$ and $|\widehat{\mu}(\xi)| \le C|\xi|^{-a}$ for some $a > 0$ and all $\xi \neq 0$. Define measures μ_j by setting $\widehat{\mu_j}(\xi) = \widehat{\mu}(2^{-j}\xi)$. Show that the operator

$$T_\mu(f)(x) = \sum_{j \in \mathbf{Z}} (f * \mu_j)(x)$$

is bounded on L^p for all $1 < p < \infty$.
[*Hint:* Use Theorem 5.3.4]

5.3.3. (*Calderón [41]/Coifman and Weiss [56]*) (a) Suppose that μ is a finite Borel measure with compact support that satisfies $|\widehat{\mu}(\xi)| \le C|\xi|^{-a}$ for some $a > 0$ and all $\xi \neq 0$. Then the maximal function

$$\mathscr{M}_\mu(f)(x) = \sup_{j \in \mathbf{Z}} \left| \int_{\mathbf{R}^n} f(x - 2^j y)\, d\mu(y) \right|$$

is bounded on L^p for all $1 < p < \infty$.
(b) Let μ be the surface measure on the sphere $\mathbf{S}^{n-1}$ when $n \ge 2$. Conclude that the *dyadic spherical maximal function* $\mathscr{M}_\mu$ is bounded on $L^p(\mathbf{R}^n)$ for all $1 < p < \infty$ whenever $n \ge 2$.
[*Hint:* Pick φ a compactly supported smooth function on $\mathbf{R}^n$ with $\widehat{\varphi}(0) = 1$. Then the measure $\sigma = \mu - \widehat{\mu}(0)\varphi$ satisfies the hypotheses of Corollary 5.3.5. But it is straightforward that

$$\mathscr{M}_\mu(f)(x) \le \Big(\sum_j |(\sigma_j * f)(x)|^2\Big)^{1/2} + |\widehat{\mu}(0)| M(f)(x),$$

from which it follows that $\mathscr{M}_\mu$ is bounded on $L^p(\mathbf{R}^n)$ whenever $1 < p < \infty$. Now let $\mu = d\sigma$ be surface measure on $\mathbf{S}^{n-1}$. It follows from the results in Appendices B.4 and B.7 that $|\widehat{d\sigma}(\xi)| \le C|\xi|^{-\frac{n-1}{2}}$.]

5.3.4. Let Ω be in $L^q(\mathbf{S}^{n-1})$ for some $1 < q < \infty$ and define the absolutely continuous measure

$$d\mu(x) = \frac{\Omega(x/|x|)}{|x|^n} \chi_{1<|x|\le 2}\, dx.$$

Show that for all $a < 1/q'$ we have that $|\widehat{\mu}(\xi)| \le C|\xi|^{-a}$. Under the additional hypothesis that Ω has mean value zero, conclude that the singular integral operator

$$T_\Omega(f)(x) = \text{p.v.} \int_{\mathbf{R}^n} \frac{\Omega(y/|y|)}{|y|^n} f(x-y)\, dy = \sum_j f * \mu_j$$

is L^p bounded for all $1 < p < \infty$. This provides an alternative proof of Theorem 4.2.10 under the hypothesis that $\Omega \in L^q(\mathbf{S}^{n-1})$.

5.3.5. For a function F on $\mathbf{R}$ define

$$u(F)(x) = \left(\int_0^\infty |F(x+t) + F(x-t) - 2F(x)|^2 \frac{dt}{t^3}\right)^{1/2}.$$

Given $f \in L^1_{\text{loc}}(\mathbf{R})$ we denote by F the indefinite integral of f, that is,

$$F(x) = \int_0^x f(t)\, dt.$$

Prove that for all $1 < p < \infty$ there exist constants c_p and C_p such that

$$c_p \big\|f\big\|_{L^p} \le \big\|u(F)\big\|_{L^p} \le C_p \big\|f\big\|_{L^p}.$$

[*Hint:* Let $\varphi = \chi_{[-1,0]} - \chi_{[0,1]}$. Then

$$(\varphi_t * f)(x) = \frac{1}{t}\big(F(x+t) + F(x-t) - 2F(x)\big)$$

and you may use Exercise 5.1.4.]

5.3.6. Let $m \in \mathscr{M}_p(\mathbf{R}^n)$. Define an operator T_t by setting $\widehat{T_t(f)}(\xi) = \widehat{f}(\xi) m(t\xi)$. Show that the maximal operator

$$\sup_{N>0} \left(\frac{1}{N}\int_0^N \big|T_t(f)(x)\big|^2\, dt\right)^{\frac{1}{2}}$$

maps $L^p(\mathbf{R}^n)$ to itself for all $1 < p < \infty$.
[*Hint:* Majorize this maximal operator by a constant multiple of the sum

$$M(f)(x) + \Big(\int_0^\infty |T_t(f)(x) - (f * \varphi_t)(x)|^2 \frac{dt}{t}\Big)^{\frac{1}{2}}$$

for a suitable function φ.]

5.3.7. (Nazarov and Seeger [206]) Let $0 < \beta < 1$ and $p_0 = (1-\beta/2)^{-1}$. Suppose that $\{f_j\}_{j\in\mathbf{Z}}$ are L^2 functions on the line with norm at most 1 that are supported in possibly different intervals of length 1. Assume that the f_j's satisfy the orthogonality relation $\big|\langle f_j \,|\, f_k\rangle\big| \le (1+|j-k|)^{-\beta}$ for all $j,k\in\mathbf{Z}$.
(a) Let $I \subseteqq \mathbf{Z}$ be such that for all $j \in I$ the functions f_j are supported in a fixed interval of length 3. Show that for all p satisfying $0 < p \le 2$ there is $C_{p,\beta} < \infty$ such that

$$\Big\|\sum_{j\in I} \varepsilon_j f_j\Big\|_{L^p} \le C_{p,\beta}\, |I|^{1-\frac{\beta}{2}}$$

whenever ε_j are complex numbers with $|\varepsilon_j| \le 1$.
(b) Under the same hypothesis as in part (a), prove that for all $0 < p < p_0$ there is a constant $C'_{p,\beta} < \infty$ such that

$$\Big\|\sum_{j\in I} c_j f_j\Big\|_{L^p} \le C'_{p,\beta}\Big(\sum_{j\in\mathbf{Z}} |c_j|^p\Big)^{\frac{1}{p}}$$

for all complex-valued sequences $\{c_j\}_j$ in ℓ^p.
(c) Derive the conclusion of part (b) without the assumption that the f_j are supported in a fixed interval of length 3.
[*Hint:* Part (a): Pass from L^p to L^2 and use the hypothesis. Part (b): Assume $\sum_{j\in\mathbf{Z}}|c_j|^p = 1$. For each $k = 0,1,\dots,$ set $I_k = \{j \in \mathbf{Z} :\, 2^{-k-1} < |c_j| \le 2^{-k}\}$. Write $\big\|\sum_{j\in\mathbf{Z}} c_j f_j\big\|_{L^p} \le \sum_{k=0}^\infty 2^{-k}\big\|\sum_{j\in I_k}(c_j 2^k) f_j\big\|_{L^p}$, use part (b), Hölder's inequality, and the fact that $\sum_{k=0}^\infty 2^{-kp}|I_k| \le 2^p$. Part (c): Write $\sum_{j\in\mathbf{Z}} c_j f_j = \sum_{m\in\mathbf{Z}} F_m$, where F_m is the sum of $c_j f_j$ over all j such that the support of f_j meets the interval $[m,m+1]$. These F_m's are supported in $[m-1,m+2]$ and are almost orthogonal.]

5.4 The Haar System, Conditional Expectation, and Martingales

There is a very strong connection between the Littlewood–Paley operators and certain notions from probability, such as conditional expectation and martingale difference operators. The conditional expectation we are concerned with is with respect to the increasing σ-algebra of all dyadic cubes on $\mathbf{R}^n$.

5.4.1 Conditional Expectation and Dyadic Martingale Differences

We recall the definition of dyadic cubes.

Definition 5.4.1. A *dyadic interval* in $\mathbf{R}$ is an interval of the form

$$\big[m2^{-k},(m+1)2^{-k}\big)$$

where m,k are integers. A *dyadic cube* in $\mathbf{R}^n$ is a product of dyadic intervals of the same length. That is, a dyadic cube is a set of the form

$$\prod_{j=1}^{n}\big[m_j2^{-k},(m_j+1)2^{-k}\big)$$

for some integers $m_1,\dots,m_n,k$.

We defined dyadic intervals to be closed on the left and open on the right, so that different dyadic intervals of the same length are always disjoint sets.

Given a cube Q in $\mathbf{R}^n$ we denote by $|Q|$ its Lebesgue measure and by $\ell(Q)$ its side length. We clearly have $|Q|=\ell(Q)^n$. We introduce some more notation.

Definition 5.4.2. For $k\in\mathbf{Z}$ we denote by $\mathscr{D}_k$ the set of all dyadic cubes in $\mathbf{R}^n$ whose side length is 2^{-k}. We also denote by $\mathscr{D}$ the set of all dyadic cubes in $\mathbf{R}^n$. Then we have

$$\mathscr{D}=\bigcup_{k\in\mathbf{Z}}\mathscr{D}_k\,,$$

and moreover, the σ-algebra $\sigma(\mathscr{D}_k)$ of measurable subsets of $\mathbf{R}^n$ formed by countable unions and complements of elements of $\mathscr{D}_k$ is increasing as k increases.

We observe the fundamental property of dyadic cubes, which clearly justifies their usefulness. Any two dyadic intervals of the same side length either are disjoint or coincide. Moreover, either two given dyadic intervals are disjoint, or one contains the other. Similarly, either two dyadic cubes are disjoint, or one contains the other.

Definition 5.4.3. Given a locally integrable function f on $\mathbf{R}^n$, we let

$$\operatorname*{Avg}_{Q} f=\frac{1}{|Q|}\int_Q f(t)\,dt$$

denote the average of f over a cube Q.

The *conditional expectation* of a locally integrable function f on $\mathbf{R}^n$ with respect to the increasing family of σ-algebras $\sigma(\mathscr{D}_k)$ generated by $\mathscr{D}_k$ is defined as

$$E_k(f)(x)=\sum_{Q\in\mathscr{D}_k}\big(\operatorname*{Avg}_{Q} f\big)\,\chi_Q(x),$$

for all $k\in\mathbf{Z}$. We also define the *dyadic martingale difference operator* D_k as follows:

$$D_k(f) = E_k(f) - E_{k-1}(f),$$

also for $k \in \mathbf{Z}$.

Next we introduce the family of Haar functions.

Definition 5.4.4. For a dyadic interval $I = [m2^{-k}, (m+1)2^{-k})$ we define $I_L = [m2^{-k}, (m+\frac{1}{2})2^{-k})$ and $I_R = [(m+\frac{1}{2})2^{-k}, (m+1)2^{-k})$ to be the left and right parts of I, respectively. The function

$$h_I(x) = |I|^{-\frac{1}{2}}\chi_{I_L} - |I|^{-\frac{1}{2}}\chi_{I_R}$$

is called the *Haar function associated with the interval I.*

We remark that Haar functions are constructed in such a way that they have L^2 norm equal to 1. Moreover, the Haar functions have the following fundamental orthogonality property:

$$\int_{\mathbf{R}} h_I(x)h_{I'}(x)\,dx = \begin{cases} 0 & \text{when } I \neq I', \\ 1 & \text{when } I = I'. \end{cases} \tag{5.4.1}$$

To see this, observe that the Haar functions have L^2 norm equal to 1 by construction. Moreover, if $I \neq I'$, then I and I' must have different lengths, say we have $|I'| < |I|$. If I and I' are not disjoint, then I' is contained either in the left or in the right half of I, on either of which h_I is constant. Thus (5.4.1) follows.

We recall the notation

$$\langle f, g\rangle = \int_{\mathbf{R}} f(x)g(x)\,dx$$

valid for square integrable functions. Under this notation, (5.4.1) can be rewritten as $\langle h_I, h_{I'}\rangle = \delta_{I,I'}$, where the latter is 1 when $I = I'$ and zero otherwise.

5.4.2 *Relation Between Dyadic Martingale Differences and Haar Functions*

We have the following result relating the Haar functions to the dyadic martingale difference operators.

Proposition 5.4.5. *For every locally integrable function f on $\mathbf{R}$ and for all $k \in \mathbf{Z}$ we have the identity*

$$D_k(f) = \sum_{I \in \mathscr{D}_{k-1}} \langle f, h_I\rangle h_I \tag{5.4.2}$$

and also

$$\big\|D_k(f)\big\|_{L^2}^2 = \sum_{I \in \mathscr{D}_{k-1}} \big|\langle f, h_I\rangle\big|^2. \tag{5.4.3}$$

Proof. We observe that every interval J in $\mathscr{D}_k$ is either an I_L or an I_R for some unique $I \in \mathscr{D}_{k-1}$. Thus we can write

$$\begin{aligned} E_k(f) &= \sum_{J\in\mathscr{D}_k} (\operatorname{Avg}_J f)\,\chi_J \\ &= \sum_{I\in\mathscr{D}_{k-1}} \left[\left(\frac{2}{|I|}\int_{I_L} f(t)\,dt\right)\chi_{I_L} + \left(\frac{2}{|I|}\int_{I_R} f(t)\,dt\right)\chi_{I_R}\right]. \end{aligned} \tag{5.4.4}$$

But we also have

$$\begin{aligned} E_{k-1}(f) &= \sum_{I\in\mathscr{D}_{k-1}} (\operatorname{Avg}_I f)\,\chi_I \\ &= \sum_{I\in\mathscr{D}_{k-1}} \left(\frac{1}{|I|}\int_{I_L} f(t)\,dt + \frac{1}{|I|}\int_{I_R} f(t)\,dt\right)(\chi_{I_L}+\chi_{I_R}). \end{aligned} \tag{5.4.5}$$

Now taking the difference between (5.4.4) and (5.4.5) we obtain

$$\begin{aligned} D_k(f) = \sum_{I\in\mathscr{D}_{k-1}} \Bigg[&\left(\frac{1}{|I|}\int_{I_L} f(t)\,dt\right)\chi_{I_L} - \left(\frac{1}{|I|}\int_{I_R} f(t)\,dt\right)\chi_{I_L} \\ &+ \left(\frac{1}{|I|}\int_{I_R} f(t)\,dt\right)\chi_{I_R} - \left(\frac{1}{|I|}\int_{I_L} f(t)\,dt\right)\chi_{I_R}\Bigg], \end{aligned}$$

which is easily checked to be equal to

$$\sum_{I\in\mathscr{D}_{k-1}} \left(\int_I f(t)h_I(t)\,dt\right)h_I = \sum_{I\in\mathscr{D}_{k-1}} \langle f,h_I\rangle\, h_I.$$

Finally, (5.4.3) is a consequence of (5.4.1). □

Theorem 5.4.6. *Every function $f \in L^2(\mathbf{R}^n)$ can be written as*

$$f = \sum_{k\in\mathbf{Z}} D_k(f), \tag{5.4.6}$$

where the series converges almost everywhere and in L^2. We also have

$$\|f\|_{L^2}^2 = \sum_{k\in\mathbf{Z}} \|D_k(f)\|_{L^2}^2. \tag{5.4.7}$$

Moreover, when $n = 1$ we have the representation

$$f = \sum_{I\in\mathscr{D}} \langle f,h_I\rangle h_I, \tag{5.4.8}$$

where the sum converges a.e. and in L^2 and also

$$\|f\|_{L^2(\mathbf{R})}^2 = \sum_{I\in\mathscr{D}} |\langle f,h_I\rangle|^2. \tag{5.4.9}$$

Proof. In view of the Lebesgue differentiation theorem, the analogue of Corollary 2.1.16 for cubes, given a function $f \in L^2(\mathbf{R}^n)$ there is a set N_f of measure zero on $\mathbf{R}^n$ such that for all $x \in \mathbf{R}^n \setminus N_f$ we have that

$$\operatorname*{Avg}_{Q_j} f \to f(x)$$

whenever Q_j is a sequence of decreasing cubes such that $\bigcap_j \overline{Q_j} = \{x\}$. Given x in $\mathbf{R}^n \setminus N_f$ there exists a unique sequence of dyadic cubes $Q_j(x) \in \mathscr{D}_j$ such that $\bigcap_{j=0}^{\infty} \overline{Q_j(x)} = \{x\}$. Then for all $x \in \mathbf{R}^n \setminus N_f$ we have

$$\lim_{j\to\infty} E_j(f)(x) = \lim_{j\to\infty} \sum_{Q\in\mathscr{D}_j} (\operatorname*{Avg}_{Q} f)\, \chi_Q(x) = \lim_{j\to\infty} \operatorname*{Avg}_{Q_j(x)} f = f(x)\,.$$

From this we conclude that $E_j(f) \to f$ a.e. as $j \to \infty$. We also observe that since $|E_j(f)| \le M_c(f)$, where M_c denotes the uncentered maximal function with respect to cubes, we have that $|E_j(f) - f| \le 2M_c(f)$, which allows us to obtain from the Lebesgue dominated convergence theorem that $E_j(f) \to f$ in L^2 as $j \to \infty$.

Next we study convergence of $E_j(f)$ as $j \to -\infty$. For a given $x \in \mathbf{R}^n$ and $Q_j(x)$ as before we have that

$$|E_j(f)(x)| = \Big| \operatorname*{Avg}_{Q_j(x)} f \Big| \le \left(\frac{1}{|Q_j(x)|} \int_{Q_j(x)} |f(t)|^2 \, dt \right)^{\frac{1}{2}} \le 2^{\frac{jn}{2}} \|f\|_{L^2},$$

which tends to zero as $j \to -\infty$, since the side length of each $Q_j(x)$ is 2^{-j}. Since $|E_j(f)| \le M_c(f)$, the Lebesgue dominated convergence theorem allows us to conclude that $E_j(f) \to 0$ in L^2 as $j \to -\infty$. To obtain the conclusion asserted in (5.4.6) we simply observe that

$$\sum_{k=M}^{N} D_k(f) = E_N(f) - E_{M-1}(f) \to f$$

in L^2 and almost everywhere as $N \to \infty$ and $M \to -\infty$.

To prove (5.4.7) we first observe that we can rewrite $D_k(f)$ as

$$\begin{aligned}
D_k(f) &= \sum_{Q\in\mathscr{D}_k} (\operatorname*{Avg}_{Q} f)\, \chi_Q - \sum_{R\in\mathscr{D}_{k-1}} (\operatorname*{Avg}_{R} f)\, \chi_R \\
&= \sum_{R\in\mathscr{D}_{k-1}} \Bigg[\sum_{\substack{Q\in\mathscr{D}_k \\ Q\subseteq R}} (\operatorname*{Avg}_{Q} f)\, \chi_Q - (\operatorname*{Avg}_{R} f)\, \chi_R \Bigg] \\
&= \sum_{R\in\mathscr{D}_{k-1}} \Bigg[\sum_{\substack{Q\in\mathscr{D}_k \\ Q\subseteq R}} (\operatorname*{Avg}_{Q} f)\, \chi_Q - \frac{1}{2^n} \sum_{\substack{Q\in\mathscr{D}_k \\ Q\subseteq R}} (\operatorname*{Avg}_{Q} f)\, \chi_R \Bigg] \\
&= \sum_{R\in\mathscr{D}_{k-1}} \sum_{\substack{Q\in\mathscr{D}_k \\ Q\subseteq R}} (\operatorname*{Avg}_{Q} f) \big(\chi_Q - 2^{-n} \chi_R \big)\,.
\end{aligned} \tag{5.4.10}$$

Using this identity we obtain that for given integers $k' > k$ we have

$$\int_{\mathbf{R}^n} D_k(f)(x)\,D_{k'}(f)(x)\,dx$$
$$= \sum_{R\in\mathscr{D}_{k-1}} \sum_{\substack{Q\in\mathscr{D}_k\\ Q\subseteq R}} (\operatorname{Avg}_Q f) \sum_{R'\in\mathscr{D}_{k'-1}} \sum_{\substack{Q'\in\mathscr{D}_{k'}\\ Q'\subseteq R'}} (\operatorname{Avg}_{Q'} f) \int (\chi_Q - 2^{-n}\chi_R)(\chi_{Q'} - 2^{-n}\chi_{R'})\,dx\,.$$

Since $k' > k$, the last integral may be nonzero only when $R' \subsetneq R$. If this is the case, then $R' \subseteq Q_{R'}$ for some dyadic cube $Q_{R'} \in \mathscr{D}_k$ with $Q_{R'} \subsetneq R$. See Figure 5.1.

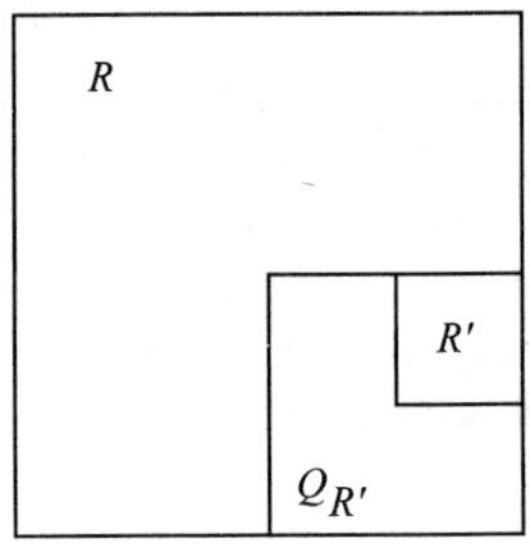

Fig. 5.1 Picture of the cubes R, R', and $Q_{R'}$.

Then the function $\chi_{Q'} - 2^{-n}\chi_{R'}$ is supported in the cube $Q_{R'}$ and the function $\chi_Q - 2^{-n}\chi_R$ is constant on any dyadic subcube Q of R (of half its side length) and in particular is constant on $Q_{R'}$. Then

$$\sum_{\substack{Q'\in\mathscr{D}_{k'}\\ Q'\subseteq R'}} (\operatorname{Avg}_{Q'} f) \int_{Q_{R'}} \chi_{Q'} - 2^{-n}\chi_{R'}\,dx = \sum_{\substack{Q'\in\mathscr{D}_{k'}\\ Q'\subseteq R'}} (\operatorname{Avg}_{Q'} f)\,(|Q'| - 2^{-n}|R'|) = 0\,,$$

since $|R'| = 2^n|Q'|$. We conclude that $\langle D_k(f), D_{k'}(f)\rangle = 0$ whenever $k \neq k'$, from which we easily derive (5.4.7).

Now observe that (5.4.8) is a direct consequence of (5.4.2), and (5.4.9) is a direct consequence of (5.4.3). □

5.4.3 The Dyadic Martingale Square Function

As a consequence of identity (5.4.7), proved in the previous subsection, we obtain that

$$\Big\| \Big(\sum_{k\in\mathbf{Z}} |D_k(f)|^2 \Big)^{\frac{1}{2}} \Big\|_{L^2(\mathbf{R}^n)} = \|f\|_{L^2(\mathbf{R}^n)}\,, \tag{5.4.11}$$

which says that the *dyadic martingale square function*

$$S(f) = \Big(\sum_{k\in\mathbf{Z}} |D_k(f)|^2\Big)^{\frac{1}{2}}$$

is L^2 bounded. It is natural to ask whether there exist L^p analogues of this result, and this is the purpose of the following theorem.

Theorem 5.4.7. *For any $1 < p < \infty$ there exists a constant $c_{p,n}$ such that for every function f in $L^p(\mathbf{R}^n)$ we have*

$$\frac{1}{c_{p',n}}\big\|f\big\|_{L^p(\mathbf{R}^n)} \le \big\|S(f)\big\|_{L^p(\mathbf{R}^n)} \le c_{p,n}\big\|f\big\|_{L^p(\mathbf{R}^n)}. \tag{5.4.12}$$

The lower inequality subsumes the fact that if $\big\|S(f)\big\|_{L^p(\mathbf{R}^n)} < \infty$, then f must be an L^p function.

Proof. Let $\{r_j\}_j$ be the Rademacher functions (see Appendix C.1) enumerated in such a way that their index set is the set of integers. We rewrite the upper estimate in (5.4.12) as

$$\int_0^1\int_{\mathbf{R}^n}\Big|\sum_{k\in\mathbf{Z}} r_k(\omega)D_k(f)(x)\Big|^p\,dx\,d\omega \le C_p^p\big\|f\big\|_{L^p}^p. \tag{5.4.13}$$

We prove a stronger estimate than (5.4.13), namely that for all $\omega\in[0,1]$ we have

$$\int_{\mathbf{R}^n}\Big|T_\omega(f)(x)\Big|^p\,dx \le C_p^p\big\|f\big\|_{L^p}^p, \tag{5.4.14}$$

where

$$T_\omega(f)(x) = \sum_{k\in\mathbf{Z}} r_k(\omega)D_k(f)(x).$$

In view of the L^2 estimate (5.4.11), we have that the operator T_ω is L^2 bounded with norm 1. We show that T_ω is weak type $(1,1)$.

To show that T_ω is of weak type $(1,1)$ we fix a function $f\in L^1$ and $\alpha>0$. We apply the Calderón–Zygmund decomposition (Theorem 4.3.1) to f at height α to write

$$f = g+b, \qquad b = \sum_j \big(f - \operatorname*{Avg}_{Q_j} f\big)\chi_{Q_j},$$

where Q_j are dyadic cubes that satisfy $\sum_j |Q_j| \le \frac{1}{\alpha}\big\|f\big\|_{L^1}$ and g has L^2 norm at most $(2^n\alpha\big\|f\big\|_{L^1})^{\frac{1}{2}}$; see (4.3.1). To achieve this decomposition, we apply the proof of Theorem 4.3.1 starting with a dyadic mesh of large cubes such that $|Q| \ge \frac{1}{\alpha}\big\|f\big\|_{L^1}$ for all Q in the mesh. Then we subdivide each Q in the mesh by halving each side, and we select those cubes for which the average of f over them is bigger than α (and thus at most $2^n\alpha$). Since the original mesh consists of dyadic cubes, the stopping-time argument of Theorem 4.3.1 ensures that each selected cube is dyadic.

We observe (and this is the key observation) that $T_\omega(b)$ is supported in $\bigcup_j Q_j$. To see this, we use identity (5.4.10) to write $T_\omega(b)$ as

$$\sum_j \left[\sum_k r_k(\omega) \sum_{R\in\mathscr{D}_{k-1}} \sum_{\substack{Q\in\mathscr{D}_k \\ Q\subseteq R}} \operatorname*{Avg}_Q[(f-\operatorname*{Avg}_{Q_j} f)\chi_{Q_j}]\,(\chi_Q - 2^{-n}\chi_R)\right]. \tag{5.4.15}$$

We consider the following three cases for the cubes Q that appear in the inner sum in (5.4.15): (i) $Q_j \subseteq Q$, (ii) $Q_j \cap Q = \emptyset$, and (iii) $Q \subsetneq Q_j$. It is simple to see that in cases (i) and (ii) we have $\operatorname{Avg}_Q[(f-\operatorname{Avg}_{Q_j} f)\chi_{Q_j}] = 0$. Therefore the inner sum in (5.4.15) is taken over all Q that satisfy $Q \subsetneq Q_j$. But then we must have that the unique dyadic parent R of Q is also contained in Q_j. It follows that the expression inside the square brackets in (5.4.15) is supported in R and therefore in Q_j. We conclude that $T_\omega(b)$ is supported in $\bigcup_j Q_j$. Using Exercise 4.3.5(a) we obtain that T_ω is weak type $(1,1)$ with norm at most

$$\frac{\alpha\big|\{|T_\omega(g)|>\frac{\alpha}{2}\}\big| + \alpha\big|\bigcup_j Q_j\big|}{\|f\|_{L^1}} \le \frac{\alpha 4\alpha^{-2}\|g\|_{L^2}^2 + \|f\|_{L^1}}{\|f\|_{L^1}} \le 2^{n+2}+1\,.$$

We have now established that T_ω is weak type $(1,1)$. Since T_ω is L^2 bounded with norm 1, it follows by interpolation that T_ω is L^p bounded for all $1<p<2$. The L^p boundedness of T_ω for the remaining $p>2$ follows by duality. (Note that the operators D_k and E_k are self-transpose.) We conclude the validity of (5.4.14), which implies that of (5.4.13). As observed, this is equivalent to the upper estimate in (5.4.12).

Finally, we notice that the lower estimate in (5.4.12) is a consequence of the upper estimate as in the case of the Littlewood–Paley operators Δ_j. Indeed, we need to observe that in view of (5.4.6) we have

$$\begin{aligned}
\big|\langle f,g\rangle\big| &= \big|\big\langle \sum_k D_k(f), \sum_{k'} D_{k'}(g)\big\rangle\big| \\
&= \Big|\sum_k\sum_{k'}\big\langle D_k(f), D_{k'}(g)\big\rangle\Big| \\
&= \Big|\sum_k \big\langle D_k(f), D_k(g)\big\rangle\Big| && \text{(Exercise 5.4.6(a))} \\
&\le \int_{\mathbf{R}^n}\sum_k |D_k(f)(x)|\,|D_k(g)(x)|\,dx \\
&\le \int_{\mathbf{R}^n} S(f)(x)\,S(g)(x)\,dx && \text{(Cauchy–Schwarz inequality)} \\
&\le \big\|S(f)\big\|_{L^p}\big\|S(g)\big\|_{L^{p'}} && \text{(Hölder's inequality)} \\
&\le \big\|S(f)\big\|_{L^p}\, c_{p',n}\big\|g\big\|_{L^{p'}}\,.
\end{aligned}$$

Taking the supremum over all functions g on $\mathbf{R}^n$ with $L^{p'}$ norm at most 1, we obtain that f gives rise to a bounded linear functional on $L^{p'}$. It follows by the Riesz representation theorem that f must be an L^p function that satisfies the lower estimate in (5.4.12). □

5.4.4 Almost Orthogonality Between the Littlewood–Paley Operators and the Dyadic Martingale Difference Operators

Next, we discuss connections between the Littlewood–Paley operators Δ_j and the dyadic martingale difference operators D_k. It turns out that these operators are almost orthogonal in the sense that the L^2 operator norm of the composition $D_k\Delta_j$ decays exponentially as the indices j and k get farther away from each other.

For the purposes of the next theorem we define the Littlewood–Paley operators Δ_j as convolution operators with the function $\Psi_{2^{-j}}$, where

$$\widehat{\Psi}(\xi) = \widehat{\Phi}(\xi) - \widehat{\Phi}(2\xi)$$

and Φ is a fixed radial Schwartz function whose Fourier transform $\widehat{\Phi}$ is real-valued, supported in the ball $|\xi| < 2$, and equal to 1 on the ball $|\xi| < 1$. In this case we clearly have the identity

$$\sum_{j\in\mathbf{Z}} \widehat{\Psi}(2^{-j}\xi) = 1, \qquad\qquad \xi \neq 0.$$

Then we have the following theorem.

Theorem 5.4.8. *There exists a constant C such that for every k,j in $\mathbf{Z}$ the following estimate on the operator norm of $D_k\Delta_j : L^2(\mathbf{R}^n) \to L^2(\mathbf{R}^n)$ is valid:*

$$\big\|D_k\Delta_j\big\|_{L^2\to L^2} = \big\|\Delta_j D_k\big\|_{L^2\to L^2} \le C2^{-\frac{1}{2}|j-k|}. \tag{5.4.16}$$

Proof. Since Ψ is a radial function, it follows that Δ_j is equal to its transpose operator on L^2. Moreover, the operator D_k is also equal to its transpose. Thus

$$(D_k\Delta_j)^t = \Delta_j D_k$$

and it therefore suffices to prove only that

$$\big\|D_k\Delta_j\big\|_{L^2\to L^2} \le C2^{-\frac{1}{2}|j-k|}. \tag{5.4.17}$$

By a simple dilation argument it suffices to prove (5.4.17) when $k = 0$. In this case we have the estimate

$$\begin{aligned}\big\|D_0\Delta_j\big\|_{L^2\to L^2} &= \big\|E_0\Delta_j - E_{-1}\Delta_j\big\|_{L^2\to L^2}\\ &\le \big\|E_0\Delta_j - \Delta_j\big\|_{L^2\to L^2} + \big\|E_{-1}\Delta_j - \Delta_j\big\|_{L^2\to L^2},\end{aligned}$$

and since the D_k's and Δ_j's are self-transposes, we have

$$\begin{aligned}\big\|D_0\Delta_j\big\|_{L^2\to L^2} = \big\|\Delta_jD_0\big\|_{L^2\to L^2} &= \big\|\Delta_jE_0-\Delta_jE_{-1}\big\|_{L^2\to L^2}\\ &\le \big\|\Delta_jE_0-E_0\big\|_{L^2\to L^2}+\big\|\Delta_jE_{-1}-E_0\big\|_{L^2\to L^2}.\end{aligned}$$

Estimate (5.4.17) when $k=0$ will be a consequence of the pair of inequalities

$$\big\|E_0\Delta_j-\Delta_j\big\|_{L^2\to L^2}+\big\|E_{-1}\Delta_j-\Delta_j\big\|_{L^2\to L^2}\le C2^j \quad \text{for } j\le 0, \tag{5.4.18}$$

$$\big\|\Delta_jE_0-E_0\big\|_{L^2\to L^2}+\big\|\Delta_jE_{-1}-E_0\big\|_{L^2\to L^2}\le C2^{-\frac{1}{2}j} \quad \text{for } j\ge 0. \tag{5.4.19}$$

We start by proving (5.4.18). We consider only the term $E_0\Delta_j-\Delta_j$, since the term $E_{-1}\Delta_j-\Delta_j$ is similar. Let $f\in L^2(\mathbf{R}^n)$. Then

$$\begin{aligned}&\big\|E_0\Delta_j(f)-\Delta_j(f)\big\|_{L^2}^2\\ &\quad=\sum_{Q\in\mathscr{D}_0}\big\|f*\Psi_{2^{-j}}-\operatorname*{Avg}_Q(f*\Psi_{2^{-j}})\big\|_{L^2(Q)}^2\\ &\quad\le\sum_{Q\in\mathscr{D}_0}\int_Q\int_Q|(f*\Psi_{2^{-j}})(x)-(f*\Psi_{2^{-j}})(t)|^2\,dt\,dx\\ &\quad\le 3\sum_{Q\in\mathscr{D}_0}\int_Q\int_Q\bigg(\int_{5\sqrt{n}Q}|f(y)||\Psi_{2^{-j}}(x-y)|\,dy\bigg)^2dt\,dx\\ &\qquad+3\sum_{Q\in\mathscr{D}_0}\int_Q\int_Q\bigg(\int_{5\sqrt{n}Q}|f(y)||\Psi_{2^{-j}}(t-y)|\,dy\bigg)^2dt\,dx\\ &\qquad+3\sum_{Q\in\mathscr{D}_0}\int_Q\int_Q\bigg(\int_{(5\sqrt{n}Q)^c}|f(y)|2^{jn+j}|\nabla\Psi(2^j(\xi_{x,t}-y))|\,dy\bigg)^2dt\,dx,\end{aligned}$$

where $\xi_{x,t}$ lies on the line segment between x and t. It is a simple fact that the sum of the last three expressions is bounded by

$$C2^{2jn}\sum_{Q\in\mathscr{D}_0}\int_{5\sqrt{n}Q}|f(y)|^2\,dy+C_M2^{2j}\sum_{Q\in\mathscr{D}_0}\int_Q\bigg(\int_{\mathbf{R}^n}\frac{2^{jn}|f(y)|\,dy}{(1+2^j|x-y|)^M}\bigg)^2dx,$$

which is clearly controlled by $C2^{2j}\big\|f\big\|_{L^2}^2$. This estimate is useful when $j\le 0$.

We now turn to the proof of (5.4.19). We set $S_j=\sum_{k\le j}\Delta_j$. Since Δ_j is the difference of two S_j's, it suffices to prove (5.4.19), where Δ_j is replaced by S_j. We work only with the term $S_jE_0-E_0$, since the other term can be treated similarly. We have

$$\begin{aligned}\big\|S_jE_0(f)-E_0(f)\big\|_{L^2}^2&=\bigg\|\sum_{Q\in\mathscr{D}_0}(\operatorname*{Avg}_Q f)\,(\Phi_{2^{-j}}*\chi_Q-\chi_Q)\bigg\|_{L^2}^2\\ &\le 2\bigg\|\sum_{Q\in\mathscr{D}_0}(\operatorname*{Avg}_Q f)\,(\Phi_{2^{-j}}*\chi_Q-\chi_Q)\chi_{5\sqrt{n}Q}\bigg\|_{L^2}^2\\ &\qquad+2\bigg\|\sum_{Q\in\mathscr{D}_0}(\operatorname*{Avg}_Q f)\,(\Phi_{2^{-j}}*\chi_Q)\chi_{(5\sqrt{n}Q)^c}\bigg\|_{L^2}^2.\end{aligned}$$

Since the functions appearing inside the sum in the first term have supports with bounded overlap, we obtain

$$\Big\| \sum_{Q\in\mathscr{D}_0} \operatorname*{Avg}_{Q} f)(\Phi_{2^{-j}} * \chi_Q - \chi_Q)\chi_{5\sqrt{n}Q}\Big\|_{L^2}^2 \le C \sum_{Q\in\mathscr{D}_0} (\operatorname*{Avg}_{Q}|f|)^2 \big\|\Phi_{2^{-j}} * \chi_Q - \chi_Q\big\|_{L^2}^2,$$

and the crucial observation is that

$$\big\|\Phi_{2^{-j}} * \chi_Q - \chi_Q\big\|_{L^2}^2 \le C 2^{-j},$$

a consequence of Plancherel's identity and the fact that $|1-\widehat{\Phi}(2^{-j}\xi)| \le \chi_{|\xi|\ge 2^j}$. Putting these observations together, we deduce

$$\Big\| \sum_{Q\in\mathscr{D}_0} (\operatorname*{Avg}_{Q} f)\,(\Phi_{2^{-j}} * \chi_Q - \chi_Q)\chi_{3Q}\Big\|_{L^2}^2 \le C \sum_{Q\in\mathscr{D}_0} (\operatorname*{Avg}_{Q}|f|)^2 2^{-j} \le C2^{-j}\|f\|_{L^2}^2,$$

and the required conclusion will be proved if we can show that

$$\Big\| \sum_{Q\in\mathscr{D}_0} (\operatorname*{Avg}_{Q} f)\,(\Phi_{2^{-j}} * \chi_Q)\chi_{(3Q)^c}\Big\|_{L^2}^2 \le C2^{-j}\|f\|_{L^2}^2. \tag{5.4.20}$$

We prove (5.4.20) by using an estimate based purely on size. Let c_Q be the center of the dyadic cube Q. For $x \notin 3Q$ we have the estimate

$$|(\Phi_{2^{-j}} * \chi_Q)(x)| \le \frac{C_M 2^{jn}}{(1+2^j|x-c_Q|)^M} \le \frac{C_M 2^{jn}}{(1+2^j)^{M/2}} \frac{1}{(1+|x-c_Q|)^{M/2}},$$

since both $2^j \ge 1$, and $|x-c_Q| \ge 1$. We now control the left-hand side of (5.4.20) by

$$\begin{aligned}
&2^{j(2n-M)} \sum_{Q\in\mathscr{D}_0}\sum_{Q'\in\mathscr{D}_0} (\operatorname*{Avg}_{Q}|f|)(\operatorname*{Avg}_{Q'}|f|) \int_{\mathbf{R}^n} \frac{C_M\, dx}{(1+|x-c_Q|)^{\frac{M}{2}}(1+|x-c_{Q'}|)^{\frac{M}{2}}}\\
&\le 2^{j(2n-M)} \sum_{Q\in\mathscr{D}_0}\sum_{Q'\in\mathscr{D}_0} \frac{(\operatorname*{Avg}_{Q}|f|)(\operatorname*{Avg}_{Q'}|f|)}{(1+|c_Q-c_{Q'}|)^{\frac{M}{4}}} \int_{\mathbf{R}^n} \frac{C_M\, dx}{(1+|x-c_Q|)^{\frac{M}{4}}(1+|x-c_{Q'}|)^{\frac{M}{4}}}\\
&\le 2^{j(2n-M)} \sum_{Q\in\mathscr{D}_0}\sum_{Q'\in\mathscr{D}_0} \frac{C_M}{(1+|c_Q-c_{Q'}|)^{\frac{M}{4}}} \left(\int_Q |f(y)|^2\,dy + \int_{Q'} |f(y)|^2\,dy\right)\\
&\le C_M 2^{j(2n-M)} \sum_{Q\in\mathscr{D}_0} \int_Q |f(y)|^2\,dy\\
&= C_M 2^{j(2n-M)} \|f\|_{L^2}^2.
\end{aligned}$$

By taking M large enough, we obtain (5.4.20) and thus (5.4.19). □

Exercises

5.4.1. (a) Prove that no dyadic cube in $\mathbf{R}^n$ contains the point 0 in its interior.
(b) Prove that every interval in $\mathbf{R}$ is contained in the union of two dyadic intervals of at most its length.
(c) Prove that every cube in $\mathbf{R}^n$ is contained in the union of 2^n dyadic cubes of at most its side length.

5.4.2. Show that the set $[m2^{-k},(m+2)2^{-k})$ is a dyadic interval if and only if m is an even integer. More generally, the set $[m2^{-k},(m+s)2^{-k})$ is a dyadic interval if and only if s is a power of 2 and m is an integer multiple of s.

5.4.3. Let Σ be the set of all $\sigma=(\sigma_1,\dots,\sigma_n)$ that satisfy $\sigma_j\in\{0,\frac{1}{2},\frac{2}{3}\}$ for all j. Show that every cube Q in $\mathbf{R}^n$ is contained in a cube of the form $\sigma+R$, where σ is in Σ and R is dyadic and has side length comparable to that of Q.

5.4.4. Show that the martingale maximal function $f\mapsto\sup_k|E_k(f)|$ is weak type $(1,1)$ with constant at most 1.

5.4.5. (a) Show that $E_N(f)\to f$ a.e. as $N\to\infty$ for all $f\in L^1_{\mathrm{loc}}(\mathbf{R}^n)$.
(b) Prove that $E_N(f)\to f$ in L^p as $N\to\infty$ for all $f\in L^p(\mathbf{R}^n)$ whenever $1<p\le\infty$.

5.4.6. (a) Show that for functions f and g, if $k\neq k'$, then we have

$$\big\langle D_k(f),D_{k'}(g)\big\rangle=0.$$

(b) Conclude that for functions f_j we have

$$\Big\|\sum_j D_j(f_j)\Big\|_{L^2}=\Big(\sum_j\|D_j(f_j)\|_{L^2}^2\Big)^{\frac{1}{2}}.$$

(c) Use Theorem 5.4.8 to show that

$$\Big\|\sum_j D_j\Delta_{j+r}D_j\Big\|_{L^2\to L^2}\le C2^{-\frac{1}{2}|r|}.$$

5.4.7. (*Grafakos and Kalton [106]*) Let D_j, Δ_j be as in Exercise 5.4.6.
(a) Prove that the operator

$$V_r=\sum_{j\in\mathbf{Z}}D_j\Delta_{j+r}$$

is L^2 bounded with norm at most a multiple of $2^{-\frac{1}{2}|r|}$.
(b) Show that V_r is L^p bounded for all $1<p<\infty$ with a constant depending only on p and n.
(c) Conclude that for each $1<p<\infty$ there is a constant $c_p>0$ such that V_r is bounded on $L^p(\mathbf{R}^n)$ with norm at most a multiple of $2^{-c_p|r|}$.
[*Hint:* Part (a): Write $\Delta_j=\Delta_j\widetilde{\Delta}_j$, where $\widetilde{\Delta}_j$ is another family of Littlewood–Paley operators and use Exercise 5.4.6(b). Part (b): Use duality and (5.1.20).]

5.5 The Spherical Maximal Function

In this section we discuss yet another consequence of the Littlewood–Paley theory, the boundedness of the spherical maximal operator.

5.5.1 Introduction of the Spherical Maximal Function

We denote throughout this section by $d\sigma$ the normalized Lebesgue measure on the sphere $\mathbf{S}^{n-1}$. For f in $L^p(\mathbf{R}^n)$, $1 \le p \le \infty$, we define the maximal operator

$$\mathscr{M}(f)(x) = \sup_{t>0} \left| \int_{\mathbf{S}^{n-1}} f(x-t\theta)\, d\sigma(\theta) \right| \tag{5.5.1}$$

and we observe that by Minkowski's integral inequality each expression inside the supremum in (5.5.1) is well defined for $f \in L^p$ for almost all $x \in \mathbf{R}^n$. The operator $\mathscr{M}$ is called the *spherical maximal function*. It is unclear at this point for which functions f we have $\mathscr{M}(f) < \infty$ a.e. and for which values of $p < \infty$ the maximal inequality

$$\big\|\mathscr{M}(f)\big\|_{L^p(\mathbf{R}^n)} \le C_p \big\|f\big\|_{L^p(\mathbf{R}^n)} \tag{5.5.2}$$

holds for all functions $f \in L^p(\mathbf{R}^n)$.

Spherical averages often make their appearance as solutions of partial differential equations. For instance, the spherical average

$$u(x,t) = \frac{1}{4\pi} \int_{\mathbf{S}^2} t\, f(x-ty)\, d\sigma(y) \tag{5.5.3}$$

is a solution of the *wave equation*

$$\begin{aligned} \Delta_x(u)(x,t) &= \frac{\partial^2 u}{\partial t^2}(x,t), \\ u(x,0) &= 0, \\ \frac{\partial u}{\partial t}(x,0) &= f(x), \end{aligned}$$

in $\mathbf{R}^3$. The introduction of the spherical maximal function is motivated by the fact that the related spherical average

$$u(x,t) = \frac{1}{4\pi} \int_{\mathbf{S}^2} f(x-ty)\, d\sigma(y) \tag{5.5.4}$$

solves *Darboux's equation*

$$\begin{aligned}\Delta_x(u)(x,t) &= \frac{\partial^2 u}{\partial t^2}(x,t)+\frac{2}{t}\frac{\partial u}{\partial t}(x,t),\\ u(x,0) &= f(x),\\ \frac{\partial u}{\partial t}(x,0) &= 0,\end{aligned}$$

in $\mathbf{R}^3$. It is rather remarkable that the Fourier transform can be used to study almost everywhere convergence for several kinds of maximal averaging operators such as the spherical averages in (5.5.4). This is achieved via the boundedness of the corresponding maximal operator; the maximal operator controlling the averages over $\mathbf{S}^{n-1}$ is given in (5.5.1).

Before we begin the analysis of the spherical maximal function, we recall that

$$\widehat{d\sigma}(\xi) = \frac{2\pi}{|\xi|^{\frac{n-2}{2}}} J_{\frac{n-2}{2}}(2\pi|\xi|),$$

as shown in Appendix B.4. Using the estimates in Appendices B.6 and B.7 and the identity

$$\frac{d}{dt}J_\nu(t) = \frac{1}{2}(J_{\nu-1}(t)-J_{\nu+1}(t))$$

derived in Appendix B.2, we deduce the crucial estimate

$$|\widehat{d\sigma}(\xi)|+|\nabla\widehat{d\sigma}(\xi)| \le \frac{C_n}{(1+|\xi|)^{\frac{n-1}{2}}}. \tag{5.5.5}$$

Theorem 5.5.1. *Let $n\ge 3$. For each $\frac{n}{n-1}<p\le\infty$, there is a constant C_p such that*

$$\big\|\mathscr{M}(f)\big\|_{L^p(\mathbf{R}^n)} \le C_p\big\|f\big\|_{L^p(\mathbf{R}^n)} \tag{5.5.6}$$

holds for all f in $L^p(\mathbf{R}^n)$. It follows that for all $\frac{n}{n-1}<p\le\infty$ and $f\in L^p(\mathbf{R}^n)$ we have

$$\lim_{t\to 0}\frac{1}{\omega_{n-1}}\int_{\mathbf{S}^{n-1}} f(x-t\theta)\,d\sigma(\theta) = f(x) \tag{5.5.7}$$

for almost all $x\in\mathbf{R}^n$. Here we set $\omega_{n-1}=|\mathbf{S}^{n-1}|$.

The proof of this theorem is presented in the rest of this section. We set $m(\xi)=\widehat{d\sigma}(\xi)$. Obviously $m(\xi)$ is a $\mathscr{C}^\infty$ function. To study the maximal multiplier operator

$$\sup_{t>0}\big|\big(\widehat{f}(\xi)m(t\xi)\big)^\vee\big|$$

we decompose the multiplier $m(\xi)$ into radial pieces as follows: We fix a radial $\mathscr{C}^\infty$ function φ_0 in $\mathbf{R}^n$ such that $\varphi_0(\xi)=1$ when $|\xi|\le 1$ and $\varphi_0(\xi)=0$ when $|\xi|\ge 2$. For $j\ge 1$ we let

$$\varphi_j(\xi) = \varphi_0(2^{-j}\xi)-\varphi_0(2^{1-j}\xi) \tag{5.5.8}$$

and we observe that $\varphi_j(\xi)$ is localized near $|\xi|\approx 2^j$. Then we have

$$\sum_{j=0}^{\infty} \varphi_j = 1 .$$

Set $m_j = \varphi_j m$ for all $j \geq 0$. The m_j's are $\mathscr{C}_0^\infty$ functions that satisfy

$$m = \sum_{j=0}^{\infty} m_j .$$

Also, the following estimate is valid:

$$\mathscr{M}(f) \leq \sum_{j=0}^{\infty} \mathscr{M}_j(f),$$

where

$$\mathscr{M}_j(f)(x) = \sup_{t>0} \big| \big(\widehat{f}(\xi)\, m_j(t\xi)\big)^{\vee}(x) \big| .$$

Since the function m_0 is $\mathscr{C}_0^\infty$, we have that $\mathscr{M}_0$ maps L^p to itself for all $1 < p \leq \infty$. (See Exercise 5.5.1.)

We define *g-functions* associated with m_j as follows:

$$G_j(f)(x) = \left(\int_0^\infty |A_{j,t}(f)(x)|^2 \frac{dt}{t} \right)^{\frac{1}{2}},$$

where $A_{j,t}(f)(x) = \big(\widehat{f}(\xi)\, m_j(t\xi)\big)^{\vee}(x)$.

5.5.2 The First Key Lemma

We have the following lemma:

Lemma 5.5.2. *There is a constant $C = C(n) < \infty$ such that for any $j \geq 1$ we have the estimate*

$$\|\mathscr{M}_j(f)\|_{L^2} \leq C 2^{(\frac{1}{2} - \frac{n-1}{2})j} \|f\|_{L^2}$$

for all functions f in $L^2(\mathbf{R}^n)$.

Proof. We define a function

$$\widetilde{m}_j(\xi) = \xi \cdot \nabla m_j(\xi),$$

we let $\widetilde{A}_{j,t}(f)(x) = \big(\widehat{f}(\xi)\, \widetilde{m}_j(t\xi)\big)^{\vee}(x)$, and we let

$$\widetilde{G}_j(f)(x) = \left(\int_0^\infty |\widetilde{A}_{j,t}(f)(x)|^2 \frac{dt}{t} \right)^{\frac{1}{2}}$$

be the associated g-function. For $f \in L^2(\mathbf{R}^n)$, the identity

$$s\frac{dA_{j,s}}{ds}(f)=\widetilde{A}_{j,s}(f)$$

is clearly valid for all j and s. Since $A_{j,s}(f)=f*(m_j^\vee)_s$ and $m_j^\vee$ has integral zero for $j\ge 1$ (here $(m_j^\vee)_s(x)=s^{-n}m_j^\vee(s^{-1}x)$), it follows from Corollary 2.1.19 that

$$\lim_{s\to 0}A_{j,s}(f)(x)=0$$

for all $x\in\mathbf{R}^n\setminus E$, where E is some set of Lebesgue measure zero. By the fundamental theorem of calculus for $x\in\mathbf{R}^n\setminus E$ we deduce that

$$\begin{aligned}(A_{j,t}(f)(x))^2&=\int_0^t\frac{d}{ds}(A_{j,s}(f)(x))^2\,ds\\&=2\int_0^t A_{j,s}(f)(x)\,s\frac{dA_{j,s}}{ds}(f)(x)\,\frac{ds}{s}\\&=2\int_0^t A_{j,s}(f)(x)\widetilde{A}_{j,s}(f)(x)\,\frac{ds}{s},\end{aligned}$$

from which we obtain the estimate

$$\big|A_{j,t}(f)(x)\big|^2\le 2\int_0^\infty\big|A_{j,s}(f)(x)\big|\,\big|\widetilde{A}_{j,s}(f)(x)\big|\,\frac{ds}{s}.\tag{5.5.9}$$

Taking the supremum over all $t>0$ on the left-hand side in (5.5.9) and integrating over $\mathbf{R}^n$, we obtain the estimate

$$\begin{aligned}\big\|\mathscr{M}_j(f)\big\|_{L^2}^2&\le 2\int_{\mathbf{R}^n}\left|\int_0^\infty A_{j,s}(f)(x)\widetilde{A}_{j,s}(f)(x)\,\frac{ds}{s}\right|^2dx\\&\le 2\big\|G_j(f)\big\|_{L^2}\big\|\widetilde{G}_j(f)\big\|_{L^2},\end{aligned}$$

where the last inequality follows by applying the Cauchy–Schwarz inequality twice. Next we claim that as a consequence of (5.5.5) we have for some $c,\widetilde{c}<\infty$,

$$\big\|m_j\big\|_{L^\infty}\le c\,2^{-j\frac{n-1}{2}}\quad\text{and}\quad\big\|\widetilde{m}_j\big\|_{L^\infty}\le\widetilde{c}\,2^{j(1-\frac{n-1}{2})}.$$

Using these facts together with the facts that the functions m_j and $\widetilde{m}_j$ are supported in the annuli $2^{j-1}\le|\xi|\le 2^{j+1}$, we obtain that the g-functions G_j and $\widetilde{G}_j$ are L^2 bounded with norms at most a constant multiple of the quantities $2^{-j\frac{n-1}{2}}$ and $2^{j(1-\frac{n-1}{2})}$, respectively; see Exercise 5.5.2. Note that since $n\ge 3$, both exponents are negative. We conclude that

$$\big\|\mathscr{M}_j(f)\big\|_{L^2}\le C2^{j(\frac{1}{2}-\frac{n-1}{2})}\big\|f\big\|_{L^2},$$

which is what we needed to prove. □

5.5.3 The Second Key Lemma

Next we need the following lemma.

Lemma 5.5.3. *There exists a constant $C = C(n) < \infty$ such that for all $j \geq 1$ and for all f in $L^1(\mathbf{R}^n)$ we have*

$$\big\|\mathscr{M}_j(f)\big\|_{L^{1,\infty}} \leq C\,2^j \big\|f\big\|_{L^1}.$$

Proof. Let $K^{(j)} = (\varphi_j)^\vee * d\sigma = \Phi_{2^{-j}} * d\sigma$, where Φ is a Schwartz function. Setting

$$(K^{(j)})_t(x) = t^{-n}K^{(j)}(t^{-1}x)$$

we have that

$$\mathscr{M}_j(f) = \sup_{t>0} |(K^{(j)})_t * f|. \tag{5.5.10}$$

The proof of the lemma is based on the estimate:

$$\mathscr{M}_j(f) \leq C\,2^j \mathcal{M}(f) \tag{5.5.11}$$

and the weak type $(1,1)$ boundedness of the Hardy–Littlewood maximal operator $\mathcal{M}$ (Theorem 2.1.6). To establish (5.5.11), it suffices to show that for any $M > n$ there is a constant $C_M < \infty$ such that

$$|K^{(j)}(x)| = |(\Phi_{2^{-j}} * d\sigma)(x)| \leq \frac{C_M\, 2^j}{(1+|x|)^M}. \tag{5.5.12}$$

Then Theorem 2.1.10 yields (5.5.11) and hence the required conclusion.

Using the fact that Φ is a Schwartz function, we have for every $N > 0$,

$$|(\Phi_{2^{-j}} * d\sigma)(x)| \leq C_N \int_{\mathbf{S}^{n-1}} \frac{2^{nj}\, d\sigma(y)}{(1+2^j|x-y|)^N}.$$

We pick an N to depend on M (5.5.12); in fact, any $N > M$ suffices for our purposes. We split the last integral into the regions

$$S_{-1}(x) = \mathbf{S}^{n-1} \cap \{y \in \mathbf{R}^n :\, 2^j|x-y| \leq 1\}$$

and for $r \geq 0$,

$$S_r(x) = \mathbf{S}^{n-1} \cap \{y \in \mathbf{R}^n :\, 2^r < 2^j|x-y| \leq 2^{r+1}\}.$$

The key observation is that whenever $B = B(x,R)$ is a ball in $\mathbf{R}^n$, then the spherical measure of the set $\mathbf{S}^{n-1} \cap B(x,R)$ is at most a dimensional constant multiple of R^{n-1}. This implies that the spherical measure of each $S_r(x)$ is at most $c_n 2^{(r+1-j)(n-1)}$, an estimate that is useful only when $r \leq j$. Using this observation, together with the fact that for $y \in S_r(x)$ we have $|x| \leq 2^{r+1-j}+1$, we obtain the following estimate for the expression $|(\Phi_{2^{-j}} * d\sigma)(x)|$:

$$
\begin{aligned}
&\sum_{r=-1}^{j} \int_{S_r(x)} \frac{C_N 2^{nj}\, d\sigma(y)}{(1+2^j|x-y|)^N} + \sum_{r=j+1}^{\infty} \int_{S_r(x)} \frac{C_N 2^{nj}\, d\sigma(y)}{(1+2^j|x-y|)^N} \\
&\quad \le C_N' 2^{nj} \left[\sum_{r=-1}^{j} \frac{d\sigma(S_r(x))\chi_{B(0,3)}(x)}{2^{rN}} + \sum_{r=j+1}^{\infty} \frac{d\sigma(S_r(x))\,\chi_{B(0,2^{r+1-j}+1)}(x)}{2^{rN}} \right] \\
&\quad \le C_N' 2^{nj} \left[\sum_{r=-1}^{j} \frac{c_n 2^{(r+1-j)(n-1)}\chi_{B(0,3)}(x)}{2^{rN}} + \sum_{r=j+1}^{\infty} \frac{\omega_{n-1}\,\chi_{B(0,2^{r+2-j})}(x)}{2^{rN}} \right] \\
&\quad \le C_{N,n} \left[2^j \chi_{B(0,3)}(x) + 2^{nj} \sum_{r=j+1}^{\infty} \frac{1}{2^{rN}} \frac{(1+2^{r+2-j})^M}{(1+|x|)^M} \right] \\
&\quad \le C_{M,n}' \frac{2^j}{(1+|x|)^M} \left[1 + \sum_{r=j+1}^{\infty} \frac{2^{(r-j)(M-N)}}{2^{j(N+1-n)}} \right] \\
&\quad \le \frac{C_{M,n}'' 2^j}{(1+|x|)^M},
\end{aligned}
$$

where we used that $N > M > n$. This establishes (5.5.12). □

5.5.4 Completion of the Proof

It remains to combine the previous ingredients to complete the proof of the theorem. Interpolating between the $L^2 \to L^2$ and $L^1 \to L^{1,\infty}$ estimates obtained in Lemmas 5.5.2 and 5.5.3, we obtain

$$
\big\| \mathscr{M}_j(f) \big\|_{L^p(\mathbf{R}^n)} \le C_p 2^{(\frac{n}{p}-(n-1))j} \big\| f \big\|_{L^p(\mathbf{R}^n)}
$$

for all $1 < p \le 2$. When $p > \frac{n}{n-1}$ the series $\sum_{j=1}^{\infty} 2^{(\frac{n}{p}-(n-1))j}$ converges and we conclude that $\mathscr{M}$ is L^p bounded for these p's. The boundedness of $\mathscr{M}$ on L^p for $p > 2$ follows by interpolation between L^q for $q < 2$ and the estimate $\mathscr{M} : L^\infty \to L^\infty$.

Exercises

5.5.1. (a) Let m be in $L^1(\mathbf{R}^n) \cap L^\infty(\mathbf{R}^n)$ that satisfies $|m^\vee(x)| \le C(1+|x|)^{-n-\delta}$ for some $\delta > 0$. Show that the maximal multiplier

$$
\mathscr{M}_m(f)(x) = \sup_{t>0} \big| \big(\widehat{f}(\xi)\, m(t\xi) \big)^\vee (x) \big|
$$

is L^p bounded for all $1 < p < \infty$.
(b) Obtain the same conclusion when $\xi^\alpha m(\xi)$ is in $L^1(\mathbf{R}^n)$ for all multi-indices α

with $|\alpha| \le [\frac{n}{2}]+1$.
[*Hint:* Control $\mathcal{M}_m$ by the Hardy–Littlewood maximal operator.]

5.5.2. Suppose that the function m is supported in the annulus $R \le |\xi| \le 2R$ and is bounded by A. Show that the g-function

$$G(f)(x) = \left(\int_0^\infty |(m(t\xi)\widehat{f}(\xi))^\vee(x)|^2 \frac{dt}{t} \right)^{\frac{1}{2}}$$

maps $L^2(\mathbf{R}^n)$ to $L^2(\mathbf{R}^n)$ with bound at most $A\sqrt{\log 2}$.

5.5.3. (*Rubio de Francia [226]*) Use the idea of Lemma 5.5.2 to show that if $m(\xi)$ satisfies $|m(\xi)| \le (1+|\xi|)^{-a}$ and $|\nabla m(\xi)| \le (1+|\xi|)^{-b}$ and $a+b>1$, then the maximal operator

$$\mathcal{M}_m(f)(x) = \sup_{t>0} \left| \left(\widehat{f}(\xi)m(t\xi)\right)^\vee(x) \right|$$

is bounded from $L^2(\mathbf{R}^n)$ to itself.
[*Hint:* Use that

$$\mathcal{M}_m \le \sum_{j=0}^\infty \mathcal{M}_{m,j},$$

where $\mathcal{M}_{m,j}$ corresponds to the multiplier $\varphi_j m$; here φ_j is as in (5.5.8). Show that

$$\left\|\mathcal{M}_{m,j}(f)\right\|_{L^2} \le C\left\|\varphi_j m\right\|_{L^\infty}^{\frac{1}{2}} \left\|\varphi_j \widetilde{m}\right\|_{L^\infty}^{\frac{1}{2}} \left\|f\right\|_{L^2} \le C 2^{j\frac{1-(a+b)}{2}} \left\|f\right\|_{L^2},$$

where $\widetilde{m}(\xi) = \xi \cdot \nabla m(\xi)$.]

5.5.4. (*Rubio de Francia [226]*) Observe that the proof of Theorem 5.5.1 gives the following more general result: If $m(\xi)$ is the Fourier transform of a compactly supported Borel measure and satisfies $|m(\xi)| \le (1+|\xi|)^{-a}$ for some $a>0$ and all $\xi \in \mathbf{R}^n$, then the maximal operator of Exercise 5.5.3 maps $L^p(\mathbf{R}^n)$ to itself when $p > \frac{2a+1}{2a}$.

5.5.5. Show that Theorem 5.5.1 is false when $n=1$, that is, show that the maximal operator

$$\mathcal{M}_1(f)(x) = \sup_{t>0} \frac{|f(x+t)+f(x-t)|}{2}$$

is unbounded on $L^p(\mathbf{R})$ for all $p<\infty$.

5.5.6. Show that Theorem 5.5.1 is false when $n \ge 2$ and $p \le \frac{n}{n-1}$.
[*Hint:* Choose a compactly supported and radial function equal to $|y|^{1-n}(-\log|y|)^{-1}$ when $|y| \le 1/2$.]

5.6 Wavelets

We are concerned with orthonormal bases of $L^2(\mathbf{R})$ generated by translations and dilations of a single function such as the Haar functions we encountered in Section 5.4. The Haar functions are generated by integer translations and dyadic dilations of the single function $\chi_{[0,\frac{1}{2})} - \chi_{[\frac{1}{2},1)}$. This function is not smooth, and the main question addressed in this section is whether there exist smooth analogues of the Haar functions.

Definition 5.6.1. A square integrable function φ on $\mathbf{R}^n$ is called a *wavelet* if the family of functions

$$\varphi_{\nu,k}(x) = 2^{\frac{\nu n}{2}}\varphi(2^\nu x - k),$$

where ν ranges over $\mathbf{Z}$ and k over $\mathbf{Z}^n$, is an orthonormal basis of $L^2(\mathbf{R}^n)$. Note that the Fourier transform of $\varphi_{\nu,k}$ is given by

$$\widehat{\varphi_{\nu,k}}(\xi) = 2^{-\frac{\nu n}{2}}\widehat{\varphi}(2^{-\nu}\xi)e^{-2\pi i 2^{-\nu}\xi\cdot k}. \tag{5.6.1}$$

Rephrasing the question posed earlier, the main issue addressed in this section is whether smooth wavelets actually exist. Before we embark on this topic, we recall that we have already encountered examples of nonsmooth wavelets.

Example 5.6.2. (The Haar wavelet) Recall the family of functions

$$h_I(x) = |I|^{-\frac{1}{2}}(\chi_{I_L} - \chi_{I_R}),$$

where I ranges over $\mathscr{D}$ (the set of all dyadic intervals) and I_L is the left part of I and I_R is the right part of I. Note that if $I = [2^{-\nu}k, 2^{-\nu}(k+1))$, then

$$h_I(x) = 2^{\frac{\nu}{2}}\varphi(2^\nu x - k),$$

where

$$\varphi(x) = \chi_{[0,\frac{1}{2})} - \chi_{[\frac{1}{2},1)}. \tag{5.6.2}$$

The single function φ in (5.6.2) therefore generates the Haar basis by taking translations and dilations. Moreover, we observed in Section 5.4 that the family $\{h_I\}_I$ is orthonormal. In Theorem 5.4.6 we obtained the representation

$$f = \sum_{I\in\mathscr{D}} \langle f, h_I\rangle h_I \qquad \text{in } L^2,$$

which proves the completeness of the system $\{h_I\}_{I\in\mathscr{D}}$ in $L^2(\mathbf{R})$.

5.6.1 Some Preliminary Facts

Before we look at more examples, we make some observations. We begin with the following useful fact.

Proposition 5.6.3. *Let $g \in L^1(\mathbf{R}^n)$. Then*

$$\widehat{g}(m) = 0 \qquad \text{for all } m \in \mathbf{Z}^n \setminus \{0\}$$

if and only if

$$\sum_{k \in \mathbf{Z}^n} g(x+k) = \int_{\mathbf{R}^n} g(t)\,dt$$

for almost all $x \in \mathbf{T}^n$.

Proof. We define the periodized function

$$G(x) = \sum_{k \in \mathbf{Z}^n} g(x+k),$$

which is easily shown to be in $L^1(\mathbf{T}^n)$. Moreover, we have

$$\widehat{G}(m) = \widehat{g}(m)$$

for all $m \in \mathbf{Z}^n$, where $\widehat{G}(m)$ denotes the mth Fourier coefficient of G and $\widehat{g}(m)$ denotes the Fourier transform of g at $\xi = m$. If $\widehat{g}(m) = 0$ for all $m \in \mathbf{Z}^n \setminus \{0\}$, then all the Fourier coefficients of G (except for $m = 0$) vanish, which means that the sequence $\{\widehat{G}\}_{m \in \mathbf{Z}^n}$ lies in $\ell^1(\mathbf{Z}^n)$ and hence Fourier inversion applies. We conclude that for almost all $x \in \mathbf{T}^n$ we have

$$G(x) = \sum_{m \in \mathbf{Z}^n} \widehat{G}(m) e^{2\pi i m \cdot x} = \widehat{G}(0) = \widehat{g}(0) = \int_{\mathbf{R}^n} g(t)\,dt\,.$$

Conversely, if G is a constant, then $\widehat{G}(m) = 0$ for all $m \in \mathbf{Z}^n \setminus \{0\}$, and so the same holds for g. $\square$

A consequence of the previous proposition is the following.

Proposition 5.6.4. *Let $\varphi \in L^2(\mathbf{R}^n)$. Then the sequence*

$$\{\varphi(x-k)\}_{k \in \mathbf{Z}^n} \tag{5.6.3}$$

forms an orthonormal set in $L^2(\mathbf{R}^n)$ if and only if

$$\sum_{k \in \mathbf{Z}^n} |\widehat{\varphi}(\xi + k)|^2 = 1 \tag{5.6.4}$$

for almost all $\xi \in \mathbf{R}^n$.

Proof. Observe that either (5.6.4) or the hypothesis that the sequence in (5.6.3) is orthonormal implies that $\|\varphi\|_{L^2} = 1$. Also the orthonormality condition

$$\int_{\mathbf{R}^n} \varphi(x-j)\overline{\varphi(x-k)}\,dx = \begin{cases} 1 & \text{when } j = k, \\ 0 & \text{when } j \neq k, \end{cases}$$

is equivalent to

$$\int_{\mathbf{R}^n} e^{-2\pi i k\cdot\xi}\widehat{\varphi}(\xi)\overline{e^{-2\pi i j\cdot\xi}\widehat{\varphi}(\xi)}\,d\xi = (|\widehat{\varphi}|^2)\widehat{\ }(k-j) = \begin{cases} 1 & \text{when } j = k, \\ 0 & \text{when } j \neq k, \end{cases}$$

in view of Parseval's identity. Proposition 5.6.3 gives that the latter is equivalent to

$$\sum_{k\in\mathbf{Z}^n} |\widehat{\varphi}(\xi+k)|^2 = \int_{\mathbf{R}^n} |\widehat{\varphi}(t)|^2\,dt = 1$$

for almost all $\xi \in \mathbf{R}^n$. □

Corollary 5.6.5. *Let $\varphi \in L^1(\mathbf{R}^n)$ and suppose that the sequence*

$$\{\varphi(x-k)\}_{k\in\mathbf{Z}^n} \tag{5.6.5}$$

forms an orthonormal set in $L^2(\mathbf{R}^n)$. Then the measure of the support of $\widehat{\varphi}$ is at least 1, that is,

$$|\operatorname{supp}\widehat{\varphi}| \geq 1\,. \tag{5.6.6}$$

If $|\operatorname{supp}\widehat{\varphi}| = 1$, then $|\widehat{\varphi}(\xi)| = 1$ for almost all $\xi \in \operatorname{supp}\widehat{\varphi}$.

Proof. It follows from (5.6.4) that $|\widehat{\varphi}| \leq 1$ for almost all $\xi \in \mathbf{R}^n$. Therefore,

$$|\operatorname{supp}\widehat{\varphi}| \geq \int_{\mathbf{R}^n} |\widehat{\varphi}(\xi)|^2\,d\xi = \int_{\mathbf{T}^n}\sum_{k\in\mathbf{Z}^n} |\widehat{\varphi}(\xi+k)|^2\,d\xi = \int_{\mathbf{T}^n} 1\,d\xi = 1\,.$$

It follows from the previous series of inequalities that if equality holds in (5.6.6), then $|\widehat{\varphi}(\xi)| = 1$ for almost all ξ in $\operatorname{supp}\widehat{\varphi}$. □

5.6.2 Construction of a Nonsmooth Wavelet

Having established these preliminary facts, we now start searching for examples of wavelets. It follows from Corollary 5.6.5 that the support of the Fourier transform of a wavelet must have measure at least 1. It is reasonable to ask whether this support can have measure exactly 1. Example 5.6.6 indicates that this can indeed happen. As dictated by the same corollary, the Fourier transform of such a wavelet must satisfy $|\widehat{\varphi}(\xi)| = 1$ for almost all $\xi \in \operatorname{supp}\widehat{\varphi}$, so it is natural to look for a wavelet φ such that $\widehat{\varphi} = \chi_A$ for some set A. We can start by asking whether the function

$$\widehat{\varphi} = \chi_{[-\frac{1}{2},\frac{1}{2}]}$$

on $\mathbf{R}$ is an appropriate Fourier transform of a wavelet, but a moment's thought shows that the functions $\varphi_{\mu,0}$ and $\varphi_{\nu,0}$ cannot be orthogonal to each other when $\mu \neq 0$. The problem here is that the Fourier transforms of the functions $\varphi_{\nu,k}$ cluster near the origin and do not allow for the needed orthogonality. We can fix this problem by considering a function whose Fourier transform vanishes near the origin. Among such functions, a natural candidate is

$$\chi_{[-1,-\frac{1}{2})} + \chi_{[\frac{1}{2},1)}, \tag{5.6.7}$$

which is indeed the Fourier transform of a wavelet.

Example 5.6.6. Let $A = [-1,-\frac{1}{2}) \bigcup [\frac{1}{2},1)$ and define a function φ on $\mathbf{R}^n$ by setting

$$\widehat{\varphi} = \chi_{A^n}.$$

Then we assert that the family of functions

$$\{2^{\nu n/2}\varphi(2^\nu x - k)\}_{k\in\mathbf{Z}^n,\nu\in\mathbf{Z}}$$

is an orthonormal basis of $L^2(\mathbf{R}^n)$ (i.e., the function φ is a wavelet). This is an example of a wavelet with *minimally supported frequency*.

To see this assertion, first note that $\{\varphi_{0,k}\}_{k\in\mathbf{Z}^n}$ is an orthonormal set, since (5.6.4) is easily seen to hold. Dilating by 2^ν, it follows that $\{\varphi_{\nu,k}\}_{k\in\mathbf{Z}^n}$ is also an orthonormal set for every fixed $\nu \in \mathbf{Z}$. Second, observe that if $\mu \neq \nu$, then

$$\operatorname{supp}\widehat{\varphi_{\nu,k}} \cap \operatorname{supp}\widehat{\varphi_{\mu,l}} = \emptyset. \tag{5.6.8}$$

This implies that the family $\{2^{\nu n/2}\varphi(2^\nu x - k)\}_{k\in\mathbf{Z}^n,\nu\in\mathbf{Z}}$ is also orthonormal.

Finally, we need to show completeness. Here we use Exercise 5.6.2 to write

$$(\varphi_{2^{-\nu}} * f)\widehat{\ }(\xi) = 2^{-\nu n}\sum_{k\in\mathbf{Z}^n}(\varphi_{2^{-\nu}} * f)(-\tfrac{k}{2^\nu})e^{2\pi i\frac{k}{2^\nu}\xi}, \quad \xi \in A^n, \tag{5.6.9}$$

where the series converges in $L^2(A^n)$. Next we observe that the following identity holds for $\widehat{\varphi}$:

$$\sum_{\nu\in\mathbf{Z}}|\widehat{\varphi}(2^\nu\xi)|^2 = 1, \qquad \xi \neq 0. \tag{5.6.10}$$

This implies that for all f in $\mathscr{S}(\mathbf{R}^n)$ we have

$$f = \sum_{\nu\in\mathbf{Z}}\varphi_{2^{-\nu}} * \varphi_{2^{-\nu}} * f = \left[\sum_{\nu\in\mathbf{Z}}\widehat{\varphi_{2^{-\nu}}}\,(\varphi_{2^{-\nu}} * f)\widehat{\ }\right]^{\vee}, \tag{5.6.11}$$

where the series converges in L^2. Inserting in (5.6.11) the value of $(\varphi_{2^{-\nu}} * f)\widehat{\ }$ given in identity (5.6.9), we obtain

$$\begin{aligned} f(x) &= 2^{-\nu n} \sum_{\nu \in \mathbf{Z}} \sum_{k \in \mathbf{Z}^n} (\varphi_{2^{-\nu}} * f)(-\tfrac{k}{2^\nu}) \big[\widehat{\varphi_{2^{-\nu}}}(\xi) e^{2\pi i \frac{k}{2^\nu} \xi}\big]^\vee (x) \\ &= 2^{-\nu n} \sum_{\nu \in \mathbf{Z}} \sum_{k \in \mathbf{Z}^n} (\varphi_{2^{-\nu}} * f)(-\tfrac{k}{2^\nu}) \varphi_{2^{-\nu}}(x + \tfrac{k}{2^\nu}) \\ &= \sum_{\nu \in \mathbf{Z}} \sum_{k \in \mathbf{Z}^n} \langle f, \varphi_{\nu,k} \rangle \varphi_{\nu,k}(x), \end{aligned}$$

where the double series converges in $L^2(\mathbf{R}^n)$. This shows that every Schwartz function can be written as an L^2 sum of $\varphi_{\nu,k}$'s, and by density the same is true for every square integrable f.

5.6.3 Construction of a Smooth Wavelet

The wavelet basis of $L^2(\mathbf{R}^n)$ constructed in Example 5.6.6 is forced to have slow decay at infinity, since the Fourier transforms of the elements of the basis are non-smooth. Smoothing out the function $\widehat{\varphi}$ but still expecting φ to be wavelet is a bit tricky, since property (5.6.8) may be violated when $\mu \neq \nu$, and moreover, (5.6.4) may be destroyed. These two obstacles are overcome by the careful construction of the next theorem.

Theorem 5.6.7. *There exists a Schwartz function φ on the real line that is a wavelet, that is, the collection of functions $\{\varphi_{\nu,k}\}_{k,\nu\in\mathbf{Z}}$ with $\varphi_{\nu,k}(x) = 2^{\frac{\nu}{2}}\varphi(2^\nu x - k)$ is an orthonormal basis of $L^2(\mathbf{R})$. Moreover, the function φ can be constructed so that its Fourier transform satisfies*

$$\operatorname{supp}\widehat{\varphi} \subseteq \big[-\tfrac{4}{3}, -\tfrac{1}{3}\big] \cup \big[\tfrac{1}{3}, \tfrac{4}{3}\big]. \tag{5.6.12}$$

Note that in view of condition (5.6.12), the function φ must have vanishing moments of all orders.

Proof. We start with an odd smooth real-valued function Θ on the real line such that $\Theta(t) = \frac{\pi}{4}$ for $t \geq \frac{1}{6} - 10^{-10}$ and such that Θ is increasing on the interval $\big[-\frac{1}{6}, \frac{1}{6}\big]$. We set

$$\alpha(t) = \sin(\Theta(t) + \tfrac{\pi}{4}), \qquad \beta(t) = \cos(\Theta(t) + \tfrac{\pi}{4}),$$

and we observe that

$$\alpha(t)^2 + \beta(t)^2 = 1$$

and that

$$\alpha(-t) = \beta(t)$$

for all real t. Next we introduce the smooth function ω defined via

$$\omega(t) = \begin{cases} \beta(-\frac{t}{2} - \frac{1}{2}) = \alpha(\frac{t}{2} + \frac{1}{2}) & \text{when } t \in \left[-\frac{4}{3}, -\frac{2}{3}\right], \\ \alpha(-t - \frac{1}{2}) = \beta(t + \frac{1}{2}) & \text{when } t \in \left[-\frac{2}{3}, -\frac{1}{3}\right], \\ \alpha(t - \frac{1}{2}) & \text{when } t \in \left[\frac{1}{3}, \frac{2}{3}\right], \\ \beta(\frac{t}{2} - \frac{1}{2}) & \text{when } t \in \left[\frac{2}{3}, \frac{4}{3}\right], \end{cases}$$

on the interval $\left[-\frac{4}{3}, -\frac{1}{3}\right] \cup \left[\frac{1}{3}, \frac{4}{3}\right]$. Note that ω is an even function. Finally we define the function φ by letting

$$\widehat{\varphi}(\xi) = e^{-\pi i \xi} \omega(\xi),$$

and we note that

$$\varphi(x) = \int_{\mathbf{R}} \omega(\xi) e^{2\pi i \xi (x - \frac{1}{2})} d\xi = 2 \int_0^\infty \omega(\xi) \cos\left(2\pi(x - \tfrac{1}{2})\xi\right) d\xi .$$

It follows that the function φ is symmetric about the number $\frac{1}{2}$, that is, we have

$$\varphi(x) = \varphi(1 - x)$$

for all $x \in \mathbf{R}$. Note that φ is a Schwartz function whose Fourier transform is supported in the set $\left[-\frac{4}{3}, -\frac{1}{3}\right] \cup \left[\frac{1}{3}, \frac{4}{3}\right]$.

Having defined φ, we proceed by showing that it is a wavelet. In view of identity (5.6.1) we have that $\widehat{\varphi_{\nu,k}}$ is supported in the set $\frac{1}{3}2^\nu \le |\xi| \le \frac{4}{3}2^\nu$, while $\widehat{\varphi_{\mu,j}}$ is supported in the set $\frac{1}{3}2^\mu \le |\xi| \le \frac{4}{3}2^\mu$. The intersection of these sets has measure zero when $|\mu - \nu| \ge 2$, which implies that such wavelets are orthogonal to each other. Therefore, it suffices to verify orthogonality between adjacent scales (i.e., when $\nu = \mu$ and $\nu = \mu + 1$).

We begin with the case $\nu = \mu$, which, by a simple dilation, is reduced to the case $\nu = \mu = 0$. Thus to obtain the orthogonality of the functions $\varphi_{0,k}(x) = \varphi(x - k)$ and $\varphi_{0,j}(x) = \varphi(x - j)$, in view of Proposition 5.6.4, it suffices to show that

$$\sum_{k \in \mathbf{Z}} |\widehat{\varphi}(\xi + k)|^2 = 1 . \tag{5.6.13}$$

Since the sum in (5.6.13) is 1-periodic, we check that is equal to 1 only for ξ in $\left[\frac{1}{3}, \frac{4}{3}\right]$. First for $\xi \in \left[\frac{1}{3}, \frac{2}{3}\right]$, the sum in (5.6.13) is equal to

$$\begin{aligned} |\widehat{\varphi}(\xi)|^2 + |\widehat{\varphi}(\xi - 1)|^2 &= \omega(\xi)^2 + \omega(\xi - 1)^2 \\ &= \alpha(\xi - \tfrac{1}{2})^2 + \beta((\xi - 1) + \tfrac{1}{2})^2 \\ &= 1 \end{aligned}$$

from the definition of ω. A similar argument also holds for $\xi \in \left[\frac{2}{3}, \frac{4}{3}\right]$, and this completes the proof of (5.6.13). As a consequence of this identity we also obtain that the functions $\varphi_{0,k}$ have L^2 norm equal to 1, and thus so have the functions $\varphi_{\nu,k}$, via a change of variables.

Next we prove the orthogonality of the functions $\varphi_{\nu,k}$ and $\varphi_{\nu+1,j}$ for general $\nu, k, j \in \mathbf{Z}$. We begin by observing the validity of the following identity:

$$\widehat{\varphi}(\xi)\overline{\widehat{\varphi}(\tfrac{\xi}{2})} = \begin{cases} e^{-\pi i\xi/2}\beta(\frac{\xi}{2}-\frac{1}{2})\alpha(\frac{\xi}{2}-\frac{1}{2}) & \text{when } \frac{2}{3}\le\xi\le\frac{4}{3}, \\ e^{-\pi i\xi/2}\alpha(\frac{\xi}{2}+\frac{1}{2})\beta(\frac{\xi}{2}+\frac{1}{2}) & \text{when } -\frac{4}{3}\le\xi\le-\frac{2}{3}. \end{cases} \tag{5.6.14}$$

Indeed, from the definition of φ, it follows that

$$\widehat{\varphi}(\xi)\overline{\widehat{\varphi}(\tfrac{\xi}{2})} = e^{-\pi i\xi/2}\omega(\xi)\omega(\tfrac{\xi}{2}).$$

This function is supported in

$$\{\xi\in\mathbf{R}:\ \tfrac{1}{3}\le|\xi|\le\tfrac{4}{3}\}\cap\{\xi\in\mathbf{R}:\ \tfrac{2}{3}\le|\xi|\le\tfrac{8}{3}\} = \{\xi\in\mathbf{R}:\ \tfrac{2}{3}\le|\xi|\le\tfrac{4}{3}\},$$

and on this set it is equal to

$$e^{-\pi i\xi/2}\begin{cases} \beta(\frac{\xi}{2}-\frac{1}{2})\alpha(\frac{\xi}{2}-\frac{1}{2}) & \text{when } \frac{2}{3}\le\xi\le\frac{4}{3}, \\ \alpha(\frac{\xi}{2}+\frac{1}{2})\beta(\frac{\xi}{2}+\frac{1}{2}) & \text{when } -\frac{4}{3}\le\xi\le-\frac{2}{3}, \end{cases}$$

by the definition of ω. This establishes (5.6.14).

We now turn to the orthogonality of the functions $\varphi_{\nu,k}$ and $\varphi_{\nu+1,j}$ for general $\nu,k,j\in\mathbf{Z}$. Using (5.6.1) and (5.6.14) we have

$$\begin{aligned}
\langle\varphi_{\nu,k}\,|\,\varphi_{\nu+1,j}\rangle &= \langle\widehat{\varphi_{\nu,k}}\,|\,\widehat{\varphi_{\nu+1,j}}\rangle \\
&= \int_{\mathbf{R}} 2^{-\frac{\nu}{2}}\widehat{\varphi}(2^{-\nu}\xi)e^{-2\pi i\frac{\xi k}{2^\nu}}\,\overline{2^{-\frac{\nu+1}{2}}\widehat{\varphi}(2^{-(\nu+1)}\xi)e^{-2\pi i\frac{\xi j}{2^{\nu+1}}}}\,d\xi \\
&= \frac{1}{\sqrt{2}}\int_{\mathbf{R}}\widehat{\varphi}(\xi)\overline{\widehat{\varphi}(\tfrac{\xi}{2})}e^{-2\pi i\xi(k-\frac{j}{2})}\,d\xi \\
&= \frac{1}{\sqrt{2}}\int_{-\frac{4}{3}}^{-\frac{2}{3}}\alpha(\tfrac{\xi}{2}+\tfrac{1}{2})\beta(\tfrac{\xi}{2}+\tfrac{1}{2})e^{-2\pi i\xi(k-\frac{j}{2}+\frac{1}{4})}\,d\xi \\
&\qquad + \frac{1}{\sqrt{2}}\int_{\frac{2}{3}}^{\frac{4}{3}}\alpha(\tfrac{\xi}{2}-\tfrac{1}{2})\beta(\tfrac{\xi}{2}-\tfrac{1}{2})e^{-2\pi i\xi(k-\frac{j}{2}+\frac{1}{4})}\,d\xi \\
&= 0,
\end{aligned}$$

where the last identity follows from the change of variables $\xi=\xi'-2$ in the second-to-last integral, which transforms its range of integration to $\left[\frac{2}{3},\frac{4}{3}\right]$ and its integrand to the negative of that of the last displayed integral.

Our final task is to show that the orthonormal system $\{\varphi_{\nu,k}\}_{\nu,k\in\mathbf{Z}}$ is complete. We show this by proving that whenever a square-integrable function f satisfies

$$\langle f\,|\,\varphi_{\nu,k}\rangle = 0 \tag{5.6.15}$$

for all $\nu,k\in\mathbf{Z}$, then f must be zero. Suppose that (5.6.15) holds. Plancherel's identity yields

$$\int_{\mathbf{R}}\widehat{f}(\xi)\overline{2^{-\frac{\nu}{2}}\widehat{\varphi}(2^{-\nu}\xi)e^{-2\pi i 2^{-\nu}\xi k}}\,d\xi = 0$$

for all ν, k and thus

$$\int_{\mathbf{R}} \widehat{f}(2^\nu \xi)\widehat{\varphi}(\xi)e^{2\pi i \xi k}\, d\xi = \big(\widehat{f}(2^\nu(\cdot))\,\widehat{\varphi}\big)\widehat{\ }(-k) = 0 \tag{5.6.16}$$

for all $\nu, k \in \mathbf{Z}$. It follows from Proposition 5.6.3 and (5.6.16) (with $k = 0$) that

$$\sum_{k\in\mathbf{Z}} \widehat{f}(2^\nu(\xi+k))\widehat{\varphi}(\xi+k) = \int_{\mathbf{R}} \widehat{f}(2^\nu \xi)\,\widehat{\varphi}(\xi)\,d\xi = \big(\widehat{f}(2^\nu(\cdot))\,\widehat{\varphi}\big)\widehat{\ }(0) = 0$$

for all $\nu \in \mathbf{Z}$.

Next, we show that the identity

$$\sum_{k\in\mathbf{Z}} \widehat{f}(2^\nu(\xi+k))\widehat{\varphi}(\xi+k) = 0 \tag{5.6.17}$$

for all $\nu \in \mathbf{Z}$ implies that $\widehat{f}$ is identically equal to zero. Suppose that $\frac{1}{3} \le \xi \le \frac{2}{3}$. In this case the support properties of $\widehat{\varphi}$ imply that the only terms in the sum in (5.6.17) that do not vanish are $k = 0$ and $k = -1$. Thus for $\frac{1}{3} \le \xi \le \frac{2}{3}$ the identity in (5.6.17) reduces to

$$\begin{aligned} 0 &= \widehat{f}(2^\nu(\xi-1))\widehat{\varphi}(\xi-1) + \widehat{f}(2^\nu \xi)\widehat{\varphi}(\xi) \\ &= \widehat{f}(2^\nu(\xi-1))e^{-\pi i(\xi-1)}\beta((\xi-1)+\tfrac{1}{2}) + \widehat{f}(2^\nu \xi)e^{-\pi i \xi}\alpha(\xi-\tfrac{1}{2}); \end{aligned}$$

hence

$$-\widehat{f}(2^\nu(\xi-1))\beta(\xi-\tfrac{1}{2}) + \widehat{f}(2^\nu \xi)\alpha(\xi-\tfrac{1}{2}) = 0, \quad \tfrac{1}{3} \le \xi \le \tfrac{2}{3}. \tag{5.6.18}$$

Next we observe that when $\frac{2}{3} \le \xi \le \frac{4}{3}$, only the terms with $k = 0$ and $k = -2$ survive in the identity in (5.6.17). This is because when $k = -1$, $\xi + k = \xi - 1 \in \big[-\frac{1}{3}, \frac{1}{3}\big]$ and this interval has null intersection with the support of $\widehat{\varphi}$. Therefore, (5.6.17) reduces to

$$\begin{aligned} 0 &= \widehat{f}(2^\nu(\xi-2))\widehat{\varphi}(\xi-2) + \widehat{f}(2^\nu \xi)\widehat{\varphi}(\xi) \\ &= \widehat{f}(2^\nu(\xi-2))e^{-\pi i(\xi-2)}\alpha(\tfrac{\xi-2}{2}+\tfrac{1}{2}) + \widehat{f}(2^\nu \xi)e^{-\pi i \xi}\beta(\tfrac{\xi}{2}-\tfrac{1}{2}); \end{aligned}$$

hence

$$\widehat{f}(2^\nu(\xi-2))\alpha(\tfrac{\xi}{2}-\tfrac{1}{2}) + \widehat{f}(2^\nu \xi)\beta(\tfrac{\xi}{2}-\tfrac{1}{2}) = 0, \quad \tfrac{2}{3} \le \xi \le \tfrac{4}{3}. \tag{5.6.19}$$

Replacing first ν by $\nu - 1$ and then $\frac{\xi}{2}$ by ξ in (5.6.19), we obtain

$$\widehat{f}(2^\nu(\xi-1))\alpha(\xi-\tfrac{1}{2}) + \widehat{f}(2^\nu \xi)\beta(\xi-\tfrac{1}{2}) = 0, \quad \tfrac{1}{3} \le \xi \le \tfrac{2}{3}. \tag{5.6.20}$$

Now consider the 2×2 system of equations given by (5.6.18) and (5.6.20) with unknown $\widehat{f}(2^\nu(\xi-1))$ and $\widehat{f}(2^\nu \xi)$. The determinant of the system is

$$\det\begin{pmatrix} -\beta(\xi-1/2) & \alpha(\xi-1/2) \\ \alpha(\xi-1/2) & \beta(\xi-1/2) \end{pmatrix} = -1 \neq 0.$$

Therefore, the system has the unique solution

$$\widehat{f}(2^\nu(\xi-1)) = \widehat{f}(2^\nu\xi) = 0,$$

which is valid for all $\nu \in \mathbf{Z}$ and all $\xi \in [\frac{1}{3}, \frac{2}{3}]$. We conclude that $\widehat{f}(\xi) = 0$ for all $\xi \in \mathbf{R}$ and thus $f = 0$. This proves the completeness of the system $\{\varphi_{\nu,k}\}$. We conclude that the function φ is a wavelet. □

5.6.4 A Sampling Theorem

We end this section by discussing how one can recover a band-limited function by its values at a countable number of points.

Definition 5.6.8. An integrable function on $\mathbf{R}^n$ is called *band limited* if its Fourier transform has compact support.

For every band-limited function there is a $B > 0$ such that its Fourier transform is supported in the cube $[-B,B]^n$. In such a case we say that the function is band limited on the cube $[-B,B]^n$.

It is an interesting observation that such functions are completely determined by their values at the points $x = k/2B$, where $k \in \mathbf{Z}^n$. We have the following result.

Theorem 5.6.9. *Let f be band limited on the cube $[-B,B]^n$. Then f can be sampled by its values at the points $x = k/2B$, where $k \in \mathbf{Z}^n$. In particular, we have*

$$f(x_1,\dots,x_n) = \sum_{k\in\mathbf{Z}^n} f\Big(\frac{k}{2B}\Big)\prod_{j=1}^{n}\frac{\sin(2\pi Bx_j - \pi k_j)}{2\pi Bx_j - \pi k_j} \tag{5.6.21}$$

for all $x \in \mathbf{R}^n$.

Proof. Since the function $\widehat{f}$ is supported in $[-B,B]^n$, we use Exercise 5.6.2 to obtain

$$\begin{aligned}\widehat{f}(\xi) &= \frac{1}{(2B)^n}\sum_{k\in\mathbf{Z}^n}\widehat{\widehat{f}}\Big(\frac{k}{2B}\Big)e^{2\pi i\frac{k}{2B}\cdot\xi}\\ &= \frac{1}{(2B)^n}\sum_{k\in\mathbf{Z}^n} f\Big(-\frac{k}{2B}\Big)e^{2\pi i\frac{k}{2B}\cdot\xi}.\end{aligned}$$

Inserting this identity in the inversion formula

$$f(x) = \int_{[-B,B]^n}\widehat{f}(\xi)e^{2\pi i x\cdot\xi}\,d\xi,$$

which holds since $\widehat{f}$ is continuous and therefore integrable over $[-B,B]^n$, we obtain

$$
\begin{aligned}
f(x) &= \int_{[-B,B]^n} \frac{1}{(2B)^n} \sum_{k\in\mathbf{Z}^n} f\Big(-\frac{k}{2B}\Big) e^{2\pi i \frac{k}{2B}\cdot\xi} e^{2\pi i x\cdot\xi}\, d\xi \\
&= \sum_{k\in\mathbf{Z}^n} f\Big(-\frac{k}{2B}\Big) \frac{1}{(2B)^n} \int_{[-B,B]^n} e^{2\pi i(\frac{k}{2B}+x)\cdot\xi}\, d\xi \\
&= \sum_{k\in\mathbf{Z}^n} f\Big(-\frac{k}{2B}\Big) \prod_{j=1}^{n} \frac{\sin(2\pi Bx_j+\pi k_j)}{2\pi Bx_j+\pi k_j}.
\end{aligned}
$$

This is exactly (5.6.21) when we change k to $-k$. □

Remark 5.6.10. Identity (5.6.21) holds for any $B' > B$. In particular, we have

$$
\sum_{k\in\mathbf{Z}^n} f\Big(\frac{k}{2B}\Big) \prod_{j=1}^{n} \frac{\sin(2\pi Bx_j-\pi k_j)}{2\pi Bx_j-\pi k_j} = \sum_{k\in\mathbf{Z}^n} f\Big(\frac{k}{2B'}\Big) \prod_{j=1}^{n} \frac{\sin(2\pi B'x_j-\pi k_j)}{2\pi B'x_j-\pi k_j}
$$

for all $x \in \mathbf{R}^n$ whenever f is band-limited in $[-B,B]^n$. In particular, band-limited functions in $[-B,B]^n$ can be sampled by their values at the points $k/2B'$ for any $B' \geq B$.

However, band-limited functions in $[-B,B]^n$ cannot be sampled by the points $k/2B'$ for any $B' < B$, as the following example indicates.

Example 5.6.11. For $0 < B' < B$, let $f(x) = g(x)\sin(2\pi B'x)$, where $\widehat{g}$ is supported in the interval $[-(B-B'), B-B']$. Then f is band limited in $[-B,B]$, but it cannot be sampled by its values at the points $k/2B'$, since it vanishes at these points and f is not identically zero if g is not the zero function.

Exercises

5.6.1. (a) Let $A = [-1,-\frac{1}{2}) \bigcup [\frac{1}{2},1)$. Show that the family $\{e^{2\pi i m x}\}_{m\in\mathbf{Z}}$ is an orthonormal basis of $L^2(A)$.
(b) Obtain the same conclusion for the family $\{e^{2\pi i m\cdot x}\}_{m\in\mathbf{Z}^n}$ in $L^2(A^n)$.
[*Hint:* To show completeness, given $f \in L^2(A)$, define h on $[0,1]$ by setting $h(x) = f(x-1)$ for $x \in [0,\frac{1}{2})$ and $h(x) = f(x)$ for $x \in [\frac{1}{2},1)$. Observe that $\widehat{h}(m) = \widehat{f}(m)$ for all $m \in \mathbf{Z}$ and expand h in Fourier series.]

5.6.2. (a) Suppose that g is supported in $[-b,b]^n$ for some $b > 0$ and that the sequence $\{\widehat{g}(k/2b)\}_{k\in\mathbf{Z}^n}$ lies in $\ell^2(\mathbf{Z}^n)$. Show that

$$
g(x) = (2b)^{-n} \sum_{k\in\mathbf{Z}^n} \widehat{g}(\tfrac{k}{2b}) e^{2\pi i \frac{k}{2b}\cdot x}
$$

when $x \in [-b,b]^n$, where the series converges in $L^2(\mathbf{R}^n)$.
(b) Suppose that g is supported in $[0,b]^n$ for some $b > 0$ and that the sequence $\{\widehat{g}(k/b)\}_{k\in\mathbf{Z}^n}$ lies in $\ell^2(\mathbf{Z}^n)$. Show that

$$g(x) = b^{-n} \sum_{k \in \mathbf{Z}^n} \widehat{g}(\tfrac{k}{b}) e^{2\pi i \frac{k}{b} \cdot x}$$

for $x \in [0,b]^n$, where the series converges in $L^2(\mathbf{R}^n)$.
(c) When $n = 1$, obtain the same as the conclusion in part (b) for $x \in [-b, -\frac{b}{2}) \bigcup [\frac{b}{2}, b)$, provided g is supported in this set.
[*Hint:* All the results follow by dilations. Part (c): Use the result in Exercise 5.6.2.]

5.6.3. Show that the sequence of functions

$$H_k(x_1, \dots, x_n) = (2B)^{\frac{n}{2}} \prod_{j=1}^{n} \frac{\sin\big(\pi(2Bx_j - k_j)\big)}{\pi(2Bx_j - k_j)}, \qquad k \in \mathbf{Z}^n,$$

is orthonormal in $L^2(\mathbf{R}^n)$.
[*Hint:* Interpret the functions H_k as the Fourier transforms of known functions.]

5.6.4. Prove the following spherical multidimensional version of Theorem 5.6.9. Suppose that $\widehat{f}$ is supported in the ball $|\xi| \le R$. Show that

$$f(x) = \sum_{k \in \mathbf{Z}^n} f\big(-\tfrac{k}{2R}\big) \frac{1}{2^n} \frac{J_{\frac{n}{2}}(2\pi |Rx + \frac{k}{2}|)}{|Rx + \frac{k}{2}|^{\frac{n}{2}}},$$

where J_a is the Bessel function of order a.

5.6.5. Let $\{a_k\}_{k \in \mathbf{Z}^n}$ be in ℓ^p for some $1 < p < \infty$. Show that the sum

$$\sum_{k \in \mathbf{Z}^n} a_k \prod_{j=1}^{n} \frac{\sin(2\pi B x_j - \pi k_j)}{2\pi B x_j - \pi k_j}$$

converges in $\mathscr{S}'(\mathbf{R}^n)$ to an L^p function A on $\mathbf{R}^n$ that is band limited in $[-B,B]^n$. Moreover, the L^p norm of A is controlled by a constant multiple of the ℓ^p norm of $\{a_k\}_k$.

5.6.6. (a) Suppose that f is a tempered distribution on $\mathbf{R}^n$ whose Fourier transform is supported in the ball $B(0, (1-\varepsilon)\frac{1}{2})$ for some $\varepsilon > 0$. Show that for all $0 < p \le \infty$ there is a constant $C_{n,p,\varepsilon}$ such that

$$\big\|f\big\|_{L^p(\mathbf{R}^n)} \le C_{n,p,\varepsilon} \big\|\{f(k)\}_k\big\|_{\ell^p(\mathbf{Z}^n)}.$$

In particular, if the values $\{f(k)\}_{k \in \mathbf{Z}^n}$ form an ℓ^p sequence, then f must coincide with an L^p function.
(b) Consider functions of the form $\sin(\pi x)/(\pi x)$ on $\mathbf{R}$ to construct a counterexample to the statement in part (a) when $\varepsilon = 0$.
[*Hint:* Take a Schwartz function Φ whose Fourier transform is supported in $B(0, \frac{1}{2})$ and that is identically equal to 1 on the support of $\widehat{f}$. Then $f = f * \Phi$. Apply Theorem 5.6.9 to the function $f * \Phi$ and use the rapid decay of Φ to sum the series.]

5.6.7. (a) Let $\psi(x)$ be a nonzero continuous integrable function on $\mathbf{R}$ that satisfies $\int_{\mathbf{R}} \psi(x)\,dx = 0$ and

$$C_\psi = 2\pi \int_{-\infty}^{+\infty} \frac{|\widehat{\psi}(t)|^2}{|t|}\,dt < \infty.$$

Define the *wavelet transform* of f in $L^2(\mathbf{R})$ by setting

$$W(f;a,b)(x) = \frac{1}{\sqrt{|a|}} \int_{-\infty}^{+\infty} f(x)\overline{\psi\Big(\frac{x-b}{a}\Big)}\,dx$$

when $a \neq 0$ and $W(f;0,b) = 0$. Show that for any $f \in L^2(\mathbf{R})$ the following inversion formula holds:

$$f(x) = \frac{1}{C_\psi} \int_{-\infty}^{+\infty} \int_{-\infty}^{+\infty} \frac{1}{|a|^{\frac{1}{2}}} \psi\Big(\frac{x-b}{a}\Big) W(f;a,b)\,db\,\frac{da}{a^2}.$$

(b) State and prove an analogous wavelet transform inversion property on $\mathbf{R}^n$.
[*Hint:* Apply Theorem 2.2.14 (5) in the b-integral to reduce matters to Fourier inversion.]

5.6.8. (*P. Casazza*) On $\mathbf{R}^n$ let e_j be the vector whose coordinates are zero everywhere except for the jth entry, which is 1. Set $q_j = e_j - \frac{1}{n}\sum_{k=1}^n e_k$ for $1 \le j \le n$ and also $q_{n+1} = \frac{1}{\sqrt{n}}\sum_{k=1}^n e_k$. Prove that

$$\sum_{j=1}^{n+1} |q_j \cdot x| = |x|^2$$

for all $x \in \mathbf{R}^n$. This provides an example of a *tight frame* on $\mathbf{R}^n$.

HISTORICAL NOTES

An early account of square functions in the context of Fourier series appears in the work of Kolmogorov [157], who proved the almost everywhere convergence of lacunary partial sums of Fourier series of periodic square-integrable functions. This result was systematically studied and extended to L^p functions, $1 < p < \infty$, by Littlewood and Paley [174], [175], [176] using complex-analysis techniques. The real-variable treatment of the Littlewood and Paley theorem was pioneered by Stein [253] and allowed the higher-dimensional extension of the theory. The use of vector-valued inequalities in the proof of Theorem 5.1.2 is contained in Benedek, Calderón, and Panzone [18]. A Littlewood–Paley theorem for lacunary sectors in $\mathbf{R}^2$ was obtained by Nagel, Stein, and Wainger [205].

An interesting Littlewood–Paley estimate holds for $2 \le p < \infty$: There exists a constant C_p such that for all families of disjoint open intervals I_j in $\mathbf{R}$ the estimate $\big\|(\sum_j |(\widehat{f}\chi_{I_j})^\vee|^2)^{\frac{1}{2}}\big\|_{L^p} \le C_p \big\|f\big\|_{L^p}$ holds for all functions $f \in L^p(\mathbf{R})$. This was proved by Rubio de Francia [225], but the special case in which $I_j = (j, j+1)$ was previously obtained by Carleson [46]. An alternative proof of Rubio de Francia's theorem was obtained by Bourgain [28]. A higher-dimensional analogue of this estimate for arbitrary disjoint open rectangles in $\mathbf{R}^n$ with sides parallel to the axes was obtained by Journé [144]. Easier proofs of the higher-dimensional result were subsequently obtained by Sjölin [246], Soria [249], and Sato [234].

Part (a) of Theorem 5.2.7 is due to Mihlin [199] and the generalization in part (b) to Hörmander [129]. Theorem 5.2.2 can be found in Marcinkiewicz's article [188] in the context of one-dimensional Fourier series. The power 6 in estimate (5.2.3) that appears in the statement of Theorem 5.2.2 is not optimal. Tao and Wright [276] proved that the best power of $(p-1)^{-1}$ in this theorem is $\frac{3}{2}$ as $p \to 1$. An improvement of the Marcinkiewicz multiplier theorem in one dimension was obtained by Coifman, Rubio de Francia, and Semmes [54]. Weighted norm estimates for Hörmander–Mihlin multipliers were obtained by Kurtz and Wheeden [166] and for Marcinkiwiecz multipliers by Kurtz [165]. Nazarov and Seeger [206] have obtained a very elegant characterization of radial L^p multipliers in large dimensions; precisely, they showed that for dimensions $n \geq 5$ and $1 < p < 2(n^2-2n-3)/(n^2-5)$, a radial function m on $\mathbf{R}^n$ is an L^p Fourier multiplier if and only if there exists a nonzero Schwartz function η such that $\sup_{t>0} t^{n/p} \big\| (m(\cdot)\eta(t\cdot))^\vee \big\|_{L^p} < \infty$. This characterization builds on and extends a previously obtained simple characterization by Garrigós and Seeger [99] of radial multipliers on the invariant subspace of radial L^p functions when $1 < p < \frac{2n}{n+1}$.

The method of proof of Theorem 5.3.4 is adapted from Duoandikoetxea and Rubio de Francia [78]. The method in this article is rather general and can be used to obtain L^p boundedness for a variety of rough singular integrals. A version of Theorem 5.3.6 was used by Christ [49] to obtain L^p smoothing estimates for Cantor–Lebesgue measures. When $p = q \neq 2$, Theorem 5.3.6 is false in general, but it is true for all r satisfying $|\frac{1}{r} - \frac{1}{2}| < |\frac{1}{p} - \frac{1}{2}|$ under the additional assumption that the m_j's are Lipschitz functions uniformly at all scales. This result was independently obtained by Carbery [43] and Seeger [238].

The probabilistic notions of conditional expectations and martingales have a strong connection with the Littlewood–Paley theory discussed in this chapter. For the purposes of this exposition we considered only the case of the sequence of σ-algebras generated by the dyadic cubes of side length 2^{-k} in $\mathbf{R}^n$. The L^p boundedness of the maximal conditional expectation (Doob [76]) is analogous to the L^p boundedness of the dyadic maximal function; likewise with the corresponding weak type $(1,1)$ estimate. The L^p boundedness of the dyadic martingale square function (Burkholder [31]) is analogous to Theorem 5.1.2. Moreover, the estimate $\big\| \sup_k |E_k(f)| \big\|_{L^p} \approx \big\| S(f) \big\|_{L^p}$, $0 < p < \infty$, obtained by Burkholder and Gundy [32] and also by Davis [70] is analogous to the square-function characterization of the H^p norm discussed in Chapter 6. For an exposition on the different and unifying aspects of Littlewood–Paley theory we refer to Stein [256]. The proof of Theorem 5.4.8, which quantitatively expresses the almost orthogonality of the Littlewood–Paley and the dyadic martingale difference operators, is taken from Grafakos and Kalton [106].

The use of quadratic expressions in the study of certain maximal operators has a long history. We refer to the article of Stein [258] for a historical survey. Theorem 5.5.1 was first proved by Stein [257]. The proof in the text is taken from an article of Rubio de Francia [226]. Another proof when $n \geq 3$ is due to Cowling and Mauceri [61]. The more difficult case $n = 2$ was settled by Bourgain [30] about 10 years later. Alternative proofs when $n = 2$ were given by Mockenhaupt, Seeger, and Sogge [200] as well as Schlag [236]. Weighted norm inequalities for the spherical maximal operator were obtained by Duoandikoetxea and Vega [79]. The discrete spherical maximal function was studied by Magyar, Stein, and Wainger [184].

Much of the theory of square functions and the ideas associated with them has analogues in the dyadic setting. A dyadic analogue of the theory discussed here can be obtained. For an introduction to the area of dyadic harmonic analysis, we refer to Pereyra [212].

The idea of expressing (or reproducing) a signal as a weighted average of translations and dilations of a single function appeared in early work of Calderón [34]. This idea is in some sense a forerunner of wavelets. An early example of a wavelet was constructed by Strömberg [270] in his search for unconditional bases for Hardy spaces. Another example of a wavelet basis was obtained by Meyer [194]. The construction of an orthonormal wavelet presented in Theorem 5.6.7 is in Lemarié and Meyer [171]. A compactly supported wavelet was constructed by Daubechies [68]. Mallat [185] introduced the notion of multiresolution analysis, which led to a systematic production of wavelets. The area of wavelets has taken off significantly since its inception, spurred by these early results. A general theory of wavelets and its use in Fourier analysis was carefully developed in the two-volume monograph of Meyer [195], [196] and its successor Meyer and Coifman [197]. For further study and a complete account on the recent developments on the subject we refer to the books of Daubechies [69], Chui [53], Wickerhauser [292], Kaiser [146], Benedetto and Frazier [19], Hérnandez and Weiss [124], Wojtaszczyk [293], Mallat [186], Meyer [198], Frazier [96], Gröchenig [115], and the references therein.

Appendix A
Gamma and Beta Functions

A.1 A Useful Formula

The following formula is valid:

$$\int_{\mathbf{R}^n} e^{-|x|^2}\,dx = \left(\sqrt{\pi}\right)^n.$$

This is an immediate consequence of the corresponding one-dimensional identity

$$\int_{-\infty}^{+\infty} e^{-x^2}\,dx = \sqrt{\pi},$$

which is usually proved from its two-dimensional version by switching to polar coordinates:

$$I^2 = \int_{-\infty}^{+\infty}\int_{-\infty}^{+\infty} e^{-x^2}e^{-y^2}\,dy\,dx = 2\pi\int_0^\infty re^{-r^2}\,dr = \pi.$$

A.2 Definitions of $\Gamma(z)$ and $B(z,w)$

For a complex number z with $\operatorname{Re} z > 0$ define

$$\Gamma(z) = \int_0^\infty t^{z-1}e^{-t}\,dt.$$

$\Gamma(z)$ is called the gamma function. It follows from its definition that $\Gamma(z)$ is analytic on the right half-plane $\operatorname{Re} z > 0$.

Two fundamental properties of the gamma function are that

$$\Gamma(z+1) = z\Gamma(z) \qquad \text{and} \qquad \Gamma(n) = (n-1)!,$$

where z is a complex number with positive real part and $n \in \mathbf{Z}^+$. Indeed, integration by parts yields

$$\Gamma(z) = \int_0^\infty t^{z-1}e^{-t}\,dt = \left[\frac{t^z e^{-t}}{z}\right]_0^\infty + \frac{1}{z}\int_0^\infty t^z e^{-t}\,dt = \frac{1}{z}\Gamma(z+1).$$

Since $\Gamma(1) = 1$, the property $\Gamma(n) = (n-1)!$ for $n \in \mathbf{Z}^+$ follows by induction. Another important fact is that

$$\Gamma(\tfrac{1}{2}) = \sqrt{\pi}.$$

This follows easily from the identity

$$\Gamma(\tfrac{1}{2}) = \int_0^\infty t^{-\frac{1}{2}} e^{-t}\, dt = 2\int_0^\infty e^{-u^2}\, du = \sqrt{\pi}.$$

Next we define the beta function. Fix z and w complex numbers with positive real parts. We define

$$B(z,w) = \int_0^1 t^{z-1}(1-t)^{w-1}\, dt = \int_0^1 t^{w-1}(1-t)^{z-1}\, dt.$$

We have the following relationship between the gamma and the beta functions:

$$B(z,w) = \frac{\Gamma(z)\Gamma(w)}{\Gamma(z+w)},$$

when z and w have positive real parts.

The proof of this fact is as follows:

$$\begin{aligned}
\Gamma(z+w)B(z,w) &= \Gamma(z+w)\int_0^1 t^{w-1}(1-t)^{z-1}\, dt \\
&= \Gamma(z+w)\int_0^\infty u^{w-1}\left(\frac{1}{1+u}\right)^{z+w} du \qquad\qquad t = u/(1+u) \\
&= \int_0^\infty \int_0^\infty u^{w-1}\left(\frac{1}{1+u}\right)^{z+w} v^{z+w-1} e^{-v}\, dv\, du \\
&= \int_0^\infty \int_0^\infty u^{w-1} s^{z+w-1} e^{-s(u+1)}\, ds\, du \qquad\qquad s = v/(1+u) \\
&= \int_0^\infty s^z e^{-s} \int_0^\infty (us)^{w-1} e^{-su}\, du\, ds \\
&= \int_0^\infty s^{z-1} e^{-s}\Gamma(w) ds \\
&= \Gamma(z)\Gamma(w).
\end{aligned}$$

A.3 Volume of the Unit Ball and Surface of the Unit Sphere

We denote by v_n the volume of the unit ball in $\mathbf{R}^n$ and by ω_{n-1} the surface area of the unit sphere $\mathbf{S}^{n-1}$. We have the following:

$$\omega_{n-1} = \frac{2\pi^{\frac{n}{2}}}{\Gamma(\frac{n}{2})}$$

and

$$v_n = \frac{\omega_{n-1}}{n} = \frac{2\pi^{\frac{n}{2}}}{n\Gamma(\frac{n}{2})} = \frac{\pi^{\frac{n}{2}}}{\Gamma(\frac{n}{2}+1)}.$$

The easy proofs are based on the formula in Appendix A.1. We have

$$(\sqrt{\pi})^n = \int_{\mathbf{R}^n} e^{-|x|^2}\,dx = \omega_{n-1}\int_0^\infty e^{-r^2} r^{n-1}\,dr,$$

by switching to polar coordinates. Now change variables $t = r^2$ to obtain that

$$\pi^{\frac{n}{2}} = \tfrac{\omega_{n-1}}{2}\int_0^\infty e^{-t} t^{\frac{n}{2}-1}\,dt = \tfrac{\omega_{n-1}}{2}\Gamma\big(\tfrac{n}{2}\big).$$

This proves the formula for the surface area of the unit sphere in $\mathbf{R}^n$.

To compute v_n, write again using polar coordinates

$$v_n = |B(0,1)| = \int_{|x|\le 1} 1\,dx = \int_{\mathbf{S}^{n-1}}\int_0^1 r^{n-1}\,dr\,d\theta = \frac{1}{n}\,\omega_{n-1}.$$

Here is another way to relate the volume to the surface area. Let $B(0,R)$ be the ball in $\mathbf{R}^n$ of radius $R > 0$ centered at the origin. Then the volume of the shell $B(0,R+h)\setminus B(0,R)$ divided by h tends to the surface area of $B(0,R)$ as $h \to 0$. In other words, the derivative of the volume of $B(0,R)$ with respect to the radius R is equal to the surface area of $B(0,R)$. Since the volume of $B(0,R)$ is $v_n R^n$, it follows that the surface area of $B(0,R)$ is $n\,v_n R^{n-1}$. Taking $R = 1$, we deduce $\omega_{n-1} = n\,v_n$.

A.4 Computation of Integrals Using Gamma Functions

Let $k_1,\dots,k_n$ be nonnegative even integers. The integral

$$\int_{\mathbf{R}^n} x_1^{k_1}\cdots x_n^{k_n} e^{-|x|^2}\,dx_1\cdots dx_n = \prod_{j=1}^n \int_{-\infty}^{+\infty} x_j^{k_j} e^{-x_j^2}\,dx_j = \prod_{j=1}^n \Gamma\Big(\frac{k_j+1}{2}\Big)$$

expressed in polar coordinates is equal to

$$\Big(\int_{\mathbf{S}^{n-1}} \theta_1^{k_1}\cdots\theta_n^{k_n}\,d\theta\Big)\int_0^\infty r^{k_1+\cdots+k_n} r^{n-1} e^{-r^2}\,dr,$$

where $\theta = (\theta_1,\dots,\theta_n)$. This leads to the identity

$$\int_{\mathbf{S}^{n-1}} \theta_1^{k_1}\cdots\theta_n^{k_n}\,d\theta = 2\Gamma\Big(\frac{k_1+\cdots+k_n+n}{2}\Big)^{-1}\prod_{j=1}^n \Gamma\Big(\frac{k_j+1}{2}\Big).$$

Another classical integral that can be computed using gamma functions is the following:

$$\int_0^{\pi/2} (\sin\varphi)^a (\cos\varphi)^b \, d\varphi = \frac{1}{2} \frac{\Gamma(\frac{a+1}{2})\Gamma(\frac{b+1}{2})}{\Gamma(\frac{a+b+2}{2})},$$

whenever a and b are complex numbers with $\operatorname{Re} a > -1$ and $\operatorname{Re} b > -1$.

Indeed, change variables $u = (\sin\varphi)^2$; then $du = 2(\sin\varphi)(\cos\varphi)d\varphi$, and the preceding integral becomes

$$\frac{1}{2}\int_0^1 u^{\frac{a-1}{2}}(1-u)^{\frac{b-1}{2}} \, du = \frac{1}{2} B\Big(\frac{a+1}{2}, \frac{b+1}{2}\Big) = \frac{1}{2} \frac{\Gamma(\frac{a+1}{2})\Gamma(\frac{b+1}{2})}{\Gamma(\frac{a+b+2}{2})}.$$

A.5 Meromorphic Extensions of $B(z,w)$ and $\Gamma(z)$

Using the identity $\Gamma(z+1) = z\Gamma(z)$, we can easily define a meromorphic extension of the gamma function on the whole complex plane starting from its known values on the right half-plane. We give an explicit description of the meromorphic extension of $\Gamma(z)$ on the whole plane. First write

$$\Gamma(z) = \int_0^1 t^{z-1} e^{-t} dt + \int_1^\infty t^{z-1} e^{-t} dt$$

and observe that the second integral is an analytic function of z for all $z \in \mathbf{C}$. Write the first integral as

$$\int_0^1 t^{z-1} \left\{ e^{-t} - \sum_{j=0}^{N} \frac{(-t)^j}{j!} \right\} dt + \sum_{j=0}^{N} \frac{(-1)^j/j!}{z+j}.$$

The last integral converges when $\operatorname{Re} z > -N-1$, since the expression inside the curly brackets is $O(t^{N+1})$ as $t \to 0$. It follows that the gamma function can be defined to be an analytic function on $\operatorname{Re} z > -N-1$ except at the points $z = -j$, $j = 0, 1, \ldots, N$, at which it has simple poles with residues $\frac{(-1)^j}{j!}$. Since N was arbitrary, it follows that the gamma function has a meromorphic extension on the whole plane.

In view of the identity

$$B(z,w) = \frac{\Gamma(z)\Gamma(w)}{\Gamma(z+w)},$$

the definition of $B(z,w)$ can be extended to $\mathbf{C} \times \mathbf{C}$. It follows that $B(z,w)$ is a meromorphic function in each argument.

A.6 Asymptotics of $\Gamma(x)$ as $x \to \infty$

We now derive *Stirling's formula*:

$$\lim_{x\to\infty}\frac{\Gamma(x+1)}{\left(\frac{x}{e}\right)^x\sqrt{2\pi x}}=1\,.$$

First change variables $t=x+sx\sqrt{\frac{2}{x}}$ to obtain

$$\Gamma(x+1)=\int_0^\infty e^{-t}t^x\,dt=\left(\frac{x}{e}\right)^x\sqrt{2x}\int_{-\sqrt{x/2}}^{+\infty}\frac{\left(1+s\sqrt{\frac{2}{x}}\right)^x}{e^{2s\sqrt{x/2}}}ds\,.$$

Setting $y=\sqrt{\frac{x}{2}}$, we obtain

$$\frac{\Gamma(x+1)}{\left(\frac{x}{e}\right)^x\sqrt{2x}}=\int_{-\infty}^{+\infty}\left(\frac{\left(1+\frac{s}{y}\right)^y}{e^s}\right)^{2y}\chi_{(-y,\infty)}(s)\,ds.$$

To show that the last integral converges to $\sqrt{\pi}$ as $y\to\infty$, we need the following:
(1) The fact that

$$\lim_{y\to\infty}\left(\frac{\left(1+s/y\right)^y}{e^s}\right)^{2y}\to e^{-s^2},$$

which follows easily by taking logarithms and applying L'Hôpital's rule twice.
(2) The estimate, valid for $y\geq 1$,

$$\left(\frac{\left(1+\frac{s}{y}\right)^y}{e^s}\right)^{2y}\leq\begin{cases}\dfrac{(1+s)^2}{e^s} & \text{when } s\geq 0,\\[2ex] e^{-s^2} & \text{when } -y<s<0,\end{cases}$$

which can be easily checked using calculus. Using these facts, the Lebesgue dominated convergence theorem, the trivial fact that $\chi_{-y<s<\infty}\to 1$ as $y\to\infty$, and the identity in Appendix A.1, we obtain that

$$\begin{aligned}\lim_{x\to\infty}\frac{\Gamma(x+1)}{\left(\frac{x}{e}\right)^x\sqrt{2x}}&=\lim_{y\to\infty}\int_{-\infty}^{+\infty}\left(\frac{\left(1+\frac{s}{y}\right)^y}{e^s}\right)^{2y}\chi_{(-y,\infty)}(s)\,ds\\&=\int_{-\infty}^{+\infty}e^{-s^2}ds\\&=\sqrt{\pi}.\end{aligned}$$

A.7 Euler's Limit Formula for the Gamma Function

For n a positive integer and $\operatorname{Re} z>0$ we consider the functions

$$\Gamma_n(z) = \int_0^n \left(1 - \frac{t}{n}\right)^n t^{z-1}\, dt$$

We show that

$$\Gamma_n(z) = \frac{n!\, n^z}{z(z+1)\cdots(z+n)}$$

and we obtain *Euler's limit formula for the gamma function*

$$\lim_{n\to\infty} \Gamma_n(z) = \Gamma(z)\,.$$

We write $\Gamma(z) - \Gamma_n(z) = I_1(z) + I_2(z) + I_3(z)$, where

$$I_1(z) = \int_n^\infty e^{-t} t^{z-1}\, dt\,,$$

$$I_2(z) = \int_{n/2}^n \left(e^{-t} - \left(1 - \frac{t}{n}\right)^n\right) t^{z-1}\, dt\,,$$

$$I_3(z) = \int_0^{n/2} \left(e^{-t} - \left(1 - \frac{t}{n}\right)^n\right) t^{z-1}\, dt\,.$$

Obviously $I_1(z)$ tends to zero as $n \to \infty$. For I_2 and I_3 we have that $0 \le t < n$, and by the Taylor expansion of the logarithm we obtain

$$\log\left(1 - \frac{t}{n}\right)^n = n\log\left(1 - \frac{t}{n}\right) = -t - L,$$

where

$$L = \frac{t^2}{n}\left(\frac{1}{2} + \frac{1}{3}\frac{t}{n} + \frac{1}{4}\frac{t^2}{n^2} + \cdots\right).$$

It follows that

$$0 < e^{-t} - \left(1 - \frac{t}{n}\right)^n = e^{-t} - e^{-L}e^{-t} \le e^{-t}\,,$$

and thus $I_2(z)$ tends to zero as $n \to \infty$. For I_3 we have $t/n \le 1/2$, which implies that

$$L \le \frac{t^2}{n} \sum_{k=0}^\infty \frac{1}{(k+1)2^{k-1}} = \frac{t^2}{n} c\,.$$

Consequently, for $t/n \le 1/2$ we have

$$0 \le e^{-t} - \left(1 - \frac{t}{n}\right)^n = e^{-t}(1 - e^{-L}) \le e^{-t} L \le e^{-t}\, \frac{ct^2}{n}\,.$$

Plugging this estimate into I_3, we deduce that

$$|I_3(z)| \le \frac{c}{n}\Gamma(\operatorname{Re} z + 2)\,,$$

which certainly tends to zero as $n \to \infty$.

Next, n integrations by parts give

$$\Gamma_n(z) = \frac{n}{nz}\frac{n-1}{n(z+1)}\frac{n-2}{n(z+2)}\cdots\frac{1}{n(z+n-1)}\int_0^n t^{z+n-1}\,dt = \frac{n!\,n^z}{z(z+1)\cdots(z+n)}.$$

This can be written as

$$1 = \Gamma_n(z)\,z\exp\left\{z\left(1+\frac{1}{2}+\frac{1}{3}+\cdots+\frac{1}{n}-\log n\right)\right\}\prod_{k=1}^{n}\left(1+\frac{z}{k}\right)e^{-z/k}.$$

Taking limits as $n \to \infty$, we obtain an *infinite product form of Euler's limit formula*,

$$1 = \Gamma(z)\,z e^{\gamma z}\prod_{k=1}^{\infty}\left(1+\frac{z}{k}\right)e^{-z/k},$$

where $\operatorname{Re} z > 0$ and γ is *Euler's constant*

$$\gamma = \lim_{n\to\infty} 1+\frac{1}{2}+\frac{1}{3}+\cdots+\frac{1}{n}-\log n.$$

The infinite product converges uniformly on compact subsets of the complex plane that excludes $z = 0, -1, -2, \dots$, and thus it represents a holomorphic function in this domain. This holomorphic function multiplied by $\Gamma(z)\,z e^{\gamma z}$ is equal to 1 on $\operatorname{Re} z > 0$ and by analytic continuation it must be equal to 1 on $\mathbf{C}\setminus\{0,-1,-2,\dots\}$. But $\Gamma(z)$ has simple poles, while the infinite product vanishes to order one at the nonpositive integers. We conclude that Euler's limit formula holds for all complex numbers z; consequently, $\Gamma(z)$ has no zeros and $\Gamma(z)^{-1}$ is entire.

An immediate consequence of Euler's limit formula is the identity

$$\frac{1}{|\Gamma(x+iy)|^2} = \frac{1}{|\Gamma(x)|^2}\prod_{k=0}^{\infty}\left(1+\frac{y^2}{(k+x)^2}\right),$$

which holds for x and y real with $x \notin \{0,-1,-2,\dots\}$. As a consequence we have that

$$|\Gamma(x+iy)| \le |\Gamma(x)|$$

and also that

$$\frac{1}{|\Gamma(x+iy)|} \le \frac{1}{|\Gamma(x)|}e^{C(x)|y|^2},$$

where

$$C(x) = \frac{1}{2}\sum_{k=0}^{\infty}\frac{1}{(k+x)^2},$$

whenever $x \in \mathbf{R}\setminus\{0,-1,-2,\dots\}$ and $y \in \mathbf{R}$.

A.8 Reflection and Duplication Formulas for the Gamma Function

The *reflection formula* relates the values of the gamma function of a complex number z and its reflection about the point $1/2$ in the following way:

$$\frac{\sin(\pi z)}{\pi} = \frac{1}{\Gamma(z)}\,\frac{1}{\Gamma(1-z)}\,.$$

The *duplication formula* relates the entire functions $\Gamma(2z)^{-1}$ and $\Gamma(z)^{-1}$ as follows:

$$\frac{1}{\Gamma(z)\Gamma(z+\frac{1}{2})} = \frac{\pi^{-\frac{1}{2}}2^{2z-1}}{\Gamma(2z)}\,.$$

Both of these could be proved using Euler's limit formula. The reflection formula also uses the identity

$$\prod_{k=1}^{\infty}\left(1-\frac{z^2}{k^2}\right) = \frac{\sin(\pi z)}{\pi z}\,,$$

while the duplication formula makes use of the fact that

$$\lim_{n\to\infty}\frac{(n!)^2\,2^{2n+1}}{(2n)!\,n^{1/2}} = 2\,\pi^{1/2}\,.$$

These and other facts related to the gamma function can be found in Olver [208].

Appendix B
Bessel Functions

B.1 Definition

We survey some basics from the theory of Bessel functions J_ν of complex order ν with $\mathrm{Re}\,\nu > -1/2$. We define the Bessel function J_ν of order ν by its *Poisson representation formula*

$$J_\nu(t) = \frac{\left(\frac{t}{2}\right)^\nu}{\Gamma(\nu+\frac{1}{2})\Gamma(\frac{1}{2})} \int_{-1}^{+1} e^{its}(1-s^2)^\nu \frac{ds}{\sqrt{1-s^2}},$$

where $\mathrm{Re}\,\nu > -1/2$ and $t \geq 0$. Although this definition is also valid when t is a complex number, for the applications we have in mind, it suffices to consider the case that t is real and nonnegative; in this case $J_\nu(t)$ is also a real number.

B.2 Some Basic Properties

Let us summarize a few properties of Bessel functions. We take $t > 0$.
(1) We have the following recurrence formula:

$$\frac{d}{dt}\left(t^{-\nu}J_\nu(t)\right) = -t^{-\nu}J_{\nu+1}(t), \qquad \mathrm{Re}\,\nu > -1/2.$$

(2) We also have the companion recurrence formula:

$$\frac{d}{dt}\left(t^{\nu}J_\nu(t)\right) = t^{\nu}J_{\nu-1}(t), \qquad \mathrm{Re}\,\nu > 1/2.$$

(3) $J_\nu(t)$ satisfies the differential equation:

$$t^2\frac{d^2}{dt^2}(J_\nu(t)) + t\frac{d}{dt}(J_\nu(t)) + (t^2-\nu^2)J_\nu(t) = 0.$$

(4) If $\nu \in \mathbf{Z}^+$, then we have the following identity, which was taken by Bessel as the definition of J_ν for integer ν:

$$J_\nu(t) = \frac{1}{2\pi}\int_0^{2\pi} e^{it\sin\theta}e^{-i\nu\theta}\,d\theta = \frac{1}{2\pi}\int_0^{2\pi}\cos(t\sin\theta - \nu\theta)\,d\theta.$$

(5) For $\operatorname{Re}\nu > -1/2$ we have the following identity:

$$J_\nu(t) = \frac{1}{\Gamma(\frac{1}{2})}\left(\frac{t}{2}\right)^\nu \sum_{j=0}^{\infty}(-1)^j \frac{\Gamma(j+\frac{1}{2})}{\Gamma(j+\nu+1)}\frac{t^{2j}}{(2j)!}.$$

(6) For $\operatorname{Re}\nu > 1/2$ the identity below is valid:

$$\frac{d}{dt}(J_\nu(t)) = \frac{1}{2}\big(J_{\nu-1}(t) - J_{\nu+1}(t)\big).$$

We first verify property (1). We have

$$\begin{aligned}\frac{d}{dt}\big(t^{-\nu}J_\nu(t)\big) &= \frac{i}{2^\nu\Gamma(\nu+\frac{1}{2})\Gamma(\frac{1}{2})}\int_{-1}^{1} se^{its}(1-s^2)^{\nu-\frac{1}{2}}\,ds \\ &= \frac{i}{2^\nu\Gamma(\nu+\frac{1}{2})\Gamma(\frac{1}{2})}\int_{-1}^{1}\frac{it}{2}e^{its}\frac{(1-s^2)^{\nu+\frac{1}{2}}}{\nu+\frac{1}{2}}\,ds \\ &= -t^{-\nu}J_{\nu+1}(t),\end{aligned}$$

where we integrated by parts and used the fact that $\Gamma(x+1) = x\Gamma(x)$. Property (2) can be proved similarly.

We proceed with the proof of property (3). A calculation using the definition of the Bessel function gives that the left-hand side of (3) is equal to

$$\frac{2^{-\nu}t^{\nu+1}}{\Gamma(\nu+\frac{1}{2})\Gamma(\frac{1}{2})}\int_{-1}^{+1} e^{ist}\Big((1-s^2)t + 2is(\nu+\tfrac{1}{2})\Big)(1-s^2)^{\nu-\frac{1}{2}}\,ds,$$

which in turn is equal to

$$-i\frac{2^{-\nu}t^{\nu+1}}{\Gamma(\nu+\frac{1}{2})\Gamma(\frac{1}{2})}\int_{-1}^{+1}\frac{d}{ds}\big(e^{ist}(1-s^2)^{\nu+\frac{1}{2}}\big)\,ds = 0.$$

Property (4) can be derived directly from (1). Define

$$G_\nu(t) = \frac{1}{2\pi}\int_0^{2\pi} e^{it\sin\theta}e^{-i\nu\theta}\,d\theta,$$

for $\nu = 0,1,2,\dots$ and $t>0$. We can show easily that $G_0 = J_0$. If we had

$$\frac{d}{dt}\big(t^{-\nu}G_\nu(t)\big) = -t^{-\nu}G_{\nu+1}(t), \qquad t>0,$$

for $\nu\in\mathbf{Z}^+$, we would immediately conclude that $G_\nu = J_\nu$ for $\nu\in\mathbf{Z}^+$. We have

$$
\begin{aligned}
\frac{d}{dt}\big(t^{-\nu}G_\nu(t)\big) &= -t^{-\nu}\left(\frac{\nu}{t}G_\nu(t) - \frac{dG_\nu}{dt}(t)\right)\\
&= -t^{-\nu}\int_0^{2\pi}\frac{\nu}{2\pi t}e^{it\sin\theta}e^{-i\nu\theta} - \frac{1}{2\pi}\left(\frac{d}{dt}e^{it\sin\theta}\right)e^{-i\nu\theta}\,d\theta\\
&= -\frac{t^{-\nu}}{2\pi}\int_0^{2\pi} i\frac{d}{d\theta}\left(\frac{e^{it\sin\theta-i\nu\theta}}{t}\right) + (\cos\theta - i\sin\theta)e^{it\sin\theta}e^{-i\nu\theta}\,d\theta\\
&= -\frac{t^{-\nu}}{2\pi}\int_0^{2\pi} e^{it\sin\theta}e^{-i(\nu+1)\theta}\,d\theta\\
&= -t^{-\nu}G_{\nu+1}(t)\,.
\end{aligned}
$$

For t real, the identity in (5) can be derived by inserting the expression

$$\sum_{j=0}^{\infty}(-1)^j\frac{(ts)^{2j}}{(2j)!} + i\sin(ts)$$

for e^{its} in the definition of the Bessel function $J_\nu(t)$ in Appendix B.1. Algebraic manipulations yield

$$
\begin{aligned}
J_\nu(t) &= \frac{(t/2)^\nu}{\Gamma(\frac{1}{2})}\sum_{j=0}^{\infty}(-1)^j\frac{1}{\Gamma(\nu+\frac{1}{2})}\frac{t^{2j}}{(2j)!}2\int_0^1 s^{2j-1}(1-s^2)^{\nu-\frac{1}{2}}s\,ds\\
&= \frac{(t/2)^\nu}{\Gamma(\frac{1}{2})}\sum_{j=0}^{\infty}(-1)^j\frac{1}{\Gamma(\nu+\frac{1}{2})}\frac{t^{2j}}{(2j)!}\frac{\Gamma(j+\frac{1}{2})\Gamma(\nu+\frac{1}{2})}{\Gamma(j+\nu+1)}\\
&= \frac{(t/2)^\nu}{\Gamma(\frac{1}{2})}\sum_{j=0}^{\infty}(-1)^j\frac{\Gamma(j+\frac{1}{2})}{\Gamma(j+\nu+1)}\frac{t^{2j}}{(2j)!}\,.
\end{aligned}
$$

To derive property (6) we first multiply (1) by t^ν and (2) by $t^{-\nu}$; then we use the product rule for differentiation and we add the resulting expressions.

For further identities on Bessel functions, one may consult Watson's monograph [288].

B.3 An Interesting Identity

Let $\operatorname{Re}\mu > -\frac{1}{2}$, $\operatorname{Re}\nu > -1$, and $t > 0$. Then the following identity is valid:

$$\int_0^1 J_\mu(ts)s^{\mu+1}(1-s^2)^\nu\,ds = \frac{\Gamma(\nu+1)2^\nu}{t^{\nu+1}}J_{\mu+\nu+1}(t)\,.$$

To prove this identity we use formula (5) in Appendix B.2. We have

$$\begin{aligned}
&\int_0^1 J_\mu(ts)s^{\mu+1}(1-s^2)^\nu\,ds \\
&\quad = \frac{\left(\frac{t}{2}\right)^\mu}{\Gamma\left(\frac{1}{2}\right)} \int_0^1 \sum_{j=0}^\infty \frac{(-1)^j\Gamma(j+\frac{1}{2})t^{2j}}{\Gamma(j+\mu+1)(2j)!} s^{2j+\mu+\mu}(1-s^2)^\nu s\,ds \\
&\quad = \frac{1}{2}\frac{\left(\frac{t}{2}\right)^\mu}{\Gamma\left(\frac{1}{2}\right)} \sum_{j=0}^\infty \frac{(-1)^j\Gamma(j+\frac{1}{2})t^{2j}}{\Gamma(j+\mu+1)(2j)!} \int_0^1 u^{j+\mu}(1-u)^\nu\,du \\
&\quad = \frac{1}{2}\frac{\left(\frac{t}{2}\right)^\mu}{\Gamma\left(\frac{1}{2}\right)} \sum_{j=0}^\infty \frac{(-1)^j\Gamma(j+\frac{1}{2})t^{2j}}{\Gamma(j+\mu+1)(2j)!} \frac{\Gamma(\mu+j+1)\Gamma(\nu+1)}{\Gamma(\mu+\nu+j+2)} \\
&\quad = \frac{2^\nu\Gamma(\nu+1)}{t^{\nu+1}} \frac{\left(\frac{t}{2}\right)^{\mu+\nu+1}}{\Gamma\left(\frac{1}{2}\right)} \sum_{j=0}^\infty \frac{(-1)^j\Gamma(j+\frac{1}{2})t^{2j}}{\Gamma(j+\mu+\nu+2)(2j)!} \\
&\quad = \frac{\Gamma(\nu+1)2^\nu}{t^{\nu+1}} J_{\mu+\nu+1}(t)\,.
\end{aligned}$$

B.4 The Fourier Transform of Surface Measure on $\mathbf{S}^{n-1}$

Let $d\sigma$ denote surface measure on $\mathbf{S}^{n-1}$ for $n \geq 2$. Then the following is true:

$$\widehat{d\sigma}(\xi) = \int_{\mathbf{S}^{n-1}} e^{-2\pi i\xi\cdot\theta}\,d\theta = \frac{2\pi}{|\xi|^{\frac{n-2}{2}}} J_{\frac{n-2}{2}}(2\pi|\xi|)\,.$$

To see this, use the result in Appendix D.3 to write

$$\begin{aligned}
\widehat{d\sigma}(\xi) &= \int_{\mathbf{S}^{n-1}} e^{-2\pi i\xi\cdot\theta}\,d\theta \\
&= \frac{2\pi^{\frac{n-1}{2}}}{\Gamma(\frac{n-1}{2})} \int_{-1}^{+1} e^{-2\pi i|\xi|s}(1-s^2)^{\frac{n-2}{2}} \frac{ds}{\sqrt{1-s^2}} \\
&= \frac{2\pi^{\frac{n-1}{2}}}{\Gamma(\frac{n-1}{2})} \frac{\Gamma(\frac{n-2}{2}+\frac{1}{2})\Gamma(\frac{1}{2})}{(\pi|\xi|)^{\frac{n-2}{2}}} J_{\frac{n-2}{2}}(2\pi|\xi|) \\
&= \frac{2\pi}{|\xi|^{\frac{n-2}{2}}} J_{\frac{n-2}{2}}(2\pi|\xi|)\,.
\end{aligned}$$

B.5 The Fourier Transform of a Radial Function on $\mathbf{R}^n$

Let $f(x) = f_0(|x|)$ be a radial function defined on $\mathbf{R}^n$, where f_0 is defined on $[0,\infty)$. Then the Fourier transform of f is given by the formula

$$\widehat{f}(\xi) = \frac{2\pi}{|\xi|^{\frac{n-2}{2}}} \int_0^\infty f_0(r) J_{\frac{n}{2}-1}(2\pi r|\xi|) r^{\frac{n}{2}} \, dr .$$

To obtain this formula, use polar coordinates to write

$$\begin{aligned}
\widehat{f}(\xi) &= \int_{\mathbf{R}^n} f(x) e^{-2\pi i \xi \cdot x} dx \\
&= \int_0^\infty \int_{\mathbf{S}^{n-1}} f_0(r) e^{-2\pi i \xi \cdot r\theta} d\theta \, r^{n-1} dr \\
&= \int_0^\infty f_0(r) \widehat{d\sigma}(r\xi) r^{n-1} dr \\
&= \int_0^\infty f_0(r) \frac{2\pi}{(r|\xi|)^{\frac{n-2}{2}}} J_{\frac{n-2}{2}}(2\pi r|\xi|) r^{n-1} dr \\
&= \frac{2\pi}{|\xi|^{\frac{n-2}{2}}} \int_0^\infty f_0(r) J_{\frac{n}{2}-1}(2\pi r|\xi|) r^{\frac{n}{2}} \, dr .
\end{aligned}$$

As an application we take $f(x) = \chi_{B(0,1)}$, where $B(0,1)$ is the unit ball in $\mathbf{R}^n$. We obtain

$$(\chi_{B(0,1)})\widehat{\ }(\xi) = \frac{2\pi}{|\xi|^{\frac{n-2}{2}}} \int_0^1 J_{\frac{n}{2}-1}(2\pi|\xi|r) r^{\frac{n}{2}} \, dr = \frac{J_{\frac{n}{2}}(2\pi|\xi|)}{|\xi|^{\frac{n}{2}}} ,$$

in view of the result in Appendix B.3. More generally, for $\operatorname{Re}\lambda > -1$, let

$$m_\lambda(\xi) = \begin{cases} (1-|\xi|^2)^\lambda & \text{for } |\xi| \le 1, \\ 0 & \text{for } |\xi| > 1. \end{cases}$$

Then

$$m_\lambda^\vee(x) = \frac{2\pi}{|x|^{\frac{n-2}{2}}} \int_0^1 J_{\frac{n}{2}-1}(2\pi|x|r) r^{\frac{n}{2}} (1-r^2)^\lambda \, dr = \frac{\Gamma(\lambda+1)}{\pi^\lambda} \frac{J_{\frac{n}{2}+\lambda}(2\pi|x|)}{|x|^{\frac{n}{2}+\lambda}} ,$$

using again the identity in Appendix B.3.

B.6 Bessel Functions of Small Arguments

We seek the behavior of $J_k(r)$ as $r \to 0+$. We fix a complex number ν with $\operatorname{Re}\nu > -\frac{1}{2}$. Then we have the identity

$$J_\nu(r) = \frac{r^\nu}{2^\nu \Gamma(\nu+1)} + S_\nu(r) ,$$

where

$$S_\nu(r) = \frac{(r/2)^\nu}{\Gamma(\nu+\frac{1}{2})\Gamma(\frac{1}{2})} \int_{-1}^{+1} (e^{irt}-1)(1-t^2)^{\nu-\frac{1}{2}} \, dt$$

and S_ν satisfies

$$|S_\nu(r)| \le \frac{2^{-\operatorname{Re}\nu} r^{\operatorname{Re}\nu+1}}{(\operatorname{Re}\nu+1)\,|\Gamma(\nu+\frac{1}{2})|\,\Gamma(\frac{1}{2})}.$$

To prove this estimate we note that

$$\begin{aligned} J_\nu(r) &= \frac{(r/2)^\nu}{\Gamma(\nu+\frac{1}{2})\Gamma(\frac{1}{2})}\int_{-1}^{+1}(1-t^2)^{\nu-\frac{1}{2}}\,dt + S_\nu(r)\\ &= \frac{(r/2)^\nu}{\Gamma(\nu+\frac{1}{2})\Gamma(\frac{1}{2})}\int_{0}^{\pi}(\sin^2\phi)^{\nu-\frac{1}{2}}(\sin\phi)\,d\phi + S_\nu(r)\\ &= \frac{(r/2)^\nu}{\Gamma(\nu+\frac{1}{2})\Gamma(\frac{1}{2})}\frac{\Gamma(\nu+\frac{1}{2})\Gamma(\frac{1}{2})}{\Gamma(\nu+1)} + S_\nu(r), \end{aligned}$$

where we evaluated the last integral using the result in Appendix A.4. Using that $|e^{irt}-1| \le r|t|$, we deduce the assertion regarding the size of $|S_\nu(r)|$.

It follows from these facts and the estimate in Appendix A.7 that for $0 < r \le 1$ and $\operatorname{Re}\nu > -1/2$ we have

$$|J_\nu(r)| \le C_0\, e^{c_0|\operatorname{Im}\nu|^2}\, r^{\operatorname{Re}\nu},$$

where C_0 and c_0 are constants depending only on $\operatorname{Re}\nu$. Note that when $\operatorname{Re}\nu \ge 0$, the constant c_0 may be taken to be absolute (such as $c_0 = \pi^2$).

B.7 Bessel Functions of Large Arguments

For $r > 0$ and complex numbers ν with $\operatorname{Re}\nu > -1/2$ we prove the identity

$$J_\nu(r) = \frac{(r/2)^\nu}{\Gamma(\nu+\frac{1}{2})\Gamma(\frac{1}{2})}\left[ie^{-ir}\int_0^\infty e^{-rt}(t^2+2it)^{\nu-\frac{1}{2}}\,dt - ie^{ir}\int_0^\infty e^{-rt}(t^2-2it)^{\nu-\frac{1}{2}}\,dt\right].$$

Fix $0 < \delta < 1/10 < 10 < R < \infty$. We consider the region $\Omega_{\delta,R}$ in the complex plane whose boundary is the set consisting of the interval $[-1+\delta, 1-\delta]$ union a quarter circle centered at 1 of radius δ from $1-\delta$ to $1+i\delta$, union the line segments from $1+i\delta$ to $1+iR$, from $1+iR$ to $-1+iR$, and from $-1+iR$ to $-1+i\delta$, union a quarter circle centered at -1 of radius δ from $-1+i\delta$ to $-1+\delta$. This is a simply connected region on the interior of which the holomorphic function $(1-z^2)$ has no zeros. Since $\Omega_{\delta,R}$ is contained in the complement of the negative imaginary axis, there is a holomorphic branch of the logarithm such that $\log(t)$ is real, $\log(-t) = \log|t| + i\pi$, and $\log(it) = \log|t| + i\pi/2$ for $t > 0$. Since the function $\log(1-z^2)$ is well defined and holomorphic in $\Omega_{\delta,R}$, we may define the holomorphic function

$$(1-z^2)^{\nu-\frac{1}{2}} = e^{(\nu-\frac{1}{2})\log(1-z^2)}$$

for $z \in \Omega_{\delta,R}$. Since $e^{irz}(1-z^2)^{\nu-\frac{1}{2}}$ has no poles in $\Omega_{\delta,R}$, Cauchy's theorem yields

$$i\int_{\delta}^{R} e^{ir(1+it)}(t^2-2it)^{\nu-\frac{1}{2}}\,dt+\int_{-1+\delta}^{1-\delta} e^{irt}(1-t^2)^{\nu-\frac{1}{2}}\,dt$$
$$+i\int_{R}^{\delta} e^{ir(-1+it)}(t^2+2it)^{\nu-\frac{1}{2}}\,dt+E(\delta,R)=0,$$

where $E(\delta,R)$ is the sum of the integrals over the two small quarter-circles of radius δ and the line segment from $1+iR$ to $-1+iR$. The first two of these integrals are bounded by constants times δ, the latter by a constant times $R^{2\operatorname{Re}\nu-1}e^{-rR}$; hence $E(\delta,R)\to 0$ as $\delta\to 0$ and $R\to\infty$. We deduce the identity

$$\int_{-1}^{+1} e^{irt}(1-t^2)^{\nu-\frac{1}{2}}\,dt = ie^{-ir}\int_0^{\infty} e^{-rt}(t^2+2it)^{\nu-\frac{1}{2}}dt - ie^{ir}\int_0^{\infty} e^{-rt}(t^2-2it)^{\nu-\frac{1}{2}}\,dt\,.$$

Estimating the two integrals on the right by putting absolute values inside and multiplying by the missing factor $r^{\nu}2^{-\nu}(\Gamma(\nu+\frac{1}{2})\Gamma(\frac{1}{2}))^{-1}$, we obtain

$$|J_\nu(r)|\le 2\frac{(r/2)^{\operatorname{Re}\nu}e^{\frac{\pi}{2}|\operatorname{Im}\nu|}}{|\Gamma(\nu+\frac{1}{2})|\Gamma(\frac{1}{2})}\int_0^{\infty} e^{-rt}t^{\operatorname{Re}\nu-\frac{1}{2}}\left(\sqrt{t^2+4}\right)^{\operatorname{Re}\nu-\frac{1}{2}}\,dt\,,$$

since the absolute value of the argument of $t^2\pm 2it$ is at most $\pi/2$. When $\operatorname{Re}\nu>1/2$, we use the inequality $(\sqrt{t^2+4})^{\operatorname{Re}\nu-\frac{1}{2}}\le 2^{\operatorname{Re}\nu-\frac{3}{2}}\left(t^{\operatorname{Re}\nu-\frac{1}{2}}+2^{\operatorname{Re}\nu-\frac{1}{2}}\right)$ to get

$$|J_\nu(r)|\le 2\frac{(r/2)^{\operatorname{Re}\nu}e^{\frac{\pi}{2}|\operatorname{Im}\nu|}}{|\Gamma(\nu+\frac{1}{2})|\Gamma(\frac{1}{2})}2^{\operatorname{Re}\nu-\frac{3}{2}}\left[\frac{\Gamma(2\operatorname{Re}\nu)}{r^{2\operatorname{Re}\nu}}+2^{\operatorname{Re}\nu}\frac{\Gamma(\operatorname{Re}\nu+\frac{1}{2})}{r^{\operatorname{Re}\nu+\frac{1}{2}}}\right].$$

When $1/2\ge\operatorname{Re}\nu>-1/2$ we use that $\left(\sqrt{t^2+4}\right)^{\operatorname{Re}\nu-\frac{1}{2}}\le 1$ to deduce that

$$|J_\nu(r)|\le 2\frac{(r/2)^{\operatorname{Re}\nu}e^{\frac{\pi}{2}|\operatorname{Im}\nu|}}{|\Gamma(\nu+\frac{1}{2})|\Gamma(\frac{1}{2})}\frac{\Gamma(\operatorname{Re}\nu+\frac{1}{2})}{r^{\operatorname{Re}\nu+\frac{1}{2}}}.$$

These estimates yield that for $\operatorname{Re}\nu>-1/2$ and $r\ge 1$ we have

$$|J_\nu(r)|\le C_0(\operatorname{Re}\nu)\,e^{\pi|\operatorname{Im}\nu|+\pi^2|\operatorname{Im}\nu|^2}\,r^{-1/2}$$

using the result in Appendix A.7. Here C_0 is a constant that depends only on $\operatorname{Re}\nu$.

B.8 Asymptotics of Bessel Functions

We obtain asymptotics for $J_\nu(r)$ as $r\to\infty$ whenever $\operatorname{Re}\nu>-1/2$. We have the following identity for $r>0$:

$$J_\nu(r) = \sqrt{\frac{2}{\pi r}} \cos\left(r - \frac{\pi\nu}{2} - \frac{\pi}{4}\right) + R_\nu(r),$$

where R_ν is given by

$$\begin{aligned} R_\nu(r) = \frac{(2\pi)^{-\frac{1}{2}} r^\nu}{\Gamma(\nu+\frac{1}{2})} e^{i(r-\frac{\pi\nu}{2}-\frac{\pi}{4})} \int_0^\infty e^{-rt} t^{\nu+\frac{1}{2}} \left[(1+\tfrac{it}{2})^{\nu-\frac{1}{2}} - 1\right] \frac{dt}{t} \\ + \frac{(2\pi)^{-\frac{1}{2}} r^\nu}{\Gamma(\nu+\frac{1}{2})} e^{-i(r-\frac{\pi\nu}{2}-\frac{\pi}{4})} \int_0^\infty e^{-rt} t^{\nu+\frac{1}{2}} \left[(1-\tfrac{it}{2})^{\nu-\frac{1}{2}} - 1\right] \frac{dt}{t} \end{aligned}$$

and satisfies $|R_\nu(r)| \le C_\nu r^{-3/2}$ whenever $r \ge 1$.

To see the validity of this identity we write

$$\begin{aligned} ie^{-ir}(t^2+2it)^{\nu-\frac{1}{2}} &= (2t)^{\nu-\frac{1}{2}} e^{-i(r-\frac{\nu\pi}{2}-\frac{\pi}{4})} (1-\tfrac{it}{2})^{\nu-\frac{1}{2}}, \\ -ie^{ir}(t^2-2it)^{\nu-\frac{1}{2}} &= (2t)^{\nu-\frac{1}{2}} e^{i(r-\frac{\nu\pi}{2}-\frac{\pi}{4})} (1+\tfrac{it}{2})^{\nu-\frac{1}{2}}. \end{aligned}$$

Inserting these expressions into the corresponding integrals in the formula proved in Appendix B.7, adding and subtracting 1 from each term $(1 \pm \frac{it}{2})^{\nu-\frac{1}{2}}$, and multiplying by the missing factor $(r/2)^\nu/\Gamma(\nu+\frac{1}{2})\Gamma(\frac{1}{2})$, we obtain the claimed identity

$$J_\nu(r) = \sqrt{\frac{2}{\pi r}} \cos\left(r - \frac{\pi\nu}{2} - \frac{\pi}{4}\right) + R_\nu(r).$$

It remains to estimate $R_\nu(r)$. We begin by noting that for a, b real with $a > -1$ we have the pair of inequalities

$$\begin{aligned} |(1\pm iy)^{a+ib} - 1| &\le 3(|a|+|b|)\left(2^{\frac{a+1}{2}} e^{\frac{\pi}{2}|b|}\right) y && \text{when } 0 < y < 1, \\ |(1\pm iy)^{a+ib} - 1| &\le (1+y^2)^{\frac{a}{2}} e^{\frac{\pi}{2}|b|} + 1 \le 2\left(2^{\frac{a+1}{2}} e^{\frac{\pi}{2}|b|}\right) y^a && \text{when } 1 \le y < \infty. \end{aligned}$$

The first inequality is proved by splitting into real and imaginary parts and applying the mean value theorem in the real part. Taking $\nu - \frac{1}{2} = a + ib$, $y = t/2$, and inserting these estimates into the integrals appearing in R_ν, we obtain

$$|R_\nu(r)| \le \frac{2^{\frac{1}{2}\mathrm{Re}\,\nu} 2^{\frac{1}{4}} e^{\frac{\pi}{2}|\mathrm{Im}\,\nu|} r^{\mathrm{Re}\,\nu}}{(2\pi)^{1/2} |\Gamma(\nu+\frac{1}{2})|} \left[\frac{3\sqrt{2}|\nu|}{2} \int_0^2 e^{-rt} t^{\mathrm{Re}\,\nu + \frac{3}{2}} \frac{dt}{t} + \frac{2\sqrt{2}}{2^{\mathrm{Re}\,\nu}} \int_2^\infty e^{-rt} t^{2\mathrm{Re}\,\nu} \frac{dt}{t}\right].$$

It follows that for all $r > 0$ we have

$$\begin{aligned} |R_\nu(r)| &\le 2 \frac{2^{\frac{1}{2}\mathrm{Re}\,\nu} e^{\frac{\pi}{2}|\mathrm{Im}\,\nu|}}{|\Gamma(\nu+\frac{1}{2})|} \left[|\nu| \frac{\Gamma(\mathrm{Re}\,\nu+\frac{3}{2})}{r^{3/2}} + \frac{r^{-\mathrm{Re}\,\nu}}{2^{\mathrm{Re}\,\nu}} \int_{2r}^\infty e^{-t} t^{2\mathrm{Re}\,\nu} \frac{dt}{t}\right] \\ &\le 2 \frac{2^{\frac{1}{2}\mathrm{Re}\,\nu} e^{\frac{\pi}{2}|\mathrm{Im}\,\nu|}}{|\Gamma(\nu+\frac{1}{2})|} \left[|\nu| \frac{\Gamma(\mathrm{Re}\,\nu+\frac{3}{2})}{r^{3/2}} + \frac{2^{\mathrm{Re}\,\nu}}{r^{\mathrm{Re}\,\nu}} \frac{\Gamma(2\mathrm{Re}\,\nu)}{e^r}\right], \end{aligned}$$

using that $e^{-t} \le e^{-t/2}e^{-r}$ for $t \ge 2r$. We conclude that for $r \ge 1$ and $\operatorname{Re} \nu > -1/2$ we have

$$|R_\nu(r)| \le C_0(\operatorname{Re} \nu) \frac{e^{\frac{\pi}{2}|\operatorname{Im} \nu|}(|\nu|+1)}{|\Gamma(\nu+\frac{1}{2})|} r^{-3/2},$$

where C_0 is a constant that depends only on $\operatorname{Re} \nu$. In view of the result in Appendix A.7, the last fraction is bounded by another constant depending on $\operatorname{Re} \nu$ times $e^{\pi^2(1+|\operatorname{Im} \nu|)^2}$.

Appendix C
Rademacher Functions

C.1 Definition of the Rademacher Functions

The Rademacher functions are defined on $[0,1]$ as follows: $r_0(t) = 1$; $r_1(t) = 1$ for $0 \le t \le 1/2$ and $r_1(t) = -1$ for $1/2 < t \le 1$; $r_2(t) = 1$ for $0 \le t \le 1/4$, $r_2(t) = -1$ for $1/4 < t \le 1/2$, $r_2(t) = 1$ for $1/2 < t \le 3/4$, and $r_2(t) = -1$ for $3/4 < t \le 1$; and so on. According to this definition, we have that $r_j(t) = \operatorname{sgn}(\sin(2^j \pi t))$ for $j = 0,1,2,\dots$. It is easy to check that the r_j's are mutually independent random variables on $[0,1]$. This means that for all functions f_j we have

$$\int_0^1 \prod_{j=0}^n f_j(r_j(t))\,dt = \prod_{j=0}^n \int_0^1 f_j(r_j(t))\,dt\,.$$

To see the validity of this identity, we write its right-hand side as

$$\begin{aligned} f_0(1)\prod_{j=1}^n \int_0^1 f_j(r_j(t))\,dt &= f_0(1)\prod_{j=1}^n \frac{f_j(1)+f_j(-1)}{2} \\ &= \frac{f_0(1)}{2^n} \sum_{S\subset\{1,2,\dots,n\}} \prod_{j\in S} f_j(1) \prod_{j\notin S} f_j(-1) \end{aligned}$$

and we observe that there is a one-to-one and onto correspondence between subsets S of $\{1,2,\dots,n\}$ and intervals $I_k = \left[\frac{k}{2^n}, \frac{k+1}{2^n}\right]$, $k = 0,1,\dots,2^n-1$, such that the restriction of the function $\prod_{j=1}^n f_j(r_j(t))$ on I_k is equal to

$$\prod_{j\in S} f_j(1) \prod_{j\notin S} f_j(-1)\,.$$

It follows that the last of the three equal displayed expressions is

$$f_0(1)\sum_{k=0}^{2^n-1} \int_{I_k} \prod_{j=1}^n f_j(r_j(t))\,dt = \int_0^1 \prod_{j=0}^n f_j(r_j(t))\,dt\,.$$

C.2 Khintchine's Inequalities

The following property of the Rademacher functions is of fundamental importance and with far-reaching consequences in analysis:

For any $0 < p < \infty$ and for any real-valued square summable sequences $\{a_j\}$ and $\{b_j\}$ we have

$$B_p\Big(\sum_j |a_j+ib_j|^2\Big)^{\frac{1}{2}} \le \Big\|\sum_j (a_j+ib_j)r_j\Big\|_{L^p([0,1])} \le A_p\Big(\sum_j |a_j+ib_j|^2\Big)^{\frac{1}{2}}$$

for some constants $0 < A_p, B_p < \infty$ that depend only on p.

These inequalities reflect the orthogonality of the Rademacher functions in L^p (especially when $p \neq 2$). Khintchine [155] was the first to prove a special form of this inequality, and he used it to estimate the asymptotic behavior of certain random walks. Later this inequality was systematically studied almost simultaneously by Littlewood [173] and by Paley and Zygmund [210], who proved the more general form stated previously. The foregoing inequalities are usually referred to by Khintchine's name.

C.3 Derivation of Khintchine's Inequalities

Both assertions in Appendix C.2 can be derived from an exponentially decaying distributional inequality for the function

$$F(t) = \sum_j (a_j+ib_j)r_j(t), \qquad t \in [0,1],$$

when a_j, b_j are square summable real numbers.

We first obtain a distributional inequality for the above function F under the following three assumptions:

(a) The sequence $\{b_j\}$ is identically zero.
(b) All but finitely many terms of the sequence $\{a_j\}$ are zero.
(c) The sequence $\{a_j\}$ satisfies $(\sum_j |a_j|^2)^{1/2} = 1$.

Let $\rho > 0$. Under assumptions (a), (b), and (c), independence gives

$$\begin{aligned}\int_0^1 e^{\rho\sum a_j r_j(t)}\,dt &= \prod_j \int_0^1 e^{\rho a_j r_j(t)}\,dt \\ &= \prod_j \frac{e^{\rho a_j}+e^{-\rho a_j}}{2} \\ &\le \prod_j e^{\frac{1}{2}\rho^2 a_j^2} = e^{\frac{1}{2}\rho^2\sum a_j^2} = e^{\frac{1}{2}\rho^2},\end{aligned}$$

where we used the inequality $\frac{1}{2}(e^x+e^{-x}) \le e^{\frac{1}{2}x^2}$ for all real x, which can be checked using power series expansions. Since the same argument is also valid for $-\sum a_j r_j(t)$, we obtain that

$$\int_0^1 e^{\rho|F(t)|}\,dt \le 2e^{\frac{1}{2}\rho^2}.$$

From this it follows that

$$e^{\rho\alpha}|\{t\in[0,1]:\,|F(t)|>\alpha\}| \le \int_0^1 e^{\rho|F(t)|}\,dt \le 2e^{\frac{1}{2}\rho^2}$$

and hence we obtain the distributional inequality

$$d_F(\alpha) = |\{t\in[0,1]:\,|F(t)|>\alpha\}| \le 2e^{\frac{1}{2}\rho^2-\rho\alpha} = 2e^{-\frac{1}{2}\alpha^2},$$

by picking $\rho=\alpha$. The L^p norm of F can now be computed easily. Formula (1.1.6) gives

$$\big\|F\big\|_{L^p}^p = \int_0^\infty p\alpha^{p-1}d_F(\alpha)\,d\alpha \le \int_0^\infty p\alpha^{p-1}2e^{-\frac{\alpha^2}{2}}\,d\alpha = 2^{\frac{p}{2}}p\Gamma(p/2).$$

We have now proved that

$$\big\|F\big\|_{L^p} \le \sqrt{2}\,(p\Gamma(p/2))^{\frac{1}{p}}\big\|F\big\|_{L^2}$$

under assumptions (a), (b), and (c).

We now dispose of assumptions (a), (b), and (c). Assumption (b) can be easily eliminated by a limiting argument and (c) by a scaling argument. To dispose of assumption (a), let a_j and b_j be real numbers. We have

$$\begin{aligned}
\Big\|\sum_j(a_j+ib_j)r_j\Big\|_{L^p} &< \Big\|\Big|\sum_j a_jr_j\Big|+\Big|\sum_j b_jr_j\Big|\Big\|_{L^p} \\
&\le \Big\|\sum_j a_jr_j\Big\|_{L^p}+\Big\|\sum_j b_jr_j\Big\|_{L^p} \\
&\le \sqrt{2}\,(p\Gamma(p/2))^{\frac{1}{p}}\Big(\Big(\sum_j|a_j|^2\Big)^{\frac{1}{2}}+\Big(\sum_j|b_j|^2\Big)^{\frac{1}{2}}\Big) \\
&\le \sqrt{2}\,(p\Gamma(p/2))^{\frac{1}{p}}\sqrt{2}\Big(\sum_j|a_j+ib_j|^2\Big)^{\frac{1}{2}}.
\end{aligned}$$

Let us now set $A_p = 2\big(p\Gamma(p/2)\big)^{1/p}$ when $p>2$. Since we have the trivial estimate $\big\|F\big\|_{L^p}\le\big\|F\big\|_{L^2}$ when $0<p\le 2$, we obtain the required inequality $\big\|F\big\|_{L^p}\le A_p\big\|F\big\|_{L^2}$ with

$$A_p = \begin{cases} 1 & \text{when } 0<p\le 2, \\ 2\,p^{\frac{1}{p}}\,\Gamma(p/2)^{\frac{1}{p}} & \text{when } 2<p<\infty. \end{cases}$$

Using Sterling's formula in Appendix A.6, we see that A_p is asymptotic to $\sqrt{p}$ as $p\to\infty$.

We now discuss the converse inequality $B_p\|F\|_{L^2} \le \|F\|_{L^p}$. It is clear that $\|F\|_{L^2} \le \|F\|_{L^p}$ when $p \ge 2$ and we may therefore take $B_p = 1$ for $p \ge 2$. Let us now consider the case $0 < p < 2$. Pick an s such that $2 < s < \infty$. Find a $0 < \theta < 1$ such that

$$\frac{1}{2} = \frac{1-\theta}{p} + \frac{\theta}{s}.$$

Then

$$\|F\|_{L^2} \le \|F\|_{L^p}^{1-\theta} \|F\|_{L^s}^{\theta} \le \|F\|_{L^p}^{1-\theta} A_s^{\theta} \|F\|_{L^2}^{\theta}.$$

It follows that

$$\|F\|_{L^2} \le A_s^{\frac{\theta}{1-\theta}} \|F\|_{L^p}.$$

We have now proved the inequality $B_p\|F\|_{L^2} \le \|F\|_{L^p}$ with

$$B_p = \begin{cases} 1 & \text{when } 2 \le p < \infty, \\ \sup\limits_{s>2} A_s^{-\frac{\frac{1}{p}-\frac{1}{2}}{\frac{1}{2}-\frac{1}{s}}} & \text{when } 0 < p < 2. \end{cases}$$

Observe that the function $s \to A_s^{-\left(\frac{1}{p}-\frac{1}{2}\right)/\left(\frac{1}{2}-\frac{1}{s}\right)}$ tends to zero as $s \to 2+$ and as $s \to \infty$. Hence it must attain its maximum for some $s = s(p)$ in the interval $(2,\infty)$. We see that $B_p \ge 16 \cdot 256^{-1/p}$ when $p < 2$ by taking $s = 4$.

It is worthwhile to mention that the best possible values of the constants A_p and B_p in Khintchine's inequality are known when $b_j = 0$. In this case Szarek [271] showed that the best possible value of B_1 is $1/\sqrt{2}$, and later Haagerup [116] found that when $b_j = 0$ the best possible values of A_p and B_p are the numbers

$$A_p = \begin{cases} 1 & \text{when } 0 < p \le 2, \\ 2^{\frac{1}{2}}\pi^{-\frac{1}{2p}}\Gamma\left(\frac{p+1}{2}\right) & \text{when } 2 < p < \infty, \end{cases}$$

and

$$B_p = \begin{cases} 2^{\frac{1}{2}-\frac{1}{p}} & \text{when } 0 < p \le p_0, \\ 2^{\frac{1}{2}}\pi^{-\frac{1}{2p}}\Gamma\left(\frac{p+1}{2}\right) & \text{when } p_0 < p < 2, \\ 1 & \text{when } 2 < p < \infty, \end{cases}$$

where $p_0 = 1.84742\ldots$ is the unique solution of the equation $2\Gamma\left(\frac{p+1}{2}\right) = \sqrt{\pi}$ in the interval $(1,2)$.

C.4 Khintchine's Inequalities for Weak Type Spaces

We note that the following weak type estimates are valid:

$$4^{-\frac{1}{p}} B_{\frac{p}{2}} \Big(\sum_j |a_j + ib_j|^2 \Big)^{\frac{1}{2}} \le \Big\| \sum_j (a_j + ib_j) r_j \Big\|_{L^{p,\infty}} \le A_p \Big(\sum_j |a_j + ib_j|^2 \Big)^{\frac{1}{2}}$$

for all $0 < p < \infty$.

Indeed, the upper estimate is a simple consequence of the fact that L^p is a subspace of $L^{p,\infty}$. For the converse inequality we use the fact that $L^{p,\infty}([0,1])$ is contained in $L^{p/2}([0,1])$ and we have (see Exercise 1.1.11)

$$\big\|F\big\|_{L^{p/2}} \le 4^{\frac{1}{p}} \big\|F\big\|_{L^{p,\infty}}.$$

Since any Lorentz space $L^{p,q}([0,1])$ can be sandwiched between $L^{2p}([0,1])$ and $L^{p/2}([0,1])$, similar inequalities hold for all Lorentz spaces $L^{p,q}([0,1])$, $0 < p < \infty$, $0 < q \le \infty$.

C.5 Extension to Several Variables

We first extend the inequality on the right in Appendix C.2 to several variables. For a positive integer n we let

$$F_n(t_1, \dots, t_n) = \sum_{j_1} \cdots \sum_{j_n} c_{j_1,\dots,j_n} r_{j_1}(t_1) \cdots r_{j_n}(t_n),$$

for $t_j \in [0,1]$, where $c_{j_1,\dots,j_n}$ is a sequence of complex numbers and F_n is a function defined on $[0,1]^n$.

For any $0 < p < \infty$ and for any complex-valued square summable sequence of n variables $\{c_{j_1,\dots,j_n}\}_{j_1,\dots,j_n}$, we have the following inequalities for F_n:

$$B_p^n \Big(\sum_{j_1} \cdots \sum_{j_n} |c_{j_1,\dots,j_n}|^2 \Big)^{\frac{1}{2}} \le \big\|F_n\big\|_{L^p} \le A_p^n \Big(\sum_{j_1} \cdots \sum_{j_n} |c_{j_1,\dots,j_n}|^2 \Big)^{\frac{1}{2}},$$

where A_p, B_p are the constants in Appendix C.2. The norms are over $[0,1]^n$.

The case $n = 2$ is indicative of the general case. For $p \ge 2$ we have

$$\begin{aligned}
\int_0^1 \int_0^1 |F_2(t_1,t_2)|^p \, dt_1 \, dt_2 &\le A_p^p \int_0^1 \Big(\sum_{j_1} \Big| \sum_{j_2} c_{j_1,j_2} r_{j_2}(t_2) \Big|^2 \Big)^{\frac{p}{2}} dt_2 \\
&\le A_p^p \Big(\sum_{j_1} \Big(\int_0^1 \Big| \sum_{j_2} c_{j_1,j_2} r_{j_2}(t_2) \Big|^p dt_2 \Big)^{\frac{2}{p}} \Big)^{\frac{p}{2}} \\
&\le A_p^{2p} \Big(\sum_{j_1} \sum_{j_2} |c_{j_1,j_n}|^2 \Big)^{\frac{p}{2}},
\end{aligned}$$

where we used Minkowski's integral inequality (with exponent $p/2 \geq 1$) in the second inequality and the result in the case $n = 1$ twice.

The case $p < 2$ follows trivially from Hölder's inequality with constant $A_p = 1$. The reverse inequalities follow exactly as in the case of one variable. Replacing A_p by A_p^n in the argument, giving the reverse inequality in the case $n = 1$, we obtain the constant B_p^n.

Likewise one may extend the weak type inequalities of Appendix C.3 in several variables.

Appendix D
Spherical Coordinates

D.1 Spherical Coordinate Formula

Switching integration from spherical coordinates to Cartesian is achieved via the following identity:

$$\int_{RS^{n-1}} f(x)\,d\sigma(x) = \int_{\varphi_1=0}^{\pi} \cdots \int_{\varphi_{n-2}=0}^{\pi} \int_{\varphi_{n-1}=0}^{2\pi} f(x(\varphi))J(n,R,\varphi)\,d\varphi_{n-1}\cdots d\varphi_1,$$

where

$$\begin{aligned}
x_1 &= R\cos\varphi_1\,,\\
x_2 &= R\sin\varphi_1\,\cos\varphi_2\,,\\
x_3 &= R\sin\varphi_1\,\sin\varphi_2\,\cos\varphi_3\,,\\
&\dots\\
x_{n-1} &= R\sin\varphi_1\,\sin\varphi_2\,\sin\varphi_3\cdots\sin\varphi_{n-2}\,\cos\varphi_{n-1}\,,\\
x_n &= R\sin\varphi_1\,\sin\varphi_2\,\sin\varphi_3\cdots\sin\varphi_{n-2}\,\sin\varphi_{n-1}\,,
\end{aligned}$$

and $0 \le \varphi_1,\dots,\varphi_{n-2} \le \pi$, $0 \le \varphi_{n-1} = \theta \le 2\pi$,

$$x(\varphi) = (x_1(\varphi_1,\dots,\varphi_{n-1}),\dots,x_n(\varphi_1,\dots,\varphi_{n-1}))\,,$$

and

$$J(n,R,\varphi) = R^{n-1}(\sin\varphi_1)^{n-2}\cdots(\sin\varphi_{n-3})^2(\sin\varphi_{n-2})$$

is the Jacobian of the transformation.

D.2 A Useful Change of Variables Formula

The following formula is useful in computing integrals over the sphere $\mathbf{S}^{n-1}$ when $n \ge 2$. Let f be a function defined on $\mathbf{S}^{n-1}$. Then we have

$$\int_{R\mathbf{S}^{n-1}} f(x)\,d\sigma(x) = \int_{-R}^{+R} \int_{\sqrt{R^2-s^2}\,\mathbf{S}^{n-2}} f(s,\theta)\,d\theta\,\frac{R\,ds}{\sqrt{R^2-s^2}}\,.$$

To prove this formula, let $\varphi' = (\varphi_2,\dots,\varphi_{n-1})$ and

$$x' = x'(\varphi') = (\cos\varphi_2, \sin\varphi_2\cos\varphi_3, \ldots, \sin\varphi_2\cdots\sin\varphi_{n-2}\sin\varphi_{n-1}).$$

Using the change of variables in Appendix D.1 we express

$$\int_{R\mathbf{S}^{n-1}} f(x)\,d\sigma(x)$$

as the iterated integral

$$\int_{\varphi_1=0}^{\pi}\left[\int_{\varphi_2=0}^{\pi}\cdots\int_{\varphi_{n-1}=0}^{2\pi} f(R\cos\varphi_1, R\sin\varphi_1\, x'(\varphi'))J(n-1,1,\varphi')\,d\varphi'\right]\frac{R\,d\varphi_1}{(R\sin\varphi_1)^{2-n}},$$

and we can realize the expression inside the square brackets as

$$\int_{\mathbf{S}^{n-2}} f(R\cos\varphi_1, R\sin\varphi_1\, x')d\sigma(x').$$

Consequently,

$$\int_{R\mathbf{S}^{n-1}} f(x)\,d\sigma(x) = \int_{\varphi_1=0}^{\pi}\int_{\mathbf{S}^{n-2}} f(R\cos\varphi_1, R\sin\varphi_1\, x')\,d\sigma(x')R^{n-1}(\sin\varphi_1)^{n-2}d\varphi_1\,,$$

and the change of variables

$$s = R\cos\varphi_1\,, \qquad \varphi_1 \in (0,\pi),$$

$$ds = -R\sin\varphi_1\,d\varphi_1\,, \qquad \sqrt{R^2-s^2} = R\sin\varphi_1\,,$$

yields

$$\int_{R\mathbf{S}^{n-1}} f(x)\,d\sigma(x) = \int_{-R}^{R}\left\{\int_{\mathbf{S}^{n-2}} f(s, \sqrt{R^2-s^2}\,\theta)\,d\theta\right\}\left(\sqrt{R^2-s^2}\right)^{n-2}\frac{R\,ds}{\sqrt{R^2-s^2}}.$$

Rescaling the sphere $\mathbf{S}^{n-2}$ to $\sqrt{R^2-s^2}\,\mathbf{S}^{n-2}$ yields the claimed identity.

D.3 Computation of an Integral over the Sphere

Let K be a function on the line. We use the result in Appendix D.2 to show that for $n \geq 2$ we have

$$\int_{\mathbf{S}^{n-1}} K(x\cdot\theta)\,d\theta = \frac{2\pi^{\frac{n-1}{2}}}{\Gamma\left(\frac{n-1}{2}\right)}\int_{-1}^{+1} K(s|x|)\left(\sqrt{1-s^2}\right)^{n-3}ds$$

when $x \in \mathbf{R}^n \setminus \{0\}$. Let $x' = x/|x|$ and pick a matrix $A \in O(n)$ such that $Ae_1 = x'$, where $e_1 = (1,0,\ldots,0)$. We have

$$\begin{aligned}
\int_{\mathbf{S}^{n-1}} K(x\cdot\theta)\,d\theta &= \int_{\mathbf{S}^{n-1}} K(|x|(x'\cdot\theta))\,d\theta \\
&= \int_{\mathbf{S}^{n-1}} K(|x|(Ae_1\cdot\theta))\,d\theta \\
&= \int_{\mathbf{S}^{n-1}} K(|x|(e_1\cdot A^{-1}\theta))\,d\theta \\
&= \int_{\mathbf{S}^{n-1}} K(|x|\theta_1)\,d\theta \\
&= \int_{-1}^{+1} K(|x|s)\,\omega_{n-2}\big(\sqrt{1-s^2}\big)^{n-2}\frac{ds}{\sqrt{1-s^2}} \\
&= \omega_{n-2}\int_{-1}^{+1} K(s|x|)\big(\sqrt{1-s^2}\big)^{n-3}\,ds\,,
\end{aligned}$$

where $\omega_{n-2}=2\pi^{\frac{n-1}{2}}\Gamma\big(\frac{n-1}{2}\big)^{-1}$ is the surface area of $\mathbf{S}^{n-2}$.

For example, we have

$$\int_{\mathbf{S}^{n-1}}\frac{d\theta}{|\xi\cdot\theta|^{\alpha}}=\omega_{n-2}\int_{-1}^{+1}\frac{1}{|s|^{\alpha}|\xi|^{\alpha}}(1-s^2)^{\frac{n-3}{2}}\,ds=\frac{1}{|\xi|^{\alpha}}\frac{2\pi^{\frac{n-1}{2}}\Gamma\big(\frac{1-\alpha}{2}\big)}{\Gamma\big(\frac{n-\alpha}{2}\big)}\,,$$

and the integral converges only when $\operatorname{Re}\alpha<1$.

D.4 The Computation of Another Integral over the Sphere

We compute the following integral for $n\geq 2$:

$$\int_{\mathbf{S}^{n-1}}\frac{d\theta}{|\theta-e_1|^{\alpha}}\,,$$

where $e_1=(1,0,\dots,0)$. Applying the formula in Appendix D.2, we obtain

$$\begin{aligned}
\int_{\mathbf{S}^{n-1}}\frac{d\theta}{|\theta-e_1|^{\alpha}} &= \int_{-1}^{+1}\int_{\theta\in\sqrt{1-s^2}\,\mathbf{S}^{n-2}}\frac{d\theta}{(|s-1|^2+|\theta|^2)^{\frac{\alpha}{2}}}\frac{ds}{\sqrt{1-s^2}} \\
&= \int_{-1}^{+1}\omega_{n-2}\frac{(1-s^2)^{\frac{n-2}{2}}}{\big((1-s)^2+1-s^2\big)^{\frac{\alpha}{2}}}\frac{ds}{\sqrt{1-s^2}} \\
&= \frac{\omega_{n-2}}{2^{\frac{\alpha}{2}}}\int_{-1}^{+1}\frac{(1-s^2)^{\frac{n-3}{2}}}{(1-s)^{\frac{\alpha}{2}}}\,ds \\
&= \frac{\omega_{n-2}}{2^{\frac{\alpha}{2}}}\int_{-1}^{+1}(1-s)^{\frac{n-3-\alpha}{2}}(1+s)^{\frac{n-3}{2}}\,ds\,,
\end{aligned}$$

which converges exactly when $\operatorname{Re}\alpha<n-1$.

D.5 Integration over a General Surface

Suppose that S is a hypersurface in $\mathbf{R}^n$ of the form $S = \{(u, \Phi(u)) : u \in D\}$, where D is an open subset of $\mathbf{R}^{n-1}$ and Φ is a continuously differentiable mapping from D to $\mathbf{R}$. Let σ be the canonical surface measure on S. If g is a function on S, then we have

$$\int_S g(y)\, d\sigma(y) = \int_D g(x, \Phi(x)) \Big(1 + \sum_{j=1}^{n} |\partial_j \Phi(x)|^2\Big)^{\frac{1}{2}} dx.$$

Specializing to the sphere, we obtain

$$\int_{\mathbf{S}^{n-1}} g(\theta)\, d\theta = \int\limits_{\substack{\xi' \in \mathbf{R}^{n-1}\\ |\xi'|<1}} \Big[g(\xi', \sqrt{1-|\xi'|^2}) + g(\xi', -\sqrt{1-|\xi'|^2})\Big] \frac{d\xi'}{\sqrt{1-|\xi'|^2}}.$$

D.6 The Stereographic Projection

Define a map $\Pi : \mathbf{R}^n \to \mathbf{S}^n$ by the formula

$$\Pi(x_1, \dots, x_n) = \left(\frac{2x_1}{1+|x|^2}, \dots, \frac{2x_n}{1+|x|^2}, \frac{|x|^2-1}{1+|x|^2}\right).$$

It is easy to see that Π is a one-to-one map from $\mathbf{R}^n$ onto the sphere $\mathbf{S}^n$ minus the north pole $e_{n+1} = (0, \dots, 0, 1)$. Its inverse is given by the formula

$$\Pi^{-1}(\theta_1, \dots, \theta_{n+1}) = \left(\frac{\theta_1}{1-\theta_{n+1}}, \dots, \frac{\theta_n}{1-\theta_{n+1}}\right).$$

The Jacobian of the map is verified to be

$$J_\Pi(x) = \left(\frac{2}{1+|x|^2}\right)^n,$$

and the following change of variables formulas are valid:

$$\int_{\mathbf{S}^n} F(\theta)\, d\theta = \int_{\mathbf{R}^n} F(\Pi(x)) J_\Pi(x)\, dx$$

and

$$\int_{\mathbf{R}^n} F(x)\, dx = \int_{\mathbf{S}^n} F(\Pi^{-1}(\theta)) J_{\Pi^{-1}}(\theta)\, d\theta,$$

where

$$J_{\Pi^{-1}}(\theta) = \frac{1}{J_\Pi(\Pi^{-1}(\theta))} = \left(\frac{|\theta_1|^2 + \cdots + |\theta_n|^2 + |1-\theta_{n+1}|^2}{2|1-\theta_{n+1}|^2}\right)^n.$$

Another interesting formula about the stereographic projection Π is

$$|\Pi(x) - \Pi(y)| = 2|x - y|(1 + |x|^2)^{-1/2}(1 + |y|^2)^{-1/2},$$

for all x, y in $\mathbf{R}^n$.

Appendix E
Some Trigonometric Identities and Inequalities

The following inequalities are valid for t real:

$$\begin{aligned}
0 < t < \frac{\pi}{2} &\implies \sin(t) < t < \tan(t)\,, \\
0 < |t| < \frac{\pi}{2} &\implies \frac{2}{\pi} < \frac{\sin(t)}{t} < 1\,, \\
-\infty < t < +\infty &\implies |\sin(t)| \le |t|\,, \\
-\infty < t < +\infty &\implies |1-\cos(t)| \le \frac{|t|^2}{2}\,, \\
-\infty < t < +\infty &\implies |1-e^{it}| \le |t|\,, \\
|t| \le \frac{\pi}{2} &\implies |\sin(t)| \ge \frac{2|t|}{\pi}\,, \\
|t| \le \pi &\implies |1-\cos(t)| \ge \frac{2|t|^2}{\pi^2}\,, \\
|t| \le \pi &\implies |1-e^{it}| \ge \frac{2|t|}{\pi}\,.
\end{aligned}$$

The following sum to product formulas are valid:

$$\begin{aligned}
\sin(a)+\sin(b) &= 2\sin\left(\frac{a+b}{2}\right)\cos\left(\frac{a-b}{2}\right), \\
\sin(a)-\sin(b) &= 2\cos\left(\frac{a+b}{2}\right)\sin\left(\frac{a-b}{2}\right), \\
\cos(a)+\cos(b) &= 2\cos\left(\frac{a+b}{2}\right)\cos\left(\frac{a-b}{2}\right), \\
\cos(a)-\cos(b) &= -2\sin\left(\frac{a+b}{2}\right)\sin\left(\frac{a-b}{2}\right).
\end{aligned}$$

The following identities are also easily proved:

$$\begin{aligned}
\sum_{k=1}^{N}\cos(kx) &= -\frac{1}{2}+\frac{\sin((N+\frac{1}{2})x)}{2\sin(\frac{x}{2})}\,, \\
\sum_{k=1}^{N}\sin(kx) &= \frac{\cos(\frac{x}{2})-\cos((N+\frac{1}{2})x)}{2\sin(\frac{x}{2})}\,.
\end{aligned}$$

Appendix F
Summation by Parts

Let $\{a_k\}_{k=0}^{\infty}$, $\{b_k\}_{k=0}^{\infty}$ be two sequences of complex numbers. Then for $N \geq 1$ we have

$$\sum_{k=0}^{N} a_k b_k = A_N b_N - \sum_{k=0}^{N-1} A_k (b_{k+1} - b_k),$$

where

$$A_k = \sum_{j=0}^{k} a_j.$$

More generally we have

$$\sum_{k=M}^{N} a_k b_k = A_N b_N - A_{M-1} b_M - \sum_{k=M}^{N-1} A_k (b_{k+1} - b_k),$$

whenever $0 \leq M \leq N$, where $A_{-1} = 0$ and

$$A_k = \sum_{j=0}^{k} a_j$$

for $k \geq 0$.

Appendix G
Basic Functional Analysis

A quasinorm is a nonnegative functional $\|\cdot\|$ on a vector space X that satisfies $\|x+y\|_X \le K(\|x\|_X + \|y\|_X)$ for some $K \ge 0$ and all $x,y \in X$ and also $\|\lambda x\|_X = |\lambda|\|x\|_X$ for all scalars λ. When $K = 1$, the quasinorm is called a norm. A quasi-Banach space is a quasinormed space that is complete with respect to the topology generated by the quasinorm. The proofs of the following theorems can be found in several books including Albiac and Kalton [1], Kalton Peck and Roberts [150], and Rudin [230].

The Hahn–Banach theorem. Let X be a normed space and X_0 a subspace. Every bounded linear functional Λ_0 on X_0 has a bounded extension Λ on X with the same norm. In addition, if Λ_0 is subordinate to a positively homogeneous subadditive functional P, then Λ may be chosen to have the same property.

Banach–Alaoglou theorem. Let X be a quasi-Banach space and let X^* be the space of all bounded linear functionals on X. Then the unit ball of X^* is weak* compact.

Open mapping theorem. Suppose that X and Y are quasi-Banach spaces and T is a bounded surjective linear map from X onto Y. Then there exists a constant $K < \infty$ such that for all $x \in X$ we have

$$\|x\|_X \le K\|T(x)\|_Y .$$

Closed graph theorem. Suppose that X and Y are quasi-Banach spaces and T is a linear map from X to Y whose graph is a closed set, i.e., whenever $x_k, x \in X$ and $(x_k, T(x_k)) \mapsto (x,y)$ in $X \times Y$ for some $y \in Y$, then $T(x) = y$. Then T is a bounded linear map from X to Y.

Uniform boundedness principle. Suppose that X is a quasi-Banach space, Y is a quasinormed space and $(T_\alpha)_{\alpha\in I}$ is a family of bounded linear maps from X to Y such that for all $x \in X$ there exists a $C_x < \infty$ such that

$$\sup_{\alpha\in I} \|T_\alpha(x)\|_Y \le C_x .$$

Then there exists a constant $K < \infty$ such that

$$\sup_{\alpha\in I} \|T_\alpha\|_{X\to Y} \le K .$$

Appendix H
The Minimax Lemma

Minimax type results are used in the theory of games and have their origin in the work of Von Neumann [286]. Much of the theory in this subject is based on convex analysis techniques. For instance, this is the case with the next proposition, which is needed in the "difficult" inequality in the proof of the minimax lemma. We refer to Fan [87] for a general account of minimax results. The following exposition is based on the simple presentation in Appendix A2 of [98].

Minimax Lemma. *Let A, B be convex subsets of certain vector spaces. Assume that a topology is defined in B for which it is a compact Hausdorff space and assume that there is a function* $\Phi : A \times B \to \mathbf{R}\bigcup\{+\infty\}$ *that satisfies the following:*
(a) $\Phi(\,.\,,b)$ *is a concave function on A for each* $b \in B$*,*
(b) $\Phi(a,\,.\,)$ *is a convex function on B for each* $a \in A$*,*
(c) $\Phi(a,\,.\,)$ *is lower semicontinuous on B for each* $a \in A$*.*
Then the following identity holds:

$$\min_{b\in B}\sup_{a\in A}\Phi(a,b) = \sup_{a\in A}\min_{b\in B}\Phi(a,b)\,.$$

To prove the lemma we need the following proposition:

Proposition. *Let B be a convex compact subset of a vector space and suppose that* $g_j : B \to \mathbf{R}\bigcup\{+\infty\}$, $j = 1,2,\ldots,n$, *are convex and lower semicontinuous. If*

$$\max_{1\leq j\leq n} g_j(b) > 0 \quad \textit{for all} \quad b \in B\,,$$

then there exist nonnegative numbers $\lambda_1,\lambda_2,\ldots,\lambda_n$ *such that*

$$\lambda_1 g_1(b) + \lambda_2 g_2(b) + \cdots + \lambda_n g_n(b) > 0 \quad \textit{for all} \quad b \in B\,.$$

Proof. We first consider the case $n = 2$. Define subsets of B

$$B_1 = \{b \in B : \; g_1(b) \leq 0\}, \quad B_2 = \{b \in B : \; g_2(b) \leq 0\}\,.$$

If $B_1 = \emptyset$, we take $\lambda_1 = 1$ and $\lambda_2 = 0$, and we similarly deal with the case $B_2 = \emptyset$. If B_1 and B_2 are nonempty, then they are closed and thus compact. The hypothesis of the proposition implies that $g_2(b) > 0 \geq g_1(b)$ for all $b \in B_1$. Therefore, the function $-g_1(b)/g_2(b)$ is well defined and upper semicontinuous on B_1 and thus attains its maximum. The same is true for $-g_2(b)/g_1(b)$ defined on B_2. We set

$$\mu_1 = \max_{b \in B_1} \frac{-g_1(b)}{g_2(b)} \geq 0, \qquad \mu_2 = \max_{b \in B_2} \frac{-g_2(b)}{g_1(b)} \geq 0.$$

We need to find $\lambda > 0$ such that $\lambda g_1(b) + g_2(b) > 0$ for all $b \in B$. This is clearly satisfied if $b \notin B_1 \bigcup B_2$, while for $b_1 \in B_1$ and $b_2 \in B_2$ we have

$$\lambda g_1(b_1) + g_2(b_1) \geq (1 - \lambda \mu_1) g_2(b_1),$$
$$\lambda g_1(b_2) + g_2(b_2) \geq (\lambda - \mu_2) g_1(b_2).$$

Therefore, it suffices to find a $\lambda > 0$ such that $1 - \lambda \mu_1 > 0$ and $\lambda - \mu_2 > 0$. Such a λ exists if and only if $\mu_1 \mu_2 < 1$. To prove that $\mu_1 \mu_2 < 1$, we can assume that $\mu_1 \neq 0$ and $\mu_2 \neq 0$. Then we take $b_1 \in B_1$ and $b_2 \in B_2$, for which the maxima μ_1 and μ_2 are attained, respectively. Then we have

$$g_1(b_1) + \mu_1 g_2(b_1) = 0,$$
$$g_1(b_2) + \frac{1}{\mu_2} g_2(b_2) = 0.$$

But $g_1(b_1) < 0 < g_1(b_2)$; thus taking $b_\theta = \theta b_1 + (1-\theta) b_2$ for some θ in $(0,1)$, we have

$$g_1(b_\theta) \leq \theta g_1(b_1) + (1-\theta) g_1(b_2) = 0.$$

Considering the same convex combination of the last displayed equations and using this identity, we obtain that

$$\mu_1 \mu_2 \theta g_2(b_1) + (1-\theta) g_2(b_2) = 0.$$

The hypothesis of the proposition implies that $g_2(b_\theta) > 0$ and the convexity of g_2:

$$\theta g_2(b_1) + (1-\theta) g_2(b_2) > 0.$$

Since $g_2(b_1) > 0$, we must have $\mu_1 \mu_2 g_2(b_1) < g_2(b_1)$, which gives $\mu_1 \mu_2 < 1$. This proves the required claim and completes the case $n = 2$.

We now use induction to prove the proposition for arbitrary n. Assume that the result has been proved for $n-1$ functions. Consider the subset of B

$$B_n = \{b \in B : \; g_n(b) \leq 0\}.$$

If $B_n = \emptyset$, we choose $\lambda_1 = \lambda_2 = \cdots = \lambda_{n-1} = 0$ and $\lambda_n = 1$. If B_n is not empty, then it is compact and convex and we can restrict $g_1, g_2, \ldots, g_{n-1}$ to B_n. Using the induction hypothesis, we can find $\lambda_1, \lambda_2, \ldots, \lambda_{n-1} \geq 0$ such that

$$g_0(b) = \lambda_1 g_1(b) + \lambda_2 g_2(b) + \cdots + \lambda_{n-1} g_{n-1}(b) > 0$$

for all $b \in B_n$. Then g_0 and g_n are convex lower semicontinuous functions on B, and $\max(g_0(b), g_n(b)) > 0$ for all $b \in B$. Using the case $n = 2$, which was first proved, we can find $\lambda_0, \lambda_n \geq 0$ such that for all $b \in B$ we have

$$\begin{aligned} 0 &< \lambda_0 g_0(b) + \lambda_n g_n(b) \\ &= \lambda_0\lambda_1 g_1(b) + \lambda_0\lambda_2 g_2(b) + \cdots + \lambda_0\lambda_{n-1} g_{n-1}(b) + \lambda_n g_n(b). \end{aligned}$$

This establishes the case of n functions and concludes the proof of the induction and hence of the proposition. □

We now turn to the proof of the minimax lemma.

Proof. The fact that the left-hand side in the required conclusion of the minimax lemma is at least as big as the right-hand side is obvious. We can therefore concentrate on the converse inequality. In doing this we may assume that the right-hand side is finite. Without loss of generality we can subtract a finite constant from $\Phi(a,b)$, and so we can also assume that

$$\sup_{a\in A}\min_{b\in B}\Phi(a,b) = 0.$$

Then, by hypothesis *(c)* of the minimax lemma, the subsets

$$B_a = \{b \in B :\ \Phi(a,b) \le 0\}, \qquad\qquad a \in A$$

of B are closed and nonempty, and we show that they satisfy the finite intersection property. Indeed, suppose that

$$B_{a_1} \cap B_{a_2} \cap \cdots \cap B_{a_n} = \emptyset$$

for some $a_1, a_2, \ldots, a_n \in A$. We write $g_j(b) = \Phi(a_j,b)$, $j = 1,2,\ldots,n$, and we observe that the conditions of the previous proposition are satisfied. Therefore we can find $\lambda_1, \lambda_2, \ldots, \lambda_n \ge 0$ such that for all $b \in B$ we have

$$\lambda_1\Phi(a_1,b) + \lambda_2\Phi(a_2,b) + \cdots + \lambda_n\Phi(a_n,b) > 0.$$

For simplicity we normalize the λ_j's by setting $\lambda_1 + \lambda_2 + \cdots + \lambda_n = 1$. If we set $a_0 = \lambda_1 a_1 + \lambda_2 a_2 + \cdots + \lambda_n a_n$, the concavity hypothesis (a) gives

$$\Phi(a_0,b) > 0$$

for all $b \in B$, contradicting the fact that $\sup_{a\in A}\min_{b\in B}\Phi(a,b) = 0$. Therefore, the family of closed subsets $\{B_a\}_{a\in A}$ of B satisfies the finite intersection property. The compactness of B now implies $\bigcap_{a\in A} B_a \neq \emptyset$. Take $b_0 \in \bigcap_{a\in A} B_a$. Then $\Phi(a,b_0) \le 0$ for every $a \in A$, and therefore

$$\min_{b\in B}\sup_{a\in A}\Phi(a,b) \le \sup_{a\in A}\Phi(a,b_0) \le 0$$

as required. □

Appendix I
The Schur Lemma

Schur's lemma provides sufficient conditions for linear operators to be bounded on L^p. Moreover, for positive operators it provides necessary and sufficient such conditions. We discuss these situations.

I.1 The Classical Schur Lemma

We begin with an easy situation. Suppose that $K(x,y)$ is a locally integrable function on a product of two σ-finite measure spaces $(X,\mu)\times(Y,\nu)$, and let T be a linear operator given by

$$T(f)(x) = \int_Y K(x,y)f(y)\,d\nu(y)$$

when f is bounded and compactly supported. It is a simple consequence of Fubini's theorem that for almost all $x\in X$ the integral defining T converges absolutely. The following lemma provides a sufficient criterion for the L^p boundedness of T.

Lemma. *Suppose that a locally integrable function $K(x,y)$ satisfies*

$$\sup_{x\in X}\int_Y |K(x,y)|\,d\nu(y) = A < \infty,$$

$$\sup_{y\in Y}\int_X |K(x,y)|\,d\mu(x) = B < \infty.$$

Then the operator T extends to a bounded operator from $L^p(Y)$ to $L^p(X)$ with norm $A^{1-\frac{1}{p}}B^{\frac{1}{p}}$ for $1\le p\le\infty$.

Proof. The second condition gives that T maps L^1 to L^1 with bound B, while the first condition gives that T maps L^∞ to L^∞ with bound A. It follows by the Riesz–Thorin interpolation theorem that T maps L^p to L^p with bound $A^{1-\frac{1}{p}}B^{\frac{1}{p}}$. □

This lemma can be improved significantly when the operators are assumed to be positive.

I.2 Schur's Lemma for Positive Operators

We have the following necessary and sufficient condition for the L^p boundedness of positive operators.

Lemma. *Let (X,μ) and (Y,ν) be two σ-finite measure spaces, where μ and ν are positive measures, and suppose that $K(x,y)$ is a nonnegative measurable function on $X\times Y$. Let $1<p<\infty$ and $0<A<\infty$. Let T be the linear operator*

$$T(f)(x)=\int_Y K(x,y)f(y)\,d\nu(y)$$

and T^t its transpose operator

$$T^t(g)(y)=\int_X K(x,y)g(x)\,d\mu(x)\,.$$

To avoid trivialities, we assume that there is a compactly supported, bounded, and positive ν-a.e. function h_1 on Y such that $T(h_1)>0$ μ-a.e. Then the following are equivalent:

(i) T maps $L^p(Y)$ to $L^p(X)$ with norm at most A.
(ii) For all $B>A$ there is a measurable function h on Y that satisfies $0<h<\infty$ ν-a.e., $0<T(h)<\infty$ μ-a.e., and such that

$$T^t\big(T(h)^{\frac{p}{p'}}\big)\le B^p\,h^{\frac{p}{p'}}\,.$$

(iii) For all $B>A$ there are measurable functions u on X and v on Y such that $0<u<\infty$ μ-a.e., $0<v<\infty$ ν-a.e., and such that

$$\begin{aligned}T(u^{p'})&\le Bv^{p'},\\ T^t(v^p)&\le Bu^p.\end{aligned}$$

Proof. First we assume (ii) and we prove (iii). Define u,v by the equations $v^{p'}=T(h)$ and $u^{p'}=Bh$ and observe that (iii) holds for this choice of u and v. Moreover, observe that $0<u,v<\infty$ a.e. with respect to the measures μ and ν, respectively.

Next we assume (iii) and we prove (i). For g in $L^{p'}(X)$ we have

$$\int_X T(f)(x)\,g(x)\,d\mu(x)=\int_X\int_Y K(x,y)f(y)g(x)\frac{v(x)}{u(y)}\frac{u(y)}{v(x)}\,d\nu(y)\,d\mu(x).$$

We now apply Hölder's inequality with exponents p and p' to the functions

$$f(y)\frac{v(x)}{u(y)}\qquad\text{and}\qquad g(x)\frac{u(y)}{v(x)}$$

with respect to the measure $K(x,y)\,d\nu(y)\,d\mu(x)$ on $X\times Y$. Since

$$\left(\int_Y\int_X f(y)^p\frac{v(x)^p}{u(y)^p}K(x,y)\,d\mu(x)\,d\nu(y)\right)^{\frac{1}{p}}\le B^{\frac{1}{p}}\big\|f\big\|_{L^p(Y)}$$

and

$$\left(\int_X \int_Y g(x)^{p'} \frac{u(y)^{p'}}{v(x)^{p'}} K(x,y)\, dv(y)\, d\mu(x)\right)^{\frac{1}{p'}} \le B^{\frac{1}{p'}} \|g\|_{L^{p'}(X)},$$

we conclude that

$$\left|\int_X T(f)(x) g(x)\, d\mu(x)\right| \le B^{\frac{1}{p}+\frac{1}{p'}} \|f\|_{L^p(Y)} \|g\|_{L^{p'}(X)}.$$

Taking the supremum over all g with $L^{p'}(X)$ norm 1, we obtain

$$\|T(f)\|_{L^p(X)} \le B\|f\|_{L^p(Y)}.$$

Since B was any number greater than A, we conclude that

$$\|T\|_{L^p(Y)\to L^p(X)} \le A,$$

which proves (i).

We finally assume (i) and we prove (ii). Without loss of generality, take here $A = 1$ and $B > 1$. Define a map $S: L^p(Y) \to L^p(Y)$ by setting

$$S(f)(y) = \left(T^t\left(T(f)^{\frac{p}{p'}}\right)\right)^{\frac{p'}{p}}(y).$$

We observe two things. First, $f_1 \le f_2$ implies $S(f_1) \le S(f_2)$, which is an easy consequence of the fact that the same monotonicity is valid for T. Next, we observe that $\|f\|_{L^p} \le 1$ implies that $\|S(f)\|_{L^p} \le 1$ as a consequence of the boundedness of T on L^p (with norm at most 1).

Construct a sequence h_n, $n = 1,2,\dots$, by induction as follows. Pick $h_1 > 0$ on Y as in the hypothesis of the theorem such that $T(h_1) > 0$ μ-a.e. and such that $\|h_1\|_{L^p} \le B^{-p'}(B^{p'} - 1)$. (The last condition can be obtained by multiplying h_1 by a small constant.) Assuming that h_n has been defined, we define

$$h_{n+1} = h_1 + \frac{1}{B^{p'}} S(h_n).$$

We check easily by induction that we have the monotonicity property $h_n \le h_{n+1}$ and the fact that $\|h_n\|_{L^p} \le 1$. We now define

$$h(x) = \sup_n h_n(x) = \lim_{n\to\infty} h_n(x).$$

Fatou's lemma gives that $\|h\|_{L^p} \le 1$, from which it follows that $h < \infty$ ν-a.e. Since $h \ge h_1 > 0$ ν-a.e., we also obtain that $h > 0$ ν-a.e.

Next we use the Lebesgue dominated convergence theorem to obtain that $h_n \to h$ in $L^p(Y)$. Since T is bounded on L^p, it follows that $T(h_n) \to T(h)$ in $L^p(X)$. It follows that $T(h_n)^{\frac{p}{p'}} \to T(h)^{\frac{p}{p'}}$ in $L^{p'}(X)$. Our hypothesis gives that T^t maps $L^{p'}(X)$

to $L^{p'}(Y)$ with norm at most 1. It follows $T^t\big(T(h_n)^{\frac{p}{p'}}\big) \to T^t\big(T(h)^{\frac{p}{p'}}\big)$ in $L^{p'}(Y)$. Raising to the power $\frac{p'}{p}$, we obtain that $S(h_n) \to S(h)$ in $L^p(Y)$.

It follows that for some subsequence n_k of the integers we have $S(h_{n_k}) \to S(h)$ a.e. in Y. Since the sequence $S(h_n)$ is increasing, we conclude that the entire sequence $S(h_n)$ converges almost everywhere to $S(h)$. We use this information in conjunction with $h_{n+1} = h_1 + \frac{1}{B^{p'}} S(h_n)$. Indeed, letting $n \to \infty$ in this identity, we obtain

$$h = h_1 + \frac{1}{B^{p'}} S(h).$$

Since $h_1 > 0$ ν-a.e. it follows that $S(h) \le B^{p'} h$ ν-a.e., which proves the required estimate in (ii).

It remains to prove that $0 < T(h) < \infty$ μ-a.e. Since $\|h\|_{L^p} \le 1$ and T is L^p bounded, it follows that $\|T(h)\|_{L^p} \le 1$, which implies that $T(h) < \infty$ μ-a.e. We also have $T(h) \ge T(h_1) > 0$ μ-a.e. □

I.3 An Example

Consider the Hilbert operator

$$T(f)(x) = \int_0^\infty \frac{f(y)}{x+y}\, dy,$$

where $x \in (0,\infty)$. The operator T takes measurable functions on $(0,\infty)$ to measurable functions on $(0,\infty)$. We claim that T maps $L^p(0,\infty)$ to itself for $1 < p < \infty$; precisely, we have the estimate

$$\int_0^\infty T(f)(x)\, g(x)\, dx \le \frac{\pi}{\sin(\pi/p)} \|f\|_{L^p(0,\infty)} \|g\|_{L^{p'}(0,\infty)}.$$

To see this we use Schur's lemma. We take

$$u(x) = v(x) = x^{-\frac{1}{pp'}}.$$

We have that

$$T(u^{p'})(x) = \int_0^\infty \frac{y^{-\frac{1}{p}}}{x+y}\, dy = x^{-\frac{1}{p}} \int_0^\infty \frac{t^{-\frac{1}{p}}}{1+t}\, dt = B(\tfrac{1}{p'}, \tfrac{1}{p})\, v(x)^{p'},$$

where B is the usual beta function and the last identity follows from the change of variables $s = (1+t)^{-1}$. Now an easy calculation yields

$$B(\tfrac{1}{p'}, \tfrac{1}{p}) = \frac{\pi}{\sin(\pi/p)},$$

so the lemma in Appendix I.2 gives that $\|T\|_{L^p \to L^p} \le \frac{\pi}{\sin(\pi/p)}$. The sharpness of this constant follows by considering the sequence of functions

$$h_\varepsilon(x) = \begin{cases} x^{-\frac{1}{p}+\varepsilon} & \text{when } x < 1, \\ x^{-\frac{1}{p}-\varepsilon} & \text{when } x \ge 1, \end{cases}$$

which satisfies

$$\lim_{\varepsilon \to 0} \frac{\|T(h_\varepsilon)\|_{L^p(0,\infty)}}{\|h_\varepsilon\|_{L^p(0,\infty)}} = \frac{\pi}{\sin(\pi/p)}.$$

We make some comments related to the history of Schur's lemma. Schur [237] first proved a matrix version of the lemma in Appendix I.1 when $p = 2$. Precisely, Schur's original version was the following: If $K(x,y)$ is a positive decreasing function in both variables and satisfies

$$\sup_m \sum_n K(m,n) + \sup_n \sum_m K(m,n) < \infty,$$

then

$$\sum_m \sum_n a_{mn} K(m,n) b_{mn} \le C \|a\|_{\ell^2} \|b\|_{\ell^2}.$$

Hardy–Littlewood and Pólya [121] extended this result to L^p for $1 < p < \infty$ and disposed of the condition that K be a decreasing function. Aronszajn, Mulla, and Szeptycki [9] proved that (iii) implies (i) in the lemma of Appendix I.2. Gagliardo in [97] proved the converse direction that (i) implies (iii) in the same lemma. The case $p = 2$ was previously obtained by Karlin [151]. Condition (ii) was introduced by Howard and Schep [131], who showed that it is equivalent to (i) and (iii). A multilinear analogue of the lemma in Appendix I.2 was obtained by Grafakos and Torres [113]; the easy direction (iii) implies (i) was independently observed by Bekollé, Bonami, Peloso, and Ricci [17]. See also Cwikel and Kerman [65] for an alternative multilinear formulation of the Schur lemma.

The case $p = p' = 2$ of the application in Appendix I.3 is a continuous version of Hilbert's double series theorem. The discrete version was first proved by Hilbert in his lectures on integral equations (published by Weyl [290]) without a determination of the exact constant. This exact constant turns out to be π, as discovered by Schur [237]. The extension to other p's (with sharp constants) is due to Hardy and M. Riesz and published by Hardy [120].

Appendix J
The Whitney Decomposition of Open Sets in $\mathbf{R}^n$

An arbitrary open set in $\mathbf{R}^n$ can be decomposed as a union of disjoint cubes whose lengths are proportional to their distance from the boundary of the open set. See, for instance, Figure J.1 when the open set is the unit disk in $\mathbf{R}^2$. For a given cube Q in $\mathbf{R}^n$, we denote by $\ell(Q)$ its length.

Proposition. *Let Ω be an open nonempty proper subset of $\mathbf{R}^n$. Then there exists a family of closed cubes $\{Q_j\}_j$ such that*

(a) $\bigcup_j Q_j = \Omega$ and the Q_j's have disjoint interiors.
(b) $\sqrt{n}\,\ell(Q_j) \le \operatorname{dist}(Q_j, \Omega^c) \le 4\sqrt{n}\,\ell(Q_j)$.
(c) If the boundaries of two cubes Q_j and Q_k touch, then

$$\frac{1}{4} \le \frac{\ell(Q_j)}{\ell(Q_k)} \le 4.$$

(d) For a given Q_j there exist at most 12^n Q_k's that touch it.

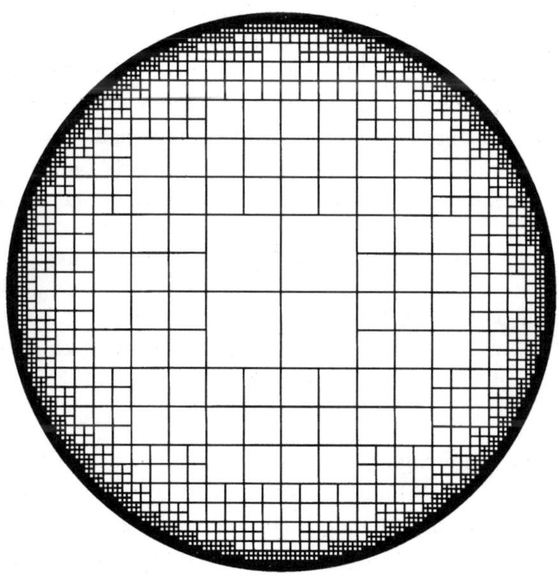

Fig. J.1 The Whitney decomposition of the unit disk.

Proof. Let $\mathscr{D}_k$ be the collection of all dyadic cubes of the form

$$\{(x_1,\dots,x_n)\in\mathbf{R}^n:\ m_j 2^{-k}\le x_j<(m_j+1)2^{-k}\},$$

where $m_j \in \mathbf{Z}$. Observe that each cube in $\mathscr{D}_k$ gives rise to 2^n cubes in $\mathscr{D}_{k+1}$ by bisecting each side.

Write the set Ω as the union of the sets

$$\Omega_k=\{x\in\Omega:\ 2\sqrt{n}2^{-k}<\operatorname{dist}(x,\Omega^c)\le 4\sqrt{n}2^{-k}\}$$

over all $k\in\mathbf{Z}$. Let $\mathscr{F}'$ be the set of all cubes Q in $\mathscr{D}_k$ for some $k\in\mathbf{Z}$ such that $Q\cap\Omega_k\neq\emptyset$. We show that the collection $\mathscr{F}'$ satisfies property (b). Let $Q\in\mathscr{F}'$ and pick $x\in\Omega_k\cap Q$ for some $k\in\mathbf{Z}$. Observe that

$$\sqrt{n}2^{-k}\le\operatorname{dist}(x,\Omega^c)-\sqrt{n}\,\ell(Q)\le\operatorname{dist}(Q,\Omega^c)\le\operatorname{dist}(x,\Omega^c)\le 4\sqrt{n}2^{-k},$$

which proves (b).

Next we observe that

$$\bigcup_{Q\in\mathscr{F}'} Q=\Omega.$$

Indeed, every Q in $\mathscr{F}'$ is contained in Ω (since it has positive distance from its complement) and every $x\in\Omega$ lies in some Ω_k and in some dyadic cube in $\mathscr{D}_k$.

The problem is that the cubes in the collection $\mathscr{F}'$ may not be disjoint. We have to refine the collection $\mathscr{F}'$ by eliminating those cubes that are contained in some other cubes in the collection. Recall that two dyadic cubes have disjoint interiors or else one contains the other. For every cube Q in $\mathscr{F}'$ we can therefore consider the unique *maximal* cube $Q^{\max}$ in $\mathscr{F}'$ that contains it. Two different such maximal cubes must have disjoint interiors by maximality. Now set $\mathscr{F}=\{Q^{\max}:\ Q\in\mathscr{F}'\}$.

The collection of cubes $\{Q_j\}_j=\mathscr{F}$ clearly satisfies (a) and (b), and we now turn our attention to the proof of (c). Observe that if Q_j and Q_k in $\mathscr{F}$ touch then

$$\sqrt{n}\,\ell(Q_j)\le\operatorname{dist}(Q_j,\Omega^c)\le\operatorname{dist}(Q_j,Q_k)+\operatorname{dist}(Q_k,\Omega^c)\le 0+4\sqrt{n}\,\ell(Q_k),$$

which proves (c). To prove (d), observe that any cube Q in $\mathscr{D}_k$ is touched by exactly 3^n-1 other cubes in $\mathscr{D}_k$. But each cube Q in $\mathscr{D}_k$ can contain at most 4^n cubes of $\mathscr{F}$ of length at least one-quarter of the length of Q. This fact combined with (c) yields (d). $\square$

The following observation is a consequence of the result just proved: Let $\mathscr{F}=\{Q_j\}_j$ be the Whitney decomposition of a proper open subset Ω of $\mathbf{R}^n$. Fix $0<\varepsilon<1/4$ and denote by Q_k^* the cube with the same center as Q_k but with side length $(1+\varepsilon)$ times that of Q_k. Then Q_k and Q_j touch if and only if Q_k^* and Q_j intersect. Consequently, every point in Ω is contained in at most 12^n cubes Q_k^*.

Appendix K
Smoothness and Vanishing Moments

K.1 The Case of No Cancellation

Let $a,b \in \mathbf{R}^n$, $\mu,\nu \in \mathbf{R}$, and $M,N > n$. Set

$$I(a,\mu,M;b,\nu,N) = \int_{\mathbf{R}^n} \frac{2^{\mu n}}{(1+2^{\mu}|x-a|)^M} \frac{2^{\nu n}}{(1+2^{\nu}|x-b|)^N}\,dx\,.$$

Then we have

$$I(a,\mu,M;b,\nu,N) \le C_0 \frac{2^{\min(\mu,\nu)n}}{\big(1+2^{\min(\mu,\nu)}|a-b|\big)^{\min(M,N)}}\,,$$

where

$$C_0 = v_n\left(\frac{M4^N}{M-n}+\frac{N4^M}{N-n}\right)$$

and v_n is the volume of the unit ball in $\mathbf{R}^n$.

To prove this estimate, first observe that

$$\int_{\mathbf{R}^n} \frac{dx}{(1+|x|)^M} \le \frac{v_n M}{M-n}\,.$$

Without loss of generality, assume that $\nu \le \mu$. Consider the cases $2^{\nu}|a-b| \le 1$ and $2^{\nu}|a-b| \ge 1$. In the case $2^{\nu}|a-b| \le 1$ we use the estimate

$$\frac{2^{\nu n}}{(1+2^{\nu}|x-a|)^N} \le 2^{\nu n} \le \frac{2^{\nu n}2^{\min(M,N)}}{(1+2^{\nu}|a-b|)^{\min(M,N)}}\,,$$

and the result is a consequence of the estimate

$$I(a,\mu,M;b,\nu,N) \le \frac{2^{\nu n}2^{\min(M,N)}}{(1+2^{\nu}|a-b|)^{\min(M,N)}} \int_{\mathbf{R}^n} \frac{2^{\mu n}}{(1+2^{\mu}|x-a|)^M}\,dx\,.$$

In the case $2^{\nu}|a-b| \ge 1$ let H_a and H_b be the two half-spaces, containing the points a and b, respectively, formed by the hyperplane perpendicular to the line segment $[a,b]$ at its midpoint. Split the integral over $\mathbf{R}^n$ as the integral over H_a and the integral over H_b. For $x \in H_a$ use that $|x-b| \ge \frac{1}{2}|a-b|$. For $x \in H_b$ use a similar inequality and the fact that $2^{\nu}|a-b| \ge 1$ to obtain

$$\frac{2^{\mu n}}{(1+2^\mu|x-a|)^M} \le \frac{2^{\mu n}}{(2^\mu \frac{1}{2}|a-b|)^M} \le \frac{4^M 2^{(\nu-\mu)(M-n)} 2^{\nu n}}{(1+2^\nu|a-b|)^M}.$$

The required estimate follows.

K.2 The Case of Cancellation

Let $a, b \in \mathbf{R}^n$, $M, N > 0$, and L a nonnegative integer. Suppose that ϕ_μ and ϕ_ν are two functions on $\mathbf{R}^n$ that satisfy

$$|(\partial_x^\alpha \phi_\mu)(x)| \le \frac{A_\alpha 2^{\mu n} 2^{\mu L}}{(1+2^\mu|x-x_\mu|)^M}, \qquad \text{for all } |\alpha| = L,$$

$$|\phi_\nu(x)| \le \frac{B 2^{\nu n}}{(1+2^\nu|x-x_\nu|)^N},$$

for some A_α and B positive, and

$$\int_{\mathbf{R}^n} \phi_\nu(x) x^\beta \, dx = 0 \qquad \text{for all } |\beta| \le L-1,$$

where the last condition is supposed to be vacuous when $L = 0$. Suppose that $N > M + L + n$ and that $\nu \ge \mu$. Then we have

$$\left| \int_{\mathbf{R}^n} \phi_\mu(x) \phi_\nu(x) \, dx \right| \le C_{00} \frac{2^{\mu n} 2^{-(\nu-\mu)L}}{(1+2^\mu|x_\mu - x_\nu|)^M},$$

where

$$C_{00} = v_n \frac{N-L-M}{N-L-M-n} B \sum_{|\alpha|=L} \frac{A_\alpha}{\alpha!}.$$

To prove this statement, we subtract the Taylor polynomial of order $L-1$ of ϕ_μ at the point x_ν from the function $\phi_\mu(x)$ and use the remainder theorem to control the required integral by

$$B \sum_{|\alpha|=L} \frac{A_\alpha}{\alpha!} \int_{\mathbf{R}^n} \frac{|x-x_\nu|^L 2^{\mu n} 2^{\mu L}}{(1+2^\mu|\xi_x - x_\mu|)^M} \frac{2^{\nu n}}{(1+2^\nu|x-x_\nu|)^N} \, dx,$$

for some ξ_x on the segment joining x_ν to x. Using $\nu \ge \mu$ and the triangle inequality, we obtain

$$\frac{1}{1+2^\mu|\xi_x - x_\mu|} \le \frac{1+2^\nu|x-x_\nu|}{1+2^\mu|x_\mu - x_\nu|}.$$

We insert this estimate in the last integral and we use that $N > L + M + n$ to deduce the required conclusion.

K.3 The Case of Three Factors

Given three numbers a,b,c we denote by $\mathrm{med}\,(a,b,c)$ the number with the property $\min(a,b,c) \le \mathrm{med}\,(a,b,c) \le \max(a,b,c)$.

Let $x_\nu, x_\mu, x_\lambda \in \mathbf{R}^n$. Suppose that ψ_ν, ψ_μ, ψ_λ are functions defined on $\mathbf{R}^n$ such that for all $N > n$ sufficiently large there exist constants $A_\nu, A_\mu, A_\lambda < \infty$ such that

$$\begin{aligned}
|\psi_\nu(x)| &\le A_\nu \frac{2^{\nu n/2}}{(1+2^\nu|x-x_\nu|)^N},\\
|\psi_\mu(x)| &\le A_\mu \frac{2^{\mu n/2}}{(1+2^\mu|x-x_\mu|)^N},\\
|\psi_\lambda(x)| &\le A_\lambda \frac{2^{\lambda n/2}}{(1+2^\lambda|x-x_\lambda|)^N},
\end{aligned}$$

for all $x \in \mathbf{R}^n$. Then the following estimate is valid:

$$\begin{aligned}
&\int_{\mathbf{R}^n} |\psi_\nu(x)|\,|\psi_\mu(x)|\,|\psi_\lambda(x)|\,dx\\
&\le \frac{C_{N,n} A_\nu A_\mu A_\lambda\, 2^{-\max(\mu,\nu,\lambda)n/2}\, 2^{\mathrm{med}\,(\mu,\nu,\lambda)n/2}\, 2^{\min(\mu,\nu,\lambda)n/2}}{((1+2^{\min(\nu,\mu)}|x_\nu-x_\mu|)(1+2^{\min(\mu,\lambda)}|x_\mu-x_\lambda|)(1+2^{\min(\lambda,\nu)}|x_\lambda-x_\nu|))^N}
\end{aligned}$$

for some constant $C_{N,n} > 0$ independent of the remaining parameters.

Analogous estimates hold if some of these factors are assumed to have cancellation and the others vanishing moments. See the article of Grafakos and Torres [114] for precise statements of these results and applications. Similar estimates with m factors, $m \in \mathbf{Z}^+$, are studied in Bényi and Tzirakis [21].

Glossary

$A \subseteq B$	A is a subset of B (not necessarily a proper subset)
$A \subsetneq B$	A is a proper subset of B
A^c	the complement of a set A
χ_E	the characteristic function of the set E
d_f	the distribution function of a function f
f^*	the decreasing rearrangement of a function f
$f_n \uparrow f$	f_n increases monotonically to a function f
$\mathbf{Z}$	the set of all integers
$\mathbf{Z}^+$	the set of all positive integers $\{1,2,3,\dots\}$
$\mathbf{Z}^n$	the n-fold product of the integers
$\mathbf{R}$	the set of real numbers
$\mathbf{R}^+$	the set of positive real numbers
$\mathbf{R}^n$	the Euclidean n-space
$\mathbf{Q}$	the set of rationals
$\mathbf{Q}^n$	the set of n-tuples with rational coordinates
$\mathbf{C}$	the set of complex numbers
$\mathbf{C}^n$	the n-fold product of complex numbers
$\mathbf{T}$	the unit circle identified with the interval $[0,1]$
$\mathbf{T}^n$	the n-dimensional torus $[0,1]^n$
$\lvert x\rvert$	$\sqrt{\lvert x_1\rvert^2+\cdots+\lvert x_n\rvert^2}$ when $x=(x_1,\dots,x_n)\in\mathbf{R}^n$
$\mathbf{S}^{n-1}$	the unit sphere $\{x\in\mathbf{R}^n:\ \lvert x\rvert=1\}$

e_j	the vector $(0,\dots,0,1,0,\dots,0)$ with 1 in the jth entry and 0 elsewhere
$\log t$	the logarithm with base e of $t>0$
$\log_a t$	the logarithm with base a of $t>0$ $(1\neq a>0)$
$\log^+ t$	$\max(0,\log t)$ for $t>0$
$[t]$	the integer part of the real number t
$x\cdot y$	the quantity $\sum_{j=1}^n x_j y_j$ when $x=(x_1,\dots,x_n)$ and $y=(y_1,\dots,y_n)$
$B(x,R)$	the ball of radius R centered at x in $\mathbf{R}^n$
ω_{n-1}	the surface area of the unit sphere $\mathbf{S}^{n-1}$
v_n	the volume of the unit ball $\{x\in\mathbf{R}^n:\ \lvert x\rvert<1\}$
$\lvert A\rvert$	the Lebesgue measure of the set $A\subseteq\mathbf{R}^n$
dx	Lebesgue measure
$\mathrm{Avg}_B f$	the average $\frac{1}{\lvert B\rvert}\int_B f(x)\,dx$ of f over the set B
$\langle f,g\rangle$	the real inner product $\int_{\mathbf{R}^n} f(x)g(x)\,dx$
$\langle f\,\vert\,g\rangle$	the complex inner product $\int_{\mathbf{R}^n} f(x)\overline{g(x)}\,dx$
$\langle u,f\rangle$	the action of a distribution u on a function f
p'	the number $p/(p-1)$, whenever $0<p\neq 1<\infty$
$1'$	the number ∞
∞'	the number 1
$f=O(g)$	means $\lvert f(x)\rvert\le M\lvert g(x)\rvert$ for some M for x near x_0
$f=o(g)$	means $\lvert f(x)\rvert\,\lvert g(x)\rvert^{-1}\to 0$ as $x\to x_0$
A^t	the transpose of the matrix A
A^*	the conjugate transpose of a complex matrix A
A^{-1}	the inverse of the matrix A
$O(n)$	the space of real matrices satisfying $A^{-1}=A^t$
$\lVert T\rVert_{X\to Y}$	the norm of the (bounded) operator $T:X\to Y$
$A\approx B$	means that there exists a $c>0$ such that $c^{-1}\le\frac{B}{A}\le c$
$\lvert\alpha\rvert$	indicates the size $\lvert\alpha_1\rvert+\dots+\lvert\alpha_n\rvert$ of a multi-index $\alpha=(\alpha_1,\dots,\alpha_n)$
$\partial_j^m f$	the mth partial derivative of $f(x_1,\dots,x_n)$ with respect to x_j
$\partial^\alpha f$	$\partial_1^{\alpha_1}\cdots\partial_n^{\alpha_n} f$
$\mathscr{C}^k$	the space of functions f with $\partial^\alpha f$ continuous for all $\lvert\alpha\rvert\le k$

$\mathscr{C}_0$	the space of continuous functions with compact support
$\mathscr{C}_{00}$	the space of continuous functions that vanish at infinity
$\mathscr{C}_0^\infty$	the space of smooth functions with compact support
$\mathscr{D}$	the space of smooth functions with compact support
$\mathscr{S}$	the space of Schwartz functions
$\mathscr{C}^\infty$	the space of smooth functions $\bigcup_{k=1}^\infty \mathscr{C}^k$
$\mathscr{D}'(\mathbf{R}^n)$	the space of distributions on $\mathbf{R}^n$
$\mathscr{S}'(\mathbf{R}^n)$	the space of tempered distributions on $\mathbf{R}^n$
$\mathscr{E}'(\mathbf{R}^n)$	the space of distributions with compact support on $\mathbf{R}^n$
$\mathscr{P}$	the set of all complex-valued polynomials of n real variables
$\mathscr{S}'(\mathbf{R}^n)/\mathscr{P}$	the space of tempered distributions on $\mathbf{R}^n$ modulo polynomials
$\ell(Q)$	the side length of a cube Q in $\mathbf{R}^n$
∂Q	the boundary of a cube Q in $\mathbf{R}^n$
$L^p(X,\mu)$	the Lebesgue space over the measure space (X,μ)
$L^p(\mathbf{R}^n)$	the space $L^p(\mathbf{R}^n,\lvert\cdot\rvert)$
$L^{p,q}(X,\mu)$	the Lorentz space over the measure space (X,μ)
$L^p_{\text{loc}}(\mathbf{R}^n)$	the space of functions that lie in $L^p(K)$ for any compact set K in $\mathbf{R}^n$
$\lvert d\mu\rvert$	the total variation of a finite Borel measure μ on $\mathbf{R}^n$
$\mathscr{M}(\mathbf{R}^n)$	the space of all finite Borel measures on $\mathbf{R}^n$
$\mathscr{M}_p(\mathbf{R}^n)$	the space of L^p Fourier multipliers, $1 \le p \le \infty$
$\mathscr{M}^{p,q}(\mathbf{R}^n)$	the space of translation-invariant operators that map $L^p(\mathbf{R}^n)$ to $L^q(\mathbf{R}^n)$
$\lVert\mu\rVert_{\mathscr{M}}$	$\int_{\mathbf{R}^n}\lvert d\mu\rvert$ the norm of a finite Borel measure μ on $\mathbf{R}^n$
$\mathcal{M}$	the centered Hardy–Littlewood maximal operator with respect to balls
M	the uncentered Hardy–Littlewood maximal operator with respect to balls
$\mathcal{M}_c$	the centered Hardy–Littlewood maximal operator with respect to cubes
M_c	the uncentered Hardy–Littlewood maximal operator with respect to cubes
$\mathcal{M}_\mu$	the centered maximal operator with respect to a measure μ
M_μ	the uncentered maximal operator with respect to a measure μ
M_s	the strong maximal operator
M_d	the dyadic maximal operator

References

1. F. Albiac and N. J. Kalton, *Topics in Banach Space Theory*, Graduate Texts in Mathematics, Vol. 233, Springer, New York, 2006.
2. S. A. Alimov, R. R. Ashurov, and A. K. Pulatov, *Multiple Fourier Series and Fourier Integrals*, Commutative Harmonic Analysis, Vol. IV, pp. 1–95, V. P. Khavin and N. K. Nikol'skiĭ (eds.), Encyclopedia of Mathematical Sciences, Vol. 42, Springer-Verlag, Berlin–Heidelberg–New York, 1992.
3. N. Y. Antonov, *Convergence of Fourier series*, Proceedings of the XXth Workshop on Function Theory (Moscow, 1995), pp. 187–196, East J. Approx. **2** (1996).
4. N. Y. Antonov, *The behavior of partial sums of trigonometric Fourier series*, PhD thesis, Institute of Mathematics and Mechanics, Ural Branch of the Russian Academy of Sciences, Ekaterinburg, 1998.
5. T. Aoki, *Locally bounded spaces*, Proc. Imp. Acad. Tokyo **18** (1942) No. 10.
6. G. I. Arhipov, A. A. Karachuba, and V. N. Čubarikov, *Trigonometric integrals*, Math. USSR Izvestija **15** (1980), 211–239.
7. J. Arias de Reyna, *Pointwise Convergence of Fourier Series*, J. London Math. Soc. **65** (2002), 139–153.
8. J. Arias de Reyna, *Pointwise Convergence of Fourier Series*, Lect. Notes in Math. 1785, Springer, Berlin–Heidelberg–New York, 2002.
9. N. Aronszajn, F. Mulla, and P. Szeptycki, *On spaces of potentials connected with L^p-spaces*, Ann. Inst. Fourier (Grenoble) **12** (1963), 211–306.
10. J. M. Ash, *Multiple trigonometric series*, Studies in harmonic analysis, pp. 76–96, MAA Stud. Math., Vol. 13, Math. Assoc. Amer., Washington, DC, 1976.
11. N. Asmar, E. Berkson, and T. A. Gillespie, *Summability methods for transferring Fourier multipliers and transference of maximal inequalities*, Analysis and Partial Differential Equations, pp. 1–34, A collection of papers dedicated to Mischa Cotlar, C. Sadosky (ed.), Marcel Dekker Inc., New York and Basel, 1990.
12. N. Asmar, E. Berkson, and T. A. Gillespie, *On Jodeit's multiplier extension theorems*, Journal d' Analyse Math. **64** (1994), 337–345.
13. K. I. Babenko, *An inequality in the theory of Fourier integrals* [in Russian], Izv. Akad. Nauk SSSR Ser. Mat. **25** (1961), 531–542.
14. A. Baernstein II, *Some sharp inequalities for conjugate functions*, Indiana Univ. Math. J. **27** (1978), 833–852.
15. N. Bary, *A Treatise on Trigonometric Series*, Pergamon Press, New York, 1964.
16. W. Beckner, *Inequalities in Fourier analysis*, Ann. of Math. **102** (1975), 159–182.
17. D. Bekollé, A. Bonami, M. Peloso, and F. Ricci, *Boundedness of Bergman projections on tube domains over light cones*, Math. Zeit. **237** (2001), 31–59.
18. A. Benedek, A. Calderón. and R. Panzone, *Convolution operators on Banach-space valued functions*, Proc. Nat. Acad. Sci. USA **48** (1962), 356–365.

19. J. Benedetto and M. Frazier (eds.), *Wavelets, Mathematics and Applications*, CRC Press, Boca Raton, FL, 1994.
20. C. Bennett and R. Sharpley, *Interpolation of Operators*, Academic Press, Series: Pure and Applied Math. 129, Orlando, FL, 1988.
21. Á. Bényi and N. Tzirakis, *Multilinear almost diagonal estimates and applications*, Studia Math. **164** (2004), 75–89.
22. J. Bergh and J. Löfström, *Interpolation Spaces, An Introduction*, Springer-Verlag, New York, 1976.
23. A. Besicovitch, *A general form of the covering principle and relative differentiation of additive functions*, Proc. of Cambridge Phil. Soc. **41** (1945), 103–110.
24. R. P. Boas and S. Bochner, *On a theorem of M. Riesz for Fourier series*, J. London Math. Soc. **14** (1939), 62–73.
25. S. Bochner, *Summation of multiple Fourier series by spherical means*, Trans. Amer. Math. Soc. **40** (1936), 175–207.
26. S. Bochner, *Lectures on Fourier integrals, (with an author's supplement on monotonic functions, Stieltjes integrals, and harmonic analysis)*, Annals of Math. Studies 42, Princeton Univ. Press, Princeton, NJ, 1959.
27. S. Bochner, *Harmonic Analysis and the Theory of Probability*, University of California Press, Berkeley, CA, 1955.
28. J. Bourgain, *On square functions on the trigonometric system*, Bull. Soc. Math. Belg. Sér B **37** (1985), 20–26.
29. J. Bourgain, *Estimations de certaines fonctions maximales, C. R. Acad. Sci. Paris Sér. I Math.* **301** (1985), 499–502.
30. J. Bourgain, *Averages in the plane over convex curves and maximal operators*, J. Analyse Math. **47** (1986), 69–85.
31. D. L. Burkholder, *Martingale transforms*, Ann. of Math. Stat. **37** (1966), 1494–1505.
32. D. L. Burkholder and R. F. Gundy, *Extrapolation and interpolation of quasilinear operators on martingales*, Acta Math. **124** (1970), 249–304.
33. A. P. Calderón, *On the theorems of M. Riesz and Zygmund*, Proc. Amer. Math. Soc. **1** (1950), 533–535.
34. A. P. Calderón, *Intermediate spaces and interpolation, the complex method*, Studia Math. **24** (1964), 113–190.
35. A. P. Calderón, *Singular integrals*, Bull. Amer. Math. Soc. **72** (1966), 427–465.
36. A. P. Calderón, *Algebras of singular integral operators*, Singular Integrals, Proc. Sympos. Pure Math. (Chicago, Ill., 1966), pp. 18–55, Amer. Math. Soc., Providence, RI, 1967.
37. A. P. Calderón and A. Zygmund, *On the existence of certain singular integrals*, Acta Math. **88** (1952), 85–139.
38. A. P. Calderón and A. Zygmund, *A note on interpolation of sublinear operators*, Amer. J. Math. **78** (1956), 282–288.
39. A. P. Calderón and A. Zygmund, *On singular integrals*, Amer. J. Math. **78** (1956), 289–309.
40. A. P. Calderón and A. Zygmund, *Algebras of certain singular integral operators*, Amer. J. Math. **78** (1956), 310–320.
41. C. Calderón, *Lacunary spherical means*, Ill. J. Math. **23** (1979), 476–484.
42. A. Carbery, *An almost-orthogonality principle with applications to maximal functions associated to convex bodies*, Bull. Amer. Soc. (N.S.) **14** (1986), 269–273.
43. A. Carbery, *Variants of the Calderón-Zygmund theory for L^p-spaces*, Rev. Mat. Iber. **2** (1986), 381–396.
44. A. Carbery, M. Christ, and J. Wright, *Multidimensional van der Corput and sublevel set estimates*, J. Amer. Math. Soc. **12** (1999), 981–1015.
45. L. Carleson, *On convergence and growth of partial sums of Fourier series*, Acta Math. **116** (1966), 135–157.
46. L. Carleson, *On the Littlewood-Paley Theorem*, Mittag-Leffler Institute Report, Djursholm, Sweden 1967.

47. M. J. Carro, *New extrapolation estimates*, J. Funct. Anal. **174** (2000), 155–166.
48. M. Christ, *On the restriction of the Fourier transform to curves: endpoint results and the degenerate case*, Trans. Amer. Math. Soc. **287** (1985), 223–238.
49. M. Christ, *A convolution inequality for Cantor-Lebesgue measures*, Rev. Mat. Iber. **1** (1985), 79–83.
50. M. Christ, *Weak type* $(1,1)$ *bounds for rough operators I*, Ann. of Math. **128** (1988), 19–42.
51. M. Christ and L. Grafakos, *Best constants for two nonconvolution inequalities*, Proc. Amer. Math. Soc. **123** (1995), 1687–1693.
52. M. Christ and J.-L. Rubio de Francia, *Weak type* $(1,1)$ *bounds for rough operators II*, Invent. Math. **93** (1988), 225–237.
53. C. Chui (ed.), *Wavelets: A Tutorial in Theory and Applications*, Academic Press, San Diego, CA, 1992.
54. R. R. Coifman, J.-L. Rubio de Francia, and S. Semmes, *Multiplicateurs de Fourier de* $L^p(\mathbb{R})$ *et estimations quadratiques*, C. R. Acad. Sci. Paris **306** (1988), 351–354.
55. R. R. Coifman and G. Weiss, *Transference Methods in Analysis*, C.B.M.S. Regional Conference Series in Math. No. 31, Amer. Math. Soc., Providence, RI, 1976.
56. R. R. Coifman and G. Weiss, *Book review of "Littlewood-Paley and multiplier theory" by R. E. Edwards and G. I. Gaudry*, Bull. Amer. Math. Soc. **84** (1978), 242–250.
57. W. C. Connett, *Singular integrals near* L^1, Harmonic analysis in Euclidean spaces, Proc. Sympos. Pure Math. (Williams Coll., Williamstown, Mass., 1978), pp. 163–165, Amer. Math. Soc., Providence, RI, 1979.
58. A. Córdoba and R. Fefferman, *A geometric proof of the strong maximal theorem*, Ann. of Math. **102** (1975), 95–100.
59. M. Cotlar, *A combinatorial inequality and its applications to* L^2 *spaces*, Rev. Mat. Cuyana, **1** (1955), 41–55.
60. M. Cotlar, *A unified theory of Hilbert transforms and ergodic theorems*, Rev. Mat. Cuyana, **1** (1955), 105–167.
61. M. Cowling and G. Mauceri, *On maximal functions*, Rend. Sem. Mat. Fis. Milano **49** (1979), 79–87.
62. M. Cwikel, *The dual of weak* L^p, Ann. Inst. Fourier (Grenoble) **25** (1975), 81–126.
63. M. Cwikel and C. Fefferman, *Maximal seminorms on Weak* L^1, Studia Math. **69** (1980), 149–154.
64. M. Cwikel and C. Fefferman, *The canonical seminorm on Weak* L^1, Studia Math. **78** (1984), 275–278.
65. M. Cwikel and R. Kerman, *Positive multilinear operators acting on weighted* L^p *spaces*, J. Funct. Anal. **106** (1992), 130–144.
66. J. E. Daly, *A necessary condition for Calderón-Zygmund singular integral operators*, J. Four. Anal. and Appl. **5** (1999), 303–308.
67. J. E. Daly and K. Phillips, *On the classification of homogeneous multipliers bounded on* $H^1(\mathbf{R}^2)$, Proc. Amer. Math. Soc. **106** (1989), 685–696.
68. I. Daubechies, *Orthonormal bases of compactly supported wavelets*, Comm. Pure Appl. Math. **61** (1988), 909–996.
69. I. Daubechies, *Ten Lectures on Wavelets*, Society for Industrial and Applied Mathematics, Philadelphia, PA, 1992.
70. B. Davis, *On the integrability of the martingale square function*, Israel J. Math. **8** (1970), 187–190.
71. B. Davis, *On the weak type* $(1,1)$ *inequality for conjugate functions*, Proc. Amer. Math. Soc. **44** (1974), 307–311.
72. M. de Guzmán, *Real-Variable Methods in Fourier Analysis*, North-Holland Math. Studies **46**, North-Holland, Amsterdam 1981.
73. M. de Guzmán, *Differentiation of Integrals in* $\mathbb{R}^n$, Lecture Notes in Math. **481**, Springer-Verlag, Berlin, 1985.
74. K. de Leeuw, *On* L_p *multipliers*, Ann. of Math. **81** (1965), 364–379.

75. J. Diestel and J. J. Uhl Jr, *Vector Measures*, Amer. Math. Soc, Providence, RI, 1977.
76. J. L. Doob, *Stochastic Processes*, Wiley, New York, 1953.
77. N. Dunford and J. T. Schwartz, *Linear Operators, I*, General Theory, Interscience, New York, 1959.
78. J. Duoandikoetxea and J.-L. Rubio de Francia, *Maximal and singular integral operators via Fourier transform estimates*, Invent. Math. **84** (1986), 541–561.
79. J. Duoandikoetxea and L. Vega, *Spherical means and weighted inequalities*, J. London Math. Soc. **53** (1996), 343–353.
80. Duong X. T. and A. McIntosh, *Singular integral operators with non-smooth kernels on irregular domains*, Rev. Mat. Iber. **15** (1999), 233–265.
81. H. Dym and H. P. McKean, *Fourier Series and Integrals*, Academic Press, New York, 1972.
82. R. E. Edwards, *Fourier Series: A Modern Introduction*, 2nd ed., Springer-Verlag, New York, 1979.
83. A. Erdélyi, *Asymptotic Expansions*, Dover Publications Inc., New York, 1956.
84. M. Essén, *A superharmonic proof of the M. Riesz conjugate function theorem*, Arkiv f. Math. **22** (1984), 241–249.
85. D. Fan, K. Guo, and Y. Pan, *A note of a rough singular integral operator*, Math. Ineq. and Appl. **2** (1999), 73–81.
86. D. Fan and Y. Pan, *Singular integral operators with rough kernels supported by subvarieties*, Amer J. Math. **119** (1997), 799–839.
87. K. Fan, *Minimax theorems*, Proc. Nat. Acad. Sci. USA **39** (1953), 42–47.
88. C. Fefferman, *On the convergence of Fourier series*, Bull. Amer. Math. Soc. **77** (1971), 744–745.
89. C. Fefferman, *The multiplier problem for the ball*, Ann. of Math. **94** (1971), 330–336.
90. C. Fefferman, *The uncertainty principle*, Bull. Amer. Math. Soc. **9** (1983), 129–206.
91. C. Fefferman and E. M. Stein, *Some maximal inequalities*, Amer. J. Math. **93** (1971), 107–115.
92. R. Fefferman, *A theory of entropy in Fourier analysis*, Adv. in Math. **30** (1978), 171–201.
93. R. Fefferman, *Strong differentiation with respect to measures*, Amer J. Math. **103** (1981), 33–40.
94. G. B. Folland and A. Sitaram, *The uncertainty principle: a mathematical survey*, J. Fourier Anal. Appl. **3** (1997), 207–238.
95. J. Fourier, *Théorie Analytique de la Chaleur*, Institut de France, Paris, 1822.
96. M. Frazier, *An Introduction to Wavelets Through Linear Algebra*, Springer-Verlag, New York, NY, 1999.
97. E. Gagliardo, *On integral transformations with positive kernel*, Proc. Amer. Math. Soc. **16** (1965), 429–434.
98. J. García-Cuerva and J.-L. Rubio de Francia, *Weighted Norm Inequalities and Related Topics*, North-Holland Math. Studies **116**, North-Holland, Amsterdam, 1985.
99. G. Garrigós and A. Seeger, *Characterizations of Hankel multipliers*, Math. Ann., to appear.
100. I. M. Gelfand and G. E. Šilov, *Generalized Functions, Vol. 1: Properties and Operations*, Academic Press, New York, London, 1964.
101. I. M. Gelfand and G. E. Šilov, *Generalized Functions, Vol. 2: Spaces of Fundamental and Generalized Functions*, Academic Press, New York, London, 1968.
102. I. Gohberg and N. Krupnik, *Norm of the Hilbert transformation in the L_p space*, Funct. Anal. Appl. **2** (1968), 180–181.
103. L. Grafakos, *Best bounds for the Hilbert transform on $L^p(\mathbb{R}^1)$*, Math. Res. Lett. **4** (1997), 469–471.
104. L. Grafakos, P. Honzík, and D. Ryabogin, *On the p-independence property of Calderón-Zygmund theory*, J. Reine und Angew. Math. **602** (2007), 227–234.
105. L. Grafakos and N. Kalton, *Some remarks on multilinear maps and interpolation*, Math. Ann. **319** (2001), 151–180.

106. L. Grafakos and N. Kalton, *The Marcinkiewicz multiplier condition for bilinear operators*, Studia Math. **146** (2001), 115–156.
107. L. Grafakos and J. Kinnunen, *Sharp inequalities for maximal functions associated to general measures*, Proc. Royal Soc. Edinb. **128** (1998), 717–723.
108. L. Grafakos and C. Morpurgo, *A Selberg integral formula and applications*, Pac. J. Math. **191** (1999), 85–94.
109. L. Grafakos and S. Montgomery-Smith, *Best constants for uncentred maximal functions*, Bull. London Math. Soc. **29** (1997), 60–64.
110. L. Grafakos and A. Stefanov, *Convolution Calderón-Zygmund singular integral operators with rough kernels*, Analysis of Divergence: Control and Management of Divergent Processes, pp. 119–143, William Bray, Časlav V. Stanojević (eds.), Birkhäuser, Boston–Basel–Berlin, 1999.
111. L. Grafakos and A. Stefanov, *L^p bounds for singular integrals and maximal singular integrals with rough kernels*, Indiana Univ. Math. J. **47** (1998), 455–469.
112. L. Grafakos and T. Tao, *Multilinear interpolation between adjoint operators*, J. Funct. Anal. **199** (2003), 379–385.
113. L. Grafakos and R. Torres, *A multilinear Schur test and multiplier operators*, J. Funct. Anal. **187** (2001), 1–24.
114. L. Grafakos and R. Torres, *Discrete decompositions for bilinear operators and almost diagonal conditions*, Trans. Amer. Math. Soc. **354** (2002), 1153–1176.
115. K. Gröhening, *Foundations of Time-Frequency Analysis*, Birkhäuser, Boston, 2001.
116. U. Haagerup, *The best constants in Khintchine's inequality*, Studia Math. **70** (1982), 231–283.
117. L.-S. Hahn, *On multipliers of p-integrable functions*, Trans. Amer. Math. Soc. **128** (1967), 321–335.
118. G. H. Hardy, *Note on a theorem of Hilbert*, Math. Zeit. **6** (1920), 314–317.
119. G. H. Hardy, *Note on some points in the integral calculus*, Messenger Math. **57** (1928), 12–16.
120. G. H. Hardy, *Note on a theorem of Hilbert concerning series of positive terms*, Proc. London Math. Soc. **23** (1925), Records of Proc. XLV–XLVI.
121. G. H. Hardy, J. E. Littlewood, and G. Pólya, *The maximum of a certain bilinear form*, Proc. London Math. Soc. **25** (1926), 265–282.
122. G. H. Hardy, J. E. Littlewood, and G. Pólya, *Inequalities*, 2nd ed., Cambridge University Press, Cambridge, UK, 1952.
123. G. H. Hardy and J. E. Littlewood, *A maximal theorem with function-theoretic applications*, Acta Math. **54** (1930), 81–116.
124. E. Hérnandez and G. Weiss, *A First Course on Wavelets*, CRC Press, Boca Raton, FL, 1996.
125. E. Hewitt and K. Ross, *Abstract Harmonic Analysis I,* 2nd ed., Grundlehren der mathematischen Wissenschaften 115, Springer-Verlag, Berlin–Heidelberg–New York, 1979.
126. I. I. Hirschman, *A convexity theorem for certain groups of transformations*, Jour. d' Analyse Math. **2** (1952), 209–218.
127. S. Hofmann, *Weak $(1,1)$ boundedness of singular integrals with nonsmooth kernels, Proc. Amer. Math. Soc.* **103** (1988), 260–264.
128. B. Hollenbeck and I. E. Verbitsky, *Best constants for the Riesz projection*, J. Funct. Anal. **175** (2000), 370–392.
129. L. Hörmander, *Estimates for translation invariant operators in L^p spaces*, Acta Math. **104** (1960), 93–140.
130. L. Hörmander, *The Analysis of Linear Partial Differential Operators I*, 2nd ed., Springer-Verlag, Berlin–Heidelberg–New York, 1990.
131. R. Howard and A. R. Schep, *Norms of positive operators on L^p spaces*, Proc. Amer Math. Soc. **109** (1990), 135–146.
132. S. Hudson, *A covering lemma for maximal operators with unbounded kernels*, Michigan Math. J. **34** (1987), 147–151.

133. R. Hunt, *An extension of the Marcinkiewicz interpolation theorem to Lorentz spaces*, Bull. Amer. Math. Soc. **70** (1964), 803–807.
134. R. Hunt, *On $L(p,q)$ spaces*, L' Einseignement Math. **12** (1966), 249–276.
135. R. Hunt, *On the convergence of Fourier series*, Orthogonal Expansions and Their Continuous Analogues (Edwardsville, Ill., 1967), pp. 235–255, D. T. Haimo (ed.), Southern Illinois Univ. Press, Carbondale IL, 1968.
136. S. Igari, *Lectures on Fourier Series of Several Variables*, Univ. Wisconsin Lecture Notes, Madison, WI, 1968.
137. R. S. Ismagilov, *On the Pauli problem*, Func. Anal. Appl. **30** (1996), 138–140.
138. T. Iwaniec and G. Martin, *Riesz transforms and related singular integrals*, J. Reine Angew. Math. **473** (1996), 25–57.
139. S. Janson, *On interpolation of multilinear operators*, Function Spaces and Applications (Lund, 1986), pp. 290–302, Lect. Notes in Math. 1302, Springer, Berlin–New York, 1988.
140. B. Jessen, J. Marcinkiewicz, and A. Zygmund, *Note on the differentiability of multiple integrals*, Fund. Math. **25** (1935), 217–234.
141. M. Jodeit, *Restrictions and extensions of Fourier multipliers*, Studia Math. **34** (1970), 215–226.
142. M. Jodeit, *A note on Fourier multipliers*, Proc. Amer. Math. Soc. **27** (1971), 423–424.
143. O. G. Jørsboe and L. Melbro, *The Carleson-Hunt Theorem on Fourier Series*, Lect. Notes in Math. 911, Springer-Verlag, Berlin, 1982.
144. J.- L. Journé *Calderón-Zygmund operators on product spaces*, Rev. Mat. Iber. **1** (1985), 55–91.
145. J. P. Kahane and Y. Katznelson, *Sur les ensembles de divergence des séries trigonometriques*, Studia Math. **26** (1966), 305–306.
146. G. Kaiser, *A Friendly Guide to Wavelets*, Birkhäuser, Boston–Basel–Berlin, 1994.
147. N. J. Kalton, *Convexity, type, and the three space problem*, Studia Math. **69** (1981), 247–287.
148. N. J. Kalton, *Plurisubharmonic functions on quasi-Banach spaces*, Studia Math. **84** (1986), 297–323.
149. N. J. Kalton, *Analytic functions in non-locally convex spaces and applications*, Studia Math. **83** (1986), 275–303.
150. N. J. Kalton, N. T. Peck, and J. W. Roberts, *An F-space sampler*, Lon. Math. Soc. Lect. Notes 89, Cambridge Univ. Press, Cambridge, UK, 1984.
151. S. Karlin, *Positive operators*, J. Math. Mech. **6** (1959), 907–937.
152. Y. Katznelson, *Sur les ensembles de divergence des séries trigonometriques*, Studia Math. **26** (1966), 301–304.
153. Y. Katznelson, *An Introduction to Harmonic Analysis*, 2nd ed., Dover Publications, Inc., New York, 1976.
154. C. Kenig and P. Tomas, *Maximal operators defined by Fourier multipliers*, Studia Math. **68** (1980), 79–83.
155. A. Khintchine, *Über dyadische Brüche*, Math. Zeit. **18** (1923), 109–116.
156. A. N. Kolmogorov, *Une série de Fourier-Lebesgue divergente presque partout*, Fund. Math. **4** (1923), 324–328.
157. A. N. Kolmogorov, *Une contribution à l' étude de la convergence des séries de Fourier*, Fund. Math. **5** (1924), 96–97.
158. A. N. Kolmogorov, *Sur les fonctions harmoniques conjuguées et les séries de Fourier*, Fund. Math. **7** (1925), 23–28.
159. A. N. Kolmogorov, *Une série de Fourier-Lebesgue divergente partout*, C. R. Acad. Sci. Paris **183** (1926), 1327–1328.
160. A. N. Kolmogorov, *Zur Normierbarkeit eines topologischen Raumes*, Studia Math. **5** (1934), 29–33.
161. S. V. Konyagin, *On everywhere divergence of trigonometric Fourier series,* Sbornik: Mathematics **191** (2000), 97–120.
162. T. Körner, *Fourier Analysis*, Cambridge Univ. Press, Cambridge, UK, 1988.

163. S. G. Krantz, *A panorama of Harmonic Analysis*, Carus Math. Monographs # 27, Mathematical Association of America, Washington, DC, 1999.
164. J. L. Krivine, *Théorèmes de factorisation dans les éspaces reticulés*, Sém. Maurey-Schwartz 1973/74, exp. XXII-XXIII, Palaiseau, France.
165. D. Kurtz, *Littlewood-Paley and multiplier theorems on weighted L^p spaces*, Trans. Amer. Math. Soc. **259** (1980), 235–254.
166. D. Kurtz and R. Wheeden, *Results on weighted norm inequalities for multipliers*, Trans. Amer. Math. Soc. **255** (1979), 343–362.
167. E. Landau, *Zur analytischen Zahlentheorie der definiten quadratischen Formen (Über die Gitterpunkte in einem mehrdimensionalen Ellipsoid),* Berl. Sitzungsber. **31** (1915), 458–476; reprinted in E. Landau "Collected Works," Vol. 6, pp. 200–218, Thales-Verlag, Essen, 1986.
168. S. Lang, *Real Analysis*, 2nd ed., Addison-Wesley, Reading, MA, 1983.
169. H. Lebesgue, *Intégrale, longeur, aire*, Annali Mat. Pura Appl. **7** (1902), 231–359.
170. H. Lebesgue, *Oeuvres Scientifiques*, Vol. I, L' Enseignement Math., pp. 201–331, Geneva, 1972.
171. P. Lemarié and Y. Meyer, *Ondelettes et bases hilbertiennes*, Rev. Mat. Iber. **2** (1986), 1–18.
172. J.-L. Lions and J. Peetre, *Sur une classe d' éspaces d' interpolation*, Inst. Hautes Etudes Sci. Publ. Math. No. **19** (1964), 5–68.
173. J. E. Littlewood, *On a certain bilinear form*, Quart. J. Math. Oxford Ser. **1** (1930), 164–174.
174. J. E. Littlewood and R. E. A. C. Paley, *Theorems on Fourier series and power series (I)*, J. London Math. Soc. **6** (1931), 230–233.
175. J. E. Littlewood and R. E. A. C. Paley, *Theorems on Fourier series and power series (II)*, Proc. London Math. Soc. **42** (1936), 52–89.
176. J. E. Littlewood and R. E. A. C. Paley, *Theorems on Fourier series and power series (III)*, Proc. London Math. Soc. **43** (1937), 105–126.
177. L. H. Loomis, *A note on Hilbert's transform*, Bull. Amer. Math. Soc. **52** (1946), 1082–1086.
178. L. H. Loomis and H. Whitney, *An inequality related to the isoperimetric inequality*, Bull. Amer. Math. Soc. **55** (1949), 961–962.
179. G. Lorentz, *Some new function spaces*, Ann. of Math. **51** (1950), 37–55.
180. G. Lorentz, *On the theory of spaces Λ*, Pacific. J. Math. **1** (1951), 411–429.
181. S. Lu, Y. Ding, and D. Yan, *Singular Integrals and Related Topics*, World Scientific Publishing, Singapore, 2007.
182. N. Luzin, *Sur la convergence des séries trigonométriques de Fourier*, C. R. Acad. Sci. Paris **156** (1913), 1655–1658.
183. A. Lyapunov, *Sur les fonctions-vecteurs complètement additives* [in Russian], Izv. Akad. Nauk SSSR Ser. Mat. **4** (1940), 465–478.
184. A. Magyar, E. M. Stein, and S. Wainger, *Discrete analogues in harmonic analysis: spherical averages*, Ann. of Math. **155** (2002), 189–208.
185. S. Mallat, *Multiresolution approximation and wavelets*, Trans. Amer. Math. Soc. **315** (1989), 69–88.
186. S. Mallat, *A Wavelet Tour of Signal Processing*, Academic Press, San Diego, CA, 1998.
187. J. Marcinkiewicz, *Sur l'interpolation d'operations*, C. R. Acad. Sci. Paris **208** (1939), 1272–1273.
188. J. Marcinkiewicz, *Sur les multiplicateurs des séries de Fourier*, Studia Math. **8** (1939), 78–91.
189. J. Marcinkiewicz and A. Zygmund, *Quelques inégalités pour les opérations linéaires*, Fund. Math. **32** (1939), 112–121.
190. J. Marcinkiewicz and A. Zygmund, *On the summability of double Fourier series*, Fund. Math. **32** (1939), 122–132.
191. J. Mateu and J. Verdera, *L^p and weak L^1 estimates for the maximal Riesz transform and the maximal Beurling transform*, Math. Res. Lett. **13** (2006), 957–966.

192. J. Mateu, J.Orobitg, and J. Verdera, *Estimates for the maximal singular integral in terms of the singular integral: the case of even kernels*, to appear.
193. A. Melas, *The best constant for the centered Hardy-Littlewood maximal inequality*, Ann. of Math. **157** (2003), 647–688.
194. Y. Meyer, *Principe d'incertitude, bases hilbertiennes et algèbres d'opérateurs*, Séminaire Bourbaki, 1985/86, No. 662.
195. Y. Meyer, *Ondelettes et Opérateurs*, Vol. I, Hermann, Paris, 1990.
196. Y. Meyer, *Ondelettes et Opérateurs*, Vol. II, Hermann, Paris, 1990.
197. Y. Meyer and R. R. Coifman, *Ondelettes et Opérateurs*, Vol. III, Hermann, Paris, 1991.
198. Y. Meyer, *Wavelets, Vibrations, and Scalings*, CRM Monograph Series Vol. 9, Amer. Math. Soc., Providence, RI, 1998.
199. S. G. Mihlin, *On the multipliers of Fourier integrals* [in Russian], Dokl. Akad. Nauk. **109** (1956), 701–703.
200. G. Mockenhaupt, A. Seeger, and C. Sogge, *Wave front sets, local smoothing and Bourgain's circular maximal theorem*, Ann. of Math. **136** (1992), 207–218.
201. K. H. Moon, *On restricted weak type* $(1,1)$, Proc. Amer. Math. Soc. **42** (1974), 148–152.
202. A. P. Morse, *Perfect blankets*, Trans. Amer. Math. Soc. **69** (1947), 418–442.
203. B. Muckenhoupt, *On certain singular integrals*, Pacific J. Math. **10** (1960), 239–261.
204. D. Müller, *A geometric bound for maximal functions associated to convex bodies*, Pacific J. Math. **142** (1990), 297–312.
205. A. Nagel, E. M. Stein, and S. Wainger, *Differentiation in lacunary directions*, Proc. Nat. Acad. Sci. USA **75** (1978), 1060–1062.
206. F. Nazarov and A. Seeger, *Radial Fourier multipliers in high dimensions*, to appear.
207. R. O' Neil, *Convolution operators and* $L(p,q)$ *spaces*, Duke Math. J. **30** (1963), 129–142.
208. F. W. J. Olver, *Asymptotics and Special Functions*, Academic Press, New York and London, 1974.
209. R. E. A. C. Paley, *A remarkable series of orthogonal functions*, Proc. London Math. Soc. **34** (1932), 241–264.
210. R. E. A. C. Paley and A. Zygmund, *On some series of functions*, Proc. Cambridge Phil. Soc. **26** (1930), 337–357.
211. J. Peetre, *Nouvelles propriétés d' éspaces d' interpolation*, C. R. Acad. Sci. Paris **256** (1963), 1424–1426.
212. M. C. Pereyra, *Lecture Notes on Dyadic Harmonic Analysis*, in "Second Summer School in Analysis and Mathematical Physics," pp. 1–60, S. Pérez-Esteva and C. Villegas-Blas (eds.), Contemp. Math. AMS, Vol. 289, Providence, RI, 2001.
213. S. Pichorides, *On the best values of the constants in the theorems of M. Riesz, Zygmund and Kolmogorov*, Studia Math. **44** (1972), 165–179.
214. M. Pinsky, N. Stanton, and P. Trapa, *Fourier series of radial functions in several variables*, J. Funct. Anal. **116** (1993), 111–132.
215. M. Reed and B. Simon, *Methods of Mathematical Physics, Vols. I, II*, Academic Press, New York, 1975.
216. F. Ricci and G. Weiss, *A characterization of* $H^1(\Sigma_{n-1})$, Harmonic analysis in Euclidean spaces, Proc. Sympos. Pure Math. (Williams Coll., Williamstown, Mass., 1978), pp. 289–294, Amer. Math. Soc., Providence, RI, 1979.
217. F. Riesz, *Untersuchungen über Systeme integrierbarer Funktionen*, Math. Ann. **69** (1910), 449–497.
218. F. Riesz, *Sur un théorème du maximum de MM. Hardy et Littlewood*, J. London Math. Soc. **7** (1932), 10–13.
219. M. Riesz, *Les fonctions conjuguées et les séries de Fourier*, C. R. Acad. Sci. Paris **178** (1924), 1464–1467.
220. M. Riesz, *Sur les maxima des formes bilinéaires et sur les fonctionnelles linéaires*, Acta Math. **49** (1927), 465–497.
221. M. Riesz, *Sur les fonctions conjuguées*, Math. Zeit. **27** (1927), 218–244.
222. M. Riesz, *L' intégrale de Riemann–Liouville et le problème de Cauchy*, Acta Math. **81** (1949), 1–223.

223. N. Riviere, *Singular integrals and multiplier operators*, Arkiv f. Math. **9** (1971), 243–278.
224. S. Rolewicz, *Metric Linear Spaces,* 2nd ed., Mathematics and Its Applications (East European Series), 20. D. Reidel Publishing Co., Dordrecht-Boston, MA; PWN, Warsaw, 1985.
225. J.-L. Rubio de Francia, *A Littlewood-Paley inequality for arbitrary intervals*, Rev. Mat. Iber. **1** (1985), 1–14.
226. J.-L. Rubio de Francia, *Maximal functions and Fourier transforms*, Duke Math. J. **53** (1986), 395–404.
227. J.-L. Rubio de Francia and J.-L. Torrea, *Vector extensions of operators in L^p spaces*, Pacific J. Math. **105** (1983), 227–235.
228. J.-L. Rubio de Francia, F. J. Ruiz, and J. L. Torrea, *Calderón-Zygmund theory for operator-valued kernels*, Adv. in Math. **62** (1986), 7–48.
229. W. Rudin, *Real and Complex Analysis*, 2nd ed., Tata McGraw-Hill Publishing Company, New Delhi, 1974.
230. W. Rudin, *Functional Analysis*, Tata McGraw-Hill Publishing Company, New Delhi, 1973.
231. D. Ryabogin and B. Rubin, *Singular integrals generated by zonal measures*, Proc. Amer. Math. Soc. **130** (2002), 745–751.
232. C. Sadosky, *Interpolation of Operators and Singular Integrals*, Marcel Dekker Inc., 1976.
233. S. Saks, *Theory of the Integral*, Hafner Publ. Co, New York, 1938.
234. S. Sato, *Note on a Littlewood-Paley operator in higher dimensions*, J. London Math. Soc. **42** (1990), 527–534.
235. L. Schwartz, *Théorie de Distributions, I, II*, Hermann, Paris, 1950-51.
236. W. Schlag, *A geometric proof of the circular maximal theorem*, Duke Math. J. **93** (1998), 505–533.
237. I. Schur, *Bemerkungen zur Theorie der beschränkten Bilinearformen mit unendlich vielen Veränderlichen*, Journal f. Math. **140** (1911), 1–28.
238. A. Seeger, *Some inequalities for singular convolution operators in L^p-spaces*, Trans. Amer. Math. Soc. **308** (1988), 259–272.
239. A. Seeger, *Singular integral operators with rough convolution kernels*, J. Amer. Math. Soc. **9** (1996), 95–105.
240. V. L. Shapiro, *Fourier series in several variables*, Bull. Amer. Math. Soc. **70** (1964), 48–93.
241. R. Sharpley, *Interpolation of n pairs and counterexamples employing indices*, J. Approx. Theory **13** (1975), 117–127.
242. R. Sharpley, *Multilinear weak type interpolation of mn-tuples with applications*, Studia Math. **60** (1977), 179–194.
243. P. Sjögren, *A remark on the maximal function for measures on $\mathbb{R}^n$*, Amer. J. Math. **105** (1983), 1231–1233
244. P. Sjögren and F. Soria, *Rough maximal operators and rough singular integral operators applied to integrable radial functions*, Rev. Math. Iber. **13** (1997), 1–18.
245. P. Sjölin, *On the convergence almost everywhere of certain singular integrals and multiple Fourier series*, Arkiv f. Math. **9** (1971), 65–90.
246. P. Sjölin, *A note on Littlewood-Paley decompositions with arbitrary intervals*, J. Approx. Theory **48** (1986), 328–334.
247. P. Sjölin and F. Soria, *Remarks on a theorem by N. Y. Antonov*, Studia Math. **158** (2003), 79–97.
248. C. Sogge, *Fourier Integrals in Classical Analysis*, Cambridge Tracts in Math. 105, Cambridge Univ. Press, Cambridge, UK, 1993.
249. F. Soria, *A note on a Littlewood-Paley inequality for arbitrary intervals in $\mathbb{R}^2$*, J. London Math. Soc. **36** (1987), 137–142.
250. F. Soria, *On an extrapolation theorem of Carleson-Sjölin with applications to a.e. convergence of Fourier series*, Studia Math. **94** (1989), 235–244.
251. E. M. Stein, *Interpolation of linear operators*, Trans. Amer. Math. Soc. **83** (1956), 482–492.

252. E. M. Stein, *Localization and summability of multiple Fourier series*, Acta Math. **100** (1958), 93–147.
253. E. M. Stein, *On the functions of Littlewood-Paley, Lusin, and Marcinkiewicz*, Trans. Amer. Math. Soc. **88** (1958), 430–466.
254. E. M. Stein, *On limits of sequences of operators*, Ann. of Math. **74** (1961), 140–170.
255. E. M. Stein, *Note on the class* $L\log L$, Studia Math. **32** (1969), 305–310.
256. E. M. Stein, *Topics in Harmonic Analysis Related to the Littlewood-Paley Theory*, Annals of Math. Studies 63, Princeton Univ. Press, Princeton, NJ, 1970.
257. E. M. Stein, *Maximal functions: Spherical means*, Proc. Nat. Acad. Sci. USA **73** (1976), 2174–2175.
258. E. M. Stein, *The development of square functions in the work of A. Zygmund*, Bull. Amer. Math. Soc. **7** (1982), 359–376.
259. E. M. Stein, *Some results in harmonic analysis in* $\mathbf{R}^n$, *for* $n \to \infty$, Bull. Amer. Math. Soc. **9** (1983), 71–73.
260. E. M. Stein, *Boundary behavior of harmonic functions on symmetric spaces: Maximal estimates for Poisson integrals*, Invent. Math. **74** (1983), 63–83.
261. E. M. Stein, *Harmonic Analysis, Real Variable Methods, Orthogonality, and Oscillatory Integrals*, Princeton Univ. Press, Princeton, NJ, 1993.
262. E. M. Stein and J. O. Strömberg, *Behavior of maximal functions in* $\mathbf{R}^n$ *for large* n, Arkiv f. Math. **21** (1983), 259–269.
263. E. M. Stein and G. Weiss, *Interpolation of operators with change of measures*, Trans. Amer. Math. Soc. **87** (1958), 159–172.
264. E. M. Stein and G. Weiss, *An extension of theorem of Marcinkiewicz and some of its applications*, J. Math. Mech. **8** (1959), 263–284.
265. E. M. Stein and G. Weiss, *Introduction to Fourier Analysis on Euclidean Spaces*, Princeton Univ. Press, Princeton, NJ, 1971.
266. E. M. Stein and N. J. Weiss, *On the convergence of Poisson integrals*, Trans. Amer. Math. Soc. **140** (1969), 34–54.
267. P. Stein, *On a theorem of M. Riesz*, J. London Math. Soc. **8** (1933), 242–247.
268. V. D. Stepanov, *On convolution integral operators*, Soviet Math. Dokl. **19** (1978), 1334–1337.
269. R. Strichartz, *A multilinear version of the Marcinkiewicz interpolation theorem*, Proc. Amer. Math. Soc. **21** (1969), 441–444.
270. J.-O. Strömberg, *A modified Franklin system and higher-order spline systems on* $\mathbb{R}^n$ *as unconditional bases for Hardy spaces*, Conference on harmonic analysis in honor of Antoni Zygmund, Vol. I, II (Chicago, Ill., 1981), pp. 475–494, Wadsworth Math. Ser., Wadsworth, Belmont, CA, 1983.
271. S. J. Szarek, *On the best constant in the Khintchine inequality*, Studia Math. **58** (1978), 197–208.
272. J. D. Tamarkin and A. Zygmund, *Proof of a theorem of Thorin*, Bull. Amer. Math. Soc. **50** (1944), 279–282.
273. T. Tao, *The weak type* $(1,1)$ *of* $L\log L$ *homogeneous convolution operators*, Indiana Univ. Math. J. **48** (1999), 1547–1584.
274. T. Tao, *A converse extrapolation theorem for translation-invariant operators*, J. Funct. Anal. **180** (2001), 1–10.
275. T. Tao and A. Seeger, *Sharp Lorentz estimates for rough operators*, Math. Ann. **320** (2001), 381–415.
276. T. Tao and J. Wright, *Endpoint multiplier theorems of Marcinkiewicz type*, Rev. Mat. Iber. **17** (2001), 521–558.
277. N. R. Tevzadze, *On the convergence of double Fourier series of quadratic summable functions*, Soobšč. Akad. Nauk Gruzin. **5** (1970), 277–279.
278. G. O. Thorin, *An extension of a convexity theorem due to M. Riesz*, Fys. Säellsk. Förh. **8** (1938), No. 14.

279. G. O. Thorin, *Convexity theorems generalizing those of M. Riesz and Hadamard with some applications*, Comm. Sem. Math. Univ. Lund [Medd. Lunds Univ. Mat. Sem.] **9** (1948), 1–58.
280. E. C. Titchmarsh, *The Theory of the Riemann Zeta Function*, Oxford at the Clarendon Press, 1951.
281. A. Torchinsky, *Real-Variable Methods in Harmonic Analysis*, Academic Press, New York, 1986.
282. J. G. van der Corput, *Zahlentheoretische Abschätzungen*, Math. Ann. **84** (1921), 53–79.
283. *On the maximal function for rotation invariant measures in* $\mathbb{R}^n$, Studia Math. **110** (1994), 9–17.
284. I. E. Verbitsky, *Estimate of the norm of a function in a Hardy space in terms of the norms of its real and imaginary part*, Amer. Math. Soc. Transl. **124** (1984), 11–15.
285. G. Vitali, *Sui gruppi di punti e sulle funzioni di variabili reali*, Atti Accad. Sci. Torino **43** (1908), 75–92.
286. J. Von Neuman, *Zur Theorie des Gesellschaftsspiele*, Math. Ann. **100** (1928), 295–320.
287. S. Wainger, *Special trigonometric series in k dimensions*, Mem. Amer. Math. Soc., No. 59, 1965.
288. G. N. Watson, *A Treatise on the Theory of Bessel Functions*, Cambridge University Press, Cambridge, UK, 1952.
289. M. Weiss and A. Zygmund, *An example in the theory of singular integrals*, Studia Math. **26** (1965), 101–111.
290. H. Weyl, *Singuläre Integralgleichungen mit besonderer Berücksichtigung des Fourierschen Integraltheorems*, Inaugural-Dissertation, Göttingen, 1908.
291. N. Wiener, *The ergodic theorem*, Duke Math. J. **5** (1939), 1–18.
292. M. V. Wickerhauser, *Adapted Wavelet Analysis from Theory to Software*, A. K. Peters, Wellesley, 1994.
293. P. Wojtaszczyk, *A Mathematical Introduction to Wavelets*, London Math. Soc. Student Texts 37, Cambridge University Press, Cambridge, UK, 1997.
294. S. Yano, *An extrapolation theorem*, J. Math. Soc. Japan **3** (1951), 296–305.
295. A. I. Yanushauskas, *Multiple trigonometric series* [in Russian], Nauka, Novosibirsk 1989.
296. K. Yosida, *Functional Analysis*, Springer-Verlag, Berlin, 1995.
297. M. Zafran, *A multilinear interpolation theorem*, Studia Math. **62** (1978), 107–124.
298. L. V. Zhizhiashvili, *Some problems in the theory of simple and multiple trigonometric and orthogonal series*, Russian Math. Surveys, **28** (1973), 65–127.
299. L. V. Zhizhiashvili, *Trigonometric Fourier series and Their Conjugates*, Mathematics and Its Applications, Kluwer Academic Publishers, Vol. 372, Dordrecht, Boston, London, 1996.
300. W. P. Ziemer, *Weakly Differentiable Functions*, Graduate Texts in Mathematics, Springer-Verlag, New York, 1989.
301. F. Zo, *A note on approximation of the identity*, Studia Math. **55** (1976), 111–122.
302. A. Zygmund, *On a theorem of Marcinkiewicz concerning interpolation of operators*, Jour. de Math. Pures et Appliquées **35** (1956), 223–248.
303. A. Zygmund, *Trigonometric Series*, Vol. I, 2nd ed., Cambridge University Press, Cambridge, UK, 1959.
304. A. Zygmund, *Trigonometric Series*, Vol. II, 2nd ed., Cambridge University Press, Cambridge, UK, 1959.
305. A. Zygmund, *Notes on the history of Fourier series,* Studies in harmonic analysis, 1–19, MAA Stud. Math., Vol. 13, Math. Assoc. Amer., Washington, DC, 1976.

Index